The EEG Handbook

Tracy Warbrick
Editor

The EEG Handbook

From Principles to Practice

Springer

Editor
Tracy Warbrick
Brain Products GmbH
Gilching, Germany

ISBN 978-3-032-20449-3 ISBN 978-3-032-20450-9 (eBook)
https://doi.org/10.1007/978-3-032-20450-9

This Springer imprint is published by the registered company Springer Nature Switzerland AG
The registered company address is: Gewerbestrasse 11, 6330 Cham, Switzerland

Preface

This book is about doing EEG research well. Our aim is to provide a trusted companion for students learning about EEG and for researchers conducting EEG studies. A key strength of this book is its emphasis on practical guidance. A book that not only explains the science behind EEG but also helps you navigate the real-world challenges of performing high-quality reproducible EEG research in the laboratory. Drawing on more than 25 years of collective experience supporting the neurophysiology research community, the authors understand where students and researchers commonly encounter difficulties and how best to support them. Alongside the fundamental scientific background and core principles of EEG measurement and analysis, we explore essential skills such as how to read EEG papers critically, how to present your own findings clearly, and how to follow good scientific practices in neuroscience. These skills form the foundation of a successful career in EEG research. Experienced academics also contribute their expertise on specific applications, adding depth and breadth to the book's content. The result is a resource designed not just to be read but to be used, whether you are preparing a course assignment, planning your next study, or troubleshooting an EEG recording in the lab.

Gilching, Germany Tracy Warbrick

Acknowledgments

I would like to thank all contributors for their hard work and commitment in bringing this project to fruition. The depth and breadth of the book's content reflect their expertise and dedication. I am especially grateful to Olga Nusser for her invaluable assistance with figures and illustrations for the chapters authored by the Brain Products team. Finally, I would like to sincerely thank Pierluigi Castellone, General Manager, Brain Products GmbH, for his support in making this book possible.

About This Book

Our aim is to offer a reliable companion for students and researchers conducting EEG studies. This book is divided into seven parts each addressing a key stage in the EEG research workflow: from basic concepts to practical lab work, data analysis, and scientific reporting. While the book is organized to guide readers step-by-step through the process of designing, conducting, and communicating high-quality EEG research, it does not have to be read linearly from cover to cover. Instead, readers are encouraged to dip into the parts most relevant to their current stage of learning, coursework, or research. The seven parts are outlined below.

Part I: Fundamentals of EEG A strong foundation in EEG research begins with understanding what the EEG signal represents and how to interpret the activity recorded. This section opens with an overview of the current state of the art in EEG, providing context for how the field has developed and where it is heading. We then introduce key concepts such as how EEG signals are generated, how they can be used in practice, and how to make full use of EEG's high temporal resolution. EEG depends on underlying biological processes; therefore it is also essential to understand the physiological and anatomical conditions under which these signals arise. For this reason, we cover the basic anatomy of the central and peripheral nervous systems, EEG in relation to human physiology, and brain states.

Part II: Experimental Design and Good Scientific Practice This part focuses on the principles of sound experimental design in EEG research. We discuss why statistical inference matters and why you should consider this in study design. We also provide practical guidance on pilot testing to optimize data quality. The section concludes with advice on establishing efficient and reproducible workflows for your studies.

Part III: EEG Data Acquisition Planning an EEG study involves numerous decisions, from selecting equipment and recording parameters to configuring triggers and managing potential artifacts. This section outlines the key considerations for effective EEG data acquisition and offers practical tips and troubleshooting strategies to help you optimize your experimental setup.

Part IV: Processing EEG Data There are many approaches to analyzing EEG data. In this section, we introduce common preprocessing steps and offer guidance on determining which steps are appropriate for your study. We also describe analytical approaches in both the time and frequency domains.

Part V: What Is EEG Used For? EEG supports a wide range of applications, and gaining an overview of these can provide inspiration and context for your own work. We begin with an outline of academic and clinical uses of EEG, followed by in-depth contributions from experts who discuss specific application areas in greater detail: BCI, focal epilepsy, neonatal and pediatric clinical applications, sleep, and mobile EEG applications.

Part VI: Multimodal EEG Applications While EEG is a powerful tool in its own right, combining it with other measurement or stimulation modalities can provide deeper insights into brain function and behavior. This section explains why specific multimodal combinations are valuable and highlights their contribution to the broader neuroimaging literature.

Part VII: Writing and Reading Research Papers Reading and writing scientific literature are essential skills for researchers. After completing a study and analyzing the data, researchers must communicate their findings clearly and transparently to the research community. This final section offers guidance on reporting EEG methods and results, including how to present data using figures, tables, and statistics. We also discuss how to read EEG literature critically and extract the information needed to evaluate and interpret published findings.

We sincerely believe that the *EEG Handbook: From Principles to Practice* will provide students and researchers with the knowledge and tools to become proficient EEG practitioners.

Contents

List of Figures

List of Tables

Editor and Contributors

About the Editor

Tracy Warbrick is Head of Education and Scientific Communication at Brain Products GmbH (Gilching, Germany). She holds a PhD in Psychology and has a multidisciplinary background spanning psychology, sport and exercise science, neuroimaging, and research methods. She has published widely in the fields of cognitive neuroscience and neuroimaging and has extensive experience in experimental design, data acquisition, and analysis of EEG and fMRI data. She is also strongly committed to teaching and learning and holds a postgraduate qualification in further and higher education. In her current role, she leads the development of high-quality educational resources for the EEG and neurophysiology research community, working closely with researchers to foster collaboration between academia and industry.

Contributors

Aime J. Aguilar-Herrera Department of Electrical and Computer Engineering, University of Houston, Houston, TX, USA

Paulo Rodrigo Bazán Brain Products GmbH, Gilching, Germany

Katherine Boere Department of Exercise Science, Physical and Health Education, The University of Victoria, Victoria, BC, Canada

David W. Carmichael School of Biomedical Engineering and Imaging Sciences, King's College London, London, UK

Silvia Casarotto Department of Biomedical and Clinical Sciences, University of Milan, Milan, Italy

IRCCS Fondazione Don Carlo Gnocchi ONLUS, Milan, Italy

Umair J. Chaudhary UCL Queen Square Institute of Neurology, University College London, London, UK

National Hospital for Neurology and Neurosurgery, University College London Hospitals NHS Foundation Trust, London, UK

Jose L. Contreras-Vidal Department of Electrical and Computer Engineering, University of Houston, Houston, TX, USA

Fernando Cross Villasana Brain Products GmbH, Gilching, Germany

Diana E. Gherman-Nagy Neuroadaptive Human-Computer Interaction, Brandenburg University of Technology Cottbus-Senftenberg, Cottbus, Germany

Mathew Rocha Hammerstrom Department of Exercise Science, Physical and Health Education, The University of Victoria, Victoria, BC, Canada

Christoph S. Herrmann Experimental Psychology Lab, Department of Psychology, Carl-von-Ossietzky Universität, Oldenburg, Germany

Cluster for Excellence "Hearing for All", Carl-von-Ossietzky Universität, Oldenburg, Germany

Research Center Neurosensory Science, Carl von Ossietzky Universitßt, Oldenburg, Germany

Michael Hoppstädter Brain Products GmbH, Gilching, Germany

Cilia Jaeger Brain Products GmbH, Gilching, Germany

David Kadlec Brain Products GmbH, Gilching, Germany

Marius Klug Neuroadaptive Human-Computer Interaction, Brandenburg University of Technology Cottbus-Senftenberg, Cottbus, Germany

Daniel S. Kluger Institute for Biomagnetism and Biosignal Analysis, University of Münster, Münster, Germany

Alex Kreilinger Brain Products GmbH, Gilching, Germany

Hannah Kreilinger Brain Products GmbH, Gilching, Germany

Olave E. Krigolson Theoretical and Applied Neuroscience Laboratory, School of Exercise Science, University of Victoria, Victoria, BC, Canada

Louis Lemieux UCL Queen Square Institute of Neurology, University College London, London, UK

Yoshua E. Lima-Carmona Department of Electrical and Computer Engineering, University of Houston, Houston, TX, USA

Marie Loescher Laboratoire de Neurosciences Cognitives et Computationnelles, Département d'Etudes Cognitives, École Normale Supérieure, Université PSL, INSERM U960, Paris, France

Yin Fen Low Brain Products GmbH, Gilching, Germany

Steven J. Luck Center for Mind & Brain and Department of Psychology, University of California, Davis, Davis, CA, USA

Ramon Martinez-Cancino Brain Products GmbH, Gilching, Germany

Jayvian Mavi Digital Health and Applied Technology Assessment (DHATA), King's College London, London, UK

Rebecca Meagher School of Biomedical Engineering and Imaging Sciences, King's College London, London, UK

Alejandro Ojeda Brain Vision LLC, Garner, NC, USA

Maxine Annel Pacheco-Ramírez Department of Electrical and Computer Engineering, University of Houston, Houston, TX, USA

Ignacio Rebollo Department of Gut Brain Interactions, German Institute of Human Nutrition, Nuthetal, Germany

Mario Rosanova Department of Biomedical and Clinical Sciences, University of Milan, Milan, Italy

Anna Sadilova School of Biomedical Engineering and Imaging Sciences, King's College London, London, UK

Lianne Sanchez-Rodriguez Department of Electrical and Computer Engineering, University of Houston, Houston, TX, USA

Boyuan Song UCL Queen Square Institute of Neurology, University College London, London, UK

Heiko I. Stecher Experimental Psychology Lab, Department of Psychology, Carl-von-Ossietzky Universität, Oldenburg, Germany

Daniel Strüber Experimental Psychology Lab, Department of Psychology, Carl-von-Ossietzky Universität, Oldenburg, Germany

Masako Tamaki RIKEN Center for Brain Science, Wako, Japan

Shivakumar Viswanathan Brain Products GmbH, Gilching, Germany

Sreekari Vogeti Experimental Psychology Lab, Department of Psychology, Carl-von-Ossietzky Universität, Oldenburg, Germany

Cluster for Excellence "Hearing for All", Carl-von-Ossietzky Universität, Oldenburg, Germany

Tracy Warbrick Brain Products GmbH, Gilching, Germany

Kimberley Whitehead Digital Health and Applied Technology Assessment (DHATA), King's College London, London, UK

Thorsten O. Zander Neuroadaptive Human-Computer Interaction, Brandenburg University of Technology Cottbus-Senftenberg, Cottbus, Germany

Part I
Fundamentals of EEG

Chapter 1
EEG in Context: Past, Present, and Future

Tracy Warbrick

Abstract Our understanding of brain function depends on the measurement and analysis techniques at our disposal. To fully understand the value of a technique, we need to consider its capabilities and limitations and be aware of its place in the wider landscape of brain imaging. Electroencephalography (EEG) offers a way to observe brain activity in near real time and is widely applicable across many research and clinical applications. It can be quantitative, objective, has good test–retest reliability, and is a rich source of information. In this chapter, we consider how EEG developed as a research tool, its place within the landscape of brain imaging, and the future of EEG.

Keywords EEG history · Brain imaging methods · EEG developments · Emerging EEG applications · Good practice · Replication · Diversity · Accessibility

1.1 EEG Technology: Past to Present

Electroencephalography (EEG) has a long history, with 2024 marking 100 years since the first EEG recordings. Its origins can be traced back to the late nineteenth century, when scientists made progress in understanding the neurophysiology and electrophysiology of the body. In the 1870s, Caton discovered that the brain also produced fluctuating electrical signals. However, it wasn't until decades later, in the 1920s, that Berger successfully recorded these signals, and we consider this the origin of EEG research. Despite this encouraging start, EEG was not widely used until the 1950s. For a comprehensive account of EEG's development from its initial discovery by Caton to its rise in the 1950s, see Niedermeyer (2005).

Since the 1950s, we have seen many technological and theoretical advances. Berger's first recordings were from a single bipolar channel using an analogue system. In contrast, today's technologies allow us to record from hundreds of

T. Warbrick (✉)
Brain Products GmbH, Gilching, Germany
e-mail: tracy.warbrick@brainproducts.com

T. Warbrick (ed.), *The EEG Handbook*,
https://doi.org/10.1007/978-3-032-20450-9_1

channels digitally. The first EEG systems used delicate electrodes, a differential amplifier for each channel, and a needle or pen to plot the signal on paper (Niedermeyer, 2005). Today, we can measure from many channels, record a larger volume of data, and use variable acquisition parameters thanks to digital signal acquisition and amplifier development.

Modern EEG systems are smaller and more portable than early EEG devices, allowing easier lab setup and enabling mobile EEG in more ecologically valid settings. Electrode technology has also developed significantly. Electrodes are more robust, and researchers have a range of electrode types, for example gel, saline, passive, and active, and application methods, for example cap-mounted, headset-mounted, and glued, to choose from. Further information on amplifier and electrode technology can be found in Chap. 12 ('Hardware for Recording EEG and Peripheral Physiology').

Analysis techniques have also advanced thanks to widespread access to computational resources and analysis methods. Early EEG research relied on visual inspection of a printed or digital trace, but we are now able to extract and quantify EEG features using sophisticated signal processing techniques. Part IV of this book is dedicated to EEG analysis and signal processing techniques. In addition, EEG can be combined with other emerging technologies such as artificial intelligence (AI), brain–computer interfaces (BCIs), and virtual reality (VR). EEG researchers are adapting to the changing technological landscape, and EEG remains a valuable method within the broader field of brain imaging and cognitive neuroscience.

The following sections will explore EEG and its relation to brain function, where EEG fits in the wider context of brain imaging methods, and possible future directions for the field.

1.2 What Do We Know About the EEG Signal?

Recording EEG from scalp electrodes provides a sensitive measure of ongoing brain activity. Neural activity generates electrical and magnetic fields, and when large populations of neurons fire synchronously, we can measure the summed activation as EEG at the scalp. These time-varying signals provide a window into the functional state of the brain and are used to understand a myriad of functions such as perception, cognition, and arousal state. A detailed account of the physiological origins of EEG is provided in Chap. 2 ('What Is EEG?').

To measure and interpret EEG signals, we need to understand what we measure and the associated limitations. Here, we focus on EEG's strengths and weaknesses in relation to other neuroimaging methods.

EEG is a direct measure of population-level neural activity and has a high temporal resolution in the order of milliseconds. This makes EEG uniquely suited to measuring the rapidly changing temporal dynamics of brain activity. However, while the temporal resolution of EEG is excellent, its spatial resolution is poor. In other words, it's difficult to determine the origin of the electrical potentials we measure at the scalp. The signal is generated in the cortical surface and must pass through multiple layers of tissue (cortex, skull, skin) to reach the scalp surface where we place our EEG electrodes. Furthermore, EEG is limited to measuring large-scale neuronal populations firing in synchrony. This means that the activity of smaller populations and asynchronous activity are difficult to measure with scalp EEG (Cohen, 2017).

We need to know how to interpret EEG signals—not only *what* EEG measures but what EEG *means*. What drives changes in EEG features? Despite much research linking the spectral, temporal, and spatial features of EEG to perceptual and cognitive processes and disease states, it's not always possible to say *how* these signals were generated by the underlying neural circuitry (Cohen, 2017). The macroscopic level of EEG doesn't allow us to measure the electrical activity of individual neurons, how this changes over time, and its connectivity with other neurons. Therefore, it's difficult to interpret a specific EEG feature in terms of corresponding features of neural activity.

Despite these limitations, EEG remains a powerful tool in the field of cognitive neuroscience. While our understanding of the drivers of the EEG signal is incomplete, its functional significance is understood well enough to provide valuable insights into electrophysiological brain dynamics. Acknowledging a technique's limitations doesn't undermine its value. Critical evaluation is necessary for its appropriate use and for understanding its contribution to the field. The remainder of this chapter considers EEG in the broader landscape of brain imaging methods and what the future of EEG might look like.

1.3 EEG and Other Neuroscience Methods

Understanding EEG's place in the wider context of brain imaging methods can help us to appreciate its unique contribution to the field. As cognitive neuroscientists, our goal is to understand how the brain works, what the measured signals mean, and how they relate to behaviour. Ideally, we would need measurements at the level of individual neurons and to track that over time. In human research this is not feasible, but we do have multiple neuroimaging tools at our disposal: electrocorticogram (ECoG), electroencephalography (EEG), magnetoencephalography (MEG), functional magnetic resonance imaging (fMRI), functional near-infrared spectroscopy (fNIRs), and positron emission tomography (PET). To use these methods effectively, we need to know what neural events (or proxies thereof) each method can detect and consider the relative spatial and temporal scales.

Each method has advantages and limitations and can make a unique contribution to our understanding of brain structure and function. We will consider what each method measures, specifically in comparison to EEG, starting at the neuron level and moving to measures of blood flow. Figure 1.1 shows the temporal and spatial resolution of common neuroimaging methods.

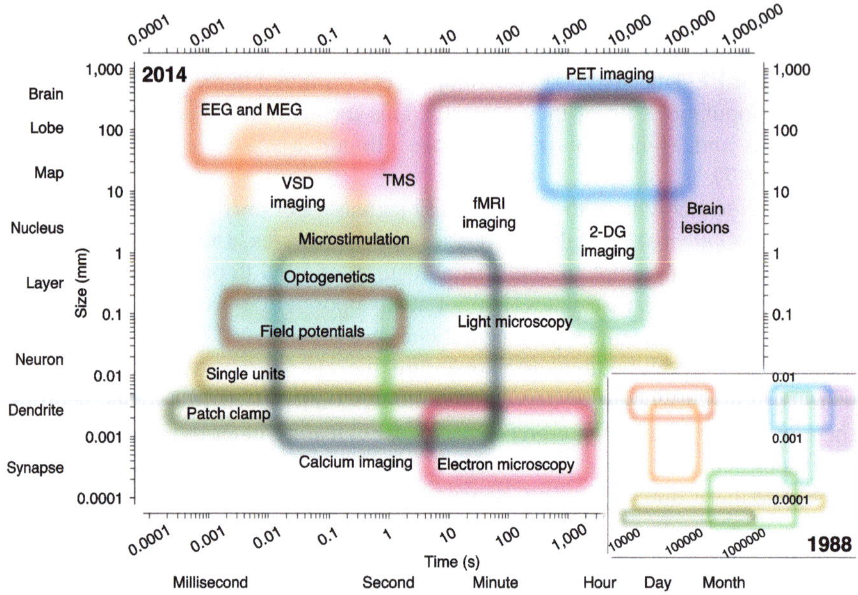

Fig. 1.1 Comparison of neuroimaging methods. (From Sejnowski et al. (2014). Reproduced with permission)

Invasive Techniques Local field potentials (LFPs) and electrocorticography (ECoG) are both invasive techniques that measure electrical signals close to their source. LFPs are an extracellular signal measured by microelectrodes implanted in the brain tissue; they reflect the highly dynamic flow of information across neural networks (Herreras, 2016). ECoG measures electrical activity using electrodes placed over the cortical surface. This provides a spatial resolution that lies somewhere between that of fully invasive (LFP) and non-invasive (EEG) neural signals (Kim et al., 2015). Both LFPs and ECoG are less susceptible to noise than EEG, due to the electrodes being closer to the source of the measured signal. Additionally, both techniques have excellent temporal resolution. However, LFPs and ECoG require surgically implanted electrodes. Therefore, in humans, their use is restricted to patients who require surgery for clinical purposes.

Magnetoencephalography (MEG) EEG and MEG are closely related and are often discussed together, as both techniques record summations of electrical activity elicited by postsynaptic potentials. While EEG measures the electrical potentials, MEG measures the magnetic fields associated with the electrical activity. Both techniques benefit from high temporal resolution, but the spatial resolution of MEG is a little better because the magnetic fields measured by MEG are less affected by tissue than electrical signals. However, MEG is limited to localising tangential sources, and since magnetic fields decay quickly, MEG is also limited to recording from the cortex.

Traditional MEG systems use superconducting quantum interference devices (SQUIDs), (Singh, 2014) which require a magnetically shielded room and a cryogenic cooling system. This makes traditional MEG more expensive than EEG and limits it to laboratory settings. Recently developed optically pumped magnetometers (OPMs) offer several advantages over traditional SQUID MEG, such as increased sensitivity and resolution, lifespan compliance, free movement, and lower cost (Schofield et al., 2024). While still an emerging technology, OPM MEG is a promising proposition for MEG research.

Functional Magnetic Resonance Imaging (fMRI) Perhaps the most common comparison among methods in recent decades is between EEG and fMRI. The neural activity that generates the EEG signal also increases metabolic demand. This triggers a vascular response known as neurovascular coupling, which leads to an increase in blood flow in the active brain region. fMRI detects these changes by using the blood oxygenation level dependent (BOLD) imaging technique (Ogawa et al., 1990), which relies on the different magnetic properties of oxygenated blood and de-oxygenated blood. The blood flow response to the increased metabolic demands, known as the haemodynamic response, is very slow in comparison to EEG. BOLD signal changes occur over several seconds (Logothetis et al., 2001), in contrast to the millisecond timescale of EEG. However, the strength of fMRI is in its sub-millimetre three-dimensional spatial resolution (Huettel et al., 2008).

With their differing temporal and spatial trade-offs, EEG and fMRI show us different aspects of neural activity (or correlates thereof), and which one is 'better' really depends on the research question. The two methods can also be complementary and are often measured simultaneously (Warbrick, 2022). We should move away from the idea that EEG is a cost-effective alternative if you can't do fMRI. They each make a unique contribution to our understanding of brain function, and it's up to us as researchers to use them appropriately.

Functional Near-Infrared Spectroscopy (fNIRS) fNIRS, like fMRI, detects changes in the BOLD signal, but using near-infrared light instead of the magnetic properties of haemoglobin. The near-infrared range (650–950 nm) reaches the cortical surface but will not penetrate it; therefore, measurement is limited to cortical regions (Li et al., 2022).

The spatial resolution of fNIRS is superior to that of EEG but is inferior to fMRI. The temporal resolution of fNIRS (approximately 100 ms) is faster than fMRI but slower than EEG. fNIRS is less susceptible to noise than EEG, and like EEG, it has the advantage of being portable, so it can be used outside of the lab. Therefore, fNIRS is well-suited to studies in naturalistic environments (Pinti et al., 2020).

Positron Emission Topography (PET) PET isn't competing in the same measurement space as EEG, but it can be a complementary method. It is considered the gold standard for metabolic imaging (Shah et al., 2013). It provides information on physiological and pathophysiological processes by measuring uptake of radioactive tracers, for example in pathologies such as brain tumours.

Typically, a tracer bolus is administered, and uptake is measured over, for example 30 min. It can be complementary to EEG, especially when also combined with fMRI in a trimodal imaging approach (Shah et al., 2013). For example, EEG can measure electrophysiological activity, fMRI can map functional connectivity, and PET can quantify energy metabolism, building a comprehensive picture of the default mode network or sleep states (Shah et al., 2017; Chen et al., 2025).

More recently, developments in functional PET (fPET) allow for better temporal resolution (Li et al., 2020). Task-related glucose uptake can be measured; however, this is sampled in the order of minutes rather than seconds, so the timescale is very different from EEG. While EEG and PET differ, being aware of the potential of fPET enhances our understanding of the scope of available imaging tools.

Multimodal EEG There is a trend towards combining EEG with other imaging and stimulation methods. Multimodal approaches exploit the complementary strengths of different techniques. The measurement methods described above use different measurement principles and vary in terms of temporal and spatial resolution. Consequently, combining them can maximise the strengths of each method to yield a more complete understanding of brain function. The technical developments in recent decades enable these multiple modal approaches, with some technologies being developed specifically for multimodal applications, for example amplifiers and EEG caps that can be used inside an MRI scanner.

EEG can also be combined with brain stimulation methods to investigate causal relationships between brain and behaviour. An exhaustive discussion of all multimodal EEG combinations is beyond the scope of the chapter, but Table 1.1 gives an overview of common combinations, the main advantages, and further reading. Some combinations also have dedicated chapters in Parts V and VI of this book.

1.4 The Future of EEG

EEG is a well-established, powerful technique and has a unique place among brain imaging methods. However, the future of EEG depends on how we choose to develop and implement EEG as a research community. While it is useful to consider the future in terms of technological advances and potential new applications, other factors must also be considered, for example availability, biases in study populations, and replicability. How we collectively choose to handle these issues will determine EEG's future.

Two recent studies surveyed or interviewed EEG practitioners on how they think the field will develop in the coming years (Hu & Yan, 2024; Mushtaq et al., 2024). These papers provide a good starting point for considering the future of EEG, and we will briefly consider the technical development, analysis, and application themes that emerged from these papers. We then consider the broader topics of availability, diversity, and replicability.

Table 1.1 Common multimodal EEG applications

Combination	Main advantage(s)	Further reading
EEG–fMRI	Measures the neural and neurovascular responses to the same stimuli Exploits the temporal resolution of EEG and the spatial resolution of fMRI Uses EEG to identify events for fMRI analysis, for example sleep and epilepsy	Warbrick (2022) Chapter 33: Combining EEG and fMRI
EEG–fNIRS	Complementary temporal and spatial resolutions Both are portable Allows validation of neurovascular coupling	Li et al. (2022)
EEG–PET/ EEG–fMRI–PET/	Investigates metabolic changes associated with brain dynamics	Shah et al. (2013)
EEG–TMS	Allows the study of cortical reactivity and connectivity at high spatiotemporal resolution Non-invasive brain stimulation	Hernandez-Pavon (2023) and Tremblay et al. (2019) Chapter 32: EEG and TMS
EEG–tACS	Modulates brain oscillations non-invasively Investigates causal links between brain oscillations and cognitive function	Herrmann et al. (2016) and Cabral-Calderin and Wilke (2020) Chapter 31: Combining EEG and Transcranial Brain Stimulation

The main advantages of each combination are listed, along with references for review papers that can be considered a good starting point for further reading

Technology Affordable hardware, artificial intelligence (AI) advances, virtual reality (VR), and brain–computer interfaces (BCI) were identified as having significant potential for advancing our understanding brain–behaviour relationships. While these technologies already exist, they are considered potential growth areas that can advance EEG research.

Data Processing The highest priorities identified were improving tools for quantitative EEG analysis and standardising protocols for data acquisition and analysis. While these seem like rather obvious things to have on our wish list, the needs remain unmet and finding solutions poses significant challenges. This chapter cannot provide detailed solutions, but highlighting the lack of robust and standardised protocols is important, and we encourage the research community to continue striving for these.

Applications Clinical and diagnostic applications were considered the most likely developments. Short term predictions were diagnosis of sleep disorders and automated processes for some clinical applications. A slightly longer-term prediction was the detection of clinical abnormalities (e.g. neurodegenerative dementia). More speculative aims were decoding dreams and reading the contents of memory. This shows that EEG is dynamic field of research with a potentially exciting future as well as a rich history.

Availability/Accessibility The benefits of EEG are its low cost and potential for widespread use. However, despite research funding, international collaboration, and technical developments, a vast majority of people don't benefit from neuroscience and technology breakthroughs, including EEG (Bringas-Vega et al., 2022). Possible reasons include the concentration of efforts in highly developed countries, the marginalisation of low- and mid-income countries due to inadequate research infrastructure, and insufficient links between research and public health needs. These issues were determined at the World Health Organization (WHO) headquarters in June 2016 in a meeting of representatives of the US BRAIN project, the European Human Brain Project, the Japan Brain/MINDS, and the Chinese, Australian, and Cuban Brain Projects (Bringas-Vega et al., 2022). EEG, as a low-cost, scalable method, would be well-suited to solving these problems, yet the problem persists. Addressing these problems should be a priority for the EEG community.

Diversity Most study participants, especially in psychology, are WEIRD meaning: *Western, Educated, Industrialised, Rich, and Democratic* (Henrich et al., 2010). This introduces a natural sampling bias in many studies conducted in universities and research institutes, where the WEIRD population is overrepresented. Besides a general sampling bias, there is a phenotypical bias in neurotechnology. For example, the exclusion of phenotypes such as skin pigmentation and hair type (Webb et al., 2022). Consequently, marginalised groups are underrepresented in the body of scientific knowledge acquired using EEG. Addressing these issues is the responsibility of researchers, institutional review boards (IRB), funding agencies, and manufacturers, and we must all work together to correct these biases.

Replicability Replicability is the cornerstone of good science and is vital to validating findings that link brain activity and cognitive function. This depends on well-defined and standardised data acquisition and analysis pipelines. However, there is limited evidence for replicability in EEG research, and this is a symptom of a wider problem in science (Baker, 2016). A contributing factor is that replication studies are rarely published because novel findings are prioritised. For example, there are more than 6000 EEG publications per year, few of which are replication studies (Pavlov et al., 2021).

The #EEGmanylabs (Pavlov et al., 2021) initiative aims to address this by replicating highly influential EEG studies across multiple labs worldwide. While findings from this study are not yet published, it's a step towards replicable, transparent EEG research.

Another reason to be optimistic is that some journals, such as Aperture Neuro, explicitly include replication studies, registered reports, and data papers within their scope. This emphasis on publishing material that supports open science is reassuring for the future of neuroscience research. This book encourages the good scientific practice that underpins transparent and reproducible research, from managing your lab to conducting your studies and reporting your research.

Ethics We could probably write a whole book on research ethics in EEG. Beyond the standard ethical requirement of informed consent and data protection, emerging technologies present new challenges, for example privacy in open data initiatives, self-diagnosis using consumer grade devices, and ethical implications of neuroenhancement. Each of these challenges will require careful attention as the field continues to evolve.

1.5 Conclusions

EEG is an effective tool for exploring the relationship between brain and behaviour. In the past 100 years, it has advanced our understanding of brain function, and EEG's future is promising as a standalone measure or in combination with other neuroimaging methods and emerging technologies. However, the future of our discipline ultimately depends on the choices we make. Mushtaq et al. (2024) conclude their paper with a call to action to 'commit to robust, ethical, inclusive, and sustainable practices'. We encourage anyone using this book as a starting point for EEG research to keep this in mind, and we aim to provide you with a strong foundation for working towards this goal.

References

Baker, M. (2016). 1,500 scientists lift the lid on reproducibility. *Nature, 533*, 452–454.

Bringas-Vega, M. L., Michel, C. M., Saxena, S., White, T., & Valdes-Sosa, P. A. (2022). Neuroimaging and global health. *NeuroImage, 260*, 119458.

Cabral-Calderin, Y., & Wilke, M. (2020). Probing the link between perception and oscillations: Lessons from transcranial alternating current stimulation. *The Neuroscientist, 26*, 57–73.

Chen, J. E., Lewis, L. D., Coursey, S. E., Catana, C., Polimeni, J. R., Fan, J., Droppa, K. S., Patel, R., Wey, H. Y., Chang, C., Manoach, D. S., Price, J. C., Sander, C. Y., & Rosen, B. R. (2025). Simultaneous EEG-PET-MRI identifies temporally coupled, spatially structured hemodynamic and metabolic dynamics across wakefulness and NREM sleep. *bioRxiv*.

Cohen, M. X. (2017). Where does EEG come from and what does it mean? *Trends in Neurosciences, 40*, 208–218.

Hernandez-Pavon, J. C., Veniero, D., Bergmann, T. O., Belardinelli, P., Bortoletto, M., Casarotto, S., Casula, E. P., Farzan, F., Fecchio, M., Julkunen, P., Kallioniemi, E., Lioumis, P., Metsomaa, J., Miniussi, C., Mutanen, T. P., Rocchi, L., Rogasch, N. C., Shafi, M. M., Siebner, H. R., Thut, G., Zrenner, C., Ziemann, U., & Ilmoniemi, R. J. (2023). TMS combined with EEG: Recommendations and open issues for data collection and analysis. *Brain Stimulation, 16*, 567–593.

Herrmann, C. S., Struber, D., Helfrich, R. F., & Engel, A. K. (2016). EEG oscillations: From correlation to causality. *International Journal of Psychophysiology, 103*, 12–21.

Herreras, O. (2016). Local field potentials: Myths and misunderstandings. *Frontiers in Neural Circuits, 10*, 101.

Henrich, J., Heine, S. J., & Norenzayan, A. (2010). The weirdest people in the world? *The Behavioral and Brain Sciences, 33*, 61–83; discussion 83-135.

Huettel, S. A., Song, A. W., & Mccarthy, G. (2008). *Functional magnetic resonance imaging*. Sinauer Assicaites Inc.

Hu, X., & Yan, C.-G. (2024). Current status and future challenges in brain imaging: Insights from leading experts. *Medicine Plus, 1*, 100049.

Kim, Y. J., Park, S. W., Yeom, H. G., Bang, M. S., Kim, J. S., Chung, C. K., & Kim, S. (2015). A study on a robot arm driven by three-dimensional trajectories predicted from non-invasive neural signals. *Biomedical Engineering Online, 14*, 81.

Li, S., Jamadar, S. D., Ward, P. G. D., Premaratne, M., Egan, G. F., & Chen, Z. (2020). Analysis of continuous infusion functional PET (fPET) in the human brain. *NeuroImage, 213*, 116720.

Li, R., Yang, D., Fang, F., Hong, K. S., Reiss, A. L., & Zhang, Y. (2022). Concurrent fNIRS and EEG for brain function investigation: A systematic, methodology-focused review. *Sensors (Basel), 22*(15), 5865.

Logothetis, N. K., Pauls, J., Augath, M., Trinath, T., & Oeltermann, A. (2001). Neurophysiological investigation of the basis of the fMRI signal. *Nature, 412*, 150–157.

Mushtaq, F., Welke, D., Gallagher, A., Pavlov, Y. G., Kouara, L., Bosch-Bayard, J., Van Den Bosch, J. J. F., Arvaneh, M., Bland, A. R., Chaumon, M., Borck, C., He, X., Luck, S. J., Machizawa, M. G., Pernet, C., Puce, A., Segalowitz, S. J., Rogers, C., Awais, M., Babiloni, C., Bailey, N. W., Baillet, S., Bendall, R. C. A., Brady, D., Bringas-Vega, M. L., Busch, N. A., Calzada-Reyes, A., Chatard, A., Clayson, P. E., Cohen, M. X., Cole, J., Constant, M., Corneyllie, A., Coyle, D., Cruse, D., Delis, I., Delorme, A., Fair, D., Falk, T. H., Gamer, M., Ganis, G., Gloy, K., Gregory, S., Hassall, C. D., Hiley, K. E., Ivry, R. B., Jerbi, K., Jenkins, M., Kaiser, J., Keil, A., Knight, R. T., Kochen, S., Kotchoubey, B., Krigolson, O. E., Langer, N., Liesefeld, H. R., Lippé, S., London, R. E., Macnamara, A., Makeig, S., Marinovic, W., Martínez-Montes, E., Marzuki, A. A., Mathew, R. K., Michel, C., Millán, J. D. R., Mon-Williams, M., Morales-Chacón, L., Naar, R., Nilsonne, G., Niso, G., Nyhus, E., Oostenveld, R., Paul, K., Paulus, W., Pfabigan, D. M., Pourtois, G., Rampp, S., Rausch, M., Robbins, K., Rossini, P. M., Ruzzoli, M., Schmidt, B., Senderecka, M., Srinivasan, N., Stegmann, Y., Thompson, P. M., Valdes-Sosa, M., Van Der Molen, M. J. W., Veniero, D., Verona, E., Voytek, B., Yao, D., Evans, A. C., & Valdes-Sosa, P. (2024). One hundred years of EEG for brain and behaviour research. *Nature Human Behaviour, 8*, 1437–1443.

Niedermeyer, E. (2005). Historical aspects of EEG. In E. Niedermeyer & F. H. Lopes Da Silva (Eds.), *Electroencephalography: Basic principles, clinical applications, and related fields* (5th ed.). Lippincott Williams & Wilkins.

Ogawa, S., Lee, T. M., Nayak, A. S., & Glynn, P. (1990). Oxygenation-sensitive contrast in magnetic resonance image of rodent brain at high magnetic fields. *Magnetic Resonance in Medicine, 14*, 68–78.

Pavlov, Y. G., Adamian, N., Appelhoff, S., Arvaneh, M., Benwell, C. S. Y., Beste, C., Bland, A. R., Bradford, D. E., Bublatzky, F., Busch, N. A., Clayson, P. E., Cruse, D., Czeszumski, A., Dreber, A., Dumas, G., Ehinger, B., Ganis, G., He, X., Hinojosa, J. A., Huber-Huber, C., Inzlicht, M., Jack, B. N., Johannesson, M., Jones, R., Kalenkovich, E., Kaltwasser, L., Karimi-Rouzbahani, H., Keil, A., Konig, P., Kouara, L., Kulke, L., Ladouceur, C. D., Langer, N., Liesefeld, H. R., Luque, D., Macnamara, A., Mudrik, L., Muthuraman, M., Neal, L. B., Nilsonne, G., Niso, G., Ocklenburg, S., Oostenveld, R., Pernet, C. R., Pourtois, G., Ruzzoli, M., Sass, S. M., Schaefer, A., Senderecka, M., Snyder, J. S., Tamnes, C. K., Tognoli, E., Van Vugt, M. K., Verona, E., Vloeberghs, R., Welke, D., Wessel, J. R., Zakharov, I., & Mushtaq, F. (2021). #EEGManyLabs: Investigating the replicability of influential EEG experiments. *Cortex, 144*, 213–229.

Pinti, P., Tachtsidis, I., Hamilton, A., Hirsch, J., Aichelburg, C., Gilbert, S., & Burgess, P. W. (2020). The present and future use of functional near-infrared spectroscopy (fNIRS) for cognitive neuroscience. *Annals of the New York Academy of Sciences, 1464*, 5–29.

Schofield, H., Hill, R. M., Feys, O., Holmes, N., Osborne, J., Doyle, C., Bobela, D., Corvilain, P., Wens, V., Rier, L., Bowtell, R., Ferez, M., Mullinger, K. J., Coleman, S., Rhodes, N., Rea, M., Tanner, Z., Boto, E., De Tiege, X., Shah, V., & Brookes, M. J. (2024). A novel, robust, and portable platform for magnetoencephalography using optically-pumped magnetometers. *Imaging Neuroscience (Cambridge, Mass.), 2*, 1–22.

Sejnowski, T. J., Churchland, P. S., & Movshon, J. A. (2014). Putting big data to good use in neuroscience. *Nature Neuroscience, 17*, 1440–1441.

Shah, N. J., Oros-Peusquens, A. M., Arrubla, J., Zhang, K., Warbrick, T., Mauler, J., Vahedipour, K., Romanzetti, S., Felder, J., Celik, A., Rota-Kops, E., Iida, H., Langen, K. J., Herzog, H., & Neuner, I. (2013). Advances in multimodal neuroimaging: Hybrid MR-PET and MR-PET-EEG at 3 T and 9.4 T. *Journal of Magnetic Resonance, 229*, 101–115.

Shah, N. J., Arrubla, J., Rajkumar, R., Farrher, E., Mauler, J., Kops, E. R., Tellmann, L., Scheins, J., Boers, F., Dammers, J., Sripad, P., Lerche, C., Langen, K. J., Herzog, H., & Neuner, I. (2017). Multimodal fingerprints of resting state networks as assessed by simultaneous trimodal MR-PET-EEG imaging. *Scientific Reports, 7*, 6452.

Singh, S. P. (2014). Magnetoencephalography: Basic principles. *Annals of Indian Academy of Neurology, 17*, S107–S112.

Tremblay, S., Rogasch, N. C., Premoli, I., Blumberger, D. M., Casarotto, S., Chen, R., Di Lazzaro, V., Farzan, F., Ferrarelli, F., Fitzgerald, P. B., Hui, J., Ilmoniemi, R. J., Kimiskidis, V. K., Kugiumtzis, D., Lioumis, P., Pascual-Leone, A., Pellicciari, M. C., Rajji, T., Thut, G., Zomorrodi, R., Ziemann, U., & Daskalakis, Z. J. (2019). Clinical utility and prospective of TMS-EEG. *Clinical Neurophysiology, 130*, 802–844.

Warbrick, T. (2022). Simultaneous EEG-fMRI: What have we learned and what does the future hold? *Sensors (Basel), 22*(6), 2262.

Webb, E. K., Etter, J. A., & Kwasa, J. A. (2022). Addressing racial and phenotypic bias in human neuroscience methods. *Nature Neuroscience, 25*, 410–414.

Chapter 2
What Is EEG?

Fernando Cross Villasana

Abstract Since the initial observations by Hans Berger in the 1920s, electroencephalography (EEG) has opened a window into studying otherwise unobservable phenomena of the brain. But how do these electrical traces come to be and moreover, what do they mean? In this chapter, we go through the physiological origins of the EEG trace recorded from the scalp, showcase the different signals that are commonly derived from it, and explore how they are used in different contexts. From sleep stages to cognitive processes, epilepsy or psychiatric conditions, the EEG has been a rich source of information on the state of the brain. Understanding these dynamics can enrich the interpretation of EEG observations, fuel new ideas, and facilitate communication with other neuroscientists.

Keywords EEG · ERP · Brain oscillation · EPSP · IPSP

2.1 Physiological Origins of the EEG

In EEG, the voltage recorded by each electrode from the scalp is mainly the result of synchronized activity from hundreds of thousands of neurons from the cortical sheet of the brain. For the most part, this activity is related to postsynaptic potentials (PSPs) from the apical dendrites of pyramidal neurons in the cortex. PSPs produce extracellular charges whose electric fields can travel through tissue and reach the electrodes on the scalp (Beniczky & Schomer, 2020; Jackson & Bolger, 2014).

PSPs are generated after a postsynaptic neuron receives excitatory or inhibitory stimulation from a presynaptic neuron. Correspondingly, PSPs can be excitatory (EPSP) or inhibitory (IPSP). In the case of EPSPs, a neuron receives excitatory neurotransmitters at the synapse, mostly into the distal side of the dendrite (Beniczky & Schomer, 2020), but can also be close to or at the soma (Jackson & Bolger, 2014). This induces postsynaptic ion channels on the membrane to allow the influx of

F. Cross Villasana (✉)
Brain Products GmbH, Gilching, Germany

T. Warbrick (ed.), *The EEG Handbook*,
https://doi.org/10.1007/978-3-032-20450-9_2

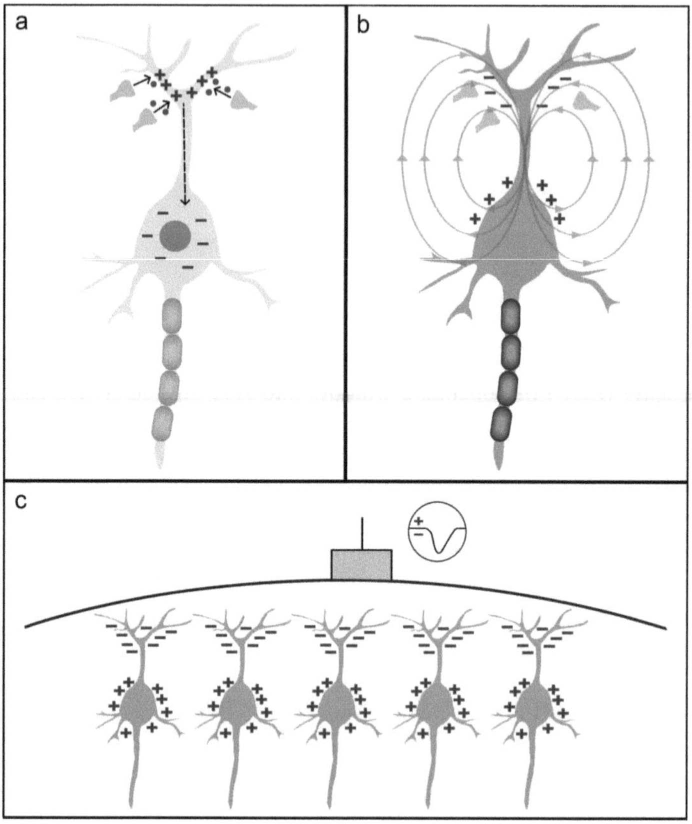

Fig. 2.1 Representation of the generation of EEG signal during EPSP. (**a**) A pyramidal neuron receives excitatory neurotransmitters at synapses on different sites of the apical dendrite's tuft. This induces a positive charge inside the neuron due to the entrance of positive sodium ions. The positive charge is propagated through the apical dendrite towards the negatively charged soma of the neuron. (**b**) Negative charges are produced in the extracellular space close to the synapses. The negatively charged region forms a dipole with the positive areas along the exterior of the neuron. (**c**) The dipoles from multiple neurons aligned in parallel summate to act as a larger dipole that can be detected on the scalp

positively charged sodium ions. Ion influx simultaneously generates a positive charge inside the neuron and a negative charge outside. The intracellular positive charge produces a differential with the negatively charged inside of the neuron, so the charge propagates towards the soma (Fig. 2.1a). Multiple EPSPs can summate at the soma to depolarize the neuron and facilitate that it generates its own action potential (Olah et al., 2025; Stuart & Sakmann, 1995), but this latter part is not normally reflected in the EEG. The extracellular negative charge, together with the relatively more positive areas along the apical dendrite, forms a dipole, that is, a negatively charged area called "sink," separated from a positively charged area called "source," creating an electric field (Fig. 2.1b). A single extracellular dipole is too small to be detected by EEG. To generate a strong enough electric field, the synchronized EPSPs of at least hundreds of thousands of neurons arranged in parallel to each other are necessary (Fig. 2.1c). The summated dipoles act as a single larger dipole whose electric field is detectable by EEG electrodes (Beniczky & Schomer, 2020; Cohen, 2017; Jackson & Bolger, 2014).

In the case of IPSPs, the process is similar to EPSP but with opposite polarities (Beniczky & Schomer, 2020), and with the site of stimulation being mostly at the proximity of the soma (Beniczky & Schomer, 2020). After receiving inhibitory neurotransmitters from a presynaptic neuron, ion channels in the postsynaptic neuron let negative chloride ions inside, creating an extracellular positive charge. The intracellular negative charge travels through the dendrite towards the soma, facilitating the polarization of the neuron and making an action potential less likely (Fricker & Miles, 2000). Meanwhile, the extracellular positive charge (source) forms a dipole with the relatively more negative region outside the neuron (sink), and the summation of multiple dipoles from neurons acting in synchrony can be detectable by EEG. EPSPs and IPSPs are in constant interplay in the neurons to regulate the brain's activity. However, the positive or negative polarity seen in EEG activity does not necessarily correspond to inhibitory or excitatory activity in the brain (Jackson & Bolger, 2014; Luck, 2014). The reason for this is that on top of the kind of PSP, the location of the PSP in the neuron affects the orientation of the extracellular dipole. In this way, an EPSP generated at a distal site of the dendrite produces the same source and sink orientation as an IPSP near the soma. Likewise, an IPSP at the distal dendrite leads to a similar dipole orientation than that from an EPSP at the soma (Beniczky & Schomer, 2020; Jackson & Bolger, 2014).

The orientation of the columns of neurons due to cortical folding also affects the orientation of the dipoles they generate and as a result affects the polarity recorded at the scalp (Beniczky & Schomer, 2020; Jackson & Bolger, 2014; Olejniczak, 2006). Radial dipoles are mainly generated over cortical gyri (Fig. 2.2) in a way that is perpendicular to the head surface (Olejniczak, 2006; Scherg et al., 2019). For radial dipoles, the recorded polarity is that of the pole facing toward the scalp. The floor of the sulci also produces radial dipoles, but they are deeper in the brain and are not normally detectable using scalp EEG. Tangential dipoles are generated within the walls of sulci and are parallel to the surface above (Fig. 2.2). They can be detected by EEG as long as they are not neutralized by another dipole from the opposing wall (Nunez & Srinivasan, 2006). In this case, electrodes on opposite sides of the dipole will detect the corresponding positive or negative pole (Olejniczak, 2006; Scherg et al., 2019). Of note, while EEG is sensitive to the electric fields from radial and tangential dipoles, the related magnetoencephalogram (MEG) technique is only sensitive to the magnetic fields from tangential dipoles (Jackson & Bolger, 2014; Nunez & Srinivasan, 2006).

For the electrical signal to reach the scalp electrodes, it must go through the extracellular medium and traverse through tissue in a process called volume conduction (Beniczky & Schomer, 2020; Jackson & Bolger, 2014). This conduction happens almost instantly as the electric field influences the ions in the medium (Luck, 2014). At the point of the meninges, skull and scalp, they act as insulating

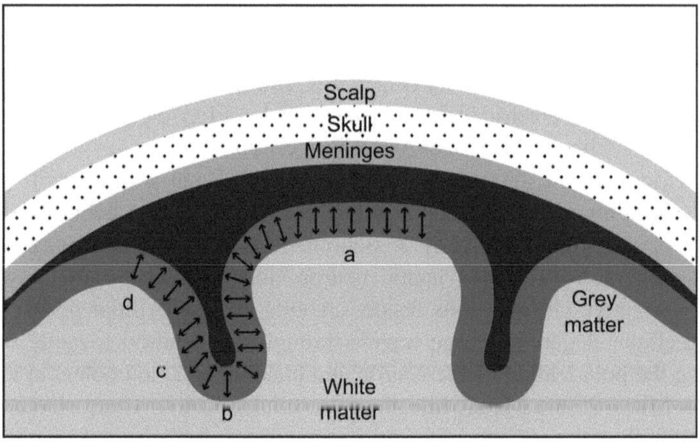

Fig. 2.2 Representation of dipoles produced in the cortex shown as double-headed arrows and their orientation with respect to the surface. (**a**) Radial dipoles produced in a gyrus are able to reach the surface, and either their positive or negative side is recorded in EEG. (**b**) Radial dipoles generated within the floor of a sulcus are deep and not normally detectable with scalp EEG. (**c**) Tangential dipoles produced at the walls of a sulcus that are faced with a dipole from the opposing wall cancel out and cannot be recorded by EEG. (**d**) A tangential dipole with the appropriate orientation that is not opposed by another dipole is able to reach the surface; the two poles would appear on opposite sides of the scalp

layers so that the electric signal further disseminates through capacitive conduction (Jackson & Bolger, 2014). In this way, when a dipole presses charged ions towards the outside of the layer, ions inside the layer will realign themselves with positive and negative charges in the same direction as the dipole. This process repeats itself across each layer of the meninges, skull, and scalp, until reaching the surface. At the surface, electroconductive gel is used to create a bridge between the scalp and the electrode. The gel and the electrode act as further capacitive layers until the charge reaches the cable and travels to the amplifier (Jackson & Bolger, 2014). The processes of volume and capacitive conduction smear and attenuate the signal (Jackson & Bolger, 2014; Nielsen et al., 2023) and, anatomical differences between individuals (e.g. skull and scalp thickness, cerebrospinal fluid) have differential effects on the recording (e.g. Klimesch, 1999; Nielsen et al., 2023; Wendel et al., 2010).

While PSPs have the largest influence on EEG, other types of brain activity are known to contribute to the EEG as well (Cohen, 2017; Olejniczak, 2006), but these are restricted to certain contexts and are less studied, for example, action potentials. Although they are not usually visible in EEG (Beniczky & Schomer, 2020), action potentials are thought to contribute to the waveforms in auditorily-evoked responses (Chertoff et al., 2010; Møller et al., 1995), and in response to somatosensory stimuli (Luck, 2014). Signals from deep sources of the brain are rare in EEG, and are studied little. Technically, to be detectable, the deeper source should produce a large enough dipole that can reach the surface, but it would still appear as a weak signal.

Investigating the possibility of recording these signals and their interpretation is still ongoing work. Special methodologies are necessary to disentangle the signal of deep sources from other brain activity and noise. In these studies, EEG is frequently accompanied by invasive intracranial sensors that were implanted in patients for clinical purposes (e.g., Fahimi Hnazaee et al., 2020).

2.2 Signals of the EEG

Understanding the origins of the EEG in the brain can explain a great deal about how neuronal activity can reach electrodes on the scalp. However, once the brain activity enters the EEG record, it shows a variety of patterns and dynamics that this model can only partially account for (Cohen, 2017). Such patterns can be considered as emerging signals in the EEG that reflect the different states and processes of the brain. These signals can be detected through different means, ranging from direct observation to advanced computational methods. But how do we understand their meaning? Over the decades, many EEG signals have been studied in relation to behavioral and neurophysiological measurements in humans and animals to clarify their significance. This work has greatly advanced the understanding of brain dynamics and generated diverse practical applications. But the knowledge is still far from complete, and research on the various EEG signals evolves continuously. With that in mind, when working with a particular signal from the EEG, it is helpful to have an overall understanding of its meaning, consider the context of the observation, and stay updated with research on that signal. The following section presents an overview of the most commonly studied EEG signals.

2.2.1 Direct Observation of EEG: Identifying States of the Brain

Since the early days of electroencephalography, direct observation of brain waves has shown that different behavioral states correspond to different wave patterns, suggesting different processes in the brain. The classic observation where eye closure led to the appearance of large oscillations termed "alpha," with 8–12 cycles per second in the visual brain areas (Britton et al., 2016b), suggests that these waves reflect a state of diminished activation in the visual cortex. This is in contrast to the pattern of mostly faster waves that is prevalent when the eyes are open (Britton et al., 2016b). More recent research combining EEG with functional magnetic resonance imaging has shown that indeed, increments in the amplitude of alpha waves correlate with a decrease in cortical neural activity (Murta et al., 2015). This notion is further reinforced when EEG is paired with transcranial magnetic stimulation, and the cortex becomes less reactive to the magnetic stimuli when alpha amplitude is higher (Taylor & Thut, 2012). However, it is important to notice that the relationship between alpha and subcortical regions or brain networks is more complex (Murta et al., 2015).

After characterizing normal brain activity, it did not take long to find abnormal discharges in patients with epilepsy around seizure episodes. This is called ictal activity and can show various patterns of spikes and waves with high amplitude (Britton et al., 2016a; Bromfield et al., 2006). Furthermore, constant monitoring of the patients' EEG reveals that they also show abnormal spikes and waves outside of seizure episodes, or interictal abnormalities, which provide valuable clinical information (Britton et al., 2016a). These types of abnormal EEG suggest episodes of over-synchronization and overspreading of activity in the neurons, which result from disruptions in the neuronal excitation and inhibition mechanisms (Bromfield et al., 2006). The scalp location of the ictal and interictal discharges can be used as a good approximation of the foci of epileptic activity in the brain (Beniczky & Schomer, 2020; Scherg et al., 2019). In the same way, the relation of EEG observations to concomitant behavior provides useful information for the diagnosis and treatment of patients with epilepsy (Britton et al., 2016a).

The discovery of sleep stages using EEG showed how sleep is not uniform and that the brain engages in different processes through the night (Aserinsky & Kleitman, 2003). This appears in the EEG as two initial stages of gradual EEG slowing, a slow-wave sleep stage with high-amplitude EEG, and a rapid-eye-movement (REM) phase where the EEG resembles the waking state (Britton et al., 2016b). These stages allow the identification of sleep patterns in healthy and clinical populations across the lifespan (Luca et al., 2015). Further research has linked sleep stages to learning and memory consolidation processes (Diekelmann & Born, 2010). Overall, the direct observation of the EEG is informative about the brain's ongoing state and its alterations. It is still a valuable tool used today by trained professionals in clinical settings for monitoring and diagnostic processes, and in research. Accumulated knowledge and the assistance of computational analyses of ongoing EEG have further refined these techniques (Aboalayon et al., 2016).

2.2.2 *Event-Related Potentials and the Brain's Reactions to Stimuli*

After observing that EEG fluctuations reflect persistent states of the brain, the next natural step was to investigate how particular stimuli affected the EEG trace under different circumstances. This led to the eventual observation of event-related potentials (ERPs) as responses in the EEG that surrounded the presentation of sensory stimuli (Woodman, 2010). ERPs are waveforms composed of a succession of positive and negative deflections that follow a particular sequence. An example ERP waveform is presented in Fig. 2.3. Sometimes, ERPs can be seen directly on the EEG, but they are more often imperceptible through direct observation and require preprocessing of the signal. Each deflection within the ERP waveform is known as a

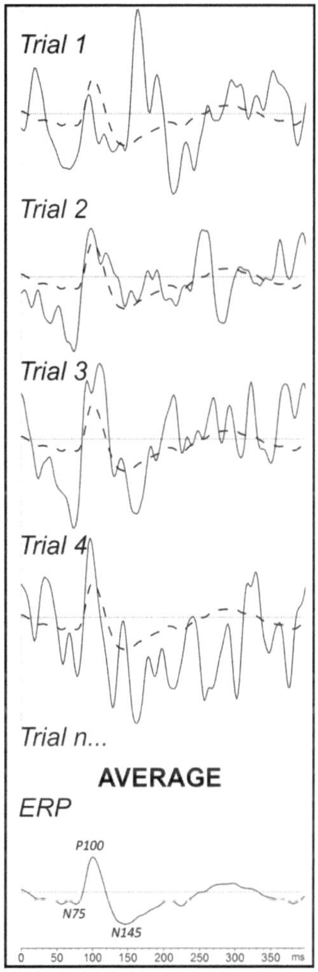

Fig. 2.3 Sample EEG of single trials during visual stimulation with a checkerboard pattern reversal at 2 Hz frequency. The corresponding ERP from the average of all trials is shown at the bottom, known in this case as the visually evoked potential (VEP), with its components N75, P100, and N145. The VEP is also shown overlaid on the single trials to highlight how each trial contains the ERP signal mixed with noise. The averaging process reduces the randomly fluctuating noise in the trials, while retaining the more consistent activity related to the brain's response to form the ERP. Modulations in the amplitude and latency of the ERP components can be used for research or clinical purposes

component. The ERP components seemingly reflect putative stages of cognitive processing by the brain, elicited by either external stimuli or internal events such as object recognition or decision-making (Luck, 2014). Changes in the latency or amplitude of these components can be used for research (e.g., induced by experimental conditions) or clinical purposes (e.g., induced by clinical conditions).

Interestingly, the first formally classified ERP component was the contingent negative variation (CNV; Luck, 2014), which happens before stimulus presentation. More precisely, the CNV shows as a slow-going negative drift that appears for a few seconds when a participant is expecting a stimulus to happen following a warning signal (Walter et al., 1964). A similar component is the "Readiness Potential," where the preparation toward the execution of a hand movement is preceded for a few seconds by a slow negative drift in the EEG, more pronounced on the side of the scalp opposite to the moving hand (Kornhuber & Deecke, 2016). Both these components show that the brain can modulate its state in anticipation of expected events or planned actions.

The ERP components that follow stimulus presentation are much faster and smaller than the CNV or Readiness Potential, corresponding to the fast pace of perception and cognition. As ERPs come wrapped within ongoing EEG activity and noise, these smaller components are much harder to observe and, for a time, were less studied. Nonetheless, with the development of techniques to average across trials, it became possible to isolate poststimulus ERPs more clearly. The idea is exemplified in Fig. 2.3: Within each trial, noise fluctuates randomly, while the event-related activity has a largely consistent pattern; therefore, by averaging across trials, the noise cancels out while the event-related activity persists in a way that the ERP can be isolated. The advent of digital computers further facilitated trial averaging and analysis of the averaged signal, leading to great leaps in ERP research and the discovery of numerous components related to sensory perception, stimulus processing, and cognitive processes. With these tools available, ERPs became the ideal partner for mental chronometry methods and became a workhorse in the field of cognitive psychology (Linden, 2007; Meyer et al., 1988). The mutual feedback between these two disciplines has, on one hand, provided a window into mental processes and, on the other, increased the understanding of what the ERP components represent in the brain and cognition.

The list of known ERP components is too long to mention here. More information can be found in Chap. 19 ("Event-Related Potentials"). Furthermore, various summaries of the components are also available in the literature (e.g., Sur & Sinha, 2009). To mention an illustrative example for now, the mismatch negativity (MMN) appears in relation to an auditory tone that differs from a stream of preceding tones (Luck, 2014), which reflects enhanced processing requirements for novel stimuli in the brain. Most strikingly, the MMN can be observed in unconscious patients in coma (Morlet & Fischer, 2014). This not only shows that the initial detection of novelty precedes conscious perception, but also turns the MMN into a prognostic tool in coma cases (Pruvost-Robieux et al., 2022). Technological advances have also allowed the use of ERPs in clinical settings, for example, to assess the integrity of auditory pathways using the sequence of components in the Auditory Brainstem Response (ABR; Stapells & Oates, 1997), or the integrity of nerve conduction based on the components in Visually Evoked Potentials (VEP; Leocani et al., 2018).

2.2.3 Frequency Analyses and the Brain's Rhythms

Among the ERP components, there is a special case that connects back to the ongoing brain waves first observed in EEG. This is the steady state visually evoked potential (SSVEP). In the SSVEP, the presentation of a visual stimulus with flickering frequencies elicits an oscillation over visual areas with the same frequency as that of the flicker (Norcia et al., 2015). This ongoing waveform is similar to the alpha oscillatory activity first observed by Berger. However, the SSVEP represents an adaptation of the brain to the flicker, rather than a naturally occurring rhythm. The SSVEP can be subjected to frequency analysis using the Fast Fourier Transform (FFT) or similar methods to quantify frequency modulations produced by the flicker on the EEG. Based on this frequency information, the SSVEP can be used for researching perception, attention, or brain dynamics (Norcia et al., 2015) or for developing practical applications such as brain–computer interfaces (Mu et al., 2024; Norcia et al., 2015).

The use of the FFT and other frequency decomposition methods has also allowed a deeper study of the brain's natural oscillations, such as the aforementioned alpha rhythm. The FFT is explained in greater detail in Chap. 18 ("Introduction of EEG Oscillations and Spectral Analysis"). For now, it suffices to mention that the FFT is able to decompose the EEG signal into various constituent frequencies, allowing the quantification of oscillations such as alpha waves, as well as other oscillations that are not necessarily visible to the naked eye. More advanced time-frequency analysis techniques are further capable of tracking the evolution of oscillations over time, either over long periods or around discrete events such as stimuli presentation or motor actions. This has vastly expanded the context in which oscillatory activity is observed and analyzed. Over time, canonical EEG frequency bands were defined (Ahmed & Cash, 2013; Arjoonsingh et al., 2024): alpha (8–12 Hz), beta (~12–30 Hz), gamma (>30 Hz), delta (1–4 Hz), and theta (4–8 Hz). Examples of these band oscillations can be seen in Fig. 2.4. However, the definition of the frequency bands may vary slightly between authors. Furthermore, bands can vary between participants (e.g. Klimesch, 1999) so that, for example, one person's alpha range may be 7.5–12 Hz, while another shows a range of 8–13 Hz. On a similar note, the frequency of oscillations varies through the life span, so that, for example, the resting EEG of young children shows oscillations in the range of adult theta, which gradually mature with age to match the frequency of adult alpha (Klimesch, 1999). Additionally, oscillations in the same band can appear in very different contexts. As an example, oscillations in the 4–8 Hz range are enhanced during concentration, but are also prominent during drowsiness, yet both cases are recognized as "theta" despite possibly reflecting different brain processes (Snipes et al., 2022). For such reasons, some authors give less priority to the canonical band classifications, preferring to report the precise frequencies that they observed in their studies (e.g. Steriade, 2000), or name the rhythms according to the context of observation (e.g., Ibarra-Lecue et al., 2022; Sterman, 1996).

Much knowledge has been gained about brain oscillations; however, despite these observations, there is no definitive account of the kind of building blocks they represent in the whole process of neurocognition, and research is still ongoing (e.g. Beste et al., 2023). As a broad summary, the waking alpha band is related to a state of quiet wakefulness, and alpha levels decrease with task engagement. As vigilance fades or tiredness sets in, alpha gradually slows in frequency and enhances its amplitude (Klimesch, 1999). In contrast, transient enhancements in alpha seem related to active inhibition over specific brain areas that are not relevant for particular tasks (Jensen & Mazaheri, 2010; Klimesch et al., 2007). Overall, as alpha amplitude increases the brain becomes less reactive to stimuli (Taylor & Thut, 2012).

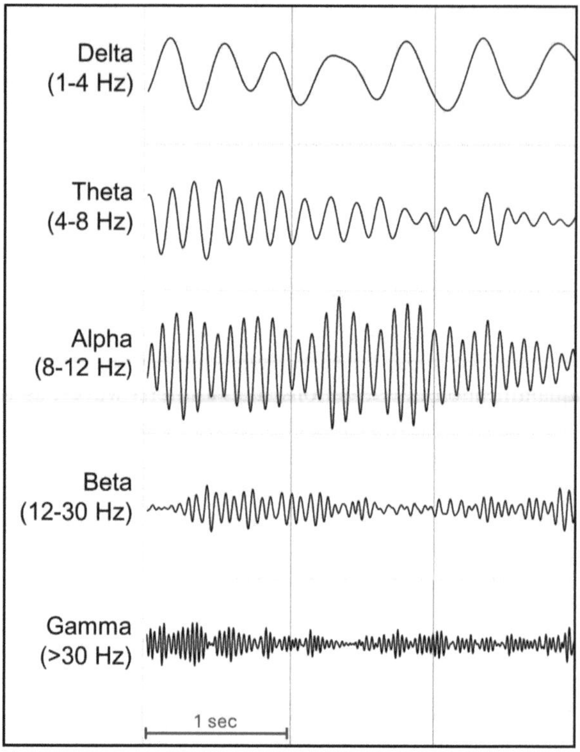

Fig. 2.4 Filtered EEG signals in the range of the different canonical bands organized in descending order. The exact band range can differ between authors

Oscillations in the alpha range may receive different names according to their location in the brain. So that names like "mu," "Rolandic," "wicket" (Niedermeyer, 1997), or "sensorimotor rhythm" (SMR; Sterman, 1996) are used for sensorimotor areas, while "tau" or "third rhythm" are used for mid-temporal auditory areas (Niedermeyer, 1997). The Beta band has been mainly observed in contexts of motor control (Pfurtscheller & Lopes da Silva, 1999), but its exact function is still unclear. Beta bursts have been proposed as a functional inhibition mechanism relevant during cognition, motor planning, and motor control (Lundqvist et al., 2024; Zich et al., 2025). Beta is also observed in working memory retrieval, and is proposed as a mechanism related to the replay of task-relevant memory contents (Ibarra-Lecue et al., 2022; Spitzer & Haegens, 2017). The gamma band is mostly observed over areas involved in active processing over the cortex, such as visual processing or memory formation (Ibarra-Lecue et al., 2022). The theta band is also related to cognitive activity, especially over central regions when conscious cognitive control is deployed (Cavanagh & Frank, 2014). Similarly, central theta is used in the assessment of cognitive workload (Chikhi et al., 2022). It has been proposed that theta facilitates the communication between different brain regions during cognitive performance (Cavanagh & Frank, 2014). Generalized Theta is enhanced during fatigue and sleep pressure (Snipes et al., 2022; Tran et al., 2020). During sleep, theta is thought to play a role in memory consolidation (Diekelmann & Born, 2010).

The Delta band is classically related to deep sleep, thought to play a role in brain restoration and plasticity (Assenza & Di Lazzaro, 2015; Hao et al., 2023). Delta's role during wakefulness is still being investigated. It has been proposed that it is involved in motivational and self-regulatory processes (Knyazev, 2012). Patients with brain injuries show persistently elevated delta levels, especially during the early stages following injury, which has shown prognostic value for clinicians (Franke et al., 2023; Sutcliffe et al., 2022). Finally, in recent years, an aperiodic component has gained increased attention as a modulator of brain activity alongside oscillations (Donoghue et al., 2020) whose exact role is still being elucidated (Brake et al., 2024).

Despite the accumulated observations and the available models of the various EEG oscillatory bands so far, it is important to stay updated with the latest research at the psychological and physiological levels. Research on brain oscillations evolves continuously in the quest for understanding the mechanisms that generate them and their functions in neurocognitive processes.

2.3 Concluding Summary

The electrical traces of brain activity recorded in EEG are a rich source of information on the brain's processes. The EEG has made important contributions to our understanding of the brain, about cognitive processes, and brings valuable information to clinical settings. Here we presented an overview of the physiological origins of EEG, the diverse signals that can be derived from it, and the insights they offer about the brain. Having an informed understanding of the different EEG signals can enrich their interpretation and facilitate the generation of new hypotheses for science or applied fields. Many more applications and analysis possibilities exist, and research on the physiology of the EEG, its physical properties, and use of its signals continues to evolve. Further chapters in this book offer a closer look at the various aspects of the acquisition, analysis, and interpretation of EEG.

References

Aboalayon, K. A. I., Faezipour, M., Almuhammadi, W. S., & Moslehpour, S. (2016). Sleep stage classification using EEG signal analysis: A comprehensive survey and new investigation. *Entropy, 18*(9), 272. https://www.mdpi.com/1099-4300/18/9/272

Ahmed, O. J., & Cash, S. S. (2013). Finding synchrony in the desynchronized EEG: The history and interpretation of gamma rhythms. *Frontiers in Integrative Neuroscience, 7*, 58. https://doi.org/10.3389/fnint.2013.00058

Arjoonsingh, A., Jamal, B. C., & Ganti, L. (2024). History and evolution of the electroencephalogram. *Cureus, 16*(8), e66385. https://doi.org/10.7759/cureus.66385

Aserinsky, E., & Kleitman, N. (2003). Regularly occurring periods of eye motility, and concomitant phenomena, during sleep. 1953. *The Journal of Neuropsychiatry and Clinical Neurosciences, 15*(4), 454–455. https://doi.org/10.1176/jnp.15.4.454

Assenza, G., & Di Lazzaro, V. (2015). A useful electroencephalography (EEG) marker of brain plasticity: Delta waves. *Neural Regeneration Research, 10*(8), 1216–1217. https://doi.org/10.4103/1673-5374.162698

Beniczky, S., & Schomer, D. L. (2020). Electroencephalography: Basic biophysical and technological aspects important for clinical applications. *Epileptic Disorders, 22*(6), 697–715. https://doi.org/10.1684/epd.2020.1217

Beste, C., Munchau, A., & Frings, C. (2023). Towards a systematization of brain oscillatory activity in actions. *Communications Biology, 6*(1), 137. https://doi.org/10.1038/s42003-023-04531-9

Brake, N., Duc, F., Rokos, A., Arseneau, F., Shahiri, S., Khadra, A., & Plourde, G. (2024). A neurophysiological basis for aperiodic EEG and the background spectral trend. *Nature Communications, 15*(1), 1514. https://doi.org/10.1038/s41467-024-45922-8

Britton, J. W., Frey, L. C., Hopp, J. L., Korb, P., Koubeissi, M. Z., Lievens, W. E., Pestana-Knight, E. M., & St. Louis, E. K. (2016a). EEG in the epilepsies. In E. K. St. Louis & L. C. Frey (Eds.), *Electroencephalography (EEG): An introductory text and atlas of normal and abnormal findings in adults, children, and infants*. American Epilepsy Society. Copyright ©2016 by American Epilepsy Society.

Britton, J. W., Frey, L. C., Hopp, J. L., Korb, P., Koubeissi, M. Z., Lievens, W. E., Pestana-Knight, E. M., & St. Louis, E. K. (2016b). The Normal EEG. In E. K. St. Louis & L. C. Frey (Eds.), *Electroencephalography (EEG): An introductory text and atlas of normal and abnormal findings in adults, children, and infants*. American Epilepsy Society. Copyright ©2016 by American Epilepsy Society. https://www.ncbi.nlm.nih.gov/pubmed/27748095

Bromfield, E. B., Cavazos, J. E., & Sirven, J. I. (2006). Basic mechanisms underlying seizures and epilepsy. In E. B. Bromfield, J. E. Cavazos, & J. I. Sirven (Eds.), *An introduction to epilepsy*. American Epilepsy Society. Copyright © 2006, American Epilepsy Society. https://www.ncbi.nlm.nih.gov/pubmed/20821849

Cavanagh, J. F., & Frank, M. J. (2014). Frontal theta as a mechanism for cognitive control. *Trends in Cognitive Sciences, 18*(8), 414–421. https://doi.org/10.1016/j.tics.2014.04.012

Chertoff, M., Lichtenhan, J., & Willis, M. (2010). Click- and chirp-evoked human compound action potentials. *The Journal of the Acoustical Society of America, 127*(5), 2992–2996. https://doi.org/10.1121/1.3372756

Chikhi, S., Matton, N., & Blanchet, S. (2022). EEG power spectral measures of cognitive workload: A meta-analysis. *Psychophysiology, 59*(6), e14009. https://doi.org/10.1111/psyp.14009

Cohen, M. X. (2017). Where does EEG come from and what does it mean? *Trends in Neurosciences, 40*(4), 208–218. https://doi.org/10.1016/j.tins.2017.02.004

Diekelmann, S., & Born, J. (2010). The memory function of sleep. *Nature Reviews. Neuroscience, 11*(2), 114–126. https://doi.org/10.1038/nrn2762

Donoghue, T., Haller, M., Peterson, E. J., Varma, P., Sebastian, P., Gao, R., Noto, T., Lara, A. H., Wallis, J. D., Knight, R. T., Shestyuk, A., & Voytek, B. (2020). Parameterizing neural power spectra into periodic and aperiodic components. *Nature Neuroscience, 23*(12), 1655–1665. https://doi.org/10.1038/s41593-020-00744-x

Fahimi Hnazaee, M., Wittevrongel, B., Khachatryan, E., Libert, A., Carrette, E., Dauwe, I., Meurs, A., Boon, P., Van Roost, D., & Van Hulle, M. M. (2020). Localization of deep brain activity with scalp and subdural EEG. *NeuroImage, 223*, 117344. https://doi.org/10.1016/j.neuroimage.2020.117344

Franke, L. M., Perera, R. A., & Sponheim, S. R. (2023). Long-term resting EEG correlates of repetitive mild traumatic brain injury and loss of consciousness: Alterations in alpha-beta power. *Frontiers in Neurology, 14*, 1241481. https://doi.org/10.3389/fneur.2023.1241481

Fricker, D., & Miles, R. (2000). EPSP amplification and the precision of spike timing in hippocampal neurons. *Neuron, 28*(2), 559–569. https://doi.org/10.1016/s0896-6273(00)00133-1

Hao, C., Li, M., Ning, Q., & Ma, N. (2023). One night of 10-h sleep restores vigilance after total sleep deprivation: The role of delta and theta power during recovery sleep. *Sleep and Biological Rhythms, 21*(2), 165–173. https://doi.org/10.1007/s41105-022-00428-y

Ibarra-Lecue, I., Haegens, S., & Harris, A. Z. (2022). Breaking down a rhythm: Dissecting the mechanisms underlying task-related neural oscillations. *Frontiers in Neural Circuits, 16*, 846905. https://doi.org/10.3389/fncir.2022.846905

Jackson, A. F., & Bolger, D. J. (2014). The neurophysiological bases of EEG and EEG measurement: A review for the rest of us. *Psychophysiology, 51*(11), 1061–1071. https://doi.org/10.1111/psyp.12283

Jensen, O., & Mazaheri, A. (2010). Shaping functional architecture by oscillatory alpha activity: Gating by inhibition. *Frontiers in Human Neuroscience, 4*, 186. https://doi.org/10.3389/fnhum.2010.00186

Klimesch, W. (1999). EEG alpha and theta oscillations reflect cognitive and memory performance: A review and analysis. *Brain Research. Brain Research Reviews, 29*(2–3), 169–195. https://doi.org/10.1016/s0165-0173(98)00056-3

Klimesch, W., Sauseng, P., & Hanslmayr, S. (2007). EEG alpha oscillations: The inhibition-timing hypothesis. *Brain Research Reviews, 53*(1), 63–88. https://doi.org/10.1016/j.brainresrev.2006.06.003

Knyazev, G. G. (2012). EEG delta oscillations as a correlate of basic homeostatic and motivational processes. *Neuroscience and Biobehavioral Reviews, 36*(1), 677–695. https://doi.org/10.1016/j.neubiorev.2011.10.002

Kornhuber, H. H., & Deecke, L. (2016). Brain potential changes in voluntary and passive movements in humans: Readiness potential and reafferent potentials. *Pflügers Archiv, 468*(7), 1115–1124. https://doi.org/10.1007/s00424-016-1852-3

Leocani, L., Guerrieri, S., & Comi, G. (2018). Visual evoked potentials as a biomarker in multiple sclerosis and associated optic neuritis. *Journal of Neuro-Ophthalmology, 38*(3), 350–357. https://doi.org/10.1097/WNO.0000000000000704

Linden, D. E. (2007). What, when, where in the brain? Exploring mental chronometry with brain imaging and electrophysiology. *Reviews in the Neurosciences, 18*(2), 159–171. https://doi.org/10.1515/revneuro.2007.18.2.159

Luca, G., Haba Rubio, J., Andries, D., Tobback, N., Vollenweider, P., Waeber, G., Marques Vidal, P., Preisig, M., Heinzer, R., & Tafti, M. (2015). Age and gender variations of sleep in subjects without sleep disorders. *Annals of Medicine, 47*(6), 482–491. https://doi.org/10.3109/07853890.2015.1074271

Luck, S. J. (2014). *An introduction to the event-related potential technique* (2nd ed.). MIT Press. https://books.google.de/books?id=SzavAwAAQBAJ

Lundqvist, M., Miller, E. K., Nordmark, J., Liljefors, J., & Herman, P. (2024). Beta: Bursts of cognition. *Trends in Cognitive Sciences, 28*(7), 662–676. https://doi.org/10.1016/j.tics.2024.03.010

Meyer, D. E., Osman, A. M., Irwin, D. E., & Yantis, S. (1988). Modern mental chronometry. *Biological Psychology, 26*(1–3), 3–67. https://doi.org/10.1016/0301-0511(88)90013-0

Møller, A. R., Jho, H. D., Yokota, M., & Jannetta, P. J. (1995). Contribution from crossed and uncrossed brainstem structures to the brainstem auditory evoked potentials: A study in humans. *Laryngoscope, 105*(6), 596–605. https://doi.org/10.1288/00005537-199506000-00007

Morlet, D., & Fischer, C. (2014). MMN and novelty P3 in coma and other altered states of consciousness: A review. *Brain Topography, 27*(4), 467–479. https://doi.org/10.1007/s10548-013-0335-5

Mu, J., Liu, S., Burkitt, A. N., & Grayden, D. B. (2024). Multi-frequency steady-state visual evoked potential dataset. *Scientific Data, 11*(1), 26. https://doi.org/10.1038/s41597-023-02841-5

Murta, T., Leite, M., Carmichael, D. W., Figueiredo, P., & Lemieux, L. (2015). Electrophysiological correlates of the BOLD signal for EEG-informed fMRI. *Human Brain Mapping, 36*(1), 391–414. https://doi.org/10.1002/hbm.22623

Niedermeyer, E. (1997). Alpha rhythms as physiological and abnormal phenomena. *International Journal of Psychophysiology, 26*(1–3), 31–49. https://doi.org/10.1016/s0167-8760(97)00754-x

Nielsen, J. D., Puonti, O., Xue, R., Thielscher, A., & Madsen, K. H. (2023). Evaluating the influence of anatomical accuracy and electrode positions on EEG forward solutions. *NeuroImage, 277*, 120259. https://doi.org/10.1016/j.neuroimage.2023.120259

Norcia, A. M., Appelbaum, L. G., Ales, J. M., Cottereau, B. R., & Rossion, B. (2015). The steady-state visual evoked potential in vision research: A review. *Journal of Vision, 15*(6), 4. https://doi.org/10.1167/15.6.4

Nunez, P. L., & Srinivasan, R. (2006). *Electric fields of the brain: The neurophysics of EEG*. Oxford University Press. https://books.google.de/books?id=fUv54as56_8C

Olah, G., Lakovics, R., Shapira, S., Leibner, Y., Szucs, A., Csajbok, E. A., Barzo, P., Molnar, G., Segev, I., & Tamas, G. (2025). Accelerated signal propagation speed in human neocortical dendrites. *eLife, 13*, RP93781. https://doi.org/10.7554/eLife.93781

Olejniczak, P. (2006). Neurophysiologic basis of EEG. *Journal of Clinical Neurophysiology, 23*(3), 186–189. https://doi.org/10.1097/01.wnp.0000220079.61973.6c

Pfurtscheller, G., & Lopes da Silva, F. H. (1999). Event-related EEG/MEG synchronization and desynchronization: Basic principles. *Clinical Neurophysiology, 110*(11), 1842–1857. https://doi.org/10.1016/s1388-2457(99)00141-8

Pruvost-Robieux, E., Marchi, A., Martinelli, I., Bouchereau, E., & Gavaret, M. (2022). Evoked and event-related potentials as biomarkers of consciousness state and recovery. *Journal of Clinical Neurophysiology, 39*(1), 22–31. https://doi.org/10.1097/WNP.0000000000000762

Scherg, M., Berg, P., Nakasato, N., & Beniczky, S. (2019). Taking the EEG back into the brain: The power of multiple discrete sources. *Frontiers in Neurology, 10*, 855. https://doi.org/10.3389/fneur.2019.00855

Snipes, S., Krugliakova, E., Meier, E., & Huber, R. (2022). The theta paradox: 4-8 Hz EEG oscillations reflect both sleep pressure and cognitive control. *The Journal of Neuroscience, 42*(45), 8569–8586. https://doi.org/10.1523/JNEUROSCI.1063-22.2022

Spitzer, B., & Haegens, S. (2017). Beyond the status quo: A role for beta oscillations in endogenous content (re)activation. *eneuro, 4*(4), ENEURO.0170-0117.2017. https://doi.org/10.1523/ENEURO.0170-17.2017

Stapells, D. R., & Oates, P. (1997). Estimation of the pure-tone audiogram by the auditory brainstem response: A review. *Audiology & Neuro-Otology, 2*(5), 257–280. https://doi.org/10.1159/000259252

Steriade, M. (2000). Corticothalamic resonance, states of vigilance and mentation. *Neuroscience, 101*(2), 243–276. https://doi.org/10.1016/s0306-4522(00)00353-5

Sterman, M. B. (1996). Physiological origins and functional correlates of EEG rhythmic activities: Implications for self-regulation. *Biofeedback and Self-Regulation, 21*(1), 3–33. https://doi.org/10.1007/BF02214147

Stuart, G., & Sakmann, B. (1995). Amplification of EPSPs by axosomatic sodium channels in neocortical pyramidal neurons. *Neuron, 15*(5), 1065–1076. https://doi.org/10.1016/0896-6273(95)90095-0

Sur, S., & Sinha, V. K. (2009). Event-related potential: An overview. *Industrial Psychiatry Journal, 18*(1), 70–73. https://doi.org/10.4103/0972-6748.57865

Sutcliffe, L., Lumley, H., Shaw, L., Francis, R., & Price, C. I. (2022). Surface electroencephalography (EEG) during the acute phase of stroke to assist with diagnosis and prediction of prognosis: A scoping review. *BMC Emergency Medicine, 22*(1), 29. https://doi.org/10.1186/s12873-022-00585-w

Taylor, P. C., & Thut, G. (2012). Brain activity underlying visual perception and attention as inferred from TMS-EEG: A review. *Brain Stimulation, 5*(2), 124–129. https://doi.org/10.1016/j.brs.2012.03.003

Tran, Y., Craig, A., Craig, R., Chai, R., & Nguyen, H. (2020). The influence of mental fatigue on brain activity: Evidence from a systematic review with meta-analyses. *Psychophysiology, 57*(5), e13554. https://doi.org/10.1111/psyp.13554

Walter, W. G., Cooper, R., Aldridge, V. J., McCallum, W. C., & Winter, A. L. (1964). Contingent negative variation: An electric sign of sensorimotor association and expectancy in the human brain. *Nature, 203*, 380–384. https://doi.org/10.1038/203380a0

Wendel, K., Vaisanen, J., Seemann, G., Hyttinen, J., & Malmivuo, J. (2010). The influence of age and skull conductivity on surface and subdermal bipolar EEG leads. *Computational Intelligence and Neuroscience, 2010*, 397272. https://doi.org/10.1155/2010/397272

Woodman, G. F. (2010). A brief introduction to the use of event-related potentials in studies of perception and attention. *Attention, Perception, & Psychophysics, 72*(8), 2031–2046. https://doi.org/10.3758/APP.72.8.2031

Zich, C., Ward, N. S., Forss, N., Bestmann, S., Quinn, A. J., Karhunen, E., & Laaksonen, K. (2025). Post-stroke changes in brain structure and function can both influence acute upper limb function and subsequent recovery. *NeuroImage: Clinical, 45*, 103754. https://doi.org/10.1016/j.nicl.2025.103754

Chapter 3
Basic Anatomy: Central Nervous System

Shivakumar Viswanathan

Abstract An integrated understanding of brain function requires both neurophysiology and neuroanatomy. Electroencephalography (EEG) is an important measure of neurophysiology. However, linking EEG measurement to neuroanatomy to understand nervous system function is often viewed as challenging and complex. In this chapter, we provide an overview of the benefits of including a neuroanatomical perspective in your EEG research. Additionally, we briefly survey relevant tools and resources to understand brain neuroanatomy using digital brain atlases. These tools can help researchers become familiar with neuroanatomy, the brain's role within the broader nervous system, and stimulate possible applications to their EEG research.

Keywords EEG · Central nervous system · Neuroanatomy · Brain atlas · 10–20 system

3.1 Introduction

Electroencephalography (EEG) is a classical measure related to *neurophysiology*, namely, the functioning and activity of the nervous system. However, linking measured EEG to its physiological relevance requires us to also consider *neuroanatomy*—the structure of the nervous system. Broadly speaking, neuroanatomy is a description of the nervous system's structure ranging from its macro-level structure down to its micro-level cellular organization. This description of what the various structures are (i.e., neuroanatomy) provides the foundation to understand what these structures do (i.e., neurophysiology).

Including neuroanatomical considerations is crucial for EEG studies. However, it can seem unfamiliar and of uncertain value for researchers who are beginning their EEG journey, especially without a background in neuroscience, medicine, or

S. Viswanathan (✉)
Brain Products GmbH, Gilching, Germany
e-mail: shivakumar.viswanathan@brainproducts.com

T. Warbrick (ed.), *The EEG Handbook*,
https://doi.org/10.1007/978-3-032-20450-9_3

biology. Therefore, it can be useful to highlight two distinct perspectives on why neuroanatomy is practically relevant for EEG.

EEG Signal Generation by the Brain The first perspective is centered on EEG as an electrophysiological measure that originates in the brain (from Buzsaki et al., 2012) (see Chap. 1: "EEG in Context: Past, Present, and Future"). Several electrical characteristics of the EEG signal depend on neuroanatomical features. For example, the EEG signal arises due to micro-anatomical features such as the laminar (i.e., layered) structure of cortical gray matter and the neuron distribution and orientation in these layers. This electrical origin determines the EEG signal's high temporal sensitivity but also its small magnitude (in the microvolt range). Furthermore, the spatial resolution of EEG is affected by features such as the convoluted shape of the cerebral gray matter relative to the various layers that separate the brain from the scalp. Therefore, an understanding of the EEG signal is incomplete without a role for anatomical features. The role of these features is an important part of advanced analyses methods such as *source analysis* that use detailed models of brain and head anatomy to estimate the neural sources of measured EEG (see Chap. 20: "EEG Source Analysis").

The Brain as Part of the Nervous System The second perspective comes from the application of EEG to gain information about nervous system function (and dysfunction) in relation to various capabilities (e.g., language, movement, perception, etc.) and physiological states (e.g., sleep, wakefulness, arousal, rest) in different populations (Cohen, 2017). In this application context, the nervous system is often treated as an extended network to detect, process, and use information about the environment and the organism's own state. For instance, the human nervous system is broadly subdivided into the central nervous system (CNS) (comprising the brain and spinal cord) and the peripheral nervous system (PNS) (Fig. 3.1a). Typical capabilities (such as perception and action) require extensive interactions within

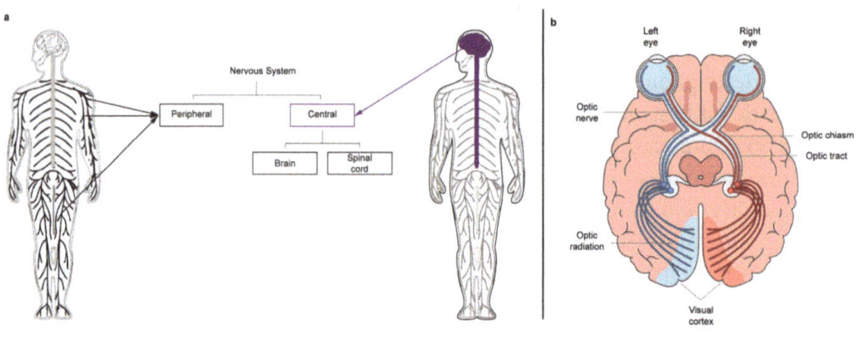

Fig. 3.1 (**a**) Organization of the central and peripheral nervous systems. The CNS consists of the brain and spinal cord. The PNS consists of an elaborate system of nerves that connect the periphery to the CNS (read more in Chap. 4: "Basic Anatomy: Peripheral Nervous System"). (**b**) Schematic of the pathway of the optic nerve (part of the PNS) that carries visual information from the retina to the occipital lobe at the posterior of the brain

and between the CNS and PNS (see example below). Therefore, understanding the relevance of EEG signals requires the broad anatomical context of the nervous system. This anatomical context is also important for multimodal approaches such as the simultaneous recording of EEG with functional magnetic resonance imaging (fMRI) (Warbrick, 2022), and when combining EEG with brain stimulation methods such as transcranial magnetic stimulation (TMS) (Lioumis & Rosanova, 2022).

As an example, consider a common scenario where EEG is recorded while participants are presented with a visual stimulus. A typical assumption is that the early EEG responses to the stimulus would be measurable at channel locations at the back of the head (e.g., Oz, O1, O2, PO1, PO2). This assumption depends on anatomical considerations. Specifically, sensory information received by the eyes is transmitted via the optic nerve (a part of the PNS) to the posterior regions of the brain (a part of the CNS) (Fig. 3.1b). Decades of functional studies have established that the posterior brain is the hub for visual sensory processing, and it is even referred to as the visual cortex. Therefore, taking anatomy into account can be valuable in planning what and where to measure, and how to interpret the results.

Motivated by the latter perspective, in this chapter (and the next), we provide a pragmatic guide on how to include a neuroanatomical perspective in your EEG research. This chapter will focus on the central nervous system (CNS). The next chapter (Chap. 4) will focus on the role of the peripheral nervous system (PNS).

Here, we provide an overview of a few useful concepts related to the CNS anatomy. Major technological advances over the past few decades have made neuroanatomical information conveniently accessible via a variety of digital tools. Familiarity with basic concepts can allow these tools to be used to include neuroanatomical considerations in an EEG study.

3.2 General Organization

The central nervous system consists of several structures. Figure 3.2a shows its hierarchical organization and its major structures. These structures vary considerably in relative size. For example, when we consider the external appearance of the brain, what is prominently visible is the cerebral cortex. A variety of structures (such as the thalamus and basal nuclei) are below the surface and referred to broadly as subcortical structures. Each of these structures contains further structural organization. We recommend that readers consult an elementary textbook on neuroanatomy for more information.

The cerebrum is of measurement-related relevance to EEG. The cerebrum (Fig. 3.2b left panel) consists of two folded sheets of gray matter (the cortex) that are organized into two hemispheres (left and right). The white matter consists of myelinated fibers that connect the neurons in the gray matter both locally and across the brain. These structural connections have an extended organization, as illustrated for the corpus callosum in Fig. 3.2b (right panel).

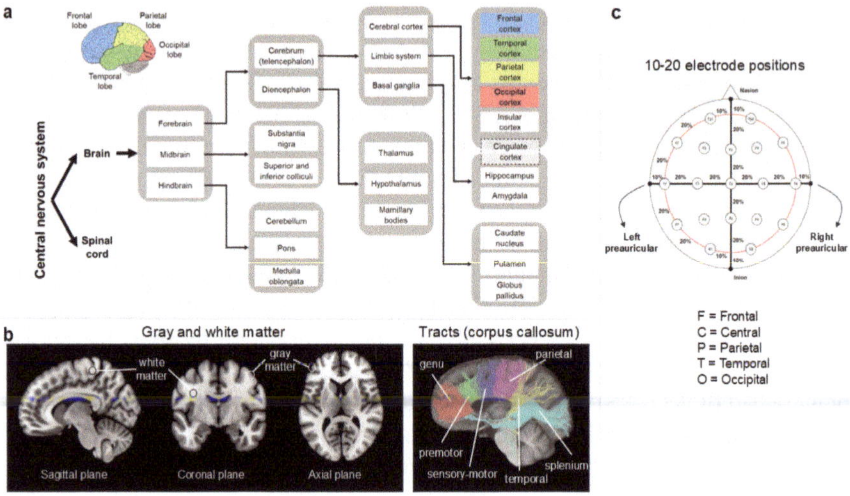

Fig. 3.2 (**a**) Chart showing the hierarchical organization of the human brain. (Adapted based on Ward, 2015). The inset shows the main lobes of the human brain. (**b**) Left: Gray-scale MRI image of an adult human brain along three standard orthogonal planes (sagittal, coronal, axial) of a coordinate system with an origin at the center of the brain. The gray matter and white matter are shown with darker and lighter shades of gray respectively. Right: Fiber tracts of the corpus callosum linking the left and right hemispheres. Tracts are colored by segment. (From Rosenbloom & Pfefferbaum, 2008). (**c**) Electrode position layout on the scalp according to the international 10–20 system

The two hemispheres are approximately mirror-symmetric along the mid-line. Therefore, many structures have a left and right version. The raised surfaces are referred to as gyri (singular: gyrus), and the lowered surfaces are referred to as sulci (singular: sulcus). Based on certain prominent sulci typically observed across individuals, each hemisphere can be further subdivided into prominent lobes (Fig. 3.2a, inset). These lobes are the frontal lobe (behind the eyes), the temporal lobe (side of the head or temples), the occipital lobe (lower back of the head), the parietal lobe (upper back of the head), and the insula (under the frontal and temporal lobes).

Although individual brains share a similar species-typical organization, there can be considerable variability between individuals in the size of the brain and the exact shape and location of the different structures. Therefore, standardized systems and tools have been developed (and continue to be developed) to precisely specify the locations of structures and features across individuals to increase the precision and reproducibility of neuroscientific findings. An important concept to describe brain locations is a *brain atlas* (discussed next).

3.3 Finding Your Way Around: Brain Atlases

3.3.1 International 10–20 System

A useful starting point to understand a brain atlas is to consider the 10–20 system used for EEG (Jasper, 1958). This convention defines standard locations on the scalp for the placement of electrodes to accommodate inter-individual variation and increase reproducibility. The 10–20 system uses distinctive landmarks on the skull (nasion, inion, pre-auricular points) that can be typically identified for every individual (Fig. 3.2c). The distances between these landmarks are then used to define locations on the scalp using relative sizes specified as percentages (i.e., 10%, 20%). The coarse correspondence between the skull and the cerebral lobes is used to name the scalp locations. Several extensions of this basic system have been developed, such as the 10–10 and 10–5 system (Oostenveld & Praamstra, 2001).

The 10–20 system (and its variants) is based on an assumed geometric scaling relationship. The assumption is that the distance between two locations on a particular head will be proportionally larger on a larger head and smaller on a smaller head. Thus, anatomically comparable locations can be identified on the scalps of two different people with different-sized heads. This geometric scaling rationale is also the core concept in relating brain locations across individuals.

3.3.2 Talairach Atlas and MNI Coordinates

Two major stereotactic systems for the brain are the Talairach coordinate system and the Montreal Neurological Institute (MNI) coordinate system. They were initially developed for neurosurgery where there is a demand for high precision in defining the location of brain structures.

Previously, a physical frame was attached to the skull based on selected landmarks. Once secured, it was possible to move precisely to specific 3D points relative to this frame. These were called *stereotactic* frames. However, due to the imprecise correspondence between the structure of the skull and that of the brain, the Talairach coordinate system was developed based on landmarks directly defined in the brain. It is defined by using two brain structures as landmarks (the anterior-commissure (AC) and posterior-commissure (PC), which are white-matter bundles that connect the right and left hemispheres) and a plane through the mid-sagittal line (https://neuroimage.usc.edu/brainstorm/CoordinateSystems, https://imaging.mrc-cbu.cam.ac.uk/imaging/MniTalairach). Using the resulting 3D coordinate system, the location of different structures within a single brain can be specified by its 3D coordinates.

As shown in Fig. 3.3a, the key idea is to place a grid defined by this coordinate system on each brain slice so that the location of structures can be precisely specified. Using this approach, Talairach and Tournoux (Harary & Cosgrove,

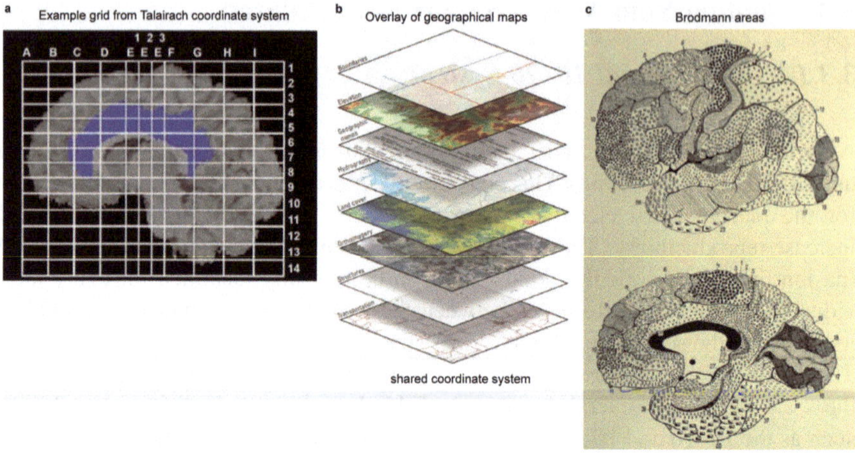

Fig. 3.3 (**a**) Example of a Talairach coordinate grid placed on a brain slice. (From Woodward et al., 2008). (**b**) Example of overlaid geographic maps with different information enabled by a shared coordinate system. (**c**) Map of cytoarchitecture properties at different locations across the brain as described by Brodmann (1909), now referred to as Brodmann areas

2019; Talairach & Tournoux, 1988) developed a detailed atlas of the entire brain using a single post-mortem brain, that is, a *template* brain. Importantly, since the procedure to define the coordinate system was based on landmarks, a comparable coordinate system could be defined for any individual brain. Then a geometric re-scaling could be used to bring the individual-specific coordinate system into correspondence with the standard template brain (i.e., similar to the scaling assumptions of the EEG 10–20 system described above). With such an approach, for example, a lesion observed on an individual's brain can be described relative to standardized coordinates on the template brain.

The advent of MRI and digital representations of the brain have dramatically increased the ease and precision of this approach. To increase its generality across individuals, the Montreal Neurological Institute (MNI) coordinates were developed using MRI images obtained from a large number of individuals (Evans et al., 1993). These template brains serve as a shared standard to help increase the reproducibility of research.

3.3.3 Enhanced Atlases: Cyto-anatomy, Connectivity-Anatomy

The coordinate framework described above has enabled the mapping of diverse types of anatomical information that can then be overlaid much like different information on a geographic map (Fig. 3.3b). Some examples of these other anatomical features are:

- *Cytoarchitecture*: The use of cellular features (cell distribution, myelin, etc.) and their variation to define "areas." A famous example is the work of Brodmann (1909), who identified variations in cellular structure across the brain (Fig. 3.3c). These Brodmann areas and gradations in myelin distribution (myeloarchitecture) define a set of locations that do not correspond to the sulcus/gyrus coarse structure (Eickhoff et al., 2018; Fan et al., 2016; Amunts et al., 2020; Abdollahi et al., 2014; Glasser et al., 2016).
- *Tractography*: The distribution and organization of white-matter tracts (Nozais et al., 2021; Salvalaggio et al., 2020; Thiebaut de Schotten et al., 2020).
- *RS-Connectivity*: Areas reliably identifiable by patterns of resting state connectivity (Eickhoff et al., 2018; Fan et al., 2016).
- *Gyral patterns* (Desikan et al., 2006).

Recommended Reading A major convenience for the practicing cognitive neuroscientist is the availability of detailed digital atlases. Several atlases are available on the web with interactive interfaces. As an example, see the Ebrain atlas: https://www.ebrains.eu/tools/human-brain-atlas. The following publications provide detailed lists of available atlases and access information

- Amunts and Zilles (2015): Architectonic Mapping of the Human Brain beyond Brodmann.
- Eickhoff et al. (2018): Imaging-Based Parcellations of the Human Brain.

3.3.4 Accessing and Using Atlases

Magnetic Resonance Imaging (MRI) has greatly simplified access to anatomical information. For example, the images in Fig. 3.2b were created using a MRIcroGL (https://www.nitrc.org/projects/mricrogl/) (Rorden & Brett, 2000) an open-source software to view MRI anatomical scans of the brain. Familiarity with viewing and navigating MRI images can provide you with access to various freely available atlases and anatomical resources.

In brief, different imaging protocols provide access to different aspects of anatomical features including connectivity structure (Magnetization-Prepared Rapid Gradient-Echo (MPRAGE), Diffusion Tensor Imaging, Diffusion Spectrum Imaging).

Representations

- *Volume*: The default is to acquire the data as a 3D-volume (x,y,z coordinates). The brain can be explored as slices passing through a particular point in this 3D volume (see Fig. 3.2b). In this view, it is possible to see the difference between gray and white matter, and to see the subcortical structures.

- *Surface*: A more advanced representation of the brain is the use of surfaces. Obtaining a surface from the acquired MRI volumes involves additional computational steps (typically performed in the background by various software). A surface representation allows features to be examined closely including by "inflating" the surface. Two points might be close together within the skull, but their neural relationship might be distant. This is of particular relevance to EEG measurement.
- *Overlay*: As described above, atlases can be placed on top of anatomical image within a common coordinate system to assess different kinds of information and features of the anatomical image.

With these basic concepts, tools, and the resources described above, it is possible to acquire a working familiarity with brain anatomy on your own.

3.4 Putting into All Together

As discussed earlier, understanding the relevance of EEG signals can be valuably informed by considering the broad anatomical context of the neural processes being investigated.

The following is a sketch of how your experimental study can be enriched by considering anatomy explicitly by considering the specific structures that might be engaged, the locations of these structures in the brain, and the manner in which they interact.

Identify During research planning, consider the cognitive processes being engaged by an experimental task (e.g., stimulus perception, attentional selection, decision-making, response selection, response execution). These functional processes can be used to estimate the expected anatomical networks and structures that might be engaged in performing this task (e.g., a lateralized motor network or a somatosensory network). This estimation step can benefit from the use of different functional atlases described above. It can be of particular value in patient populations where further anatomical information might be available.

The full anatomical scope includes the role of the peripheral nervous system. For example, a "simple" task of viewing a stimulus and responding to it, requires the engagement of the CNS with the PNS to receive sensory information and to generate movements of a peripheral effector (e.g., a finger).

Formulate The estimated anatomical networks provide a prior expectation of when and where activity in the brain is likely to be engaged. This information can be used to ensure that the experiment is designed to acquire relevant information. For example, if the networks are engaged differently between experimental conditions would the acquisition design (e.g., number of electrodes, their placement) be capable of detecting these differences.

Peripheral Sensors The expected anatomical networks in the experiment could indicate a role for peripheral physiological measures (e.g., electromyography (EMG), electrocardiography (ECG), eye-tracking) depending on the phenomena being investigated. In simplistic terms, these sensors can help track the timing and magnitude of information arriving to the brain, and the output of information from the brain to the periphery.

Apply These considerations can inform the selection of appropriate EEG measures to analyze the effects of interest: their electrode locations, lateralization, and timing, especially in coordination with peripheral sensors. It can also inform the interpretation of the obtained findings.

3.5 Conclusion

Here we have sought to highlight the value of including a neuroanatomical perspective to your EEG study. Although neuroanatomy is a large and complex topic, there are several tools and resources that can help you increase your familiarity with this topic. This can have many concrete benefits in better design and analysis strategies for your EEG studies. Finally, considering the nervous system as an extended interacting network that includes the brain as well as the peripheral nervous system can enrich how a study is formulated. The companion chapter on the peripheral nervous system provides further information about how the CNS and PNS interact to achieve various functional capabilities.

References

Abdollahi, R. O., Kolster, H., Glasser, M. F., Robinson, E. C., Coalson, T. S., Dierker, D., Jenkinson, M., Van Essen, D. C., & Orban, G. A. (2014). Correspondences between retinotopic areas and myelin maps in human visual cortex. *NeuroImage, 99*, 509–524.

Amunts, K., & Zilles, K. (2015). Architectonic mapping of the human brain beyond Brodmann. *Neuron, 88*, 1086–1107.

Amunts, K., Mohlberg, H., Bludau, S., & Zilles, K. (2020). Julich-brain: A 3D probabilistic atlas of the human brain's cytoarchitecture. *Science, 369*, 988–992.

Brodmann, K. (1909). *Vergleichende Lokalisationslehre der Grosshirnrinde in ihren Prinzipien dargestellt auf Grund des Zellenbaues*. Barth.

Buzsaki, G., Anastassiou, C. A., & Koch, C. (2012). The origin of extracellular fields and currents—EEG, ECoG, LFP and spikes. *Nature Reviews. Neuroscience, 13*, 407–420.

Cohen, M. X. (2017). Where does EEG come from and what does it mean? *Trends in Neurosciences, 40*, 208–218.

Desikan, R. S., Segonne, F., Fischl, B., Quinn, B. T., Dickerson, B. C., Blacker, D., Buckner, R. L., Dale, A. M., Maguire, R. P., Hyman, B. T., Albert, M. S., & Killiany, R. J. (2006). An automated labeling system for subdividing the human cerebral cortex on MRI scans into gyral based regions of interest. *NeuroImage, 31*, 968–980.

Eickhoff, S. B., Yeo, B. T. T., & Genon, S. (2018). Imaging-based parcellations of the human brain. *Nature Reviews. Neuroscience, 19*, 672–686.

Evans, A. C., Collins, D. L., Mills, S. R., Brown, E. D., Kelly, R. L., & Peters, T. M. (1993). 3D statistical neuroanatomical models from 305 MRI volumes. In *1993 IEEE conference record nuclear science symposium and medical imaging conference, San Francisco, CA, USA* (pp. 1813–1817). IEEE.

Fan, L., Li, H., Zhuo, J., Zhang, Y., Wang, J., Chen, L., Yang, Z., Chu, C., Xie, S., Laird, A. R., Fox, P. T., Eickhoff, S. B., Yu, C., & Jiang, T. (2016). The human brainnetome atlas: A new brain atlas based on connectional architecture. *Cerebral Cortex, 26*, 3508–3526.

Glasser, M. F., Coalson, T. S., Robinson, E. C., Hacker, C. D., Harwell, J., Yacoub, E., Ugurbil, K., Andersson, J., Beckmann, C. F., Jenkinson, M., Smith, S. M., & Van Essen, D. C. (2016). A multi-modal parcellation of human cerebral cortex. *Nature, 536*, 171–178.

Harary, M., & Cosgrove, G. R. (2019). Jean Talairach: A cerebral cartographer. *Neurosurgical Focus, 47*, E12.

Jasper, H. H. (1958). The ten-twenty electrode system of the international federation. *Electroencephalography and Clinical Neurophysiology, 10*, 371–375.

Lioumis, P., & Rosanova, M. (2022). The role of neuronavigation in TMS-EEG studies: Current applications and future perspectives. *Journal of Neuroscience Methods, 380*, 109677.

Nozais, V., Forkel, S. J., Foulon, C., Petit, L., & Thiebaut De Schotten, M. (2021). Functionnectome as a framework to analyse the contribution of brain circuits to fMRI. *Communications Biology, 4*, 1–12.

Oostenveld, R., & Praamstra, P. (2001). The five percent electrode system for high-resolution EEG and ERP measurements. *Clinical Neurophysiology, 112*, 713–719.

Rorden, C., & Brett, M. (2000). Stereotaxic display of brain lesions. *Behavioural Neurology, 12*, 191–200.

Rosenbloom, M. J., & Pfefferbaum, A. (2008). Magnetic resonance imaging of the living brain: Evidence for brain degeneration among alcoholics and recovery with abstinence. *Alcohol Research & Health, 31*, 362–376.

Salvalaggio, A., De Filippo De Grazia, M., Zorzi, M., Thiebaut De Schotten, M., & Corbetta, M. (2020). Post-stroke deficit prediction from lesion and indirect structural and functional disconnection. *Brain: A Journal of Neurology, 143*, 2173–2188.

Talairach, J., & Tournoux, P. (1988). *Co-planar stereotaxic atlas of the human brain*. Thieme Medical Publishers.

Thiebaut De Schotten, M., Foulon, C., & Nachev, P. (2020). Brain disconnections link structural connectivity with function and behaviour. *Nature Communications, 11*, 5094.

Warbrick, T. (2022). Simultaneous EEG-fMRI: What have we learned and what does the future hold? *Sensors (Basel), 22*(6), 2262.

Ward, J. (2015). *The student's guide to cognitive neuroscience*. Psychology Press.

Woodward, S. H., Kaloupek, D. G., Schaer, M., Martinez, C., & Eliez, S. (2008). Right anterior cingulate cortical volume covaries with respiratory sinus arrhythmia magnitude in combat veterans. *Journal of Rehabilitation Research and Development, 45*, 451–463.

Chapter 4
Basic Anatomy: Peripheral Nervous System

Cilia Jaeger

Abstract This chapter provides an overview of the neuroanatomy of the peripheral nervous system. The peripheral nervous system (PNS) consists of a complex anatomical system that enables information exchange between the brain and the body. It transmits information from sensory receptors to the brain, relays motor commands from the brain to the body, and regulates homeostasis. This chapter introduces the anatomy and function of the peripheral nervous system and outlines its two main domains: the somatic nervous system and the autonomic nervous system. The chapter illustrates how sensory information is detected, relayed to the brain, and how this communication produces specific body functions or responses.

Keywords Peripheral nervous system · Somatic nervous system · Autonomic nervous system · Receptors

4.1 Introduction

Chapter 3 ("Basic Anatomy: Central Nervous System") introduced the underlying neuroanatomy that forms the basis of the brain activity recorded in electroencephalography (EEG). Much of the neural activity that we record in EEG is connected to processing sensory information that is received from the peripheral nervous system (PNS). For example, a visual evoked response originates from visual information being detected by special receptors in the eye and relayed to the brain. Similarly, the brain can send information to the peripheries to elicit a behavioral response, such as voluntary muscle contraction. While EEG does not directly measure the underlying physiology of the peripheral nervous system, it is important to understand how these systems function together to respond to stimuli in the environment and elicit behavior. This chapter provides an overview of the anatomy and function of the

C. Jaeger (✉)
Brain Products GmbH, Gilching, Germany
e-mail: cilia.jaeger@brainproducts.com

T. Warbrick (ed.), *The EEG Handbook*,
https://doi.org/10.1007/978-3-032-20450-9_4

peripheral nervous system, highlighting how information about the body and the external environment is detected by specialized receptors and relayed to and from the central nervous system (CNS).

4.2 Overview of the Peripheral Nervous System

The peripheral nervous system comprises the parts of the nervous system outside of the brain and the spinal cord (Fig. 4.1). The PNS is composed of the somatic and autonomic nervous systems. The somatic nervous system consists of the sensory nervous system, which provides information about the surrounding environment, and the motor system, which enacts voluntary and involuntary muscle function. The autonomic nervous system governs bodily functions that occur involuntarily and are important for maintaining homeostasis. These include heart rate, breathing rate, and digestion. The autonomic nervous system can be further divided into the sympathetic, parasympathetic, and enteric nervous systems. The cranial nerves are also considered part of the PNS. The cranial nerves innervate the face, neck, and tongue. There are 12 cranial nerves, with some classified as part of the somatic nervous system and the rest classified as part of the autonomic nervous system (Table 4.1).

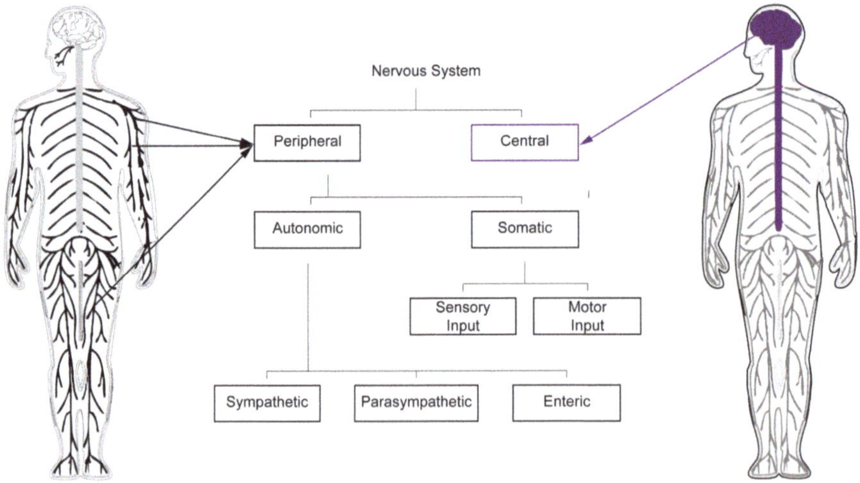

Fig. 4.1 Overview of the nervous system divisions. The nervous system is composed of the CNS (purple) and the PNS (black). The PNS can be further divided into the somatic nervous system, consisting of sensory input and motor outputs, and the autonomic nervous system, which is composed of the sympathetic, parasympathetic, and enteric nervous systems

Table 4.1 Cranial nerves

Number	Name	Function	CNS connection	Type
I	Olfactory	Smell	Olfactory bulb	Sensory
II	Optic	Vision	Thalamus	Sensory
III	Oculomotor	Eye movement, pupil contraction	Brainstem	Motor
IV	Trochlear	Eye movement	Brainstem	Motor
V	Trigeminal	Sensation in the face, chewing	Brainstem	Both
VI	Abducens	Eye movement	Brainstem	Motor
VII	Facial	Contraction of facial muscles, taste, salivation, lacrimation	Brainstem	Both
VIII	Vestibulochoclear	Balance, hearing	Brainstem	Sensory
IX	Glossopharyngeal	Taste, swallowing, salivation, blood pressure regulation	Brainstem	Both
X	Vagus	Speech, swallowing, regulation of heart, lungs, digestion	Brainstem	Both
XI	Spinal accessory	Neck and shoulder movement	Brainstem/ spinal cord	Motor
XII	Hypoglossal	Tongue movement, throat movement	Brainstem	Motor

4.3 Basic Anatomical Unit of the PNS: Ganglia and Nerves

The different PNS domains share common anatomical components. Similar to the CNS, the basic unit in the PNS is the neuron. Neurons make up the structures called the ganglia and peripheral nerves. Within the PNS, collections of neuronal cell bodies form ganglia. Their axons extend outward, traveling together in bundles that make up peripheral nerves, eventually reaching target tissues in the periphery. These bundles of axons are called fascicles and are surrounded by connective tissue called the perineurium, which provides structural support. The blood vessels surrounding the fascicles nourish axons with oxygen. Several fascicles are further bundled together by connective tissue to form a peripheral nerve. The signal transmitted along peripheral nerves needs to travel long distances to and from the peripheries to the CNS. To facilitate this, Schwann cells form a myelin sheath around axons in the PNS. Myelination speeds up electrical signal transmission along the nerves, allowing faster communication.

There are two categories of peripheral nerves: afferent and efferent. Afferent nerves transmit signal from the peripheries to the brain. These nerves are also called sensory nerves as they detect environmental or peripheral stimuli and relay specific information about the stimuli to the brain. Efferent nerves relay motor commands from the CNS to the muscles, organs, and glands in the body. These nerves are also called motor nerves as they carry commands for voluntary and involuntary motor action.

4.4 Anatomy of Somatic Nervous System

The somatic nervous system connects the central nervous system to the periphery and performs voluntary movements. The somatic nervous system is also known as a single-cell nervous system, as the motor and sensory neuron bodies synapse onto the central nervous system directly.

4.4.1 Receptors

The sensory nerves have specialized receptors at the end of the nerve that respond to different types of stimuli and enable the electrochemical conversion of the signal. The response to a sensory stimulus causes a change within a receptor cell or transmembrane protein that changes the ion composition within the cell by opening ion channels. This elicits an action potential within the sensory axon and relays the action potential toward the central nervous system. The body has many specialized types of sensory receptors that respond to very different environmental stimuli. A detailed description of the composition and function of each receptor type is outside the scope of this book. Here, we will focus on some of the sensory receptors that are involved in sensory modalities most prominently studied in conjunction with EEG. A more in-depth overview of receptors is provided in Leclerc and Wray (2023). The most recognized senses, such as vision, audition, olfaction, and gustation, are called specialized sensory modalities, which have a specialized organ that processes and transmits the sensory information to the brain; below we cover vision and audition. In addition, we cover balance and general sensory modalities.

4.4.1.1 Vision

The sense of vision is detected by specialized receptors located within the retina of the eye. These specialized receptors are called photoreceptors, which change their membrane potential when stimulated by light energy or photons. The photoreceptors regulate neurotransmitter release of the bipolar cells, which synapse onto the retinal ganglion cell. The retinal ganglion cells converge into the optic nerve, which transmits the visual input as an electrical signal to the thalamus. There are two types of photoreceptors classified as cones and rods. The cones are further divided into three types of specialized photoreceptors that respond to different wavelengths in the light spectrum: red, green, and blue. Their combined activity enables the perception of a wide variety of colors. The rods are highly sensitive to light but lack color sensitivity. They are located on the periphery of the retina, transmit gray-scale light, and assist with low-light vision (Openstax, 2022).

4.4.1.2 Audition

Another specialized sense is audition, which mediates the transduction of sound waves into an electrical signal within the ear. The cartilaginous outer structure of the ear, known as the auricle, funnels sound waves into the auditory canal. At the end of the auditory canal, the tympanic membrane vibrates in response to incoming sound waves. The vibrations from the tympanic membrane become amplified by three small bones, known as ossicles, that are located in the inner ear. The enhanced vibrations are transmitted to the oval window of the inner ear. Within the inner ear, the cochlea is attached to the oval window. The cochlea contains a fluid-filled tube. The pressure from the vibrating ossicles moves the fluid within the cochlea in a wave-like pattern. The cochlear duct is surrounded by the basilar membrane that contains hair cells, which have hair-like projections known as stereocilia. As the fluid in the cochlear duct moves, it pushes the stereocilia to bend. As the stereocilia bend, ion channels open, changing the ion composition of the hair cells. This causes the hair cell to depolarize. This type of receptor is also known as a mechanoreceptor. A mechanical change in the receptor cell causes the opening of ion channels. Depolarization of the hair cells releases neurotransmitter, which stimulates the cochlear nerve. The cochlear nerve projects to the brainstem, where the sense of audition is then further transmitted to the auditory cortex.

4.4.1.3 Vestibular System and Balance

The ear also contains the vestibular system, which helps with the sense of balance. The mechanoreceptors responsible for balance are located within the vestibule of the inner ear. The utricle and saccule sense head position. They are also composed of specialized hair cells with stereocilia that extend into a viscous gel known as the otolithic membrane. As the head tilts forward and back, the otolithic membrane slides over the hair cells, causing the stereocilia to bend and some hair cells to either depolarize or hyperpolarize. The position of the head is then determined by the brain, depending on the composition of depolarized hair cells. Head rotation is sensed by the semicircular canals, which consist of three ringlike structures oriented in different planes of the body. At the base of each semicircular canal, a structure known as the ampulla contains hair cells. These hair cells extend into the gelatinous structure, the cupula. As the head rotates in a direction that is parallel to the plane of a given semicircular canal, the fluid within the cupula causes the stereocilia on the hair cells to bend, which again opens ion channels and causes depolarization of the hair cells. The information is then transmitted to the vestibular ganglions and through the vestibulocochlear nerve to the brain stem and cerebellum.

4.4.1.4 General Sensory Modalities

General sensory modalities have specialized receptor cells distributed throughout the body. An example of such a sensory modality is somatosensation. Somatosensation is a general term for proprioception, interoception, and touch and consists of different sensory receptors that respond to vibration, temperature, pressure, and pain. Somatosensory receptors can be located in the skin, muscles, and within visceral organs. They also consist of free nerve endings that have dendrites directly embedded in the tissue or encapsulated nerve endings, which are covered in connective tissue. Two types of free nerve ending receptor cells include thermoreceptors and nociceptors, which transduce temperature and pain. Temperature receptors detect temperature that differs from body temperature. Nociceptors respond to either thermal, chemical, or mechanical stimuli. A common example of a chemical ligand that initiates pain is capsaicin, which is found in spicy foods. Capsaicin binds to an ion channel regulating temperature and induces a sensation of heat. There are several encapsulated mechanoreceptors, stretch receptors, or receptors that detect vibration.

4.4.2 Somatic Sensory System

The sensory signal ascends from the receptors along the peripheral process of the sensory nerve along the dorsal root to the cell body, which sits in the dorsal root ganglion just outside of the spinal cord. Figure 4.2 highlights the two different pathways by which sensory information from the dorsal root ganglion travels to the CNS. The two pathways are classified as the dorsal column system and the spinothalamic tract and relay different information from the environment. Proprioception and fine touch sensory input is primarily relayed along the dorsal column system (Koop & Tadi, 2025). In the dorsal column system, a second central process extends from the cell body in the dorsal root ganglion and ascends ipsilaterally through the dorsal column and synapses onto a secondary neuron located in the medulla. Temperature and nociception are primarily relayed along the spinothalamic tract. In the spinothalamic pathway, the central process, which extends from the cell body in the dorsal root ganglion, synapses contralaterally to a second neuron whose cell body is located in the spinal cord grey matter and terminates in the thalamus. In both sensory ascending pathways, the secondary neuron is considered part of the CNS and synapses with the thalamus, which relays peripheral information to the somatosensory cortex.

The efferent counterpart of the afferent sensory pathways is the motor peripheral nerves. Voluntary movement is controlled by the motor cortex in the CNS along the corticospinal tract. This pathway is also known as the primary descending motor pathway. Upper motor neurons descend from the primary motor cortex through the brainstem along the spinal cord and synapse with the lower motor neurons. The cell bodies of lower motor neurons are located in the ventral root, and the axons from the lower motor neurons project to peripheral skeletal muscles via the ventral horn. The

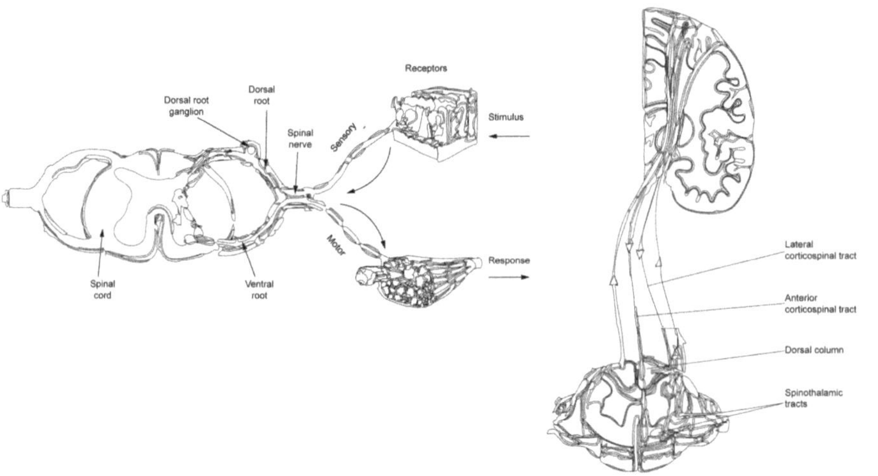

Fig. 4.2 Somatic nervous system. (**a**) Receptors from the peripheries convert a stimulus into an electrical signal that propagates along the myelinated sensory neuron to the dorsal root of the vertebrae. From there, the signal can be directly transmitted to the efferent motor neuron that innervates the effector muscle. The signal from the afferent sensory neuron also synapses onto a secondary sensory neuron in the spinal cord that transmits the signal to the brain. (**b**) Afferent pathways consist of either the dorsal column or the spinothalamic tracts. Efferent pathways are divided into the anterior and lateral corticospinal tracts

electrical signal from the motor neurons is then translated into mechanical muscle contractions at the neuromuscular junction. The neuromuscular junction is in the center of the muscle fiber, also referred to as the innervation zone. Electromyography (EMG) can be used to measure the summation of electrical activity generated by muscle fibers contracting in response to motor unit stimulation at a muscle of interest. The placement of EMG electrodes is covered in more detail in Chap. 12 ("Hardware for Recording EEG and Peripheral Physiology").

4.4.3 *Reflex Arc in the Somatic Nervous System*

While the somatic nervous system is under voluntary control of the CNS, the body must be able to respond quickly to dangerous external stimuli. There is an adaptation of the body, known as a reflex, to briefly circumvent the CNS and mediate an automatic, involuntary response to certain stimuli. Reflexes are present in both the somatic and autonomic nervous systems. Reflexes consist of the reflex arc, which is the neural pathway that receives stimuli from a sensory receptor at a specific location and initiates an effector to elicit a functional response at the original site of sensory input, such as a muscle contraction. Reflex arcs can be monosynaptic, which involves a sensory neuron synapsing directly with a motor neuron to elicit an immediate response, such as the patellar knee-jerk response. Reflex arcs can also be polysynaptic, in which the sensory neuron synapses with an interneuron within the dorsal horn, which transmits the signal to the efferent motor neuron to elicit a response (OpenStax, 2022).

4.5 Anatomy of the Autonomic Nervous System

The autonomic nervous system (ANS) consists of nerves innervating glands, blood vessels, and internal organs. It is also called the involuntary or visceral nervous system. The ANS is divided into the parasympathetic and the sympathetic nervous systems, which are anatomically and functionally distinct. Most tissues are innervated by both systems, as the two systems work together with opposing effects. Figure 4.3 highlights the function of the parasympathetic and sympathetic nervous

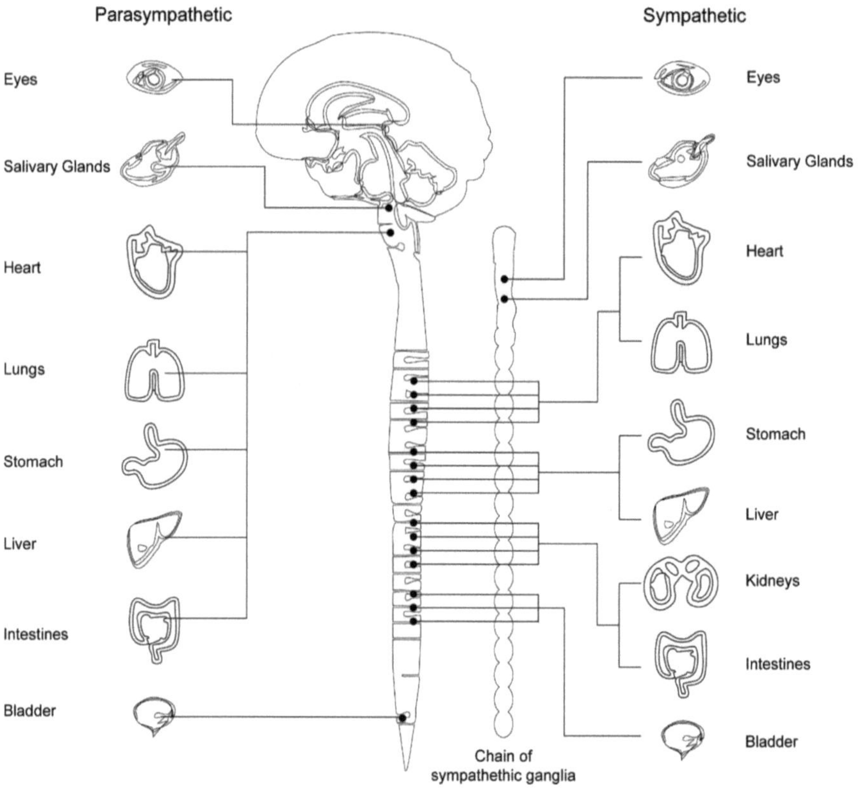

Fig. 4.3 Autonomic nervous system. The autonomic nervous system consists of the sympathetic and parasympathetic nervous systems that innervate different organs of the body. The efferent pathways of the autonomic nervous system consist of two neurons: the preganglionic and the postganglionic neurons. The preganglia synapse onto the postganglia, whose cell bodies are located in the paravertebral sympathetic ganglion chain

systems and how they work in an opposing manner. The ANS is a disynaptic pathway in which the efferent pathways consist of two neurons, a preganglionic neuron and a postganglionic neuron, that transmit signals from the CNS to the periphery. The synapse between the postganglionic neuron and effector tissue is called the neuroeffector junction, which is organized into varicosities that release neurotransmitter over a large surface area (McCorry, 2007).

4.5.1 Sympathetic Nervous System

The sympathetic nervous system prepares the body to react to stressors in its environment by initiating a rapid physiological response, often termed the "fight or flight" response. Visceral sensory receptor cells detect stimuli, which are relayed along afferent nerve fibers through the sympathetic chain ganglia to the dorsal root ganglion. The sensory nerves then synapse to secondary nerves in the dorsal horn that relay information to the CNS or to efferent connections via reflex arcs.

The efferent preganglionic neurons originate from the CNS, specifically from the thoracic and lumbar regions (McCorry, 2007). The efferent preganglionic sympathetic connections are also referred to as the thoracolumbar system. Within the thoracolumbar system, the preganglionic cell body is located in the lateral horn of the vertebral column, where it projects to a postganglion located in a network of sympathetic chain ganglia that runs parallel to the vertebral column. A preganglion can project to several postganglionic fibers. Projections from the postganglion neuron project to the effector tissue in the periphery. The preganglionic neurons of the sympathetic nervous system release acetylcholine, and most postganglionic neurons of the sympathetic nervous system secrete norepinephrine. The sympathetic nervous system regulates heart rate, respiration, and pupil responses.

4.5.2 Parasympathetic Nervous System

The parasympathetic nervous system stimulates "rest and digest" functions of the ANS and regulates visceral organs. The parasympathetic nervous system is organized similarly to the sympathetic nervous system, except that some of the afferent and efferent pathways are categorized as cranial nerves and have a direct connection with the brainstem. The efferent preganglionic nerve fibers from the cranial division also arise from the brainstem and synapse with the postganglionic nerves near the effector tissue. The other division is located in the sacral region (McCorry, 2007). The sensory receptors located in the visceral organs transmit electrical signals along the sensory afferent neurons to the sacral dorsal root ganglia, where they synapse onto a secondary neuron that relays information from the spinal cord to the brain. The preganglionic neurons pass through the lateral horn of the sacral spinal cord. The parasympathetic preganglia target the terminal ganglia, which are located near or within the target tissue. The postganglia project a short distance to the target organ or tissue. The primary neurotransmitter of the preganglionic and the postganglionic neurons of the parasympathetic nervous system is acetylcholine.

4.5.3 Autonomic Reflex Example

Autonomic reflexes play a large role in maintaining homeostatic balance within an organism. The ANS is largely regulated by autonomic reflexes. Sensory input from glands and organs is relayed to the hypothalamus and brainstem, which are the control centers for bodily processes such as heart rate, blood pressure, body temperature, glandular secretion, and digestion. An example of an autonomic reflex is the chemoreceptor reflex, which regulates respiration rate based on the concentration of oxygen and carbon dioxide in the blood. Specialized chemoreceptors in the aortic arch near the heart and in the carotid artery in the neck monitor partial pressure of gases in the blood, sensing fluctuations in oxygen and carbon dioxide levels. The chemoreceptors can also detect changes in pH or acidity of the blood, which is correlated to the amount of dissolved carbon dioxide and hydrogen ions in the blood. A decrease in oxygen inhibits oxygen-sensitive potassium ion channels in glomus cells (type I cells) located in the carotid and aortic bodies. The inhibition of potassium channels causes a buildup of potassium and subsequently leads to cell depolarization. The glomus cell depolarization leads to neurotransmitter releases by opening voltage-gated calcium channels. The release of neurotransmitters activates afferent nerve endings of either the glossopharyngeal nerve within the carotid body or the vagus nerve within the aortic body. The peripheral afferent nerves relay the signal to the brainstem. The brainstem then regulates efferent autonomic responses that increase breathing rate to enhance oxygen intake and increase the heart rate to enhance oxygen delivery. When carbon dioxide dissolves in the blood, it reacts with water molecules and forms carbonic acid, which lowers the blood pH. Type I glomus cells are also sensitive to a decrease in pH levels. When the pH level of the blood drops, potassium channels are inhibited, and a similar signal transmission pathway as described for oxygen levels is activated, increasing breathing rate (Iturriaga, 2021). The autonomic chemoreceptor reflex can also be indirectly measured by recording respiration rate and heart rate changes, which can be measured with peripheral sensors, as explained in Chap. 12 ("Hardware for Recording EEG and Peripheral Physiology").

4.6 Cranial Nerves

The 12 cranial nerves consist of afferent and efferent connections whose nuclei originate from the brainstem or cerebrum. The cranial nerves can be classified as sensory, motor, or both and mostly innervate the head and neck area. Some of the cranial nerves are part of the autonomic nervous system, and some cranial nerves are part of the somatic nervous system.

Table I lists the twelve cranial nerves, their function, the type of peripheral nerves they are classified as, and their connection within the CNS. The olfactory and optic nerves are sensory nerves that allow smell and vision. The oculomotor, trochlear, and abducens nerves are all responsible for controlling eye movements; however, they innervate different extraocular muscles. The trigeminal nerve is responsible for the sensation of the skin on the face. It also innervates the muscles required for chewing food. The vestibulocochlear nerve transmits sensory information to the brain for hearing and balance. The glossopharyngeal nerve controls muscle movements in the mouth and upper throat. It also transmits sensory signals for the sense of taste and regulates the production of saliva. The vagus nerve regulates homeostatic function of the organs in the thoracic and upper abdominal cavities and regulates breathing rate, heart rate, and digestion. The spinal accessory nerve is a motor nerve responsible for controlling the muscles of the neck and upper shoulders. The last cranial nerve, the hypoglossal nerve, controls muscles of the tongue and throat.

Most of the cranial nerves belong to the somatic nervous system. The vagus nerve is considered part of the autonomic nervous system, specifically the parasympathetic division. Three cranial nerves perform functions that belong to both the somatic and the autonomic nervous systems. The oculomotor nerve fibers both innervate the eye, considered an organ, and the muscle within the eye, the iris. The facial and glossopharyngeal nerves innervate glands involved in salivation.

4.7 Function of the PNS and CNS as a Unit

A key example of the function of the PNS working together with the CNS is the processing of pain. To demonstrate how the two nervous systems work together in the context of electrophysiological research, we present an example of neurofeedback and pain processing and perception.

Figure 4.4 shows the physiological response to touching a hot surface. Thermal changes activate nociceptors in the skin, which convert the stimulus into electrical impulses that travel along sensory neurons to the dorsal horn of the spinal cord. The sensory signal is then transmitted along a reflex arc to the interneurons within the dorsal horn, which relays the signal to the motor neurons. At the neuromuscular junction, the signal is then converted to muscular contraction and results in a muscular response, such as pulling the arm away from the hot stimulus by causing muscle contraction in the arm. Simultaneously, the pain signal is further transmitted from primary to secondary sensory neurons that travel to the brainstem and thalamus, where pain is further transmitted and processed within the somatosensory cortex.

The CNS can also override the reflex response. The somatosensory cortex also transmits the signal generated by the painful stimulus to the prefrontal cortex and anterior cingulate cortex. These areas are involved in decision-making and emotional regulation, respectively, and allow for conscious perception of pain. If the stimulus is perceived as tolerable, the CNS sends inhibitory signals via the

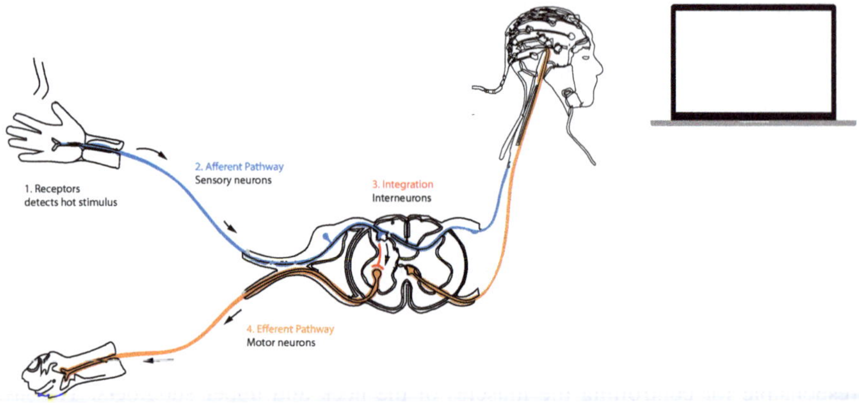

Fig. 4.4 Reflex arc and neurofeedback in electrophysiology. Nociceptors in the skin of the hand detect sudden temperature changes. This is converted to an electrical signal that is transmitted along the afferent sensory pathway to the dorsal horn. Interneurons in the vertebrae then transmit the electrical signal to the efferent motor pathway. The signal is transmitted to the neuromuscular junction, which causes contraction of the muscles in the forearm, effectively pulling the hand away from a hot source. Electrodes on the forearm can be used to measure muscle contraction in response to the nociceptive stimulus. The painful stimulus is also transmitted to the CNS, where it is consciously perceived and can be modulated. For example, in neurofeedback training, the pain response can be mindfully downregulated. The CNS transmits an inhibitory signal along efferent pathways that activate inhibitory neurons in the ventral horn and inhibit muscle contraction of the forearm

descending pathway in the spinal cord to inhibitory interneurons in the ventral horn that synapse onto the motor neurons of the hand and suppress their excitability.

In an experimental setting, neurofeedback can be used in combination with EEG and EMG to modulate the response to a painful stimulus. The response to pain can be measured with EMG to measure the muscular response of pulling the hand away and the reaction time between administering a painful heat stimulus and muscle contraction. Pain perception can also be detected with EEG electrodes placed over the sensorimotor cortex or prefrontal cortex. Figure 4.4 shows a participant receiving real-time feedback of their perceived pain perception. They are asked to suppress their pain-related brain activity and subsequently suppress their hand pulling away from the hot, painful stimulus. Alongside the EEG, feedback from EMG can be used to see how well the participant is able to downregulate their pain response and muscle activity in the forearm. This example of a paradigm is often used in studies investigating chronic pain (Kern et al., 2024).

4.8 Concluding Remarks

This chapter covers the basic anatomy of the peripheral nervous system and connects the nervous system to other systems in the body, such as the muscular system and endocrine system. It is essential to understand how the CNS and PNS work together to understand how function, cognition, and behavior are linked. Peripheral physiology can interact with central electrophysiology. Autonomic activity, such as changes in skin conduction or heart rate have shown to be coupled to brain states (Huang et al., 2018), which are covered in detail in Chap. 6 ("Brain States"). Incorporating peripheral measures into EEG research can provide contextual information on body and brain interactions and behavior.

References

Huang, J., Ulke, C., Sander, C., Jawinski, P., Spada, J., Hegerl, U., & Hensch, T., 2018. Impact of brain arousal and time-on-task on autonomic nervous system activity in the wake-sleep transition. *BMC Neuroscience, 19*(1), 18. Available at: https://www.ncbi.nlm.nih.gov/pmc/articles/PMC5896037/. Accessed 27 Sept 2025.

Iturriaga, R., Alcayaga, J., Chapleau, M. W., & Somers, V. K. (2021). Carotid body chemoreceptors: Physiology, pathology, and implications for health and disease. *Physiological Reviews, 101*(3), 1177–1235. Available at: https://www.ncbi.nlm.nih.gov/pmc/articles/PMC8526340/. Accessed 27 Sept 2025

Kern, M., Sperlich, B., & Rief, W. (2024). Evaluating the effectiveness of neurofeedback in chronic pain management: A narrative review. *Frontiers in Psychology, 15*, 1369487. Available at: https://doi.org/10.3389/fpsyg.2024.1369487. Accessed 27 Sept 2025

Koop, L. K. & Tadi, P. (2025). Neuroanatomy, sensory nerves. In *StatPearls* [Internet]. StatPearls Publishing. Available at: https://www.ncbi.nlm.nih.gov/books/NBK539846/. Accessed 27 Sept 2025.

Leclerc, A., & Wray, A. (2023). Sensory Receptors. In *StatPearls* [Internet]. StatPearls Publishing. Available at: https://www.ncbi.nlm.nih.gov/books/NBK539861/. Accessed 23 June 2025.

McCorry, L. K. (2007). Physiology of the autonomic nervous system. *American Journal of Pharmaceutical Education, 71*(4), 78. Available at: https://www.ncbi.nlm.nih.gov/pmc/articles/PMC1959222/. Accessed 27 Sept 2025.

OpenStax. (2022). The Somatic Nervous System. In *Anatomy and physiology 2e*. OpenStax. Available at: https://openstax.org/books/anatomy-and-physiology-2e/pages/14-introduction. Accessed 27 Sept 2025.

Chapter 5
EEG in the Context of Human Physiology

Michael Hoppstädter

Abstract Electroencephalography (EEG) as a noninvasive measurement technique is recorded from the scalp. Hence, it does not measure any signal directly at cell level but rather the signal that reaches the electrodes. Therefore, EEG should be considered in the context of human physiology which interacts with and influences the EEG signal. This chapter discusses aspects of macroscopic head anatomy and physiology that affect the EEG signal. The focus will be on ocular, muscular and skin potentials, and vascular activity, and how they influence the EEG signal.

Keywords Ocular potentials · Muscle artifacts · Sweating · Blood circulation · Volume conduction

5.1 Introduction

What does physiology mean? As with many technical terms, physiology comes from old Greek, and it roughly translates to the study of nature. Compared to anatomy as the study of the structure of the body, physiology refers to the processes occurring in the cells and tissues of the living organism.

EEG and Physiology In a nutshell, the electroencephalography (EEG) signal originates mostly from synchronous excitatory postsynaptic potentials. These are largely generated by pyramidal neurons in the cortex (see Chap. 2: "What Is EEG?"). Thus, EEG itself is a consequence of physiological processes at the cell level. The signal then travels from the cortex through different tissues of the head until it is eventually picked up by the sensors on the scalp. As other physiological processes happen at the same time and in the same place, it makes sense to look at the bigger picture. Consider that the EEG signal can be influenced by the physical properties of

M. Hoppstädter (✉)
Brain Products GmbH, Gilching, Germany
e-mail: michael.hoppstaedter@brainproducts.com

T. Warbrick (ed.), *The EEG Handbook*,
https://doi.org/10.1007/978-3-032-20450-9_5

the tissues that it permeates. Also, the human body generates other electromagnetic signals which will interact with the EEG and have an influence on what is finally measured.

We should therefore think about the wider context in which EEG arises when we interpret the signal. This will also help us to differentiate between what in our recording is true EEG and what is due to other physiological signals. In the following sections, we will consider some specific physiological processes and how they manifest in the EEG data.

5.2 Head Anatomy and Signal Propagation

A Brief and Macroscopic Anatomy of the Head Starting from the inside out, the brain is a conglomerate of billions of nerve cells (the *grey matter*) and their processes (i.e., axons and dendrites, the *white matter*), while the whole structure is supported through *glia cells*. Clusters of varying cell compositions form different organs in the brain that serve specialized functions. The outer boundary of the brain is the *cerebral cortex*, a heavily folded structure made up of multiple layers (for a more detailed account of brain anatomy, please see Chap. 3: "Basic Anatomy: Central Nervous System"). The brain is encased by several layers of skin, the so-called *meninges*. There are also spaces between the meninges and within the brain which are not filled with air but *liquor*, for example, the four *ventricles*. As brain cells need to be supplied with oxygen, the brain is permeated with many blood vessels and larger blood-filled cavities (the *sinuses*). Travelling further toward the outside, we reach the skull which is covered in muscles that enable movement of the head and face. Finally, the head is covered in skin. It is not a uniform organ, but consists of multiple layers, whereby the *epidermis* (the outermost layer) creates a physical and chemical barrier. The skin also includes other structures (so-called *skin appendages*) such as sweat and sebaceous glands and hairs (see Fig. 5.1).

Signal Propagation and Volume Conduction As an electromagnetic signal, the EEG propagates through conductive materials and is measured from a distance (Rutkove, 2007). This is usually referred to as *volume conduction* (for a more detailed description, see Chap. 2: "What Is EEG?"). The different tissues are not equal in their composition and neither in their physical properties. For instance, liquor has a conductivity that is roughly five times that of grey matter (McCann et al., 2019). Another crucial point is that the EEG does not travel in a direct way from its source to the head surface, but it spreads evenly from the cortex in all directions. Since electrical activity is not just emitted from one source but from many, the signals also mix on their way. This means that an EEG sensor placed at any scalp location will not only record the EEG signal from the neuron populations in the cortical patch directly below it but also from the surrounding area. The signals from nearby neurons will certainly make up a large portion of the EEG at that electrode, but signals from more distant parts of the brain will also contribute. Thus, overall, the

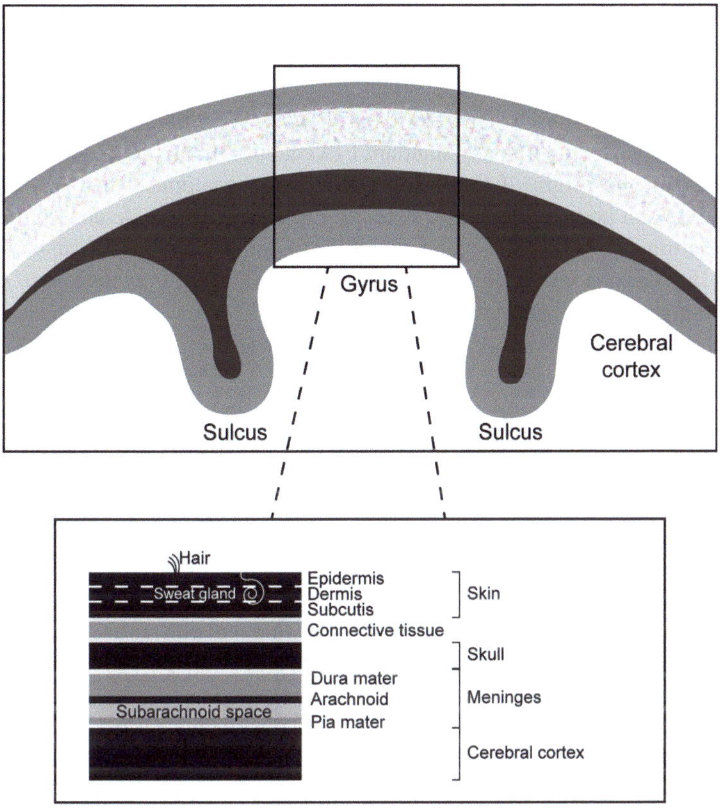

Fig. 5.1 Simplified schematic of the tissues surrounding the brain

location of the source, its orientation, and the conductivity of the different tissues between the source and the electrode will have an effect on what EEG signal is measured at the scalp (Jochmann et al., 2011; Vorwerk et al., 2024).

Volume conduction is also the reason why EEG effects, like event-related potentials (ERPs), do not only appear at single electrodes but show instead rather widespread distributions over the scalp. Neighboring electrodes will display a similar effect, with amplitudes being largest in the vicinity of the origin of the effect (e.g., visual evoked potentials are largest at lateral occipital electrodes, which sit above the visual cortex). And since the generators of EEG usually form an electrical dipole, it is also common to see the inverse effect on the opposite side of the head, that is, a left hemisphere negativity accompanied by a right hemisphere positivity (Scherg et al., 2019).

The cells that emit the EEG signal are not directly in contact with our recording electrodes, but they are part of this complex context of neuroanatomy. This means that the EEG signal will have traversed all the aforementioned tissues via volume conduction on its way from the cortex to the electrode. To fully understand what signal is recorded, it is thus necessary to consider all the elements that interact and potentially influence the EEG signal along its path.

5.3 How EEG Is Influenced by Other Physiological Signals

There are a few physiological signals that play a bigger role in EEG recordings, and you should consider these when working with EEG. We will look at signals that you will most likely encounter in any standard EEG dataset. In this chapter, we will focus on discussing the origin of those signals (from an anatomical and physiological perspective) and how they interfere with the EEG signal. More detailed information on these artifacts and how they can be handled in your recorded data can be found in Chaps. 15 ("Getting Clean EEG Data: Artifacts and How to Prevent Them") and 17 ("EEG Pre-processing and Artifact Handling").

5.3.1 The Eyes and Ocular Potentials

Let us first consider the different parts of the human eye. The largest part is the eyeball, which lets the light pass through the lens (in the front) toward the retina (in the back). On top of the lens sits the *cornea*, which refracts the light and focuses it on the retina. The eyelid is meant to protect the eyeball and to distribute tear fluid evenly through blinking. Six eye muscles enable fine movements and rotations of the eyeball (Trepel, 2008). This allows us to flexibly shift the gaze to locations in our environment without being reliant on head movements.

The eyeball forms an electrical dipole that is positively charged in the front and negatively charged in the back (Malmivuo & Plonsey, 1995). Therefore, any eye movement will cause changes in the electric field of the eyeball. When the eyelid is closed during blinking, the contact between the eyelid and the cornea creates a current flow toward the scalp (Croft & Barry, 2000). Blinking leads to an electrical potential in the range of 0.4–1 mV (Malmivuo & Plonsey, 1995), which is much larger than what we usually observe with EEG. As with EEG, these ocular potentials reach the EEG electrodes on the scalp through volume conduction (Lins et al., 1993). The largest effects are visible in electrodes in the vicinity of the eyes, such as frontopolar (e.g., Fp1 and Fp2) and lateral frontal channels (e.g., F9 and F10). However, ocular potentials also travel to more distant sensors, whereby they become smaller with increasing distance. It is still possible to see ocular potentials at central electrodes and sometimes beyond.

As blinking cannot be suppressed over longer time periods, eye blinks will inevitably show up in any EEG recording of awake participants. In contrast, eye movements can be avoided more easily than blinks (e.g., when participants are guided to focus on the center of a presentation screen), but occasional small movements are still likely to happen. Since their magnitude is much larger than more subtle effects like ERPs, ocular potentials would likely conceal the EEG activity of interest. Therefore, they must be handled as part of data processing.

5.3.2 *Facial Muscles and EMG*

The human skull is lined with many muscles which allow us to perform a variety of smaller and larger movements. These are most notably the *muscles of mastication* (four muscles) and the facial or *mimetic muscles* (more than 20 muscles across five groups: ocular, nasal, oral, auricular, scalp, and neck). The muscles of mastication connect to the lower jaw to support the opening and closing of the jaw and chewing. Whereas the mimetic muscles connect to the skin and allow fine-grained facial expressions such as smiling, wrinkling the nose, or closing the eyelids.

Activity in facial muscles can have small to detrimental effects on concurrent EEG depending on which muscles are involved. Generally, facial muscle activity generates electrical potentials in a wide frequency range, roughly from 20 to 200 Hz (Muthukumaraswamy, 2013). Hence, the lower end of the muscle activity spectrum overlaps with EEG frequencies of interest in the beta and gamma frequency bands. Strong artifacts can occur in the case of teeth clenching or chewing, which can easily affect all scalp channels and obscure EEG activity entirely (Luck, 2014). Some artifacts are likely to be localized to specific electrodes or regions. For example, a rather common issue is tension in neck muscles that leads to persistent high-frequency noise in single, mostly temporal electrodes. Smaller facial movements such as smiling or moving the jaw can also lead to artifacts across lateral electrodes, while movements of the nose or the forehead usually affect more frontally located channels.

As strong muscle activity can obscure the underlying EEG activity, it is a good idea to instruct participants to minimize body and facial movements during EEG recordings. However, it is difficult to avoid this completely, and thus, muscle artifacts need to be addressed during data processing.

5.3.3 *Sweat Glands and Skin Potentials*

The skin is made up of several layers, embedded skin organs, and hair. It has both protective and sensory functions (mechanical, thermal, and nociceptive). Skin and hair form a physical barrier that a researcher has to overcome through thorough cap preparation to collect high-quality EEG data.

Another issue arises from sweat glands, which sit in deeper skin layers (the *dermis* and *subcutis*) across the whole body, and drain sweat via ducts to the surface (Trepel, 2008). Their main function is thermoregulation. In an EEG recording situation, sweating might increase due to high temperature in the environment or physical activity (e.g., measurement during physical exercise). Since sweat secretion is mediated via the sympathetic nervous system (Benedek & Kaernbach, 2010; see also Chap. 4: "Basic Anatomy: Peripheral Nervous System"), psychological stress can also lead to increased sweat secretion, for example, when participants are presented with emotionally arousing stimuli.

Sweat is mostly made up of water, but it also contains trace amounts of minerals such as sodium and chlorine. Therefore, sweating changes the conductivity of the skin (Boucsein, 2012), and this becomes a problem if it happens under the EEG electrodes. Electrodes on hairless and thus more exposed areas, such as the forehead, are at a higher risk of being affected by sweating. The result is a slow drift that produces low-frequency oscillations with large amplitude variations (Kappenman & Luck, 2010). Thus, sweat artifacts can pose a serious problem when they contaminate the EEG since they cannot be easily removed or attenuated. They are best avoided in the first place by optimizing recording conditions.

The effect of sweat glands on skin conductivity can also be a signal of interest when *electrodermal activity* (EDA) is measured as a marker of emotional arousal (Boucsein, 2012). This signal is usually not recorded from the scalp but from skin areas on the hand or foot (Boucsein et al., 2012). EDA measurements are often combined with EEG recordings in emotion research.

5.3.4 Blood Vessels and Heartbeat

The head's main blood supply is bilaterally driven via two arteries, the *arteria carotis interna* and *arteria vertebralis*, which then branch out again into three further arteries (the *arteria cerebri anterior*, *media*, and *posterior*, again left and right) to transport oxygenated blood to dedicated brain areas. In contrast to other internal organs where large blood vessels enter the organ at specific points, these large afferent vessels run on the brain surface along the sulci. De-oxygenated blood is transported back via a separate system of veins that discharge into larger sinuses leading (via the *vena jugularis interna*) back to the heart (Trepel, 2008).

As the heart itself is an electrical organ, the effects of the cardiac cycle are propagated throughout the body and can also be picked up by EEG. This *cardiac field artifact* coincides with the *R-peak* of the cardiac cycle. However, at the same time, the EEG can be impacted in a more direct way by blood circulation. Due to the rhythmic movement of blood being pumped through the body, it is possible to pick up this signal with EEG electrodes placed on top or close to the pulsating arteries running below the scalp surface. These so-called *pulsatility artifacts* will be visible in the recorded EEG activity as a wave- or spike-like artifact at the frequency of the heartbeat, but with a roughly constant delay of 200 ms (Kern et al., 2013). These artifacts can affect the signal at single electrodes, but it is also possible that they enter all EEG sensors if the contaminated channel is part of the reference. This can happen, for instance, when you are using a mastoid reference as these electrodes are usually placed above the *arteria auricularis posterior* (a smaller artery branching off the *arteria carotis externa*), which runs between the mastoid process and the ear canal (Luck, 2014). Higher blood pressure and thinner skin can increase this effect, and therefore, older participants are especially prone to showing pulse artifact contamination in the EEG.

5.4 Conclusion

EEG provides a measure of the electrical activity of the brain. However, we do not measure directly from the pyramidal cells (as with invasive depth electrodes) or from the cortex (as with *electrocorticography*) but from the scalp surface. Therefore, we need to consider what else will influence the signal on the way from its source to the sensor.

This chapter has put the EEG signal into context with human anatomy and physiology. We have learned about the different tissues through which the EEG signals must travel via volume conduction, and the effects of other physiological signals that can overlap with EEG at the sensor. Some of the artifacts that we discussed can be prevented to some extent. Those artifacts that we cannot avoid will have to be treated during data processing. Some can be attenuated nicely so that we can keep the data for analysis. However, we might also lose data due to more severe artifacts. Thus, reading this chapter should form the basis for a better understanding of artifact prevention and handling.

References

Benedek, M., & Kaernbach, C. (2010). Decomposition of skin conductance data by means of nonnegative deconvolution. *Psychophysiology, 47*, 647–658.

Boucsein, W. (2012). *Electrodermal activity*. Springer.

Boucsein, W., Fowles, D. C., Grimnes, S., Ben-Shakhar, G., Roth, W. T., Dawson, M. E., Filion, D. L., & Society for Psychophysiological Research Ad Hoc Committee on Electrodermal Measures (2012). Publication recommendations for electrodermal measurements. *Psychophysiology, 49*, 1017–1034.

Croft, R. J., & Barry, R. J. (2000). Removal of ocular artifact from the EEG: A review. *Neurophysiologie Clinique, 30*, 5–19.

Jochmann, T., Gullmar, D., Haueisen, J., & Reichenbach, J. R. (2011). Influence of tissue conductivity changes on the EEG signal in the human brain: A simulation study. *Zeitschrift für Medizinische Physik, 21*, 102–112.

Kappenman, E. S., & Luck, S. J. (2010). The effects of electrode impedance on data quality and statistical significance in ERP recordings. *Psychophysiology, 47*(5), 888–904.

Kern, M., Aertsen, A., Schulze-Bonhage, A., & Ball, T. (2013). Heart cycle-related effects on event-related potentials, spectral power changes, and connectivity patterns in the human ECoG. *NeuroImage, 81*, 178–190.

Lins, O. G., Picton, T. W., Berg, P., & Scherg, M. (1993). Ocular artifacts in EEG and event-related potentials. I: Scalp topography. *Brain Topography, 6*, 51–63.

Luck, S. J. (2014). *An introduction to the event-related potential technique*. MIT Press.

Malmivuo, J., & Plonsey, R. (1995). *Bioelectromagnetism: Principles and applications of bioelectric and biomagnetic fields*. Oxford University Press.

McCann, H., Pisano, G., & Beltrachini, L. (2019). Variation in reported human head tissue electrical conductivity values. *Brain Topography, 32*, 825–858.

Muthukumaraswamy, S. D. (2013). High-frequency brain activity and muscle artifacts in MEG/EEG: A review and recommendations. *Frontiers in Human Neuroscience, 7*, 138.

Rutkove, S. B. (2007). Introduction to volume conduction. In A. S. Blum & S. B. Rutkove (Eds.), *The clinical neurophysiology primer*. Humana Press.

Scherg, M., Berg, P., Nakasato, N., & Beniczky, S. (2019). Taking the EEG back into the brain: The power of multiple discrete sources. *Frontiers in Neurology, 10*, 855.
Trepel, M. (2008). *Neuroanatomie: Struktur und Funktion*. Elsevier/Urban & Fischer Verlag.
Vorwerk, J., Wolters, C. H., & Baumgarten, D. (2024). Global sensitivity of EEG source analysis to tissue conductivity uncertainties. *Frontiers in Human Neuroscience, 18*, 1335212.

Chapter 6
Brain States

Cilia Jaeger

Abstract Brain states are patterns of neural activity that correlate with behavior and reflect underlying neural processing. Within the electroencephalography (EEG) recording, brain states can be classified by observing the temporary patterns or fluctuations within the electrical signal over time that can be observed across individuals. Technological advancements in recording and analysis methods have greatly enhanced our ability to study brain dynamics, which has in turn given rise to a better understanding of the structure and function of the brain. In this chapter, we define brain states in the context of EEG and provide examples of research fields studying brain states. We highlight how classification of brain states within specific applications has helped us to understand the function of both the healthy brain and the disruption of function in disease.

Keywords Wakefulness · Sleep states · Anesthesia · Neural synchrony · Oscillatory rhythms · Brain networks

6.1 Introduction

The raw electroencephalography (EEG) signal is a mix of electrical activity generated by different sources of underlying neural populations and other physiological signals (see Chap. 2: "What Is EEG?" for further information on the EEG signal). Yet how exactly can we relate EEG to a neuroscientific question? Neural processing and subsequent brain activity are not static. Rather, brain activity changes with arousal, cognitive tasks, and pathology. Similarly, EEG recordings fluctuate over time. These fluctuations are not only random noise but can instead reflect meaningful patterns of neural activity that can correspond to a functional role. This chapter introduces the concept of brain states—dynamic patterns of neural activity—and explains how EEG can be used to study them.

C. Jaeger (✉)
Brain Products GmbH, Gilching, Germany
e-mail: cilia.jaeger@brainproducts.com

T. Warbrick (ed.), *The EEG Handbook*,
https://doi.org/10.1007/978-3-032-20450-9_6

6.2 Definition of Brain States in the Context of EEG

6.2.1 What Is a Brain State?

A brain state is defined as a recurring, reliable pattern of neural activity observed across large-scale brain networks. These patterns are temporary and serve a functional role in supporting cognitive and behavioral processes (Tang et al., 2012; Kringelbach & Deco, 2020; Lee et al., 2024). Brain states are identified by their neural patterns observed in EEG, functional magnetic resonance imaging (fMRI), or other whole-brain neuroimaging methods (Greene et al., 2023).

While brain states emerge with cognitive or physiological processes, they are distinct from physiological or cognitive states. Brain states reflect the neural configuration that enables or constrains cognitive processes. A cognitive state refers to a specific mental operation, such as attention, memory, meditation, or mental imagery. A physiological state consists of bodily or homeostatic processes, including arousal-related states, sleep, hunger, and stress (Greene et al., 2023).

Brain states are influenced by neuromodulation of neurotransmitter levels and subsequent neural activity. Neuromodulation shapes network coordination and shifts between brain states (Lee & Dan, 2012). Brain states are also affected by environmental factors such as sensory input, task demand, and cognitive workload, which can cause changes in arousal, attention, and emotional responses (Chang et al., 2021; López et al., 2019; Deco et al., 2022).

Studying the transition between brain states provides insight into how the brain adapts information processing to support cognition and behavior and links neural dynamics to mental and physiological functions.

6.2.2 Brain States Measured with EEG

When measuring scalp EEG, dynamic patterns distributed across distinct spatial regions or electrodes can be observed over time in response to specific behavior or cognitive feature within a population of individuals. The distinct and recurring configurations of electrical activity that emerge with a behavioral response or cognitive process correspond to underlying neural activity and function (Buzsaki & Watson, 2012). EEG can be a useful tool to measure these so-called brain states. Patterns observed in either the EEG time-domain or frequency-domain can be used to classify and study the relevance of brain states.

One of the first distinct patterns observed with EEG was an increase in the amplitude of the posterior alpha rhythm that occurs in response to closing the eyes. Since this observation, patterns within the alpha response have been studied within the context of vigilance and attention, sensory processing, and working memory (Siegel et al., 2012). Fluctuations of neural oscillations within distinct

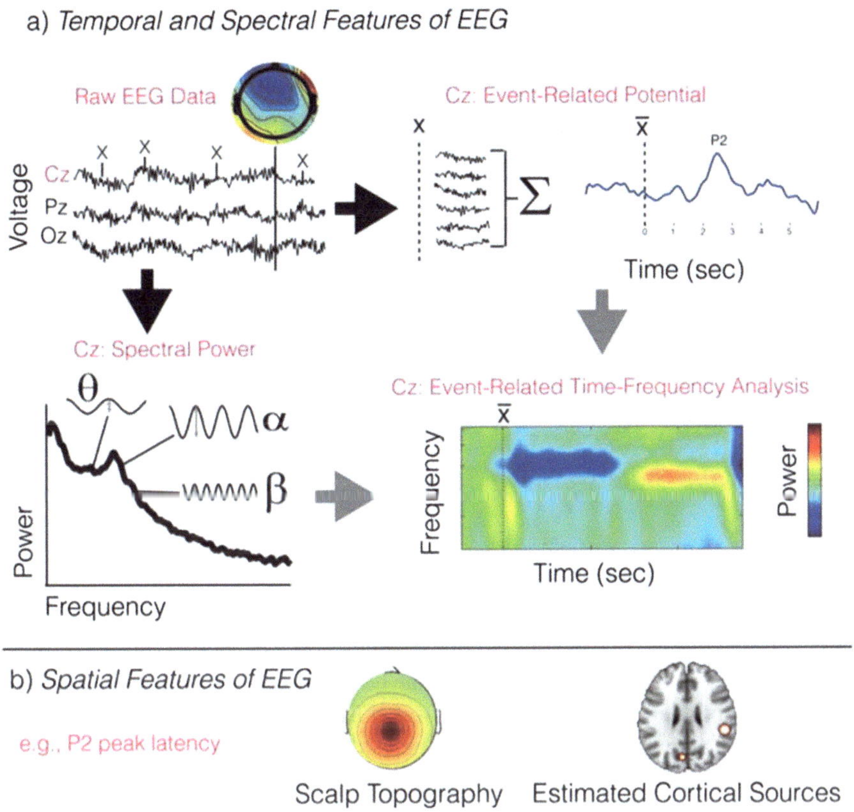

Fig. 6.1 EEG features to characterize brain states. (**a**) Temporal or spectral features can be extracted from the raw EEG data. In the time-domain, the voltage changes evoked in response to a stimulus are averaged together across several repeated trials to measure the event-related potential evoked by the stimulus. The raw EEG data can also be broken down into its component frequencies to investigate activity across neural oscillations. An event-related time-frequency analysis can also be applied to study how the brain's oscillatory activity changes over time in response to a stimulus. (**b**) EEG features can also be extracted spatially by either looking at the spatial distribution of EEG signal properties across scalp topography maps or by source localizing specific EEG features. (Reproduced from Lenartowicz and Loo (2014), with permission

frequency bands are one of the prominent features used to characterize brain states and their link to behavior and cognition (Van Bree et al., 2025; Babiloni et al., 2020).

Frequency-based measures are one example of an EEG feature that can be used to characterize brain states. Time-domain features such as event-related potentials (Woodman, 2010) and connectivity measures such as phase locking value or coherence (Sadaghiani et al., 2022) are other common EEG features used to study brain states. Figure 6.1 provides a brief overview of common EEG features used to study brain states.

The study of brain states not only plays an important role in understanding the function of the brain during different states of consciousness, cognition, and behavior. It also helps study aberrant brain states, in which atypical patterns of neural activity are associated with neurological or psychiatric conditions (Voytek & Knight, 2015; Buzsaki & Watson, 2012). EEG has been a crucial tool to measure dysregulated patterns of electrical activity and the underlying neural dynamics. For example, the onset of epileptic seizures induces a specific pattern of high-amplitude spikes that can be observed across a subset of EEG channels (Andrade-Valenca et al., 2011). The characteristic spiking features and high temporal resolution of EEG have made it an important tool for studying epilepsy.

The following sections highlight examples of brain states that can be recorded with EEG, their role in understanding the function of the brain, and how the classification of brain states allows the study of pathological brain function.

6.3 Examples of Brain States

6.3.1 Awake State ⇔ Sleep State

EEG has been a fundamental tool for linking the circadian rhythm to neural activity patterns and distinct sleep and wakefulness states. The circadian rhythm results from oscillatory gene transcription that drives rhythmic changes in hormone expression and sleep-wake cycles in response to the 24-h day–night cycle (Wright et al., 2012; Mendoza, 2025). Distinct EEG patterns have helped characterize the awake and sleep states and their transitions. During wakeful states, EEG recordings are dominated by high frequencies and low amplitudes that change in response to external stimuli and cognitive demand (Scammell et al., 2017). Sleep can be categorized into two main stages based on a characteristic behavior of rapid eye movements (REM) that is present during REM sleep and lacking during non-REM sleep. During REM, eye movements elicit polarity changes in ocular electrodes and frontal electrodes and desynchronous, low-amplitude, mixed-frequency neural oscillations within EEG recordings (Girardeau & Lopes-Dos-Santos, 2021). Whereas in non-REM sleep, high-amplitude, slow-frequency oscillations dominate the EEG signal (Clawson et al., 2016). During non-REM sleep, transient states of cortical coupling of slow waves and sleep spindles have been linked to memory consolidation (Fogel & Smith, 2011).

Along with a combination of other imaging techniques, such as the structural imaging of nerve tracts, the characterization of neural patterns during the transition from wake to sleep has highlighted key brain structures. It has also revealed the communication patterns that drive circadian rhythms in the body and the brain. Specifically, it has helped clarify the intricate link between the suprachiasmatic nucleus (SCN) in the hypothalamus, thought to be the master clock driving rhythmic gene transcription and hormone regulation, and rhythmic signaling along the hypothalamic nerve cells with other brain structures via neurotransmitter regulation (Mendoza, 2025). Figure 6.2 highlights signaling pathways between the structures in the hypothalamus, the brainstem, thalamus, and cortex that are thought to create a complex feedback loop that regulates arousal, wakefulness, and consciousness.

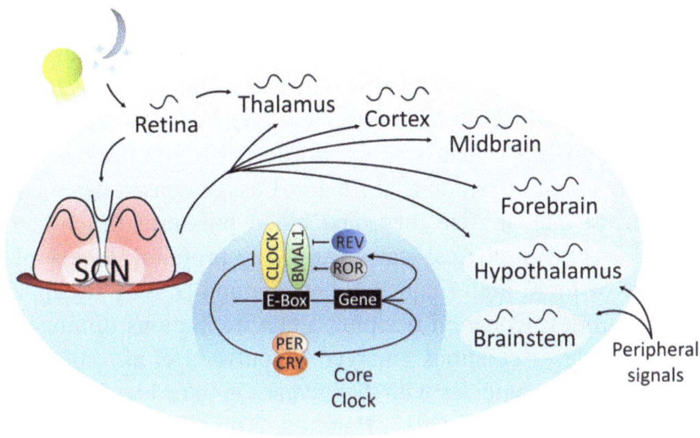

Fig. 6.2 The circadian rhythm. The suprachiasmatic nucleus (SCN) of the hypothalamus acts as the principal clock that regulates rhythmic bodily processes such as sleep–wake cycles and appetite, as well as rhythmic processes of the autonomic nervous system that are important for homeostasis. The SCN neurons have a 24-h rhythm that is generated through a feedback loop of clock gene transcription and protein inhibition. The SCN then transmits electrical signals to other brain regions, which are involved in cognition, mood, and other neural processes. (The figure is reproduced from Mendoza (2025), with permission)

The transition from sleep to wakefulness is characterized by the shift in the prominence of synchronous slow-wave activity to desynchronous high-frequency activity in the EEG. The degree of attenuation of high-amplitude, low-frequency activity has also been linked to varying degrees of arousal and attention (Xu et al., 2021; Wright et al., 2012; von Gall, 2022).

While it is still not fully understood which underlying processes modulate frequency-specific oscillatory activity, one key structure involved in wakefulness and arousal is the reticular activating system (RAS) in the brainstem (Garcia-Rill et al., 2013). Electrical stimulation of the RAS has been shown to cause a shift in EEG signal from the prominence of slow wave synchronicity to an EEG pattern of desynchronous, low-amplitude, high-frequency activity resembling wakefulness (Moruzzi & Magoun, 1949). These EEG patterns of decreased power across high frequencies and decreased complexity in the EEG patterns correlate with a comatose state as compared to an awake state responsive to external stimulus.

6.3.2 *Consciousness States: Presence ⇔ Loss (Anesthesia)*

The RAS is thought to neuromodulate cortical activity via ascending projections to the thalamus, hypothalamus, and cerebral cortex (Kinomura et al., 1996). Measuring differences in neural activity along the ascending RAS pathways has become a main target for studies of consciousness and consciousness disorders. Lesion studies and pharmacological modulation of activity along these pathways have helped define the causal mechanisms behind behavior, brain states, and levels of consciousness

(Edlow et al., 2012; Garcia-Rill et al., 2013; Scammell et al., 2017). Anesthesia is a prime example of how transitions between brain states can be linked to different levels of consciousness. To assess levels of consciousness, general anesthesia can be administered during a cognitive task with more and less salient stimuli. Generally, the responsiveness to less salient stimuli will decrease first, followed by a loss of response to salient stimuli (Purdon et al., 2013). Loss of consciousness is also linked to changes in heart rate, muscle tone, and blood pressure, which can further be monitored to assess physiological states. Applying propofol, a general anesthetic, results in a shift of more high-frequency beta-gamma activity to alpha activity in frontal EEG electrodes. Coherent occipital alpha oscillations diminished, whereas coherent frontal alpha oscillations emerged (Sepulveda et al., 2020). The loss of response to stimuli also coincides with the increase in ultra-low-frequency (less than 1 Hz) EEG power (Purdon et al., 2013; Hagihira, 2015). These EEG patterns reverse when regaining consciousness. By observing EEG patterns under anesthesia, researchers gain controlled, reproducible states that allow them to study the neural dynamics underlying the shifts between wakefulness and unconscious states.

In this section, we have highlighted brain states that exhibit distinct neural patterns and elicit specific behavioral responses, which can be easily studied, such as the transition between wakefulness and sleep or the loss of consciousness induced by anesthesia. The next section explores how defining brain states in healthy individuals can help identify deviations linked to neurological or physiological conditions.

6.4 Pathological Brain States

Brain states have the potential to provide insight into how normal brain functions differ from abnormal states that may be associated with various neurological and psychiatric conditions. This section provides a brief insight into specific electrical patterns recorded in EEG that have been linked to certain pathologies.

6.4.1 Traumatic Brain Injury

EEG may serve as a valuable tool to identify pathological brain states linked to traumatic brain injury. For example, epileptiform, high-amplitude, high-frequency activity has been observed in EEG shortly after a concussion. These high activity patterns are followed by distributed suppression of cortical activity and slowing of oscillatory activity recorded in EEG (Ianof & Anghinah, 2017). Alterations in the frequency of the posterior alpha rhythm have also been observed subacutely within EEG recordings following mild traumatic brain injury (Kadri & Apriani, 2022). While it is difficult to quantify these changes in EEG features within specific individuals experiencing traumatic brain injury, classifying aberrant brain states linked to traumatic events may help identify the impact of brain injury on cognition.

6.4.2 ADHD

Population comparisons between participants diagnosed with and without attention-deficit hyperactivity disorder have revealed elevated theta power and decreased beta power in participants with attention deficit hyperactivity disorder (ADHD) (Newson & Thiagarajan, 2019). Yet the reproducibility of these results has proven difficult in part due to ADHD being a spectrum disorder, and individuals diagnosed with it may exhibit a variety of symptoms. Other EEG features, such as event-related potentials, may be more advantageous for linking deficits in sensory processing and sensory discrimination, a common behavioral outcome in ADHD, with a specific brain state. For example, early event-related components such as N1, P1, and P2 have been linked to early sensory processing and have been found to be altered in ADHD (Kaiser et al., 2020). A study by Kaiser et al. (2021) demonstrated that during a cued continuous performance task, the amplitude of the parietal P3 component was lower in children with ADHD than the control group. This study illustrates how the P3 component, an EEG feature that has been linked with sustained attention (Johnstone et al., 2013) differs between two study populations in response to a given task.

These examples underscore the importance of EEG in understanding various pathologies by highlighting how specific electrical activity patterns are linked to conditions such as epilepsy, concussion, traumatic brain injury (TBI), ADHD, and schizophrenia. Since many neurological and psychiatric disorders present heterogeneously within clinical populations, it has proven challenging to define brain states linked to specific disorders. Yet identifying neural activation patterns corresponding to distinct symptoms in combination offers promising insight into using brain states for diagnosis and treatment options. Some in-depth examples are covered in Chap. 24 ("Clinical Applications").

6.5 Framework of Brain States

So why does the characterization of brain states remain such an integral part of neuroscience research? The brain is highly organized and dynamic, and it is constantly coordinating neural processes linked to complex behavior and cognition. Brain states representing distributed patterns of activity elicited by physiological or cognitive processes also have a functional relevance for future behavioral and physiological responses (Lopez et al. 2019; Tang et al., 2012). So far, this chapter has highlighted the value of identifying distinct brain states. Yet defining brain states linked to more complex cognitive tasks is challenging due to the distributed and less well-defined nature of the underlying neural activity. Within this framework, it is possible that more than one brain state can be present at a given instant of time and can be present across different cognitive tasks (Greene et al., 2023; van Bree et al., 2025; Spitzer & Haegens, 2017). Furthermore, cognitive function is rarely localized in one specific brain region but rather emerges from large-scale networks (Voytek &

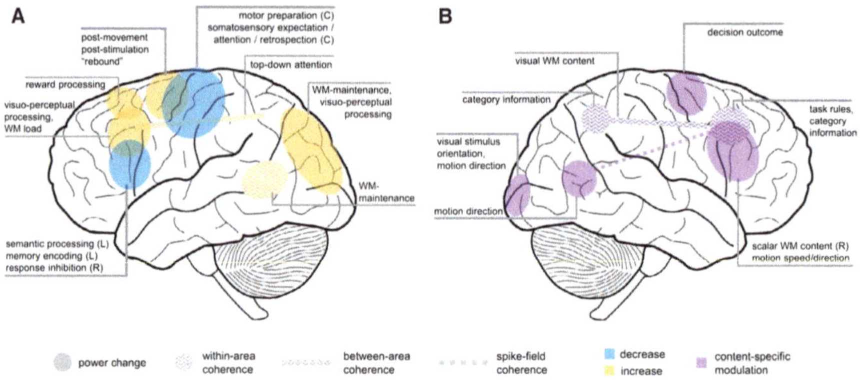

Fig. 6.3 Brain states characterized by differences in neural activity patterns in the beta frequency. (**a**) Changes in beta power, within-area, and between-area coherence have been observed in different brain regions and are also linked with different cognitive processes. (**b**) The location and type of beta-related activity change with respect to content-specific information processing, for example, during working memory. This figure highlights how activity patterns across brain regions can overlap for different cognitive processes, therefore emphasizing the importance of studying brain states and dynamic neural activity patterns. (This figure is reproduced from Spitzer and Haegens (2017), with permission)

Knight, 2015; Siegel et al., 2012). Figure 6.3 depicts different brain regions that have been shown to be activated in multiple cognitive tasks. This makes linking brain states to specific cognitive functions challenging.

A variety of techniques are available for recording whole-brain activity and studying brain function, including positron emission tomography (PET), magneto-encephalography (MEG), functional near-infrared spectroscopy (fNIRS), EEG, and fMRI. There are also tools for modulating brain activity, such as transcranial magnetic stimulation (TMS) or focused ultrasound. Yet these techniques differ with respect to the type of neural signal recorded, as well as the temporal and spatial scales of the signal. Due to the number of imaging techniques and differences in spatial and temporal resolution, brain states may present differently when recorded with different tools (Greene et al., 2023). This can make it challenging to define brain states and has left an incomplete picture of how the dynamics of neural activity from the individual cell level to whole brain activity relate to behavior and function. Advancements in combining neural acquisition techniques and computational modelling provide an opportunity for conceptualizing the relationship between different signal properties, neural dynamics, and brain function in health and disease and have helped decompose complex and dynamic cognitive processes.

For example, within EEG, patterns of frequency-specific oscillatory activity that change in line with a measurable behavioral response in a cognitive task have shaped theories for their role in neural communication. During sensory processing, localized narrow-band high-frequency gamma oscillations correlate with distinct stimuli features such as luminance and color in visual processing (Han et al., 2022). Based on such correlations, localized high-frequency activity is thought to play a role in feature binding and object representation during sensory perceptual processing (Hermes et al., 2015). Whereas correlations between fluctuations in perceptual performance have been linked to fluctuations in the amplitude or phase of low-frequency oscillations (Voytek et al., 2010; Siegel et al., 2012). Furthermore, task performance and attention have also been correlated with synchronization of low-frequency oscillations across distributed brain regions (Canolty & Knight, 2010). These findings suggest low-frequency oscillations may modulate cognitive processes such as attention, which are thought to occur across distributed brain regions.

Dynamic cross-frequency coupling between frequency-specific oscillations in correspondence with cognitive tasks suggests that cross-frequency coupling may coordinate neural activity organized across different temporal scales across different brain regions. For instance, low-frequency alpha coupling with high-frequency gamma oscillations may set the temporal framework for which sensory information is encoded within a specific region (Voytek et al., 2010). Thus, a shift in the cross-frequency alpha-gamma coupling may represent a causal shift in a brain state corresponding to a transition in attentional demand (Canolty & Knight, 2010). Examining changes in cross-frequency coupling across different brain states is just one example of how information obtained from EEG can be interpreted to build a complex model of how the brain can maintain hierarchical organization and flexibility. Further, classifying brain states in combination with other neuroscience techniques can help neuroscientists build a better representation of brain biology and function in a normal state.

More recently, there is growing interest in using EEG to study state-related connectivity, complementing more traditional fMRI connectivity research. While resting state fMRI has been instrumental in mapping large-scale brain networks, the high temporal resolution of EEG is advantageous for studying rapid, real-time brain state dynamics that cannot be captured with fMRI. Studying the fast reconfiguration of neural activity enhances understanding of processes such as cognitive flexibility and adaptability (Sadaghiani et al., 2022). Task-switching studies have found that the reorganization of low-frequency oscillatory networks in preparation for switching tasks appears to support the brain's ability to proactively prepare for and maintain task-relevant information (Lopez et al., 2019).

Historically, large-scale functional connectivity of the brain is more typically studied in MRI. However, EEG connectivity also offers key insights into polyrhythmic brain networks, which are defined as patterns of coordinated communication across multiple frequency bands. Sadaghiani et al. (2022) review evidence showing that slower neural rhythms can organize the timing of faster, local activity. Thereby suggesting that fast, transient, and electrophysiological patterns are linked to large-scale network dynamics more traditionally studied in fMRI. This interplay of connectivity strengthens the argument that rhythmic interactions measured across different neuroimaging techniques coordinate complex neural communication within anatomically and functionally defined networks.

6.6 Concluding Summary

EEG can be used to measure patterns of neural activity corresponding to some behavioral outcome. Studying dynamic patterns in EEG recordings, such as the composition of specific frequency bands within oscillatory activity, has led to a better understanding of how the brain can adaptively process information in relation to behavior and cognition. However, brain states remain incompletely understood because they are dynamic, often overlapping, and can share similar neural signatures across different cognitive or physiological conditions. Furthermore, there is no universal framework for classifying brain states. This is due to different measuring techniques available for measuring brain activity that require special signal processing and are prone to artifacts. These factors can also influence how the neural activity is interpreted.

References

Andrade-Valenca, L. P., Dubeau, F., Mari, F., Zelmann, R., & Gotman, J. (2011). Interictal scalp fast oscillations as a marker of the seizure onset zone. *Neurology, 77*(6), 524–531. https://doi.org/10.1212/WNL.0b013e318228bee2

Babiloni, C., Barry, R. J., Başar, E., Blinowska, K. J., Cichocki, A., Drinkenburg, W. H. I. M., Klimesch, W., Knight, R. T., Lopes Da Silva, F., Nunez, P., Oostenveld, R., Jeong, J., Pascual-Marqui, R., Valdes-Sosa, P., & Hallett, M. (2020). International Federation of Clinical Neurophysiology (IFCN)—EEG research workgroup: Recommendations on frequency and topographic analysis of resting state EEG rhythms. Part 1: Applications in clinical research studies. *Clinical Neurophysiology, 131*(1), 285–307. https://doi.org/10.1016/j.clinph.2019.06.234

Buzsáki, G., & Watson, B. O. (2012). Brain rhythms and neural syntax: Implications for efficient coding of cognitive content and neuropsychiatric disease. *Dialogues in Clinical Neuroscience, 14*(4), 345–367.

Canolty, R. T., & Knight, R. T. (2010). The functional role of cross-frequency coupling. *Trends in Cognitive Sciences, 14*(11), 506–515. https://doi.org/10.1016/j.tics.2010.09.001

Chang, Y., He, C., Tsai, B. Y., & Ko, L. W. (2021). Multi-parameter physiological state monitoring in target detection under real-world settings. *Frontiers in Human Neuroscience, 15*, 785562. https://doi.org/10.3389/fnhum.2021.785562

Clawson, B. C., Durkin, J., & Aton, S. J. (2016). Form and function of sleep spindles across the lifespan. *Neural Plasticity, 2016*, 6936381. https://doi.org/10.1155/2016/6936381

Deco, G., Sanz Perl, Y., Bocaccio, H., Tagliazucchi, E., & Kringelbach, M. L. (2022). The INSIDEOUT framework provides precise signatures of the balance of intrinsic and extrinsic dynamics in brain states. *Communications Biology, 5*(1), 572. https://doi.org/10.1038/s42003-022-03505-7

Edlow, B. L., Takahashi, E., Wu, O., Benner, T., Dai, G., Bu, L., Grant, P. E., Greer, D. M., Greenberg, S. M., Kinney, H. C., & Folkerth, R. D. (2012). Neuroanatomic connectivity of the human ascending arousal system critical to consciousness and its disorders. *Journal of Neuropathology & Experimental Neurology, 71*(6), 531–546. https://doi.org/10.1097/NEN.0b013e3182588293

Fogel, S. M., & Smith, C. T. (2011). The function of the sleep spindle: A physiological index of intelligence and a mechanism for sleep-dependent memory consolidation. *Neuroscience & Biobehavioral Reviews, 35*(5), 1154–1165. https://doi.org/10.1016/j.neubiorev.2010.12.003

Garcia-Rill, E., Kezunovic, N., Hyde, J., Simon, C., Beck, P., & Urbano, F. J. (2013). Coherence and frequency in the reticular activating system (RAS). *Sleep Medicine Reviews, 17*(3), 227–238. https://doi.org/10.1016/j.smrv.2012.06.002

Girardeau, G., & Lopes-Dos-Santos, V. (2021). Brain neural patterns and the memory function of sleep. *Science, 374*(6567), 560–564.

Greene, A.S., Horien, C., Barson, D., Scheinost, D. & Constable, R.T. (2023). Why is everyone talking about brain state? Trends in Neurosciences, 46(7), pp.508–524. https://doi.org/10.1016/j.tins.2023.04.001.

Hagihira, S. (2015). Changes in the electroencephalogram during anaesthesia and their physiological basis. *British Journal of Anaesthesia, 115*(Suppl 1), i27–i31. https://doi.org/10.1093/bja/aev212

Han, C., Shapley, R., & Xing, D. (2022). Gamma rhythms in the visual cortex: Functions and mechanisms. *Cognitive Neurodynamics, 16*(4), 745–756. https://doi.org/10.1007/s11571-021-09767-x

Hermes, D., Miller, K. J., Wandell, B. A., & Winawer, J. (2015). Stimulus dependence of gamma oscillations in human visual cortex. *Cerebral Cortex, 25*(9), 2951–2959. https://doi.org/10.1093/cercor/bhu091

Ianof, J. N., & Anghinah, R. (2017). Traumatic brain injury: An EEG point of view. *Dementia & Neuropsychologia, 11*(1), 3–5. https://doi.org/10.1590/1980-57642016dn11-010002

Johnstone, S. J., Barry, R. J., & Clarke, A. R. (2013). Ten years on: A follow-up review of ERP research in attention-deficit/hyperactivity disorder (1872–8952 (Electronic)). *Clinical Neurophysiology, 124*, 644–657.

Kadri, A., & Apriani, N. (2022). Electroencephalography findings in traumatic brain injury. *The Open Neurology Journal, 16*. https://doi.org/10.2174/1874205X-v16-e2206100

Kaiser, A., Aggensteiner, P. M., Baumeister, S., Holz, N. E., Banaschewski, T., & Brandeis, D. (2020). Earlier versus later cognitive event-related potentials (ERPs) in attention-deficit/hyperactivity disorder (ADHD): A meta-analysis. *Neuroscience and Biobehavioral Reviews, 112*, 117–134. https://doi.org/10.1016/j.neubiorev.2020.01.019

Kaiser, A., Aggensteiner, P. M., Holtmann, M., Fallgatter, A., Romanos, M., Abenova, K., Alm, B., Becker, K., Dopfner, M., Ethofer, T., Freitag, C. M., Geissler, J., Hebebrand, J., Huss, M., Jans, T., Jendreizik, L. T., Ketter, J., Legenbauer, T., Philipsen, A., et al. (2021). EEG data quality: Determinants and impact in a multicenter study of children, adolescents, and adults with attention-deficit/hyperactivity disorder (ADHD). *Brain Sciences, 11*(2). https://doi.org/10.3390/brainsci11020214

Kinomura, S., Larsson, J., Gulyás, B., & Roland, P. E. (1996). Activation by attention of the human reticular formation and thalamic intralaminar nuclei. *Science, 271*(5248), 512–515. https://doi.org/10.1126/science.271.5248.512

Kringelbach, M. L., & Deco, G. (2020). Brain states and transitions: Insights from computational neuroscience. *Cell Reports, 32*(10), 108128. https://doi.org/10.1016/j.celrep.2020.108128

Lee, S. H., & Dan, Y. (2012). Neuromodulation of brain states. *Neuron, 76*(1), 209–222. https://doi.org/10.1016/j.neuron.2012.09.012

Lee, K., Ji, J. L., Fonteneau, C., Berkovitch, L., Rahmati, M., Pan, L., Repovs, G., Krystal, J. H., Murray, J. D., & Anticevic, A. (2024). Human brain state dynamics are highly reproducible and associated with neural and behavioral features. *PLoS Biology, 22*(9), e3002808. https://doi.org/10.1371/journal.pbio.3002808

Lenartowicz, A., & Loo, S. K. (2014). Use of EEG to diagnose ADHD. *Current Psychiatry Reports, 16*(11), 498. https://doi.org/10.1007/s11920-014-0498-0

López, M. E., Pusil, S., Pereda, E., Maestú, F., & Barceló, F. (2019). Dynamic low frequency EEG phase synchronization patterns during proactive control of task switching. *NeuroImage, 186*, 70–82. https://doi.org/10.1016/j.neuroimage.2018.10.068

Mendoza, J. (2025). Brain circadian clocks timing the 24h rhythms of behavior. *npj Biological Timing & Sleep, 2*, 13. https://doi.org/10.1038/s44323-025-00030-8

Moruzzi, G., & Magoun, H. W. (1949). Brain stem reticular formation and activation of the EEG. *Electroencephalography and Clinical Neurophysiology, 1*(1–4), 455–473. https://doi.org/10.1016/0013-4694(49)90219-9

Newson, J.J. & Thiagarajan, T.C. (2019). EEG frequency bands in psychiatric disorders: a review of resting state studies. Frontiers in Human Neuroscience, 12, p.521. https://doi.org/10.3389/fnhum.2018.00521

Purdon, P. L., Pierce, E. T., Mukamel, E. A., Prerau, M. J., Walsh, J. L., Wong, K. F., Salazar-Gomez, A. F., Harrell, P. G., Sampson, A. L., Cimenser, A., Ching, S., Kopell, N. J., Tavares-Stoeckel, C., Habeeb, K., Merhar, R., & Brown, E. N. (2013). Electroencephalogram signatures of loss and recovery of consciousness from propofol. *Proceedings of the National Academy of Sciences of the USA, 110*(12), E1142–E1151. https://doi.org/10.1073/pnas.1221180110

Sadaghiani, S., Brookes, M. J., & Baillet, S. (2022). Connectomics of human electrophysiology. *NeuroImage, 247*, 118788. https://doi.org/10.1016/j.neuroimage.2021.118788

Scammell, T. E., Arrigoni, E., & Lipton, J. O. (2017). Neural circuitry of wakefulness and sleep. *Neuron, 93*(4), 747–765. https://doi.org/10.1016/j.neuron.2017.01.014

Sepulveda, P., Cortinez, L. I., Irani, M., Egana, J. I., Contreras, V., Sanchez Corzo, A., Acosta, I., & Sitaram, R. (2020). Differential frontal alpha oscillations and mechanisms underlying loss of consciousness: A comparison between slow and fast propofol infusion rates. *Anaesthesia, 75*(2), 196–201. https://doi.org/10.1111/anae.14885

Siegel, M., Donner, T., & Engel, A. (2012). Spectral fingerprints of large-scale neuronal interactions. *Nature Reviews Neuroscience, 13*, 121–134. https://doi.org/10.1038/nrn3137

Spitzer, B., & Haegens, S. (2017). Beyond the status quo: A role for beta oscillations in endogenous content (re)activation. *eNeuro, 4*(4), ENEURO.0170-17.2017. https://doi.org/10.1523/ENEURO.0170-17.2017

Tang, Y. Y., Rothbart, M. K., & Posner, M. I. (2012). Neural correlates of establishing, maintaining, and switching brain states. *Trends in Cognitive Sciences, 16*(6), 330–337. https://doi.org/10.1016/j.tics.2012.05.001

Van Bree, S., Levenstein, D., Krause, M. R., Voytek, B., & Gao, R. (2025). Processes and measurements: A framework for understanding neural oscillations in field potentials. *Trends in Cognitive Sciences, 29*(5), 448–466. https://doi.org/10.1016/j.tics.2024.12.003

von Gall, C. (2022). The effects of light and the circadian system on rhythmic brain function. International Journal of Molecular Sciences, 23(5). https://doi.org/10.3390/ijms23052778

Voytek, B., & Knight, R. T. (2015). Dynamic network communication as a unifying neural basis for cognition, development, aging, and disease. *Biological Psychiatry, 77*(12), 1089–1097. https://doi.org/10.1016/j.biopsych.2015.04.016

Voytek, B., Canolty, R. T., Shestyuk, A., Crone, N. E., Parvizi, J., & Knight, R. T. (2010). Shifts in gamma phase-amplitude coupling frequency from theta to alpha over posterior cortex during visual tasks. *Frontiers in Human Neuroscience, 4*, 191. https://doi.org/10.3389/fnhum.2010.00191

Woodman, G. F. (2010). A brief introduction to the use of event-related potentials in studies of perception and attention. *Attention, Perception & Psychophysics, 72*(8), 2031–2046. https://doi.org/10.3758/APP.72.8.203

Wright, K. P., Lowry, C. A., & Lebourgeois, M. K. (2012). Circadian and wakefulness-sleep modulation of cognition in humans. *Frontiers in Molecular Neuroscience, 5*, 50. https://doi.org/10.3389/fnmol.2012.00050

Xu, S., Akioma, M., & Yuan, Z. (2021). Relationship between circadian rhythm and brain cognitive functions. *Frontiers in Optoelectronics, 14*(3), 278–287. https://doi.org/10.1007/s12200-021-1090-y

Chapter 7
Basic Time Concepts in EEG Practice

Shivakumar Viswanathan

Abstract A valuable strategy to understand human brain function is to record neural activity dynamics while a person is engaged in specific cognitive/mental states. However, noninvasive functional recordings are often indirect as with electroencephalography (EEG), where brain activity is recorded from the scalp. Using indirect recordings effectively relies on the assumed relationship between brain activity (what, when, and where it happens) and its corresponding signature in the functional recording. These assumptions guide practical decisions ranging from experiment design to the analysis of the recordings. In this chapter, we discuss basic relational assumptions related to time in EEG practice. Using an example of a trial-based experiment, we illustrate how assumptions relating neural activity to the EEG timeline help to answer three common questions during data analysis: (1) Where is a trial's neural activity in the EEG recording? (2) Where in a trial's activity timeline is a task-relevant process? (3) What does this activity look like? We highlight how these analysis approaches are uniquely supported by the high time-sensitivity of EEG measurements.

Keywords EEG analysis · Time-resolution · Segmentation · Time-locking · Time domain · Frequency domain

7.1 Introduction

The functioning of the brain involves the dynamic activity and interactions of neural populations across the brain. Understanding how this activity enables different capabilities (e.g., memory, perception, cognition) is a major challenge for neuroscience, both in laboratory and clinical settings. One powerful strategy to address this challenge involves recording ongoing brain activity *while* a person is engaged in the capabilities and states of interest. Electroencephalography (EEG) is a widely used

S. Viswanathan (✉)
Brain Products GmbH, Gilching, Germany
e-mail: shivakumar.viswanathan@brainproducts.com

T. Warbrick (ed.), *The EEG Handbook*,
https://doi.org/10.1007/978-3-032-20450-9_7

technology to obtain functional recordings (see Chap. 1: "EEG in Context: Past, Present, and Future"). EEG is the measurement of electrical potentials on the scalp that are generated by the brain's neuro-electrical activity (Buzsáki et al., 2012; Cohen, 2017). Despite being an indirect measure, a distinguishing feature of EEG is its high time-sensitivity. Specifically, rapid changes in neural activity on the millisecond scale can be detected and recorded with scalp EEG measurements, aided by the high time-resolution of modern EEG amplifiers.

Due to EEG's high time-sensitivity, time-related considerations influence how EEG recordings are analyzed and linked to cognitive/neural functions of interest. However, this influence might not always be obvious, especially for researchers who are new to EEG. In this chapter, we seek to make basic time-related considerations more explicit to help readers in their research decisions. The time-related concepts discussed here are general and not specific to a particular experimental design or analysis approach. However, in the interest of simplicity, we focus here on task-related experiments where timing considerations are often crucial. In this context, we discuss a set of practical challenges that arise during analysis and how time-related concepts guide commonly used solutions.

To provide a concrete scenario for our discussion, consider an experiment where participants perform an instructed visual perception task on multiple trials with the simultaneous acquisition of EEG. Each trial begins with the presentation of a visual stimulus, for example, dots that can move either to the left or to the right (Williams & Sekuler, 1984; Kelly & O'Connell, 2013). The participant's task is to make a perceptual decision about the stimulus, for example, are the dots moving to the left or right? This decision is reported with a behavioral response, for example, pressing a designated left or right button. This response completes the trial. The next trial follows after a brief intertrial interval. The simultaneous EEG measurement enables the participant's brain activity to be recorded while they perform the task on trial after trial. We encourage readers to keep this experiment in mind as we will refer to it throughout this chapter.

Table 7.1 illustrates the trial-wise information obtained from a participant in the above task. The stimulus direction (i.e., left or right) defines the experimental

Table 7.1 Example of trial-specific structured information from a participant in a hypothetical visual perception task (see text)

Trial number	Condition (stimulus direction)	Behavioral measures Reported judgment	 Accuracy	 RT(ms)	Neural activity measure (from EEG)
1	Left	Left	Correct	890 ms	?
2	Right	Right	Correct	1340 ms	?
3	Right	Left	Wrong	930 ms	?
4	Left	Left	Correct	1101 ms	?
…	…			…	
200	…			…	?

The final column indicates neural activity measures that need to be obtained from the EEG recording (indicated by "?")

conditions of interest. The behavioral response to the stimulus on a trial provides a measure of response accuracy (was the motion direction judged correctly?) and response time (RT) (elapsed time between stimulus and response). The recorded EEG carries information about the cognitive and neural processes engaged while performing the task. Therefore, the objective is to extract relevant measures of neural activity specific to each trial (last column of Table 7.1). Meeting this objective, however, presents a practical difficulty.

Unlike the single behavioral response to each trial, the EEG recordings produce a large amount of data. Each recording is a structured dataset of voltages over time from multiple electrodes at different scalp locations. Therefore, a strategy is needed to find the appropriate data specific to each trial within the large amount of recorded data. This is a necessary requirement to be able to extract trial-specific neural activity measures. Time-related concepts play a valuable role in addressing this difficulty, as discussed next.

7.2 Where in the Recording Is a Trial's Activity?

An important goal of measuring EEG in a task-based experiment is to obtain brain activity measures *specific to every trial* (as shown in Table 7.1). However, in typical experiments. EEG data is recorded *continuously* over extended durations, namely, across multiple trials without interruption, rather than per trial. Therefore, a basic question has to be answered for every trial *T* (column 1, Table 7.1): *Where in the EEG recording is the activity specific to that trial T?*

A reliable, practical answer to this question is crucial for multiple reasons.

First, the timing and identity of a trial can vary from participant to participant. In typical task-based experiments (as in the example above), trials of different conditions are presented in randomized order across participants. Furthermore, the duration of trials and the intertrial intervals can vary over the experiment and between participants.

Second, EEG is recorded with a high time-resolution. Modern EEG recording systems can have sampling frequencies in the kilohertz range (e.g., 1000 Hz, 2500 Hz, 5000 Hz). For example, a continuous 10-min EEG recording from a single channel with a 1000 Hz sampling rate would produce 600,000 consecutive snapshots (or samples) of the EEG potential at that channel, namely, one sample every 1 ms. Therefore, it is crucial to accurately identify the EEG samples that are specific to each trial.

In EEG research, a common strategy to link each trial to its associated EEG data exploits a general assumption about how EEG is related to neural activity. To explain this strategy, it can be useful to describe the experiment using three timelines, as shown schematically in Fig. 7.1. Unlike the listing of trials only by their order of presentation (as in Table 7.1), the actual time of occurrence of each trial (and its related events) is crucial.

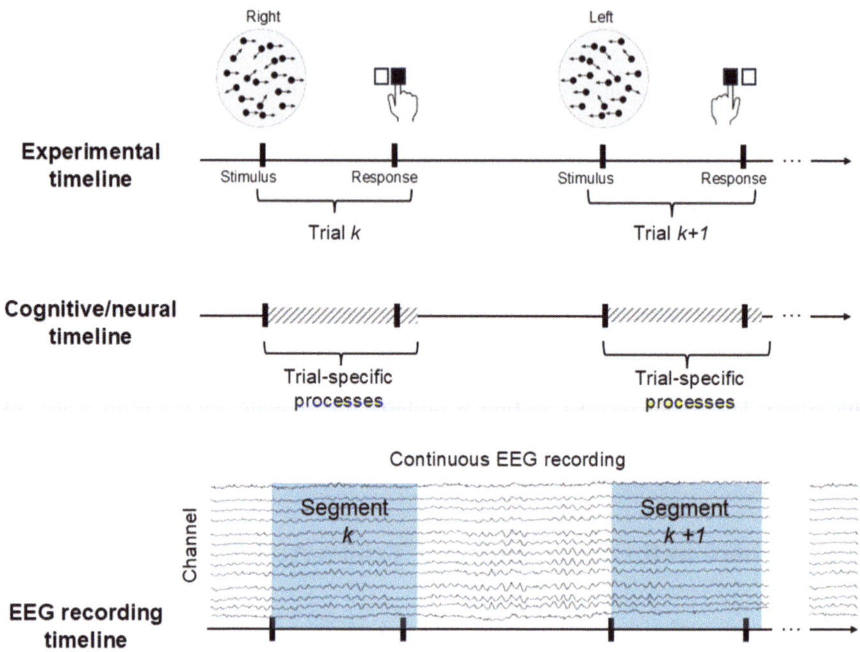

Fig. 7.1 Schematic of three related timelines in a trial-based EEG experiment. The *experimental timeline* (top) shows example events from two consecutive trials of an experiment. Event times are shown by short vertical lines. The *cognitive/neural timeline* (middle) shows intervals of neural activity (hatched lines) related to each trial in the upper timeline. The *EEG recording timeline* (bottom) shows a continuous EEG recording with one voltage time-series (wavy line) per channel. Segments related to trials k and $k+1$ are highlighted

Experimental Timeline This timeline describes the timing of all trials across the experiment (including each trial's related events such as the stimulus onset and response).

Cognitive/Neural Timeline This is the hypothesized timeline of the cognitive/neural processes engaged in performing the task on each trial. In its simplest form, the task-specific neural activity begins with the onset of that trial's stimulus and ends following the behavioral response (shown with hatched lines).

EEG Recording Timeline The continuous EEG recording produces one time-series per recording channel.

The link between each trial (on the experimental timeline) and its EEG activity (on the EEG recording timeline) depends on how the cognitive/neural timeline is linked to the EEG timeline.

Scalp EEG voltages capture the effects of electrical dipoles that are formed by the synchronized activity of neural populations in the brain's cerebral cortex (Buzsáki et al., 2012). These dipoles create electrical fields that are detectable by the scalp electrodes almost immediately (i.e., without timing delays), even though these fields are spatially distorted by the layers between the cortex and the scalp. Specifically, a cortical dipole that forms at time t would be expected to produce a detectable EEG signature at $t + \Delta t$, where Δt is bounded by the sampling period of the EEG recording system. For convenience, we will refer to this as EEG's *immediacy* property.

This immediacy property has useful practical consequences. As shown in Fig. 7.1, the neural activity specific to a trial begins with the onset of that trial's stimulus and ends following the behavioral response. Due to the immediacy property, the trial's EEG signatures would also occur over the same interval. In practical terms, the event timings that define a trial on the experimental timeline also define the interval of the EEG recording with trial-specific activity.

The above rationale is implemented as a routine data-processing operation called *segmentation* by common EEG analysis software. In this operation, timing information related to trial-specific events is used to cut the continuous recording into trial-specific segments. These segments are also referred to as *epochs*, due to their time-based definition. Once obtained, these segments/epochs can then be further analyzed to extract trial-specific neural activity measures.

Accurate trial-specific segmentation depends on the precise recording of when experimental events occur on the EEG timeline. This is an important practical priority during EEG acquisition and is supported by a combination of hardware and software strategies (also see Chap. 14: "Triggers").

In summary, timing-based segmentation enables each trial in an experiment to be associated with a corresponding segment of the EEG recording. This strategy is made possible by EEG's *immediacy* property, that is, the immediate effect of (certain) neural activity patterns on the recorded EEG. Importantly, immediacy is a property of the EEG modality and is not linked to a particular experimental design. For example, this property is *not* shared by popular neuroimaging modalities such as functional magnetic resonance imaging (fMRI), where there is a delay of ~10–15 s between neural activity on the cognitive/neural timeline and its detection on the fMRI recording timeline (Arthurs O, 2002; Logothetis et al., 2001).

7.3 Identifying Task Processes Within a Trial Segment

The term "activity" is an important term in functional neuroimaging. Informally, it can broadly refer to the dynamic state of an entity (e.g., a neural population or network or even the brain) when it is engaged in "doing" something (e.g., a functionally relevant process or computation) in contrast to being disengaged or inactive.

The activity at different time-points during a trial might be associated with different functional processes related to the task. For example, in our hypothetical motion perception experiment, performing the task on each trial involves several cognitive operations between the onset of the stimulus and the response. These operations include the sensory detection of the presented visual stimulus, the perceptual identification of the stimulus (e.g., whether the dots are moving to the left or to the right), the selection of an appropriate response (e.g., press left or right), and operations to physically execute the selected response.

Therefore, in a particular segment, the EEG data at different time-points might be related to different cognitive operations (Cisek, 2007; Pastor-Bernier & Cisek, 2011; Viswanathan et al., 2020). This scenario poses the question: *Where in a trial-specific segment is activity related to a specific cognitive/neural operation?*

In EEG research, one common strategy to address this question is referred to as *time-locking.*

Recall that segmentation involves "cutting" the continuous EEG recording into trial-specific segments. This uses trial-specific times that are defined on the EEG recording timeline (Fig. 7.1). For instance, the segment for trial 48 might start at 320 s and end at 322.5 s on the EEG recording timeline, that is, relative to the start of the entire recording. *Time-locking* is the definition of a new timeline *specific to each segment.* Typically, a characteristic event of the trial, for example, the stimulus onset, is treated as a reference time-point, assigning it the time $t = 0$. Each time-point of a segment is now assigned a time stamp that is "time-locked" to this reference time-point, for example, +240 ms relative to the reference event.

Due to the immediacy property, these segment-specific timelines provide a powerful tool for analyses to identify different underlying processes (discussed below). Therefore, time-locking is often a critical precursor step for event-related analysis approaches, such as event-related potential (ERPs) (Chap. 19: "Event-Related Potentials"), time-frequency analysis of oscillations (Keil et al., 2022; Donoghue et al., 2021), and machine learning approaches (King & Dehaene, 2014; Wolff et al., 2017; Blom et al., 2020). Two common uses of time-locking are discussed below, namely, baseline correction and cross-trial comparison.

7.3.1 Baselines to Distinguish Task Relevant and Irrelevant Activity

The brain is active even when the task of interest is *not* being performed, for example, to maintain and perform various functions, such as breathing, maintaining wakefulness, and holding body posture (Gratton et al., 2018; Raichle, 2015; Hommelsen et al., 2022). Therefore, it is crucial to distinguish task-specific activity from background ongoing activity.

Time-locking provides a convenient solution in many trial-based experimental designs. The first informative event of each trial (e.g., the stimulus in our task example) can be used as the reference event. Therefore, an interval *before* the reference event (i.e., −300 ms to 0 ms) serves as a baseline. Activity during this pre-event baseline interval is uninformed by the details of the reference event. For example, at −100 ms the participant has yet to see the stimulus and does not know its identity. Therefore, EEG differences between the task-relevant interval (i.e., $t > 0$) and the baseline interval (i.e., $t < 0$) are a convenient way to distinguish task-specific activity from background activity. Importantly, this baseline can be defined for each segment.

7.3.2 *Cross-Trial Comparability and Flexible Time-Locking*

Segment-specific timelines enable EEG activity measured at a particular time on one segment (e.g., +220 ms following stimulus on segment #3) to be compared to the corresponding time-point on another relevant segment (e.g., +220 ms following stimulus on segment #78). This comparability can help identify neural processes with a characteristic timing across segments (i.e., time-locked neural processes (Keil et al., 2022)) and whether these processes differ between experimental conditions. Importantly, this cross-segment comparability extends to segments from different participants performing the same task.

However, timestamps are not always sufficient to ensure cross-trial comparability. This is a concern when each trial involves multiple events (e.g., stimulus and response) with a variable duration between these events across trials. For instance, Response Times to the same stimulus can differ between trials due to natural variability, as illustrated in Table 7.1 for trials 1, 3, and 4. As a consequence, time-locking the segment to one event can increase comparability near that event but reduce it for time-points related to the other event.

Figure 7.2a shows the timelines of multiple segments that are time-locked to the stimulus event (green vertical line). Time-points near the stimulus (e.g., $t = +30$ ms) have a comparable context across segments suitable to study cognitive/neural processes that immediately followed the stimulus, for example, related to sensory and perceptual processing. However, near the response event (black vertical line), the same timestamp (e.g., $t = +120$ ms) is associated with activity *before* the response in some segments and with activity *after* the response on other segments. Therefore, time-points around the response are not suited to study pre-response (or post-response) cognitive/neural processes across segments. However, as shown in Fig. 7.2b, this response-related ambiguity can be resolved by time-locking the segments to the response event. Now, activity near the response event (e.g., $t = -30$ ms) has a comparable context across segments, although the tradeoff is that comparability is now lost near the stimulus event (e.g., $t = -120$ ms).

Thus, time-locking EEG data from the same trial to different events provides a flexible strategy to distinguish activity related to different neural processes and their ordering (e.g., before or after an event) (Luck, 2005; Wessel, 2012; Pfurtscheller & Lopes da Silva, 1999; Falkenstein et al., 2000; Gehring et al., 2018; Kornhuber & Deecke, 1964; Viswanathan et al., 2019).

In summary, by using segment-specific timelines, the trial-specific activity can be analyzed further to identify elements that are specific to the cognitive/neural processes being studied. The simplicity and flexibility of this approach are deceptive as this approach depends on EEG's immediacy property. For instance, the flexibility to time-lock the same activity to different events to resolve ambiguity (as shown in Fig. 7.2) is not available to low immediacy methods such as fMRI (Viswanathan et al., 2020; Ploran et al., 2007).

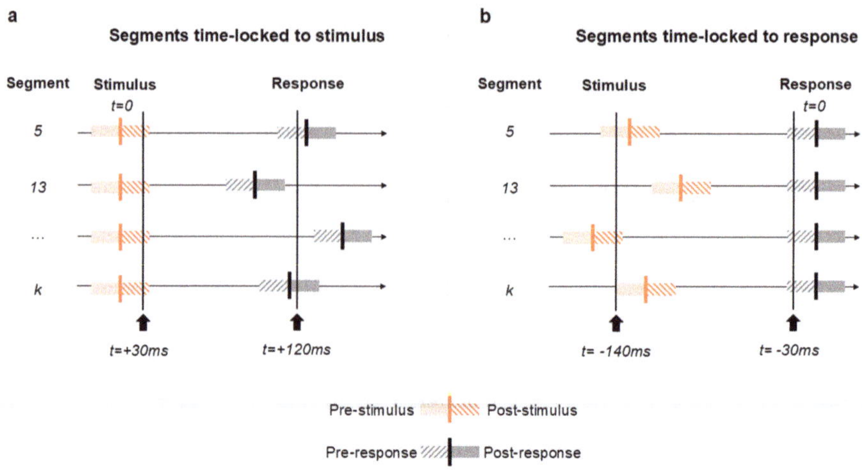

Fig. 7.2 (**a**) Example segment timelines that are time-locked to the stimulus (short vertical line at $t = 0$). On each segment, activity immediately before (pre-) and after (post-) each event is shown with differing hatching patterns. Time-points near the stimulus event have comparable activity across segments, as shown for $t = +30$ ms (poststimulus activity). However, this is not true near the response event, as shown for $t = +120$ ms; some segments have pre-response activity and others post-response activity. (**b**) Timelines of segments time-locked to the response event (short vertical line at $t = 0$). Time-points close to the response event have comparable activity across segments ($t = -30$ ms), but this is not true around the stimulus event ($t = -120$ ms)

7.4 What Does Activity Look Like on the Timeline?

Thus far, we have discussed two timeline-based strategies. The first is to isolate trial-specific activity in the continuous EEG recording (key idea: timeline segmentation). The second is to help isolate task-relevant activity within trial-related segments of the EEG recording (key idea: time-locking). Both of these strategies are based on assumed regularities in the timing relationship between the EEG recording and underlying neural activity. However, these strategies do not make assumptions about the specific form of this activity. In this part, we briefly discuss two distinct formulations of activity based on their relation to time.

Neural activity can fluctuate over time. The issue is how these activity fluctuations might be revealed in the fluctuations of the measured EEG signal. In one formulation, the EEG signal at each time-point is assumed to be an indicator of underlying activity. However, in the second formulation, the assumption is that activity can take different forms, and certain forms of activity are only revealed by signal changes over *time-intervals* (rather than single time-points). The difference between these two formulations is shown in Fig. 7.3 using hypothetical EEG signal changes from a single trial.

In Fig. 7.3a, the magnitude of the signal at each time-point is initially close to zero. This magnitude then deviates away from zero at multiple time-points starting

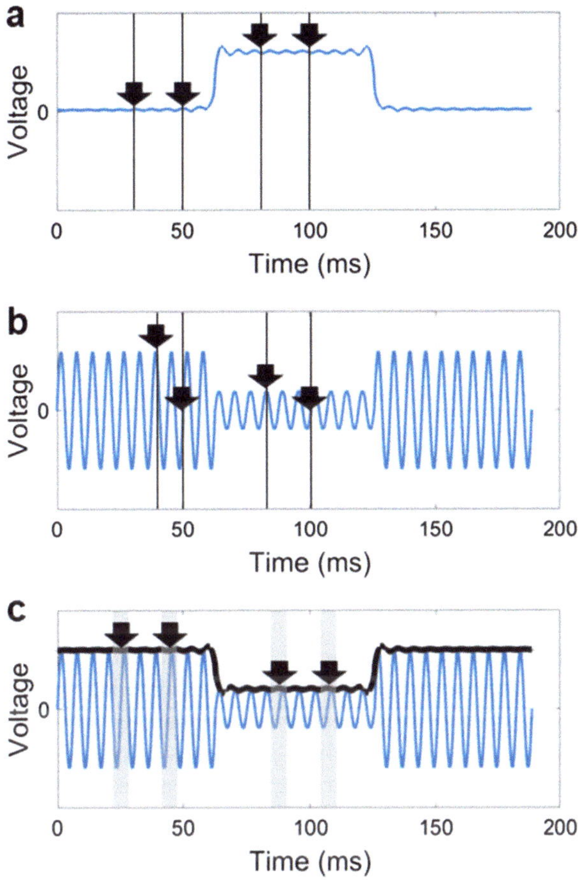

Fig. 7.3 (**a**) Activity based on signal magnitude at different time-points. At time-points before ~60 ms, the signal magnitude is close to zero (examples shown by arrows) but deviates away from zero from ~60 to 130 ms (examples shown by arrows). (**b**) The signal magnitude at single time-points is close to zero and deviates away from zero in quick succession (examples shown with arrows). (**c**) The maximum signal magnitude over short time-intervals (examples highlighted in gray) reveals a coherent pattern across the entire signal duration (thick black line)

from ~60 to ~130 ms before returning to zero. This pattern supports a general interpretation that there is higher activity between ~60 and ~130 ms than at other time-points.

The above interpretation based on time-points is challenged when we consider Fig. 7.3b. The signal deviates away from zero and returns close to zero at multiple time-points over the entire range. The size of these deviations also varies repeatedly. This scenario does not provide a coherent view of activity as in Fig. 7.3a. However, rather than considering the signal values at single time-points, consider the signal changes over time-intervals, that is, across multiple neighboring time-points (highlighted areas in Fig. 7.3c). When we consider the maximum deviation in each of these intervals over the entire signal duration (thick black line), we observe that the activity shows a coherent pattern that is first high before ~60 ms, with a decrease from ~60 to ~130 ms, followed by an increase again.

These two views of activity changes over time are not mutually exclusive, where one is correct and the other incorrect. Instead, they both expand how the concept of activity is practically analyzed. A time-point specific formulation can be appropriate in many scenarios and is the basis for the event-related potentials approach (Chap. 19, event-related potentials). However, an interval-based formulation can be powerful when the signal of interest is an *oscillation*, as in Fig. 7.3b, where the signal periodically completes a full transition from positive to negative (i.e., one cycle) and completes k cycles every second (a frequency of k Hertz (Hz)) (see Chap. 18: "Introduction to EEG Oscillations and Spectral Analysis").

In summary, the flexibility in formulating activity changes over time is possible due to the high time-resolution of EEG. This high time-resolution supports analyses based on single time-points (e.g., every 4 ms with a sampling frequency of 250 Hz) as well as over time-intervals ranging from tens of milliseconds to several seconds.

7.5 Conclusion

In this chapter, we have provided you with an overview of basic time concepts in EEG with a focus on their practical implications. EEG is often referred to as having a high time-resolution. However, the full impact of this feature can be underappreciated. A key takeaway from this chapter is how the high immediacy of EEG measurements can influence practical decisions about how EEG is formulated and analyzed, such as segmentation and time-locking. Importantly, immediacy is a property of the measurement modality and is not linked to a particular experimental design. fMRI also involves continuous recordings and has to address similar concerns about extracting trial-specific measures. However, due to fMRI's low immediacy, options such as segmentation and time-locking are unavailable. Therefore, in conclusion, considering the role of time carefully in your EEG study can maximize the value provided by EEG as an approach to study brain function.

References

Arthurs O, B. S. (2002). Neural origins of the fMRI BOLD signal. *Trends in Neurosciences, 25*, 27–31.

Blom, T., Feuerriegel, D., Johnson, P., Bode, S., & Hogendoorn, H. (2020). Predictions drive neural representations of visual events ahead of incoming sensory information. *Proceedings of the National Academy of Sciences of the United States of America, 117*, 7510–7515.

Buzsáki, G., Anastassiou, C. A., & Koch, C. (2012). The origin of extracellular fields and currents—EEG, ECoG, LFP and spikes. *Nature Reviews. Neuroscience, 13*, 407–420.

Cisek, P. (2007). Cortical mechanisms of action selection: The affordance competition hypothesis. *Philosophical Transactions of the Royal Society of London. Series B, Biological Sciences, 362*, 1585–1599.

Cohen, M. X. (2017). Where does EEG come from and what does it mean? *Trends in Neurosciences, 40*, 208–218.

Donoghue, T., Schaworonkow, N., & Voytek, B. (2021). Methodological considerations for studying neural oscillations. *European Journal of Neuroscience, 55*, 3502–3527.

Falkenstein, M., Hoormann, J. R., Christ, S., & Hohnsbein, J. (2000). ERP components on reaction errors and their functional significance: A tutorial. *Biological Psychology, 51*, 87–107.

Gehring, W. J., Goss, B., Coles, M. G. H., Meyer, D. E., & Donchin, E. (2018). The error-related negativity. *Perspectives on Psychological Science, 13*, 200–204.

Gratton, C., Laumann, T. O., Nielsen, A. N., Greene, D. J., Gordon, E. M., Gilmore, A. W., Nelson, S. M., Coalson, R. S., Snyder, A. Z., Schlaggar, B. L., Dosenbach, N. U. F., & Petersen, S. E. (2018). Functional brain networks are dominated by stable group and individual factors, not cognitive or daily variation. *Neuron, 98*, 439–452.e5.

Hommelsen, M., Viswanathan, S., & Daun, S. (2022). Robustness of individualized inferences from longitudinal resting state EEG dynamics. *European Journal of Neuroscience, 56*, 1–32.

Keil, A., Bernat, E. M., Cohen, M. X., Ding, M., Fabiani, M., Gratton, G., Kappenman, E. S., Maris, E., Mathewson, K. E., Ward, R. T., & Weisz, N. (2022). Recommendations and publication guidelines for studies using frequency domain and time-frequency domain analyses of neural time series. *Psychophysiology, 59*, e14052.

Kelly, S. P., & O'connell, R. G. (2013). Internal and external influences on the rate of sensory evidence accumulation in the human brain. *The Journal of Neuroscience, 33*, 19434–19441.

King, J. R., & Dehaene, S. (2014). Characterizing the dynamics of mental representations: The temporal generalization method. *Trends in Cognitive Sciences, 18*, 203–210.

Kornhuber, H. H., & Deecke, L. (1964). Hirnpotentialänderungen beim Menschen vor und nach Willkurbewegungen, dargestellt mit Magnetband-Speiche- rung und Ruckwartsanalyse. *Pflügers Archiv, 281*, 52–52.

Logothetis, N. K., Pauls, J., Augath, M., Trinath, T., & Oeltermann, A. (2001). Neurophysiological investigation of the basis of the fMRI signal. *Nature, 412*, 150–157.

Luck, S. J. (2005). *An introduction to the event-related potential technique*. MIT Press.

Pastor-Bernier, A., & Cisek, P. (2011). Neural correlates of biased competition in premotor cortex. *Journal of Neuroscience, 31*, 7083–7088.

Pfurtscheller, G., & Lopes Da Silva, F. H. (1999). Event-related EEG/MEG synchronization and desynchronization: Basic principles. *Clinical Neurophysiology, 110*, 1842–1857.

Ploran, E. J., Nelson, S. M., Velanova, K., Donaldson, D. I., Petersen, S. E., & Wheeler, M. E. (2007). Evidence accumulation and the moment of recognition: Dissociating perceptual recognition processes using fMRI. *The Journal of Neuroscience, 27*, 11912–11924.

Raichle, M. E. (2015). The brain's default mode network. *Annual Review of Neuroscience, 38*, 433–447.

Viswanathan, S., Wang, B. A., Abdollahi, R. O., Daun, S., Grefkes, C., & Fink, G. R. (2019). Freely chosen and instructed actions are terminated by different neural mechanisms revealed by kinematics-informed EEG. *NeuroImage, 188*, 26–42.

Viswanathan, S., Abdollahi, R. O., Wang, B. A., Grefkes, C., Fink, G. R., & Daun, S. (2020). A response-locking protocol to boost sensitivity for fMRI-based neurochronometry. *Human Brain Mapping, 41*, 1–19.

Wessel, J. R. (2012). Error awareness and the error-related negativity: Evaluating the first decade of evidence. *Frontiers in Human Neuroscience, 6*, 1–16.

Williams, D. W., & Sekuler, R. (1984). Coherent global motion percepts from stochastic local motions. *Vision Research, 24*, 55–62.

Wolff, M. J., Jochim, J., Akyürek, E. G., & Stokes, M. G. (2017). Dynamic hidden states underlying working-memory-guided behavior. *Nature Neuroscience, 20*(6), 864–871.

Part II
Experimental Design and Good Scientific Practice

Chapter 8
Designing Your EEG Study

Fernando Cross Villasana

Abstract Numerous factors influence the design and implementation of a new research project. The motivations for doing a study may vary: for example, answering a question left from a previous experiment, exploring the variables involved in a novel field, or perhaps devising a practical application. These different aims have profound effects on the way a research question is posed, how the hypothesis is derived, the experimental protocol, and the methods used for data analysis. In EEG studies, designing a study also requires consideration of resource availability, participant welfare, and precision of the measurements. Together, these factors guide the design of studies that aim to generate new and reliable knowledge. This chapter aims to present an integral perspective on study design, while highlighting the various nuances to consider at each stage of the process.

Keywords Experimental design · Study design · Experiment · EEG · ERP

8.1 Introduction

This chapter presents a comprehensive overview to guide the reader through the various stages involved in designing an electroencephalography (EEG) research study. However, each of the topics presented is much more extensive than can be covered here. Readers who require a deeper review of each topic can refer to the literature cited for further detail. References to other chapters in this book are also included. A summary of the design process can be visualized in the following Fig. 8.1.

F. Cross Villasana (✉)
Brain Products GmbH, Gilching, Germany

T. Warbrick (ed.), *The EEG Handbook*,
https://doi.org/10.1007/978-3-032-20450-9_8

1. RESEARCH QUESTION
Narrow the research scope

- Identify gap in knowledge
  - Previous results
  - Practical problems
  - Own experience
- Build theoretical framework
  - Reviews
  - Original research
  - Expert interviews
  - Field investigation
- Evaluate state of the art
- Follow question setting guidelines

2. HYPOTHESIS
Define a testable statement

- Derive one or more from the research question
  - Relationship between variables
- Alternative hypothesis (H_A)
  - Positive statement
  - Independent variable
  - Dependent variable
- Null hypothesis (H_0)
  - Opposite of H_A
  - Gather data against it
  - If not supported: investigate H_A
- Correlational studies
  - Co-occurring variables
- Exploratory studies
  - No prior hypothesis
  - Stricter statistical criteria

3. STUDY DESIGN
Define the data collection strategy

- Choose task or paradigm
- Choose EEG index
- Avoid confounds
- Sample/Group(s)
  - Sufficient size
  - Within subjects design
  - Between subjects design
  - Mixed design
- Gender balance
- Representation/Diversity
- Group comparability
- Statistical power
  - Group design
  - Sample size
  - Effect size
  - Data quality

4. STUDY IMPLEMENTATION
Optimize data collection

- Paradigm implementation
  - Trial number/Amount of data
  - Trial length
    - Baseline
    - Trial timing
  - Randomization
  - Experimental blocks
    - Block length/breaks
    - Counter-balancing
- Measurement precision
  - Adjust hardware
  - Train personnel
- Experiment protocol
- Pilot test

Fig. 8.1 Overview of the study design process

8.2 Setting a Research Question

The aim of research is to address gaps or uncertainties in the current state of knowledge in a particular subject. Phrasing the gap in knowledge as a question is a first step toward narrowing the scope of a research project, defining testable hypotheses, and identifying the appropriate resources to implement the study (Malik & Amin, 2017; Ratan et al., 2019). Therefore, the proper setting of a research question is crucial to the success of a study (Farrugia et al., 2010; Thabane et al., 2009).

The source of a research question is influenced by the context and experience of the researcher. In basic research, the question often emerges from the results of a previous study or from currently open questions within a field (Malik & Amin, 2017). In applied or clinical research, a problem is first identified, then refined research questions are devised with the goal of improving practical outputs (Ratan et al., 2019; Thabane et al., 2009). The research question can also result from an idea that emerged from the researcher's own experience, personal interests, or from letting imagination roam (Thabane et al., 2009).

Deep knowledge of the topic at hand is crucial for forming an informed research question (Farrugia et al., 2010). This knowledge can be gathered from different sources and put together into a *theoretical framework* that will specify the concepts behind the current project (Kivunja, 2018). An important initial step is to perform a review of the literature to identify the current trends, methods, technologies, and open questions in the field (Fandino, 2019; Farrugia et al., 2010). Chapter 36 ("How to Evaluate an EEG Research Paper") contains insights on how to extract the most relevant information from scientific literature. Systematic reviews and overviews of reviews (meta-reviews) are excellent sources to learn the field's broad perspective and current state of the art (Hunt et al., 2018). Reviews can also point toward the original studies that are relevant to the aims of the current research project. It is important to consider one's own limitations and biases when collecting and organizing information (Niso et al., 2022). In this sense, exchange with peers (Niso et al., 2022) and interviews with experts or mentors can be very helpful for refining the research question (Farrugia et al., 2010; Thabane et al., 2009). Experts can provide orientation through the literature and share their firsthand experience on the practical details when researching a particular subject. In applied and clinical contexts, field investigation with the population is particularly useful to identify relevant issues from daily practice that research can address (Farrugia et al., 2010; Thabane et al., 2009).

When building the theoretical framework, a critical analysis of the state of the art will help to refine the research question. How complete is the knowledge to date? Is there an existing theory that clarifies the current research question, or are there perhaps competing theories? Are there only isolated observations and disconnected ideas? What are the current uncertainties? Likewise, when checking singular research reports, consider methodological concerns to further inform the research question: Was a particular sample too small? Were there confounding factors? Were there issues during data collection or processing? Do you detect logical fallacies in the interpretation of results? Further insights into the critical reading of scientific literature can be found in Chap. 36.

The setting of the research question is a dynamic process where an initial question can be adapted as more information is gathered (Doody & Bailey, 2016; Haynes, 2006). Besides topic-specific information, elements like the feasibility of implementation, resource availability, institutional support, or ethical concerns also have influence in shaping the research question (Fandino, 2019; Ratan et al., 2019). Formulating the research question can be challenging. To aid the question setting process, various guidelines have been proposed (e.g. Fandino, 2019; Farrugia et al.,

2010; Ratan et al., 2019; Thabane et al., 2009). For example, the FINER framework provides criteria for assuring the quality of a research question, so that it is Feasible, Interesting, Novel, Ethical, and Relevant (Farrugia et al., 2010; Thabane et al., 2009). Guidelines often include checklists to ensure that the research questions comply with the criteria (e.g. Fandino, 2019). In this way, question-setting guidelines can be a helpful tool for clarifying the research question. It is worth searching for a suitable guideline that applies to the particular research field at hand.

8.3 Setting a Hypothesis

8.3.1 Defining Hypotheses

Once a research question is in place, it is further narrowed into a series of hypotheses. A hypothesis can be defined as a statement that proposes an explanation for a phenomenon. It is a proposed answer to the research question and serves as a basis for empirical testing (Bulajic et al., 2012; Jeong & Kwon, 2006). In EEG, a hypothesis mostly involves a prediction about how a particular electromagnetic index relates to a specific condition, such as an experimental treatment, group membership, or site of recording (Keil et al., 2014). As an example, for the question "Does the resting state alpha power affect subsequent task performance?" a hypothesis can be derived: "Greater resting alpha power correlates with increased performance in a following task." During the research study, evidence is collected that can be in favor of or against the hypothesis. Therefore, the hypothesis is the main guide for what data will be collected and how it will be analyzed (Keil et al., 2014; Thompson & Skau, 2023).

Hypotheses necessarily contemplate a relationship between variables (Thompson & Skau, 2023). In *experimental research*, one variable(s) is manipulated to produce an effect on another variable(s) that is measured by the experimenter. The manipulated variable is known as the *independent variable (IV)*, while the affected variable is the *dependent variable (DV)*. For example, in the hypothesis "Greater task difficulty produces greater mid-line theta amplitude," task difficulty is the independent variable that is manipulated by the experimenter. Meanwhile, the amplitude of midline theta induced by the task is the dependent variable.

Not all studies are experimental, and other approaches to research can be implemented. In *correlational research*, no variable is manipulated; rather, the relationship between co-occurring variables is investigated. A hypothesis of this type can be: "Alpha peak frequency increases with age between childhood and adolescence". In this case, neither alpha peak frequency nor the ageing of the participants is deliberately manipulated, but both are measured in search of a correlation between them.

In *exploratory studies*, measurements can be first performed without a specific hypothesis with the aim of identifying relevant variables. Since exploratory studies involve testing multiple variables without a hypothesis, stricter statistical controls are necessary to prevent outputs produced by chance from being perceived as meaningful observations (Luck & Gaspelin, 2017; Szucs & Ioannidis, 2017). Sometimes, incidental findings are encountered during experimental or correlational studies that were not predicted by the hypothesis. Any new hypothesis derived from observations during the analysis phase should be regarded as exploratory and receive the corresponding statistical controls. This prevents making an interpretation of observations that potentially resulted from chance (Luck & Gaspelin, 2017; Puce & Hamalainen, 2017).

8.3.2 Testing the Hypothesis

To test a hypothesis using statistics, the most common approach is *null hypothesis testing*. In this approach, the experimental hypothesis must be accompanied by a *null hypothesis* (H_0), which proposes that no relationship exists between the given variables. When paired with the null hypothesis, the experimental hypothesis becomes the *alternative hypothesis* (H_A). Testing the null hypothesis requires gathering evidence against the proposal that there is no relationship between the experimental variables (Pernet, 2015). Rejecting the null hypothesis does not automatically mean that the alternative hypothesis is true (Pernet, 2015; Szucs & Ioannidis, 2017). However, it indicates that a deeper investigation is needed into the relationship between the variables. This should be based on the theoretical framework and statistical analyses (Luck & Gaspelin, 2017; Szucs & Ioannidis, 2017).

Null hypothesis testing is a widely used method for scientific inference in neuroscience and related fields. However it has limitations and has received criticism, mostly regarding misunderstandings about the meaning of rejecting the null hypothesis, and the statistical criteria used for rejection (Szucs & Ioannidis, 2017). This has led to a series of recommendations and complementary approaches to strengthen null hypothesis testing. From an experimental design perspective, recent years have seen an emphasis on pre-registration of experiments (Niso et al., 2022). In this way, the public can identify the original hypotheses, the theoretical background, and the intended methodology. Great emphasis has also been placed on doing replication studies to strengthen previous observations and facilitate new predictions (Luck & Gaspelin, 2017; Nebe et al., 2023; Niso et al., 2022). Data sharing allows other researchers to confirm and aggregate results from previous studies (Nebe et al., 2023). Having a solid theoretical background with clear predictions is also important to support arguments in favor of the experimental effects after rejecting the null hypothesis (Luck & Gaspelin, 2017; Szucs & Ioannidis, 2017). The former proposals are accompanied by recommendations for statistics, such as using power estimation to find an optimal sample size, and reporting the effect size of the results (Keil et al., 2014; Niso et al., 2022; Szucs & Ioannidis, 2017).

8.4 Design of the Study

With the alternative and null hypotheses set in place, it is time to design a study where they can be put to test. The aim is to fine-tune all factors involved in data collection to obtain the clearest possible evidence for testing the hypotheses. Importantly, all procedures decided upon must comply with ethical guidelines (e.g., country regulations) and be approved by the local ethics committee.

The nuances of experimental design can be complex. However, the key point is to make it possible to attribute the effects observed on the dependent variable to the manipulation of the independent variable. In other words, the potential effects of other variables, such as noise in the recordings, length of the experiment, participant age, or random variation, should be minimized through effective experimental design. These additional variables that affect the measurement are known as *confounds*.

In EEG studies, the factors to fine-tune are many indeed. These include elements such as the criteria for participant selection, sample size, number of trials, hardware settings, data processing methods, or participant well-being. Often, a decision regarding one factor will affect other factors, which is why fine-tuning them frequently involves trade-offs. Each research study is unique, requiring its own set of decisions during its design. A good initial source of information when designing a study is found in previous publications using similar methods. Additionally, consulting sources on good scientific practice (e.g. Niso et al., 2022), technical guidelines (e.g. Keil et al., 2014), and literature about the EEG index used in the study can guide well-informed decisions during experimental design. In the following, we will cover the main aspects to consider.

8.4.1 Contextualization of the Hypothesis

The main questions at this point are how to generate the evidence necessary to test the hypothesis, what strategies to follow to obtain the data, and how the data generated will be analyzed.

8.4.1.1 Experimental Paradigm

The choice of experimental task or paradigm, and the situation in which the data will be recorded, should be suitable to assess the neurocognitive process under investigation. There are two main kinds of paradigms used in human neuroscience. In neurocognitive studies, the most common kind by far is the *event-related-design*, where brain activity is analyzed around the occurrence of a discrete event, such as the presentation of a sensory stimulus, the execution of a response, or other actions. This design is useful to track the fast pace of brain responses during perception and cognition. Multiple trials of such events are required to obtain a stable response and detect experimental effects. The other kind is the *block-design*. In this approach, brain activity is tracked over longer periods of time, allowing the study of persistent states of the brain over minutes or hours, for example, vigilance levels or resting state connectivity. Though most studies focus on one of these approaches, event-related and blocked designs are compatible with each other.

Multiple standard experimental paradigms have been developed to test faculties such as memory, perception, attention, vigilance, or motor control. A repository with freely available scripts for diverse paradigms is the ERP-Core database (Kappenman et al., 2021). Besides standard experimental paradigms, some studies require a specific paradigm to be designed from scratch. Examples include current efforts for experiments to approximate naturalistic stimuli (e.g., video clips), with the aim of increasing the experiment's ecological validity. In the same direction, out-of-lab recordings with real-world tasks have become increasingly common, enabled by mobile EEG devices (Cruz-Garza et al., 2017). Some examples are sporting activities (e.g. Cheron et al., 2016) or machinery operation (e.g. Borghini et al., 2014). These paradigms should be well justified in the theoretical framework to support their validity for investigating the neurocognitive processes of interest.

8.4.1.2 EEG Index

The EEG index of choice should reflect the psychological or physiological processes under investigation. EEG has the particularity that multiple indexes for analysis can be derived from a single measurement. For example, multiple components can be analyzed from an ERP waveform, or multiple frequency bands from spectral analysis, all derived from multiple channels (Cohen, 2014; Luck & Gaspelin, 2017). It is generally advised to base the hypothesis on fewer variables that are well supported by the theoretical framework (Luck & Gaspelin, 2017). Adding more EEG indexes to the study means adding more variables to the analysis, hence increasing the likelihood of spurious findings in one of them by chance. When having more variables, more stringent statistical criteria would be required during the analysis phase, which entails the risk of dismissing a relevant observation. Of course, besides the main hypothesis, further explorative analyses are possible, always using suitable statistical techniques (Luck & Gaspelin, 2017; Puce & Hamalainen, 2017).

8.4.1.3 Group/Sample

The sample from the population must be suitable for obtaining the necessary data. In basic neurocognitive research, participants are most often young, healthy adults, partly because of their accessibility, but also their suitability to research basic cognitive phenomena. To reduce influence from external variables, participants are screened regarding their health and their cognitive and perceptual status (Keil et al., 2014). Age, gender, and social status of the participants need to be accounted for, keeping participants in a comparable age range and with balanced gender representation, while also reporting where and how the participants were recruited (Keil et al., 2014).

In recent years, numerous calls have been made to increase the diversity of participants in research samples and improve gender, ethnic, and social representation (Niso et al., 2022). It is further argued that since participants most often come from communities near universities and research centers, most samples have W.E.I.R.D. characteristics, standing for Western, Educated, Industrialized, Rich, and Democratic (Nebe et al., 2023; Niso et al., 2022). Such biased samples can limit the generalizability of the observations to the overall population. Therefore, calls have been made for greater efforts to increase diversity in the samples and to reach out to further communities to do research (Niso et al., 2022).

The number of groups and experimental conditions to test is an important factor to consider as it affects the kind of statistical tests and the number of trials required to enhance statistical sensibility (Boudewyn et al., 2018; Puce & Hamalainen, 2017). The different designs can be visualized in Fig. 8.2. A common framework is to test two or more experimental conditions in a single group, called a *within-subjects design*. When two or more groups are compared in a single condition, it is referred to as a *between-subjects design*. A combination where more than one group is tested in more than one condition represents a *mixed design*. In basic neurocognitive research, within-subjects designs are generally preferred as they reduce intersubject variability. In other words, within-subjects designs have greater statistical power. Within-subjects designs can further enhance their statistical power by increasing the number of trials from each participant (Boudewyn et al., 2018). While between-subjects and mixed designs also profit from having more trials (or otherwise more data), it has been observed that between-subjects mostly enhances statistical power with a greater number of participants, likely by compensating for intersubject variability (Boudewyn et al., 2018).

In between-subjects and mixed designs, it is important that the overall demographic characteristics of groups such as age, education, and gender balance are comparable. This can be visualized in Fig. 8.2. In clinical studies, efforts are made to match the participants of the healthy control group with those of the clinical group (e.g., an educated 30-year-old male patient is paired with an educated male of similar age in the control group). It must also be reported that the groups do not show statistically significant differences in these variables to ensure they do not act as a confound.

The number of participants in the groups, together with the strength of the experimental effect, is determinant of statistical power. Power calculators are available to estimate the number of participants required for reliable hypothesis testing according to an expected effect size seen in previous reports. If no previous effect size is available, power estimation can be based on a conservative estimate of the expected effect size. Importantly, if EEG is being combined with other techniques or measures, the effect size to consider for power estimation should be based on the measure with the weakest effect (Nebe et al., 2023).

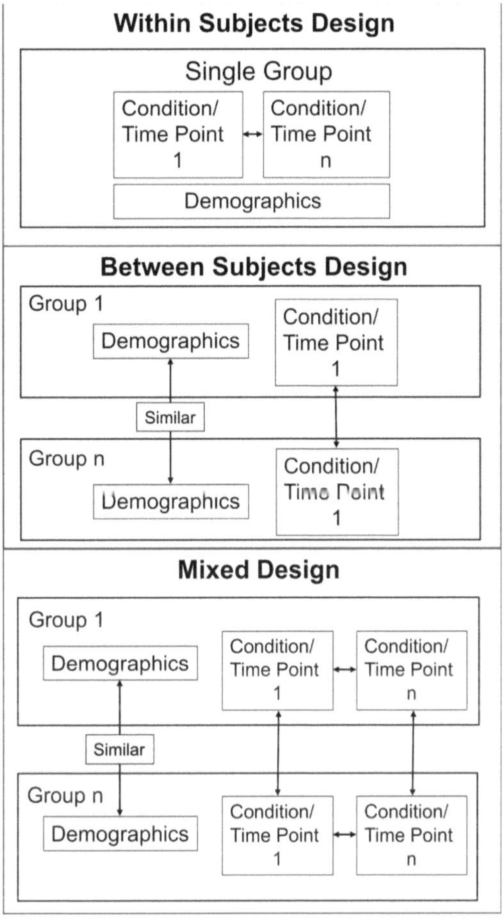

Fig. 8.2 Experimental designs considering groups and experimental conditions and/or points of measurement

8.4.2 Implementation of the Study

Having identified the strategy and main elements required for testing the hypothesis, the next step is to implement the strategy efficiently to obtain data of the highest possible quality. Besides the number of participants, the precision of measurements is an important contributor to statistical power, especially when the ideal sample size cannot be reached for practical reasons (Nebe et al., 2023). This requires the fine-tuning of the experimental paradigm and of all settings and procedures for data collection. Likewise, it is important to keep in mind the requirements of the planned data processing and statistical analysis, to adjust data collection accordingly (e.g., having a suitable baseline length, collecting enough trials).

8.4.2.1 Paradigm/Task Implementation

The chosen experimental paradigm needs to be optimized to the current requirements. The aim is to gather sufficient data for analysis while controlling for confounds. The different parameters that play a role in this are summarized in Table 8.1 and described in more detail below.

In event-related designs, this optimization mainly entails the number of trials. In principle, more trials improve the EEG signal-to-noise ratio and stabilize random variability. However, too many trials may exhaust participants and alter the brain's responses, introducing a confound. Hence, a balance must be found. One must also consider that some trials will be lost due to artifacts or due to participant error. The number of trials per condition is a delicate decision. Useful references for deciding the number of trials are previous studies with similar designs, as well as specific literature about the EEG index being used (e.g., P300 component, theta oscillation, connectivity measures). Some studies have been specifically conducted to assess the impact of the number of trials on different EEG signals. For example, Boudewyn et al. (2018) show complex interactions between the ERP component used, the use of between- or within-subjects design, and the effect size to define adequate trial numbers. But overall, increasing the number of trials improved statistical power. Borras et al. (2022), with various motor signals, suggest 50 trials as a reasonable standard, while cautioning that adjustments may be needed for different circumstances (e.g., small effect sizes, clinical populations). Both authors advise having a balance between trial numbers and participant well-being.

The order in which trials from different conditions are presented is also important to prevent confounds: If all trials from Condition A are presented first, and all trials from Condition B come second, it could happen that the state of the participant

Table 8.1 Trial and block checklist

Trial number/amount of data
Sufficient to reduce variability and improve signal quality, compensate for loss of trials/data
Not excessive, to avoid overly long sessions
Trial timing
Baseline length: Sufficient for the signal of interest
Trial length: Captures signal of interest
Intertrial interval: Enough to recede trial aftereffects and jittered to prevent habituation and expectation
Order of trials
Randomized: To counter learning and exhaustion confounds
Sequential: If justified by the design: E.g., easy condition first, difficult second
Blocks
Block length: Appropriate to keep focus
Number of blocks: Suitable to collect the required trials or sufficient data
Counterbalancing: Shuffle starting block between-subjects (prevent confounds)
Breaks between blocks: Long enough for the participant to recover

varies between them, affecting task performance. For example, participants might not yet be used to the task during Condition A, while they may be tired for Condition B. Through *randomization* of the trial order, a balance is achieved where any learning, exhaustion, or repetition effects affect all experimental conditions equally.

The length of each individual trial should be determined by the task requirements and the EEG index of interest. This involves the duration of stimulus presentation, time for the participant to respond, and time for EEG modulations to show. Between trials or within trials with sequences of stimuli (e.g., cue and target), the timing between stimuli needs to be considered. The stimulus-onset-asynchrony (SOA) is the period between the onset points of two stimuli, and the inter-stimulus-interval (ISI) is the period between the offset of one stimulus presentation and the onset of the following stimulus. Both parameters are calibrated to prevent the overlap of neural responses to subsequent stimuli and adjust task difficulty (Cohen, 2014; Luck, 2014). The timing between trials is another factor that affects the measurement and the length of the experiment. An intertrial interval (ITI) is necessary for any aftereffects to recede, and avoid the overlap of neural processes between trials. Adding a jitter on top of a standard ITI is helpful to prevent habituation and expectation effects. The precise length of these timing parameters depends on the paradigm in question, and it is recommended to base these settings on the literature from the field in question.

In most cases each trial contains a baseline period before the stimulus or event, which facilitates quantifying the EEG modulations with respect to a reference period. The length of the baseline varies depending on the EEG signal. In ERP studies, baselines of hundreds of milliseconds are common (e.g. Feuerriegel & Bode, 2022), but the ideal baseline length is still debated (e.g. Delorme, 2023; Feuerriegel & Bode, 2022). For time-frequency analyses longer baselines are required, which are adjusted according to the wave-length of the oscillation in question, so that baselines can reach lengths of seconds (Cohen, 2014).

Most neurocognitive experiments are long; therefore, it is necessary to break the stimuli presentation into separate *blocks* with equal trial numbers and introduce breaks between blocks for participants to rest and recover. During breaks, participants can stop attending to the stimuli, relax, and move freely. Besides individual trials, blocks also serve as a way to introduce experimental conditions, for example, in block designs with continuous measurements, but also in event-related designs. *Counterbalancing* refers to a systematic variation in the way participants are exposed to the experiment's settings and conditions most often related to the blocks. For example, in some experimental tasks, the response hand is switched between blocks as a control measure. Counterbalancing in this case implies alternating the initial response hand across participants and switching the response in each successive block. Counterbalancing prevents the response hand itself from becoming a confound. This measure can be used on top of randomization for better variable control. But in certain designs where randomization is not possible, counterbalancing can still be used to prevent confounds. For example, an experiment may require all trials within a block to be Condition A or Condition B without randomizing, but counterbalancing can still be used to alternate which is the initial condition for each participant.

8.4.2.2 Measurement Precision

Recording conditions should be arranged to ensure that noise is reduced to a minimum, events during the recording are timed with precision, and that the data contains the signal of interest. This requires setting up the hardware to optimize the measurement and reduce artifacts, which is described in Chap. 15 ("Getting Clean Data: Artifacts and How to Prevent Them). However, the personnel who operate the devices are just as important. Operators should be well-trained in setting up the EEG and other devices accurately and in an adequate time frame, so that suitable data are obtained without exhausting participants (Nebe et al., 2023). Operators also need to monitor all devices during acquisition and make adjustments when necessary, for example, check the cap position and electrode impedance during block breaks.

The artifact handling strategy must be considered during the study design, including the signal of interest, recording situation, and EEG preprocessing pipeline. To illustrate, one may expect different artifacts from healthy participants in the laboratory, patients in a clinic, or mobile recordings. According to the situation, one may consider requesting healthy young participants to try to reduce their blinking to avoid this artifact. In contrast, in a clinical study, let patients blink freely to assure their well-being, and later use blink correction techniques on the data. Meanwhile, for mobile or patient data, where it is known that the data will be noisier, one may opt to record more data/trials compared to standard lab studies, and use advanced data processing techniques.

Some EEG signals require special controls regarding artifacts, and this must be considered during design. For example, accurate detection of eye movements is required when investigating the gamma oscillation, because activation of ocular muscles generates spikes with energy in the same frequency range (Niso et al., 2022). Also consider the processing techniques in relation to the signal of interest and recording situation, taking care that the processing applied to the data does not drastically affect the signal (Niso et al., 2022; Pernet et al., 2020).

8.4.2.3 Experimental Protocol

An experimental protocol specifies the procedures that the operators must follow for collecting data. It indicates the number of required sessions, and within each session, it specifies the duration and steps to follow (e.g., welcoming participants, signing informed consent, recording data, debriefing).

For counterbalancing purposes, the protocol must identify factors such as the experimental conditions that apply to each participant, which block or response mapping is applied first, or the block order to follow. Requirements for participants must also be specified. This involves control of substance use, such as caffeine, alcohol, or cigarettes (e.g., avoid caffeine 2 h before the experiment, in some laboratories), and accounting for medication status (e.g., on-medication, off-medication). Other common requirements are having normal or corrected to normal vision/perception, washing hair, avoiding the use of hair products, and having good sleep the night before. The time of the day is another important factor, as circadian rhythms and participants' routines affect the brain state and performance. As a general rule, working hours between 09:00 and 17:00 are suitable so participants are not tired or hungry. But optimal timing must be confirmed with each individual participant, considering their individual routines. Chapter 11 ("Study Workflow and Lab Management") includes a detailed overview of the nuances of implementing an EEG experiment and the day-to-day administration of an EEG laboratory.

8.4.2.4 Pilot Testing

Running a pilot test that involves all aspects from participant recruitment to data analysis is the best way to ensure that everything works as intended before the real experiment. If a problem was detected at any stage, it is possible to correct and adjust settings. You can consult Chap. 10 ("Pilot Testing") for a detailed review of the piloting process.

8.5 Concluding Summary

Careful study design is critical for the success of a research project. It is always worth spending sufficient time on ensuring that all elements are in place to guarantee a successful study. In this chapter, we provided an overview of the most important factors to consider while designing EEG studies. However, it is always advisable to review similar studies to the planned study design to get the most detailed information on the requirements specific to the area of study.

References

Borghini, G., Astolfi, L., Vecchiato, G., Mattia, D., & Babiloni, F. (2014). Measuring neurophysiological signals in aircraft pilots and car drivers for the assessment of mental workload, fatigue and drowsiness. *Neuroscience and Biobehavioral Reviews, 44*, 58–75. https://doi.org/10.1016/j.neubiorev.2012.10.003

Borras, M., Romero, S., Alonso, J. F., Bachiller, A., Serna, L. Y., Migliorelli, C., & Mananas, M. A. (2022). Influence of the number of trials on evoked motor cortical activity in EEG recordings. *Journal of Neural Engineering, 19*(4), 046050. https://doi.org/10.1088/1741-2552/ac86f5

Boudewyn, M. A., Luck, S. J., Farrens, J. L., & Kappenman, E. S. (2018). How many trials does it take to get a significant ERP effect? It depends. *Psychophysiology, 55*(6), e13049. https://doi.org/10.1111/psyp.13049

Bulajic, A., Stamatovic, M., & Cvetanovic, S. (2012). The importance of defining the hypothesis in scientific research. *International Journal of Education Administration and Policy Studies, 4*(8), 170–176.

Cheron, G., Petit, G., Cheron, J., Leroy, A., Cebolla, A., Cevallos, C., Petieau, M., Hoellinger, T., Zarka, D., Clarinval, A. M., & Dan, B. (2016). Brain oscillations in sport: Toward EEG biomarkers of performance. *Frontiers in Psychology, 7*, 246. https://doi.org/10.3389/fpsyg.2016.00246

Cohen, M. (2014). *Analyzing neural time series data: Theory and practice*. MIT Press. https://doi.org/10.7551/mitpress/9609.001.0001

Cruz-Garza, J. G., Brantley, J. A., Nakagome, S., Kontson, K., Megjhani, M., Robleto, D., & Contreras-Vidal, J. L. (2017). Deployment of mobile EEG technology in an art museum setting: Evaluation of signal quality and usability. *Frontiers in Human Neuroscience, 11*, 527. https://doi.org/10.3389/fnhum.2017.00527

Delorme, A. (2023). EEG is better left alone. *Scientific Reports, 13*(1), 2372. https://doi.org/10.1038/s41598-023-27528-0

Doody, O., & Bailey, M. E. (2016). Setting a research question, aim and objective. *Nurse Researcher, 23*(4), 19–23. https://doi.org/10.7748/nr.23.4.19.s5

Fandino, W. (2019). Formulating a good research question: Pearls and pitfalls. *Indian Journal of Anaesthesia, 63*(8), 611–616. https://doi.org/10.4103/ija.IJA_198_19

Farrugia, P., Petrisor, B. A., Farrokhyar, F., & Bhandari, M. (2010). Practical tips for surgical research: Research questions, hypotheses and objectives. *Canadian Journal of Surgery, 53*(4), 278–281. https://www.ncbi.nlm.nih.gov/pubmed/20646403

Feuerriegel, D., & Bode, S. (2022). Bring a map when exploring the ERP data processing multiverse: A commentary on Clayson et al. 2021. *NeuroImage, 259*, 119443. https://doi.org/10.1016/j.neuroimage.2022.119443

Haynes, B. (2006). Forming research questions. *Journal of Clinical Epidemiology, 59*(9), 881–886. https://doi.org/10.1016/j.jclinepi.2006.06.006

Hunt, H., Pollock, A., Campbell, P., Estcourt, L., & Brunton, G. (2018). An introduction to overviews of reviews: Planning a relevant research question and objective for an overview. *Systematic Reviews, 7*(1), 39. https://doi.org/10.1186/s13643-018-0695-8

Jeong, J.-S., & Kwon, Y.-J. (2006). Definition of scientific hypothesis: A generalization or a causal explanation? *Journal of the Korean Association for Research in Science Education, 26*, 637–645.

Kappenman, E. S., Farrens, J. L., Zhang, W., Stewart, A. X., & Luck, S. J. (2021). ERP CORE: An open resource for human event-related potential research. *NeuroImage, 225*, 117465. https://doi.org/10.1016/j.neuroimage.2020.117465

Keil, A., Debener, S., Gratton, G., Junghofer, M., Kappenman, E. S., Luck, S. J., Luu, P., Miller, G. A., & Yee, C. M. (2014). Committee report: Publication guidelines and recommendations for studies using electroencephalography and magnetoencephalography. *Psychophysiology, 51*(1), 1–21. https://doi.org/10.1111/psyp.12147

Kivunja, C. (2018). Distinguishing between theory, theoretical framework, and conceptual framework: A systematic review of lessons from the field. *International Journal of Higher Education, 7*(6), 14566–14566. https://doi.org/10.5430/ijhe.v7n6p44

Luck, S. J. (2014). *An introduction to the event-related potential technique, second edition*. MIT Press. https://books.google.de/books?id=SzavAwAAQBAJ

Luck, S. J., & Gaspelin, N. (2017). How to get statistically significant effects in any ERP experiment (and why you shouldn't). *Psychophysiology, 54*(1), 146–157. https://doi.org/10.1111/psyp.12639

Malik, A., & Amin, H. U. (2017). *Designing EEG experiments for studying the brain 1st edition design code and example datasets*. Academic.

Nebe, S., Reutter, M., Baker, D. H., Bölte, J., Domes, G., Gamer, M., Gärtner, A., Gießing, C., Gurr, C., Hilger, K., Jawinski, P., Kulke, L., Lischke, A., Markett, S., Meier, M., Merz, C. J.,

Popov, T., Puhlmann, L. M. C., Quintana, D. S., et al. (2023). Enhancing precision in human neuroscience. *eLife, 12*. https://doi.org/10.7554/eLife.85980

Niso, G., Krol, L. R., Combrisson, E., Dubarry, A. S., Elliott, M. A., Francois, C., Hejja-Brichard, Y., Herbst, S. K., Jerbi, K., Kovic, V., Lehongre, K., Luck, S. J., Mercier, M., Mosher, J. C., Pavlov, Y. G., Puce, A., Schettino, A., Schon, D., Sinnott-Armstrong, W., et al. (2022). Good scientific practice in EEG and MEG research: Progress and perspectives. *NeuroImage, 257*, 119056. https://doi.org/10.1016/j.neuroimage.2022.119056

Pernet, C. (2015). Null hypothesis significance testing: A short tutorial. *F1000Res, 4*, 621. https://doi.org/10.12688/f1000research.6963.3

Pernet, C., Garrido, M. I., Gramfort, A., Maurits, N., Michel, C. M., Pang, E., Salmelin, R., Schoffelen, J. M., Valdes-Sosa, P. A., & Puce, A. (2020). Issues and recommendations from the OHBM COBIDAS MEEG committee for reproducible EEG and MEG research. *Nature Neuroscience, 23*(12), 1473–1483. https://doi.org/10.1038/s41593-020-00709-0

Puce, A., & Hamalainen, M. S. (2017). A review of issues related to data acquisition and analysis in EEG/MEG studies. *Brain Sciences, 7*(6), 58. https://doi.org/10.3390/brainsci7060058

Ratan, S. K., Anand, T., & Ratan, J. (2019). Formulation of research question—Stepwise approach. *Journal of Indian Association of Pediatric Surgeons, 24*(1), 15–20. https://doi.org/10.4103/jiaps.JIAPS_76_18

Szucs, D., & Ioannidis, J. P. A. (2017). When null hypothesis significance testing is unsuitable for research: A reassessment. *Frontiers in Human Neuroscience, 11*, 390. https://doi.org/10.3389/fnhum.2017.00390

Thabane, L., Thomas, T., Ye, C., & Paul, J. (2009). Posing the research question: Not so simple. *Canadian Journal of Anaesthesia, 56*(1), 71–79. https://doi.org/10.1007/s12630-008-9007-4

Thompson, W. H., & Skau, S. (2023). On the scope of scientific hypotheses. *Royal Society Open Science, 10*(8), 230607. https://doi.org/10.1098/rsos.230607

Chapter 9
Applying Statistics in Your EEG Research

Shivakumar Viswanathan

Abstract Electroencephalography (EEG) is used to measure weak electrical potentials generated on the scalp by the brain's activity. However, EEG recordings have several sources of measurement variability due to interindividual differences and the specifics of measurement. Therefore, the application of statistical inference has a crucial role in EEG data analysis to detect experimental effects. In this chapter, we provide an introductory overview of the role of statistical inference in an EEG study. We discuss the sources of uncertainty in EEG measurements and how EEG analyses are organized to reduce measurement-related error. Finally, we highlight the importance of reducing measurement error, both during acquisition and analysis, and the value of well-defined hypotheses specific to the EEG experiment.

Keywords EEG · Statistics · EEG analysis · Measurement error · Hypothesis testing

9.1 Introduction

A major goal of experimental research with electroencephalography (EEG) is to evaluate hypotheses about brain function. A key element to achieving this objective involves applying *statistics*.

This chapter is for readers who are beginning their EEG research journeys and are faced with the question: *How do I apply statistics in my EEG study?* This is an important question as statistics has multiple roles in a typical EEG study. While planning a study, statistical considerations help to turn a scientific question into a structured experiment (see Chap. 8: "Designing Your EEG Study"). While analyzing EEG data, statistical methods and tests are applied, often extensively, to understand observed data patterns and their relevance. The influence of statistics does not end there. When a study's findings are published in a scientific article, other researchers often use the reported statistical details and their graphical visualizations to judge the

S. Viswanathan (✉)
Brain Products GmbH, Gilching, Germany
e-mail: shivakumar.viswanathan@brainproducts.com

T. Warbrick (ed.), *The EEG Handbook*,
https://doi.org/10.1007/978-3-032-20450-9_9

findings (Picton et al., 2000; Keil et al., 2014, 2022) (see Chap. 36: "Writing Up Your EEG Research" and Chap. 37: "How to Evaluate an EEG Research Paper").

In this chapter, we focus on the role of statistics for EEG data analysis. Statistics provides a sophisticated toolkit to address a broad class of challenges (related to uncertainty). However, selecting the right tools to apply requires an understanding of the specific challenges to be solved. We provide an overview of the demands that EEG data present for statistical analysis. This is intended to help readers craft a statistical analysis strategy that is customized to their own research.

9.2 Why Is Statistics Needed in EEG Research?

When conducting EEG research, a first (and often overlooked) issue to address is *why* statistical analysis is needed for a study. In this section, we discuss how uncertainties in an EEG experiment can make it necessary to apply statistical analysis.

The term *statistics* derives from the Italian phrase *ragione di stato* (i.e., science of the state), referring to the use of quantitative data for statecraft (Ostasiewicz, 2014). In its modern usage, statistics refers to a general mathematical approach (and related numerical procedures) to address the role of *uncertainty* in empirical studies. The relationship between uncertainty and statistics, as described nicely by the psychologist S.S. Stevens (Stevens, 1968): "In those disciplines where measurement is noisy, uncertain, and difficult, it is only natural that statistics should flourish. [...] At the other extreme, if accurate measurement were achieved in every inquiry, many of the needs for statistics would vanish."

EEG measurement involves several uncertainties. A careful assessment of these uncertainties can be valuable before we consider possible statistical solutions.

EEG measurements are used to detect and record fluctuations in electrical potentials on the scalp that are generated by the brain's neuroelectric activity (also see Chap. 2: "What is EEG?" and Chap. 20: "EEG Source Analysis"). An EEG recording thus provides a structured set of numerical values (i.e., data) that are associated with a person's brain activity. Due to this association, EEG recordings are a valuable *dependent measure* for experimental studies of brain function. This use of EEG presents a set of typical uncertainties, as described below.

Experimental EEG studies are often conducted to resolve unknowns about how the brain implements a particular capability or function (e.g., visual attention, motor control). Although experiments vary greatly in their specifics, the rationale is often as follows:

1. *Experimental conditions*: Experimenters use controlled strategies to modulate/manipulate the cognitive function of interest while EEG is being measured. These manipulation strategies can take various forms, such as instructed tasks, external stimulation, pharmacological interventions, and so on. It can also involve distinct groups of individuals with relevant characteristics, for example, clinical patient

groups or groups defined by age. With these manipulations, the objective is to measure a person's EEG under distinct conditions or treatments (e.g., condition 1: resting with eyes open, condition 2: resting with eyes closed).
2. *Neural activity*: Each experimental condition is assumed to produce neural activity related to the cognitive/mental function of interest.
3. *EEG recording:* The condition-specific neural activity is, in turn, assumed to produce condition-specific scalp potentials that are recorded by EEG measurement.

To summarize, the assumed relationships are: (1) each experimental condition influences neural activity, and (2) this neural activity influences the recorded EEG. Therefore, comparing the effects of the experimental conditions (the independent variables) on the EEG data (the dependent variables) can be informative about the neural processes of interest.

Detecting these effects quantitatively in the EEG data presents two critical uncertainties: effect uncertainty and measurement uncertainty.

Effect Uncertainty Experiments are conducted to understand unknowns about brain function. Therefore, the effect of the experimental conditions can be uncertain until the EEG data are analyzed. For example, in a particular study, presenting visual images of human faces (condition 1) might be expected to evoke EEG responses that are absent for images of houses (condition 2). The researchers might hypothesize that the perceptual processing of face stimuli engages a distinctive brain network that is inactive for house stimuli (Kanwisher, 2010). However, this hypothesis might involve potentially incorrect assumptions about the brain. Therefore, the presence/absence of these effects in the actual data can provide novel information to either confirm or revise prior scientific assumptions. This uncertainty is compounded by a second source of uncertainty—EEG measurements can be noisy.

Measurement Uncertainty Apart from signals related to neural activity, the measured EEG includes noise and random variation from various other sources (see Chap. 15: "Getting Clean EEG Data: Artifacts and How to Prevent Them"). The net effect for the measured EEG can be summarized simplistically as: *Measured value* = *True value* + *Error*. EEG measures a weak bioelectric signal in the microvolt range. Therefore, a major source of measurement error is electrical interference from the measuring environment (e.g., power line noise) and other bioelectric phenomena (e.g., muscle activity, eye movements). Several additional factors can produce inter-recording differences both within and between individuals, which can range from individual variability in scalp-to-electrode contact (e.g., related to hair thickness, skin properties) to neural activity itself (e.g., related to individual neuroanatomy/physiology) (Hommelsen et al., 2022).

In summary, experiments are conducted to resolve *effect uncertainty*. Evaluating the presence/absence of an experimental effect (i.e., the true difference between experimental conditions) and its properties is part of scientific research. However, this objective is confounded by *measurement uncertainty* from various sources. Specifically, in an EEG recording, the relative contribution of relevant brain activity (i.e., *True value*) and measurement variability (i.e., *Error*) can be uncertain, especially when the *True value* is unknown (i.e., effect uncertainty). Therefore, addressing these overlapping uncertainties is critical.

Statistics provides a quantitative framework to navigate these uncertainties, namely, by estimating the true effect despite the presence of measurement error.

Applying statistics does not eliminate the sources of the above uncertainties. Therefore, practical steps to reduce or even eliminate measurement error should always be a high priority during EEG acquisition and data analysis. This priority has an influence on when statistics is applied during EEG data analysis, as discussed next.

9.3 When Is Statistics Applied During EEG Data Analysis?

Figure 9.1 is a schematic of a typical organization of EEG data analysis and where statistics has a role. This is meant as a coarse overview as specifics can differ from study to study. We describe the steps in detail below.

9.3.1 Raw EEG Data

In a typical EEG study, recordings are acquired from a sample of individuals using a particular experimental protocol. These individuals are selected from one or more well-defined *populations* (discussed in more detail below). The raw EEG recording

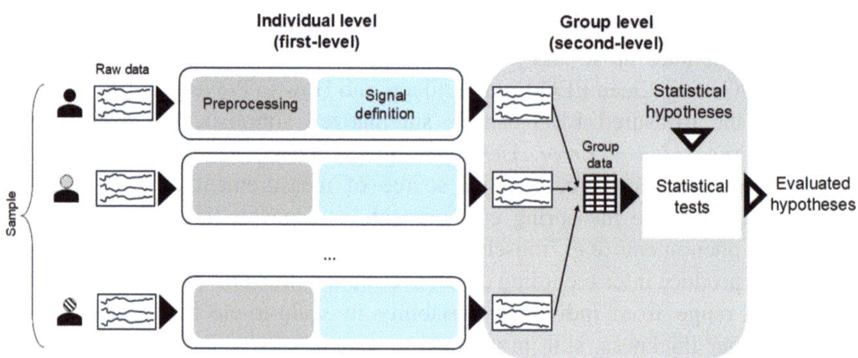

Fig. 9.1 Schematic of a typical EEG data analysis organization for hypothesis testing. Raw data are obtained from individual participants who are sampled from the population (left). They undergo an individual-level or first-level analysis, which consists of a preprocessing step followed by signal definition steps (see text for details). The processed data from individual participants are then pooled together for group-level or second-level analysis. The group data are then used to evaluate statistical hypotheses by applying statistical tests. The outcomes are numerical measures evaluating these hypotheses

from each person usually consists of the following: (1) a structured set of continuous time-series of voltages from multiple electrode locations on the person's scalp (i.e., channel × time) often also with auxiliary data from peripheral physiology sensors (such as electrooculography (EOG) and electromyography (EMG)); (2) markers with information about the person's state and relevant events during the recording (e.g., related to instructions, stimuli, behaviors).

The raw data cannot be interpreted as is and requires further analysis. This analysis usually has two sequential stages (see Fig. 9.1). The first involves data analyses performed independently on individual datasets (referred to as individual-level analysis or first-level analysis). This is followed by an analysis performed on data pooled from all participants (referred to as group-level or second-level analysis) with the application of statistics.

9.3.2 *Individual-Level (First-Level) Analysis*

The individual-level analysis has two sequential (but interrelated) steps.

Preprocessing The raw data are often acquired with parameter settings optimized for accurate measurement and digitization (e.g., choice of sampling rate, filters). Furthermore, the raw data can include extraneous (i.e., non-brain) information, also known as artifacts. Several of these artifacts have distinctive signatures (e.g., related to blinking, muscle contractions). Preprocessing refers to the various transformations applied to the raw data to convert it into a form suitable for analysis. This includes the removal of artifacts of known origin and the correction of biases. The reader can consult Chap. 17 ("EEG Pre-processing and Artifact Handling") for more details on preprocessing and artifact handling.

Signal Definition The clean preprocessed data is then used to extract EEG activity specific to each experimental condition. This might involve linking the clean data to condition-specific details, for example, using the recorded markers (see Chap. 7: "Basic Time Concepts in EEG Practice"). These data are processed further to define the EEG signal of interest using one or more approaches, for example, event-related potential analysis (Chap. 19: "Event-Related Potentials"), spectral and time-frequency analysis (Chap. 18: "Introduction to EEG Oscillations and Spectral Analysis"), source analysis (Chap. 20: "EEG Source Analysis"), connectivity analysis (Lachaux et al., 1999) (Chap. 20: "EEG Source Analysis"), and so on. This step can include the application of statistics, for example, to model intertrial variability (Pernet et al., 2011) or when applying machine learning (King & Dehaene, 2014; Hommelsen et al., 2022).

The outcome of this stage is a summary of each individual's EEG activity in the different experimental conditions. Importantly, this activity has a lower contribution of measurement error as compared to the raw data.

9.3.3 *Group-Level (Second-Level) Analysis*

The individual data from the first-level analysis are pooled together into groups for second-level or group analysis. This grouping is based on the individual's population membership (e.g., patient group, healthy control group) and the goal of the statistical analysis.

Unlike the first-level analysis, where the focus is on the available data, the focus of the second-level analysis is on *statistical hypotheses* (i.e., predictions about the data). The main objective of this stage is to evaluate hypotheses on the group data by the application of statistical methods, discussed in more detail below.

9.3.3.1 Statistical Hypotheses

A statistical hypothesis is a quantitative prediction of the effect of selected independent variables on selected dependent variables in the experiment. This hypothesis is needed for statistical calculations. An example of a hypothesis might be: *The mean beta power at channel Cz is higher in the rest condition than in the motor imagery condition.*

Statistical hypotheses are formulated as general statements relevant to the populations of interest rather than specific to the sample of participants in the study. Apart from being a numerical prediction, a statistical hypothesis in the EEG context involves the selection of a relevant dependent variable. The above example hypothesis is a prediction for one frequency band (i.e., beta band [13–33 Hz] from a large range of available oscillatory frequencies) at one particular channel (i.e., Cz of the several recorded channels). However, EEG data are typically *multivariate*, namely, involving more than one dependent variable. EEG is recorded continuously over time from multiple scalp locations. Therefore, even a 2-second segment of an EEG recording with 64-channels and 500 Hz sampling rate is associated with 64,000 voltage values (Channel (64) × Time (2 s × 500 Hz)).

Therefore, EEG analyses can benefit from statistical hypotheses that include *a well-formulated rationale for variable selection* (Picton et al., 2000; Luck & Gaspelin, 2017). These hypotheses can be derived from broader scientific hypotheses, neuroanatomical considerations (see Chaps. 3: "Basic Anatomy: Central Nervous System" and 4: "Basic Anatomy: Peripheral Nervous System") and details of the experiment design. Without variable selection, the large number of dependent variables can dramatically increase effect uncertainty, which can require complex analyses to address (Pernet et al., 2015; Groppe et al., 2011; Maris & Oostenveld, 2007).

The major hypotheses are defined *before* the data are acquired and guide experiment design and analysis planning. Multiple alternative (or competing) hypotheses can be defined for the same combination of independent and dependent variables based on competing scientific considerations.

9.3.4 Application of Statistical Inference

Statistical inference procedures are applied to evaluate each statistical hypothesis with the actual corresponding data. This is the step where the confounding effect of measurement uncertainty is numerically addressed.

A large variety of statistical procedures (or tests) could be used here as they are not specific to EEG, for example, t-test, Analysis of Variance (ANOVA), and general linear model (GLM). For details on specific statistical tests and how to choose a particular test, we recommend the reader consult a standard statistical textbook for further information. In general, before performing a test, the data's consistency with the test's assumptions is evaluated. Typical requirements are that the data are normally distributed, and that there are no extreme outliers. Major violations of a test's assumptions require a re-evaluation of how to proceed, for example, re-evaluating the data quality, or the application of a better-suited test.

This step can be practically implemented with any standard statistical package (for example, R (https://www.r-project.org/), SPSS (https://www.ibm.com/products/spss), JASP (https://jasp-stats.org/), NumPy (https://numpy.org/). Furthermore, this step requires organizing the data into a suitable format for the statistical tests, often referred to as data "wrangling" (Wickham, 2014). These formats can vary in complexity from simple tables for t-test (Table 9.1) to more involved formats when using, for instance, a multiple-factor ANOVA or a linear mixed model. The difficulties of data wrangling should not be underestimated since errors during this step can directly impact the results. To reduce errors, many statistical programs are integrated with programming environments that enable data wrangling to be conveniently and reproducibly scripted.

Running a statistical procedure produces multiple test-dependent numerical outputs, such as p-values, effect sizes, and confidence intervals.

Table 9.1 Illustrative example of a numerical table prepared for a within-subject t-test to evaluate a statistical hypothesis about the mean difference in ERP magnitudes between two conditions (e.g., left, right) at channel Pz at time + 200 ms

	Voltage		
Participant	condition = left (Pz, +200 ms)	Condition = right (Pz, +200 ms)	Difference: left − right (Pz, +200 ms)
1	0.3285	−0.2143	0.5428
2	1.3332	0.0573	−1.3905
….			….
30	−0.6386	0.9211	−1.5597

9.3.5 Interpretation and Inference

Despite being numerical input-to-output calculations, statistical procedures involve several complex assumptions about the data and uncertainty (Wasserstein & Lazar, 2016; Wasserstein et al., 2019) (see Fig. 9.2 and next section). Therefore, the numerical outputs of the statistical tests are used along with other considerations (e.g., data plots, contextual information) to evaluate whether the hypotheses are supported by the data or suggest rejection. This reasoning is typically reported in the results section of a scientific article. Furthermore, this step can trigger the definition of further post hoc hypotheses and statistical testing.

In summary, the application of statistical calculations is performed near the end of EEG data analysis and relies on the previous steps. The first-level analysis helps reduce the extent of measurement error in the analyzed data. While the first-level analysis is strongly focused on data processing, at the group-level, the main focus is

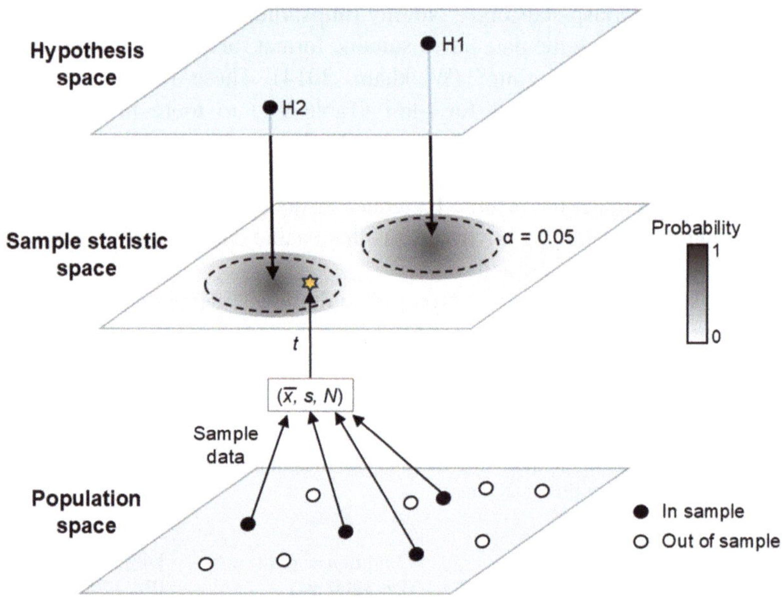

Fig. 9.2 Schematic of statistical inference for a single variable. The hypothesis space (upper plane) is the space of possible statistical hypotheses where *H*1 and *H*2 are examples (black dots). For each hypothesis, a probability is assigned to every possible dataset that could be obtained under the effects of random chance if that hypothesis were true. Each dataset is represented by a sample statistic value, and the space of all possible sample statistic values is shown in the middle plane. The shading indicates the probability assigned to each sample statistic (dataset) by a hypothesis. The alpha threshold for each hypothesis is shown as a dotted line. The colored star indicates a sample statistic obtained from the measured sample (bottom plane). It is assigned a different probability by different hypotheses (probability < alpha for *H*1 but > alpha for *H*2). The measured sample (bottom plane) is a subset of the population. The sample mean ($\bar{x}$), sample standard deviation (s), and sample size (N) are used to calculate the statistic for the sample

on statistical hypotheses, which are crucial inputs to this stage. Furthermore, the outputs at the second-level are numerical evaluations of these hypotheses, rather than data transformations. Therefore, applying statistics in EEG analysis benefits from well-specified hypotheses.

9.4 How Does Statistics Account for Measurement Uncertainty?

The relative ease of running statistical procedures using modern software tools can underestimate the mathematical concepts and assumptions behind them (Gigerenzer, 2004, 2018; Nieuwenhuis et al., 2011; Wasserstein et al., 2019). A general understanding of how statistical methods address EEG-specific uncertainty can be valuable to the application of statistics.

Figure 9.2 provides a simplified Venn diagram schematic of statistical evaluation of a single dependent variable V in the EEG recording. As described earlier, the measured value of a variable (V_{meas}) is assumed to be a mixture of its true value (V_{true}) and error (e.g., related to; measurement). For simplicity, $V_{\text{meas}} = V_{\text{true}}$ + error. Both V_{true} and the error are unknown.

9.4.1 Hypotheses (Upper Plane of Fig. 9.2)

A first step in statistical analysis (e.g., with a t-test) is to make a guess about the possible value of mean V_{true}, namely, a statistical hypothesis. A hypothesis (shown as a black dot, upper plane in Fig. 9.2) is one of many possible alternatives. For example, hypothesis $H1$ might be that mean $V_{\text{true}} > 0$, while an alternative $H2$ might be that mean $V_{\text{true}} < 0$. By evaluating different hypotheses (i.e., guesses) against the measured data, the objective is to arrive at better estimates of mean V_{true}, which is unknown and might never be exactly knowable.

The hypothesis is defined as mean V_{true} rather than per individual. The mean is a measure of the central tendency across a collection of individuals—it can be evaluated on the entire population as well as on samples of different sizes N.

9.4.2 Population and Sample Data (Bottom Plane of Fig. 9.2)

The bottom plane shows the values of V_{meas} of N individuals in the sample. This is data obtained from the first-level analysis. Importantly, the value of N (i.e., the sample size) and the specific individuals included in the analyzed sample are decided well before the group-level analysis.

A critical idea in statistical inference is that properties observed in a sample are informative about properties of the population. Without this assumption, statistical inference would be unable to reason beyond the sample. For this to be possible, the sample is assumed to be *representative* of the population, namely, sharing properties of the entire population. This can be illustrated by considering an urn filled with 1000 colored marbles (the population), where 70% are colored red and the remaining 30% are green. If the marbles are thoroughly mixed in the urn, then a randomly selected sample of $N = 20$ marbles is representative of the full population as it would be expected to have color proportions similar to the entire urn, even if somewhat higher or lower. However, if all the red marbles were concentrated at the top of the urn and all the green marbles were concentrated at the bottom of the urn, then a sample from the urn (even if selected randomly) would be biased, leading to erroneous conclusions about the population. Along similar lines, the selection of a *representative* sample is a critical requirement during experiment design.

Although the population is a mathematical assumption, it is the set of individuals that we seek to meaningfully learn about in a research study. Therefore, defining a suitable population is critical for the value of the study. The population is often defined using a combination of inclusion and exclusion criteria, for example, age, demographics, and clinical profile.

9.4.3 Sample Statistic (Middle Plane of Fig. 9.2)

How is the hypothesis linked to the acquired data? The key idea is to use hypothesis H to generate a model of all possible ways in which random variation can influence mean V_{meas}, if the hypothesis H were actually true.

In Fig. 9.2, this is shown by the arrow from H to the middle plane. Each point in this plane is a sample statistic value, e.g., t-statistic. For a sample of size N, a sample can be described by a unique t-value (Eq. 9.1)

$$t = \frac{\bar{x} - \mu 0}{(s)/\sqrt{N}} \tag{9.1}$$

where $\bar{x}$ is the sample mean, s indicates the sample standard deviation and $\mu 0$ is the statistical hypothesis. This plane represents the statistical values for all possible samples (of size N), actually measured as well as hypothetically possible.

The model for H assigns a probability value to every possible sample statistic value. This is shown by the gradations in gray color, where dark gray indicates a higher probability and lighter colors indicate lower probability. In parametric statistics, this model is derived mathematically, i.e., independent of the actual data. This probability is often denoted as $P(d|H)$, namely, the probability of obtaining data d given that the hypothesis H is true.

Different hypotheses assign different probabilities to the sample statistic values. Therefore, the same sample statistic can be assigned a probability $p_{H1} = P(d|H1)$ according to hypothesis $H1$ and a different probability $p_{H2} = P(d|H2)$ according to hypothesis $H2$. Hence, d might have a higher probability if $H1$ were true but a lower probability if $H2$ were true. However, this does not imply that $H2$ is more probable than $H1$ since $P(H1|d)$ is not equal to $P(d|H1)$. The reason is due to Bayes Theorem (van de Schoot et al., 2021), which we do not discuss further here.

The key step is to evaluate the actual sample data (colored star). Specifically, if H is true, is the data consistent with this hypothesis? One approach has been to set a threshold probability below which the data is deemed to be inconsistent with H (shown as a dotted line). This threshold is denoted as α (alpha). It is set by the researcher and is typically 0.05. So, an extreme p-value less than 0.05 is interpreted as having a low consistency with hypothesis H, if H is true.

With this sketch of the key concepts, we see that statistical inference can be a powerful aid to reasoning but also involves many interacting assumptions. Meeting these assumptions involves decisions made before data acquisition (e.g., sample size, sample selection). It can also be seen that this procedure cannot tell us what is "true" but evaluates the consistency of data if a particular hypothesis were to be true. Therefore, how these procedures are applied and their interpretation require careful attention to the EEG context. Finally, these considerations highlight that a decision about a hypothesis with statistical inference (e.g., $p < 0.05$) is incomplete without the details of the evaluation (Curran-Everett & Benos, 2004, 2007; Lang & Altman, 2015).

9.5 Conclusion

In this chapter, we have provided you with an overview of the uncertainties that arise in EEG measurement that motivate the application of statistics. A careful diagnosis of these uncertainties in your study can help you select a suitable statistical approach. Before applying statistical analyses, you should consider minimizing the contribution of measurement error during data acquisition and other analysis stages. Finally, a major recommendation is to formulate well-specified statistical hypotheses. Well-specified hypotheses with a rationale for variable selection can benefit statistical analysis and enable a clear interpretation of the results.

References

Curran-Everett, D., & Benos, D. J. (2004). Guidelines for reporting statistics in journals published by the American Physiological Society. *American Journal of Physiology. Regulatory, Integrative and Comparative Physiology, 287*, R247–R249.

Curran-Everett, D., & Benos, D. J. (2007). Guidelines for reporting statistics in journals published by the American Physiological Society: The sequel. *Advances in Physiology Education, 31*, 295–298.

Gigerenzer, G. (2004). Mindless statistics. *The Journal of Socio-Economics, 33*, 587–606.
Gigerenzer, G. (2018). Statistical rituals: The replication delusion and how we got there. *Advances in Methods and Practices in Psychological Science, 1*, 198–218.
Groppe, D. M., Urbach, T. P., & Kutas, M. (2011). Mass univariate analysis of event-related brain potentials/fields I: A critical tutorial review. *Psychophysiology, 48*, 1711–1725.
Hommelsen, M., Viswanathan, S., & Daun, S. (2022). Robustness of individualized inferences from longitudinal resting state EEG dynamics. *European Journal of Neuroscience, 56*, 1–32.
Kanwisher, N. (2010). Functional specificity in the human brain: A window into the functional architecture of the mind. *Proceedings of the National Academy of Sciences of the United States of America, 107*, 11163–11170.
Keil, A., Debener, S., Gratton, G., Junghofer, M., Kappenman, E. S., Luck, S. J., Luu, P., Miller, G. A., & Yee, C. M. (2014). Committee report: Publication guidelines and recommendations for studies using electroencephalography and magnetoencephalography. *Psychophysiology, 51*, 1–21.
Keil, A., Bernat, E. M., Cohen, M. X., Ding, M., Fabiani, M., Gratton, G., Kappenman, E. S., Maris, E., Mathewson, K. E., Ward, R. T., & Weisz, N. (2022). Recommendations and publication guidelines for studies using frequency domain and time-frequency domain analyses of neural time series. *Psychophysiology, 59*, e14052.
King, J. R., & Dehaene, S. (2014). Characterizing the dynamics of mental representations: The temporal generalization method. *Trends in Cognitive Sciences, 18*, 203–210.
Lachaux, J.-P., Rodriguez, E., Martinerie, J., & Varela, F. J. (1999). Measuring phase synchrony in brain signals. *Human Brain Mapping, 208*, 194–208.
Lang, T. A., & Altman, D. G. (2015). Basic statistical reporting for articles published in biomedical journals: The "statistical analyses and methods in the published literature" or the SAMPL guidelines. *International Journal of Nursing Studies, 52*, 5–9.
Luck, S. J., & Gaspelin, N. (2017). How to get statistically significant effects in any ERP experiment (and why you shouldn't). *Psychophysiology, 54*, 146–157.
Maris, E., & Oostenveld, R. (2007). Nonparametric statistical testing of EEG- and MEG-data. *Journal of Neuroscience Methods, 164*, 177–190.
Nieuwenhuis, S., Forstmann, B. U., & Wagenmakers, E. J. (2011). Erroneous analyses of interactions in neuroscience: A problem of significance. *Nature Neuroscience, 14*, 1105–1107.
Ostasiewicz, W. (2014). The emergence of statistical science. *Śląski Przegląd Statystyczny, 12*, 76–77.
Pernet, C. R., Chauveau, N., Gaspar, C., & Rousselet, G. A. (2011). LIMO EEG: A toolbox for hierarchical LInear MOdeling of ElectroEncephaloGraphic data. *Computational Intelligence and Neuroscience, 2011*, 831409.
Pernet, C. R., Latinus, M., Nichols, T. E., & Rousselet, G. A. (2015). Cluster-based computational methods for mass univariate analyses of event-related brain potentials/fields: A simulation study. *Journal of Neuroscience Methods, 250*, 85–93.
Picton, T. W., Bentin, S., Berg, P., Donchin, E., Hillyard, S. A., Johnson, R., Miller, G. A., Ritter, W., Ruchkin, D. S., Rugg, M. D., & Taylor, M. J. (2000). Guidelines for using human event-related potentials to study cognition: Recording standards and publication criteria. *Psychophysiology, 37*, 127–152.
Stevens, S. S. (1968). Measurement, statistics and the schemapiric view. *Science, 161*, 849–856.
Van de Schoot, R., Depaoli, S., King, R., Kramer, B., Märtens, K., Tadesse, M. G., Vannucci, M., Gelman, A., Veen, D., Willemsen, J., & Yau, C. (2021). Bayesian statistics and modelling. *Nature Reviews Methods Primers, 1*, 1.
Wasserstein, R. L., & Lazar, N. A. (2016). The ASA statement on p-values: Context, process, and purpose. *The American Statistician, 70*, 129–133.
Wasserstein, R. L., Schirm, A. L., & Lazar, N. A. (2019). Moving to a world beyond "$p < 0.05$". *American Statistician, 73*, 1–19.
Wickham, H. (2014). Tidy data. *Journal of Statistical Software, 59*, 1–23.

Chapter 10
Pilot Testing

Paulo Rodrigo Bazán

Abstract Pilot testing is a crucial preliminary phase in electroencephalography (EEG) studies, aimed at optimizing experimental design, data acquisition, and analysis pipelines before full-scale data collection. This chapter discusses the significance of pilot testing in ensuring high-quality EEG data. First, we define what pilot testing is and clarify what it is not. We then explore the various study characteristics that can be assessed during the pilot phase, such as task parameters, equipment setup, signal quality, and experimental procedures. The chapter further examines the preparation required for the pilot phase, emphasizing the definition of criteria to assess the experiment. Finally, we discuss the iterative nature of pilot testing and its integration into the study preparation process.

Keywords Pilot testing · EEG studies · Experimental design · EEG data acquisition · Equipment setup · Experimental procedures

10.1 What Pilot Testing Is

Pilot testing refers to the preliminary phase of a study; its purpose is to evaluate several aspects of the study before fully investing in and proceeding with data acquisition. It involves the internal checks and fine-tuning of the experiment, which should be part of every research study (Barbosa et al., 2022; Boudewyn et al., 2023). In a broader sense, it evaluates the feasibility of the proposed study and provides helpful information for future studies with similar tasks or techniques. Therefore, it can be relevant as a study on its own, and there are even journals dedicated to pilot studies (Lancaster, 2015).

Although it is common to find published articles with "pilot" or "feasibility" in their titles, the exact definition of these terms is not universally agreed. Some authors use the terms as synonyms (Ruel et al., 2016), and others suggest that one is a subset

P. R. Bazán (✉)
Brain Products GmbH, Gilching, Germany
e-mail: paulo-rodrigo.bazan@brainproducts.com

T. Warbrick (ed.), *The EEG Handbook*,
https://doi.org/10.1007/978-3-032-20450-9_10

of the other (Eldridge et al., 2016), and yet others suggest specific and mutually exclusive definitions (Thabane et al., 2010; O'Cathain et al., 2015). Additionally, pilot testing can be seen as a pretesting of a specific task, questionnaire, or data acquisition instrument (van Teijlingen & Hundley, 2002). This pretesting focuses on the validation of such an instrument (Ruel et al., 2016).

In electroencephalography (EEG) studies, the pilot phase is used to optimize the experiment, from the design and EEG data acquisition procedures to the analysis pipeline. The aim is to ensure high-quality EEG data and adequate control of the experimental environment in a safe and efficient way (as covered in Chap. 16: "Practical Aspects of EEG Data Acquisition"). Here, piloting can be very helpful to check if all the needed triggers are working, and if the needed synchrony between equipment is being achieved (Boudewyn et al., 2023). It also allows testing and improvement of the analysis methods and codes. Furthermore, it also allows an initial check to confirm whether it is possible to study the research question with the planned analysis method and experimental design. It is also possible to use pilot testing to validate new EEG equipment or technologies, but this is beyond the scope of this chapter. We will focus on pilot testing as part of every EEG study, as the initial check and optimization of the planned experimental protocol.

10.1.1 Why Pilot Testing Matters

Imagine doing hundreds of data acquisition sessions over a few years, only to realize that the EEG signal was too noisy. Or that the data cannot be analyzed because there are significant jitters in the markers, improper task instructions, or incorrect parameters. This would be a disappointing outcome and a waste of valuable time and resources. Although this is an extreme scenario, it illustrates how important pilot testing can be, as it could easily detect and eventually solve such issues.

The pilot test is a tool to improve the quality of the research project: different characteristics can be assessed, and the best parameters can be selected. Additionally, pilot testing can provide a better estimate of logistical requirements and of the resources needed, which are helpful for funding requests (van Teijlingen & Hundley, 2002). It can also evaluate other aspects, such as the recruitment steps, consent and retention rates, and randomization procedures when preparing clinical trials (Leon et al., 2011; Lancaster, 2015).

All this information is very relevant for preregistering your study. The pilot test can confirm whether the EEG feature of interest is elicited by the experiment design and detected by the analysis pipeline. This increases the chance of being able to fully execute the study as preregistered (Paul et al., 2021).

10.1.2 What Pilot Testing Is Not, and Important Considerations

Pilot tests are not shortcuts, and they do not replace the necessary steps in preparing a research study, such as reviewing the literature, defining a research question, designing the experiment, and planning the analysis to answer the question (Barbosa et al., 2022). To properly plan and perform the pilot phase, an initial proposition of the study with a clear hypothesis is necessary. Piloting will help to refine the experiment design and the analysis. It can also help to determine whether the research question can be assessed with the proposed methodology.

Pilot testing is not just a study with a small sample size. Its goals are different from those of the main study. For example, pilots should not be used for inferences related to the main hypothesis to be tested in the main study (Leon et al., 2011; Lancaster, 2015). Variable estimates taken from pilot data will not have the required precision due to the small sample. For example, it is generally no longer recommended to use pilot data to define the sample size for the study. As this relies on estimating the effect size of the main outcome, this can yield an inappropriate sample size calculation (Leon et al., 2011; In, 2017). Estimation of other parameters, such as the standard deviation of the variable of interest, will also be suboptimal (Sim, 2019). However, there is still some debate as to whether some outcomes could be considered to estimate sample size (Lewis et al., 2021).

Is it appropriate to include the pilot data in the final study analysis? There are two issues to be considered here. First, it is likely that the piloting phase will result in adjustments to the experimental procedure. Therefore, data collected in the pilot tests will be different from the data from the main study. This variation can have an impact on the outcomes (Leon et al., 2011; In, 2017). Second, to be able to include the pilot sample into the final study, the preliminary analysis must be unrelated (orthogonal) to the study hypotheses (Barbosa et al., 2022). Although inferences and statistical testing of the hypothesis with the pilot data are not recommended, the preliminary analysis of the pilot data involves checking the presence of the expected EEG signature in the data and testing the desired analysis pipeline. Therefore, a dedicated sample is needed for the pilot phase in most cases.

10.2 How to Prepare and Run the Pilot Testing

Now that we know what pilot testing can and cannot do, we can discuss practical tips for properly planning and executing the pilot phase. First, we need to define the main goals of the pilot phase. We must consider whether the plan is to publish it or to use it to define the parameters to be reported in the preregistration of the main study. Alternatively, it could be executed only as an internal step. In any case, we must define the characteristics to be checked and optimized for the experiment to work. Once these are defined, we can choose measurable outcomes to assess them. This

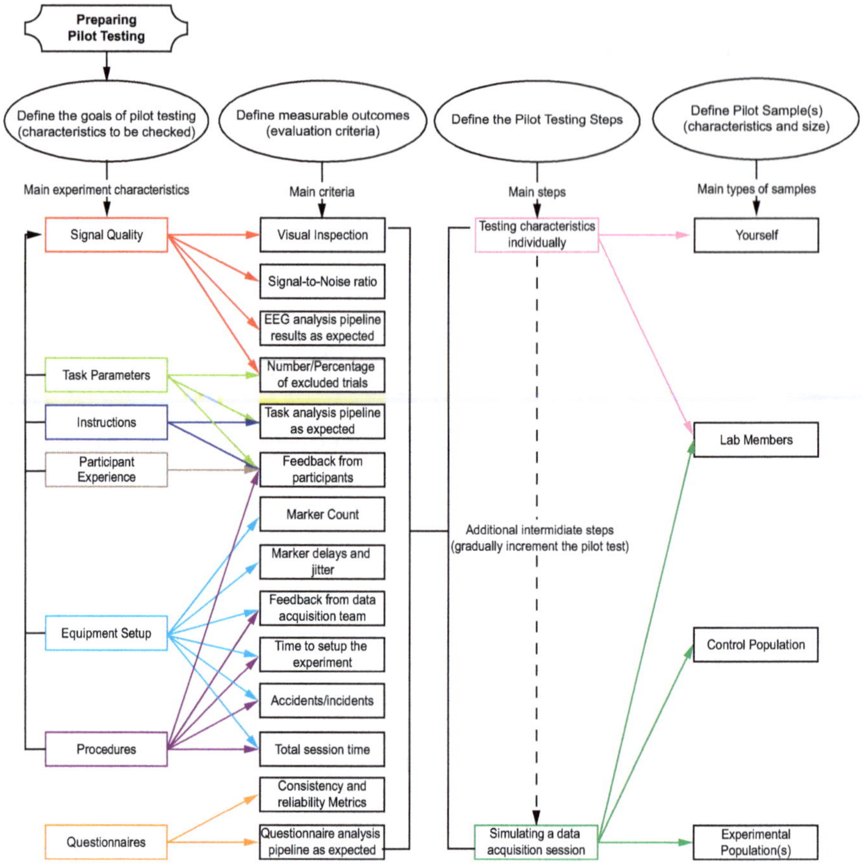

Fig. 10.1 Preparing pilot testing flowchart. The flowchart illustrates the process of preparing pilot testing, including defining the study's goals, measurable outcomes, testing steps, and pilot sample characteristics. The key experiment characteristics can be optimized (from signal quality to questionnaires) by assessing the suggested criteria (arrow connections). Almost all these goals can have an impact on signal quality (except for the questionnaires). The goals and outcomes help define the steps and the samples used in pilot testing. Different phases are associated with different sample groups, including lab members, control populations, and experimental participants. These phases go from testing the characteristics individually to simulating a data acquisition session for thorough preparation

also requires clear evaluation criteria to confirm we can continue with our experiment. All these points will help design the pilot testing steps and the required pilot samples. An overview of the piloting phase preparation is given in Fig. 10.1. Its main steps are further discussed below.

10.2.1 *Define Characteristics and Evaluation Criteria*

The pilot phase can involve variations of the experimental task or of the equipment setup, which can be compared to decide which is more suitable. For this, we need to know which characteristics we want to test. There is some room for exploration as well, as one goal of the pilot tests is to evaluate factors that were not accounted for during the design of the experiment. However, having a pilot phase without clear outcomes will likely result in difficulties in interpreting the pilot data. It is important to define the criteria that will be used to conclude the pilot and proceed with the next research phases. These criteria will prevent both proceeding without a proper experimental protocol and being trapped in an optimization loop (there will always be points to be improved in scientific progress). Here we highlight some of the main characteristics to be assessed in the pilot phase of EEG studies (see Fig. 10.1) and suggest criteria to assess them.

10.2.1.1 Signal Quality

This is a key point for all EEG studies, as the signal quality can be affected by many factors. We want to ensure a low noise level in the data acquisition environment, which can be evaluated first with a visual inspection of the signal. It is helpful to monitor the signal during the pilot test to identify the source of the noise. Monitoring the filtered signal can provide an initial idea of whether the frequency filters can handle the noise. Using Fast-Fourier Transform will also be helpful to identify the source, as the main noise frequencies will be highlighted. The signal quality evaluation should also check whether the expected EEG feature of interest is observed. For example, when doing an oddball task, a P300 evoked potential should be detected; when using visual stimuli, visually evoked potentials should be present, such as P100.

An objective way to assess EEG signal quality is the signal-to-noise ratio. The threshold for acceptable signal quality depends on the signal of interest and on the planned analysis. For example, if a study is focused on the alpha and beta frequency range and there is some 50 Hz line noise, it may not impact the analysis. Therefore, running the desired analysis pipeline is important as it will show if it can properly handle the observed noise. Further, by running the pipeline, we evaluate how many trials were excluded in preprocessing. Both the number and the percentage are relevant metrics to assess signal quality. We want to make sure that a minimum number of trials is available for analysis, but we also want to make the experiment efficient. This is useful when evaluating the feasibility of data acquisition in a noisy environment, or during a task that is expected to elicit more noise due to movement.

Most aspects of the experiment can have an impact on the EEG signal. Therefore, the signal quality must also be assessed during the optimization of such characteristics, as discussed below.

10.2.1.2 Task Parameters

Many characteristics of the stimuli and task parameters can be adjusted during pilot testing. For example, task duration, number of trials, stimulus size or volume, interval between trials, difficulty, the sequence of the stimuli, and the interval between stimuli and between trials (not just the length but the distribution of the interval). We must define what will be carefully tuned. As an initial step, it is important to confirm that the participants can understand and perform the task. Specific experimental tasks can be programmed and used in the pilot phase to optimize these parameters. For example, the size of a cue stimulus can be varied, and the performance associated with each size can be evaluated. This will determine the optimal size to use in the main study.

One way to evaluate the task parameters is by assessing behavioral performance. The number and the percentage of excluded trials are again relevant. Task results are also important, such as reaction times and the number and percentage of correct responses. These metrics help confirm that the outcomes of the task are consistently and properly recorded. For example, a big variation over time or between experimental blocks may be related to fatigue or learning effects, or it may indicate improper coding of the task responses (Barbosa et al., 2022).

One helpful tool to optimize the task, as well as other parts of the experiment, is a structured questionnaire for the pilot phase. The questionnaire should include open questions to allow participants to express their overall perception and experience of the study (Barbosa et al., 2022). Open questions can assess the strategy used by the participants, as sometimes they can find an alternative and easier way to do the task, such as blurring their vision during a Stroop test. The questionnaire should also contain objective and scale rating questions to quantify and compare the tested experimental variations.

The EEG signal quality should also be evaluated during the adjustment of task parameters. For example, fatigue, learning, and the strategy used by the participant can also affect the event-related potentials and the EEG results overall. Further, the EEG signal-to-noise ratio and the analysis pipeline provide important information for defining the number of trials.

10.2.1.3 Instructions

How well participants understand the task will have an impact on their performance. Therefore, how the task is explained to them is crucial and must be optimized. Failures in task comprehension or adopting strategies other than the one intended for the task can usually be detected in the task results, which will be outside of the expected range. Dedicated questionnaires are also effective for detecting communication issues. For example, we can ask the participant if the instructions were clear and to rate the clarity of the instructions. It is important to check the EEG signal and expected results, as miscommunication can lead to different cognitive components being involved in the task than originally planned in the study design (Amaro & Barker, 2006).

10.2.1.4 Participant Experience

It is important to make the experiment as pleasant as possible and always provide a kind and respectful environment (Barbosa et al., 2022). This can be measured via a dedicated questionnaire, asking the participants about their comfort during or their satisfaction with the experiment. The EEG data can be inspected for indicators of emotional state or alertness, for example.

10.2.1.5 Equipment Setup

Several aspects of equipment setup play a role in a successful experiment, from the position of the equipment to the recording software parameters. These should be considered during pilot testing.

In general, it is important to ensure correct and precise triggers (as discussed in Chap. 14: “Triggers”). For this, the total marker count and the jitter in the markers are effective metrics. The count must match exactly the intended number (e.g., number of trials). The maximum acceptable jitter depends on the EEG analysis pipeline, although it should be very small overall. If these metrics are outside of the expected levels, the trigger connections should be checked, as well as the task programming.

The setup should also be practical and efficient. We can evaluate it by measuring the time it takes to set up the experiment, as well as the total session time. Once optimized, these will be informative for scheduling the data acquisition of the study. The input from the research group can help identify what needs improvement. The research group can use a questionnaire (e.g., grade the equipment setup) or a checklist of characteristics to discuss together if adjustments are needed.

The equipment positioning and the data acquisition setup can have an impact on the data quality, and therefore on the study outcomes. Visual inspection of the data quality and signal-to-noise ratio can be used as outcomes. If there is significant noise in the data, try to adjust and identify the source, either by checking nearby electrical equipment or confirming if the noise is specific to the room. Also check if the data acquisition notebook is connected to the power supply (this can, in some cases, generate noise). Further tips to optimize equipment positioning are presented in Chap. 16: “Practical Aspects of EEG Data Acquisition.”

Further, it is important to consider the safety of the setup. It is expected that no accident or incident will happen, but we must control and register them properly. In case something happens, having detailed information will allow more effective adjustments to the setup.

10.2.1.6 Procedures

The overall protocol or study workflow can be optimized during pilot testing. A helpful tool is feedback from the participants about their experience during the experiment. The goal is to evaluate the full procedure, including the instructions given to the participant before the day of the experiment, the routine on the data acquisition day, and the entire experience of the participant during the experiment.

Similarly, the feedback from the research staff during the tests is also very relevant. Confirm that the role of each person during the data acquisition is clear, and that each lab member is comfortable with their role—all the people involved in the experiment should feel respected, not only the participant. Aside from adjustments to the protocol, consider increasing the personnel involved in data acquisition if necessary. Additional training can be provided to lab members if needed. The initial pilot tests, possibly with lab members as participants, can be a good training opportunity. However, it is important to simulate the real data acquisition and avoid different procedures due to knowing the participant/colleague.

As additional outcomes, you can assess:

- Data loss during the pilot testing
- Quality of the acquired data
- Consistency of experimental procedures between different pilot test data acquisitions
- The duration of the experimental session and of each step separately.

10.2.1.7 Questionnaires

Although questionnaires do not directly influence the EEG signal quality, they can complement the neurophysiological measures and the task performance. For example, it can help during participant screening to control variables that can influence the study outcomes. A dedicated short, structured interview or questionnaire can be used to confirm that the questions are clear and properly understood by the participants (van Teijlingen & Hundley, 2002). If necessary, the study questionnaire can be adjusted and the testing reiterated. Pilot studies can also be used to assess the reliability of the questionnaire as part of the validation process (Bujang et al., 2022). It is also recommended to have different questions to evaluate the same feature. This will allow consistency and reliability of the questionnaire to be assessed to see if the desired aspect is really being measured by the questionnaire.

10.2.2 Define the Different Steps of Your Pilot Tests

The pilot tests are best optimized when organized in several steps (Fig. 10.2). For example, an initial step would be to check the main characteristics of the experiment separately, such as the individual instruments used in the experiment, the equipment setup, and the tasks and stimuli being used. We can check the noise level in the EEG recorded in the experiment room. We can, in parallel, check the task parameters considering only the behavior. The preliminary testing steps are especially valuable when either the task or the instruments used are not validated within the study

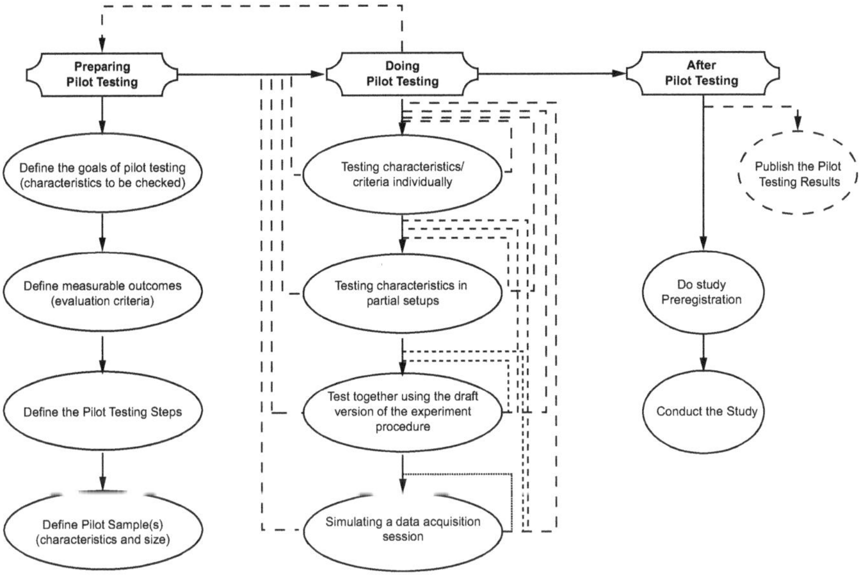

Fig. 10.2 Flowchart of the main phases in pilot testing. Three phases of pilot testing are outlined: preparation, execution, and post-testing steps. The diagram presents the steps within each phase and their order. The preparation phase (preparing) includes defining study goals, evaluation criteria, steps of the testing, and samples used. The execution (doing) requires testing individual characteristics, partial setups, testing the experiment procedure, and then simulating data acquisition session. It can require a return to previous steps or even to the previous phase to prepare new pilot tests. The steps after pilot testing involve publishing the results (optional), preregistering the study, and then conducting the full study. Optional steps and returns to previous steps are presented in dashed lines

population (Nebe et al., 2023). Once each main characteristic is tested, combined partial setups can be made. For example, testing the task with the EEG to check the triggers and the task parameters based on the EEG signal. Then the steps progress by incorporating more parts of the experiment and testing the initial versions of the protocol. The last step is simulating a complete data acquisition session, to confirm the experiment is ready.

Pilot testing steps are executed iteratively until the desired optimization is achieved (Barbosa et al., 2022). The results from one step can lead to a return to previous steps. For example, after a partial setup assessing the EEG signal during the task, we may need to go back to adjusting the task by itself. The pilot results can even suggest a return to prior study phases, such as refining the research question, if the pilot testing results suggest it cannot be properly assessed.

Please keep in mind that the steps can be adjusted to each type of study. For example, when doing EEG-fMRI, it is recommended to divide the signal quality check into three steps: First, check the signal outside the scanner room, to test that the triggers and clock synchrony are correct; then test inside the scanner room without an fMRI sequence, to evaluate the artifacts generated by the MR; and then test the signal with a sequence (Störmer et al., 2016).

10.2.3 Define the Samples Characteristics and Size for the Pilot Testing

As an initial suggestion, perform your experiment yourself as a participant (Barbosa et al., 2022). This will provide helpful information and insights. For example, it can be the case that you imagined that the task would elicit a specific cognitive process; however, by doing the task, you realize that this is not the case. Also, it is helpful to consider the experiment from the participant's perspective, to understand how the experiment is perceived, and what emotions it can induce. You can get an idea of the participant's comfort during the task. Check your comfort when having the EEG prepared on you, executing the experiment with it, and cleaning up afterwards.

This is just a first step; your personal impression may be very different from that of others doing the same experiment. It is therefore useful to have feedback from research colleagues as they can further discuss the details of the experiment, considering their scientific proficiency (Ruel et al., 2016). However, it is important to avoid peer pressure. Your colleagues shouldn't feel obliged to participate in the pilot phase; this should always be voluntary (Barbosa et al., 2022).

You and your lab members are usually more suitable as participants for the initial steps. This is valid both for testing the characteristics separately and the first round of subsequent, more complete steps. After these internal adjustments, the pilot tests should also involve a small sample of the control and of the experimental populations to consider their specific context and needs. Then one common question arises: What should be the sample size for pilot tests? Here it is important to differentiate a pilot that is meant as a feasibility study of a clinical trial, for example, and the iterative pilot testing we are targeting for EEG studies—although in neither case, it is recommended to have power calculations. In clinical trials, the pilot sample size can be defined either as a percentage of the sample to be included in the final study (Eldridge et al., 2016) or by using previous studies as reference (Kunselman, 2024). However, for internal EEG pilot tests, there isn't an exact recommendation for sample size. One reason is their iterative nature. The key point is to confirm that all the characteristics of the experiment were tested and to be confident that the experiment is properly designed. When in doubt, you likely need another pilot session.

10.2.4 What to Do After the Pilot Testing (Publication and Preregistration)

Once the study has properly been adjusted and optimized during the iterative pilot testing, it is time to prepare the publication of the pilot testing results, if it was designed with that goal. Otherwise, it is a good time to do the preregistration of your study (Barbosa et al., 2022; Paul et al., 2021). Preregistration is recommended to ensure scientific reproducibility. Once the study is preregistered, we are ready to start the data acquisition of our study.

10.3 Concluding Summary

In summary, pilot testing is a key phase in any EEG study to improve its quality. It reduces the risk of improper data acquisition, which would result in useless data. It is important to bear in mind that pilot testing does not replace other steps in a research project, but it can be used to optimize the study workflow. Further, a pilot study is not a small sample version of the main study as its goals are different. The pilot aims to tune the experiment rather than test hypotheses. Specifically in the context of EEG studies, pilot testing offers the opportunity to optimize parameters and EEG signal quality, to ensure that the desired EEG feature can be measured. To do effective pilot testing, it is important to prepare by considering the characteristics that you want to test. Aside from checking that the experiments are feasible, it is possible to test a few different parameters to select the best one. Further, it is important to have clear criteria to evaluate these aspects. The pilot testing should be seen as an iterative phase: If an issue is detected, we can go back, adjust the experiment, and test again. Once the pilot phase is concluded, you will have the tools to do a detailed preregistration of your study, which is highly recommended.

References

Amaro, E., & Barker, G. J. (2006). Study design in fMRI: Basic principles. *Brain and Cognition, 60*, 220–232.

Barbosa, J., Stein, H., Zorowitz, S., Niv, Y., Summerfield, C., Soto-Faraco, S., & Hyafil, A. (2022). A practical guide for studying human behavior in the lab. *Behavior Research Methods, 55*, 58–76.

Boudewyn, M. A., Erickson, M. A., Winsler, K., Ragland, J. D., Yonelinas, A., Frank, M., Silverstein, S. M., Gold, J., Macdonald, A. W., Carter, C. S., Barch, D. M., & Luck, S. J. (2023). Managing EEG studies: How to prepare and what to do once data collection has begun. *Psychophysiology, 60*, e14365.

Bujang, M. A., Khee, H. Y., & Yee, L. K. (2022). *A step-by-step guide to questionnaire validation research*. Institute for Clinical Research (ICR), National Institutes of Health.

Eldridge, S. M., Lancaster, G. A., Campbell, M. J., Thabane, L., Hopewell, S., Coleman, C. L., & Bond, C. M. (2016). Defining feasibility and pilot studies in preparation for randomised controlled trials: Development of a conceptual framework. *PLoS One, 11*, e0150205.

In, J. (2017). Introduction of a pilot study. *Korean Journal of Anesthesiology, 70*, 601–601.

Kunselman, A. R. (2024). A brief overview of pilot studies and their sample size justification. *Fertility and Sterility, 121*, 899–901.

Lancaster, G. A. (2015). Pilot and feasibility studies come of age! *Pilot and Feasibility Studies, 1*, 1–4.

Leon, A. C., Davis, L. L., & Kraemer, H. C. (2011). The role and interpretation of pilot studies in clinical research. *Journal of Psychiatric Research, 45*, 626–629.

Lewis, M., Bromley, K., Sutton, C. J., Mccray, G., Myers, H. L., & Lancaster, G. A. (2021). Determining sample size for progression criteria for pragmatic pilot RCTs: The hypothesis test strikes back! *Pilot and Feasibility Studies, 7*, 40–40.

Nebe, S., Reutter, M., Baker, D. H., Bölte, J., Domes, G., Gamer, M., Gärtner, A., Gießing, C., Gurr, C., Hilger, K., Jawinski, P., Kulke, L., Lischke, A., Markett, S., Meier, M., Merz, C. J., Popov, T., Puhlmann, L. M. C., Quintana, D. S., Schäfer, T., Schubert, A. L., Sperl, M. F. J.,

Vehlen, A., Lonsdorf, T. B., & Feld, G. B. (2023). Enhancing precision in human neuroscience. *eLife, 12*, e85980.
O'cathain, A., Hoddinott, P., Lewin, S., Thomas, K. J., Young, B., Adamson, J., Jansen, Y. J. F. M., Mills, N., Moore, G., & Donovan, J. L. (2015). Maximising the impact of qualitative research in feasibility studies for randomised controlled trials: Guidance for researchers. *Pilot and Feasibility Studies, 1*, 1–13.
Paul, M., Govaart, G. H., & Schettino, A. (2021). Making ERP research more transparent: Guidelines for preregistration. *International Journal of Psychophysiology, 164*, 52–63.
Ruel, E., Wagner, W. E., III, & Gillespie, B. J. (2016). Pretesting and pilot testing. In *The practice of survey research: Theory and applications*. SAGE.
Sim, J. (2019). Should treatment effects be estimated in pilot and feasibility studies? *Pilot and Feasibility Studies, 5*, 107–107.
Störmer, R., Bártolo, M., Geraci, S., & Warbrick, T. (2016). *Simultaneous EEG and BOLD fMRI: Best setup practice in a nutshell* [Online]. Brain Products Press Release: Brain Products GmbH. Available: https://pressrelease.brainproducts.com/eeg-fmri/. Accessed 05 Sept 2025.
Thabane, L., Ma, J., Chu, R., Cheng, J., Ismaila, A., Rios, L. P., Robson, R., Thabane, M., Giangregorio, L., & Goldsmith, C. H. (2010). A tutorial on pilot studies: The what, why and how. *BMC Medical Research Methodology, 10*, 1–1.
Van Teijlingen, E., & Hundley, V. (2002). The importance of pilot studies. *Nursing standard (Royal College of Nursing (Great Britain): 1987), 16*, 33–36.

Chapter 11
Study Workflow and Lab Management

Tracy Warbrick and David Kadlec

Abstract Running a successful EEG lab relies on clear communication, documentation, and sharing of resources. This chapter outlines strategies for establishing a lab management plan to ensure the smooth operation of your lab and that it produces high-quality research. On an individual level, a researcher needs to be organised, consistent, and thorough in their approach. We provide tips for planning workflows for before, during, and after a measurement. We also highlight the importance of keeping your own lab notebook to accurately and reliably document your work.

Keywords Measurement checklist · Consistent procedures · Replicability · Transparency · Lab notebook · Lab logbook

11.1 Introduction

The scientific community has a responsibility to conduct transparent and reproducible research. Effective lab management and consistent study workflows are essential to high-quality EEG studies. Establishing repeatable study procedures, well-documented protocols, and consistently monitoring equipment use benefits not only individual researchers but also research groups, and ultimately the published scientific knowledge base. This chapter covers tips for lab management, keeping your own personal lab notebook, and planning study workflows.

T. Warbrick (✉) · D. Kadlec
Brain Products GmbH, Gilching, Germany
e-mail: tracy.warbrick@brainproducts.com; david.kadlec@brainproducts.com

T. Warbrick (ed.), *The EEG Handbook*,
https://doi.org/10.1007/978-3-032-20450-9_11

11.2 Lab Management

To support the effective operation of your lab, it is essential to implement a lab management plan and to designate a responsible person. This is particularly important when multiple experiments are running and when facilities are shared by many users. This not only helps general organisation and scheduling, but also facilitates equipment monitoring, identifying equipment overuse, underuse, and devices prone to problems.

Effective lab management is also integral to transparent and reproducible research. Yet these practices are often overlooked in the scientific community (Baker, 2016) and as researchers we must strive to improve. While commitment is needed from the whole community for systemic change, as individuals we can contribute by establishing clear protocols for our own research. This will enable better record keeping, project management, data sharing, and information dissemination both within research groups and across the wider scientific community (Monaghan et al., 2023).

If resources permit, appointing a lab manager can optimise the productivity and efficiency of the lab. In the absence of an official lab manager position, assigning a responsible person is a good use of resources. Lab users are more likely to follow procedures if someone is made accountable for running the lab. It also provides a contact point for all lab users, making them more likely to follow the correct procedures and report problems.

Below, we outline key points for establishing policies and procedures that will help your lab run smoothly.

11.2.1 Admin and Organisation

- Use a transparent booking system, including who has priority (if appropriate).
- Manage the budget: who is responsible for this, what is the procedure for requesting equipment?
- Establish methods for sharing information among lab users, e.g. wikis, logbooks, mailing list, chat group.
- Consumables (e.g. gel, syringes, tape): How to monitor use and decide when they need ordering. Who orders them?
- Establish data backup procedures.
- Ensure that all users understand and comply with established procedures and policies.
- Have clear emergency procedures, e.g. what happens if the fire alarm sounds, who is responsible for participants, and making sure the lab is cleared?
- Users should promptly report any system malfunctions or incidents without hesitation. For example, a broken electrode, interruption of data recording, or allergic reactions to gel or tape (rare).

- Establish a procedure for training new people. New lab users need to be familiar with the lab procedures as well as how to use the equipment appropriately.
- Encourage all lab users to have a written study protocol for all studies.
- Have a recommended pilot testing procedure for new studies.

11.2.2 Hardware and Software Maintenance

- Keep a usage log to track how often the equipment is being used.
- Establish a procedure for reporting or logging problems and damage. Also consider associated procedures for initiating repairs.
- Schedule regular function checks, especially for equipment that is frequently used or used by multiple users.
- Keep software up to date. Also consider when to do this in relation to running studies. If in doubt check with the manufacturer.
- Have standard cleaning and disinfection procedures.

11.2.3 Lab Logbook

Keep a lab book to record all measurements. This should be clear and easy to complete, e.g. a table of basic information for each measurement. See Table 11.1 for an example of information to include.

Electronic lab notebooks (ELNs) might be more suited to some studies than paper versions, e.g. those that require increasingly large data volumes, data complexity,

Table 11.1 Example of simple information that could be included in a lab logbook

Date and time	
Project name	
Principal investigator's name	
Experimenter name and contact details	
Recording number in the present study	To track how many recordings per study/research group
Equipment used	Amplifier: *(serial number or lab label)*
	Battery: *(serial number or lab label)*
	EEG cap: *(serial number or lab label)*
Equipment status notes	Battery charged after use? Any malfunctions or damage?
Consumables used	
Consumables status notes	Is anything running out? Order initiated?
Has the data been archived?	
Additional notes	

Fig. 11.1 The role an electronic lab notebook (ELN) can play in the research data life cycle. (Vandendorpe et al., 2024. Reproduced under Creative Commons CC BY 4.0. Unaltered)

accessibility, and traceability (Higgins et al., 2022). An ELN can be a very useful tool for the whole research process (see Fig. 11.1). However, implementing one is not trivial, and you should invest some time in choosing one that meets your requirements (Vandendorpe et al., 2024).

Regardless of the method chosen, it should be easily accessible for all users; people are more likely to comply with simple, transparent procedures. Solutions can be as easy as having an online booking calendar, a paper lab book, and displaying procedures in the lab or as complex as implementing an ELN for the whole study process. The essential requirements are to establish clear working procedures and to assign a person responsible for their implementation.

11.3 Keep Your Own Lab Notebook

Most labs will have a shared lab book or measurement logbook (see Sect. 1.2), but it is also considered good scientific practice to keep your own lab notebook. A lab notebook can serve as tool for planning and organising your work, recording your work, and helping to troubleshoot or diagnose recurring problems, especially when communicating with the manufacturer's support team. It can even have a role in protecting any intellectual property that comes from your research.

Traditionally, scientists kept paper notebooks which allow flexibility in content and organisation. However, paper notebooks have some drawbacks in terms of data retrieval and data sharing; they can also be misplaced leading to a complete loss of the research record. Electronic lab notebooks can avoid these potential problems through searchable indices, easier data sharing, and automatic backup and archiving (Wright, 2009). It could be that your university or lab has implemented an ELN, researchers are advised to check whether such a resource is available. If not, you can establish a paper or electronic approach that fits your needs and preferences.

The specific content and structure of your lab book will depend on your requirements. You could consider including the following:

- Pilot testing plans, results, and adjustments.
- Ideas for your next studies and what stimulated them.
- Possible improvements to the current procedures and workflows.
- Notes from discussions with advisors and peers.
- Measurement information
  - Similar information to the lab notebook plus additional details that could be useful for you during analysis, e.g. how did the cap preparation go (impedance okay? noisy channels?), anything unusual during recording (e.g. excessive movement), anything of note during the debrief?
  - Much of this information overlaps with that recorded in your study workflow checklist (see below), but those checklists are usually stored with each participant's data. Your lab notebook provides a central record that you have close at hand when you're back at your desk analysing data or writing your paper.

11.4 Study Workflow

Although listing tasks and their order might seem obvious, it can play a crucial role at multiple stages of your study (Boudewyn et al., 2023). At the lab level having a study workflow establishes good scientific practice, keeps everyone informed, and promotes accountability for running a study. At the individual level, a study workflow or checklist supports effective planning and execution. During planning, it ensures that all necessary steps are covered and helps to estimate how long each measurement will take. It can also help you identify what can be done before your participant arrives to save you and them time. When executing the study, having a workflow ensures that no steps are forgotten. This is particularly useful when multiple researchers are involved; it brings consistency to the measurements by making sure everyone follows a standard procedure. After the measurement, the workflow serves as a reminder to ensure that paperwork is complete, equipment is cleaned and ready for the next recording, and data are stored appropriately.

Below is an example of a workflow checklist that is organised into pre, during, and post measurement phases. This framework can serve as a starting point for your own study workflow. We recommend completing a checklist every measurement and storing with the participant's data (digital or paper).

11.4.1 Pre-measurement

The pre-measurement part of the study is as important as the measurement itself. With a little planning you can reduce the likelihood of problems occurring and ensure a smooth measurement for you and your participant.

- During participant recruitment
  - Send screening forms ahead of scheduling a measurement, especially if you have any exclusion criteria. It would be unfortunate to have to exclude a participant once they have arrived at the lab.
  - It's useful to ask for the circumference of the participant's head so you can prepare a cap before they arrive. You should provide instructions for how to do this, so they measure it at the correct point. A short document or graphic will be sufficient.
  - Provide full information about the study. Having someone drop out on the day of the study because they weren't fully informed is inconvenient. The participant should know what will happen on the day, what they will have to do, and how long it will take.
- Day before measurement
  - Charge batteries (where applicable).
  - Send a reminder email to the participant, including where and when to arrive, how to prepare for their visit to the lab, e.g. no hair products, no coffee or other substances (if relevant for your study).
  - Prepare consumables: gel, tape, syringes, etc. If anything is running low, let your lab manager know or order some more.
- Day of measurement—before the participant arrives
  - Check temperature and humidity in the recording room and adjust when needed. Sweating can influence your EEG data and being too hot or too cold can influence your participant's comfort and perhaps their ability to focus on the task.
  - Prepare digital or print versions of the consent form and questionnaires.

  - Prepare cap (if size is known). If you are using a sponge based-electrode system soak your net. If your cap has removable electrodes, pre-populate cap.
  - Start your recording software and open the correct recording workspace.
  - Start your experimental software and open your task (where needed).
  - Consider having a dummy version of your experiment that you can run to check triggers.

11.4.2 Measurement

It's essential that you follow a standard procedure for your measurements. You want to avoid forgetting any steps or introducing confounds to the experiment because you did something different for some of the measurements.

- Participant preparation
  - Complete the informed consent procedure.
  - Complete pre-test questionnaires (if relevant).
  - Put the EEG cap/net/headset on the participant and place any additional sensors.
  - Do an impedance check and reduce impedance to the target level.
  - Check data quality.
  - It can be helpful to show the participant the effect of their behaviour on the data. For example, ask them to blink and clench their jaw and point out how the signal changes.
  - Explain the task. Alternatively, this could be done prior to the cap preparation.
  - Perform a practice run of the task if appropriate for your study. You can also use this as an opportunity to check the triggers.
- During the measurement
  - Start the recording! This sounds obvious, but it's a mistake people do make.
  - Check that triggers are arriving when your paradigm is running.
  - Monitor participant's state and whether breaks might be needed.
    - Ask how they're doing in between runs.
    - Keep an eye on the data quality, e.g. are there movement artifacts that could indicate the participant is uncomfortable?
  - Monitor the data and make a note of anything that might affect your analysis or that you might want to address immediately, e.g. if an electrode has become very noisy, perhaps you could check the impedance at the next suitable time.
  - Document anything unusual. For example, the participant blinked a lot, there was an unusual amount of movement artifact, channel X had a bad signal from a certain time point.

11.4.3 Post Measurement

- Remove the EEG cap/net/headset and sensors.
- Offer the participant options for washing their hair.
- Complete post-test questionnaires (where needed).
- Debrief: explain the study, especially if some aspects couldn't be explained before the experiment. Ask if the participant has any questions. Also, ask the participant for their feedback on how the study was, it could be useful for planning future studies.
- Clean the EEG cap/net/headset and sensors and store appropriately for drying.
- Tidy the lab and leave it ready for the next measurement.
- Store your data appropriately.
- Complete the lab book and document any issues with hardware or software that need to be addressed or could be useful for other lab users (see Sect. 1.2).
- Analyse the recorded data. You don't have to do this immediately after the session but it's a good idea to keep on top of data analysis as you go rather than leave it all until the end of the study. By exploring your data, you can spot any problems with your recording setup, e.g. trigger codes that are incorrect or see whether your paradigm is working, e.g. expected behavioural effects are seen. If anything is wrong, you can fix it for the next participants rather than find out at the end of your study.

11.5 Conclusion

In this chapter, we outlined strategies for establishing study workflows and implementing effective lab management. Thorough planning and systematic organisation are essential for running a successful EEG experiment and maintaining an efficient EEG lab. The paper by Boudewyn et al. (2023) provides a valuable overview of factors to consider when planning and running an EEG study. Although the paper is primarily aimed at large-scale multicentre studies, many principles apply to studies of any scale.

Other chapters in this book offer complementary guidance on planning and executing your studies: Chap. 10 Pilot Testing, Chap. 15 Getting Clean EEG Data: Artifacts and How to Prevent them, and Chap. 16 Practical aspects of EEG Data Analysis.

References

Baker, M. (2016). 1,500 scientists lift the lid on reproducibility. *Nature, 533*, 452–454.

Boudewyn, M. A., Erickson, M. A., Winsler, K., Ragland, J. D., Yonelinas, A., Frank, M., Silverstein, S. M., Gold, J., Macdonald, A. W., 3rd, Carter, C. S., Barch, D. M., & Luck, S. J. (2023). Managing EEG studies: How to prepare and what to do once data collection has begun. *Psychophysiology, 60*, e14365.

Higgins, S. G., Nogiwa-Valdez, A. A., & Stevens, M. M. (2022). Considerations for implementing electronic laboratory notebooks in an academic research environment. *Nature Protocols, 17*, 179–189.
Monaghan, J., Brady, S. M., Haswell, E. S., Roy, S., Schwessinger, B., & Mcfarlane, H. E. (2023). Running a research group in the next generation: Combining sustainable and reproducible research with values-driven leadership. *Journal of Experimental Botany, 74*, 1–6.
Vandendorpe, J., Adam, B., Wilbrandt, J., Lindstadt, B., & Forstner, K. U. (2024). Ten simple rules for implementing electronic lab notebooks (ELNs). *PLoS Computational Biology, 20*, e1012170.
Wright, J. M. (2009). Make it better but don't change anything. *Automated Experimentation, 1*, 5.

Part III
EEG Data Acquisition

Chapter 12
Hardware for Recording EEG and Peripheral Physiology

Tracy Warbrick and Cilia Jaeger

Abstract In this chapter, we describe the components of a system for recording EEG and peripheral physiology, how they work, and how they influence the recorded data. We cover electrode types and features that affect data quality, preparation time, and participant comfort. We introduce some examples of peripheral physiology sensors and the technology behind them. We consider the role of the amplifier and what parameters are important to the recorded data. Finally, we consider adding triggers to the setup for recording event markers. At the end of the chapter, you should be able to identify the components of your system, explain what they do, and make decisions about your own setup.

Keywords EEG amplifier · Electrodes · Peripheral physiology · Sensors · EEG measurement principles · EEG hardware · Peripheral physiology hardware · EEG recording principles

12.1 Introduction

As an EEG researcher, knowing how your system works and how it influences the data you acquire is essential. It is also crucial to accurately report the equipment used in your studies (Keil et al., 2014), and to do this you need to be aware of relevant parameters and their roles. In this chapter, we will cover the components of a system for recording EEG and peripheral physiology, the different technologies available, and the important parameters associated with each of them. Our aim is to help you make the right decisions about your hardware and recording parameters. A good place to start is to think about the requirements of your study; for example, what's your signal of interest, who are your participants, do you have any special recording conditions (e.g., mobile or multimodal recordings)? We encourage you to think about these questions and keep your requirements in mind throughout the chapter.

T. Warbrick (✉) · C. Jaeger
Brain Products GmbH, Gilching, Germany
e-mail: tracy.warbrick@brainproducts.com

T. Warbrick (ed.), *The EEG Handbook*,
https://doi.org/10.1007/978-3-032-20450-9_12

12.2 Components of the System

To record EEG data, we need something to detect the signal, something to measure it, a way to record it, and, when needed, a way to integrate an experimental paradigm (Fig. 12.1). Here we consider the role of electrodes, peripheral physiology sensors, and amplifiers and the technical features that influence your data. Note that recording software is covered in Chap. 13: Software for Recording EEG and Peripheral Physiology.

12.2.1 Detecting the Signal: EEG Electrode Technology

The variety of electrode options available can seem overwhelming: gel, salt water, dry, active, passive, cap, headset, low/high density. The choice can be made easier by thinking about the desired data quality, participants, and recording environment for your study. These factors will influence acceptable electrode impedance, comfort, duration of preparation, and duration of recording.

The purpose of the electrode is to detect the electrical signal generated by the underlying neural populations. Here, we will focus on surface EEG electrodes placed on the scalp. To record a signal, we need to make a connection between the scalp and

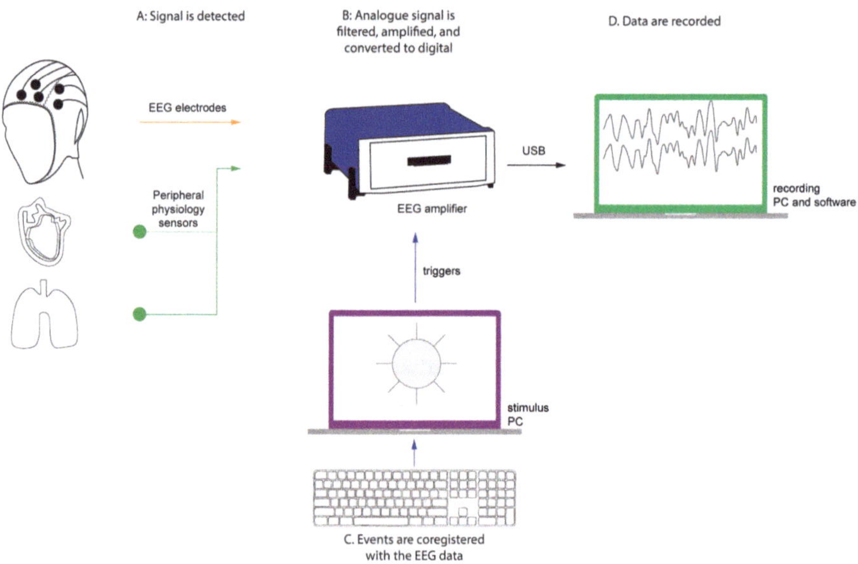

Fig. 12.1 The components of an EEG recording setup. (**a**) EEG electrodes and peripheral physiology sensors are used to detect the signal. (**b**) The amplifier converts the signal from analog to digital, amplifies, and filters the signal. (**c**) Event markers can be co-registered with the data using triggers. (**d**) Data (and triggers) are recorded using dedicated software

the electrode surface. To achieve this, we need a conductive medium or to apply light pressure. To appreciate the different electrode types and associated conductive medium, we need to cover the concept of impedance.

Impedance is the opposition to alternating current, in other words, how easily can a current pass through. When we talk about electrode impedance, we mean impedance between the electrode and the skin. We aim for a low impedance value because we want the EEG signal to 'pass through' with as little opposition as possible. A low impedance also helps to reduce susceptibility to environmental electromagnetic noise. This is important because the electrode and lead wire act as an antenna and pick up environmental electromagnetic noise. Therefore, the signal reaching the amplifier is a mix of EEG signal and noise. Much of the noise arriving at the amplifier will be attenuated by amplifier features such as input impedance and common mode rejection (see Sect. 12.2.3). However, anything we can do to reduce the amount of noise will improve our signal-to-noise ratio and enhance the signal. We can of course reduce the amount of noise in the environment (see Chap. 16: Practical Aspects of EEG Data Acquisition) but we cannot eliminate it all, therefore reducing electrode impedance is important.

It's good practice to have a target impedance for your study and to stay under this threshold consistently across your measurements. To measure impedance a low voltage alternating current is applied and resistance to this current is measured. It's important to note that in most systems where impedance values are displayed, the value is an approximation within the impedance measurement range rather than an accurate, specific value. So having a target threshold or range rather than a specific target value is recommended.

Electrode impedance is influenced by the electrode type (active versus passive), the conductive medium used (gel, salt water, direct contact), and the human head (skin, sweat, hair, moisture).

Human Factors Effective electrode preparation can help to reduce the influence of the human head. The optimal preparation strategy will depend on the electrode type that you use. Practical tips for different electrodes and applications are provided in Chap. 16: Practical Aspects of EEG Data Acquisition.

Electrode Type Active electrodes include a small electronic circuit that performs impedance conversion. This makes active electrodes more robust against cable motion and noisier environments than passive electrodes. It also means that higher electrode impedance can be tolerated with active compared to passive electrodes. This reduces preparation time, which might be beneficial under some circumstances or in certain populations.

Connection Between Electrode and Scalp We will cover three ways of connecting the electrode to the scalp: electrolyte gel, saline solution, and direct skin contact.

For each combination of electrode type and contact method, we will consider target impedance, preparation time, and recording duration. It is also important to consider data quality expectations, the population being studied, and the study environment.

12.2.1.1 Passive Electrode Plus Gel

A low impedance is needed for passive electrodes (see Table 12.1), and an abrasive gel is usually used to help reduce impedance. Reducing impedance can be time-consuming and is less well tolerated by some groups such as children and some clinical populations. Furthermore, care must be taken to avoid breaking the skin (Ferree et al., 2001). The low impedance and conductive gel mean that excellent SNR can be achieved, and complex analyses on small signals of interest are possible. This combination is suitable for standard lab environments and electromagnetically quiet locations as well as special applications where active electrode technology is not appropriate, e.g., for EEG-fMRI.

12.2.1.2 Active Electrodes Plus Gel

For active gelled electrodes we can tolerate a higher impedance (see Table 12.1), so we don't need an abrasive gel, and we don't need to work as hard as for passive electrodes. Therefore, we can save time with the preparation. It is also possible to achieve excellent SNR and do complex analyses on small signals of interest with this combination. However, some caution is advised for very fast changes in the signal (Laszlo et al., 2014). This combination of electrode and gel also allows recordings of more than 8 h and is suitable for lab-based studies, especially noisier environments, mobile studies, and studies involving movement.

12.2.1.3 Passive Electrode and Saline Soaked Sponges

Sponge-based electrode systems use saline solution to conduct the EEG signal. The whole electrode net is soaked in saline solution and this allows an even quicker preparation because you don't need to put gel in each individual electrode. Therefore, this combination is suitable for those less tolerant of long preparations and any study where quick preparation is required. However, it isn't possible to achieve impedances as low as with gel-based electrodes (see Table 12.1). Furthermore, saltwater sponges are usually used with passive electrodes, consequently, there is a risk of picking up more environmental noise and having a lower signal-to-noise

Table 12.1 Target impedance, estimated preparation time, and approximate recording duration for commonly used combinations of electrode type and conduction method

Electrode and conduction method	Target impedance (kΩ)	Preparation time for 32 channels (minutes)	Recording duration
Gelled passive	5–10	30	>8 h
Gelled active	25	5–10	>8 h
Saline solution passive	60–100	5	60–90 min
Dry active	500–2500	5	45 min

ratio. Good SNR is possible, and you can expect to reliably measure ERPs and frequencies up to 100 Hz. Saline nets are suitable for the lab and environments where active electrodes are not possible, e.g., EEG-fMRI.

12.2.1.4 Dry Electrodes

Dry electrode technologies don't use any conductive gel or saline. The most common type of dry EEG electrodes relies on direct contact between the electrode and the scalp (for further information on contact and non-contact dry electrodes, see Shad et al., 2020). Contact between the scalp and electrode surface is achieved using a small amount of mechanical pressure. Preparation is usually very quick, making it suitable for those less tolerant of long preparation. Impedance remains high due to the lack of conductive medium to stabilize the connection between scalp and electrode (see Table 12.1). For this reason, dry electrodes are often combined with active electrode technology. Acceptable SNR is possible, you can expect to measure ERPs and lower EEG frequencies, e.g., up to 45 Hz. Dry electrode technologies are often used in EEG headsets popular for mobile EEG applications due to the need for quick preparation time in non-standard measurement environments.

12.2.1.5 Electrode Positions

While the choice of electrode technology is important, so is where you put the electrodes. A standard positioning system allows comparison across measurements and studies. The most commonly used is called the 10–20 system (Jasper, 1958). Electrode positions are defined by percentages of distances between anatomical landmarks on the head and named with a letter representing the lobe of the brain (e.g., F = frontal) and a number (odd numbers left, even numbers right), see Fig. 12.2. Due to an increasing number of electrodes being used in EEG research, Oostenfeld and Praamstra (2001) extended the 10–20 system to the 10–5 system. Electrodes can be positioned individually by measuring the distance and using an adhesive to fix the electrodes in place. A cap or net can also be used where the electrode positions are predefined. Of course, head shapes and sizes differ so this is to some extent an approximation, but you can improve accuracy by using an appropriately sized cap or net and fitting it properly (see Chap. 16: Practical Aspects of EEG Data Acquisition).

12.2.2 Detecting the Signal: Sensors for Other Measures

Peripheral physiology can be a useful addition to an EEG study or a valuable measure on its own. A variety of measures are possible, and some applications are described in Chap. 34: Bridging Brain and Body: Complementing EEG with

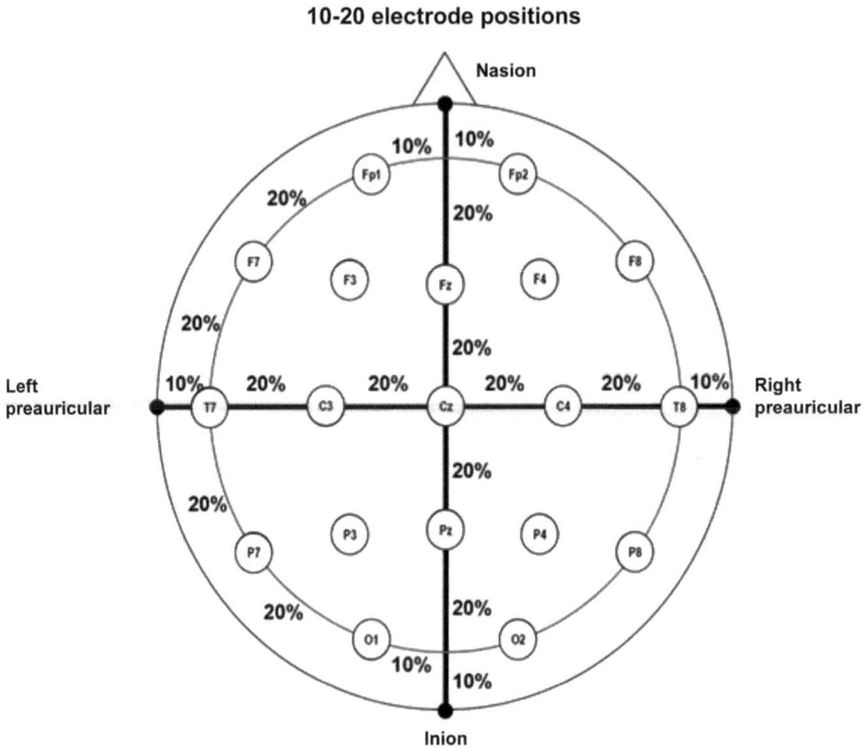

Fig. 12.2 The 10–20 system for naming EEG electrodes (Jasper, 1958). Electrode positions are defined by percentages of distances between anatomical landmarks on the head. They are named with a letter representing the lobe of the brain (*F* frontal, *P* parietal, *C* central, *T* temporal, *O* occipital) and a number (odd numbers left, even numbers right). The lines connecting anatomical landmarks, as well as the circumference, are shown. These lengths are defined as 100%. Electrodes are placed at distances of 10% or 20% of these lines

Peripheral Physiological Signals. Some have different measurement principles from EEG and require specific sensors. It is also possible to record non-physiological signals such as force and acceleration alongside your physiological measure. This section explains the measurement principles used to record some commonly used physiological signals.

12.2.2.1 Bipolar Peripheral Electrophysiology

Let's begin with the peripheral signals that use similar measurement principles as EEG to record electrical activity. These include measuring activity either from the heart (electrocardiography, ECG), the muscles (electromyography, EMG), or the eyes (electrooculography, EOG). The principal measurement technique involves a bipolar recording principle, in which the differential electrical signal is taken between two electrodes. Bipolar measurements are described in Sect. 12.2.3. The types of electrodes that can be used are the same as for EEG. This includes gel-based and dry electrodes as well as active or passive electrodes. Depending on the type of activity recorded, the electrodes should be placed in certain locations to obtain an optimal difference wave to be recorded by the electrodes.

EOG For vertical eye movement one electrode is placed above the eyebrow on the forehead and one electrode is placed below the lower eyelid. For horizontal eye activity, one electrode is placed on the outer canthus of the eye and one electrode is placed on the outer canthus of the opposite eye (Fig. 12.3a). The EOG electrodes measure polarity changes of the eye as the eye moves vertically and horizontally (López et al., 2022).

EMG For recordings of skeletal muscle, one electrode pair should be placed on the muscle belly, which entails the motor unit (Fig. 12.3c). The other electrode should be placed on midline of the tendinous insert of the muscle (Zaheer et al., 2012).

ECG There are several electrode placement layouts for ECG recordings that range from a 3-lead layout to a 12-lead layout. Here we will cover the simpler 3-lead layout, which places electrodes based on the Einthoven Triangle (Barold, 2003). This involves a bipolar electrode pair, with one electrode on the left arm and the other on the right arm. A third electrode, that often functions as the ground electrode, is placed on the left ankle (Fig. 12.3b).

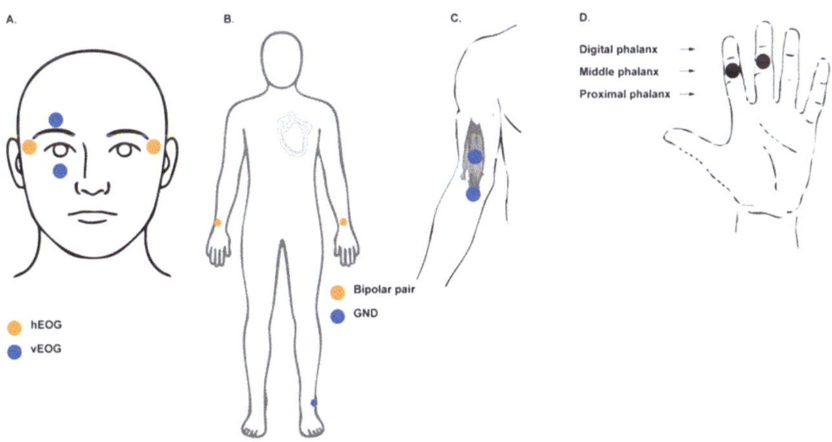

Fig. 12.3 Placement of electrodes for peripheral electrophysiology recordings. (**a**) Indicates placement of bipolar pair of electrodes for vertical EOG, vEOG, recordings and horizontal EOG, hEOG; (**b**) Placement of bipolar electrodes and ground (GND) electrode for ECG recordings; (**c**) indicates muscle body tendon placement for EMG recordings. The example shows the bicep muscle of the arm; and (**d**) electrode placement for skin conductance with GSR sensor

For all bipolar muscle electrodes, a ground electrode is needed to account for external electrical noise. The purpose of the ground electrode is covered in Sect. 12.2.3. The ground electrode should be placed on a bony protrusion or cartilage with minimal electrical activity.

12.2.2.2 Peripheral Physiological Sensors

Data from other types of sensors can also be recorded alongside EEG that can provide additional information on physiological responses. These sensors often utilize different recording principles and require a conversion of the signal to an electrical signal such that the peripheral signal can be measured alongside the EEG recordings.

12.2.2.3 GSR

A sensor that utilizes electrical conductance modulated by physiological responses is the GSR sensor which measures galvanic skin response (GSR) or electrodermal activity. The GSR sensor measures the skin's electrical conductance, which is modulated by sweat gland activity and the electrolyte concentration within sweat. The conductance of skin increases during sweating. The sweat glands are regulated by the autonomic nervous system, which was described in Chap. 4: Basic Anatomy: Peripheral Nervous System. The GSR sensors can measure skin conductance levels, which is the baseline of skin conductance established over time. It can also measure rapid skin conductance changes due to fear- or stress-induced stimuli, which is known as the skin conductance response (Boucsein et al., 2012). A GSR sensor consists of a pair of electrodes, placed on the middle phalanx of the pointer and middle finger (Fig. 12.3d). A small constant voltage is applied across the two electrodes to establish skin conductance, which changes over time or due to an external stimulus. The electrical conductance is then converted to voltages and recorded by an amplifier.

12.2.2.4 Respiration

A respiration belt can be used to monitor changes in breathing rates, which is also associated with autonomic nervous system functions (Liu et al., 2017). The respiration belt is a pneumatic sensor that measures pressure changes that correspond to the rhythmic expansion and contraction of the chest during breathing. During inhalation, the chest expands, and a positive flow of air induces pressure on the pneumatic sensor. During exhalation, the chest contracts and air is expelled from the lungs, which results in a negative airflow or vacuum within the respiration belt. A transducer is also required to convert pressure changes into an electrical signal. The respiration belt can measure both breathing rate and amplitude of respiration, correlating to the amount of airflow (Stern et al. (2001).

12.2.2.5 Photoplethysmography (PPG)

PPG is often used to detect systemic changes to heart rate, blood pressure, and vascular tone. These peripheral responses are under autonomic nervous system control and often correlate with physiological measurements linked to homeostasis and stress responses (Allen, 2007). PPG is a volumetric measurement that detects local perfusion changes in oxygenated hemoglobin within the blood by using infrared light. Infrared light is absorbed differently by oxygenated and deoxygenated hemoglobin and thus can be used to measure blood oxygenation changes linked to the cardiac cycle and blood pressure changes. The sensor is placed on a finger. One side of the sensor emits infrared light through the finger. The blood vessels within the finger then absorb the infrared light proportionally to the amount of oxygen saturation within the blood. A photo diode on the opposite side of the finger then records the amount of unabsorbed infrared light. Over time the PPG signal shows fluctuations in oxygen levels within the blood. In combination with ECG recordings, the PPG sensor is a useful tool for measuring heart rate variability linked to different physiological conditions.

A combination of peripheral measurements alongside EEG recordings can be useful to help characterize behavior linked to simultaneous neurophysiological changes recorded with EEG.

12.2.3 Measuring the Signal: Amplifiers

There are some basic concepts and parameters associated with EEG amplifiers that influence the recorded data. We will cover the main features of an EEG amplifier to describe how your signal is recorded.

The role of the amplifier is to amplify the signal, convert it from an analog to a digital signal, and transfer it to a recording device. EEG is usually recorded using a differential amplifier; the measured signal is the difference between two electrodes. In the case of EEG, this is usually a reference electrode and the EEG signal electrode (see Fig. 12.4). For peripheral physiology such as ECG and EMG (Sect. 12.2.2) a bipolar amplifier is usually used, where each recorded signal is the difference between pairs of recording electrodes (see Fig. 12.4). We will focus on EEG signals in this section, but the same principles apply to peripheral physiology measurements.

The ground electrode plays an important role in EEG recording, specifically, in common mode rejection (CMR). The term common mode refers to identical signals that appear in phase in all electrodes (e.g., electrical mains noise). Common mode *rejection* refers to the attenuation of common mode noise. The ground electrode is connected to the ground circuit of the amplifier and picks up common-mode electrical noise. This noise is subtracted from the reference and signal EEG channels. Rather than measuring a direct difference between the reference and EEG electrodes, the difference between (*EEG—ground*) and (*reference—ground*) is measured.

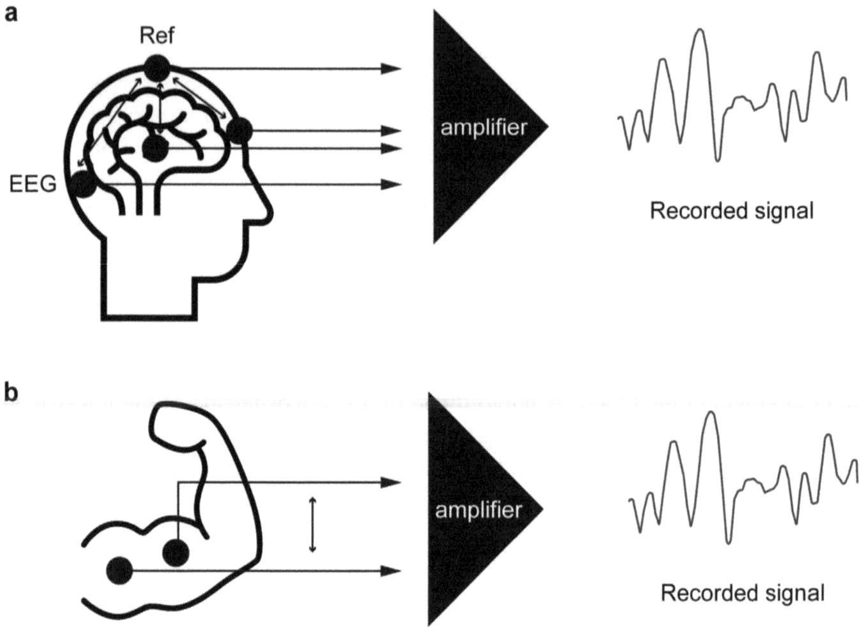

Fig. 12.4 Referential and bipolar recording schemes for EEG and peripheral physiology. (**Part a**) Shows a referential recording scheme where the signal at each EEG electrode is recorded relative to a reference electrode. Note that the distance differs between the reference electrode and each recording electrode, this will influence the signal recorded. The position of the reference electrode relative to the EEG electrodes is therefore important. (**Part b**) Shows a bipolar recording scheme for EMG. The position of the recording electrodes relative to each other (rather than to a refence electrode as in **part a**) will influence the recorded signal

Amplifier input impedance also plays a role in noise suppression. Amplifier input impedance and electrode impedance (Sect. 12.2.1) should not be confused. For electrode impedance we want the signal to pass through, so a low impedance is preferred. For amplifier input impedance we want to *measure* the signal, so a high impedance is required. It may seem counterintuitive to want a low impedance in one place and high impedance in another, but we have different goals for the signal at the electrode and at the amplifier. The ratio of electrode impedance and amplifier impedance is important (Shad et al., 2020). The EEG signal is small therefore it's important to retain as much of the signal as possible and to minimize the effect of noise. Modern amplifiers have very high input impedance; this means that higher electrode impedance can be tolerated than with early EEG systems. However, it is worth remembering that not all types of noise are attenuated by the high input impedance and low electrode impedance is generally recommended. For a detailed explanation of amplifier impedance and its relation to electrode impedance please see Jackson and Bolger (2014).

Once the signal reaches the amplifier it passes through hardware filters to remove frequencies outside of the desired range. The signal is then passed to the analog to digital converter (ADC). An analog signal is perfectly resolved in time, but a digital signal is a sampled signal. How often and how many of those digital samples are taken is referred to as the sampling rate. This determines how true to the original analog signal the digital signal is and determines the temporal resolution (Fig. 12.5).

The analog signal also gets sampled at regular amplitude intervals (see Fig. 12.6). The voltage intervals at which the continuous signal is sampled (or digitized) are defined by the bit depth and amplitude resolution of the ADC. The bit depth determines how many possible digital values a signal can have, and the amplitude resolution determines the size of steps between values. The more possible values, the more fine-grained the differentiation between voltage steps and the closer to the original analog signal the digital signal will be (Fig. 12.6). The bit depth and the amplitude resolution determine another important parameter: measurement range. This is the maximum and minimum voltage values that can be recorded. For example, a 24-bit amplifier has 2^{24} (16,777,216) possible steps. With a resolution of 0.0487 μV/bit, the measurement range is ±409.6 mV.

At this point you might be thinking that we should always use the highest sampling rate and highest amplitude resolution possible. But at some point, higher

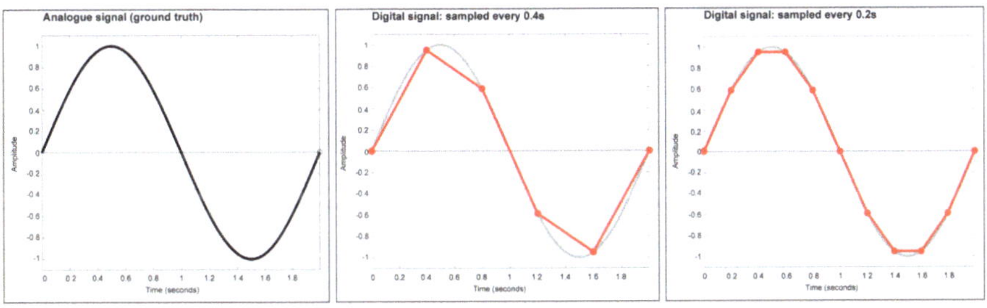

Fig. 12.5 Analog to digital conversion: temporal resolution. An analog signal perfectly resolved in time is shown in the left panel. A digital signal is a *sampled* signal, obtained by sampling the analog signal at discrete points in time. The higher the sampling rate, the better the time resolution. The middle panel shows a signal sampled every 0.4 s and the right panel shows a signal sampled every 0.2 s. The signal in the right panel is closer to the original analog signal than the digital

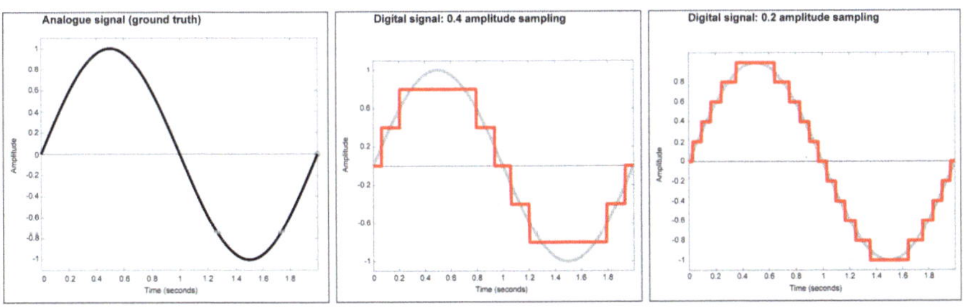

Fig. 12.6 Analog to digital conversion: amplitude resolution. A continuous signal gets recorded (& amplified) at regular amplitude intervals. The higher the bits per volt, the higher the amplitude resolution. The left panel shows the original analog signal. The middle panel shows a signal sampled with an amplitude of 0.4. The right panel shows a signal sampled with an amplitude of 0.2. The signal in the right panel has a more fine-grained resolution and is closer to the original analog signal

resolution doesn't bring much return with respect to the signal and the consequences of going higher might not be acceptable: increases in file size and decreases in processing speed. Consequently, we don't want to push temporal and amplitude resolution as high as possible. But what is enough and what is too low?

It might help to consider the consequences of too low resolution. If our sampling rate is too low, we can potentially lose useful information and we run the risk of aliasing. Aliasing is when different signals become indistinguishable when sampled at an insufficient sampling rate. You can see a signal in the data that isn't real, it's an alias. To prevent aliasing, we can implement anti-aliasing filters. This is done on the hardware level and is often a fixed feature of the hardware. Even if we don't choose our antialiasing filters, we do need to choose a sampling rate and for this we need to consider the Nyquist frequency, or folding frequency. This is the theoretical maximum frequency that can be recorded without aliasing artifacts. For a given sampling rate, up to half of that frequency can still be represented in the recording, e.g., for a sampling rate of 1000 Hz, you could expect usable data up to 500 Hz. In practice, anti-aliasing filters are often considerably lower than the Nyquist frequency and it is recommended that data are sampled at 5–10 times the frequency of interest (Weiergraeber et al., 2016).

If our amplitude resolution is too low, we run the risk of saturation, or clipping, in the data. In other words, the amplitude of the signal exceeds the measurement range of the amplifier. For most applications a standard EEG amplifier offers a sufficient measurement range. However, if you are working in an application where very high amplitudes are expected, for example, scanner related artifacts in the MR environment or during TMS stimulation, you need to be aware of the measurement range of your amplifier. In both cases you want to record the full range of the artifacts and these are much larger in amplitude than EEG, therefore you need to make sure your amplifier is capable of the required measurement range.

Amplifier technology has progressed in recent decades, and it is now possible to record EEG in many different environments. In addition to standard, stationary, laboratory amplifiers, small mobile amplifiers and amplifiers integrated into electrode headsets are available. This means you can record high-quality data in real-world settings outside of the laboratory. It is also possible to combine EEG with other measurement (e.g., fMRI) or stimulation (e.g., TMS) modalities, making multimodal recording possible. These developments are significant in terms of what EEG can be used for. Parts V and VI of this book cover EEG applications and multimodal recordings, respectively, and showcase what applications benefit from advanced EEG amplifier technology.

12.3 Conclusion

We've introduced EEG electrodes, peripheral physiology sensors, and EEG amplifiers. We've also covered the main features and explained how they affect your data. It is important to keep in mind the aims of your study, the study population, and the environment you will measure in when deciding what equipment to use. You should also consider your planned data analysis when planning your measurement setup because the recording decisions you make will influence the data you will analyze later. The information in this chapter will help you to make informed decisions about electrodes, sensors, and amplifiers that are suitable for your application and study population.

References

Allen, J. (2007). Photoplethysmography and its application in clinical physiological measurement. *Physiological Measurement, 28*(3), R1–R39. https://doi.org/10.1088/0967-3334/28/3/R01

Barold, S. S. (2003). Willem Einthoven and the birth of clinical electrocardiography a hundred years ago. *Cardiac Electrophysiology Review, 7*(1), 99–104. https://doi.org/10.1023/a:1023667812925

Boucsein, W., Fowles, D. C., Grimnes, S., Ben-Shakhar, G., Roth, W. T., Dawson, M. E., Filion, D. L., & Society for Psychophysiological Research Ad Hoc Committee on Electrodermal Measures. (2012). Publication recommendations for electrodermal measurements. *Psychophysiology, 49*(8), 1017–1034. https://doi.org/10.1111/j.1469-8986.2012.01384.x

Ferree, T. C., Luu, P., Russell, G. S., & Tucker, D. M. (2001). Scalp electrode impedance, infection risk, and EEG data quality. *Clinical Neurophysiology, 112*, 536–544.

Jackson, A. F., & Bolger, D. J. (2014). The neurophysiological bases of EEG and EEG measurement: A review for the rest of us. *Psychophysiology, 51*, 1061–1071.

Jasper, H. H. (1958). The ten twenty electrode system of the international federation. *Electroencephalography and Clinical Neurophysiology, 10*, 371–375.

Keil, A., Debener, S., Gratton, G., Junghofer, M., Kappenman, E. S., Luck, S. J., Luu, P., Miller, G. A., & Yee, C. M. (2014). Committee report: Publication guidelines and recommendations for studies using electroencephalography and magnetoencephalography. *Psychophysiology, 51*, 1–21.

Laszlo, S., Ruiz-Blondet, M., Khalifian, N., Chu, F., & Jin, Z. (2014). A direct comparison of active and passive amplification electrodes in the same amplifier system. *Journal of Neuroscience Methods, 235*, 298–307.

Liu, T. T., Nalci, A., & Falahspour, M. (2017). The global signal in fMRI: Nuisance or information? *NeuroImage, 150*, 213–229. https://doi.org/10.1016/j.neuroimage.2017.02.036

López, A., Ferrero, F. J., Llopis, M. V., & Campo, J. C. (2022). Reference electrode placement in EOG-based systems design. In *2022 IEEE international symposium on medical measurements and applications (MeMeA)* (pp. 1–5). Messina, Italy. https://doi.org/10.1109/MeMeA54994.2022.9856469

Oostenveld, R., & Praamstra, P. (2001). The five percent electrode system for high-resolution EEG and ERP measurements. *Clinical Neurophysiology, 112*, 713–719.

Shad, E., Molinas, M., & Ytterdal, T. (2020). Impedance and noise of passive and active dry EEG electrodes: A review. *IEEE Sensors Journal, PP*(99), 1–1.

Stern, R.M., Ray, W.J. And Quigley, K.S. (2001). Respiratory system. In: Psychophysiological recording. 2nd ed.: Oxford University Press, pp.142–156.

Weiergräber, M., Papazoglou, A., Broich, K., & Müller, R. (2016). Sampling rate, signal bandwidth and related pitfalls in EEG analysis. *Journal of Neuroscience Methods, 268*, 53–55.

Zaheer, F., De Luca, C. J., & Roy, S. H. (2012). Preferred sensor sites for surface EMG signal decomposition. *Journal of Neuroscience Methods, 207*(2), 204–212. https://doi.org/10.1016/j.jneumeth.2012.03.017

Chapter 13
Software for Recording EEG and Peripheral Physiology

Tracy Warbrick and David Kadlec

Abstract In addition to hardware considerations, the choices made in the recording software will influence the quality of the acquired data. This chapter covers the main features of EEG recording software, important recording parameters, monitoring and recording your data effectively, and troubleshooting tips for common recording software problems.

Keywords EEG data format · EEG recording parameters · EEG software configuration · EEG data monitoring

13.1 Purpose and Features

The main purpose of your recording software is to create a stored version of your data. After analog to digital conversion, the data are transferred to the recording device, which may be a desktop PC, laptop, tablet, or mobile phone.

Data Format EEG and peripheral physiology data are typically stored in binary or text format. Other relevant information, such as recording parameters, metadata, and markers, can be stored alongside the EEG data. Because raw data and metadata formats vary across the EEG community, the specific structure of the saved data depends on the software and settings. To encourage standardisation of data storage and description across neuroimaging studies, the Brain Imaging Data Structure (BIDS) data format was implemented. BIDS facilitates sharing and reusing data, the application of automatic pipelines, and using quality assurance protocols. It was initially implemented for MRI Data (Gorgolewski et al., 2016) and later extended to EEG data (Pernet et al., 2019). The EEG-BIDS specification recommends that one of two data formats be used: The European Data Format (EDF) (an ongoing international effort to standardise EEG data format), and the BrainVision Core Data Format (developed by Brain Products GmbH) (Pernet et al., 2019). It is also recommended

T. Warbrick (✉) · D. Kadlec
Brain Products GmbH, Gilching, Germany
e-mail: tracy.warbrick@brainproducts.com; david.kadlec@brainproducts.com

T. Warbrick (ed.), *The EEG Handbook*,
https://doi.org/10.1007/978-3-032-20450-9_13

that additional metadata extracted from the manufacturer-specific data files are stored in the 'sidecar JSON' file. Further details can be found in Pernet et al. (2019). Researchers are encouraged to use the BIDS format; it will bring order and structure to your data, and you can play your part in standardising EEG data formats.

Data Handling Once recorded, the EEG can be data stored locally or in cloud-based storage systems. Any personal data stored in the cloud must comply with GDPR rules and it's the researcher's responsibility to know how their data are handled. Also consider whether the raw data can be accessed or some processing is applied before it's displayed. Generally, it's preferable to access the raw data so you can assess the quality and apply your own processing pipeline. If the data are processed before you receive them, it's crucial to know what specific steps have been applied.

Basic Features In addition to recording data, most EEG software includes impedance measurement and data monitoring. Impedance measurement allows you to establish good contact between the electrode and the scalp. Chapter 16, Practical Aspects of EEG Data Acquisition provides advice on improving impedance. Note that some systems don't include impedance measurements and, in this case, you should rely on signal quality to decide whether to proceed with measurements. Data monitoring allows you to check the signal before starting the recording and to observe signal quality during the recording.

Advanced Features Some software packages offer advanced features that can enhance your experimental setup. For example, experimental control, synchronised video recording, real-time access to the data, or Lab Streaming Layer (LSL) connection. Real-time access to the data and a straightforward LSL connection can be helpful when you need to merge data streams from different systems. Not every system is equipped with interfaces for sharing event markers or dedicated time stamps or for registering these in the data stream. If this functionality is important to your study, verify that the system supports LSL for synchronising signals and event information.

Open Source Versus Proprietary Software EEG software can be open source or proprietary; each has advantages and limitations, and their suitability depends on the circumstances.

Open-source software (OSS) is released under a license that allows personal or commercial use. Users have access to the source code and can modify it and redistribute the original and/or modified versions, as OSS is typically categorised as free software. However, be aware that not all free software is open source. In the neuroscience community, OSS is usually developed and maintained by scientists and academic institutions. It is developed to meet a scientific need, rather than a commercial requirement. While the concept of OSS is positive and increasingly relied upon in neuroscience (Poldrack et al., 2019), maintenance and support for open-source packages are uncertain (Westner, 2024). Users should also verify that free OSS is standalone, or whether it requires another programme to run. For example, some free EEG software requires MATLAB, which is a commercial software that requires a license.

Proprietary, or commercial, software is protected by legal measures that restrict its use, distribution, and modification. There will be an End-user License Agreement (EULA) or Terms of Service (TOS) that outlines the terms and conditions under which the software can be used and accessed. End users cannot access the source code of commercial software, and it remains the intellectual property of the company. Commercial software development is typically managed by a team of developers who take care of modifications, upgrades, and bug fixes. The software provider is also responsible for ensuring that the implemented processes are correct. In addition to the customer support that usually comes with commercial software, this maintenance and accountability can justify the purchase price.

Operating Systems It is also necessary to consider the operating system on which the software will run, and whether that is compatible with practices in your lab (e.g. Windows Operating System, Linux, Apple's macOS, Android, or iOS). It is sometimes possible to emulate the required operating system within another operating system, but this can influence the functionality of the software and is generally not supported by the software provider. Regardless of the operating system, the recording software should be installed on a separate computer to other parts of the experiment (e.g. stimulus programme) to avoid unwanted interactions.

13.2 Before Starting Your Study

The stored EEG data are directly influenced by hardware and software configuration. This section covers key parameters that should be considered and their effects on the data. Some parameters are related to the hardware concepts discussed in Chap. 12, Hardware for Recording EEG and Peripheral Physiology; it is recommended that you also read that chapters as it provides complementary advice.

Even when the recording software is bound to the hardware, researchers should be familiar with the parameters available and how they relate to the recorded data. This ensures an informed choice of recoding parameters based on the study requirements.

13.2.1 General Parameters

The options available will vary depending on the hardware and software combination used, but there are some general parameters that should be considered.

Number of Channels Make sure the correct number of channels is specified and that they match the physical channels of the amplifier. For example, if your amplifier is capable of recording 64 channels but your study is only using 32, confirm that the correct ones are selected.

Channel Types If you are recording different types of data, for example, peripheral physiology in addition to your EEG, ensure that your channels are set up correctly. Signals such as electromyography (EMG) typically require a bipolar recording setup while some measures, such as respiration, are recorded from sensors with different parameters. Consult the manufacturer's guidelines for optimal recording with different sensors.

Sampling Rate The sampling rate should be 5–10 times larger than the frequency of interest (Weiergräber et al., 2016). However, extremely high sampling rates create large data files, especially for longer recordings, so choosing the highest available is not always the most sensible option. If multiple sampling rates are available choose the most appropriate for your study based on the planned analyses and recommendations in the literature.

Amplitude Resolution The measurement range of an amplifier is determined by how many bits it has and the amplitude resolution: in other words, how many possible digital values the signal can have and the size of steps between values. The number of bits is a fixed hardware feature but sometimes the amplitude resolution is configurable. A smaller amplitude resolution will give you a more fine-grained representation of the original signal but a smaller measurement range for a given number of bits. For example, a 16-bit amplifier has $2^{16} = 65{,}536$ possible steps. With a resolution of 0.1 μV/bit the measurement range is ±3.28 mV and at 0.5 μV/bit it is ±16.384 mV.

Filters It's important to make the distinction between hardware filters, software filters, and display-only filters.

The hardware filters determine the frequency range reaching the analog to digital converter and include an antialiasing filter (see Chap. 12, Hardware for Recording EEG and Peripheral Physiology).

Software filters are applied to your data after they have been digitised but before they are saved to disk. It is generally recommended to avoid using software filters as the filtered frequencies are permanently removed. Instead, it's better to record the whole range determined by the hardware filter and then filter offline when needed.

Display filters are applied for visualisation purposes so you can view a 'clean' signal during recording, but the full range of data determined by your hardware filters is preserved. For example, if you're not able to remove all sources of mains electrical noise from your recording environment, you will see a 50 Hz or 60 Hz signal in your data. A display filter can be used to remove this noise so you can view a cleaner signal without affecting the stored data.

Triggers and Event Markers The recording software should be configured to receive triggers and record event markers. There are several ways to register the timing information of the main events of your experiments along with your EEG data, for example using hardware triggers and software markers. Chapter 14, Triggers, covers the different options in detail.

13.2.2 Special Applications

Some applications require specific recording parameters. For example, the high-amplitude artifacts seen in EEG-fMRI and EEG-TMS require a large measurement range. Very small signals such as an auditory brain stem response require a very high sampling rate in comparison to regular EEG. Covering all special applications is beyond the scope of this chapter. The important message is to be aware of any specific conditions and what can be done to accommodate these conditions and maximise your data quality. Manufacturer guidelines and literature in your field of study are a good place to start.

13.2.3 Real-Time Processing

Some applications require real-time processing. For example, in closed-loop brain stimulation EEG signal features are used to trigger a stimulation device such as transcranial magnetic stimulation (TMS). In Brain Computer Interface (BCI) applications, immediate feedback is provided to the participant based on an EEG feature. Some sleep recording software can perform real-time sleep scoring and requires access to the real-time data. This typically requires a connection between the recording software and another client (e.g. remote data access). Input from an institution's IT department might be necessary, e.g. if permissions are restricted. Therefore, it is best practice to set up and test the real-time processing pipeline during the pilot testing phase.

13.3 During a Measurement Session

It's tempting to just press record and assume that everything is running as expected. However, it is essential to perform some initial checks before starting the recording and to monitor the signal during recording. It's important to pay attention to detail and to be consistent across measurements.

Impedance Check In Chap. 12, Hardware for Recording EEG and Peripheral Physiology, the role of impedance at the electrode scalp interface was covered. Researchers should establish a target impedance range for a study, this will depend on the electrode type and the application. Impedance at all recording electrodes should be checked prior to starting a measurement. The options available will depend on the software but you should be aware of the scale that you are using and whether you can switch between scales. Some systems might offer an LED impedance indication system on the electrodes, in which case it's crucial to know what levels are set for the light colours. Chapter 16, Practical Aspects of EEG Data Acquisition, provides some tips and tricks for achieving your target impedance with different amplifier and electrode configurations.

Monitoring the Data The display should be set up to optimally view the signal. Use a scale that is meaningful and will allow you to assess signal quality or to spot any problems. For high-density recordings, e.g. 256 channels, it's not practical to view all channels at once, but rather to view fewer channels in subsets. For signals with different amplitudes or temporal properties, the scaling should be adjusted to view each signal optimally. For example, the galvanic skin response (GSR) is much slower than EEG and requires a different temporal scale for optimal viewing. If display filters are available, it is important to know whether this is enabled and how to en−/disable this feature. Furthermore, it might be useful to view data in a different montage, for example re-referenced. This functionality isn't available in all software, but if available it should be configured before the recording is started.

Triggers and Markers Triggers should be verified before the recording, especially in shared labs where settings might change between recordings. In addition to trial markers, it might be possible to add on-the-fly annotations to the data to indicate noteworthy events, e.g. excessive movement or interruptions.

Save the Data! Make sure that the data are being recorded. For example, some programs require you to save a data file at the end of recording, others start saving to disk automatically once you start. Note that your software might have a monitoring mode that allows you to view the data but it is not yet recording, it's important to switch from monitoring to saving. This sounds obvious, but it's not an uncommon error.

13.4 Troubleshooting

It's important to be prepared for potential problems when recording EEG data. While specific problems will differ across software programmes, there are some typical issues that can be addressed with standard troubleshooting strategies. Below we list a few common problems and where to start troubleshooting them.

License Issues It can be frustrating if your software won't open due to licensing restrictions. Researchers should know where to find the license information: license number, expiry date, how to activate it, and how to renew it. This information will also be needed if you need to contact the software support service. It is good practice to check the software expiry date during pilot testing to ensure that it won't expire during the study.

Communication Between the Hardware and Software A common problem is the failure to establish communication between the hardware and software. System or software manuals from the manufacturer are a good place to start for troubleshooting advice. Communication problems will most likely be solved during pilot testing; nonetheless, it's good to know where to start troubleshooting if problems occur during measurements.

Unexpected Interruptions Avoid using the same computer for recording the data and for other aspects of the study, such as your stimulation paradigm. This will prevent data overflow problems (Windows), screen freezes, and other interruptions. It's also important to make sure that the recording computer won't go into sleep mode.

Maintenance Software should be kept up to date. Regularly check for updates and patches from the manufacturer. When updating the software, consider whether an update will have an impact on ongoing studies. It's good practice to record all data in a study with the same version of the recording software.

Support Remember that most commercial software comes with a support service. When contacting a support team it is helpful to provide the version number of the software, your license information, and a thorough description of the problem, including error messages (screenshots are helpful). If you are using OSS there is often a forum, sometimes moderated by the developers, that can be a useful place to start.

13.5 Conclusion

This chapter has introduced the key features of EEG recording software and the parameters you need to consider when setting up your recording environment. It is important to understand how software choices influence the data, and we recommend that you spend some time setting up your recording software. It is also helpful to pilot test your choices before beginning data acquisition for your study. We have also outlined good practice for monitoring data quality during a measurement and provided troubleshooting tips for common software-related problems. We recommend familiarising yourself with these during the pilot testing phase of your study.

References

Gorgolewski, K. J., Auer, T., Calhoun, V. D., Craddock, R. C., Das, S., Duff, E. P., Flandin, G., Ghosh, S. S., Glatard, T., Halchenko, Y. O., Handwerker, D. A., Hanke, M., Keator, D., Li, X., Michael, Z., Maumet, C., Nichols, B. N., Nichols, T. E., Pellman, J., Poline, J. B., Rokem, A., Schaefer, G., Sochat, V., Triplett, W., Turner, J. A., Varoquaux, G., & Poldrack, R. A. (2016). The brain imaging data structure, a format for organizing and describing outputs of neuroimaging experiments. *Scientific Data, 3*, 160044.

Pernet, C. R., Appelhoff, S., Gorgolewski, K. J., Flandin, G., Phillips, C., Delorme, A., & Oostenveld, R. (2019). EEG-BIDS, an extension to the brain imaging data structure for electroencephalography. *Scientific Data, 6*, 103.

Poldrack, R. A., Gorgolewski, K. J., & Varoquaux, G. (2019). Computational and informatic advances for reproducible data analysis in neuroimaging. *Annual Review of Biomedical Data Science, 2*, 119–138.

Weiergräber, M., Papazoglou, A., Broich, K., & Müller, R. (2016). Sampling rate, signal bandwidth and related pitfalls in EEG analysis. *Journal of Neuroscience Methods, 268*, 53–55.

Westner, B. U. (2024). *Cycling on the freeway: The perilous state of open source neuroscience software*. 2403.19394. Available: https://arxiv.org/abs/2403.19394.

Chapter 14
Triggers

Alex Kreilinger and Paulo Rodrigo Bazán

Abstract In this chapter, we will cover the importance of triggers and how they can be used to add contextual information to data recordings. We will also demonstrate different common ways to generate and record triggers, along with suggestions for alternative approaches if triggers are not an option. Finally, we will demonstrate strategies that are considered good practice.

Keywords Synchronization · Triggers · Events · Markers

14.1 Introduction

Triggers are one way to synchronize external events with EEG recordings. They are necessary to allow the analysis of time-locked signals. To maximize the quality of the results it is vital that the triggers are recorded with the EEG data as precisely as possible.

First, some clarification of the terminology used around triggering is needed to understand the difference between events, markers, and triggers (Fig. 14.1):

- *Event:* describes something that is happening which is relevant in the context of the recording. For example, an event could be a behavior from the participant, such as a motor response (e.g., a button press). Or the appearance of a stimulus in an experimental paradigm, e.g., visual, acoustic, or tactile.
- *Marker:* describes such an event in the software. It is the information about *what and when* and is provided either directly in the recording, or in an additional supporting file. For example, a marker file may contain a matrix of types and corresponding time stamps. This information is particularly important for EEG data analysis.

A. Kreilinger (✉) · P. R. Bazán
Brain Products GmbH, Gilching, Germany
e-mail: alex.kreilinger@brainproducts.com; paulo-rodrigo.bazan@brainproducts.com

T. Warbrick (ed.), *The EEG Handbook*,
https://doi.org/10.1007/978-3-032-20450-9_14

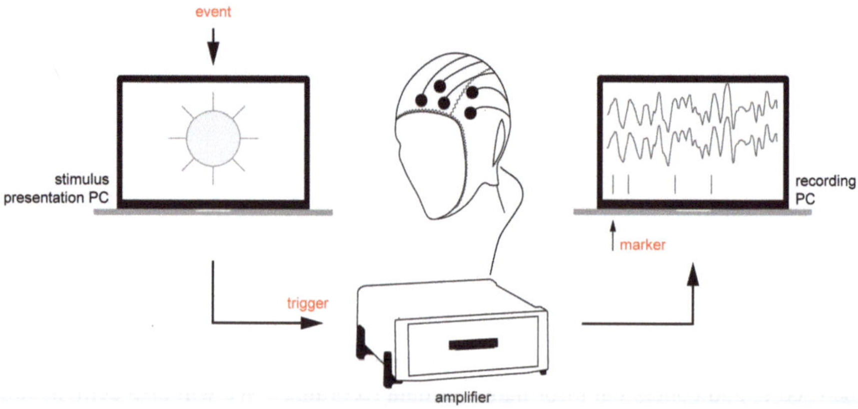

Fig. 14.1 Events, markers, and triggers in a visual paradigm. In this case, a visual cue appearing on the screen is the event. The stimulus presentation computer sends a trigger to the EEG amplifier at the same time. The trigger gets recorded together with the EEG data. The triggers are represented as markers in the recorded file

- *Trigger:* is the hardware embodiment of an event, represented by an electric or optical signal. Usually, these triggers are sampled and recorded directly with the EEG to provide optimal timing precision.

14.2 Importance of Triggers

EEG has a very high temporal resolution (Chap. 2, What is EEG). Therefore, to properly analyze EEG signals in relation to a task or stimulus, the time of the task or stimulus needs to be recorded with high precision. In this context, having triggers is a reliable form of relating and synchronizing events to EEG data. Triggers are also helpful for multimodal data acquisition, as they can be used to synchronize the data acquired from different devices. For example, the EEG system can mirror or forward the triggers it receives from a stimulation source to another device, such as a functional near-infrared spectroscopy (fNIRS) system. This also works the other way around: for example, an eye-tracking system can send triggers to the EEG amplifier. This will then generate common markers in the multimodal recording which can be used to merge the data or analyze it together. Another option is to replicate triggers with a y-cable or a one-to-many trigger replicator.

The required precision for triggers will depend on the research questions and desired analysis. For example, if the analysis focuses on brain activity sustained over several seconds, it is likely that a smaller precision is good enough. In contrast, for event-related potential (ERP) analysis, where an average of trials will be generated in a time-locked window of a few hundred milliseconds, even variations of a few milliseconds can significantly affect the results. In these cases, millisecond precision is required, and high precision triggers are especially important.

The precision is affected by latency and jitter:

- *Latency:* is related to the delays between the actual onset of the event and the trigger being received or the marker being generated in the EEG recording. It is recommended to check these delays, as latency can be adjusted during the analysis, if it is known and constant. However, for near real-time applications, latency needs to be very short, such as in transcranial magnetic stimulation (TMS)-EEG, brain-computer interface (BCI), or neurofeedback.
- *Jitter:* is the variation in the delays from trial to trial. These delays are more problematic because they are variable. Therefore, we need to reduce them to the precision level that we want to assure in our experiment.

Figure 14.2 demonstrates the effects of jitter and latency on the waveform of an ERP.

14.3 Advantages of Triggers

While aligning the timing between EEG signals and events may seem straightforward, it is not guaranteed in every setup. There are many factors that can disrupt this alignment, but by encoding the event with a hardware trigger, the exact time of the event is recorded precisely when it really happens. This is because the hardware input directly synchronizes the trigger signal with the EEG signal. This means that potential delays caused by separate processing pipelines of markers and EEG data can be avoided. For example, different stimulus presentation software tools could otherwise have a negative effect on precision. There are also differences in how fast data can be read from the amplifier to the PC, tablet, or smartphone. Therefore, the main advantage of the triggers is the high precision due to the simultaneous sampling together with EEG signals. Additionally, in some cases triggers can be generated directly from the stimulus or participant response, reducing the need for additional programming when compared to software solutions.

As an example, a common scenario is the presentation of a visual stimulus. The aim of an experiment is to record the brain's response to a visual stimulus, or to a variety of visual stimuli. These stimuli can be created in a software environment and consist of pictures that appear and disappear at specific moments. Now, to analyze the ERP associated with these visual stimuli, it is important that the synchronization between event and signal is precise. We can guarantee this if every individual visual stimulus is accompanied by a corresponding trigger input that gets sampled simultaneously with the EEG. Ideally, the visual stimulus can be used itself for creating the trigger input. This can be done by recording the visual stimulus via a photo sensor and sending a trigger as soon as a certain threshold is exceeded. This way, the true appearance of the stimulus can be verified directly in the recording. Otherwise, one can never be sure that a stimulus happens when it is supposed to. For example, a

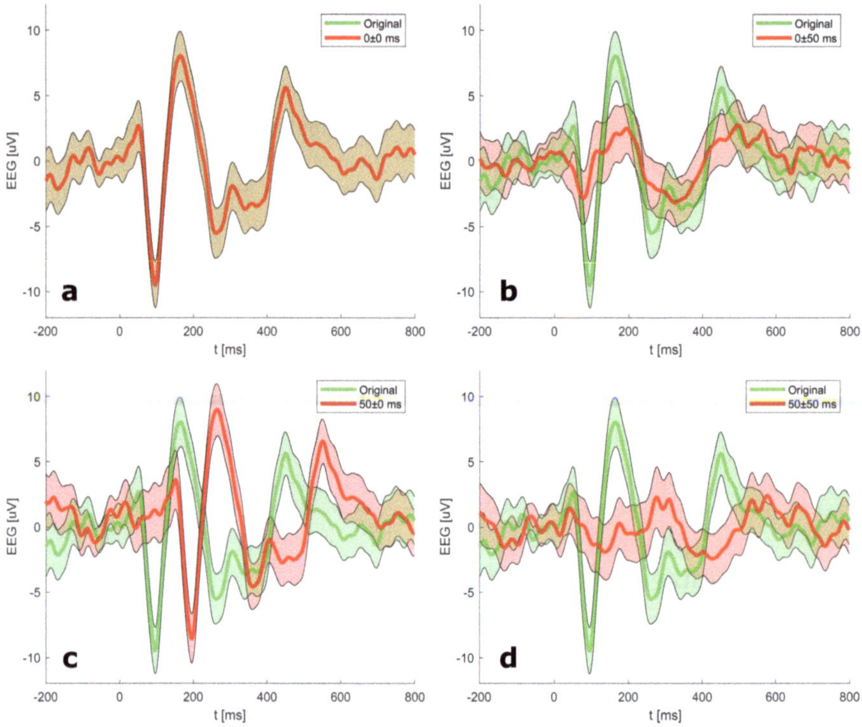

Fig. 14.2 Effects of jitter and latency on an ERP ± confidence interval. A stable latency changes the time of the ERPs' appearance in relation to the event (in this case at $t = 0$ ms). Increasingly high jitter directly affects the waveform. The higher the jitter, the more of the original waveform is getting lost. The four examples demonstrate the effects of latency and jitter, simulated with normally distributed delays based on mean latency ± jitter in milliseconds. (**a**) No latency and no jitter: the original and simulated ERPs overlap perfectly. (**b**) Here, the mean latency is zero, but the jitter has a high standard deviation of 50 ms. The waveform of the ERP is noticeably different. (**c**) Constant, jitter-free latency: the ERP is preserved but shifted on the time axis. The amplitude can also be affected if baseline correction is used. (**d**) High latency and high jitter significantly distort the original waveform and timing information of the ERP

visual stimulus might be delayed because of an additional screen refresh cycle that was not accounted for, or an acoustic stimulus might require some additional time to load the sound into a buffer. Not accounting for these possibilities can easily lead to jitter and latency issues in the order of tens of milliseconds, which will have an impact on the ERP analysis. At the end of this chapter, we will provide more details on how to set up an experiment to verify the timing accordingly.

If possible, connecting triggers to the EEG recording hardware is a great way to make sure that important external events are synchronized with the other recorded signals without any potential interference.

14.4 Disadvantages of Triggers

Adding triggers to the setup typically requires additional accessories and cables that can be cumbersome to connect, can make the setup more complicated, and require more preparation time. In certain cases, it may be beneficial to have a very fast and uncomplicated design to keep the recording time to a minimum. It could be that the recording itself is already extensive or that the participants need to be kept motivated to not lose their concentration. There are also recording setups that do not foresee the inclusion of triggers, either by not allowing the physical connection or by not needing triggers in the first place. For example, there are cases in which all the relevant context is provided directly within the acquisition software itself, e.g., in sleep EEG.

14.5 Alternatives to Triggers

As already mentioned, triggers may not always be the preferred method, and there may even be circumstances where triggers are not necessary. A generalization is difficult as this depends on the respective environment of the selected EEG amplifier, the research question, and planned analysis. It might be the case that markers can be created and stored directly within the acquisition software. The amplifier might also be connected to a platform which creates context by means of a marker or an auxiliary signal and saves it directly with the data (e.g., OpenViBE (Renard et al., 2010)). In these cases, it is highly recommended to perform thorough testing during the pilot phase to make sure signals and markers are properly synchronized (Chap. 10, Pilot Testing).

A common alternative to triggers is to use Lab Streaming Layer (LSL) (Kothe et al., 2025). LSL is a protocol that manages the synchronization of multiple signal sources in a shared local network. It works with regular streams that have a specific sampling rate, but also with irregular streams where data points only appear occasionally, or at irregular intervals. Instead of triggers, one can create such an irregular marker stream that is available in the network. Any LSL client can then connect to the data and marker streams, allowing signals to be processed and/or saved together with contextual information. A study using TMS-EEG to compare hardware triggers and software markers via LSL is presented in Miziara et al. (2025).

One drawback of combining such LSL marker streams and LSL data streams is the time that passes between the sampling of a data point and its availability in the stream. Whereas an LSL marker can be sent to the network almost immediately, the sample from the EEG amplifier is subject to delays due to several factors: communication on the driver level, transmission time via Bluetooth® or WiFi, or potentially some internal buffering. Many factors can add up to create a latency that is no longer negligible, especially when dealing with time-critical ERPs.

14.6 Good Practice for Using Triggers

Once the decision to use triggers has been made, it is good practice to adhere to a few principles that will ensure that no information is lost, and that the information is presented in a clear and precise manner.

- All triggers must be encoded in a way that they do not overlap with other triggers or result in misinterpretations at the receiver's end. For example, in a typical EEG experiment, triggers can be created by a computer that is running a stimulus presentation software, such as Presentation® (Neurobehavioral Systems, Inc., Berkeley, CA, www.neurobs.com), E-Prime® (Psychology Software Tools, Pittsburgh, PA, https://support.pstnet.com/), PsychoPy (Peirce, et al., 2019), MATLAB® (The MathWorks Inc. MATLAB, Natick, Massachusetts: The MathWorks Inc. https://www.mathworks.com), or similar. These software tools interact with a device that can translate the software commands to triggers. This can be done via USB, serial, or parallel port, with the USB port connected to a trigger device becoming more and more the standard. Regardless of how, the user must ensure the device sends signals in a format that fits to the requirements of the EEG amplifier. For example, consider the possible ways that the trigger could be sent via the interface. The corresponding command might require a string, an integer, or it might even be necessary to encode the message as a binary or hex code number: the following dummy code snippets all encode the number 42 but might be interpreted differently: 'send('42')', 'send(42)', 'send('00101010')', or 'send('0x2A')'. Therefore, users are advised to inform themselves about how the triggers are sent and received. Users should also find out if trigger codes need to be actively reset, or if the software tool already sends pulses with a pre-defined length, after which this happens automatically. If such a reset does not happen, sending another trigger value might have no effect because the lines' values have not changed. Worse, it could even be the case that sending different codes after each other without a reset in between leads to completely wrong numbers at the receiver, formed by the combination of codes. This will impact ERP analysis. If it is necessary to send additional information in parallel, a viable option is to separate the available bits into groups (Table 14.1). If the EEG amplifier trigger input consists of 8 bits, the input can be divided into eight groups at maximum. A more likely division is using two groups of 4 bits, or one 7-bit group with the option of a 1-bit trigger input that can be used as an on/off input.
- The pulse width of a sent trigger code needs to be carefully considered. On one hand, it should not be too long because otherwise it can block sending subsequent triggers. On the other hand, it should not be too short because very short pulses can easily be missed if the onset and offset occur within a single sampling interval. The minimum pulse width, therefore, should always be at least 2 sampling intervals long to be on the safe side. If the length of the pulse width cannot be adjusted, losing triggers can be mitigated by either increasing the sampling rate of the amplifier, or by using a hardware solution that extends the pulse duration to an adequate length.

Table 14.1 Three examples of how to divide and group 8 bits: One single group of 8 bits to decode 0–255 values, two groups of 4 bits to decode values of 0–15 each, and a 7-bit group with a single bit to decode 0–127 values and an on/off state

Bit number	Bit weight	Decimal value	Value range	Bit number	Bit weight	Decimal value	Value range	Bit number	Bit weight	Decimal value	Value range
0	2^0	1	0–255	0	2^0	1	0–15	0	2^0	1	0–127
1	2^1	2		1	2^1	2		1	2^1	2	
2	2^2	4		2	2^2	4		2	2^2	4	
3	2^3	8		3	2^3	8		3	2^3	8	
4	2^4	16		0	2^0	1	0–15	4	2^4	16	
5	2^5	32		1	2^1	2		5	2^5	32	
6	2^6	64		2	2^2	4		6	2^6	64	
7	2^7	128		3	2^3	8		0	2^0	1	0–1
Single 8-bit group				Two 4-bit groups				7-bit and 1-bit group			

With more groups, more independent trigger codes can be sent without interfering with each other, but the reduced number of bits results in smaller value ranges per group

- Sometimes it is not enough to get triggers from one source only. In a typical experiment, the software tool is used to provide the stimuli and the participant's task is to respond. This can be done via a button press on the computer's keyboard, but also with a button that can directly create a trigger. The response can also be directly caused by the brain (neurofeedback or BCI). Alternatively, changes in the environment can be used as responses as well. For example, after exceeding a predefined pressure threshold on a force sensor or a brightness threshold on a photo sensor. As mentioned previously, it is crucial to take care that these different trigger modalities do not interfere with each other. This has to be done either on a software or hardware level, or on both.
- Another important decision is how to deliver the triggers on the hardware side. Triggers can be sent via a dedicated cable. For example, from the parallel output of a computer or from a dedicated trigger device to the trigger-in port of the EEG amplifier system or a dedicated device used for receiving trigger signals. Connecting cables comes with the burden of less mobility, however, this way the transmission is most reliable. Triggers can also be sent wirelessly. Here, triggers can be lost due to the transmitter and receiver going out of range, or that heavy environmental noise causes interference and bad reception. In this case, it is good practice to use redundancy when sending trigger codes: instead of sending only a single trigger, it is better to send the trigger multiple times or in predefined sequences. This way, if one or more triggers are lost, the information can be recovered by analyzing the context.
- In some cases, it can become necessary to deal with the underlying characteristics of the trigger signal itself. When using an analog signal, it should be verified that the signal fulfills the requirements for causing a state change at the input if it toggles between high and low states. When using transistor-transistor logic (TTL), for example, there are certain thresholds above or below which an input signal is classified as low or high, but there are variations of TTL where thresholds may differ. Of course, sending electrical signals is not the only way to send triggers. Optical signals are also an alternative, for example in environments where long electric cables are not desirable or even prohibited (e.g., in magnetic resonance (MR) environments). In other words, the user may have to consider such things when the recording solution does not come as a ready-to-go complete package.
- One should also think about how the nature of the trigger signal itself is used to encode information: the onset of a trigger can be linked to the falling edge when the previous state is high and then becomes low during the trigger pulse (when the event happens), and the state returns to high after the pulse width; the idle state could be low and a trigger could be encoded at the rising edge of a short high pulse; or, the trigger mode could be configured as a toggle. In this case, each rising or falling edge is used as a trigger, see Fig. 14.3.

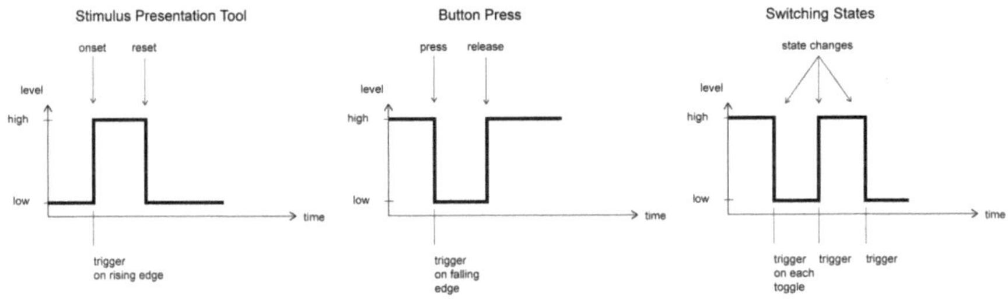

Fig. 14.3 Trigger pulses. From left to right, the triggers shall be generated on: the rising edge, caused by a stimulus presentation software tool that sets the idle level to low and sends a short high pulse to indicate an event; falling edge, where a button is pressed, pulling a high level to low for a short time; on both edges, where each level change is linked to an event

14.7 Using Triggers for Timing Verification

Based on the analysis of the exact timeline of an experiment, we can gain a lot of information that is directly relevant for assessing ERPs and for making sure that the setup is working as intended. In this final part of the chapter, we will describe an experiment that can be used to verify the timing between triggers and actual events.

The example can be expanded to include checking timing parameters between hardware and software triggering (hardware triggers versus software markers). Such a verification test can be useful if the experiment includes both hardware triggers and software markers, in which cases the measured latencies can be corrected. It can also help to decide between hardware and software options when the choice has not yet been finalized. While we describe an actual experiment, the setup can be adjusted where necessary to suit your own requirements and the details can be considered as pure suggestions.

The two main characteristics we want to evaluate in this setup are the latency and jitter (see Fig. 14.2).

14.7.1 Setup

A stimulus presentation software tool is running on a computer. A participant is sitting in front of the computer screen and is observing the display. A photo sensor is placed on the screen to acquire brightness levels. The onsets of events can be encoded by programming a flashing pattern at the position of the photo sensor. The sensor can either be used as a trigger source online by generating triggers based on a brightness threshold, or by analyzing the analog signal later offline. The analog signal and/or the triggers are recorded with an EEG amplifier, along with the EEG of the participant, Fig. 14.4.

A simple ERP paradigm can be used as the foundation of the experiment (just a flashing pattern or a visual oddball paradigm). The only requirement is that there is at

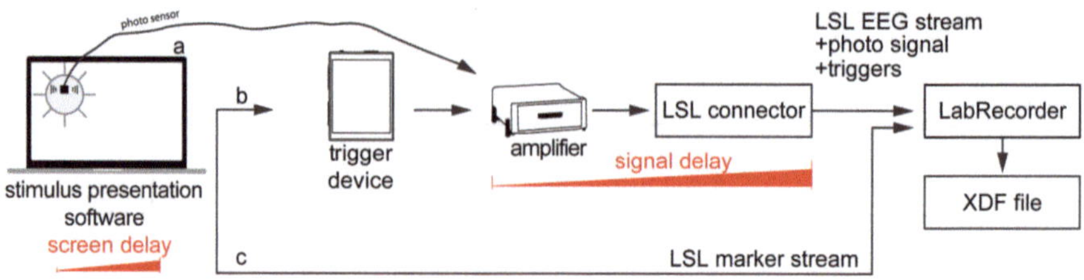

Fig. 14.4 Schematic of a timing verification experiment. A stimulus presentation software is used to: generate a visual event on a display (**a**), send a trigger to an EEG amplifier (**b**), and send an LSL marker to the network (**c**). Ideally, all of these events should happen simultaneously. The EEG data stream, including the brain signals from the participant and the photo signal from a photo sensor, is also sent via LSL and can be synchronized with the LSL marker stream to analyze the timing between the events. The test can be used with or without the LSL option

least one stimulus related to a visual event, i.e., a specific visual event observable on the display. Every time the stimulus in question is generated, a trigger is also created from the stimulus presentation software by interfacing a trigger device via USB or sending signals directly via the parallel port. When we also want to assess the correlation between software markers and hardware triggers, we can additionally send an LSL marker simultaneously. When choosing to do this, all LSL data and marker streams need to be synchronized and recorded in the network (e.g., in LabRecorder[1]).

14.7.2 Analysis

The experiment is started, and all data and marker streams are recorded in one file. This file now contains the *what and when* for up to four components:

1. The time when the stimulus is supposed to happen encoded by a trigger.
2. The time when the stimulus really happened encoded by the photo signal, either as an analog signal or already encoded as a trigger by applying a specific threshold.
3. The brain signal recorded from the participant. Either a simple visual ERP or even a P300 if using an oddball paradigm.
4. The time when the stimulus is supposed to happen encoded by a software marker (if going with the LSL option).

[1] https://github.com/labstreaminglayer/App-LabRecorder

14.7.3 Interpretation

The timing relations between these three or four components can have important implications:

- If the software markers (LSL) arrive earlier than the other components, most importantly the triggers, it means that the signals need more time to pass through the hardware until they are available in the data stream. This is only relevant when markers (generated by a software) should be used to mark events that are recorded with an amplifier (hardware). It may be necessary to adjust the latency in this case. The latency can depend on several factors: the amplifier model, wireless versus wired, the number of channels, the sampling rate, or the local network. When changing any of these factors, it is recommended to run the verification again.
- If the actual onset of the stimulus is delayed significantly, this can hint at a problem in the computer setup. For example, if a visual stimulus is delayed for multiples of the screen refresh cycle time (e.g., 16.67 ms for a 60-Hz display), it can indicate that the graphics settings are not optimized. Typical reasons can be connecting too many displays, activating filter settings that buffer multiple cycles, not using full screen, or simply running too many programs in the background.
- If there is not only a stable latency, but also high jitter, this can point to potential problems in the setup. For example, network difficulties, issues with the Bluetooth® transmission caused by interfering environment, or simply an error in the programming.
- Sometimes it is simply good to know how different stimulus presentation software tools work in specific situations. As these different solutions have different strengths and weaknesses, it is good to have the tools at hand to evaluate when to use which software and how to mitigate potential issues.

14.8 Conclusion

In this chapter, we highlighted the advantages of triggers over software markers and provided examples for how to use them to mark important events relevant for analysis. We showed examples where other options can work as good alternatives but also indicated potential drawbacks. In addition, we showed an example of how triggers can be used to verify timing characteristics of an experiment setup.

A clear distinction can be made between events, markers, and triggers, each serving a specific function in experimental synchronization. While hardware triggers remain a gold standard for precise timing, alternatives such as software markers (e.g., via LSL) offer flexibility and can be appropriate depending on the experimental context. Combining different marker, data, and trigger streams can introduce timing discrepancies, highlighting the need for systematic testing and validation to ensure temporal accuracy.

References

Kothe, C., Shirazi, S. Y., Stenner, T., Medine, D., Boulay, C., Grivich, M. I., et al. (2025). The lab streaming layer for synchronized multimodal recording. *Imaging Neuroscience*. https://doi.org/10.1162/IMAG.a.136

Miziara, I. M., Fallon, N., Marshall, A., & Lakany, H. (2025). A comparative study to assess synchronisation methods for combined simultaneous EEG and TMS acquisition. *Scientific Reports, 15*, 12816. https://doi.org/10.1038/s41598-025-97225-7

Peirce, J., Gray, J. R., Simpson, S., MacAskill, M., Höchenberger, R., Sogo, H., et al. (2019). PsychoPy2: Experiments in behavior made easy. *Behavior Research, 51*, 195–203. https://doi.org/10.3758/s13428-018-01193-y

Renard, Y., Lotte, F., Gibert, G., Congedo, M., Maby, E., Delannoy, V., et al. (2010). OpenViBE: An open-source software platform to design, test, and use brain–computer interfaces in real and virtual environments. *Presence Teleoperators and Virtual Environments, 19*(1), 35–53. https://doi.org/10.1162/pres.19.1.35

Chapter 15
Getting Clean Data: Artifacts and How to Prevent Them

David Kadlec, Shivakumar Viswanathan, and Hannah Kreilinger

Abstract Recording high-quality electroencephalographic (EEG) data is fundamental for achieving reliable results in research and clinical applications. However, the EEG signal is highly vulnerable to various artifacts that can arise from physiological sources, such as eye blinks, muscle activity, and cardiac rhythms. In addition, environmental and technical influences, including electrode impedance, cable movement, or electromagnetic interference, can negatively influence the signal. This chapter offers a comprehensive overview of approaches to minimize such artifacts during data acquisition. Central considerations include optimal preparation of the electrodes, lowering the impedance values on the participant's head, as well as maintaining consistent recording conditions. Understanding the source of artifacts, executing preventive measures, and being able to detect different kinds of artifacts as they occur are often more effective than relying only on post-processing corrections. This chapter serves as a practical guide to recording clean data for more effective and reliable analysis.

Keywords Signal quality · Artifact reduction · Physiological artifacts · Technical artifacts · Best practice · Data acquisition · Lowering impedance

15.1 Introduction

One unwanted element when recording EEG is the presence of artifacts. As EEG researchers, the goal is to acquire the highest-quality data possible, and to achieve this, we must minimize artifacts. A completely artifact-free EEG signal is very unlikely, so our best approach is to be aware of the sources of artifacts and take every possible step to prevent or minimize them. This chapter begins by explaining the concepts of signal and noise, and why their ratio is essential, especially for smaller artifacts. We will then take a look at some common artifacts found in EEG

D. Kadlec (✉) · S. Viswanathan · H. Kreilinger
Brain Products GmbH, Gilching, Germany
e-mail: david.kadlec@brainproducts.com; shivakumar.viswanathan@brainproducts.com; hannah.kreilinger@brainproducts.com

T. Warbrick (ed.), *The EEG Handbook*,
https://doi.org/10.1007/978-3-032-20450-9_15

signals and consider whether there are any steps that can be taken to avoid or reduce them.

Some artifacts can have devastating effects on the recorded data and should be prevented in the first place (e.g., signal saturation, electrode detachment). It is best to address these issues during the recording process. In contrast, other artifacts are unavoidable (e.g., blinks) and need to be tackled with appropriate offline processing strategies. For each artifact example, we will explain its origin and what the EEG researcher needs to do to minimize its impact on the recorded EEG data, either during recording or during offline data processing. Handling artifacts offline will be covered in detail in Chap. 17: EEG Pre-processing and Artifact Handling, here we just provide tips on where to start.

15.2 The Idea of Signal-to-Noise Ratio (SNR)

To better understand signal quality, it is an advantage to be familiar with the signal-to-noise ratio (SNR). The term *signal-to-noise ratio* is relevant in almost every technical field (electronics, audio, imaging, etc.), the field of EEG is no different. The signal-to-noise ratio indicates how strong the desired EEG signal is relative to the background noise (particularly electromagnetic interference) and how apparent the cortical activity is relative to unwanted interference (artifacts). A high SNR means the desired EEG signal is much stronger than the noise, and the EEG is more straightforward to record and interpret. In contrast, a lower SNR means there is more noise, and it will be more challenging to interpret the desired signal. A good example comes from the audio field: if you watch and listen to a movie during a flight and have the 'noise canceling' functionality turned off on your headphones, it is difficult to understand the characters because of the humming background noise of the engines, chatter from passengers, etc. —(low SNR). But once you turn on the 'noise canceling' functionality, the SNR becomes higher, and you understand the desired dialogue in the movie much better.

The interindividual EEG signal varies and so do the artifacts and the background noise in a given recording session, therefore we can also expect the SNR to vary. Ideally, we would have a standard number that represents a good SNR, however, this has not been established for EEG data. So how do we know if the SNR is high or low, and whether it's sufficient for our experiment?

Acceptable SNR might differ for different types of signals and/or data analysis. In spontaneous EEG recordings, brain signals, such as theta, alpha, beta, or gamma rhythms, are of relatively low amplitude. This makes the signals of interest more susceptible to environmental noise, so a high signal-to-noise ratio (SNR) is essential. In event-related potential (ERP) studies, where averaging across multiple trials to isolate short time-locked responses to stimuli, it is also crucial to maintain a good SNR. However, the averaging process improves the SNR because the random noise is canceled out while the ERP accumulates. It is important to consider your signal of interest when establishing an acceptable SNR for your study.

Visual inspection of the EEG signal remains the most commonly used method in laboratory settings, especially before and during the EEG recordings.

It is possible to do this in a more quantitative way and some EEG analysis software allows you to calculate SNR, often simply defined as *SNR = power of signal/power of noise*. Additional tools assessing signal quality are discussed in Chap. 22: Quantifying EEG and ERP Data Quality. One such metric is the standardized measurement error (SME), which helps to quantify the data quality of an ERP (Luck et al., 2021). This method can be applied after trial recordings, during data analysis, to assess whether further refinements to the paradigm or laboratory setup are needed.

In general, the SNR improves through good electrode preparation and control of the recording environment, both of which enhance the quality of spontaneous EEG and ERP data. Visual inspection is the most common approach in labs and this is generally what we look for:

- Good/High SNR: Clean, smooth visible brain activity during resting (i.e., visible alpha activity and peaks at occipital region).
- Poor/Low SNR: Results in spikes, line noise, and unusual patterns that cannot be explained by typical brain waves.

15.3 Sources of Artifact

There are two major categories of artifacts: physiological and technical.

Physiological artifacts in EEG originate from the participant's own biological activity, such as eye movements (EOG), muscle activity (EMG), cardiac signals (ECG), or skin potentials.

Technical artifacts arise from non-biological sources, such as power line interference, electrode cable movements, or malfunctioning equipment, as well as electromagnetic fields from nearby devices.

Both artifact categories arise independently of the brain but can be mistakenly interpreted as cerebral signals. The recorded waves from the cortex are blended with physiological and technical artifacts and being able to distinguish artifacts from signal is crucial. To help you with this, throughout this chapter artifacts are presented with the aid of images and explained in the corresponding tables with solutions to minimize these artifacts.

15.4 Common Physiological Artifacts

15.4.1 Eye Artifacts

Eye movement artifacts are caused by the movement of the eyeball. The cornea is positively charged, while the fundus (the inside back surface of the eye) is negatively charged. These circumstances create a static potential field (dipole). When the eyeball rotates, the eye dipole rotates with it, causing a potential change at the nearby electrodes (corneo-retinal dipole artifact) (Lins et al., 1993).

Vertical eye movements result from the rotation of the eyeball along the vertical axis, generating EOG (VEOG) potential primarily visible in the frontal EEG channels (Fp1, Fp2).

Horizontal eye movements, caused by rotation along the horizontal axis (HEOG), lead to characteristic saccades detectable at lateral frontal sites (e.g., F7, F8). Especially in polysomnographic EEG recordings, electrodes are typically positioned at the outer canthi of the eyes to capture horizontal eye movements with high spatial sensitivity, facilitating accurate identification of sleep stages, particularly rapid eye movement (REM) sleep.

Saccades and smooth pursuit movements are eye movements in any direction (horizontal, vertical, or oblique) that shift the gaze and produce artifacts in the EEG signal as illustrated in the following Fig. 15.1.

Eyeblinks are predominantly caused by eyelid closure, and slight eyeball rotation, generating short-duration, high-amplitude artifacts in the EEG.

Blinking is a natural reflex that helps keep the eyes refreshed and moisturized. This artifact is unavoidable but can be made predictable.

During a blink, the eyelids act like sliding electrodes, producing an electric field and temporarily increasing the conductivity between the electrodes and the corneo-retinal dipole. This phenomenon is often referred to as the "riding" or "rider artifact" (Matsuo et al., 1975).

One can further differentiate between spontaneous (natural) and voluntary (intentional) blinks. Spontaneous blinks are typically shorter in duration and lower in

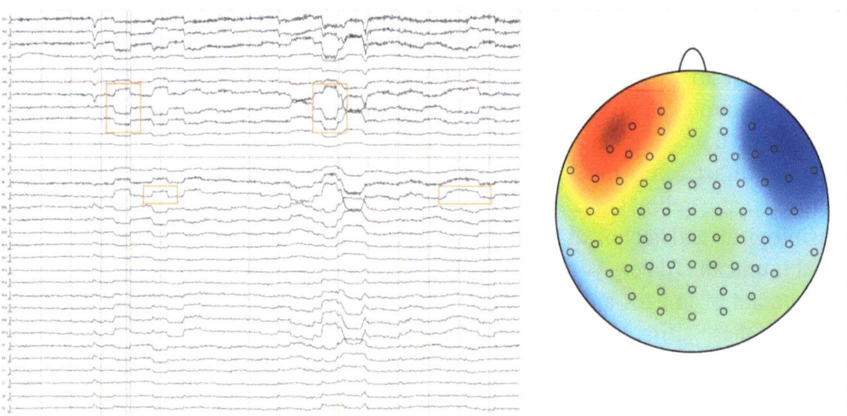

Fig. 15.1 EEG recording illustrating artifacts from eye blinks and horizontal eye movements. The signal is referenced to Fz, and channel order is from frontal to posterior regions. The data are high-pass filtered at 0.1–40 Hz to enhance visibility of these characteristic slow deflections. The topography of a representative saccade exposes the typical spatial distribution and amplitude of the related artifact, with noticeable activity in frontal and temporal regions due to extraocular muscle engagement

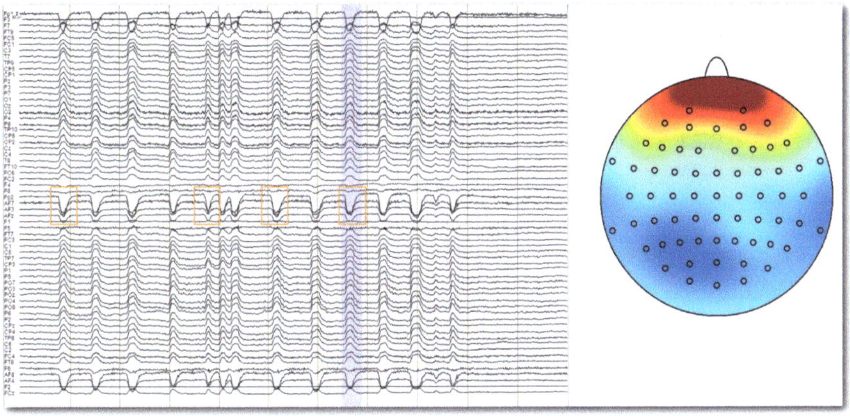

Fig. 15.2 EEG data illustrating a sequence of voluntary and spontaneous eye blinks, recorded across 64 channels using the 10–20 electrode montage. The signal was high-pass filtered at 0.1 Hz and low-pass filtered at 40 Hz. The blink sequence consists of four voluntary blinks, followed by three spontaneous blinks, and then five additional voluntary blinks. A marked segment of the EEG trace is shown on the right in a topographic mapping view, highlighting frontal activity associated with blink artifacts. The topography of a representative blink peak reveals the characteristic spatial distribution and amplitude of the artifact, with prominent activation in frontal regions

amplitude, while voluntary blinks tend to last longer and produce higher amplitude signals. However, in terms of their effect on the EEG, both types of blinks generate similar artifacts, a prominent positive deflection in the frontal electrodes (Fp1 and Fp2). In the Fig. 15.2, you can distinguish between the first 4 voluntary blinks followed by three spontaneous blinks, and then five further voluntary blinks.

Some EEG labs, especially those with a focus on polysomnography, visual tasks, or co-registering eye-tracking, generally place an electrode in vertical alignment with one of the frontopolar (Fp1; Fp2) electrodes, which is best suited for co-registering the vertical electrooculogram (EOG) and blink activity.

Eye flutter or *twitching* (Eye myoclonia) doesn't happen often, but this involuntary eyelid twitch leads to irregular and fast waves on the EEG. To get an idea of what it looks like you can imagine rapidly occurring eyeblink artifacts, which subside within a short time.

Rapid Eye movement (REM) is relevant in sleep EEG research. At this specific sleep stage, the eyes move predominantly in the horizontal direction at a higher frequency, but lower voltage amplitude compared to intended horizontal eye movements during wakefulness. These eye movements, while technically considered artifacts in EEG recordings, are both unavoidable and informative. Rather than being dismissed as mere noise, they serve as an indicator of sleep stage and provide meaningful insight into sleep studies.

Table 15.1 summarizes recommendations for minimizing physiological artifacts, including strategies for preparation before recordings, guidance for instructing participants, and procedures for managing artifacts during breaks or between trials.

Table 15.1 Eye artifacts and tips for reducing their occurrence

Artifacts	Solutions		
	Before recording (instruction)	Before recording (setup)	During (monitoring)
Eyeblinks	Blink with a natural pattern	Check already during the pilot testing if blinking occurs frequently and appears time-locked to the stimuli, which can happen in visually driven tasks. If so, reduce the contrast, luminance, and match the stimuli of different conditions Brief pauses in the paradigm can be beneficial.	Annotate and note in a log file unnatural blinking behavior. For example, if the participant is blinking each time after a visual stimulus on the screen.
Vertical eye movement	Keep your gaze fixed at the center of the screen and avoid looking up or down, or around the screen.	During visual tasks, maintaining focus on the display prevents unwanted eye movements. To further reduce distractions, ensure background behind the stimulation screen remains neutral.	Document any atypical eye movement artifacts, such as instances where the participant shifts their gaze or appears to redirect attention.
Horizontal eye movement	Maintain visual focus on the stimulus or target object. Avoid eye movements unless the task requires gaze shifts	Gaze shift can appear more often when the stimuli are presented at variable spatial locations. Depending on the paradigm, providing a fixation point at the center of the screen will help keep the participant focussed.	Use the same approach as for vertical eye movement

One can demonstrate many of these eye artifacts and their effect on the EEG before the recordings. Inviting the participant to perform specific actions (i.e., intentional blinking and eye movements in vertical and horizontal directions) induces the artifacts. These tasks raise the participants' awareness of their impact on EEG recordings. However, it is important to emphasize that maintaining natural behavior is essential.

15.4.2 ECG Artifacts

ECG and pulse artifacts are rarely seen in standard lab EEG recordings, but when they do occur, recognizing their rhythmic patterns and peaks can be helpful for offline analysis. There is not much one can do before and during the EEG recordings.

These artifacts are more frequent and visible in special applications, such as the simultaneous recording of EEG and fMRI due to the magnetic field(s).

The *ECG artifact* originates from the heart's electrical activity. The rhythmic sharp waveform can spread across multiple electrodes, particularly on the left side of the head, such as the lower temporal or mastoid electrodes, due to the heart's position and its relatively high voltage output. The artifact typically appears in phase with the R-peak of the heartbeat and if an electrocardiogram signal is co-registered with the EEG you can clearly see this.

In contrast, the *pulse artifact* shown in Fig. 15.3 is mechanical in origin, resulting from blood pulsing through the arteries beneath the scalp. If it is present at all, it typically affects a single electrode, often in the mastoid or neck area, and appears as slow fluctuations, time-locked to the heartbeat. This artifact indicates that the electrode is likely placed directly above an artery, such as the temporal or carotid artery (Amin et al., 2023).

To reduce the impact of pulse artifact, depending on the flexibility of the cap or net, the electrode can be shifted slightly in either direction to see if that lowers or eliminates the artifact. Adjusting one individual electrode without moving the EEG cap is generally challenging. Therefore, offline processing is usually the right approach here (see Chap. 17: Pre-processing and Artifact Handling).

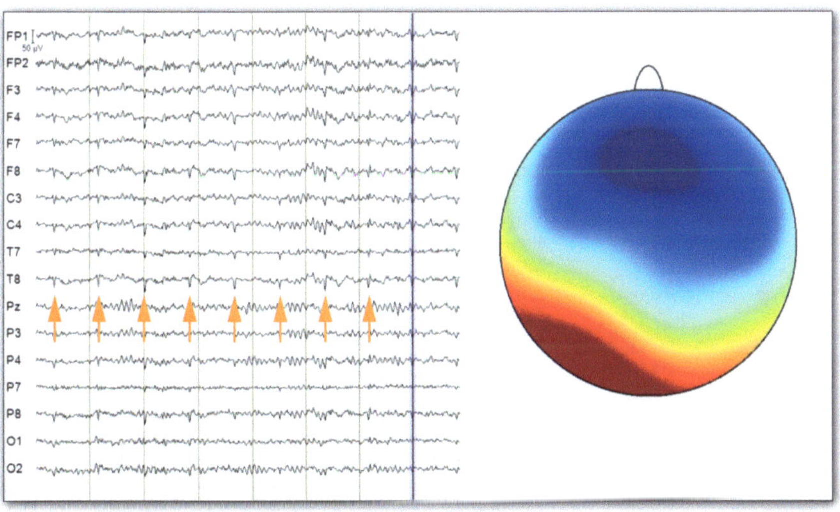

Fig. 15.3 Pulse artifact in EEG, every second or so, at nearly the same moment across many of the channels, a sharp vertical spike suddenly appears. These spikes are very brief, narrow, and point upward or downward depending on the EEG's polarity, like tall thin peaks. They are much sharper and taller than the slower rolling background waves, which makes them stand out clearly. The topographic map of a single pulsation peak, displayed to the right, illustrates the typical spatial distribution of pulse-related artifacts

15.4.3 Electromyographic (EMG)—Muscular Artifacts

Muscle activity causes *EMG artifacts* in EEG, generating high-frequency signals that can overlap with the entire EEG frequency range. The artifact amplitude depends on the intensity of muscle activation and the specific muscle groups involved. The closer the active muscles are to the EEG electrodes, the more they affect the signal. For example, head movements or facial muscle activity produce much stronger artifacts than isolated muscle contractions in distant areas like the legs. Luckily, *EMG artifacts* are the most easily prevented. Avoiding the muscular artifacts is recommended because the frequency bandwidth overlaps with the entire EEG spectrum, but whether you can achieve this will depend on your paradigm and the task performed by participants.

Talking and chewing artifacts are generated through the muscles located mainly under the temporal electrodes but can spread over the whole EEG map. These actions usually happen consciously, however, in some cases the participants chew their tongue or perform mouth movement unintentionally.

Clenching teeth artifacts, visible in Fig. 15.4, originate from the face and jaw-muscle groups, mainly under the temporal part of the EEG caps. They have higher frequencies with larger amplitudes and can spread across the entire scalp. Participants may begin to experience slight muscle cramping or jaw clenching when intensely focused on a task. Additionally, an overly tight chinstrap can contribute to discomfort or trigger unintended movement artifacts.

Table 15.2 outlines further recommendations for minimizing artifacts similar to those caused by chewing or talking.

Glossokinetic artifact, or tongue artifact, originates from the muscle group within your oral cavity and can be registered with the electrodes on the participant's scalp.

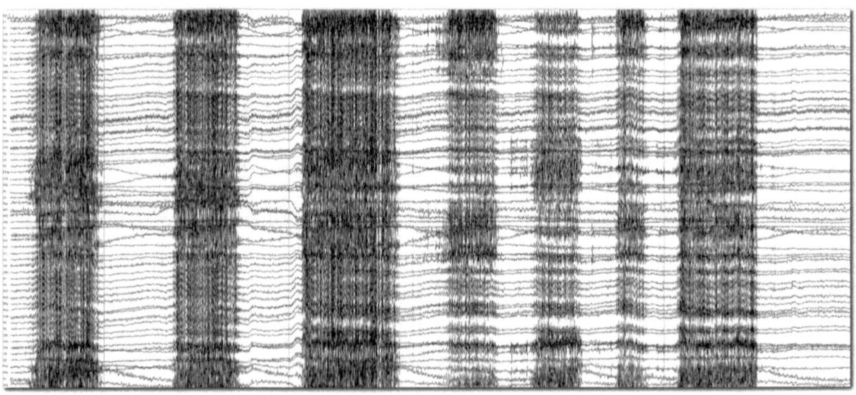

Fig. 15.4 EMG artifacts in EEG—Participant was asked to clench the teeth and mimic chewing behavior. Overall, the picture shows how muscle contractions can dominate the recording, masking the underlying brain activity with strong, sharp, and irregular patterns of EMG. It clearly illustrates why muscle relaxation and minimizing facial movements are important for obtaining clean EEG data

Table 15.2 EMG artifacts and suggested solutions

Artifacts	Solutions		
	Before recording (instruction)	Before recording (setup)	During (monitoring)
Talking/ chewing	Don't talk during the paradigm, and don't chew chewing gum or tongue.	Brief pauses in the paradigm can be beneficial for participant to ask questions or for the investigator to provide further instructions. If the paradigm is focused on speech patterns loosen the chin strap or maybe select chest strap instead.	Use annotations and notes in a log file if EMG artifacts occur frequently.
Clenching teeth	Stay relaxed especially the jaw and facial muscles. Don't press your lips together during the recording.	Loosen the chinstrap slightly to reduce tension around the jaw.	Build in breaks of sufficient length and suggest to reduce tension.
Ear wiggling/ smiling & frowning/ eyebrow movement	Keep the face neutral over the course of recordings.	Correct size of the EEG cap and adapted recording duration can avoid face muscle activation.	If required, a reminder during pauses will solve it.
Shoulder and neck tensions	Find a comfortable seated position and remain still. Minimize face and head movement to ensure clean recordings.	Depending on the paradigm, consider incorporating short breaks and allowing participants to adjust their seating or lying positions. Ensuring a comfortable setup (such as a chair with armrests and a headrest) can help minimize movement and reduce artifacts.	Monitor for frequent posture shifts. If observed, schedule breaks and modify setup, adjust chair height or armrests as needed.

Some participants might move their tongue unintentionally when focused on the task. For demonstration purposes, it can be helpful to ask participants to say "La-la-la". This simple vocalization allows observation of muscle activity while minimizing complex articulation. To ensure clean EEG data, participants should avoid unnecessary tongue movements during recordings.

Fascial muscle artifacts in EEG recordings are largely avoidable. While participants may attempt to relax, habitual expressions such as *smiling, wiggling the ears, raising the eyebrows, or frowning* can still occur unconsciously. Even low-intensity activation of these muscle groups can interfere with the EEG signal (see Figs. 15.5 and 15.6).

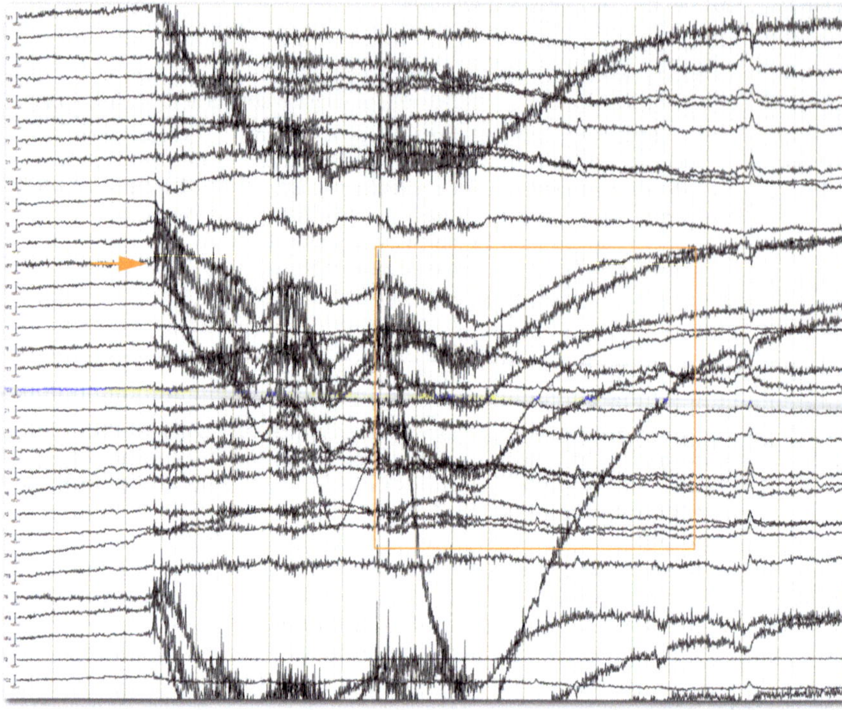

Fig. 15.5 Facial muscle activity, such as frowning or raising the eyebrows, generates high frequency artifacts in the EEG-data. These movements are typically reflected as slower signal drifts, as shown in the figure

A brief demonstration of how facial movements affect the EEG can help the participants understand the importance of maintaining a neutral and inactive facial expression during the recordings.

Due to human anatomy, the EEG electrodes can even pick up *shoulder* and *neck tension*, typically in the occipital channels of the EEG cap or net. It isn't uncommon for participants to adapt their posture and, with it, increase muscle tension once the participant is wearing an EEG cap. Participants can start to shrug, roll, and move their shoulders for relaxation purpose, or adjust their sitting or lying position. These muscle-related artifacts can be minimized with a well-designed lab setup, a properly fitted EEG cap, and sufficient slack in the cables. For longer recordings, providing adequate breaks, a comfortable seating position, and cushioning at pressure points can further help reduce tension and improve data quality.

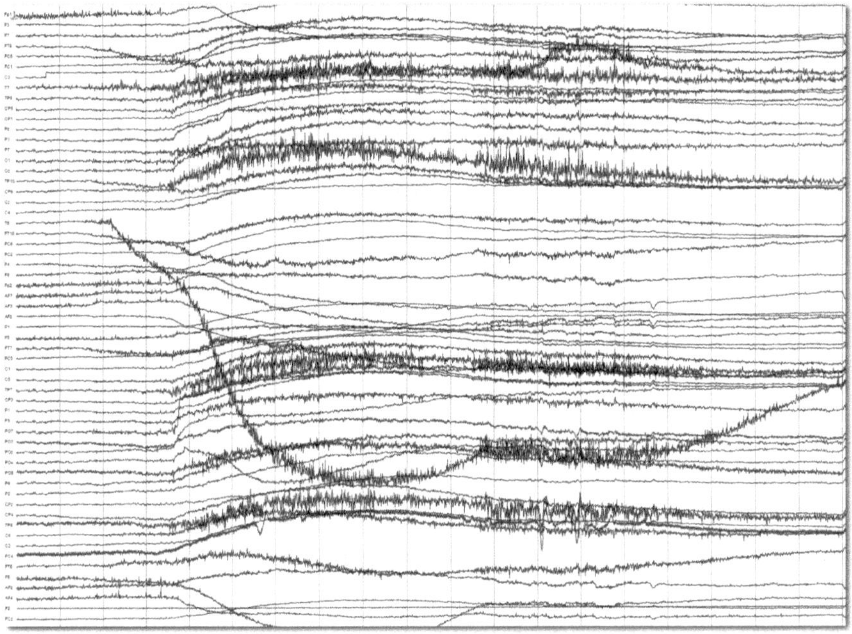

Fig. 15.6 Facial muscle activity associated with smiling and laughing produces characteristic artifacts in EEG recordings. Typically showing as slow signal drifts. The figure illustrates these distortions across a subset of 64 EEG channels, highlighting their widespread impact on scalp recordings. The artifacts are most prominent in frontal and temporal regions, where the muscle activity interferes with the EEG bandwidth

15.4.4 Other Physiological Artifacts

Sweat artifacts, even mild sweating, can cause a change in skin conductivity, which in turn translates into slow voltage drifts in the EEG. Sweat and perspiration artifacts are more common in warm, humid lab environments and can introduce slow signal drifts in EEG recordings. As outlined in Table 15.3, such effects can be minimized by maintaining a slightly cooler room temperature during the session. Participants should wear clothing appropriate for the lab environment and the expected duration of the recording.

Respiratory artifact is a combination of sources; the deep inhalation and exhalation cycles transfer to a mechanical movement of the EEG electrodes, which results in rhythmic amplitudes. These can be observed when the caps are fastened with chest belts and/or the head is lying on the electrodes, or a headrest is in use.

Swallowing artifacts appear when the chinstraps and fixations of the EEG caps and nets are either too tight or the strap bumps against the participant's Adam's apple. In this case, the whole cap is moving, affecting most of the electrodes; the tongue muscle also contributes to the artifact.

Table 15.3 Other physiological artifacts and optional solutions

Artifacts	Solutions Before recording (instruction)	Before recording (setup)	During (monitoring)
Sweat and perspiration	Check with the participant to confirm their comfort level.	Check the room temperatures and if required cool it down before the EEG recordings	Monitor and check with the participant about a comfortable temperature.
Respiration	Breathe naturally and not deeply. Ensure proper cap placement and fit.	Place a towel or suitable foam beneath the participant's head to avoid pressure points by the electrodes and possible friction.	Should the artifact result from mechanical influence, ask the participant to lift their head from the headrest.
Swallowing	Assess whether swallowing or taking a sip of water produces a visible artifact. If so, adjusting the chinstrap may resolve the issue.	Make sure the cap fits snug but still comfortable.	If the chinstrap is too tight, ask the participant to adjust it.

15.5 Common Technical Artifacts

15.5.1 Technical Artifacts

Non-physiological signal distortion can originate from a wide range of sources, including some that interact and overlap with the desired brain signals. These artifacts can manifest in various forms, such as sharp spikes, high-frequency noise, high-voltage fluctuations, or slow drifts, overlaying and distorting the EEG signal.

Here we categorize technical artifacts as environmental, electrical interference, and mechanical:

Environmental noise/electrical interference

Electromagnetic interference is one of the most common electrical artifacts in EEG recordings, also called *line noise.* The power line noise typically occurs at 50 or 60 Hz, depending on the country's power grid. It is often present from the very start of an EEG session. Any electronic device powered by alternating current, such as power strips, laptop chargers, adapters, projectors, or treadmills, can introduce this interference into the EEG recordings. Even devices like monitors or dimmable lights can cause localized artifacts, especially in electrodes positioned near the source (e.g., frontal channels).

Interference from external devices

External devices can operate at frequencies other than 50/60 Hz, so it is a good idea to look for noise in other frequency bands.

While modern amplifiers do a terrific job of reducing noise, external devices can still interfere with EEG recordings. The sources mentioned previously, such as dimmable lights, wireless devices, and medical or laboratory equipment (including stimulation devices and ventilators), may exhibit frequency-specific noise ranges that vary in amplitude and frequency over time. These artifacts are often localized to specific electrodes near the source of interference.

Static discharge

Static discharges can cause brief, high-amplitude spikes in the EEG signal that may look like biological events (i.e., epileptiform seizure spikes) and can be misleading. This artifact can originate from hair movement, synthetic materials (such as plastic chairs and carpets), or even clothing that builds up and releases a static charge. In the EEG, it is visible as a sharp wave that can affect multiple channels simultaneously.

These examples are summarized in Table 15.4, which outlines key sources, characteristics, and improvement strategies for common technical artifacts in EEG recordings.

15.5.2 Electrode Artifacts

Electrode artifacts, as illustrated in Fig. 15.7, result from physical or electrical disturbances at the interface between the electrode's conductive part and the skin. These artifacts may mimic physiological signals or create false waveforms that distort the EEG. Familiar sources include conductive gel/paste drying out and loose or shifting electrodes due to head movement. Also, a poorly fitting electrode cap can lead to electrode shifts. In Fig. 15.7, several randomly distributed electrodes exhibit interference at higher frequencies, highlighting the impact of such disturbance.

Furthermore, it's important to take care of your EEG equipment because corroded electrodes and damaged electrode leads can also cause electrode artifacts. Malfunctioning components may cause random spikes, signal dropouts, or flatlines in individual EEG channels, compromising data quality.

Loose electrode contact can lead to slow drifts of the signal at individual electrode positions, which can suddenly cause a phenomenon called "electrode pop". The loose electrode's electrochemical instability and the alteration in conductance result in a fast voltage change. This may be triggered by a loose-fitted cap, quick body movement, dried gel contacts, or hair pushing the cap away. The ground and reference electrodes can also be affected by this type of artifact which would consequently affect all other channels, so it is important to pay attention to these electrodes during preparation.

Mechanical influence on electrodes caused by touching or applying pressure (e.g., by additional equipment), tugging on the cable, etc., can all cause artifacts in the signal. Accidental electrode touching during recordings (i.e., if the participant spontaneously needs to scratch the head) can lead to electrical discharge and cause electrode pop, electrical noise, or simply moving the electrode. A dry lab environment, especially during the wintertime, when heaters are running, can facilitate static electricity (described above) and lead to frequent electrostatic discharges.

Table 15.4 Cable and electrode artifacts with troubleshooting strategies

Artifacts	Solutions		
	Before recording	Before recording (setup)	During (monitoring)
Cable movement	Gently move the cables after the cap preparation to observe the effect	Ensure cables are properly routed and strain reliefs are in place. The participant should be able to move their head freely while wearing the EEG cap, net, or headset.	Pause the recording, check that cables are securely arranged with proper strain relief and instruct the participant to minimize head movement.
Loose electrode	Always check impedance values and visually confirm proper electrode contact before starting EEG recording.	Ensure that the EEG cap fits securely. If necessary, apply an additional fixation. Check that each electrode has sufficient gel and inspect for any movement-related artifacts that may indicate poor contact or electrode displacement.	If loose electrodes or poor contact are detected, pause the session, reapply conductive medium, adjust the cap for a secure fit, and assess whether the issue can be resolved during a brief recording break.
Line noise	Turn off and disconnect all unnecessary electrical equipment.	Eliminate unnecessary electrical devices. Avoid coiled cables around metallic table frames. The participant should be as far as possible from electronical sources like multi power strips, power lines, etc.	Monitor for line noise during recordings and keep track of which electronic devices were powered on in the environment or by the participant. It may also be helpful to apply a *Display Filter* set to the specific interfering frequency (e.g., 50 or 60 Hz) to visualize and assess the impact of line noise in the EEG signal more clearly.
Defective or malfunctioning electrodes, leads, and connectors	Regularly inspect and test electrodes, cables, and connectors before recordings.	Replace any damaged or unreliable components to prevent unnecessary signal loss or artifacts.	Use annotations when random spikes, signal dropouts, or flatlines appear in individual EEG channels

15.5.3 Gel-Related Artifacts

These artifacts can occur when insufficient gel is applied under an electrode or when sponges have not absorbed enough saline to ensure proper skin-electrode contact. On the contrary, applying excessive amounts of gel or paste may lead to so-called gel bridges between adjacent electrodes, particularly when the conductive medium is less viscous and runs along the scalp. These bridges create an electrical shortcut between the electrodes. Signals from bridged electrodes may appear identical, making data interpretation unreliable.

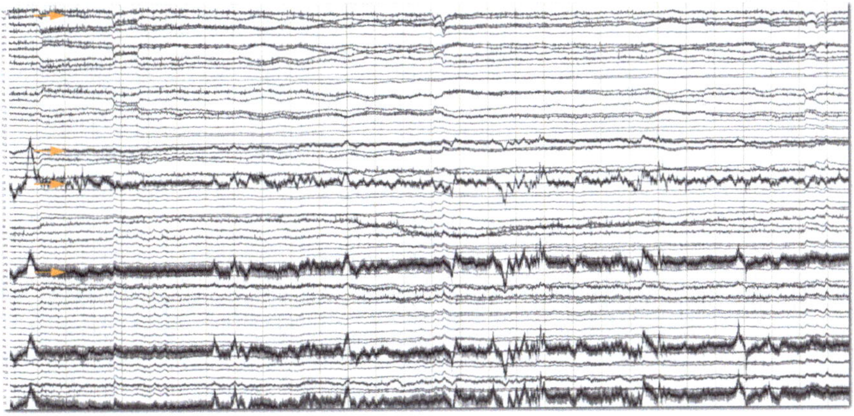

Fig. 15.7 EEG signal distortion caused by higher impedance values, manifesting in increased sensitivity to high-frequency interference. Electrode and cable movements due to whole body movement introduce random waves of higher amplitudes, which are further masked by 50 Hz line noise. A few EEG channels with these described artifacts are marked

15.5.4 Movement Artifacts

Although commonly categorized as physiological artifacts, movement-related disturbances often involve a combination of physiological (e.g., EMG) and technical factors, such as electrode, cap, or cable movement and displacement. This double impact makes EEG recordings particularly challenging during activities involving considerable motion, such as walking, dancing, running, or cycling.

15.5.5 Body/Head Movements

As noted previously, different types of body movements produce corresponding changes in the EEG signal, as illustrated in Fig. 15.8. The intensity and acceleration of movement play a critical role in artifact severity.

For example, running on a treadmill can produce rhythmic artifacts in the EEG signal (Gwin et al., 2010). The artifact is often caused by the repetitive, slight lifting of electrodes from the skin, which disturbs the contact between the skin and the conductive gel.

It is generally recommended to prioritize thorough preparation of the electrode cap and to ensure the secure placement of the caps, nets, headsets, and electrodes. The precise placement and stable attachment of EEG caps and electrodes form the foundation for all EEG recording scenarios.

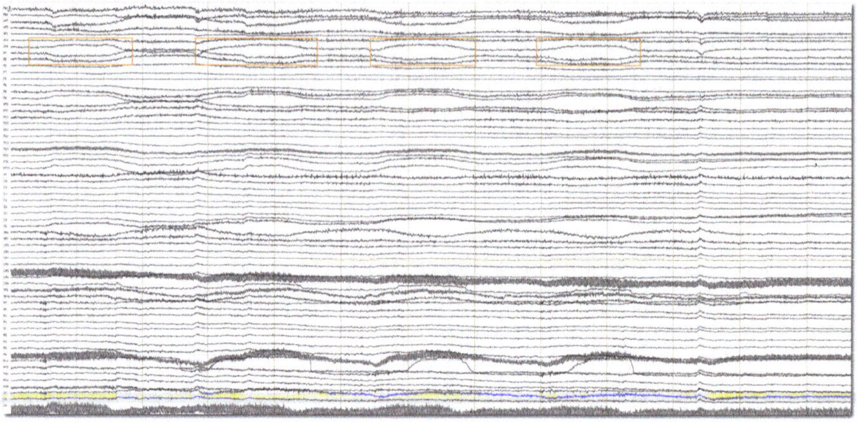

Fig. 15.8 Regular bumps in EEG recordings during running reflect head motion-induced artifacts, with their amplitude and frequency closely tied to the intensity and acceleration of movement. Channels are arranged from frontal to posterior regions, progressing from the left to the right hemisphere. The artifact is especially visible in the second half of the window on the frontal electrodes

15.5.6 Cable Movement Artifacts

Artifacts caused by *cable motion* are reflected in the EEG at the frequency of the cable's swinging and introduce oscillations that can overlap with the EEG bandwidth of interest. Cable movement can also be mechanically transferred to the electrodes themselves, as illustrated in the following picture (see Fig. 15.9). Such artifacts are more visible in passive electrode and dry electrode technology systems, where mechanical stability and skin contact are more susceptible to disruption.

15.6 Artifacts in Advanced Applications and Multi-modal Recordings

Some special applications give rise to specific artifacts in addition to those commonly seen in lab environments. It's beyond the scope of this chapter to cover all of them in detail; we provide some examples and some references for further reading.

15.6.1 EEG and Functional MRI

In EEG-fMRI, the EEG system is exposed to strong magnetic fields, which can cause artifacts in the EEG data. These artifacts originate from electromagnetic induction,

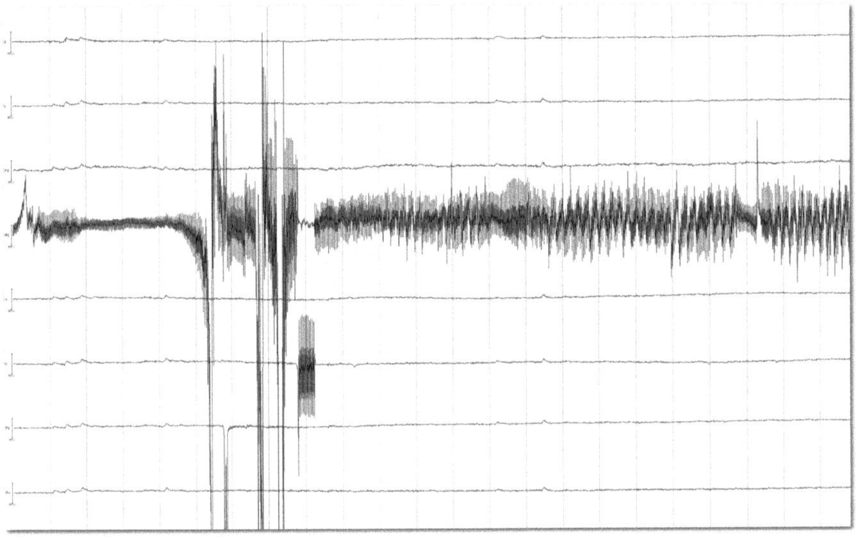

Fig. 15.9 Cable movement artifacts, compounded by 50 Hz line noise, often result from higher electrode impedance. Mechanical disturbances, like vibrations, tugging, or loose connections, can introduce voltage shifts

where changing magnetic fields induce voltages in the EEG electrodes and leads, according to Faraday's law, for more details see Chap. 33: Combining EEG and fMRI. The most prominent and unavoidable artifact is the gradient artifact caused by rapid magnetic gradient switching used for spatial encoding in MRI. The artifact appears in large, periodic, and high-amplitude oscillations that reach thousands of microvolts and completely overlay the EEG signal (Chowdhury et al., 2019). The previously mentioned ECG-related artifacts become more visible in the MR environment and can affect all channels. The cardio-ballistic artifact is prominent in magnetic fields. It is caused by a combination of subtle head and scalp movements, as well as blood flow in synchrony with the cardiac pulse. In addition to well-known sources of interference, other forms of motion can significantly contribute to artifacts in the EEG signal, particularly during simultaneous EEG-fMRI recordings. For instance, spontaneous head movements alter the position of electrodes within the magnetic field, leading to changes in the signal.

Moreover, vibrations can also be problematic in the MR environment. These arise from several sources, like the scanner's vibration during sequences, cold heads associated with the helium pump cooling the MR scanner's superconducting magnets, or the ventilation system. Good preparation and pilot testing are key to minimizing the effects of the MR environment on your EEG data.

15.6.2 EEG and Non-invasive Brain Stimulation

It is becoming increasingly common to measure EEG during brain stimulation. The stimulation devices can induce artifacts in EEG data. For example, when recording EEG with concurrent TMS pulses, considerable spikes in the EEG occur, lasting for a few milliseconds until the signal recovers to baseline (Wischnewski et al., 2024).

Electrical stimulation can induce a current artifact in the EEG data. The most common forms of stimulation are tDCS (transcranial direct current stimulation), tACS (transcranial alternating current stimulation), and ECT (electroconvulsive therapy). These electrical artifacts exhibit DC voltage shifts that scale with the current intensity during tDCS. With tACS, a dominant oscillatory artifact, with amplitudes often several orders higher than the EEG spikes (Feher & Morishima, 2016). This application is described in more detail in Chap. 31: Combining EEG and Transcranial Brain Stimulation.

ECT typically produces massive voltage spikes during the stimulation, often causing signal saturation or flatlining in EEG channels. As a result, EEG is commonly recorded only before and after stimulation, with the EEG cap or net often removed entirely during the procedure.

15.7 Optimizing the EEG Recording Quality

A clean EEG signal allows the researcher to analyze and process true brain activity, leading to reliable, accurate, and meaningful interpretations. An artifact-polluted EEG will obscure the brain activity signal, which can lead to avoidable artifact correction processing and filtering steps and difficulties in interpreting the results appropriately. It is worth considering that artifact attenuation techniques can also influence your signal of interest and unnecessary processing steps should be avoided. Therefore, recording clean data by preventing artifacts in the first place is always a priority. However, some artifacts– such as eye blinks are often unavoidable, and an appropriate artifact handling strategy should be implemented. Below we provide some tips for how to record a clean EEG signal. For tips on pre-processing steps for removing unavoidable artifacts, such as eye blinks, refer to Chap. 17: EEG Pre-processing and Artifact Handling.

15.7.1 Focus on the Cap Preparation

Select an EEG cap or net that fits the participant properly. A well-fitting cap ensures stable EEG signal quality over the course of recordings and provides a comfortable experience for the participant.

After applying the EEG cap, net, or headset, aim to achieve impedance levels appropriate for the specific electrode technology in use. Elevated electrode impedance can significantly degrade data quality and signal fidelity (Kappenman & Luck, 2010).

- For standard passive electrodes, it is generally recommended that impedance values remain below 10 kOhm to ensure optimal signal quality.
- With advanced active electrodes, impedance values can exceed 20 kOhm without compromising the signal clarity.
- Dry and sponge-based electrodes typically operate at higher impedance levels. While they offer faster preparation and more convenience, these are more vulnerable to signal artifacts.

That said, the acceptable impedance level also depends on the recording environment. In low-noise environments, EEG recordings may tolerate higher electrode impedance without compromising the signal quality.

In addition to the impedance measurement, you should check the signal quality. In fact, some systems do not provide an impedance measurement mode, and the electrodes' contact quality needs to be determined according to the signal during the monitoring mode, based on waveform stability and noise characteristics.

15.7.2 Optimize the Recording Environment

Once the cap preparation is complete, ensure the recording environment is set up for the best EEG data collection. The general recording environment should be optimized during pilot testing but there are some things you should pay attention to for each recording to ensure good, clean data. For detailed guidance on lab setup and environmental considerations, see Chap. 16: Practical Aspects of EEG Data Collection.

Here are key steps to follow:

- Ensure the participant is seated comfortably, with proper back and head support if possible. Adjust armrests, distance to keyboards, response pads, monitor, or any other equipment involved in the study.
- Verify that all cables are securely routed and allow the participant sufficient freedom of movement.
- Use strain reliefs for cables and adjust the amplifier's position as needed to minimize tugging or tension. Avoid any squeezing of the cables.
- Turn off or unplug any unnecessary electronic devices (e.g., mobile phones, Wi-Fi routers and devices, power strips, etc.) to reduce electromagnetic interference.
- Keep electrical equipment as far away from the participant as possible.
- Guide the participant through simple tasks such as eye blinking. Consider displaying the EEG signal in real time to illustrate the influences on the signal.

15.7.3 During the Recording

Cautiously observing the participant during data acquisition and making notes can save you or the data analyst some time in interpreting and explaining the artifact interferences.

- Closely monitor the signal stream, and the participant's behavior.
- Use online annotations in the recordings software to document any uncommon artifact or behavior.
- Ask for the participant's feedback during breaks—are they comfortable, are they becoming fatigued?
- Adjust the position of the participant if required.
- Instruct the participants to relax during the breaks if needed.
- Keep detailed notes in your lab log, including any observations or deviations from the standard procedure.

15.7.4 Post Recordings

On completion of the EEG session disconnect the participant from the system and carefully remove the electrode cap or net from their head. This marks the start of post-session procedures and preparation for the next recordings.

- Ask for participant's feedback—is there anything you can act on to reduce artifacts in subsequent sessions?
- Maintaining your equipment well ensures proper functioning and reduces the likelihood of artifacts due to, e.g., electrode degradation. Be sure to:
  - Clean the electrode caps/nets or headsets according to the manufacturer's instructions.
  - Perform maintenance checks as specified by the manufacturer.
  - Store the equipment, especially the EEG caps and electrodes, according to the manufacturer's guidelines.
  - Inspect the recorded EEG data and note down especially artifacts of technical nature.
  - Check the equipment (i.e., electrode surfaces, cables, and connectors) according to the notes.

Most EEG system vendors provide dedicated materials, documentation, and support to help optimize lab setup. Utilizing these resources can enhance signal quality and minimize artifacts. Support teams can also assist with setup, trouble-shooting, recommend best practices, and apply system-specific adjustments. So, check whether your EEG system's manufacturer offers this service.

15.8 Conclusion

This chapter provided a comprehensive overview of common EEG artifacts, explaining both their nature and sources, which is critical for accurate identification and interpretation of the EEG signal. For most artifact types, practical solutions were offered to help prevent, reduce, or correct interferences. Being aware of these artifacts is crucial for ensuring optimal data quality, as it enables EEG researchers to take appropriate action, whether through preventive measures such as pilot testing or real-time adjustments during an experiment (e.g., addressing signal loss or saturation). While some artifacts are unavoidable, their impact can often be reduced through careful planning, lab setup, and participant preparation, including the use of offline processing strategies tailored to the specific challenges anticipated in the study.

References

Amin, U., Nascimento, F. A., Karakis, I., Schomer, D., & Benbadis, S. R. (2023). Normal variants and artifacts: Importance in EEG interpretation. *Epileptic Disorders, 25*(5), 591–648. https://doi.org/10.1002/epd2.20040. Epub 2023 Jul 27. PMID: 36938895.

Chowdhury, M. E. H., Khandakar, A., Mullinger, K. J., Al-Emadi, N., & Bowtell, R. (2019). Simultaneous EEG-fMRI: Evaluating the effect of the EEG cap-cabling configuration on the gradient artifact. *Frontiers in Neuroscience, 13*, 690. https://doi.org/10.3389/fnins.2019.00690

Fehér, K. D., & Morishima, Y. (2016). Concurrent electroencephalography recording during transcranial alternating current stimulation (tACS). *Journal of Visualized Experiments, 107*, e53527. https://doi.org/10.3791/53527. PMID: 26862814; PMCID: PMC4828151.

Gwin, J. T., Gramann, K., Makeig, S., & Ferris, D. P. (2010). Removal of movement artifact from high-density EEG recorded during walking and running. *Journal of Neurophysiology, 103*(6), 3526–3534. https://doi.org/10.1152/jn.00105.2010. Epub 2010 Apr 21. PMID: 20410364; PMCID: PMC3774587.

Kappenman, E. S., & Luck, S. J. (2010). The effects of electrode impedance on data quality and statistical significance in ERP recordings. *Psychophysiology, 47*(5), 888–904. https://doi.org/10.1111/j.1469-8986.2010.01009.x. Epub 2010 Mar 29. PMID: 20374541; PMCID: PMC2902592.

Lins, O. G., Picton, T. W., Berg, P., & Scherg, M. (1993). Ocular artifacts in EEG and event-related potentials I: Scalp topography. *Brain Topography, 6*(1), 51–63. https://doi.org/10.1007/BF01234127

Luck, S. J., Stewart, A. X., Simmons, A. M., & Rhemtulla, M. (2021). Standardized measurement error: A universal metric of data quality for averaged event-related potentials. *Psychophysiology, 58*(6), e13793. https://doi.org/10.1111/psyp.13793. Epub 2021 Mar 29. PMID: 33782996; PMCID: PMC8169536.

Matsuo, F., Peters, J. F., & Reilly, E. L. (1975). Electrical phenomena associated with movements of the eyelid. *Electroencephalography and Clinical Neurophysiology, 38*(5), 507–511. https://doi.org/10.1016/0013-4694(75)90191-1. PMID: 50174.

Wischnewski, M., Shirinpour, S., Alekseichuk, I., Lapid, M. I., Nahas, Z., Lim, K. O., Croarkin, P. E., & Opitz, A. (2024). Real-time TMS-EEG for brain state-controlled research and precision treatment: A narrative review and guide. *Journal of Neural Engineering, 21*(6), 061001. https://doi.org/10.1088/1741-2552/ad8a8e. PMID: 39442548; PMCID: PMC11528152.

Chapter 16
Practical Aspects of EEG Data Acquisition

Paulo Rodrigo Bazán

Abstract How your lab is set up and how you prepare your participant for the experiment can have an impact on the quality of your EEG recording. In this chapter we provide tips and tricks regarding the infrastructure and equipment organization in your lab, as well as data acquisition optimization in different EEG setups. Our recommendations stand on four pillars: signal quality, experimental control, operational efficiency, and safety. Both stationary and mobile EEG setups are covered. We focus on examples of different electrode technologies and on considerations for special populations, using children as an example, to provide some guidelines that you can follow in your daily research work.

Keywords EEG data acquisition · EEG laboratory setup · Participant preparation · Signal quality optimization · Experimental control · Data acquisition efficiency · EEG Safety · Mobile EEG · Stationary EEG · EEG guidelines

16.1 Introduction

It can be challenging to set up an EEG lab or implement a new EEG study. To help you achieve such goals, we will first provide advice on the lab infrastructure, equipment position, and procedures. Then we will discuss practical aspects of data acquisition using examples of specific EEG setups. Our aim is to maximize the following:

- *Signal quality*—We want to make sure that we can properly measure the signal of interest, which is the brain activity related to our experimental task. However, the measured signal can also include other effects, such as brain activity unrelated to the task, other physiological activity, and signals from other sources (not directly from the participant). Therefore, it is very important to reduce the effects of these

P. R. Bazán (✉)
Brain Products GmbH, Gilching, Germany
e-mail: paulo-rodrigo.bazan@brainproducts.com

T. Warbrick (ed.), *The EEG Handbook*,
https://doi.org/10.1007/978-3-032-20450-9_16

sources of noise (for details on artifacts see Chap. 15, Getting Clean EEG data: Artifacts and How to Prevent Them).

- *Control over the experimental environment*—We must consider conditions that affect the behavior of the participant and, consequently, the measured signal. The goal is to make sure the effects we observe in our experiment are due to the characteristics of the experimental task and the population being evaluated, and not to uncontrolled variables. Therefore, it is important to reduce distractions and behaviors that are not related to the study hypothesis. This also strengthens the reproducibility of the experiment.
- *Efficiency*—The goal here is to make data acquisition both simpler and faster, optimizing the resources of the lab.
- *Safety*—Of course, we must consider safety aspects to prevent risk to the participants, the research team, and the equipment.

16.2 Lab Infrastructure

The specific application will determine the infrastructure requirement such as room size, number of rooms, or location of rooms. For example, when working with young participants, it is recommended to have a separate welcoming room with toys and to let them feel comfortable in the lab environment before moving to the data acquisition room (Hervé et al., 2022). Generally, the data acquisition room should be welcoming, but without adding distractions or noise sources. The lab setups presented by Luck (2014) and Ledwidge et al. (2018) are good starting points for planning the layout of an EEG lab. Below, we provide some general recommendations for optimizing key aspects of an EEG lab.

16.2.1 Signal Quality

The goal is to optimize the signal-to-noise ratio for the brain signal being measured. However, the data acquisition room can have several sources of electrical noise, which can impact data quality. In particular, lighting and temperature control systems can induce artifacts in the data. This is due to the mains powerline and to AC-DC (as these devices are AC powered) and DC-DC converters (Luck, 2014; Koyanagi et al., 2017). These will induce artifacts at the AC frequency (which varies between 50 and 60 Hz, depending on the country) and its harmonics, and possibly aliasing from higher frequency noise, depending on the EEG hardware filters (see Chap. 12, Hardware for Recording EEG and Peripheral Physiology). If the desired EEG signal is close to the AC frequency, it is recommended to reduce their impact by using DC-powered lights and/or placing the converters away from the EEG electrodes and amplifier.

Similarly, potential noise from additional devices in multimodal settings must be considered. It is helpful to limit the electrical devices within the room to those strictly needed for data acquisition. A good way to avoid noise is to have additional physiological measures directly connected to the EEG amplifier, avoiding the use of additional power sources. This is often possible for electromyography, electrocardiography, and electrodermal skin activity, for example.

The room's electrical installation should be properly grounded to prevent other noise coming through the mains connections of the connected devices. Mains noise can be avoided by using a battery-powered EEG system, as well as having notebooks running on internal batteries for data acquisition. Avoiding furniture made of metal or conductive material is also helpful to prevent signal changes in case the participant touches the desk, for example.

Susceptibility to artifacts depends on the type of electrodes used. For example, gel-based active electrodes are less impacted by line noise, because active electrodes usually have an impedance conversion on the electrode, reducing the antenna effect on the cables. If the frequency of the EEG signal of interest does not include the AC frequency, and/or active electrodes are used, a general, unshielded room should be enough.

Alternatively, a Faraday cage could be used to filter external noise coming from the surrounding environment. For example, rooms with a large concentration of equipment or big machinery (elevators, motors, treadmills, IT servers) can impact the EEG data. It is best to have the data acquisition room well away from such noise sources.

Aside from the external influences on signal quality, it is also important to consider other sources of artifacts that could be influenced by the lab environment. For example, sweat will change the conductive properties of the skin and cause low-frequency drifts in the EEG signal. Therefore, it is worth considering a temperature control system. However, make sure it is not close to the EEG system to avoid introducing a source of noise.

16.2.2 Control Over the Experimental Environment

Distractions can impact task performance and should be minimized. For instance, if the experiment is carried out in a dark room to maximize visual effects, the small "power on" light of devices should be covered, and the equipment should be positioned behind the participant.

As researchers we can also be a source of distraction, e.g., by communicating with colleagues or typing observations during the session. If possible, we recommend having a dedicated room for data acquisition/for the participant, and an adjacent room for monitoring and controlling data acquisition. To allow proper observation of the participant's behavior, the room should have a camera system and/or a window behind the participant. Furthermore, a communication system will be needed in this case.

Acoustic noise can also be a distraction. It is important to turn off smartphones or put them in silent mode (vibrations should also be avoided). Further, having the experiment room close to loud areas (kitchen, bathrooms, a parking lot, areas with a high circulation of people) will likely impact the data acquisition. If this can't be avoided, acoustic isolation around the room is an option. Additionally, it is a good idea to have a sign on the lab door to remind people to be quiet and to not enter during a measurement.

Participants can be uncomfortable if the room is either too hot or too cold, or if their chair is uncomfortable. This should be considered when furnishing the lab and when deciding whether heating or air conditioning is needed.

16.2.3 Efficiency

The lab infrastructure can also help optimize data acquisition:

- *Equipment setup and cap preparation*—The room size and layout along with its furniture should allow for easy access to needed equipment and consumables, as well as freedom of motion around the participant to prepare the electrodes.
- *Transition between participants*—As scheduling several participants in sequence can be an effective strategy for data acquisition, it is helpful to have a separate room for pre- and post-experiment activities, such as completing the consent form and questionnaires. This allows the preparation of the next participant, while the previous is finishing the experiment.
- *Clean up*—It is necessary to have a place close by for cleaning the electrodes after the experiment, such as a sink. Similarly, if gel-based electrodes are used, the participants will need to wash their hair after the experiment. Having facilities such as a hair washing chair, a large sink, or a shower close to the lab is a good idea.
- *Data storage and backup*—The lab infrastructure should consider the proper resources for data storage and backup, either by having an internal backup server or a cloud system for data storage.

16.2.4 Safety

Data privacy and protection must be considered when defining the method for data storage and backup. Please consider your local regulations, especially when using cloud storage services for backup or for data processing.

We also need to ensure the physical safety of the experiment participants. When planning the lab infrastructure, aside from the common safety concerns for buildings, like having an emergency exit route and clear safety protocols, there are some tips related specifically to the EEG. For example, proper grounding of the electrical installation is important to avoid electrical discharges to the participant, especially if the EEG system is not battery powered (has a mains connection). Additionally, if possible, the electrical installation of the room should be planned to provide enough outlets in appropriate locations, to avoid the use of cord extensions.

From the perspective of EEG maintenance, it is important to avoid metal furniture. If the electrode touches metal surfaces, this can lead to the formation of an alloy. Therefore, avoid metal sinks, or have a dedicated plastic bucket to clean the electrodes. Also, make sure to have proper storage for the equipment, regarding temperature and humidity, as indicated by the EEG manufacturer.

16.3 Position of the Equipment and Accessories

Once we have the proper infrastructure, we can focus on how to optimize the position of the equipment in the lab and the electrode cable routing. When attempting to optimize the position of the equipment, there are two common approaches:

- *Fixed position of equipment*—In other words, the equipment is stored in the position where it is going to be used. This can be faster and simpler as the equipment is always ready. This is usually recommended for stationary setups with bigger and heavier equipment. However, this requires having a single application for the data acquisition room, which sometimes is not the case. Using cable organizers, such as wraps and sleeves, is a good option for fixed position setups.
- *Flexible positions*—In this case, the equipment has a dedicated position for storage but has to be set up for each data acquisition. This setup is more suitable for data acquisition in different locations, even outside the lab (e.g., naturalistic or ecological studies). Further, it can be used when the acquisition rooms are not dedicated to a single device or experiment type. Furniture with wheels that allow the system and its accessories to be easily moved is very useful in this kind of setup.

Regardless of whether you have fixed or flexible positions, there are some things you should pay attention to. Here we will again consider our four optimization points:

- *Signal quality*—We should position the EEG amplifier away from other devices that are needed for the study but that can generate noise. This is more relevant when working with dry or passive electrodes, as active electrodes are less susceptible to this type of noise. Proper cable routing also helps minimize artifacts related to the motion of the electrodes, as entangled cables can pull and move the electrodes, therefore, the cables should be as stable as possible. Cable routing is especially important in applications where a significant noise induction occurs, such as EEG-TMS (Chap. 32, EEG-TMS) and EEG-fMRI (Chap. 33, Combining EEG and fMRI).

- *Control over the experimental environment*—Ideally, the equipment and cables should be outside of the field of view of the participant, for example, being directed to the back of the participant, as we want to avoid status lights from the devices to become distractors. Therefore, having a table, a cart, or a cabinet behind the participant is helpful to place the equipment. Alternatively, the equipment can also be positioned to the side of the participants, outside of their field of view.
- *Efficiency*—The equipment position needs to allow proper movement of the researcher around the participant throughout the experiment. Aside from the main equipment, it is helpful to plan the positioning, organization, and use of common accessories. For example, when using gel-based or sponge-based electrodes, paper and cloth towels will be helpful, so this should also be within the reach of the researcher. A recommended alternative for flexible setups is to have a cart or moving cabinet for each main device, with drawers for its accessories.
- *Safety*—Being able to move around the participant is also important for safety. We want to minimize the chance of accidents related to bumping, pulling, or tripping on the equipment and cables. Proper cable organization should leave some margin for motion for the participant and should not impede the movement of the researcher. Further, the risk of damage to the cables can be reduced by avoiding cable entanglement.

The best position can be evaluated during the pilot testing phase (Chap. 10, Pilot Testing).

16.4 Lab Procedures

Often, there are simultaneous projects running in parallel and different users working with the same EEG equipment in a lab. Consequently, it is very important to have clear procedures that everyone can follow. These procedures should cover the safety aspects of the building overall (e.g., emergency procedures), the specific aspects of handling the equipment, and data storage and backup. One option is to have quick guides as posters regarding these procedures, but it is also important to have structured training for the lab members (Boudewyn et al., 2023).

For further details on how to establish good lab management procedures and study protocols, see Chap. 12, Study Workflow and Lab Management. Below, we provide suggestions for specific setups.

16.5 Practical Aspects Associated with Specific Setups

As some EEG setups can present specific challenges, we consider different scenarios and offer tips to deal with these challenges.

16.5.1 Mobile Setups

The previous considerations apply to stationary setups and, to some extent, mobile setups. But mobile setups have special characteristics that must be considered. There are usually two main options when working with mobile EEG: headsets or caps connected to portable EEG amplifiers. When working with headsets, it is important to consider their weight and comfort according to the length of the experiment. Additionally, depending on head motion associated with the task, the headset may move out of position, so it is important to ensure stability if stronger movements are expected. Therefore, headsets are a nice alternative for shorter experiments that require a fast setup and do not have a lot of head movement.

Caps are usually more stable in this sense; however, they require special attention to the lead wires. Using cable guides and Velcro straps attached to the cap can be helpful to organize and stabilize the cables. Using overcaps might help stabilize the cables, but they can increase sweating, which also causes artifacts. Better alternatives are tubular elastic net dressing retainers or sports pre-wrap tapes (Emmerling & Kreilinger, 2018). Gel-based active electrodes will also be a good option, as they minimize artifacts from cable movement. The portable EEG amplifier should be placed in a comfortable and stable position as well. For this, a harness can be used; alternatively, some caps also have a pouch to carry the EEG amplifier. Several mobile EEG amplifiers have built-in acceleration sensors, which can provide important information for dealing with motion artefacts. However, these will be limited to the position of the EEG amplifier—consider this when defining the amplifier position in your study. Alternatively, some systems allow additional acceleration sensors to be connected, to capture motion information from other parts of the body.

Wireless data transmission is also an important factor in mobile EEG. To avoid missing data points during mobile recordings, test and adjust the range during the pilot phase (Chap. 10, Pilot Testing). Having a clear line of sight between the receiver and the transmitter will help if longer ranges are needed. To avoid interferences in the data transmission, reduce the number of devices that use wireless data transmission to the minimum necessary—check if cellphones and computers are using similar communication frequencies. Reducing the sampling rate can also increase stability. To ensure wireless trigger reliability, send a pattern of triggers instead of a single trigger for an event. This will allow you to recover the information if one of the triggers is lost (Chap. 14, Triggers).

16.5.2 Electrode Types

As there are several types of electrodes, it is important to follow the manufacturer's guidelines to achieve the best signal and protect the equipment. The main points are the target impedance values, the procedures to lower the impedances, and how to clean and maintain the electrodes. Before we present specific tips, the following apply to all electrode technologies:

- Ground and reference electrodes (if available) must have a low impedance, as they affect all channels.
- After lowering the impedances, check the signal quality by looking for standard signals (e.g., blinks, alpha waves).
- Confirm which parts of the system can be wet for cleaning.
- Avoid strong disinfectants.

16.5.2.1 Passive Sponge-Based Electrodes

Sponge-based electrodes offer a very fast setup, and they leave almost no residuals for the participant to clean after the experiment; this makes them very useful for children and some clinical populations. However, as the sponge can dry over time, they are more suitable for shorter experiments (up to around 60–90 min). To optimize the length of the experiment, control the room's ventilation, temperature, and humidity, and plan short pauses to wet the sponges with a pipette. Further, special attention to environmental and motion noise is required, as these electrodes work with higher impedances (60–100 kOhms).

When preparing the participant, a towel can be used to gently absorb the excess water in the electrodes after removing them from the saline solution. To help lower the impedances, a pipette can be used to move the hair away from the sponges. Considering safety, prepare towels to protect the water-sensitive parts of the system. Similarly, a towel can be placed over the shoulders of the participants, to avoid getting their clothes wet. To ensure the durability of the sponges and electrodes, use salt with high purity when preparing the saltwater solution, following the manufacturer's recommendations.

16.5.2.2 Gel-Based Electrodes

The main advantage of gel-based electrodes is the high signal quality, as gel allows for better contact with the scalp. One thing to consider is that the impedances continue to improve for a few minutes before stabilizing. Therefore, instead of working on an electrode until the target impedance is reached, it is better to go over all electrodes applying the gel; then go back and work more carefully on those that still have higher impedances. There are two types of gel-based electrodes:

Passive electrodes—offer very good signal quality and allow longer experiments (e.g., sleep studies) but require lower impedance values (5–10 kOhms). To achieve low impedance values, a light abrasion of the scalp to clean it is necessary. This can be done either as a prior step, or directly with an abrasive gel. It is good to show the participant the gel and explain the abrasion procedure. Working continuously or too long on a single electrode can cause discomfort or even harm the participant. Ask the participant for feedback to avoid excessive abrasion of the skin. As the preparation with passive electrodes can take a little longer it is good to keep the participant relaxed and entertained, with a conversation or a short movie, for example.

Active electrodes—offer both relatively fast preparation time and high signal quality. Impedances around 25 kOhms should provide a good signal. However, specific applications can require very low impedances even with active electrodes, such as EEG-TMS. When using active gel-based electrodes, it is common to apply gel using syringes with blunt needles. To reassure the participant it is safe, open the needle in front of them and touch the back of their hand with the blunted tip.

16.5.2.3 Dry Electrodes

Dry electrodes require neither gel nor saltwater. Therefore, they make EEG recording easier for experiments outside of the lab: fewer accessories for preparing and cleaning are needed. However, as the electrodes need to be in contact with the scalp, the pressure needed for a good contact can cause discomfort over time. This technology is mainly for shorter experiments. It works with higher impedances (up to 2500 kOhms), therefore, the bandwidth of the usable EEG signal is usually limited—check if this is suitable for your research. When preparing the electrodes, make sure the participant is comfortable with the electrode pressure. Any small discomfort will tend to increase and become an issue over time. Some dry electrodes and their components (headsets) cannot get wet, so double check before cleaning—isopropyl alcohol swabs can be a good option.

16.5.3 Special Populations

When working with specific populations, their needs must be evaluated, and the lab environment and EEG procedures should be adapted accordingly (Kappenman & Luck, 2015). Here, children will be our example population, but similar considerations could be made for other populations.

16.5.3.1 Children

When working with children, providing a welcoming environment is important (Hervé et al., 2022). The staff should be properly trained to communicate with appropriate language to the children and to their parents, paying attention to any needs they may have (Cotter et al., 2002). For example, when working with infants, it is important to check if they need to be fed before the experiment, and make sure that they feel comfortable with the lab environment, so consider having extra time for the children to get used to the data acquisition room (Turk et al., 2022). With infants the chance of a device acting as a distractor is bigger, and the child may want to reach and grab it, which could cause accidents, which highlights the need for the proper equipment positioning (out of the field of view of the participant). Further, the lead wires should also be organized to avoid accidental grabbing by the infant (Hinz & Pletti, 2021). Additionally, consider using electrode technologies with faster setup that also provide comfort such as sponge-based and active gel-based electrodes. Video data synchronized with the EEG data can be very useful, as some of the behaviors can be better identified with the video, especially in interaction studies (Turk et al., 2022). Eye tracking can also add valuable information to EEG with children (Kulke, 2024).

16.6 Concluding Summary

In summary, recording high-quality EEG data requires optimizing the signal quality and controlling the experimental environment. Additionally, this should be achieved in a safe and efficient way. One important aspect is the lab infrastructure, from the planning of the room and its surroundings to optimizing the devices present in the data acquisition room. The room should have proper electrical installation and only include the necessary equipment, providing comfort without offering distractions. Adjusting the position of the equipment in the lab can also improve EEG data acquisition, by placing the EEG further away from noise sources and out of the participant's sight, in an accessible and safe way. Further, the lab should have well-defined procedures for data acquisition and for training members to do it. Although each setup will have its own challenges and opportunities, the tips provided here can be a starting point for optimizing your own experiment.

References

Boudewyn, M. A., Erickson, M. A., Winsler, K., Ragland, J. D., Yonelinas, A., Frank, M., Silverstein, S. M., Gold, J., Macdonald, A. W., Carter, C. S., Barch, D. M., & Luck, S. J. (2023). Managing EEG studies: How to prepare and what to do once data collection has begun. *Psychophysiology, 60*, e14365.

Cotter, R. B., Burke, J. D., Loeber, R., & Navratil, J. L. (2002). Innovative retention methods in longitudinal research: A case study of the developmental trends study. *Journal of Child and Family Studies, 11*, 485–498.

Emmerling, T., & Kreilinger, A. (2018). *actiCAP slim active electrodes walkthrough* [Online]. Available: https://pressrelease.brainproducts.com/active-electrodes-walkthrough-slim. Accessed 02 Sept 2025.

Hervé, E., Mento, G., Desnous, B., & François, C. (2022). Challenges and new perspectives of developmental cognitive EEG studies. *NeuroImage, 260*, 119508–119508.

Hinz, N. -A., & Pletti, C. (2021). *R-Nets with infants: A walkthrough* [Online]. Available: https://pressrelease.brainproducts.com/rnet-infants-walkthrough/. Accessed 02 Sept 2025.

Kappenman, E. S., & Luck, S. J. (2015). Best practices for event-related potential research in clinical populations. *Biological Psychiatry: Cognitive Neuroscience and Neuroimaging, 1*, 110–110.

Koyanagi, S., Saito, K., Muroi, K., & Tanaka, M. (2017). A study on magnetic near-field noise and EMI of LED light bulb. In *2017 international conference on electronics packaging, ICEP 2017* (pp. 590–593).

Kulke, L. (2024). Coregistration of EEG and eye-tracking in infants and developing populations. *Attention, Perception, & Psychophysics, 87*, 228–237.

Ledwidge, P., Foust, J., & Ramsey, A. (2018). Recommendations for developing an EEG laboratory at a primarily undergraduate institution. *Journal of Undergraduate Neuroscience Education, 17*, A10–A10.

Luck, S. J. (2014). Setting up and running an ERP lab. In J. L. Steven (Ed.), *An introduction to the event-related potential technique* (2nd ed.). The MIT Press.

Turk, E., Endevelt-Shapira, Y., Feldman, R., Van Den Heuvel, M. I., & Levy, J. (2022). Brains in sync: Practical guideline for parent–infant EEG during natural interaction. *Frontiers in Psychology, 13*, 833112.

Part IV
Processing EEG Data

Chapter 17
EEG Preprocessing and Artifact Handling

Yin Fen Low and Ramon Martinez-Cancino

Abstract Transforming raw EEG recordings into interpretable data is a complex process that demands careful preprocessing. While published studies often outline their methods, these can be difficult for researchers new to EEG analysis to fully grasp. This chapter aims to bridge that knowledge gap by providing a clear overview of essential preprocessing steps.

We begin by discussing the core components of EEG preprocessing, including filtering, re-referencing, and artifact handling. These steps are foundational for improving data quality and ensuring the validity of subsequent analyses. Preprocessing is typically considered complete when the data is ready for segmentation, which is the process of dividing continuous EEG signals into meaningful time windows or epochs for analysis.

Rather than advocating for a single, standardized pipeline, this chapter equips you with the essential knowledge needed to design and adapt preprocessing workflows suited to specific research goals. While not exhaustive, this overview is supported by references to key resources and recent literature to guide deeper exploration.

To facilitate practical implementation, the final section presents a compilation of established EEG processing toolboxes and software to support the initiation and refinement of preprocessing pipelines.

Keywords Preprocessing · Data transformation · Artifact handling · Artifact attenuation · Artifact rejection

Y. F. Low (✉) · R. Martinez-Cancino
Brain Products GmbH, Gilching, Germany
e-mail: yinfen.low@brainproducts.com; ramon.martinez@brainproducts.com

T. Warbrick (ed.), *The EEG Handbook*,
https://doi.org/10.1007/978-3-032-20450-9_17

17.1 Introduction

The two main objectives of preprocessing are:

- *Data transformation:* Raw EEG data is often unsuitable for analysis in its original form. Preprocessing transforms it into a format that is appropriate for further computations.
- *Artifact handling:* EEG signals are frequently contaminated with non-neural signals, such as eye movements, muscle activity, or electrical noise. These artifacts need to be removed or minimized as much as possible to obtain clearer and more accurate neural signals.

The following sections outline common preprocessing procedures; however, the order in which they are presented does not necessarily reflect the sequence in which they should be applied.

17.2 Common Preprocessing Steps: Data Transformation

17.2.1 Inspecting Data

Once the data is loaded, conducting a quick inspection of its appearance and properties is often beneficial. Things you should or can do are as follows:

- *Ensure that all EEG channels and their corresponding coordinates are correctly imported.* Accurate channel locations are essential for various preprocessing steps and for generating meaningful visualizations. For a detailed discussion on electrode placement and coordinate systems, refer to Robert Oostenveld's blog on the topic (Oostenveld, n.d.).
- *Inspect the data for anomalies,* such as missing event markers, incorrect marker timings, or flat (non-functioning) channels. If any irregularities are found, cross-reference them with the experimental design or recording setup to identify potential causes.
- *Standardize event markers* by renaming them for clarity and consistency. Remove anything that is unnecessary for your analysis.

17.2.2 Changing the Sampling Frequency

Adjusting the sampling frequency involves either decreasing (downsampling) or increasing (upsampling) the number of data points recorded per second. When data are collected at a high sampling rate (e.g., >1000 Hz), it is often beneficial to downsample the data offline to a lower frequency that is more suitable for analysis. This can significantly reduce file size and improve processing efficiency.

In contrast, upsampling is less commonly used but can be helpful in specific situations, for example to improve temporal precision when estimating ERP peak latencies. Although it does not add new information, it allows more accurate estimation of the timing of waveform peaks.

For analyses focused on frequency ranges up to the Beta band (approximately 30 Hz), a sampling rate between 150 and 300 Hz is typically sufficient. However, certain preprocessing steps, such as EMG analysis, fMRI gradient artifact correction, or TMS pulse artifact removal may require higher sampling rates. After these steps are completed, the data can usually be downsampled without loss of relevant information.

Importantly, low-pass filtering must be applied before downsampling to avoid aliasing. While this step is critical, it is often handled automatically within built-in functions, which may lead users to overlook its importance.

17.2.3 Re-referencing

Re-referencing implies altering the "baseline" of the recorded data offline using different schemes to suit various purposes. Every re-reference scheme possesses distinct benefits and limitations (Luck, 2014b; Yao et al., 2019). Generally, re-referencing may be required for any of the following reasons:

- *Impact on signal amplitude:* If the online reference is near the channel of interest, it can affect the signal's amplitude, thereby reducing the desired effect. In this case, re-referencing the data to a channel not too close to the region of interest is recommended.
- *Artifact reduction:* If all scalp channels exhibit common artifacts (e.g., line noise, drifts), re-referencing helps mitigate these spatially widespread signals. The *Average Reference* scheme, in particular, improves signal-to-noise ratio (Bigdely-Shamlo et al., 2015; Tsuchimoto et al., 2021), provided there are sufficient channels (>64 channels) for even head coverage (Hu et al., 2018) and no bad channels are included. This makes the *Average Reference* a common choice in mobile EEG studies (Arad et al., 2018; Klug & Gramann, 2020).
- *Hemispheric bias:* EEG recordings often utilize a single electrode as the online reference. When referencing is limited to either the left or right mastoid, the resulting voltage distribution may exhibit hemispheric asymmetry.

  To mitigate this bias, re-referencing to the average of the left and right mastoids is recommended. An alternative is the previously mentioned *Average Reference* scheme, which is most effective with high-density montages (typically 64 or more electrodes) (Nunez & Srinivasan, 2006) and at least 50% uniform scalp coverage (Luck, 2014c).
- *Comparability with literature:* The choice of reference site can also be field-dependent. So, it would make sense to cross-check with other published research papers in your area of research to know the reference they used. To facilitate

result comparison, it may be necessary to re-reference the data using the referencing scheme employed in the literature being compared.

It is important to ensure that the new reference site is clean; otherwise, noise will propagate to other channels after referencing. Overall, re-referencing provides the opportunity to explore different schemes and choose the most suitable one for a specific research question or hypothesis.

17.2.4 Interpolating Channels or Data Portions

Channel interpolation refers to estimating the signal at specific electrode locations based on data from surrounding channels. This technique is commonly used to replace EEG channels that are affected by persistent artifacts.

Many EEG processing tools use *spherical spline* interpolation by default, as it generally yields high-quality results. However, not all scalp channels are necessarily included in the interpolation process. The specific channels used often depend on the chosen algorithm, user-defined parameters, and the characteristics of the dataset.

While there are no strict guidelines regarding the maximum number of channels that can be interpolated, it is recommended to apply interpolation sparingly and only when necessary. Overuse of interpolation can potentially distort the true underlying brain activity. As such, interpretations based on interpolated data should be made with caution.

In specific applications, such as EEG combined with TMS, interpolation can help mitigate artifacts caused by the TMS pulse. A common approach is to interpolate the data portion within the pulse time window before proceeding with further preprocessing or analysis. For more detailed strategies on managing EEG-TMS data, see Hernandez-Pavon et al. (Hernandez-Pavon et al., 2023).

17.2.5 Segmenting Data

Segmentation involves breaking down continuous EEG data into shorter, manageable segments or epochs. In studies focusing on event-related changes, continuous data is divided into segments around events, defined by pre- and post-stimulus intervals. These intervals should be selected based on the experimental paradigm and the expected effects. Typical steps for generating ERPs are discussed in Chap. 19: Event-Related Potentials.

Other types of analyses require specific segmentation considerations:

- *Wavelet* analysis: The segment length should include a safety margin at both the left and right borders to avoid border and smearing effects (Herrmann et al., 2014; Roach & Mathalon, 2008).

- *FFT* or *Welch's* method: Data are divided into short, non-overlapping or overlapping segments, which should be long enough to contain at least one cycle of the lowest frequency of interest. Moreover, for computational efficiency, the number of data points per segment is often chosen as a power of two.

To learn more about spectral analysis, you are referred to Chap. 18: Introduction to EEG Oscillations and Spectral Analysis.

17.3 Common Preprocessing Steps: Artifact Handling

In Chap. 15: Getting Clean EEG Data, strategies to avoid or minimize artifacts during data acquisition are presented. Despite these efforts, recorded data may still be contaminated with various artifacts and noise. It is essential to remove or minimize these artifacts to obtain neural signals that more accurately reflect the underlying brain activity.

17.3.1 Filtering

Offline filtering attempts to modify the frequency characteristics of data, aimed at reducing unwanted frequencies by removing noise from the data. Non-causal zero-phase filters are often preferred because they preserve the phase of all frequency components that remain in the signal, which is important for many analyses. Filters generally fall into four categories:

- *High-pass filter:* Attenuates frequencies below the low-cutoff. Common cutoffs in EEG are between 0.1 and 0.5 Hz to reduce drifts such as body sway or skin potentials.
- *Low-pass filter:* Reduces frequencies above the high-cutoff. As most EEG frequencies that are usually studied lie below 40 Hz, this is a common cutoff frequency. This helps minimize the impact of muscular artifacts and other high-frequency noise.
- *Band-pass filter:* The combination of high and low pass filters.
- *Notch filter:* A band rejection filter used to attenuate line noise at 50 or 60 Hz and its harmonics.

Furthermore, in publications, you might encounter two distinct types of filters: finite impulse response (FIR) and infinite impulse response (IIR), each based on different design principles, but both are suitable for filtering EEG data. Note that, a higher filter order produces a steeper magnitude response, resulting in stronger suppression beyond the cutoff frequency and thus improved frequency selectivity. However, this increased selectivity comes at the expenses of reduced temporal precision.

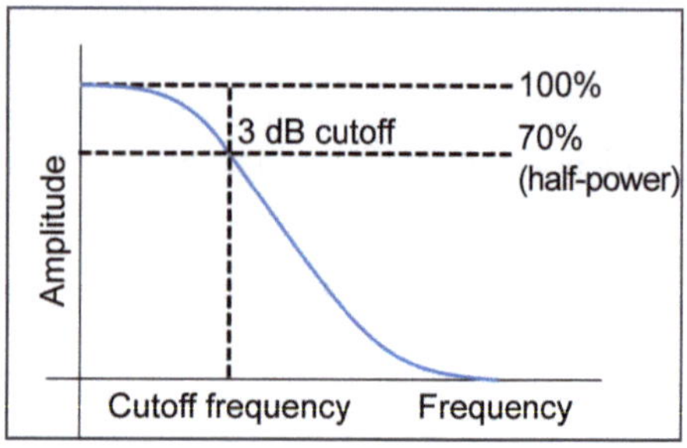

Fig. 17.1 Magnitude response plot of a 4th-order IIR Butterworth low-pass filter, based on the half-power cutoff definition. About 70.7% of the original amplitude is retained at the cutoff frequency

When discussing cutoff frequencies, the two commonly used terms are *half-power* cutoff and *half-amplitude* cutoff. The half-power cutoff, also known as the −3 dB point, refers to the frequency at which the power of the signal is reduced by half (approximately 70.7% of the original amplitude), as illustrated in Fig. 17.1. In contrast, the half-amplitude cutoff is defined by a −6 dB reduction, which corresponds to 50% of the original amplitude.

Filtering is a widely reported preprocessing step in publications. However, it is advisable to minimize the extent of filtering and apply it only, when necessary, as filtering can unpredictably alter the data. Interpreting the results with caution is essential. This topic has been thoroughly discussed, and practical guidelines for applying filters to EEG data are published (de Cheveigne & Nelken, 2019; Widmann et al., 2015; Zhang et al., 2024a, b).

17.3.2 Attenuating Artifacts

Stereotypical artifacts such as blinks, lateral eye movements, and muscle activity often exhibit consistent shapes and distributions. They can be reduced or removed through *correction* methods *without losing data points.*

17.3.2.1 Independent Component Analysis (ICA)

ICA is a blind source separation (BSS) method which involves decomposing the recorded EEG signal into components that represent brain activity and artifacts. These artifacts have characteristic topographies, shapes, and time courses, such as blinks and eye movements, as illustrated in Fig. 17.2. By reconstructing the signal after removing the artifact-related components, the influence of artifacts is effectively reduced.

Various ICA-based methods have been proposed, including *Infomax ICA* (Makeig et al., 1996), *Fast ICA* (Hyvärinen & Oja, 2000), *RELICA* (Artoni et al., 2014), and *AMICA* (Klug et al., 2024; Palmer et al., 2010). The choice of the ICA algorithm depends on the specific characteristics of your data and the type of artifacts

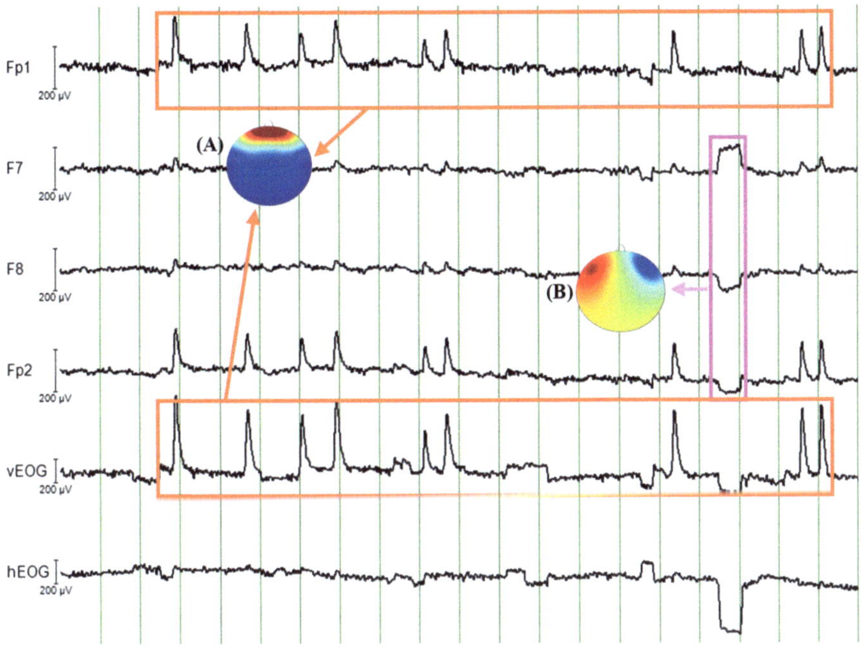

Fig. 17.2 Topographic maps of blinks and eye movements. (**a**) Blinks are prominent bell-shaped deflections over the data. They show a predominantly frontal topography. (**b**) Lateral eye movements show a box-shaped morphology, appear strongest in frontopolar and frontotemporal channels, and typically show opposite polarity on each side of the scalp. Note: the interval between the vertical green lines is 1 s

you are dealing with (Delorme et al., 2012; Dimigen, 2020; Frølich & Dowding, 2018; Gorjan et al., 2022; Hoffmann, 2009; Klug et al., 2024; Leutheuser et al., 2013; Stergiadis et al., 2022). Among these, *Infomax ICA* is perhaps the most used and frequently cited in publications.

The quality of ICA can be affected by several factors, which should be carefully considered:

- *Number of EEG channels:* The maximum number of components that can be isolated in EEG data corresponds to the number of electrodes used. While decomposition can be performed with low-density recordings, it is generally recommended to use at least 64 EEG channels for optimal results (Cohen, 2014; Klug & Gramann, 2020). Increasing the number of channels enhances spatial resolution and improves the accuracy of independent component analysis.
- *Amount of training data:* A minimum of (number of channel)^2 × 20 (for fewer than 64 channels) or (number of channel)^2 × 30 (for more than 64 channels) data points is recommended to perform ICA (Makeig & Onton, 2011). It is also important that the training data contain sufficient representative neuronal activity, including artifacts intended to be remove.

- *Quality of training data:* Slow drifts in the data can significantly deteriorate ICA performance. To mitigate this, applying a high-pass filter is strongly recommended. The cutoff frequency should be selected in a balanced way that is neither too low nor too high (i.e., usually not more than 1 Hz). Some literature suggests a cutoff between 1 and 2 Hz for clean decompositions, while continuing the analyses on less or unfiltered data afterwards to avoid distorting the time course of ERPs (Debener et al., 2010; Klug & Gramann, 2020; Winkler et al., 2015). Additionally, applying a low-pass filter reduces high-frequency noise, preventing it from dominating ICA components and improving the extraction of meaningful signals.

  Furthermore, the following points should be kept in mind when applying ICA:

- *Over-correction:* ICA is purely data-driven, meaning that the decomposed components are based on statistical properties of the signal and do not map precisely onto physiological processes. When using ICA for artifact removal, it is important to carefully examine all decomposed components to make well-informed decisions about whether a component should be discarded. Over-correction, which occurs when the components removed to correct artifacts also contain valid EEG, should be avoided.
- *Rank-deficiency:* ICA assumes that the signal is a linear mixture of independent sources, with the maximum number of extractable source signals limited by the number of channels. However, when channels are interpolated or re-referenced to a common average (or mean of mastoids), the principle of independence is violated. This results in rank-deficiency, which must be addressed to ensure a proper computation of the inverse ICA weight matrix. Mitigation of this issue involves running ICA with a reduced number of components or performing ICA before interpolation or re-referencing.

17.3.2.2 Regression Techniques

Regression-based methods, such as the *Gratton and Coles* algorithm (Gratton et al., 1983), can be utilized for the identification and correction of ocular artifacts. These methods leverage the EOG channels to estimate the influence of blinks and eye movements on the EEG channels via regression. Ocular artifacts are then subtracted based on a calculated correction factor. This approach is computationally efficient, though optimal performance necessitates EOG recordings. For a comprehensive overview of current methods for handling ocular artifacts, see (Jiang et al., 2019; Ronca et al., 2024).

Another notable method is the *Carbon Wire Loop (CWL)* regression technique, designed to reduce motion artifacts in brain signal recordings, particularly in simultaneous EEG-fMRI studies (van der Meer et al., 2016). CWLs are flexible carbon wires placed near the electrodes that capture motion-related signals without detecting neural activity. These motion signals are used as regressors in a linear model to estimate and remove motion-induced noise from the recorded brain data, significantly enhancing signal quality.

17.3.2.3 Template Subtraction Methods

Methods such as *Average Artifact Subtraction* (AAS) (Allen et al., 2000) can be used to remove gradient and cardioballistic artifacts in EEG data collected within an MR environment. In this method, artifact templates are calculated by averaging over adjacent artifact epochs and subsequently the templates are subtracted to reduce the artifacts. More details on handling artifacts for EEG-fMRI data can be found in Warbrick's review paper (Warbrick, 2022).

17.3.3 Rejecting Artifacts

Non-stereotypical artifacts can sporadically appear on various electrodes and exhibit different shapes. Examples include sudden electrode pops across multiple channels, significant movements by the participant affecting several channels, or transient muscular tensions impacting different regions of the head. These artifacts can be carefully addressed through interpolation or completely rejected. *Rejection* involves *removing* portions of the affected data or specific channels, or in extreme cases, *discarding* the entire dataset from a participant, to prevent compromising the group analysis and study results.

Artifacts must be detected before they can be rejected. This can be achieved either manually by using expert judgment to mark the data, or by applying specific criteria for automatic identification. There are no standardized parameters for artifact detection as it varies depending on the unique characteristics of the data in each experiment. Therefore, it's crucial to identify parameters that best suit the current dataset. Ideally, these parameters should be consistently applicable to all datasets within a study. However, exceptions may arise; for instance, if some datasets exhibit higher levels of noise, it may be sensible to adjust the criteria accordingly.

It is important to note that relying solely on artifact rejection without incorporating attenuation methods could lead to an unrepresentative sample of trials, particularly in mobile environments. Therefore, rather than using only one approach, a more effective strategy is to combine correction with rejection (Zhang et al., 2024c). As a general practice, give priority to removing sporadic artifacts and correcting predictable patterns, such as blinks.

17.4 What to Consider When Developing a Preprocessing Pipeline?

The recent recommendation to either avoid or minimally preprocess EEG data, based on the assertion that most preprocessing techniques do not improve data quality (Delorme, 2023), has sparked controversy. This view has been challenged by several EEG experts (e.g., de Cheveigné, 2023; Larsen & Versace, 2024). These experts argue that while such an approach may be applicable to relatively clean EEG recordings, preprocessing can be crucial for improving the quality and interpretability of noisier or more complex EEG data.

Therefore, it is generally imprudent to completely disregard data preprocessing. To support informed decision-making in designing a valid and effective EEG preprocessing pipeline for a specific dataset, several key factors have been outlined for consideration:

- *Data characteristics:* Not all preprocessing steps are always relevant; their necessity depends on the quality of the acquired data. Noisy datasets often demand more comprehensive preprocessing and artifact rejection. Consequently, this may involve interpolating more channels, applying more stringent filtering, and adjusting the criteria for artifact detection.
- *Intended analysis and research goals:* The desired outcome measure can dictate the necessary processing steps and their sequence. For example, using a Laplacian filter (also known as Current Source Density, CSD) as a preprocessing step before performing connectivity analysis can mitigate the effect of volume conduction through the scalp, leading to a more precise calculation of connectivity. If the research objective is to compare your findings with a previous study, it would make sense to replicate the processing pipeline reported in that publication, rather than creating a different one, to ensure comparability.
- *Knowledge of preprocessing steps:* Understanding the prerequisites for specific analysis steps is crucial. For instance, attenuating slow drifts and high-frequency noise, as well as rejecting non-stereotypical artifacts, can enhance the effectiveness of ICA. Additionally, interpolating bad channels prior to applying an average reference helps to maintain a balanced spatial representation across the scalp, which is important for preserving the accuracy of topographical and source-level analyses.
- *Suitability:* Some steps are better suited for continuous or for segmented data, while others can be applied to both, depending on requirements. For example, editing the markers or re-referencing can be performed on both segmented and continuous data. However, high-pass filtering is best applied to long continuous data to avoid edge artifacts that might contaminate the region of interest on shorter segments. Similarly, ICA is more efficient when applied to continuous data, as larger, uninterrupted data allow for better separation of components.
- *Order:* Some steps within a pipeline can be interchangeable. However, others can significantly influence the final output, depending on whether they are executed

earlier or later and whether they are linear or non-linear operations. Luck comprehensively explains these concepts in his book and also provides practical guidance for developing processing pipelines for ERP analysis (Luck, 2014a). Equally important are the parameters used at each step, which should be documented in detail to ensure proper evaluation and reproducibility (Keil et al., 2014).

As you can see, constructing an EEG processing pipeline involves many interconnected decisions. Therefore, dedicating sufficient time to thoughtfully plan and refine the workflow will help ensure reliable and interpretable results.

17.5 Tools for Processing and Analyzing EEG

A variety of software can be utilized to process and analyze EEG data, encompassing both open-source and proprietary options. The choice of software largely depends on the research project's requirements and the user's programming proficiency. Finding a single application that meets all needs is often challenging. Rather than focusing on one particular tool, it may be more effective to adopt a platform that can interact with multiple software to some extend and allow, where possible, the creation of custom scripts. Below is a list of some software options (the order is arbitrary and the list is not exhaustive). Some other tools are mentioned in the review paper by Das et al. (Das et al., 2023):

- *EEGLAB* (Delorme & Makeig, 2004): An open-source *MATLAB* toolbox offering a comprehensive environment for processing EEG data, including functions for preprocessing, visualization, and analyzing ERPs and time-frequency data. Several automated pipelines have been developed based on this platform, such as the *PREP* (Bigdely-Shamlo et al., 2015), *HAPPE* (Gabard-Durnam et al., 2018; Monachino et al., 2022), *RELAX* (Bailey et al., 2023a, b; Hill et al., 2024), *ADJUST* (Mognon et al., 2011), and *DISCOVER-EEG* (Gil Ávila et al., 2023).
- *ERPLAB Toolbox:* A free, open source Matlab package for analyzing ERP data, closely intergrated with EEGLAB (Lopez-Calderon & Luck, 2014).
- *BrainVision Analyzer:* A commercial software package that provides tools for the analysis of EEG, ERP, and other neurophysiological data. The software is known for its user-friendly interface, intuitive features, and reliable scientific methods. It also allows interfacing with *MATLAB* and extending processing functionalities via macro scripting.
- *FieldTrip:* Another open-source *MATLAB* toolbox designed for analyzing MEG, EEG, and other neurophysiological data. It offers advanced functions for time-frequency analysis, source reconstruction, and statistical analysis.
- *CURRY:* A commercial software offering tools for multi-modal neuroimaging and neurophysiological data analysis. It's widely used in both research and clinical settings, particularly for epilepsy evaluation.

- *MNE-Python:* An open-source software package for processing MEG and EEG data in Python. It includes tools for preprocessing, source localization, and statistical analysis, making it a versatile choice for researchers comfortable with Python.
- *OpenVibe:* An open-source software platform designed for brain-computer interface (BCI) applications, which includes functionalities for processing and analyzing EEG signals.
- *BESA:* A commercial software that is particularly well-suited for source analysis and connectivity analysis of EEG and MEG data.

17.6 Concluding Remarks

We have explored several common preprocessing steps and shed light on the essential considerations for developing an effective preprocessing pipeline. To conclude, it is crucial to underscore that preprocessing techniques are not magical solutions capable of transforming poor-quality data into high-quality data. Rather, they serve to refine and enhance already good data, facilitating more accurate and meaningful analysis. It is important to remember that the foundation of any successful EEG analysis lies in the quality of the raw signal—there is simply no substitute for a clean, well-recorded signal. Therefore, while preprocessing is an indispensable step, it should always be complemented by rigorous data acquisition practices to ensure the best possible outcomes.

References

Allen, P. J., Josephs, O., & Turner, R. (2000). A method for removing imaging artifact from continuous EEG recorded during functional MRI. *NeuroImage, 12*(2), 230–239. https://doi.org/10.1006/nimg.2000.0599

Arad, E., Bartsch, R. P., Kantelhardt, J. W., & Plotnik, M. (2018). Performance-based approach for movement artifact removal from electroencephalographic data recorded during locomotion. *PLoS One, 13*(5), e0197153. https://doi.org/10.1371/journal.pone.0197153

Artoni, F., Menicucci, D., Delorme, A., Makeig, S., & Micera, S. (2014). RELICA: A method for estimating the reliability of independent components. *NeuroImage, 103*, 391–400. https://doi.org/10.1016/j.neuroimage.2014.09.010

Bailey, N. W., Biabani, M., Hill, A. T., Miljevic, A., Rogasch, N. C., McQueen, B., Murphy, O. W., & Fitzgerald, P. B. (2023a). Introducing RELAX: An automated pre-processing pipeline for cleaning EEG data—Part 1: Algorithm and application to oscillations. *Clinical Neurophysiology, 149*, 178–201. https://doi.org/10.1016/j.clinph.2023.01.017

Bailey, N. W., Hill, A. T., Biabani, M., Murphy, O. W., Rogasch, N. C., McQueen, B., Miljevic, A., & Fitzgerald, P. B. (2023b). RELAX part 2: A fully automated EEG data cleaning algorithm that is applicable to event-related-potentials. *Clinical Neurophysiology, 149*, 202–222. https://doi.org/10.1016/j.clinph.2023.01.018

Bigdely-Shamlo, N., Mullen, T., Kothe, C., Su, K. M., & Robbins, K. A. (2015). The PREP pipeline: Standardized preprocessing for large-scale EEG analysis. *Frontiers in Neuroinformatics, 9*, 16. https://doi.org/10.3389/fninf.2015.00016

Cohen, M. X. (2014). 6.5 How many electrodes you will need. In *Analyzing neural time series data: Theory and practice* (pp. 65–66). The MIT Press. https://doi.org/10.7551/mitpress/9609.003.0023

Das, R. K., Martin, A., Zurales, T., Dowling, D., & Khan, A. (2023). A survey on EEG data analysis software. *The Sciences, 5*(2), 23. https://www.mdpi.com/2413-4155/5/2/23

de Cheveigné, A. (2023). Is EEG is better left alone? *bioRxiv*, 2023.2006.2019.545602. https://doi.org/10.1101/2023.06.19.545602

de Cheveigne, A., & Nelken, I. (2019). Filters: When, why, and how (not) to use them. *Neuron, 102*(2), 280–293. https://doi.org/10.1016/j.neuron.2019.02.039

Debener, S., Thorne, J., Schneider, T. R., & Viola, F. C. (2010). 3.1 Using ICA for the analysis of multi-channel EEG data. In *Simultaneous EEG and fMRI: Recording, analysis, and application*. Oxford University Press. https://doi.org/10.1093/acprof:oso/9780195372731.003.0008

Delorme, A. (2023). EEG is better left alone. *Scientific Reports, 13*(1), 2372. https://doi.org/10.1038/s41598-023-27528-0

Delorme, A., & Makeig, S. (2004). EEGLAB: An open source toolbox for analysis of single-trial EEG dynamics including independent component analysis. *Journal of Neuroscience Methods 134*(1), 9–21. https://doi.org/10.1016/j.jneumeth.2003.10.009

Delorme, A., Palmer, J., Onton, J., Oostenveld, R., & Makeig, S. (2012). Independent EEG sources are dipolar. *PLoS One, 7*(2), e30135. https://doi.org/10.1371/journal.pone.0030135

Dimigen, O. (2020). Optimizing the ICA-based removal of ocular EEG artifacts from free viewing experiments. *NeuroImage, 207*, 116117. https://doi.org/10.1016/j.neuroimage.2019.116117

Frølich, L., & Dowding, I. (2018). Removal of muscular artifacts in EEG signals: A comparison of linear decomposition methods. *Brain Informatics, 5*(1), 13–22. https://doi.org/10.1007/s40708-017-0074-6

Gabard-Durnam, L. J., Mendez Leal, A. S., Wilkinson, C. L., & Levin, A. R. (2018). The Harvard automated processing pipeline for electroencephalography (HAPPE): Standardized processing software for developmental and high-artifact data. *Frontiers in Neuroscience, 12*. https://doi.org/10.3389/fnins.2018.00097

Gil Ávila, C., Bott, F. S., Tiemann, L., Hohn, V. D., May, E. S., Nickel, M. M., Zebhauser, P. T., Gross, J., & Ploner, M. (2023). DISCOVER-EEG: An open, fully automated EEG pipeline for biomarker discovery in clinical neuroscience. *Scientific Data, 10*(1), 613. https://doi.org/10.1038/s41597-023-02525-0

Gorjan, D., Gramann, K., De Pauw, K., & Marusic, U. (2022). Removal of movement-induced EEG artifacts: Current state of the art and guidelines. *Journal of Neural Engineering, 19*(1), 011004. https://doi.org/10.1088/1741-2552/ac542c

Gratton, G., Coles, M. G. H., & Donchin, E. (1983). A new method for off-line removal of ocular artifact. *Electroencephalogrography and Clinical Neurophysiology, 55*(4), 468–484. https://doi.org/10.1016/0013-4694(83)90135-9

Hernandez-Pavon, J. C., Veniero, D., Bergmann, T. O., Belardinelli, P., Bortoletto, M., Casarotto, S., Casula, E. P., Farzan, F., Fecchio, M., Julkunen, P., Kallioniemi, E., Lioumis, P., Metsomaa, J., Miniussi, C., Mutanen, T. P., Rocchi, L., Rogasch, N. C., Shafi, M. M., Siebner, H. R., et al. (2023). TMS combined with EEG: Recommendations and open issues for data collection and analysis. *Brain Stimulation, 16*(2), 567–593. https://doi.org/10.1016/j.brs.2023.02.009

Herrmann, C. S., Rach, S., Vosskuhl, J., & Strüber, D. (2014). Time–frequency analysis of event-related potentials: A brief tutorial. *Brain Topography, 27*(4), 438–450. https://doi.org/10.1007/s10548-013-0327-5

Hill, A. T., Enticott, P. G., Fitzgerald, P. B., & Bailey, N. W. (2024). RELAX-Jr: An automated pre-processing pipeline for developmental EEG recordings. *Human Brain Mapping, 45*(14), e70034. https://doi.org/10.1002/hbm.70034

Hoffmann, S. (2009). *Independent component analysis of ocular artifacts.* https://hss-opus.ub.ruhr-uni-bochum.de/opus4/frontdoor/index/index/docId/2223

Hu, S., Lai, Y., Valdes-Sosa, P. A., Bringas-Vega, M. L., & Yao, D. (2018). How do reference montage and electrodes setup affect the measured scalp EEG potentials? *Journal of Neural Engineering, 15*(2), 026013. https://doi.org/10.1088/1741-2552/aaa13f

Hyvärinen, A., & Oja, E. (2000). Independent component analysis: Algorithms and applications. *Neural Networks, 13*(4), 411–430. https://doi.org/10.1016/S0893-6080(00)00026-5

Jiang, X., Bian, G.-B., & Tian, Z. (2019). Removal of artifacts from EEG signals: A review. *Sensors, 19*(5), 987. https://www.mdpi.com/1424-8220/19/5/987

Keil, A., Debener, S., Gratton, G., Junghofer, M., Kappenman, E. S., Luck, S. J., Luu, P., Miller, G. A., & Yee, C. M. (2014). Committee report: Publication guidelines and recommendations for studies using electroencephalography and magnetoencephalography. *Psychophysiology, 51*(1), 1–21. https://doi.org/10.1111/psyp.12147

Klug, M., & Gramann, K. (2020). Identifying key factors for improving ICA-based decomposition of EEG data in mobile and stationary experiments. *The European Journal of Neuroscience.* https://doi.org/10.1111/ejn.14992

Klug, M., Berg, T., & Gramann, K. (2024). Optimizing EEG ICA decomposition with data cleaning in stationary and mobile experiments. *Scientific Reports, 14*(1), 14119. https://doi.org/10.1038/s41598-024-64919-3

Larsen, B. A., & Versace, F. (2024). EEG might be better left alone, but ERPs must be attended to: Optimizing the late positive potential preprocessing pipeline. *International Journal of Psychophysiology, 205*, 112441. https://doi.org/10.1016/j.ijpsycho.2024.112441

Leutheuser, H., Gabsteiger, F., Hebenstreit, F., Reis, P., Lochmann, M., & Eskofier, B. (2013). Comparison of the AMICA and the InfoMax algorithm for the reduction of electromyogenic artifacts in EEG data. *2013 35th Annual International Conference of the IEEE Engineering in Medicine and Biology Society (EMBC)*, Osaka, Japan, 2013, pp. 6804-6807. https://doi.org/10.1109/EMBC.2013.6611119

Lopez-Calderon, J., & Luck, S. J. (2014). ERPLAB: an open-source toolbox for the analysis of event-related potentials. *Frontiers in Human Neuroscience*, 8, 213. https://doi.org/10.3389/fnhum.2014.00213

Luck, S. J. (2014a). Apendix: Linear operations, nonlinear operations, and the order of processing steps. In *An introduction to the event-related potential technique* (pp. 341–350). MIT Press.

Luck, S. J. (2014b). Basic principles of ERP recording. In *An introduction to the event-related potential technique* (pp. 147–183). MIT Press.

Luck, S. J. (2014c). *An introduction to the event-related potential technique.* MIT Press.

Makeig, S., & Onton, J. (2011). ERP features and EEG dynamics: An ICA perspective. In *The Oxford handbook of event-related potential components.* Oxford University Press. https://doi.org/10.1093/oxfordhb/9780195374148.013.0035

Makeig, S., Bell, A. J., Jung, T.-P., & Sejnowski, T. J. (1996). Independent component analysis of electroencephalographic data. In D. S. Touretzky, M. C. Mozer, & M. E. Hasselmo (Eds.), *Advances in neural information processing systems 8: Proceedings of the 1995 conference* (Vol. 8, pp. 145–151). The MIT Press. https://dl.acm.org/doi/10.5555/2998828.2998849

Mognon, A., Jovicich, J., Bruzzone, L., & Buiatti, M. (2011). ADJUST: An automatic EEG artifact detector based on the joint use of spatial and temporal features. *Psychophysiology, 48*(2), 229–240. https://doi.org/10.1111/j.1469-8986.2010.01061.x

Monachino, A. D., Lopez, K. L., Pierce, L. J., & Gabard-Durnam, L. J. (2022). The HAPPE plus Event-Related (HAPPE+ER) software: A standardized preprocessing pipeline for event-related potential analyses. *Developmental Cognitive Neuroscience, 57*, 101140. https://doi.org/10.1016/j.dcn.2022.101140

Nunez, P. L., & Srinivasan, R. (2006). *Electric fields of the brain: The neurophysics of EEG.* Oxford University Press.

Oostenveld, R. (n.d.). High-density EEG electrode placement. *Robert Oostenveld's blog*. https://robertoostenveld.nl/electrode/#oostenveld2001

Palmer, J., Kreutz-Delgado, K., & Makeig, S. (2010). *AMICA: An adaptive mixture of independent component analyzers with shared components*. https://sccn.ucsd.edu/~jason/amica_a.pdf

Roach, B. J., & Mathalon, D. H. (2008). Event-related EEG time-frequency analysis: An overview of measures and an analysis of early gamma band phase locking in schizophrenia. *Schizophrenia Bulletin, 34*(5), 907–926. https://doi.org/10.1093/schbul/sbn093

Ronca, V., Capotorto, R., Di Flumeri, G., Giorgi, A., Vozzi, A., Germano, D., Virgilio, V. D., Borghini, G., Cartocci, G., Rossi, D., Inguscio, B. M. S., Babiloni, F., & Aricò, P. (2024). Optimizing EEG signal integrity: A comprehensive guide to ocular artifact correction. *Bioengineering, 11*(10), 1018. https://www.mdpi.com/2306-5354/11/10/1018

Stergiadis, C., Kostaridou, V.-D., & Klados, M. A. (2022). Which BSS method separates better the EEG signals? A comparison of five different algorithms. *Biomedical Signal Processing and Control, 72*, 103292. https://doi.org/10.1016/j.bspc.2021.103292

Tsuchimoto, S., Shibusawa, S., Iwama, S., Hayashi, M., Okuyama, K., Mizuguchi, N., Kato, K., & Ushiba, J. (2021). Use of common average reference and large-Laplacian spatial-filters enhances EEG signal-to-noise ratios in intrinsic sensorimotor activity. *Journal of Neuroscience Methods, 353*, 109089. https://doi.org/10.1016/j.jneumeth.2021.109089

van der Meer, J. N., Pampel, A., Van Someren, E. J. W., Ramautar, J. R., van der Werf, Y. D., Gomez-Herrero, G., Lepsien, J., Hellrung, L., Hinrichs, H., Moller, H. E., & Walter, M. (2016). Carbon-wire loop based artifact correction outperforms post processing EEG/fMRI corrections—A validation of a real-time simultaneous EEG/fMRI correction method. *NeuroImage, 125*, 880–894. https://doi.org/10.1016/j.neuroimage.2015.10.064

Warbrick, T. (2022). Simultaneous EEG-fMRI: What have we learned and what does the future hold? *Sensors, 22*(6), 2262. https://doi.org/10.3390/s22062262

Widmann, A., Schroger, E., & Maess, B. (2015). Digital filter design for electrophysiological data—A practical approach. *Journal of Neuroscience Methods, 250*, 34–46. https://doi.org/10.1016/j.jneumeth.2014.08.002

Winkler, I., Debener, S., Müller, K. R., & Tangermann, M. (2015, August 25–29). On the influence of high-pass filtering on ICA-based artifact reduction in EEG-ERP. In *2015 37th annual international conference of the IEEE Engineering in Medicine and Biology Society (EMBC)*.

Yao, D., Qin, Y., Hu, S., Dong, L., Bringas Vega, M. L., & Valdes Sosa, P. A. (2019). Which reference should we use for EEG and ERP practice? *Brain Topography, 32*(4), 530–549. https://doi.org/10.1007/s10548-019-00707-x

Zhang, G., Garrett, D. R., & Luck, S. J. (2024a). Optimal filters for ERP research I: A general approach for selecting filter settings. *Psychophysiology, 61*. https://doi.org/10.1111/psyp.14531

Zhang, G., Garrett, D. R., & Luck, S. J. (2024b). Optimal filters for ERP research II: Recommended settings for seven common ERP components. *Psychophysiology, 61*. https://doi.org/10.1111/psyp.14530

Zhang, G., Garrett, D. R., Simmons, A. M., Kiat, J. E., & Luck, S. J. (2024c). Evaluating the effectiveness of artifact correction and rejection in event-related potential research. *Psychophysiology, 61*(5), e14511. https://doi.org/10.1111/psyp.14511

Chapter 18
Introduction to EEG Oscillations and Spectral Analysis

Ramon Martinez-Cancino and Yin Fen Low

Abstract Oscillatory brain activity reflects the dynamic coordination of neural processes. Understanding these oscillations requires methods beyond the time domain to reveal their frequency-domain information. This chapter introduces spectral analysis as the foundation for studying EEG oscillations. Beginning with the mathematical description of oscillatory processes, we build intuition around the core concepts of amplitude, frequency, and phase. We then show how the dot product provides the mathematical, geometric, and intuitive basis for Fourier analysis. The discrete Fourier transform (DFT) is presented as a practical tool for decomposing EEG time series into the frequency domain, enabling precise characterization of oscillatory activity. Finally, we discuss the limitations of Fourier analysis and outline ways to address them. Rather than thoroughly presenting advanced methods in frequency analysis, this chapter aims to equip readers with the conceptual and analytical framework needed to begin studying EEG time series from a frequency-domain perspective.

Keywords Brain oscillations · Neural rhythms · Spectral analysis · Fourier analysis · Dot product · Discrete Fourier transform · Power spectrum · Signal processing

18.1 Introduction

As you may have discovered in earlier chapters, electroencephalography (EEG) is a powerful tool for looking into the brain through its electrical activity. One of the most distinctive features of EEG activity is its rhythmic nature – an attribute that extends beyond the human brain. In the 1870s, physician Richard Caton first noted the presence of rhythmic electrical activity in the brains of animals (Caton, 1875). A few decades later, Hans Berger observed similar phenomena in humans (Berger,

R. Martinez-Cancino (✉) · Y. F. Low
Brain Products GmbH, Gilching, Germany
e-mail: ramon.martinez@brainproducts.com; yinfen.low@brainproducts.com

T. Warbrick (ed.), *The EEG Handbook*,
https://doi.org/10.1007/978-3-032-20450-9_18

1929). Over time, researchers realized that these neural "rhythms" are not random fluctuations but reflect structured patterns of neural activity. Many theories propose that these oscillations help organize and synchronize neural activity across various spatial and temporal scales (Varela et al., 2001; Ward, 2003; Buzsaki & Draguhn, 2004). Such oscillatory coordination has been linked to processes ranging from perception and cognition to consciousness (Ward, 2003; Buzsaki & Draguhn, 2004).

Although a fascinating topic, this chapter does not deal with the historical origins or biophysical mechanisms responsible for these rhythmic patterns. Instead, its goal is to provide a theoretical base for interpreting the information they convey. *Rhythms* have been our term of choice thus far, but moving forward, we will intentionally use the term *oscillations* to emphasize the cyclic, repetitive nature of brain electrical activity. Although *brain rhythms* and *brain oscillations* are often used interchangeably in neuroscience, a gradual conceptual evolution – fueled in part by interdisciplinary collaboration and the advent of refined analytical techniques – has prompted a shift toward the more precise term *oscillations*.

This chapter explores the foundations of oscillation analysis: spectral analysis. This approach becomes more intuitive when we consider the brain's electrical signals to be inherently oscillatory. Given the strongly oscillatory nature of EEG recordings, they have been characterized by the speed of these oscillations since their early days. For instance, in his first observation of human EEG, Hans Berger described what we know today as Alpha and Beta waves, which have frequency ranges of 7–12 Hz and 12–30 Hz, respectively. Since that initial observation, rhythmic activity in EEG has been widely reported and studied due to its ability to characterize both normal and pathological brain states.

It is worth noting that embracing this oscillation-centric perspective is not just a technical convenience. As Cohen (2014) points out, focusing on EEG oscillations means acknowledging that it captures a dynamic, multidimensional domain of brain activity. In this view, traditional event-related potentials (ERPs) represent only a fraction of the information available in the EEG signal (Cohen, 2014). Conversely, oscillation-based frequency analyses can reveal additional dimensions of brain dynamics that a purely time-domain approach might overlook (Herrmann et al., 2014). This premise fundamentally shapes the content of the pages ahead.

In the first part of the chapter, we introduce a mathematical description to characterize oscillatory processes using a simple experimental setup. We discuss the concepts of frequency, phase, and amplitude. Next, we build the intuition that will allow us to justify the use of Fourier analysis, a key tool for frequency-domain analysis of brain oscillations. Following, we introduce the dot product as a foundation for Fourier analysis. We then present the Discrete Fourier Transform (DFT) as a practical implementation of this analysis. Throughout the chapter, the aim is to equip you with the conceptual outlook and intuition needed to understand the spectral analysis of EEG oscillations and unlock the rich information they contain in their frequency domain. Finally, we present some of the DFT's limitations and suggest alternative approaches. A list of further readings is also provided at the end of the chapter to complement the discussions.

18.2 Characterizing an Oscillatory Process

18.2.1 Fundamental Characteristics of an Oscillatory Process

Until now, we have been talking about frequency and oscillation without a proper introduction. We will address these concepts now. Let us consider a mass – the bob – suspended by a string from a fixed point, free to swing in a vertical plane. This setup is known as a pendulum (see Fig. 18.1a). The bob's position at any given time can be described by its coordinates along the x and y axes. However, let us focus on its motion along the x-axis.

From this perspective, as the pendulum swings, the projection (imagine its shadow) of the bob onto the x-axis traces a line between $-A$ and $+A$. The mass travels from one extreme to the other with variable speed: it moves fastest at the midpoint and slows to zero at the extremes, precisely at the instant it reverses direction. We can model the bob's position at time t, denoted $x(t)$, with the following sine-wave expression (Eq. 18.1):

$$x(t) = A \sin(2\pi ft + \theta) \tag{18.1}$$

Let us break this expression into smaller elements:

- A represents the maximum displacement of the motion around the midpoint.
- sin is the sine function, which takes as its argument $2\pi ft + \theta$.
- π (pi) is a fundamental constant approximately equal to 3.14159, that is, the ratio of a circle's circumference to its diameter (Wikipedia, 2025). In angle measurement, π radians is equal to 180 degrees.
- f is the *frequency*, indicating how many full swings the bob makes in a unit of time.

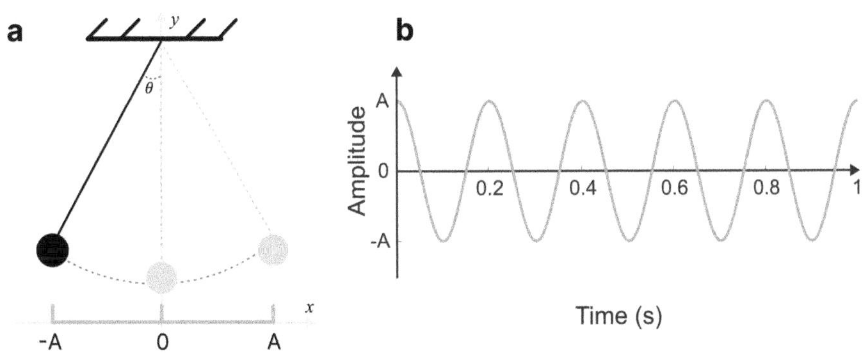

Fig. 18.1 Pendulum motion and corresponding sinusoidal oscillation. (**a**) Pendulum oscillating in a vertical plane, with its bob's displacement on the x-axis restricted to the range $-A$ to A. (**b**) A sinusoidal function describes the pendulum's movement along the x-axis within this range

- θ is the *phase* (or phase angle offset), specifying where in its cycle the motion starts at $t = 0$.

Figure 18.1b shows the waveform generated by this function, and this is the position of the bob over time. We have four important concepts to highlight here:

- *Oscillation*: Repetitive or cyclic variation in a system state over time. In our pendulum example, a full swing of the bob – from one extreme to the other and back – exemplifies an oscillation. This periodic motion can be described mathematically using sine or cosine functions.
- *Amplitude*: The maximum displacement from the midpoint of the oscillation. This value, A, gives us an immediate idea of the "size" or "strength" of the oscillation. This magnitude is related to the power or energy in each frequency – or frequency band – through the squared amplitude of the oscillation.
- *Frequency*: Indicates how many full oscillations occur in a unit of time, measured in hertz (Hz), where 1 Hz equals one complete cycle per second. In our pendulum example, the frequency f captures the speed at which the pendulum swings back and forth.
- *Phase*: Specifies the starting point of the oscillation at $t = 0$. It determines where in the cycle the pendulum begins its motion. A non-zero phase implies the bob may already be going through a swing when we start observing it.

18.2.2 From Time to Frequency and Back

Consider now three sinusoids with different amplitudes and frequencies (Fig. 18.2a). When added together, they produce a more complex waveform that still contains all the original frequency components (Fig. 18.2b). This complex signal in the time domain can also be represented in the frequency domain by finding the signal's components distribution of amplitudes over frequencies (Fig. 18.2c). This representation conveys critical information on the power contribution to a signal from its individual frequency components, and it is important to characterize the signal in this domain.

A neat theoretical demonstration, you might say – but not exactly how things play out in real-world scenarios. In practice, we are usually presented with the opposite challenge: Given a complex signal like the EEG, how can we identify and quantify its underlying frequency components? This task is the core of spectral or frequency analysis and is made possible by well-established signal processing methods like the Fourier analysis.

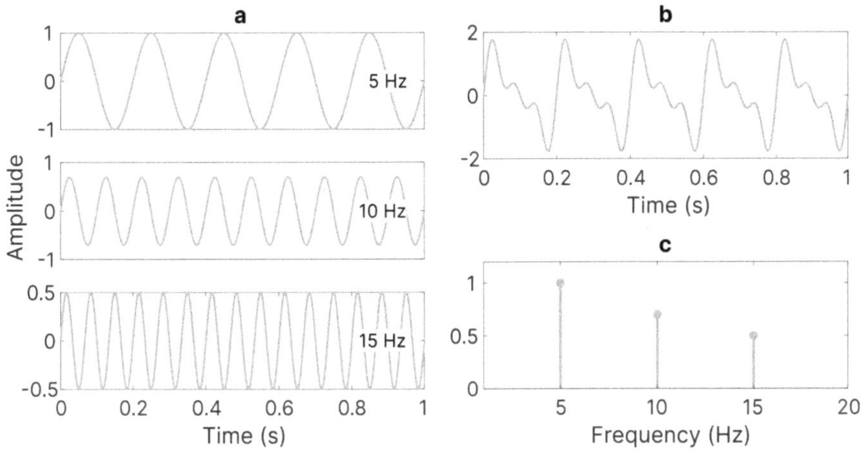

Fig. 18.2 Composite of sinusoids at 5, 10, and 15 Hz and corresponding frequency representation (**a**). Sinusoids at three different frequencies: 5, 10, and 15 Hz. (**b**) Composite signal obtained by summing the three sinusoids in panel **a**. (**c**) Power spectrum of the composite signal shown in panel **b**, showing peaks at the corresponding frequencies

18.3 Foundation for Spectral Analysis: The Dot Product

The dot product is a mathematical operation on two vectors – think of two equal-length sequences of numbers – that returns a single number. This operation is at the core of the Fourier transform.

Computing the dot product is relatively simple; given the vectors $A = [a_1, a_2, \ldots, a_n]$ and $B = [b_1, b_2, \ldots, b_n]$ where n is the number of elements of the vectors, to compute the dot product between A and B, defined as $A \cdot B$, you multiply each A vector element by their corresponding component of the vector B, and then sum up all the values from each product (Eq. 18.2).

$$A \cdot B = \sum_{i=1}^{n} a_i b_i = a_1 b_1 + a_2 b_2 + \cdots + a_n b_n \tag{18.2}$$

The result of the dot product between two vectors can be interpreted in various ways depending on the context, although its definition remains consistent. For our signal processing application, a particularly useful perspective is its geometric interpretation. The dot product between the vectors A and B can also be defined by Eq. 18.3.

$$A \cdot B = |A||B| \cos \theta \tag{18.3}$$

where θ is the angle between the vectors A and B, and | | represents the length of the vector. This relationship is depicted in Fig. 18.3a, where we have assumed that the

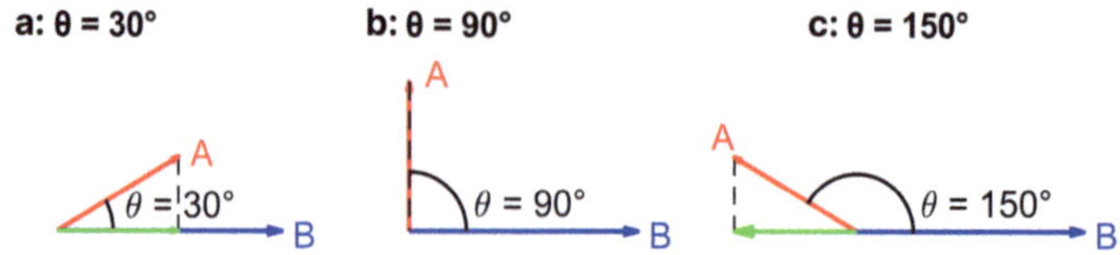

Fig. 18.3 Dot product between two vectors at different angles. Each panel illustrates the dot product between vectors A (red) and B (blue). The green vector represents the projection of A onto B, quantifying the component of A that lies in the direction of B. As the angle θ increases, the projection shortens and can become negative, as seen in panel **c**. (**a**) $\theta = 30°$ ($\pi/6$) (**b**) $\theta = 90°$ ($\pi/2$) (**c**) $\theta = 150°$ ($5\pi/6$)

vectors in question can be represented in a bi-dimensional space, e.g., $A = [2, 1]$, $B = [4, 0]$. As the dimensions, or number of elements of the vectors grow, it may be more complicated or even impossible to picture the graphical representation of the vectors, but essentially, the interpretation of the dot product will remain as in our 2D example. From this geometrical perspective, two common interpretations arise, associated with aligning and projecting one vector onto the other. From the projection perspective, the dot product is the length of the projection of A into the direction of B when the two vectors are placed so that their origins coincide (Weisstein, 1999). That is, how much of A "*lies along*" the direction of B. From the alignment perspective, the dot product may act as an alignment score between these two vectors by indicating whether they point in similar directions or not. For example, see Fig. 18.3a, if $\theta < \pi/2$ the $\cos\theta > 0$ and $A \cdot B > 0$, here the two vectors point mainly in the same direction, and the dot product is a score of how much. In Fig. 18.3b $\theta = \pi/2$ the $\cos\theta = 0$ therefore $A \cdot B = 0$, this indicates that these two vectors have orthogonal directions, like a perfect cross. The last case, in Fig. 18.3c, shows the case when $\theta > \pi/2$ the $\cos\theta < 0$ and $A \cdot B < 0$ which indicates that these two vectors have somehow opposite directions. It may already be obvious that, despite the interpretation followed, the dot product indicates how similar the vectors involved in the operation are.

18.4 Fourier Analysis

18.4.1 From Vectors to Sinusoids: The Fourier Connection

Nevertheless, why is there an insistence on discussing the dot product and its geometrical interpretation? This is because it creates a powerful intuition that goes beyond simple vectors. The same principle behind the dot product can be used to decompose complex signals into sinusoids using the Fourier analysis. Here is a hint: imagine a complex signal, a segment from an EEG recording, for example. This is equivalent to a large vector with dimension equal to the number of sample points. Then, imagine a pure sinusoid – a sinusoid with a single frequency. How would we determine how much of that sinusoid is present in our EEG signal? Of course, it can be done by taking a dot product!

Conceptually, the Fourier analysis operates by projecting the signal onto a whole set of sinusoidal "basis vectors" with different frequencies, and each time asking: *How well does my signal match this specific sine wave?* By doing this for all relevant basis vectors, we decompose the signal into several sinusoidal components that, when weighted and summed, can describe the original signal exactly. In this sense, the dot product serves as the engine driving Fourier decomposition; therefore, developing a strong intuition for this operation is highly relevant. In the following sections, we will dive more into the Fourier analysis.

18.4.2 The Fourier Family

While many individuals have contributed to the development of Fourier analysis, the name itself honors Jean-Baptiste Joseph Fourier (1768–1830), a French mathematician and physicist, for his mathematical insights into the practical applications of these techniques. The origins of Fourier analysis can be traced back to his first paper on the Theory of Heat published in 1807 and expanded in 1822 (Fourier, 1822), where he discussed the expansion of functions in trigonometric series.

Fourier proposed that any continuous and smooth function could be expressed in terms of trigonometric or exponential functions. Following his work, many others expanded on this field, making the study of Fourier analysis seem daunting today due to the variety of names and classifications used.

Signals, for example, can be classified as either continuous, where values are defined at every instant in time, or discrete, where values are only defined at specific time points. Additionally, signals can be periodic, featuring repeating patterns, or aperiodic, which do not show such patterns. The combination of these two characteristics creates four distinct categories of signals, each requiring different approaches and terminology within Fourier analysis (Smith, 1997).

In our context of EEG analysis, some assumptions can be made to help us decide which approach is more appropriate. For example, our signals are recorded digitally over limited periods of time. This means we do not have infinite recordings and do not measure signals continuously. Instead, we sample them at regular time intervals, resulting in a finite number of data points. However, no such Fourier Transform is designed to deal with finite-length signals. Furthermore, sines and cosines are defined as infinite signals, from minus to plus infinity, and they cannot be used to build up a finite signal. The trick is imagining the signals in question have infinite samples on the left and right of the actual points with value zero. Then, the resultant signal looks discrete and aperiodic, and the Discrete Fourier Transform (DFT) can be applied (Smith, 1997).

18.4.3 Discrete Fourier Transform

We now know that the DFT fits the type of data we deal with when analyzing the EEG in digital signal processing (DSP) settings: sampled at fixed intervals and limited in duration. Now let us get into the details: imagine we have a signal sampled regularly at time intervals Δt during the recording process into N discrete points $x[n]$, where $n = 0, 1, 2, \cdots, N - 1$. This means our signal has a finite length, with a time duration $N\Delta t$, and a sampling frequency $f_s = 1/\Delta t$. The DFT allows us to transform this discrete-time signal into a discrete-frequency representation using the transformation in Eq. 18.4.

$$X[k] = \sum_{n=0}^{N-1} x[n] e^{-i2\pi kn/N}, k = 0, 1, 2, \cdots, N - 1 \tag{18.4}$$

Here, $X[k]$ is the – complex – Fourier coefficient of the time series x at frequency k. In this expression, you can also identify the main terms within the sum operator; one is the signal to decompose, $x[n]$, and the other one is an exponential function $e^{-i2\pi kn/N}$ with a complex argument. Here, we say complex because the term $i = \sqrt{-1}$ is present, as in the coefficient $X[k]$. This may look intimidating at first, but it is another way to represent trigonometric functions in an exponential notation through Euler's formula (Eq. 18.5).

$$e^{i\theta} = \cos\theta + i\sin\theta \tag{18.5}$$

Introducing the Euler notation into Eq. 18.4 lead us to Eq. 18.6, a more familiar expression in line with our previous discussions (remember our dot product!).

$$\begin{aligned} X[k] &= \sum_{n=0}^{N-1} x[n]\cos(2\pi kn/N) - i\,x[n]\ \sin(2\pi kn/N), \quad k \\ &= 0, 1, 2, \cdots, N - 1 \end{aligned} \tag{18.6}$$

This clearly illustrates how the DFT quantifies the alignment of our original signal with the corresponding cosine and sine waves at specific frequencies k.

A valid question at this point might be: How is this set of frequencies k determined? To answer this, we need to understand that the highest frequency that can be extracted from a time series is given by its Nyquist frequency, which is half its sampling frequency (Eq. 18.7).

$$f_N = \frac{f_s}{2} \tag{18.7}$$

And that the number of unique frequencies that can be extracted from an N-length time series – known as the frequency resolution of the signal – equals one-half of the number of data points plus the zero frequency, $N/2 + 1$ (Cohen, 2014). The k frequencies in Eq. 18.6 correspond then to linearly spaced $N/2 + 1$ frequencies between zero and f_N. Here, we have assumed N to be even, which is the most common case in practice.

It's important to mention that the DFT delivers an exact reconstruction of the signal in its frequency domain and is fully reversible without any loss of information. This is done by the inverse Fourier transform (Eq. 18.8).

$$x[n] = \frac{1}{N}\sum_{k=0}^{N-1} X[k]e^{i2\pi kn/N}, n = 0, 1, 2, \cdots, N-1 \tag{18.8}$$

This inverse reconstruction can be particularly useful, for instance, when a signal has been transformed into its frequency domain; some frequency components are rejected, and then the signal is reconstructed. This exemplifies a typical filtering use case.

It is particularly important to know that the DFT, as presented here, only serves didactic purposes. Yes, it delivers exact results that allow the representation of a signal in the frequency space and its reconstruction back to the time domain with zero loss. However, it is hugely inefficient computationally for practical use. There are at least three methods used to implement the DFT for practical applications. First, the problem can be approached as a set of simultaneous equations. The second method, DFT by correlation, is based on detecting a known signal within another – think of correlating the signal with sinusoidal basis functions, and the last, and probably the most popular, is the Fast Fourier Transform (FFT). The FFT is an ingenious algorithm that decomposes a DFT with N points into N DFTs, each with a single point (Smith, 1997). Computationally, the speed of the DFT is proportional to the square of the number of sample points in the signal, N^2, but in the FFT case it is $N \log N$. This could be the difference between 1 min and 1000 min (17 h) (Cohen, 2014)!

18.4.4 Power Spectrum

After performing the Fourier analysis, the next step is to present and interpret its results. As mentioned previously, the Fourier coefficients are complex magnitudes with real and imaginary parts; therefore, they can be represented as in Eq. 18.9.

$$X[k] = X_{Re}[k] + i\, X_{Im}[k] \tag{18.9}$$

From here, we can easily compute the magnitude of each frequency component as in Eq. 18.10; the power is then obtained by taking the square of this value.

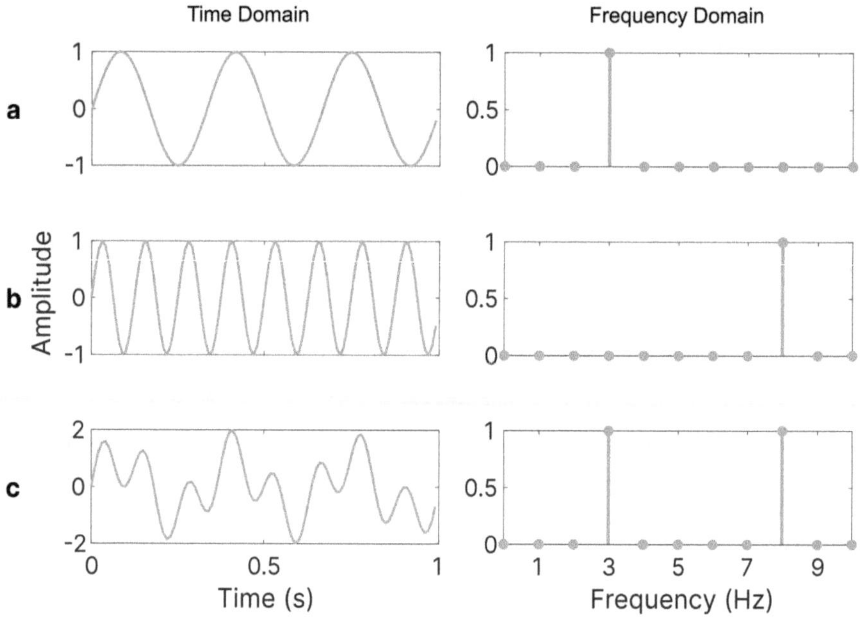

Fig. 18.4 Representation of the Fourier analysis of sinusoids. (**a**) A 3 Hz sine wave (left) and its spectrum (right), showing a single peak at 3 Hz. (**b**) An 8 Hz sine wave (left) and its spectrum (right), showing a single peak at 8 Hz. (**c**) The sum of the 3 Hz and 8 Hz sine waves (left). The corresponding spectrum (right) displays two peaks, reflecting the presence of both frequency components

$$|X[k]| = \sqrt{X_{Re}[k]^2 + X_{Im}[k]^2} \tag{18.10}$$

A common way of visualizing the results from the Fourier analysis is through a 2-D plot showing the values of Eq. 18.10: frequency on the x-axis and power or amplitude on the y-axis (Cohen, 2014) as in Fig. 18.4.

18.4.5 Limitations of Fourier Analysis and Advanced Methods

Fourier analysis is widely regarded as one of the most powerful tools for performing and understanding frequency and time-frequency analysis in signal processing. Nevertheless, it does have its limitations. In this section, we will briefly discuss these limitations and suggest alternative methods that can help mitigate their drawbacks. To complement the discussions in this chapter, we provide a list of further readings to enhance your understanding of the concepts covered.

Fourier analysis assumes that the signal is stationary throughout the analysis window. This means that the statistical properties of the data remain constant over time, and the power representation from the Fourier perspective does not change over time. For EEG signals, this limitation is critical when we analyze processes that change over time, such as the response to external stimuli or due to endogenous processes like the development of an epileptic seizure. This limitation has two practical consequences for our results: first, it reduces the "resolution" of the Fourier analysis, since to represent a more complex frequency structure on a non-stationary signal, more closely spaced frequency components need to be added. Second, we implicitly give away all the temporal information in our signal; there is no description of how the spectrum evolves over time. Of course, this can be overcome if we analyze shorter data segments where we can assume the signal is stationary, and by making this a moving window, we can achieve a dynamic description of the frequency domain over time. This is indeed one of the main rationales behind the short-time Fourier transform (Oppenheim, 1999). However, here we are presented with the trade-off between time and frequency resolution. A shorter window yields better time localization but poor frequency resolution, while a longer window improves frequency discrimination at the cost of smearing timing information (Daubechies, 1992; Mallat, 1999; Samar et al., 1999, Rosenblatt et al., 2014).

The limitations discussed can be addressed by using wavelet transforms (Oppenheim, 1999), which inherently handle non-stationary signals by analyzing EEG data in both time and frequency domains. Unlike the Discrete Fourier Transform (DFT), wavelet analysis does not assume that signals are stationary, allowing the spectral content to change over time. Another alternative approach to the DFT is the Multitaper Analysis (Thomson, 2000), which improves spectral estimates by reducing variance, making it particularly useful for analyzing short or noisy time series.

Further Readings

Addison, P. S. (2002). *The illustrated wavelet transform handbook*. CRC Press.
Cohen, M. X. (2014). *Analyzing neural time series data: Theory and practice*. MIT Press.
Cohen, M. X. (2014). *Fundamentals of time-frequency analyses in Matlab/Octave*. MIT Press.
Smith, S. W. (1997). *The scientist and engineer's guide to digital signal processing*. California Technical.
van Drongelen, W. (2007). *Signal processing for neuroscientists. Introduction to the analysis of physiological signals*. Academic.

References

Berger, H. (1929). Über das elektroenkephalogramm des menschen. *Archiv für Psychiatrie und Nervenkrankheiten, 87*(1), 527–570.

Buzsaki, G., & Draguhn, A. (2004). Neuronal oscillations in cortical networks. *Science, 304*(5679), 1926–1929.

Caton, R. (1875). Electrical currents of the brain. *The Journal of Nervous and Mental Disease, 2*(4), 610.

Cohen, M. X. (2014). *Analyzing neural time series data: Theory and practice*. MIT Press.

Daubechies, I. (1992). *Ten lectures on wavelets*. SIAM.

Fourier, J. (1822). *Theorie analytique de la chaleur*. par M. Fourier, Chez Firmin Didot, père et fils.

Herrmann, C. S., Rach, S., Vosskuhl, J., & Strüber, D. (2014). Time–frequency analysis of event-related potentials: A brief tutorial. *Brain Topography, 27*, 438–450.

Mallat, S. (1999). *A wavelet tour of signal processing*. Elsevier.

Oppenheim, A. V. (1999). *Discrete-time signal processing*. Pearson Education India.

Rosenblatt, M., Figliola, A., Paccosi, G., Serrano, E., & Rosso, O. A. (2014). A quantitative analysis of an EEG epileptic record based on multiresolution wavelet coefficients. *Entropy, 16*(11), 5976–6005.

Samar, V. J., Bopardikar, A., Rao, R., & Swartz, K. (1999). Wavelet analysis of neuroelectric waveforms: A conceptual tutorial. *Brain and Language, 66*(1), 7–60.

Smith, S. W. (1997). *The scientist and engineer's guide to digital signal processing*. California Technical Pub.

Thomson, D. J. (2000). Multitaper analysis of nonstationary and nonlinear time series data. In W. Fitzgerald, R. Smith, A. Walden, & P. Young (Eds.), *Nonlinear and nonstationary signal processing* (pp. 317–394). Cambridge University Press.

Varela, F., Lachaux, J.-P., Rodriguez, E., & Martinerie, J. (2001). The brainweb: Phase synchronization and large-scale integration. *Nature Reviews Neuroscience, 2*(4), 229–239.

Ward, L. M. (2003). Synchronous neural oscillations and cognitive processes. *Trends in Cognitive Sciences, 7*(12), 553–559.

Weisstein, E. W. Dot product. MathWorld—A Wolfram Web Resource. https://mathworld.wolfram.com/DotProduct.html

Wikipedia contributors. (2025, June 21). Pi. Wikipedia. Retrieved June 27, 2025, from https://en.wikipedia.org/w/index.php?title=Pi&oldid=1296692125

Chapter 19
Event-Related Potentials

Michael Hoppstädter

Abstract Event-related potentials (ERPs) are one of the most common applications for EEG. Many ERPs are linked to different sensory or cognitive processes, and they are investigated whenever researchers are interested in brain responses time-locked to an event.

This chapter will enable you to create your own ERPs from pre-processed EEG data and provide you with some background information on the basic principles behind ERPs and how to interpret them. The characteristic shape of the ERP waveform will be discussed, and some typical examples of ERP components will be covered in brief. The chapter will close with a discussion of typical ERP features that can be extracted for statistical analysis.

Keywords ERP · Evoked response · Sensory component · Cognitive component · Epoching · Baseline correction

19.1 Introduction

EEG is a well-established measure and has been around for more than 100 years. While ERPs were also investigated early in the 1930s (Luck, 2014), their rise to prominence only started in the 1960s with publications on the contingent negative variation or CNV (Walter et al., 1964) and the P300 (Sutton et al., 1965).

What Is an ERP The explanation is already in the name—Event-Related Potential. Thus, the focus of interest is an *electrical potential* (since the recording technique is EEG) that is *related to an event*. The term *event* usually refers to stimulation of any sensory modality, for instance, visual, auditory, haptic, olfactory, or even pain. However, the responses evoked by these stimuli are so small that they do not stand out from the ongoing oscillatory activity, and it would be hard to detect and analyze them. Overcoming this issue is rather simple: the response to either the same

M. Hoppstädter (✉)
Brain Products GmbH, Gilching, Germany
e-mail: michael.hoppstaedter@brainproducts.com

T. Warbrick (ed.), *The EEG Handbook*,
https://doi.org/10.1007/978-3-032-20450-9_19

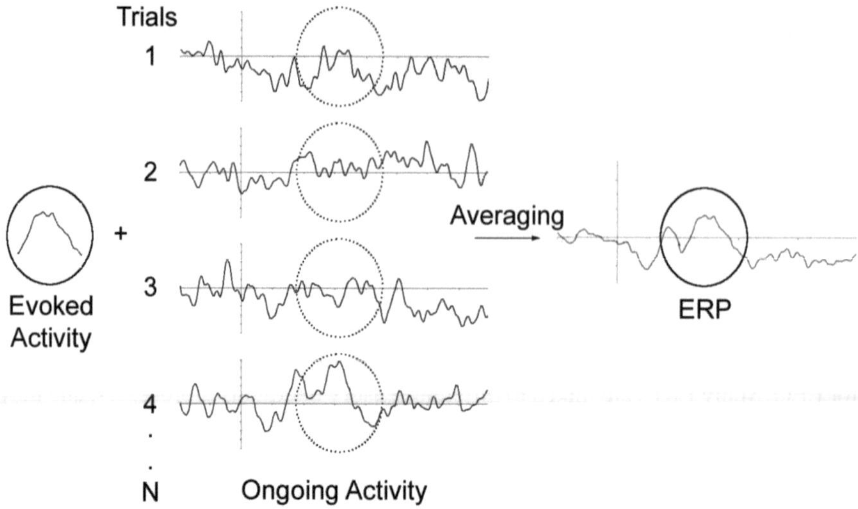

Fig. 19.1 Schematic of the ERP logic. Each trial contains ongoing task-unrelated background activity and evoked activity that is due to the stimulation. Averaging over many repetitions reduces the ongoing activity and amplifies the evoked activity in the final ERP

or similar stimuli is not only measured once but many times, and then all responses are averaged to yield the mean activity related to that type of event—the ERP.

The evoked response to the stimulus is very similar across repetitions as the EEG oscillations are roughly showing the same deflections at the same time. They are *time* and *phase-locked*. Therefore, averaging these time and phase-locked responses amplifies the evoked activity. The ongoing oscillatory background activity, however, will be random across trials as it is neither time nor phase-locked. Such random activity will mostly cancel out through averaging. Hence, the stimulus-unrelated background noise will be attenuated (see Fig. 19.1). From a signal detection perspective, this means that the *signal-to-noise ratio* (SNR) increases by adding more repetitions, as the noise (i.e., task-unrelated variability) is continuously reduced (Luck, 2014). Whether the ERPs are generated by many evoked responses that sum up or through a phase resetting of ongoing background oscillations (Sauseng et al., 2007) is a matter of debate.

19.2 How to Get from EEG to ERPs

A Simple ERP Experiment Since ERPs have been investigated for decades, many classical paradigms are known in the EEG community. An easy and widely used example is the oddball paradigm which has been investigated in many variations in thousands of studies. In its simplest form, you have 2 stimuli, they can be for example different shapes (like X's and O's) or different tones. Imagine the same tone is repeated many times and the participant only needs to listen. Sometimes the sequence will be interrupted with another tone of a different pitch. This second tone occurs more rarely and it cannot be predicted. The data segments around the tones are then averaged separately for frequent and rare tones to get the ERPs for the two conditions. Finally, the two averages are contrasted to reveal any differences in the ERP waveform between conditions. This very easy paradigm is a great example for explaining the logic of ERP experiments, and I will come back to it at different points of this chapter.

How to Measure the Right Responses Two factors are key to this process: precise timing and correct labeling. Timing is important because the brain reacts extremely quickly to external stimuli. To accurately detect the neuronal responses that occur within a few hundred milliseconds, you need to know exactly when a stimulus is presented, or when a response is given. This is achieved by sending *triggers* to the EEG amplifier (see Chap. 14: Triggers). Any time delays (especially unsystematic jitter) must be avoided, and therefore it is essential to properly set up and test your recording pipeline (see Chap. 10: Pilot Testing). It is also crucial to know what has been presented. Coming back to the oddball example, you need to know which of the two tones was presented when. For most experiments, there will be more than two stimuli, thus it is important to know to which category, or *condition,* each stimulus belongs. Otherwise, it would be impossible to average all stimuli of the same condition. Thus, appropriate labeling of the triggers is the second key factor.

19.2.1 How to Process Your ERP Data

19.2.1.1 Pre-processing

It is necessary to treat the raw, continuous EEG data with adequate pre-processing techniques to clean and prepare it for ERP analysis. There is no general standard pre-processing pipeline for ERPs, but some steps are commonly used across many ERP studies. I want to mention a few key factors that play a big role in pre-processing and should be considered when planning the analysis:

Sampling Rate Consider before recording which sampling rate you need to accomplish the planned analyses. It is not necessary to keep a very high rate for ERPs, but do not down-sample too far.

Filtering You need to determine the right filter band that does not distort your ERPs. Consider especially the low cutoff which removes slow frequencies from the data. Also think about whether you need a notch filter (to remove mains noise). See also Chap. 22: Quantifying EEG and ERP Data Quality for a detailed discussion of ERP components and appropriate filter parameters.

Optimal Reference While the online reference is often preset by the hardware, you can always re-reference offline. The final reference for the analysis should be chosen based on the spatial distribution of the ERP effect of interest. Consider the choice of reference in publications using similar paradigms to ensure comparability of your findings.

Artifact Handling In order to avoid contamination of the ERP waveform, it is often necessary to attenuate or reject large artifacts. You should focus on ocular, muscle, and cardiac artifacts and large head or body movements.

Generally, pre-processing should always be adapted to your experiment and analysis goals. See Chap. 17: EEG Pre-processing and Artifact Handling for a detailed discussion of this topic.

19.2.1.2 Trial Selection

Once the continuous EEG data has been cleaned, it is time to select the time intervals that you want to further investigate. A few different names are used to refer to the intervals around experimental stimuli: trials, epochs, and segments are commonly used but they mean the same thing. The procedure is often called epoching or segmentation. For simplicity, I will stick to the term *trial*. If the recording contains markers (as a result of sending triggers during the experiment) that tag the time point and the condition of each stimulation, these can be used to segment the data into smaller chunks, only keeping the experimental trials.

In addition to identifying the right data portions in your recording, you also need to decide how long the trials should be. Generally, the trials should be long enough to capture the full ERP response. In most cases, you will keep a section of data around the stimulation marker, so that you have a *pre-stimulus* and a *post-stimulus period*. The pre-stimulus period is usually used for baseline correction (see Sect. 19.2.1.3) while the post-stimulus period is the interval in which you observe the reaction to the stimulus, i.e., the ERPs. Note that there are exceptions: some ERPs occur prior to the stimulation while others are locked to the time of a response to the stimulus (see Sect. 19.4.3 for examples). Importantly, the timeline of a trial should be the same for all trials that will be averaged together, and also across all conditions that you will compare against each other. The frequency of the stimulus presentation should also be considered when you design the experiment. The inter-stimulus interval (ISI) should be long enough so that successive trials will not overlap. However, this is not always possible, for example, in paradigms with fast stimulation.

19.2.1.3 Baseline Correction

As EEG is constantly fluctuating due to ongoing oscillations, the voltage level is changing over the course of the recording. Hence, the mean voltage between trials can vary considerably. As soon as you want to compare the amplitudes between trials, this becomes a problem. Thus, removing this offset in amplitude level between trials is considered a prerequisite for ERPs, so that all trials share the same *baseline* (i.e., mean amplitude level). The usual procedure is to take the mean of a dedicated baseline interval and subtract that value from all data points in the trial. This will shift the baseline of each trial to a level around zero. The pre-stimulus period is the obvious choice of baseline interval as activity in the time before the stimulation should be relatively low. This is another reason why the ISI should be long enough so that the reaction to the preceding stimulus has waned. In cases when increased activity in the pre-stimulus period is expected, the full trial could be used instead as a baseline interval.

As for the trial length, you also need to decide the length of the baseline interval. It should not be too short so that extreme values could bias the calculation of the mean. But it should also not be too long to keep it limited to a period of little variability. Many published ERP studies use an interval of 200 ms. The reasoning is that this is equal to two periods of an alpha oscillation which is the dominant oscillation in human background EEG. Using multiples of an alpha cycle might therefore reduce the influence of alpha on the ERPs as positive and negative parts of a cycle should cancel out (Luck, 2014).

19.2.1.4 Averaging

The last step in calculating ERPs is averaging the trials. All included trials should be free of artifacts. Otherwise, if you average only a small number of trials, even a single larger artifact can compromise the ERP. Conversely, this is less of an issue if you have many trials that are largely artifact-free. Usually, all trials of one condition are averaged together. However, in some cases, multiple averages could be created for subsections of the experiment. For example, if you have a very long experiment, and you want to compare the responses for early and late trials. Other common approaches to check the reliability of an ERP could be to contrast odd and even trials, or to sample trials at random.

The processing steps discussed so far are required for an ERP analysis. However, there are further optional steps. A common example is artifact handling on the trial level. In fact, it is not unusual to do a coarse artifact screening early on to remove only large-scale artifacts that would otherwise interfere with other pre-processing steps. In this case, it is likely that some subtler artifacts will survive, and a more fine-grained artifact rejection can be done before averaging trials.

19.2.2 Interpreting ERPs

Now that you know how to compute ERPs, let's take a step back and think about what you might want to investigate with an ERP experiment. A classical research question could be that you have two conditions A and B, and you expect a difference in the ERP waveform between the conditions due to some experimental manipulation. Coming back to the oddball example, the two conditions could be frequent and rare tones. Another approach is to explore differences in the same experimental condition between experimental groups, for instance, between healthy controls and patients. Thus, the next step after creating the ERPs would be to compare them to assess the research question.

A common way to explore differences between conditions is to simply overlay both waveforms and compare the differences visually. This can be extended to computing their *difference wave*. If you subtract two waveforms from each other, the result shows how the difference between two conditions (or groups) varies over time. This can help you determine when exactly an ERP effect of interest emerges. You might not only be interested in when but also where an ERP appears. To inspect this, you can plot the difference wave as a *topography*. This can help you locate the spatial focus of an effect (see Sect. 19.3 for more information).

19.2.3 Group Analysis

The average ERP response across all datasets of a study is often called the *grand average* ERP. This is usually calculated for each condition separately. Through the additional averaging step, the SNR will again improve with every added dataset. The grand average ERP can be investigated between conditions and between groups in the same way as the single-subject ERPs (see Sect. 19.2.2). The last step to complete the analysis is to extract features for statistics but this will be discussed later in this chapter (see Sect. 19.5).

19.2.4 Single-Trial Analysis

An alternative to a classical ERP analysis is to stop before averaging the trials and investigate the single trials instead. This can be useful, for example, when you are interested in the variability of a signal over time. However, the consequence of looking at single trials is that they are based on only one observation and therefore the SNR will be much lower due to the higher noise level.

19.3 Characteristics of the ERP and Its Components

Since EEG is generally an oscillating signal, the ERP also presents itself as a waveform pattern of alternating positive and negative-going deflections (see Fig. 19.2). These individual deflections are usually referred to as ERP *components*. However, not every peak is necessarily a component if it has not been related to a specific physiological or cognitive response.

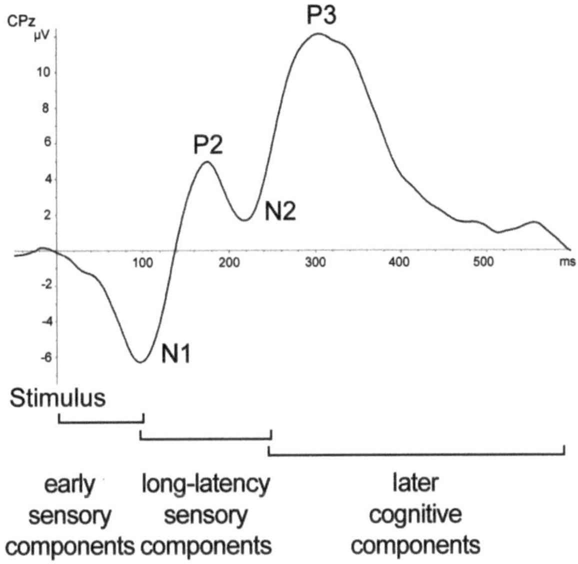

Fig. 19.2 A simplified overview of the waveform pattern of ERP components. The figure is based on an auditory oddball paradigm and the ERP at CPz is displayed. Sensory components and a P3 are visible. Note that negative is plotted downward. The approximate time ranges in which the different classes of ERP components fall are indicated at the bottom

Each time point of the ERP waveform can be described by several parameters. While not every point in the ERP is of equal interest, the start and the peak of specific deflections usually are. They can be characterized by their *onset* and *peak latency* with respect to the stimulus presentation. The *polarity* refers to the direction of the deflection, meaning if it is positive or negative going. Finally, the *amplitude* (i.e., the current value of voltage) and *phase* (i.e., the position in the ongoing oscillation) can be extracted at each time point.

Naming of ERP Components The characteristic shape of alternating positive and negative deflections also inspired the nomenclature of ERP components. The logic is simple: if the component is going in a negative direction the name starts with N, or if positive-going with P. Then the time around which the component usually peaks is added. Thus, a positive-going potential peaking around 300 ms after a stimulus would be called *P300*. Many ERP components are named following this scheme, though sometimes a short form may be used (i.e., P3 instead of P300). This focuses more on the position of the component in the overall sequence of deflections rather than on its peak time which can vary. However, some components have been named by using more phenomenological or functional descriptions, like the *contingent negative variation* or the *mismatch negativity* (see Sect. 19.4 for examples). It makes sense to add a word of caution here: you should always check if an ERP is plotted with negative polarity pointing upward or downward as both options are used in the ERP literature.

Topographies Another characteristic of ERP components is that they usually have a typical localization. Hence, it is very common to look at topographies which map the voltage distribution on the scalp. These can be created for single points, like the onset or the peak of a component, and you can follow the change in spatial distribution over time. You can also map the mean amplitude of an ERP component. This type of plot is commonly found in the results section of any ERP paper (for more information on typical EEG figures, see Chap. 36: How to Evaluate an EEG Research Paper, and Fig. 36.2 for an example of a topography).

Variability Importantly, these characteristics are not absolutely fixed. While an ERP component has a *typical* onset, peak latency, and topography, those parameters can vary with many experimental variables. The specifics of the paradigm, the stimulus timing, the type of stimuli, and sensory modality are just a few examples. Also, the characteristics of the tested participant group can affect ERPs, for example, their age or if they have a psychiatric condition. Some aspects of data processing such as the referencing scheme or the number of trials contributing to the ERP can influence the appearance of the final ERPs as well.

19.4 Commonly Investigated ERP Components

This section provides some examples of commonly investigated ERP components. The list is not exhaustive, but it mentions many ERP components that you are likely to encounter in the ERP literature. If you are looking for more in-depth information about a component, there are some great books focused on ERP components (see for instance Luck, 2014; Handy, 2005; Luck & Kappenman, 2011).

One way of classifying ERP components is by looking at the time window in which they occur. First, there are the *early sensory components* occurring within roughly the first 100 ms after the stimulus that are elicited in response to auditory and visual stimuli. Then follows a range of *long-latency sensory components* within the first 200 ms. These react to the stimulus characteristics, and they vary in their shape depending on sensory domain. Later potentials can be described as more *cognitive components*. They are task-dependent and correlate with higher cognitive functions. These components can also be labeled as *endogenous,* while those components that are automatically triggered through a sensory stimulus are *exogenous* (Luck, 2014). Earlier components have a rather small amplitude but a high frequency, while the later components usually show larger amplitudes at a much slower frequency.

19.4.1 Early Sensory Components

Auditory brainstem responses (ABRs) are the earliest auditory component and will be automatically and reliably evoked when auditory stimulation with clicks is used. They reflect information processing on the auditory pathway from the cochlea through the brainstem to the thalamus. There are usually up to seven peaks in roughly millisecond distance labeled I to VII.

Table 19.1 Examples of early sensory ERP components

Component	Timing w.r.t. stimulus onset	Distribution	Common experimental observations
ABR	Within 10 ms.	Usually recorded from vertex with mastoid reference.	Measured to assess integrity of auditory pathway. Usually sampled at 20 kHz. Shape depends on participant and stimulus characteristics, e.g., higher intensity leads to shorter peak latencies, higher frequency stimulation results in longer peak latencies (Pratt, 2012).
C1	Peaks ~100 ms. Might overlap with P1 if C1 is positive.	Strongest at posterior midline sites (e.g., Oz, Pz).	Originates in V1. Elicited by any visual stimulus. Can be affected by sample characteristics, e.g., smaller amplitude and longer latency in schizophrenic patients (Schechter et al., 2005).

The earliest visual response is the *C1* wave which originates from processing in the primary visual cortex (V1). It can be of either polarity (i.e., negative or positive) depending on where in the visual field a stimulus is presented (Jeffreys & Axford, 1972).

Refer to Table 19.1 for a summary of early sensory components.

19.4.2 Long-Latency Sensory Components

These components are related to the processing of stimulus characteristics. They are not strictly specific to a sensory domain, but I focus on examples from auditory and visual stimulation since they are most commonly used.

The *P100* relates to early visual processing (like the C1) of a stimulus, but further downstream of V1 in the visual pathway. Both the visual and auditory *N100* are widely studied in early sensory processing and have been linked to the detection of change.

The *N170* can be considered a special version of a visual N100 linked to the processing of human faces, or to other stimuli for which the participant has acquired expert knowledge, e.g., birds (Tanaka & Curran, 2001).

The *P200* has been linked to processing basic visual target features of a stimulus, and it reacts similarly to a P300 (see Sect. 19.4.3).

In the N200 range, you can find several sub-components, like the *N2a* which is also known as *Mismatch Negativity (MMN)* and the *N2b***,** which has been linked to response inhibition.

Refer to Table 19.2 for a summary of long-latency sensory components.

Table 19.2 Examples of long-latency sensory ERP components

Component	Timing w.r.t. stimulus onset	Distribution	Common experimental observations
P100	Peaks 100–130 ms. Might overlap with C1 (if C1 is positive).	Lateral occipital maximum.	Elicited through any visual stimulation. Sensitive to stimulus characteristics.
N100	Peaks ~100 ms. Anterior peak precedes posterior peak.	Widespread with anterior and posterior peaks.	Investigated with basic stimuli, like pure tones or a flickering checkerboard. Suffers from refractory effects, i.e., smaller amplitudes for shorter ISIs.
N170	Peaks 150–200 ms.	Parieto-occipital maximum and bilateral.	Increased amplitude to the perception of faces (Eimer, 2011; Rossion & Jacques, 2012). Usually accompanied by central positivity (*vertex positive potential*). Originates from activity in the fusiform and occipital face areas (Haxby et al., 2000).
P200	Peaks ~200 ms.	Fronto-central maximum.	Can be investigated with oddball or visual search paradigms. More pronounced for target features in infrequent visual stimuli.
N2a/MMN	Peaks ~200 ms.	Strongest at fronto-central midline sites.	Mismatch detection in a sequence of standard stimuli, usually tested with pure tones. Elicited automatically without the need to consciously attend the stimulus. Amplitude increases with stronger deviation from standard stimulus (Näätänen & Kreegipuu, 2012).
N2b	Peaks ~200 ms.	Anterior maximum.	Measure of response inhibition, classically investigated with Go/No-Go paradigm. More negative amplitude with increasing frequency of Go stimulus (Bruin & Wijers, 2002).

19.4.3 Later Cognitive Components

Numerous components fall into this category as ERPs have been used to investigate many different cognitive processes. I have selected a few prominent examples.

The *P300* (or more precisely the P3b component) is used similarly to the MMN as a marker of change detection in stimulus sequences, however, the task has to be relevant (contrary to the MMN).

Two components have been investigated intensively in language research. While the *N400* is considered a proxy for semantic congruency, the *P600* serves as a marker of syntactic violations.

ERPs have also been related to the recruitment of resources for processing emotional stimuli. The *LPP (late positive potential)* is thought to reflect the allocation of additional attention to emotionally relevant stimuli (Nordstrom & Wiens, 2012).

In the learning phase of long-term memory experiments, one can investigate the *Dm effect* (*difference due to memory*, also known as *subsequent memory effect*). This involves contrasting the ERP in response to items that were later successfully remembered minus the ERP in response to later forgotten items and yields a widespread positivity.

There are some ERPs related to motor responses, which are typically slow waves that evolve over longer time intervals. As an example, the *LRP* (*lateralized readiness potential*, also known by its German name *Bereitschaftspotential*) is a slowly varying negative deflection in anticipation of a motor response.

The *ERN* (*error-related negativity*) is different from all aforementioned ERP components because it is related to a response rather than the stimulation. Hence, it is a *response-locked potential*. The ERN is thought to reflect error monitoring and cognitive control (see Gehring et al. (2012) for an overview).

Refer to Table 19.3 for a summary of later cognitive components.

19.4.4 ERP Components in Multimodal Recording Scenarios

ERPs are not always recorded by themselves, but also in multimodal recording scenarios. This enables you to create ERPs based on events other than experimental triggers. Many scenarios are possible, but I want to mention three examples:

EEG and ECG If you add an electrocardiogram (ECG) to the EEG recording, you can investigate *heartbeat-evoked potentials,* a marker of neural interoceptive processing of cardiac activity (Schandry et al., 1986). The idea is to detect each R-peak of the QRS complex and use this to align the EEG signal with the heartbeat. This ERP peaks between 250 and 450 ms after the heartbeat in frontal regions.

EEG and TMS With transcranial magnetic stimulation (TMS) of the motor cortex, you can provoke a motor response and use the EEG to record *motor-evoked potentials*. However, you can also align the EEG with the trigger of the TMS pulse to generate *transcranial-evoked potentials*. These produce a complex waveform, and they are considered a measure of cortical reactivity (Hernandez-Pavon et al., 2023).

EEG and Eye Tracking Eye trackers allow you to check where exactly your participant is looking. You can use this information to exclude trials in which a stimulus was not attended, thus improving the validity of your ERP. It is also

Table 19.3 Examples of later cognitive ERP components

Component	Timing w.r.t. stimulus onset	Distribution	Common experimental observations
P300/P3b	Peaks ~300 ms.	Widespread centro-parietal maximum.	Classically investigated with the oddball paradigm, both in auditory (e.g., pure tones) and visual (e.g., X's and O's) domain. Larger amplitude with increasing deviancy and decreasing frequency of targets compared to standards (Polich, 2004).
N400	Peaks ~400 ms.	Centro-parietal maximum, usually shifted to the right hemisphere.	Typically examined using linguistic materials (written or spoken or combined). The effect can also be demonstrated with non-linguistic materials like meaningful line drawings (Kutas & Federmeier, 2000). Amplitude scales with the distance of the target word from the semantic context.
P600	Peaks ~600 ms.	Posterior maximum.	Investigated with either written or spoken language materials, usually full sentences. Larger for sentences with non-grammatical structure, e.g., words appearing in incorrect position or use of the wrong verb form.
LPP	Similar time window to P300 but more extended.	Centro-parietal maximum.	Observed in response to emotionally arousing stimuli (usually complex visual scenes). Responses to both positive or negative emotional scenes show increased amplitudes compared to neutral ones (Liu et al., 2012).
Dm-effect	Wider positivity between 400 and 800 ms.	Centro-parietal maximum, influenced by stimulus category (e.g., pictures vs. words).	Connected to successful encoding of items to long-term memory (Wilding & Ranganath, 2012). Replicated for a variety of stimuli (e.g., spoken or written words, visual objects).
LRP	Starts ~1 s before the motor response.	Fronto-central maximum.	Lateralization and polarity depend on which motor response is carried out, e.g., left vs. right hand, or

(continued)

Table 19.3 (continued)

Component	Timing w.r.t. stimulus onset	Distribution	Common experimental observations
			hand vs. foot (Brunia et al., 2012). Considered a marker of movement preparation but it can also be observed when the movement is inhibited (Smulders & Miller, 2012).
ERN	Starts before the response. Peaks ~100 ms after the response.	Midline fronto-central maximum.	Time-locked to the response. Usually observed in difference wave between correct responses minus incorrect responses. For partial errors (e.g., wrong button press inhibited and corrected), the amplitude falls in between full errors and correct trials. Corrected errors show a larger amplitude than uncorrected ones.

possible to obtain *fixation-related potentials*. This means to align the EEG with the fixation marker from the eye tracker instead of using the stimulus trigger. This provides a better time estimate of the beginning of stimulus processing (Hutzler et al., 2007).

19.5 Extraction of ERP Features

Once your ERP analysis is complete, you usually want to investigate the effects statistically. This means checking for significant differences between two or more conditions, groups, or EEG channel locations. Features of interest differ across studies, and I will discuss a few common examples below (see also Fig. 19.3).

Peak-Related Features It is possible to base the comparison on the peak of an ERP component, i.e., the single point with the maximum amplitude of the deflection. This *peak amplitude* (Fig. 19.3a) can be extracted for all conditions, groups, and sensors of interest and statistically tested. Note that the peak of an ERP component is likely to fall into a similar time range across electrodes or compared datasets, but it is unlikely to coincide perfectly. The *peak latency* (Fig. 19.3b) can therefore also be a variable of interest.

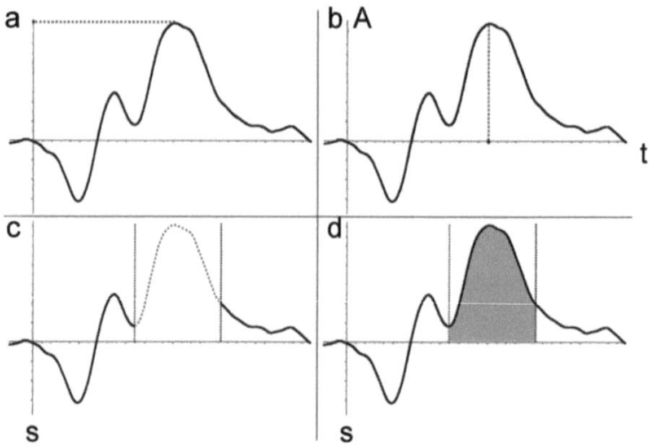

Fig. 19.3 Overview of typical measures extracted from ERPs based on a P3 component. From the peak of a component, both the peak amplitude (**a**) and the peak latency (**b**) can be determined. Aggregated measures like the mean amplitude (**c**) or the area under the curve (**d**) can be computed within a defined time window around the component. *S* Stimulus, *A* Amplitude, *t* Time

ERPs can still contain some higher frequency oscillations, which means that the data can fluctuate around the peak. To avoid picking just one of many small "peaks", it is common to average across neighboring data points around the maximum (see also next section). This filters out the high-frequency noise but still does not guarantee that you selected the "true" peak among many small fluctuations (Luck, 2014; see also Chap. 22: Quantifying EEG and ERP Data Quality, specifically Sect. 22.2.5).

Sometimes, the distance or amplitude difference between two points could be of interest, for instance between two neighboring peaks (*peak-to-peak*) or between a peak and the preceding depression (*trough-to-peak*).

Mean Amplitude and Area Measures Another common way to extract features of an ERP is to aggregate values over time. This could be the *mean amplitude* (Fig. 19.3c) of a specific time interval related to an ERP component. However, this works best for shorter and well-defined components (thus not optimal for slow waves). Ideally, the time range should not be selected based on the visually apparent difference in the data itself as this would be circular reasoning (Keil et al., 2014). You could define it based on prior observation, for instance, by running a pilot study, or consider other studies on the same topic. Another way to optimize this is to apply data-driven methods. For instance, you can use non-parametric permutation tests to find time windows and locations to probe afterwards with more classical statistical tests (Maris & Oostenveld, 2007).

The benefit of using mean amplitudes is that you counteract the high-frequency noise issue mentioned above. However, a mean amplitude can be biased by extreme values within the time window. Also, if the time window contains both negative and positive amplitudes, they will cancel each other out when computing the mean. A way around this issue is to use area measures, like the *area under the curve* (Fig. 19.3d). This refers to the area between the curve and the x-axis which can be considered as always being positive so that negative and positive parts of the curve would not cancel out.

19.6 Conclusion

This chapter introduced the concept of ERPs, their characteristic waveform, and a selection of commonly investigated components. ERPs have been studied extensively for more than half a century and have proven to be a powerful tool to explore sensory and cognitive processing. At the same time, the logic behind the analysis is easy to grasp: cutting the data into segments around the events, removing offsets, and averaging over segments. The results can be communicated through a variety of output measures and compelling figures. Thus, ERPs provide a simple yet elegant solution to investigate evoked brain responses with high temporal resolution. The ERP method is still widely popular in EEG research as reflected by the prolific publication of new studies. Hence, ERPs should be an essential part of the EEG researcher's analysis toolkit.

References

Bruin, K. J., & Wijers, A. A. (2002). Inhibition, response mode, and stimulus probability: A comparative event-related potential study. *Clinical Neurophysiology, 113*, 1172–1182.

Brunia, C. H. M., Van Boxtel, G. J. M., & Böcker, K. B. E. (2012). Negative slow waves as indices of anticipation: The Bereitschaftspotential, the contingent negative variation, and the stimulus-preceding negativity. In E. S. Kappenman & S. J. Luck (Eds.), *The Oxford handbook of event-related potential components*. Oxford University Press.

Eimer, M. (2011). The face-sensitive N170 component of the event-related brain potential. In *The Oxford handbook of face perception* (Vol. 28, pp. 329–344). Oxford University Press.

Gehring, W. J., Liu, Y., Orr, J. M., & Carp, J. (2012). The error-related negativity (Ern/Ne). In E. S. Kappenman & S. J. Luck (Eds.), *The Oxford handbook of event-related potential components*. Oxford University Press.

Handy, T. C. (2005). *Event-related potentials: A methods handbook*. MIT Press.

Haxby, J. V., Hoffman, E. A., & Gobbini, M. I. (2000). The distributed human neural system for face perception. *Trends in Cognitive Sciences, 4*, 223–233.

Hernandez-Pavon, J. C., Veniero, D., Bergmann, T. O., Belardinelli, P., Bortoletto, M., Casarotto, S., Casula, E. P., Farzan, F., Fecchio, M., Julkunen, P., Kallioniemi, E., Lioumis, P., Metsomaa, J., Miniussi, C., Mutanen, T. P., Rocchi, L., Rogasch, N. C., Shafi, M. M., Siebner, H. R., Thut, G., Zrenner, C., Ziemann, U., & Ilmoniemi, R. J. (2023). TMS combined with EEG: Recommendations and open issues for data collection and analysis. *Brain Stimulation, 16*, 567–593.

Hutzler, F., Braun, M., Võ, M. L. H., Engl, V., Hofmann, M., Dambacher, M., Leder, H., & Jacobs, A. M. (2007). Welcome to the real world: Validating fixation-related brain potentials for ecologically valid settings. *Brain Research, 1172*, 124–129.

Jeffreys, D. A., & Axford, J. G. (1972). Source locations of pattern-specific components of human visual evoked potentials. I. Component of striate cortical origin. *Experimental Brain Research, 16*, 1–21.

Keil, A., Debener, S., Gratton, G., Junghofer, M., Kappenman, E. S., Luck, S. J., Luu, P., Miller, G. A., & Yee, C. M. (2014). Committee report: Publication guidelines and recommendations for studies using electroencephalography and magnetoencephalography. *Psychophysiology, 51*, 1–21.

Kutas, M., & Federmeier, K. D. (2000). Electrophysiology reveals semantic memory use in language comprehension. *Trends in Cognitive Sciences, 4*, 463–470.

Liu, Y., Huang, H., McGinnis-Deweese, M., Keil, A., & Ding, M. (2012). Neural substrate of the late positive potential in emotional processing. *The Journal of Neuroscience, 32*, 14563–14572.

Luck, S. J. (2014). *An introduction to the event-related potential technique*. MIT Press.

Luck, S. J., & Kappenman, E. S. (2011). *The Oxford handbook of event-related potential components*. Oxford University Press.

Maris, E., & Oostenveld, R. (2007). Nonparametric statistical testing of EEG- and MEG-data. *Journal of Neuroscience Methods, 164*, 177–190.

Näätänen, R., & Kreegipuu, K. (2012). The mismatch negativity (MMN). In E. S. Kappenman & S. J. Luck (Eds.), *The Oxford handbook of event-related potential components*. Oxford University Press.

Nordstrom, H., & Wiens, S. (2012). Emotional event-related potentials are larger to figures than scenes but are similarly reduced by inattention. *BMC Neuroscience, 13*, 49.

Polich, J. (2004). Clinical application of the P300 event-related brain potential. *Physical Medicine and Rehabilitation Clinics of North America, 15*, 133–161.

Pratt, H. (2012). Sensory ERP components. In E. S. Kappenman & S. J. Luck (Eds.), *The Oxford handbook of event-related potential components*. Oxford University Press.

Rossion, B., & Jacques, C. (2012). The N170: Understanding the time course of face perception in the human brain. In E. S. Kappenman & S. J. Luck (Eds.), *The Oxford handbook of event-related potential components*. Oxford University Press.

Sauseng, P., Klimesch, W., Gruber, W. R., Hanslmayr, S., Freunberger, R., & Doppelmayr, M. (2007). Are event-related potential components generated by phase resetting of brain oscillations? A critical discussion. *Neuroscience, 146*, 1435–1444.

Schandry, R., Sparrer, B., & Weitkunat, R. (1986). From the heart to the brain: A study of heartbeat contingent scalp potentials. *The International Journal of Neuroscience, 30*, 261–275.

Schechter, I., Butler, P. D., Zemon, V. M., Revheim, N., Saperstein, A. M., Jalbrzikowski, M., Pasternak, R., Silipo, G., & Javitt, D. C. (2005). Impairments in generation of early-stage transient visual evoked potentials to magno- and parvocellular-selective stimuli in schizophrenia. *Clinical Neurophysiology, 116*, 2204–2215.

Smulders, F. T. Y., & Miller, J. O. (2012). The lateralized readiness potential. In E. S. Kappenman & S. J. Luck (Eds.), *The Oxford handbook of event-related potential components*. Oxford University Press.

Sutton, S., Braren, M., Zubin, J., & John, E. R. (1965). Evoked-potential correlates of stimulus uncertainty. *Science, 150*, 1187–1188.

Tanaka, J. W., & Curran, T. (2001). A neural basis for expert object recognition. *Psychological Science, 12*, 43–47.

Walter, W. G., Cooper, R., Aldridge, V., McCallum, W., & Winter, A. (1964). Contingent negative variation: An electric sign of sensori-motor association and expectancy in the human brain. *Nature, 203*, 380–384.

Wilding, E. L., & Ranganath, C. (2012). Electrophysiological correlates of episodic memory processes. In E. S. Kappenman & S. J. Luck (Eds.), *The Oxford handbook of event-related potential components*. Oxford University Press.

Chapter 20
EEG Source Analysis

Alejandro Ojeda

Abstract EEG source analysis enables the study of cognitive and physiological processes closer to their origin within the brain. To that end, it combines anatomical information, usually derived from participant-specific or template MRI segmentations of the head and inner layers of tissue, with EEG signals using mathematical methods that transform the latter into a functional brain imaging modality. This chapter will touch on the fundamentals of EEG source analysis, including its modeling assumptions, representative source localization algorithms, and source-level statistical inference and connectivity.

Keywords Source imaging · Head model · Forward and inverse problems · Regularization · Constraints · Statistical analysis and connectivity in the source space · Multiple comparison correction · Source summarization

20.1 Introduction

As mentioned in Chap. 2, What is EEG?, the EEG is a voltage signal that can be measured on the scalp surface and originates from the electrical activity of cortical neurons. The layer of pyramidal neurons in the cortex is thought to be the primary contributor to the EEG (and MEG) signal because the spatial orientation of these neurons resembles equivalent current dipoles. When enough localized populations of these neurons fire synchronously, they can produce extracellular currents large enough to spread across different layers of conductive tissue (e.g., cerebrospinal fluid (CSF), bone, and skin) and can be measured with voltage sensors on the scalp. These conductive layers of tissue form the volume conductor, which has the effect of mixing currents produced by different neural populations. As a result, an EEG sensor on the scalp picks up a voltage signal composed of contributions from multiple

A. Ojeda (✉)
Brain Vision LLC, Garner, NC, USA
e-mail: alejandro.ojeda@brainvision.com

T. Warbrick (ed.), *The EEG Handbook*,
https://doi.org/10.1007/978-3-032-20450-9_20

neural sources scattered over the cortical surface (and not only those directly underneath the sensor).[1]

Typically, we use electrical models of the head to describe how neural signals propagate and mix on their way from their source to the EEG sensors. Depending on the assumptions, these models can furnish different variants of Maxwell's equations, which we can solve for simulated, unitary source activities (see Table 20.1). This is known as solving the forward problem of the EEG. The solution of which is the so-called lead field matrix, which encodes the contribution of a unitary current dipole normal to the cortical surface at the *j*th location to the *i*th scalp sensor. If source direction is included in the model, the contributions along the *x*, *y*, and *z* axes are also solved for and included in the lead field.

Terminology

Volume conductor	Electrically conductive tissue within the head through which neural currents propagate (e.g., brain, cerebrospinal fluid, bone, and skin).
Volume conductor effect	It is the mixing of extracellular currents produced by distant as well as nearby neural populations; therefore, a voltage sensor on the scalp picks up a signal composed of contributions from multiple neural sources.
Head model	Electrical model of the head that describes how neural currents propagate from their sources to the sensors. The model usually includes information and assumptions about the geometry and conductivity of the tissue.
Sensor space	Component of the head model that defines the sensor positions with respect to the head. For example, sensor positions could be digitized and co-registered with a skin segmentation derived from the participant's structural MRI.
Source space	Component of the head model where neural sources are defined. For example, in realistic head models, one can use grey matter segmentation derived from structural MRI to constrain sources to the grey matter. Another constraint could be to confine the sources to the cortical surface only.
Forward problem	Given the equations derived from the head model, one finds the voltage projection on the sensor space produced by simulated unitary current dipoles placed on the source space.
Lead field matrix	Matrix that contains the solution to the forward problem for a dense array of sources. Each row in this matrix represents the contribution of every unitary source to the given sensor. For example, the element $k_{i,j}$ represents the voltage induced by the *j*th unitary current dipole onto the *i*th sensor.

[1] In addition to cortical sources, there is evidence that subcortical sources contributed to the EEG signals (Seeber et al., 2019).

Inverse problem	Estimate the sources that generated a set of scalp voltage measurements.
EEG source imaging (ESI)	The solution of the inverse problem of the EEG in a dense source space. In this case, the output of the inverse algorithm is a map of estimated current source density values. The resulting map can be rendered into an image (electrical tomography) or used for subsequent statistical analyses.
Equivalent current dipole (ECD)	A dipole model that summarizes the electrical activity of an underlying source configuration of unknown spatial extent. Dipole fitting algorithms utilize this model to determine the location of a few equivalent sources that best explain a given scalp topography.
Constraints	Set of mathematical restrictions used to bias the solution of an inverse algorithm toward biologically plausible source configurations.
Regularization parameter	A parameter used to obtain a unique inverse solution. It can be interpreted as a trade-off between the reconstruction error and the weight assigned to the constraints.
Euclidean norm	Square root of the sum of squares of a vector's components. This norm is typically used to represent the reconstruction error of an inverse algorithm when solving for a single source vector.
Frobenius norm	It is an extension of the Euclidean norm to matrices, and it is used to represent the reconstruction error of an inverse algorithm when solving for multiple source vectors simultaneously.
Type I error	When statistically testing a hypothesis, this is the error of rejecting the null hypothesis when it is true. This is also known as a false positive.
Multiple comparisons problem	When performing inference on the source space, the larger the number of sources, the more likely it is to find erroneous effects, therefore, incurring a Type I error.
Anatomical connectivity	Indicates connectivity at the structural level (axons and bundles of axons connecting different brain regions).
Functional connectivity	Indicates an instantaneous relationship (correlation) between brain regions.
Effective connectivity	Indicates a causal relationship between brain regions.

As long as the geometrical and electrical properties of the participant's brain don't change significantly (due to aging, injury, illness, etc.), the lead field matrix can be calculated once and then used for estimating the neural sources that most likely generated an observed brain topography for an experimental condition of interest. This is known as solving the inverse problem of the EEG. The solution to this problem turns the EEG into a functional brain imaging modality by mapping the signals measured on the scalp onto anatomically relevant brain structures (see Fig. 20.1). Thereby allowing researchers to study behavioral, physiological, and cognitive responses linked to brain function and anatomy.

Table 20.1 Common modeling approaches to solve the forward problem of the EEG

Method	Summary	Pros	Cons	References
Spherical	The sources are distributed over a sphere, typically fitted to the cortical surface extracted from participant's MRI segmentation. The layers representing the skull and scalp are modeled accordingly, increasing the radius parameter and using a ratio of fixed conductivities for each layer.	Has analytical solution. Serves as benchmark for more sophisticated methods.	Inaccurate.	Hallez et al. (2007)
BEM	Boundary Element Method (BEM): MRI segmentations are used to extract cortex, skull, and scalp surfaces and define compartments of homogeneous isotropic conductivity. The sources are modeled as dipoles distributed over the cortical surface. Potential distributions are solved numerically on the boundary of each compartment.	Captures the geometry of the head better than the spherical model.	Does not model inhomogeneous and anisotropic conductivities and can be challenging to apply on surfaces with holes.	Stenroos and Sarvas (2012) and Gramfort et al. (2010)
FEM	Finite Element Method (FEM): MRI is used to discretize the entire volume into small elements. The anisotropy and inhomogeneity of the tissue can be modeled if Diffusion Tensor imaging (DTI) data is available. The potentials are solved numerically for each element.	Can accurately model any head shape and manage heterogeneous and anisotropic conductivities.	The method has singularities that need to be handled by the numeric solver.	Wolters et al. (2007)

Having an EEG source signal representation in a well-defined anatomical space opens the door to statistical analyses of individuals and groups in the source space. Furthermore, ERP, frequency-domain, correlation, and causality studies can be

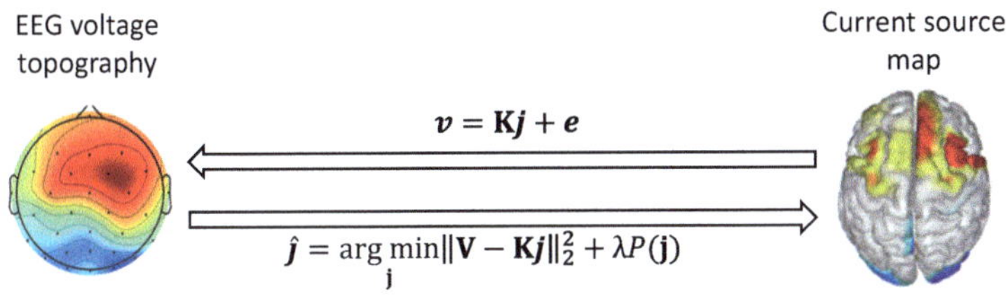

Fig. 20.1 Schematic representation of EEG forward and inverse problems. The right-to-left arrow represents the EEG voltage generation model; source dipole currents on the right spread throughout the tissue, producing a voltage pattern on the left. The left-to-right arrow indicates the inverse problem: the estimation of the most likely current source pattern that could have generated the observed EEG voltage scalp pattern

carried out entirely in the source space. In the rest of the chapter, we will delve further into (1) the fundamentals of EEG source imaging, (2) statistical inference in the source space, and (3) source connectivity.

20.2 Fundamentals of EEG Source Imaging

In this chapter, we distinguish between source imaging and source localization. Both families of algorithms start with a scalp topography typically representing EEG voltages (or EEG voltage components if a decomposition such as ICA was previously applied (Debener et al., 2010)). Then, by source imaging, we mean that the algorithm's output is an image (or map) of current source density values that can be rendered in the source space. By source localization, however, we mean that the output is a discrete set of points indicating the center of the equivalent current dipoles that generated the topography. The key difference between the two is that for imaging, the source model is defined using a dense set of points distributed on a cortical surface or 3D grid. While for localization, the source model is sparse, and we identify the location of a handful of equivalent current dipoles that provide a plausible explanation for the scalp topography. See Fig. 20.2.

The discrete source model has been successfully applied to localize auditory and visual ERPs (Makeig et al., 1997, 2002; Debener et al., 2008). However, despite its simplicity, the sparse nature of this source space complicates further statistical analysis because equivalent dipoles may not be found in the same locations across conditions or subjects. Usually, it requires working on clusters of sources and other techniques (Bigdely-Shamlo et al., 2013). Dealing with this type of analysis is beyond the scope of this chapter, so from now on we will focus on distributed EEG source Imaging (ESI).

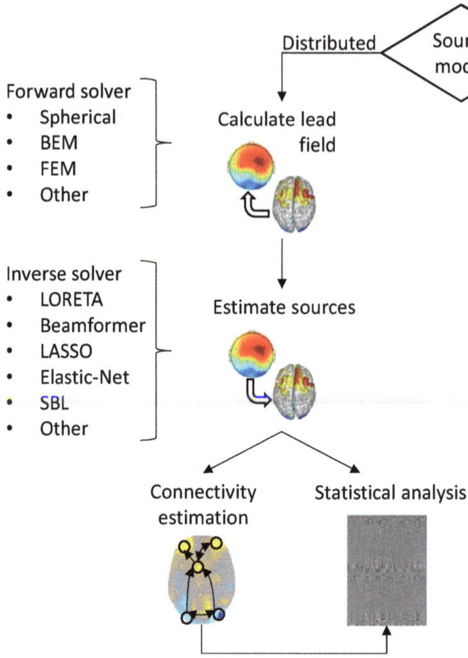

Fig. 20.2 Typical EEG source analysis flowchart. The source model determines the subsequent pipeline. If the model is discrete, we need to go the route of dipole modeling and clustering and decide how to perform subsequent single-participant and group analyses on dipoles or clusters of dipoles. If the source model is distributed, we build a participant-specific or template head model and calculate the lead field using a forward solver (see Table 20.1). Next, we use an inverse solver to estimate the source maps for selected time points of interest in the EEG (see Table 20.2). Then, we can perform connectivity analysis on the source time series. Statistical analyses can be performed on sources or connectivity features at the participant level or in groups. An important caveat for group analyses is that unless we use the same template head model for all participants, we need to co-register participant source spaces to ensure that the features on which we perform statistics correspond to the same anatomical structures

20.3 Forward Problem

The typical ESI pipeline starts by defining a source space. If participant-specific MRI segmentations are available, a personalized head model is constructed. Otherwise, a generic one can be used for all participants based on a population template MRI. We will revisit the distinction between participant-specific and template-based head models later when we discuss the statistical analysis in the source space. The head model is built according to the forward solver of choice. For BEM, we need to provide the cortex, in-skull, out-skull, and skin surfaces, and we solve for sources placed on the vertices of the cortical surface. For FEM, grey matter (usually cortex only), CSF, and head segmentations are needed, and we solve for a 3D grid of sources placed on the grey matter. A spherical model solver can work on a surface and a 3D grid source space. Next, we need to co-register the MRI-derived head model with the position of the sensors where voltages were measured. Notwithstanding that every approximation can introduce errors, if participant-specific sensor positions are missing, we can use a standard montage (that includes our sensors as a subset) placed on the head surface. Then we proceed to calculate the lead field matrix.

20.4 Source Estimation

The relationship between scalp voltages and current sources in the brain is referred to as the measurement model, and can be expressed by a linear equation (Dale & Sereno, 1993),

$$v = Kj + e \tag{20.1}$$

where v represents a vector of sensor voltages at a given time point, K is the lead field matrix, j is a vector of current density values on the source space, and e is the measurement noise usually assumed normally distributed. The normality of the sensor noise rests on the assumption that gross artifacts were removed in pre-processing. So, to estimate the source vector j, we need to invert this equation (solve the inverse problem of the EEG, see Fig. 20.1). A key aspect of this inverse problem is that there are many more sources than sensors, so without further assumptions, there are infinite source configurations that can render the same scalp topography. To tackle this problem, we seek to estimate a source vector $\widehat{j}$ that best explains the measurements subject to additional constraints,

$$\widehat{j} = \arg\min_{j} \|v - Kj\|_2^2 + \lambda P(j) \tag{20.2}$$

where the first term ($\|v - Kj\|_2^2$) is the data fit, taken as the square of the $l2$-norm (Euclidean norm) of the reconstruction error, the second is a penalty function that biases the solution toward a source configuration that exhibits some desired features, and λ is a regularization parameter that regulates the weight given to the penalty with respect to the data fit term. The regularization parameter can be determined by minimizing the generalized cross-validation curve (Golub et al., 1979), the L-curve method (Hansen, 1992), or by grid search. Equation 20.2 belongs to a class of problems called Regularized Least Squares (RLS), and many popular inverse algorithms can be written in this framework, some of which we summarize in Table 20.2. See Michel and Brunet (2019) for an extended list of popular EEG source analysis tools.

Thanks to the normality of the measurement noise, even many Bayesian inversion methods can be represented in the RLS framework and vice versa. In the Bayesian framework, the error term is derived from a data likelihood function, the penalty

Table 20.2 Shows a non-exhaustive collection of popular inverse methods from the RLS family

Method	Summary	Pros	Cons	References
MNE	Minimum norm estimation (MNE) is obtained with $P(j) = \|\|j\|\|_2^2$.	It has a closed-form solution.	It is biased toward superficial sources and sensitive to noise.	Hamalainen and Ilmoniemi (1994)
wMNE	Weighted MNE is obtained with $P(j) = \|\|Wj\|\|_2^2$, where the matrix W acts as a depth normalization operator.	It has a closed-form solution and can recover superficial and deeper sources.	It is sensitive to noise.	Lin et al. (2006)
LORETA	Low-resolution electro-magnetic tomography (LORETA) is obtained with $P(j) = \|\|Lj\|\|_2^2$, where L is a discrete Laplacian operator that enforces correlation among nearby sources.	It has a closed-form solution and can recover superficial and deeper sources. Enforces smooth continuous solutions. It is robust to noise.	It is less sensitive to focal sources, and it tends to overestimate the spatial extent of the source activations.	Pascual-Marqui et al. (1994)
LASSO	Least absolute shrinkage selection operator (LASSO) is obtained with $P(j) = \|\|j\|\|_1$, where the $l1$-norm enforces sparse solutions.	It is good for recovering focal and localized solutions.	It has an iterative solution, thus usually more expensive to compute than other closed-form solution algorithms. It is sensitive to noise.	Tibshirani (1996)
ENET	Elastic net (ENET) is obtained with $P(j) = \|\|M_1 j\|\|_2^2 + \|\|M_2 j\|\|_1$, where M_1 and M_2 are linear operator. ENET is also found in the literature as MxNE (Mixed-norm Estimate).	Combines $l1$ and $l2$ norms to promote solutions that are smooth and non-overlapping, therefore limiting the excessive smoothness sometimes exhibit by LORETA while enforcing local correlations.	Like Lasso, it is solved iteratively.	Vega-Hernández et al. (2008) and Gramfort et al. (2012)

See Paz-Linares et al. (2017) and Vega-Hernández et al. (2008) for a more in-depth exposition

comes from a prior density, and the regularization parameters, known as hyperparameters, are estimated within the same probabilistic framework. Although an exposition of Bayesian approaches is beyond the scope of this chapter, it is worth briefly mentioning that the Sparse Bayesian Learning (SBL) family of algorithms has been very successful in ESI (Friston et al., 2008; Owen et al., 2012; Henson et al., 2011).

SBL algorithms use Automatic Relevance Determination (ARD) priors (Wipf & Nagarajan, 2008) to estimate source maps in a very flexible and data-driven manner. Without explicitly designing for smooth or sparse solutions (the typical opposite ends of the spectrum), the algorithms automatically select the optimal (in the Bayesian evidence sense) sparsity and smoothness required in every region of the source space so that the resulting source maps best explain the data without any user intervention. Furthermore, it has been shown that this framework can be extended to model and identify non-brain artifactual sources and that the recovered sources are maximally independent, thereby bridging EEG imaging and ICA (Ojeda et al., 2021).

20.5 Statistical Inference in the Source Space

So far, we have discussed the solution of the inverse problem for a single measurement (v and j are vectors representing one topography and a corresponding source map). Given the linearity of our measurement model (Eq. 20.1), with minimal modifications, we can use the same inversion algorithms above to solve for multiple measurements simultaneously. Note that for the purpose of performing statistical inference, by multiple measurements we mean independent trials or even subjects (rather than autocorrelated sequential data).

For example, let us suppose we are interested in identifying the generators of an ERP component. As shown in Fig. 20.3, we first select the single-trial EEG topographies at the desired component latency to form the matrix V, which has dimensions channels by trials. The data matrix V is the input to our inverse mapping algorithm. Beforehand, we constructed a head model, from which we calculated the lead field matrix K. With these ingredients we estimate the matrix J, of dimensions sources by trials. In this example, the null hypothesis (H_0) can be that the value of a source at a given location is not significantly different from zero. A simple way to test this hypothesis is to calculate a t-test for each source.

We note that in ESI, the number of trials (data points) is usually far smaller than the number of variables (sources); therefore, the p-value at which we threshold our statistical maps (the t-statistic in this example) needs to be corrected for multiple comparisons to guard against false positives (Type 1 error). Three popular ways to correct for multiple comparisons are:

- Bonferroni correction: divides the p-value by the number of sources tested. This method is very stringent because since the sources are usually in the thousands, the corrected p-value tends to zero.
- False discovery rate (FDR): is a method to control the Type 1 error less stringent than Bonferroni (Benjamini & Hochberg, 1995).
- Resampling: this is a non-parametric method that consists of resampling the data and constructing the min/max (empirical) distributions on the output statistic,

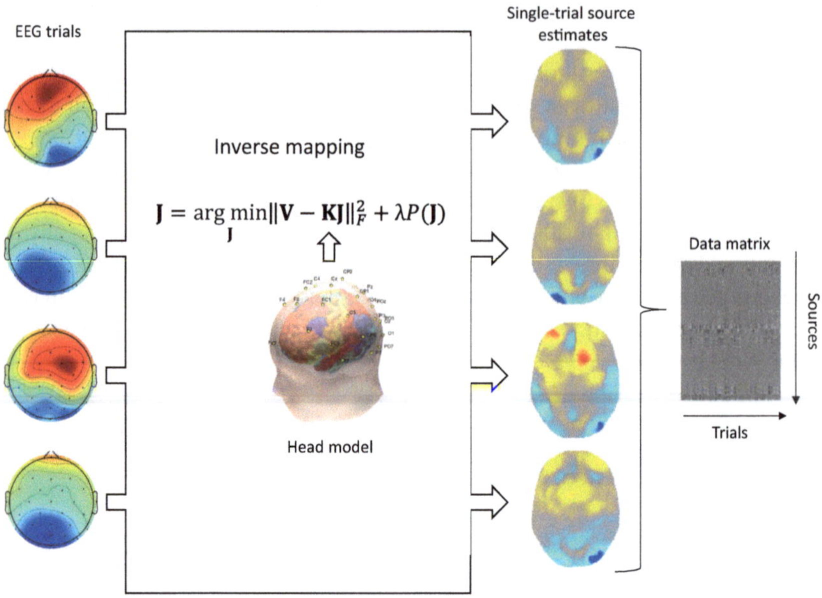

Fig. 20.3 Statistical inference in the source space. From left to right, the figure displays a collection of EEG topographies corresponding to the same latency but a different trial. Then we use the inverse solver to estimate single-trial source maps. Afterwards, the estimated source maps are collected into a matrix of sources by trials, which can be used for statistical inference. Note that in this case we group the single-trial EEG topographies into the columns of matrix V; consequently, the error term uses the Frobenius norm ($\|V - KJ\|_F$) instead of the Euclidean norm, and we solve for the matrix J, where each column represents the corresponding single-trial source estimate

then we take the p-value threshold from these empirical distributions (Nichols & Holmes, 2002). If min/max distributions are considered too stringent (i.e., driven by outliers), 5/95 percentiles can be used as an alternative (Valdes-Hernandez et al., 2010).

20.6 Source Connectivity

We would not want to finish this chapter without saying a few words about EEG source connectivity. Rather than reproducing comprehensive reviews already available (Cao et al., 2021; He et al., 2019), we will focus on practical considerations for performing source connectivity analysis.

First, it is useful to distinguish between three subdivisions of brain connectivity:

- Anatomical: this is at the structural level, axons and bundles of axons connecting different brain regions. Anatomical connectivity can be estimated with DTI.

- Functional: instantaneous relationship (correlation) between brain regions.
- Effective: causal (directed) dependence between brain regions.

Both functional and effective connectivity can be estimated with EEG. Crucially, effective connectivity allows the creation of mechanistic models of brain function and responses to different experimental manipulations. This allows us to explain brain mechanisms in terms of causal networks (Friston, 2011).

Early work on EEG-based brain connectivity used multichannel sensor data (Kaminski & Blinowska, 1991). However, since channel-level connectivity is largely blind to the mixing effect of the volume conduction, it is expected to lead to misleading interpretations. Therefore, it is generally accepted that EEG connectivity should be estimated on the source space (He et al., 2019).

EEG source connectivity is usually estimated in a two-step approach: (1) resolve the time series of the sources and (2) calculate connectivity metrics on those time series. With enough structural modeling it is also possible to simultaneously estimate source activity and coupling parameters (Kiebel et al., 2008). Two approaches to estimating source time series are:

- Piecewise estimation: we estimate sources based on current data without considering the past. We can either fix the regularization parameters or update them regularly to provide some degree of inverse model adaptation (Ojeda et al., 2018).
- Sequential filtering: here we use a Kalman filter-like approach in which source estimates are based on the current and past data points (Ojeda et al., 2021; Ghumare et al., 2018; Cheung et al., 2010). This approach is usually more computationally demanding.

As noted in the previous section, there are many more sources than data points, so to keep our problem tractable, it is desirable to calculate connectivity in a reduced space. Although nearby sources are usually highly correlated, extracting a source time series that summarizes the activity within a region of interest (ROI) remains an area of active research (Brkic et al., 2023; Bruña et al., 2023). If source orientations are not estimated (only amplitudes), a popular approach is to average all the time series within the ROI (Xie et al., 2022; Mullen et al., 2015) and use one of the existing algorithms (Cao et al., 2021; He et al., 2019) to calculate brain source connectivity on that reduced space.

Extra caution must be taken when summarizing orientation-free sources (source activity with x, y, z components). Note that if we reduce source components to their magnitude and then average within an ROI, the resulting time series undergoes a nonlinear transformation that fundamentally changes the dynamics of the original components. To sidestep the drawbacks of ROI summarization, several connectivity metrics have been extended to a multidimensional context, allowing us to assess the functional relationship between pairs of ROIs with different numbers of sources without any collapsing (Basti et al., 2020; Geerligs et al., 2016).

20.7 Conclusion

Source analysis is a powerful tool that comes to our aid whenever it is important to explain EEG scalp patterns in terms of the underlying anatomical structures. A source analysis pipeline begins by constructing a head model and calculating the lead field matrix using an off-the-shelf forward solver. This is typically done before any EEG measurements are taken. Once we have collected the participant's EEG data, we determine the latencies of the scalp topographies that we want to map into the cortical space. Depending on the data, experiment design, and hypothesis, we pick the inverse solver that best fits our needs. For instance, if we want to locate the source of an epileptic focus, a solver with a sparse penalty, such as LASSO or ENET, makes sense; if speed is important because maybe we want to give source-derived feedback to the user in real-time, a method with analytical solution such as MNE, wMNE, or LORETA is called for; otherwise, combining sparsity and smoothness in a data-driven manner as done in ENET is always a good idea.

When it comes to performing statistical analysis in the source space, we have seen that it is important to correct the output statistics for multiple comparisons to control for false positives. Furthermore, when it comes to group analysis, we must ensure that the head models of all participants are in anatomical correspondence. Otherwise, we must co-register them in a common space so that we can make statistical inferences on the same anatomical locations across the group.

We mentioned two common approaches to estimating source time series prior to calculating interactions: piecewise (instantaneous) and sequential filtering. While sequential filtering is usually more accurate, it is also more computationally expensive. We also gave a word of caution regarding the problem of summarizing source activity within regions of interest.

We finish the chapter by noting that each of the topics mentioned here is an active field of research in and of itself. Nevertheless, we hope this introductory material can guide readers interested in pursuing EEG source imaging studies, and we encourage them to follow their learning journey by delving into the references given here and beyond.

References

Basti, A., Nili, H., Hauk, O., Marzetti, L., & Henson, R. N. (2020). Multi-dimensional connectivity: A conceptual and mathematical review. *NeuroImage, 221*, 117179.

Benjamini, Y., & Hochberg, Y. (1995). Controlling the false discovery rate: A practical and powerful approach to multiple testing. *Journal of the Royal Statistical Society: Series B: Methodological, 57*, 289–300.

Bigdely-Shamlo, N., Mullen, T., Kreutz-Delgado, K., & Makeig, S. (2013). Measure projection analysis: A probabilistic approach to EEG source comparison and multi-subject inference. *NeuroImage, 72*, 287–303.

Brkic, D., Sommariva, S., Schuler, A. L., Pascarella, A., Belardinelli, P., Isabella, S. L., Pino, G. D., Zago, S., Ferrazzi, G., Rasero, J., Arcara, G., Marinazzo, D., & Pellegrino, G. (2023). The impact of ROI extraction method for MEG connectivity estimation: Practical recommendations for the study of resting state data. *NeuroImage, 2023*, 120424.

Bruña, R., Cuesta, P., Ramírez-Toraño, F., Suárez-Méndez, I., & Pereda, E. (2023). The more, the merrier: Multivariate phase synchronization methods excel pairwise ones in estimating functional brain connectivity from reconstructed neural sources. *bioRxiv*, 2023.01.19.524740.

Cao, J., Zhao, Y., Shan, X., Wei, H. L., Guo, Y., Chen, L., Erkoyuncu, J. A., & Sarrigiannis, P. G. (2021). Brain functional and effective connectivity based on electroencephalography recordings: A review. *Human Brain Mapping, 43*, 860–879.

Cheung, B. L., Riedner, B. A., Tononi, G., & Van Veen, B. D. (2010). Estimation of cortical connectivity from EEG using state-space models. *IEEE Transactions on Biomedical Engineering, 57*, 2122–2134.

Dale, A. M., & Sereno, M. I. (1993). Improved localizadon of cortical activity by combining EEG and MEG with MRI cortical surface reconstruction: A linear approach. *Journal of Cognitive Neuroscience, 5*, 162–176.

Debener, S., Hine, J., Bleeck, S., & Eyles, J. (2008). Source localization of auditory evoked potentials after cochlear implantation. *Psychophysiology, 45*, 20–24.

Debener, S., Thorne, J., Schneider, T. R., & Viola, F. C. (2010). 1213.1 Using ICA for the analysis of multi-channel EEG data. In M. Ullsperger & S. Debener (Eds.), *Simultaneous EEG and fMRI: Recording, analysis, and application*. Oxford University Press.

Friston, K. J. (2011). Functional and effective connectivity: A review. *Brain Connectivity, 1*, 13–36.

Friston, K., Harrison, L., Daunizeau, J., Kiebel, S., Phillips, C., Trujillo-Barreto, N., Henson, R., Flandin, G., & Mattout, J. (2008). Multiple sparse priors for the M/EEG inverse problem. *NeuroImage, 39*, 1104–1120.

Geerligs, L., Cam, C., & Henson, R. N. (2016). Functional connectivity and structural covariance between regions of interest can be measured more accurately using multivariate distance correlation. *NeuroImage, 135*, 16–31.

Ghumare, E. G., Schrooten, M., Vandenberghe, R., & Dupont, P. (2018). A time-varying connectivity analysis from distributed EEG sources: A simulation study. *Brain Topography, 31*, 721–737.

Golub, G. H., Heath, M., & Wahba, G. (1979). Generalized cross-validation as a method for choosing a good ridge parameter. *Technometrics, 21*, 215–223.

Gramfort, A., Papadopoulo, T., Olivi, E., & Clerc, M. (2010). OpenMEEG: Opensource software for quasistatic bioelectromagnetics. *Biomedical Engineering Online, 9*, 45.

Gramfort, A., Kowalski, M., & Hamalainen, M. (2012). Mixed-norm estimates for the M/EEG inverse problem using accelerated gradient methods. *Physics in Medicine and Biology, 57*, 1937–1961.

Hallez, H., Vanrumste, B., Grech, R., Muscat, J., De Clercq, W., Vergult, A., D'Asseler, Y., Camilleri, K. P., Fabri, S. G., Van Huffel, S., & Lemahieu, I. (2007). Review on solving the forward problem in EEG source analysis. *Journal of Neuroengineering and Rehabilitation, 4*, 46.

Hamalainen, M. S., & Ilmoniemi, R. J. (1994). Interpreting magnetic fields of the brain: Minimum norm estimates. *Medical & Biological Engineering & Computing, 32*, 35–42.

Hansen, P. C. (1992). Analysis of discrete ill-posed problems by means of the L-curve. *SIAM Review, 34*, 561–580.

He, B., Astolfi, L., Valdes-Sosa, P. A., Marinazzo, D., Palva, S., Benar, C. G., Michel, C. M., & Koenig, T. (2019). Electrophysiological brain connectivity: Theory and implementation. *IEEE Transactions on Biomedical Engineering*. https://doi.org/10.1109/TBME.2019.2913928

Henson, R. N., Wakeman, D. G., Litvak, V., & Friston, K. J. (2011). A parametric empirical Bayesian framework for the EEG/MEG inverse problem: Generative models for multi-subject and multi-modal integration. *Frontiers in Human Neuroscience, 5*, 76.

Kaminski, M. J., & Blinowska, K. J. (1991). A new method of the description of the information flow in the brain structures. *Biological Cybernetics, 65*, 203–210.

Kiebel, S. J., Garrido, M. I., Moran, R. J., & Friston, K. J. (2008). Dynamic causal modelling for EEG and MEG. *Cognitive Neurodynamics, 2*, 121–136.

Lin, F. H., Witzel, T., Ahlfors, S. P., Stufflebeam, S. M., Belliveau, J. W., & Hamalainen, M. S. (2006). Assessing and improving the spatial accuracy in MEG source localization by depth-weighted minimum-norm estimates. *NeuroImage, 31*, 160–171.

Makeig, S., Jung, T. P., Bell, A. J., Ghahremani, D., & Sejnowski, T. J. (1997). Blind separation of auditory event-related brain responses into independent components. *Proceedings of the National Academy of Sciences of the United States of America, 94*, 10979–10984.

Makeig, S., Westerfield, M., Jung, T. P., Enghoff, S., Townsend, J., Courchesne, E., & Sejnowski, T. J. (2002). Dynamic brain sources of visual evoked responses. *Science, 295*, 690–694.

Michel, C. M., & Brunet, D. (2019). EEG source imaging: A practical review of the analysis steps. *Frontiers in Neurology, 10*, 325.

Mullen, T. R., Kothe, C. A., Chi, Y. M., Ojeda, A., Kerth, T., Makeig, S., Jung, T. P., & Cauwenberghs, G. (2015). Real-time neuroimaging and cognitive monitoring using wearable dry EEG. *IEEE Transactions on Biomedical Engineering, 62*, 2553–2567.

Nichols, T. E., & Holmes, A. P. (2002). Nonparametric permutation tests for functional neuroimaging: A primer with examples. *Human Brain Mapping, 15*, 1–25.

Ojeda, A., Kreutz-Delgado, K., & Mullen, T. (2018). Fast and robust Block-Sparse Bayesian learning for EEG source imaging. *NeuroImage, 174*, 449–462.

Ojeda, A., Kreutz-Delgado, K., & Mishra, J. (2021). Bridging M/EEG source imaging and independent component analysis frameworks using biologically inspired sparsity priors. *Neural Computation, 33*, 2408–2438.

Owen, J. P., Wipf, D. P., Attias, H. T., Sekihara, K., & Nagarajan, S. S. (2012). Performance evaluation of the Champagne source reconstruction algorithm on simulated and real M/EEG data. *NeuroImage, 60*, 305–323.

Pascual-Marqui, R. D., Michel, C. M., & Lehmann, D. (1994). Low resolution electromagnetic tomography: A new method for localizing electrical activity in the brain. *International Journal of Psychophysiology, 18*, 49–65.

Paz-Linares, D., Vega-Hernandez, M., Rojas-Lopez, P. A., Valdes-Hernandez, P. A., Martinez-Montes, E., & Valdes-Sosa, P. A. (2017). Spatio temporal EEG source imaging with the hierarchical Bayesian elastic net and elitist Lasso models. *Frontiers in Neuroscience, 11*, 635.

Seeber, M., Cantonas, L. M., Hoevels, M., Sesia, T., Visser-Vandewalle, V., & Michel, C. M. (2019). Subcortical electrophysiological activity is detectable with high-density EEG source imaging. *Nature Communications, 10*, 753.

Stenroos, M., & Sarvas, J. (2012). Bioelectromagnetic forward problem: Isolated source approach revis(it)ed. *Physics in Medicine and Biology, 57*, 3517–3535.

Tibshirani, R. (1996). Regression shrinkage and selection via the Lasso. *Journal of the Royal Statistical Society: Series B: Methodological, 58*, 267–288.

Valdes-Hernandez, P. A., Ojeda-Gonzalez, A., Martinez-Montes, E., Lage-Castellanos, A., Virues-Alba, T., Valdes-Urrutia, L., & Valdes-Sosa, P. A. (2010). White matter architecture rather than cortical surface area correlates with the EEG alpha rhythm. *NeuroImage, 49*, 2328–2339.

Vega-Hernández, M., Martínez-Montes, E., José Sánchez-Bornot, M., Lage-Castellanos, A., & Pedro Valdés-Sosa, A. (2008). Penalized least squares methods for solving the EEG inverse problem. *Statistica Sinica, 18*, 1535–1551.

Wipf, D. P., & Srikantan Nagarajan, S. (2008). A new view of automatic relevance determination. In *Proceedings of the twenty-first annual conference on neural information processing systems, Vancouver, British Columbia, Canada* (pp. 1625–1632). Curran Associates, Inc.

Wolters, C., Köstler, H., Möller, C., Härdtlein, J., & Anwander, A. (2007). Numerical approaches for dipole modeling in finite element method based source analysis. *International Congress Series, 1300*, 189–192.

Xie, W., Toll, R. T., & Nelson, C. A. (2022). EEG functional connectivity analysis in the source space. *Developmental Cognitive Neuroscience, 56*, 101119.

Chapter 21
Online Processing

Alex Kreilinger and Alejandro Ojeda

Abstract This chapter introduces online processing of EEG signals, as opposed to the usual offline analysis of previously recorded data. We will explain where it is necessary to use online processing, how to get access to and deal with raw data, and point out the main differences between online and offline analysis. We conclude the chapter with an example of an online experiment to provide a foundation for our readers.

Keywords Online · (Near) real-time · LSL · BCI

21.1 Introduction

A large portion of scientific research involves recording data in different scenarios and analyzing them afterward to examine cause and effect, to determine the impact of a given factor, or to investigate basic mechanisms of the brain.

In some cases, it is not enough to have access to only saved data, it is necessary to use current measurements to estimate brain state changes online, i.e., directly when they occur. These estimations are used to provide immediate feedback that can either positively or negatively affect future behavior.

To have a meaningful impact, such feedback needs to be linked to certain events and arbitrary delays in relation to these events are not acceptable. These timing constraints can vary depending on the behavior or the underlying mechanism that needs to be affected. For example, in a brain-computer interface (BCI) that is controlled with a rather long imagination task (in the range of a few seconds), the delay of the feedback may not be that critical. However, when the aim is to stimulate

A. Kreilinger (✉)
Brain Products GmbH, Gilching, Germany
e-mail: alex.kreilinger@brainproducts.com

A. Ojeda
Brain Vision LLC, Garner, NC, USA
e-mail: alejandro.ojeda@brainvision.com

T. Warbrick (ed.), *The EEG Handbook*,
https://doi.org/10.1007/978-3-032-20450-9_21

with a transcranial magnetic stimulation (TMS) pulse directly at the onset of an alpha wave, the delay needs to be significantly lower than a wavelength of the alpha frequency. Moreover, fast and precise feedback can be helpful for reinforcing successful responses (Enriquez-Geppert et al., 2017; Arvaneh et al., 2015; Sherlin et al., 2011).

Terminology
In this chapter we use the term 'online'. In this context, 'online' means that something is happening right now, as opposed to 'offline', where data is first stored in a file and only later analyzed. Frequently, the term 'real-time' is used instead, although technically 'real-time' refers to the guaranteed processing of a task within a specific, pre-defined time frame. Actual 'real-time' requires a real-time operating system, which is often not available. The more correct term would be 'near real-time' when the word 'online' should be avoided, as it may cause a misunderstanding with the meaning in the internet context. We acknowledge that 'real-time' has more or less been accepted by a large part of the EEG research community and suggest that the reader can make up their own mind.

The most simple and straightforward kind of online processing is visualization. Displaying the EEG traces during recording is a vital part of EEG data collection and something that most researchers do by default. A clear recommendation is, where possible, to be able to monitor acquired signals constantly before and during a recording. This way, the operator can react immediately in case of deteriorating signal quality. Artifacts, such as line noise, heavy sweating, or muscular activity, can easily be observed, and counteractions can be initiated. In addition, simple display filters can help to visualize the underlying signals in cases where heavy artifacts cannot be avoided during the measurement.

A step further would be to map the signals to something more meaningful to the participant. For example, the magnitude of alpha band power can be directly mapped onto a visual element—such as the size of an object or the saturation of a color—allowing individuals to modulate their behavior accordingly.

Advancing further, classification techniques can be introduced, including simple thresholding, linear classifiers, support vector machines, and neural networks. The range of available options continues to grow rapidly, especially with the ongoing advancements in artificial intelligence. See Fig. 21.1 for a basic example.

The crux of online processing is that the analyses need to be fast and that they need to be able to work with present and past data only. This not only means that not all the data is available at once, but also that processing time needs to be accounted for. A good example is how to use filters: while in a typical offline analysis it is not a problem to create zero-phase filters by running the filter in both directions (forward and backward), such a filter is not possible online, unless huge buffers are used as epochs, thereby introducing significant delays.

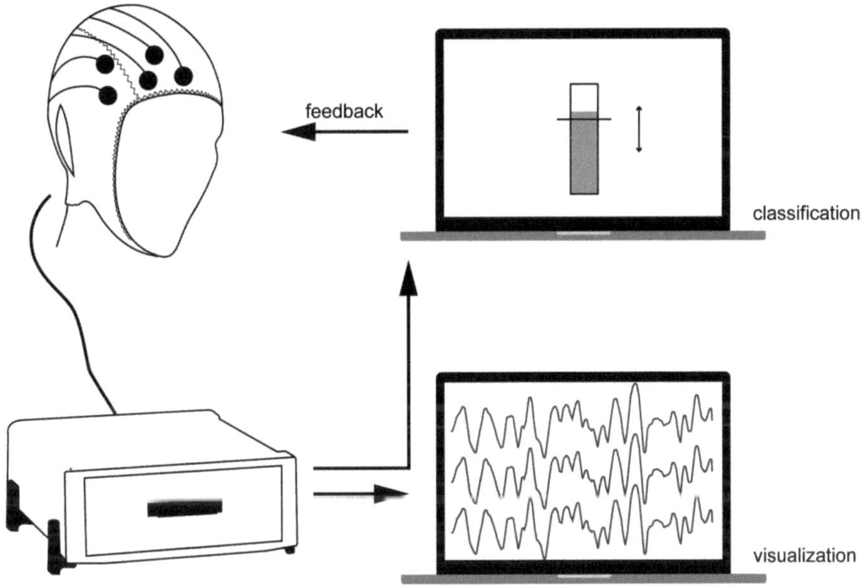

Fig. 21.1 Schematic of a basic online visualization of the acquired EEG signals. In parallel, a classifier translates the brain activity into a bar graph that the participant is trying to control. Successful (and unsuccessful) control attempts are immediately shown as feedback, which can in turn alter brain activity

In addition, when acquiring signals online, one cannot simply exclude bad trials due to artifacts or make the choice to remove a bad channel from the data when noticing a bad electrode contact, for example. These potential issues need to be well planned ahead of time, for example, by applying online artifact detection methods that can react to these occurrences on the fly. In general, many of these potential problems can be mitigated, but they do require additional planning. It is also always a good strategy to at least simulate the data processing pipeline before asking a participant to take part in an experiment. For more information refer to Chap. 10: Pilot Testing.

21.2 Raw Data Access

Before beginning to plan for an online processing pipeline, the first step is to determine if and how it is possible to get access to the raw EEG data from the amplifier.

21.2.1 How to Get Raw Data

Depending on the manufacturer and the type of amplifier, there may be different options for accessing raw data. Depending on the requirements, the operator can make the right choice.

- Data access comes with the device's own software. Some manufacturers provide their own data acquisition software with the amplifier. In those cases, it might be possible to access data online. The communication through these data outlets may follow a common format or internal standards. Notable examples are the Remote Data Access (RDA) feature of BrainVision Recorder (Brain Products GmbH, Gilching, Germany) or Lab Streaming Layer (LSL) that is becoming more and more the standard among many companies.

Lab Streaming Layer (LSL)
Lab Streaming Layer (LSL): LSL is a software framework designed for the streaming and receiving of data and marker streams over a local network. One of its main features is that the synchronization of multiple streams is handled directly by the framework itself. It was developed in 2010 by Christian Kothe and has since become a widely used tool in neurophysiological research. It is supported by many tools and languages, such as Python, MATLAB® (The MathWorks Inc. MATLAB, Natick, Massachusetts: The MathWorks Inc. https://www.mathworks.com), C++, Unity (*Unity Game Engine*. Unity Technologies. https://unity.com), and more. For many EEG amplifiers, LSL outlets are either available through the open-source community or directly from the manufacturer. More information on LSL can be found in the publication from the creators (Kothe et al., 2025) and on the official documentation website.[1]

- There are alternative tools available—often provided by the open-source community—which work with specific amplifiers and generate data streams. Many of those tools are using the LSL protocol. These so-called LSL connectors communicate directly with the amplifier to read data and stream the signals to the local network. Other tools or protocols that are dedicated to providing data access from amplifiers are the TOBI interface A (Breitwieser et al., 2012) or BrainFlow.[2]
- Another option would be to look into readily available standalone platforms that aim to provide a full package for solutions such as BCIs, neuroimaging, or neural signal processing. These platforms often provide drivers, sources, plugins (however they may be called in the respective platform) that can directly access data from the amplifier. In addition, they can provide a whole selection of different processing modules, such as visualization, recording, artifact detection, and

[1] https://labstreaminglayer.readthedocs.io/
[2] https://brainflow.org

classification. Depending on the platform, the options can be numerous. Available platforms include OpenViBE (Renard et al., 2010), NeuroPype (Intheon, La Jolla, CA), or BCI2000 (Schalk et al., 2004).
- If none of the above options provide the right composition, the ultimate option would be to ask the manufacturer for a software development kit (SDK) or application programming interface (API) that allows a knowledgeable software developer to create their own applications. This facilitates configuring the EEG amplifier, acquiring data from the amplifier, and processing data immediately in the application to generate feedback online. A typical example of how such an SDK can be used might just be to create an LSL connector.

21.2.2 How to Work with Raw Data Online

The chosen method for getting raw data might already pre-determine how you work with the data. However, there might still be options available to you.

- If the manufacturer's own software package is used for accessing the raw data, chances are high that there is also a tool available for online data processing. For example, when using BrainVision Recorder with RDA, the complementary BrainVision RecView can be used to work with the data online and to select different filters for further processing.
- When using one of the openly available protocols, such as LSL, there are often plugins available for the major programming or scripting languages. LSL, for example, can be easily incorporated in Python, MATLAB®, C++, or C#. Using these languages, it is then possible to write signal processing code or rely on readily available code from the community. The specifics of the individual programming languages should be considered with respect to the planned processing steps. For example, if a very high processing speed is required, C++ might be better suited than Python or MATLAB®, but of course there may be other factors playing a role in the final decision.
- Stand-alone platforms often provide signal processing nodes and may even include options for running user's custom code in case the provided algorithms are not enough.
- Developers who are using the SDK typically also have the means to write their own code following data acquisition.

21.2.3 What Factors to Consider for Online Processing

Depending on the requirements, different factors can play a role in deciding which kind of hardware and/or software to use. Here, we focus mainly on factors that influence online processing rather than providing a recipe that could bias the reader toward an EEG amplifier, accessory, or software to use. To learn more about hardware and software, refer to Chap. 12: Hardware for Recording EEG and Peripheral Physiology and Chap. 13: Software for Recording EEG and Peripheral Physiology. Asking yourself the following questions will help you determine what is right for your study:

- *Is it a mobile or stationary recording?* This affects the decision whether to choose a wireless or cable-bound recording setup. While it is possible for participants to carry around portable computers, it is more convenient to go for a wireless connection in such cases. Factors relevant to the online processing are the potential additional delay that is caused by using wireless transmission, such as Bluetooth®. An obvious drawback is the potential loss of the connection between transmitter and receiver. Wireless transmission also limits the capacity of data that can be sent. This can be dictated by the technology itself (e.g., Bluetooth® low energy with a potential data rate of ~1 Mbps) or by the necessary amount of power to increase the WiFi data rate, which would drain the batteries of a mobile amplifier quickly. That said, high channel counts with high sampling rates are usually only provided by stationary, cable-bound amplifiers.
- *How fast and how precise are the timing requirements?* Before deciding on a piece of equipment, it is important to know what kind of latencies and jitters are to be expected when accessing the data streams online. Depending on the drivers, the use of cables versus wireless transmission, and data protocol, these parameters can have severe effects on the further signal processing pipeline and feedback loop. For offline analyses, it usually does not matter if the signal is recorded with a large latency, as long as the relevant event markers are synchronized and come with the same latency. For online processing, this can become a big problem depending on the goals of the experiment. For example, for closed-loop systems, where the feedback must be provided within a time-critical constraint, it is important to have access to the data as fast as possible. Here, the word real-time can become relevant if a real-time operating system or environment is available. If that is not the case, the alternative must be that data can be typically accessed within a very short amount of time. Examples are the BrainVision TurboLink that sends data from an actiCHamp Plus EEG-amplifier in under 1.5 ms. This can be linked to a bossdevice from sync2brain (Lieb et al., 2023) that is running on a real-time operating system and uses an estimate of the next alpha wave cycle to trigger TMS pulses accordingly. Requirements can be a bit looser for less timing-sensitive scenarios, but typical criteria include a recommendation for low latency and low jitter. It is good practice to test such parameters before running measurements or at least to acquire the information beforehand.
- *What is the nature of the acquired signal?* In some cases, signals can only be accessed in an altered manner, for example, with mandatory temporal and/or spatial filters. Depending on the signal of interest, this can be problematic when important signal components are already attenuated by the filters. Ideally, signals can be accessed as raw as possible.

- *How well can contextual information and other measurements of interest be synchronized with the brain signals online?* This question is a bit dependent on previous questions and on how the raw data is accessed in the first place. For example, when event markers need to be stored together with the EEG, a trigger solution that connects on a hardware level might already be good enough, but as soon as other modalities or markers need to be supplied on a software level, other considerations might be necessary. For example, it is worth checking whether there are LSL connectors already available for the other modality (e.g., reading just the mouse cursor or even other biosignals such as functional near-infrared spectroscopy (fNIRS)). For more information on how to synchronize signals and events, refer to Chap. 14: Triggers.

21.3 Designing an Online Processing Experiment, an Example

We summarize the chapter by presenting considerations for a hypothetical experiment conceived to use online processing. In this experiment, the user will try to control a BCI based on motor imagery using EEG. The BCI shall use two classes and be able to control different actions in a computer game depending on the current context. In the steps that follow, we provide guidance on selecting appropriate hardware and software, as well as general considerations for conducting the BCI experiment. See also Fig. 21.2 for more information.

1. Choosing the hardware: in this case, the criteria include channel count, coverage of the relevant area of the brain, data quality, and sampling rate, and importantly, how raw data can be accessed online. In this specific case, the requirements are not very exclusive. One should just make sure to use a solution that provides good quality data, a full coverage of the whole head (~32 channels) or the possibility of a flexible montage for covering the motor area required for motor imagery. Make sure you have a high enough sampling rate. This should not be an issue for motor imagery where a typical 500 Hz sampling rate is high enough to record all frequency bands of interest. The raw data access feature is also relevant for the next part.
2. Choosing the software. Often, the selected hardware comes with different options for accessing the raw data. Depending on the further steps and how the feedback needs to be created, these details become relevant here. The software not only has to be able to read the data but also needs to analyze and classify the signals before any kind of feedback can be reported in any modality (visual, acoustic, or tactile). Nowadays, one can rely on a large community that provides an abundance of information on signal processing for all kinds of different purposes. For example, code packages can be used based on Python or MATLAB®, but also other software platforms provide many options for online processing with the focus lying more on the feedback part, such as Unreal Engine (Safikhani et al., 2024) or

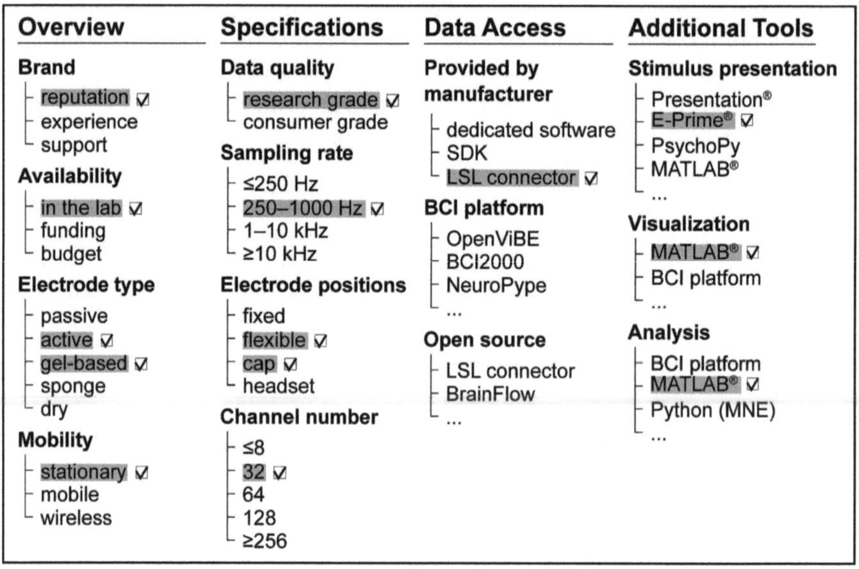

Fig. 21.2 Selection of possible choices when designing an experiment and deciding what kind of hardware and software to use. The choices can be affected by many different factors (e.g., project-related requirements, personal experience, or simply by having access to specific equipment). In this example, the selected setup for the experiment is driven by practical considerations such as reputation and availability in the lab, but also the fulfillment of certain requirements, such as the electrode and amplifier type, data quality, sampling rate, and number of channels. Raw data can be accessed via an LSL connector provided by the manufacturer, and a mix of E-Prime® and MATLAB® is used to train, visualize, and analyze data. Refer to Chap. 12: Hardware for Recording EEG and Peripheral Physiology and Chap. 13: Software for Recording EEG and Peripheral Physiology for more information

Unity (for example, via LSL[3]). Of course, it is also an option to use an all-in-one solution such as OpenViBE.

3. Once these decisions are finalized, the preparation of the experiment can begin. Ideally, the first training sessions are simulated and analyzed. Then, participants can be brought in for the first pilot sessions. If classifiers need to be created, it must be considered that these cannot be applied to the data in the same way as offline. Again, it might be necessary to change filter settings and make sure to be aware of the constraints when only short amounts of data can be analyzed and that, ideally, the processing can be repeated frequently. If the processing power is insufficient to do this at every sample, it may be necessary to run the process in frames where data is only analyzed in steps of multiple samples.

[3]https://github.com/labstreaminglayer/LSL4Unity

21.4 Conclusion

In this chapter, we introduced the concept of online EEG data processing and discussed its relevance in scenarios where offline analysis is insufficient. We outlined various methods for accessing raw EEG data during acquisition, including manufacturer-specific solutions, open-source data streaming tools, and integrated platforms that combine data access with processing, visualization, and storage capabilities. Key considerations for selecting appropriate hardware and software components were highlighted, particularly in relation to the demands of (near) real-time experimental setups. Throughout, we emphasized the importance of precise terminology—clarifying distinctions between *online*, *real-time*, and *near real-time* processing—and discussed how these terms relate to system capabilities and constraints. Furthermore, we examined the limitations imposed by online processing on analysis pipelines, such as the inability to apply bidirectional filters or to reference future data. These considerations were illustrated through the design of a BCI paradigm, demonstrating the practical steps and trade-offs involved in planning an online experiment.

References

Arvaneh, M., Ward, T. E., & Robertson, I. H. (2015). Effects of feedback latency on P300-based brain-computer interface. In *2015 37th annual international conference of the IEEE Engineering in Medicine and Biology Society (EMBC)* (pp. 2315–2318). https://doi.org/10.1109/EMBC.2015.7318856

Breitwieser, C., Daly, I., Neuper, C., & Müller-Putz, G. R. (2012). Proposing a standardized protocol for raw biosignal transmission. *IEEE Transactions on Biomedical Engineering, 59*(3), 852–859. https://doi.org/10.1109/TBME.2011.2174637

Enriquez-Geppert, S., Huster, R. J., & Herrmann, C. S. (2017). EEG-neurofeedback as a tool to modulate cognition and behavior: A review tutorial. *Frontiers in Human Neuroscience, 11*. https://doi.org/10.3389/fnhum.2017.00051

Kothe, C., Shirazi, S. Y., Stenner, T., Medine, D., Boulay, C., Grivich, M. I., et al. (2025). The lab streaming layer for synchronized multimodal recording. *Imaging Neuroscience*. https://doi.org/10.1162/IMAG.a.136

Lieb, A., Zrenner, B., Zrenner, C., Kozák, G., Martus, P., Grefkes, C., & Ziemann, U. (2023). Brain-oscillation-synchronized stimulation. *BMC Neurology, 23*(204). https://doi.org/10.1186/s12883-023-03235-1

Renard, Y., Lotte, F., Gibert, G., Congedo, M., Maby, E., Delannoy, V., et al. (2010). OpenViBE: An open-source software platform to design, test, and use brain–computer interfaces in real and virtual environments. *Presence Teleoperators and Virtual Environments, 19*(1), 35–53. https://doi.org/10.1162/pres.19.1.35

Safikhani, S., Gattringer, V., Schmied, M., Pirker, J., & Wriessnegger, S. C. (2024). The influence of realism on the sense of presence in virtual reality: Neurophysiological insights using EEG. *Multimodal Technologies and Interaction, 8*(11), 104. https://doi.org/10.3390/mti8110104

Schalk, G., McFarland, D. J., Hinterberger, T., Birbaumer, N., & Wolpaw, J. R. (2004). BCI2000: A general-purpose brain-computer interface (BCI) system. *IEEE Transactions on Biomedical Engineering, 51*(6), 1034–1043. https://doi.org/10.1109/TBME.2004.827072

Sherlin, L. H., Arns, M., Lubar, J., Heinrich, H., Kerson, C., Strehl, U., & Sterman, M. B. (2011). Neurofeedback and basic learning theory: Implications for research and practice. *Journal of Neurotherapy, 15*(4). https://doi.org/10.1080/10874208.2011.623089

Chapter 22
Quantifying EEG and ERP Data Quality

Steven J. Luck

Abstract We can record brain activity (the electroencephalogram or EEG) from electrodes on the scalp, even though the skull and scalp separate the electrodes from the brain. However, the signals we record this way are often contaminated by noise and artifacts that are an order of magnitude larger than the EEG signals of interest. It is therefore essential to pay close attention to the quality of the data in order to obtain stable, meaningful, and statistically significant effects. Until recently, however, there was no widely accepted method for quantifying data quality for EEG signals, especially for the event-related potentials (ERPs) embedded within the EEG. In this chapter, I describe a new metric of ERP data quality, called the standardized measurement error (SME). I provide an intuitive description of how it works and then I provide examples of how it has been used to quantify data quality across several common ERP paradigms and to determine which EEG and ERP processing operations do the best job of maximizing data quality. I also describe how the SME can be extended to time-frequency analyses and how psychometric measures of reliability can be used to assess data quality in studies that focus on individual differences.

Keywords ERPs · Event-related potentials · Analysis methods

22.1 Introduction

It seems like a miracle that we can record meaningful brain activity from electrodes placed on the scalp, with the meninges, skull, fat, and skin between the neurons and our electrodes. The brain signals of interest are often only a microvolt or two on the scalp, and these signals are embedded within noise that is typically an order of magnitude larger. For example, Fig. 22.1 shows the averaged ERP waveform from a single participant in an oddball experiment (Fig. 22.1a) along with the 20 single-trial

---

S. J. Luck (✉)
Center for Mind & Brain and Department of Psychology, University of California, Davis, Davis, CA, USA
e-mail: sjluck@ucdavis.edu

T. Warbrick (ed.), *The EEG Handbook*,
https://doi.org/10.1007/978-3-032-20450-9_22

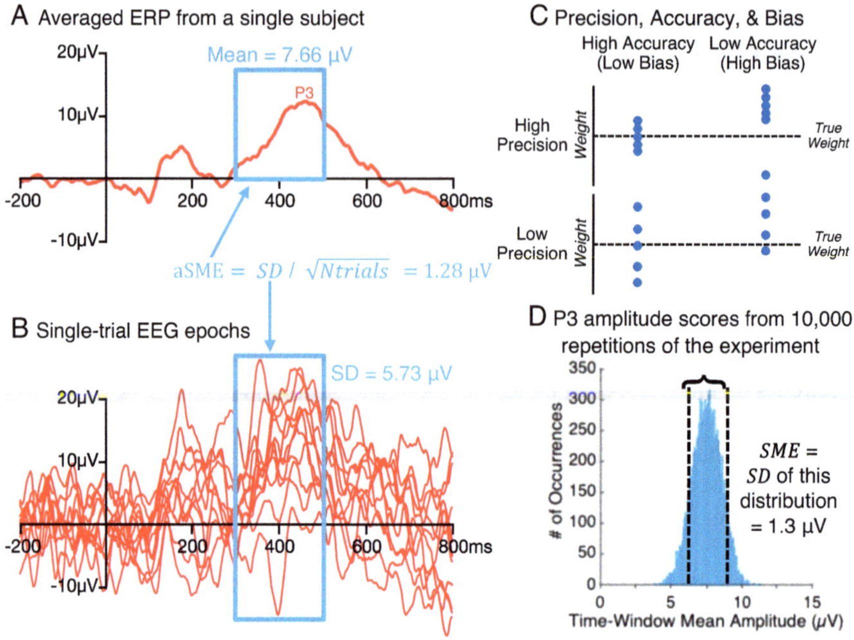

Fig. 22.1 (**a**) Averaged ERPs for the oddball trials for one participant in an oddball experiment in which each participant pressed one button for frequently occurring digits and another button for rarely occurring letters (or vice versa). Time zero is stimulus onset. (**b**) Single-trial EEG epochs from the oddball trials that were averaged together to create the waveform in panel A. If the amplitude of the P3 wave is scored as the mean voltage between 300 and 500 ms in the averaged ERP waveform, the analytic standardized measurement error (aSME) can be computed by measuring the mean voltage between 300 and 500 ms in the single-trial EEG epochs and applying the equation shown between panels A and B. (**c**) Illustration of the concepts of precision, accuracy, and bias for measuring the weight of an object. (**d**) Conceptual approach for understanding the SME, in which the same participant is tested in 10,000 different replications of the oddball experiment (assuming no fatigue or learning). For each replication, the oddball trials are averaged together, and the mean voltage between 300 and 500 ms is measured from the averaged ERP waveform. The histogram shows the P3 amplitude scores obtained from each of these 10,000 replications. The SME is the standard deviation (SD) of these 10,000 scores.

EEG epochs that were averaged together to create the averaged ERP waveform for the oddballs (Fig. 22.1b). There is clearly tremendous trial-to-trial variability in the EEG that is not obvious from the averaged ERP waveform, with a clear P3 component in the averaged ERP that is difficult to see in the single-trial EEG epochs.

When I first started running EEG/ERP experiments about 40 years ago, my data looked a lot like the waveforms shown in Fig. 22.1, but I had no idea whether the amount of trial-to-trial EEG variability I was seeing was "normal." And I had no idea whether I had enough trials to obtain a stable averaged ERP waveform given this amount of EEG variability. I desperately wished that I had a metric of data quality that would allow me to determine whether my EEG and ERP waveforms were sufficiently clean. I kept looking in the EEG/ERP literature for data quality metrics, but I never found one that seemed very useful. After several decades, I gave up on looking for an existing metric of data quality, and I worked with a set of collaborators to develop one (Luck et al., 2021) and implement it in ERPLAB Toolbox (Lopez-Calderon & Luck, 2014). In the process, I learned about additional data quality metrics that can be used for ERPs, time-frequency analyses, and other psychophysiological measures.

The goal of this chapter is to explain these metrics of data quality so that you can use them in your own research. A good metric of data quality can quantify how much statistical power you have lost because of the noise in your single-trial EEG epochs and how much power you could gain by increasing the number of trials. It can help you decide which trials to reject and which channels to interpolate. It can tell you which preprocessing techniques work best to increase your data quality and improve your statistical power. And it can provide you with information about whether the data from a given participant or a given dataset has too much noise to be believable.

22.1.1 Defining Data Quality

Before we can discuss metrics of data quality, we need to spend a moment considering what we mean by the term *data quality*. When we measure a simple physical quantity, such as the weight of an object, the quality of the measurement is classically defined in terms of *accuracy* (or *bias*) and *precision*. This is illustrated graphically in Fig. 22.1c for the measurement of weight; we assume that the true weight does not change across measurement attempts. A measurement is accurate if the average measurement across many measurement attempts is close to the true value (e.g., the true weight in Fig. 22.1c). Bias is the converse of accuracy: a measurement is biased if it is systematically shifted in a specific direction away from the true value. Precision reflects the spread of values across measurement attempts. A measurement method is precise if we get similar values on each measurement attempt. A measurement method can be accurate without being precise (e.g., a broad spread of values, but with an average near the true value), and it can be precise without being accurate (e.g., a narrow spread of values, all of which are systematically shifted away from the true value). Or it can be both accurate and precise, or both inaccurate and imprecise.

In this chapter, I will define data quality in terms of precision, not accuracy or bias, because precision is more related to the concept of how "noisy" a dataset is. Accuracy and bias are important, but they usually become an issue during the process of scoring the amplitude or latency of an ERP component. Rather than trying to quantify accuracy/bias, a common approach is to use amplitude or latency quantification algorithms that are mathematically guaranteed to be unbiased (see Chapter 9 in Luck, 2014).

I would also like to distinguish between a *measurement* and a *score*. I like to define a *measurement* as a quantification of a simple physical quantity, such as the raw EEG voltage at a particular moment in time. I then define a *score* as a value that we compute by applying an algorithm to a set of measurements, such as the peak amplitude of the voltage between 300 and 500 ms in the averaged ERP waveform or the power between 8 and 12 Hz in a Fourier transform of a single-trial EEG epoch. In this chapter, we will focus on the data quality of scores, not measurements, because the scores are what you use as the dependent variables in your statistical analyses and use to test your scientific hypotheses. The precision of the measurements that are used to compute a score contributes to the precision of the score, so the precision of a score reflects measurement precision as well as higher-order factors such as artifacts. Thus, for the purposes of this chapter, *data quality is defined as the precision of a score*.

22.1.2 Chapter Overview

Now that I have defined what I mean by data quality, we can move on to discuss how data quality can be quantified. The next section will provide a detailed description of the metric of data quality that my lab developed for averaged ERPs, called the *standardized measurement error* or *SME* (Luck et al., 2021). This will be followed by examples of SME values from actual experiments, a discussion of how SME is related to statistical power, and concrete examples of how the SME can be used to improve ERP research. The SME is appropriate only when scores are obtained from averaged ERPs, but later sections of the chapter will describe how the SME could be readily extended to time-frequency analysis and other psychophysiological measures based on averages, and how metrics of *reliability* can be used to quantify data quality in research on individual differences.

22.2 Quantifying Data Quality in Averaged ERPs with the Standardized Measurement Error (SME)

22.2.1 A Simple Case: Analytic SME for Time-Window Mean Amplitude Scores

I'm going to start with a simple but common case, in which the amplitude of an ERP component is scored from an averaged ERP waveform as the mean voltage within a measurement window, as illustrated in Fig. 22.1a, b.

The data in the figure were taken from the oddball experiment of Kappenman and Luck (2010), with 80 *standard* trials and 20 *oddball* trials. Panel A shows the averaged ERP waveform for the oddballs in one participant, and Panel B shows the 20 EEG epochs that were averaged together to compute this averaged waveform. In this experiment, the amplitude of the P3 wave was scored as the mean voltage between 300 and 500 ms in the averaged ERP waveform, yielding a value of 7.66 μV.

When we score an amplitude from an averaged ERP waveform as the mean voltage across a time window, this is typically called the *mean amplitude* scoring method. However, this term can be confusing, because it involves both a mean across trials (when creating the averaged ERP waveform) and a mean across time points. As a result, I like to refer to it as the *time-window mean amplitude* scoring method to emphasize that it is a mean across time points.

This scoring method is unbiased (Luck et al., 2021). Thus, if we averaged together an infinite number of trials and measured the time-window mean amplitude from this averaged ERP waveform, we would obtain the true value for this participant. But how close to the true value is the score of 7.66 μV that we obtained from an average of only 20 trials? Informally, we can think of data quality as being an indicator of how much confidence we have that the score obtained from a given participant is close to the true score (i.e., the score we would obtain if we had an infinite number of trials).

Consequently, the data quality is inversely related to the amount of trial-to-trial variability in the single-trial EEG epochs. If the time-window mean amplitude in the single-trial epochs is similar from trial to trial, then the value we get from an average of 20 trials should be pretty close to the value we would get from average of an infinite number of trials. But if the time-window mean amplitude varies a great deal from trial to trial, then the value obtained from an average of 20 trials might be quite different from the true value. So, our first step in quantifying the data quality for the time-window mean amplitude from the averaged ERP waveform is to obtain time-window mean amplitude from each single-trial epoch and quantify the variability as the standard deviation (SD) of these single-trial values.

I did this for the 20 single-trial EEG epochs in Fig. 22.1b, and I obtained an SD of 5.73 μV. This is a useful value that tells me something important about the data quality of the EEG, but my ultimate goal is to quantify the precision of the time-window mean amplitude score I obtained from the averaged ERP waveform. I would have a perfectly precise score if the averaged ERP was computed from an infinite number of trials, and it seems obvious that the precision will decline as I have fewer and fewer trials. Thus, we need to take into account the number of trials that are being averaged together (which we denote N). Mathematically, the precision of a time-window mean amplitude value obtained from an averaged ERP waveform is directly related to the square root of N.

This brings us to how we compute our measure of data quality for averaged ERPs, the standardized measurement error (SME). The SME for a time-window mean amplitude score obtained from an averaged ERP waveform is simply the SD of the time-window mean amplitude values obtained from the single-trial EEG epochs divided by the square root of N. For reasons I will describe later, we call this the *analytic SME or aSME*, and it is defined as

$$aSME = SD \div \sqrt{N} \tag{22.1}$$

For the data shown in Fig. 22.1a, b, the SD was 5.73 μV and N was 20, so the aSME is $5.73 \div \sqrt{20}$ or 1.28 μV. The SME is an estimate of the expected *error* in the score, so it is the inverse of the precision (i.e., a greater SME indicates poorer data quality). To decide whether the SME for a given score is bad (large) or good (small), you can compare it to the score itself. For example, the score in this case was a time-window mean amplitude of 7.66 μV, and the estimated error of this score (the SME) was 1.28 μV. An error of 1.28 μV for a score of 7.66 μV isn't great, but it isn't terrible. I will have more to say about interpreting SME values later.

Note that the aSME will increase if there is more trial-to-trial variability in the scores obtained from the single-trial EEG epochs or if we have fewer trials in our averaged ERP waveform (all else being equal). Both of those things make perfect sense for a metric of data quality for averaged ERPs.

The aSME is automatically computed by ERPLAB Toolbox (version 8 and higher) for a set of default time windows when you create averaged ERP waveforms, and you can also specify custom time windows. The aSME can also be computed by the MNE-Python package (Gramfort et al., 2013). It should also be available soon as a *solution* in BrainVision Analyzer (BrainVision Analyzer, Brain Products GmbH, Gilching, Germany). And it is straightforward to use Eq. 22.1 to compute the aSME in custom code; Matlab examples are provided at https://doi.org/10.18115/D58G91. I hope that it will eventually be a standard part of all ERP analysis software, because everyone should be concerned with data quality.

22.2.2 Exactly What the SME Means

Now that we've seen how the aSME is computed, let's consider exactly what it means. You may have noted that Eq. 22.1 is the same as the familiar equation for the *standard error of the mean* or *SEM*. As an example of the SEM, imagine that you measured the height of 16 participants, and you found that the mean of these 16 single-participant heights was 180 cm with a SD of 32 cm. The standard error of this mean would be calculated by dividing the SD by the square root of the number of participants, which would be $32 \div \sqrt{16} = 8$ cm. You might then create a bar graph showing the mean of 180 cm with error bars of ±8 cm. You probably have some intuitions about whether an SEM of 8 cm is large or small relative to a mean of 180 cm. You can use those same intuitions when thinking about whether an aSME is large or small relative to an ERP amplitude score.

The aSME is just like the SEM, except that it is the standard error of the mean of the single-trial time-window mean amplitudes from a single participant. In other words, if you score the time-window mean amplitude from each individual trial for a participant, and take the mean of these values, the aSME is the standard error of this mean. It turns out that the mean of the single-trial time-window mean amplitudes is exactly equal to the time-window mean amplitude obtained from the averaged ERP waveform. So, we can use the standard error of the single-trial mean amplitudes to estimate the standard error of the time-window mean amplitude obtained from the averaged ERP waveform.

This is a key point, but it can be a little confusing, so it's worth repeating in a different way. If we obtain the time-window mean amplitude from 300 to 500 ms for each of the 20 single-trial EEG epochs in Fig. 22.1b, and we take the mean of these 20 values, we will get a mean of 7.66 μV. This is exactly the same value I obtained when I measured the time-window mean amplitude from 300 to 500 ms in the averaged ERP waveform. This was not an accident: You will always get the same value by scoring the time-window mean amplitude from the single-trial epochs and then averaging these values together or by first averaging the single-trial waveforms and then scoring the time-window mean amplitude from this averaged ERP waveform. If we take the SD of the 20 single-trial values (5.73 μV) and divide by the square root of 20 (4.47), we will get an SEM of 1.28 μV. This is the SEM of the 20 single-trial values, but it is also the standard error of the time-window mean amplitude obtained from the averaged ERP waveform, because this value obtained from the averaged ERP waveform is necessarily the same as the mean of the single-trial values. In other words, the standard error of the mean of the single-trial values is the *standard error of estimate* for the score obtained from the averaged ERP waveform. When we use the standard error of the mean (SEM) in this way, we call it the standardized measurement error (SME).

But what does this standard error actually mean? Roughly speaking, the standard error of a score tells you how consistent the score would be if you could repeat the same experiment many times in a given participant (without any learning or fatigue) and obtain the score for each repetition of the experiment. For example, imagine that we took the participant whose data are shown in Fig. 22.1a, b and repeated the same oddball experiment 10,000 times (without any learning or fatigue, but with different random noise in each repetition). Each time, we would get 20 oddball trials, create an averaged ERP waveform from these 20 trials, and obtain the time-window mean amplitude from 300 to 500 ms in this averaged ERP waveform. We might get a score of 8.2 μV for the first repetition of the experiment, 6.9 μV for the second repetition, 6.3 μV for the third repetition, etc. This is illustrated in Fig. 22.1d.

We would end up with 10,000 P3 amplitude scores for this participant (one for each of the 10,000 repetitions of the experiment). We could then quantify the variability of these 10,000 scores. If the scores are very consistent, this means that our score has high precision (i.e., that we have good data quality). But if the scores vary widely from repetition to repetition, this means that our score has low precision (i.e., that we have poor data quality). Figure 22.1d shows a simulation of this imaginary procedure in which the true score was 7.5 μV. You can see that the score obtained from most experiments was within ±1.5 μV of the true value of 7.5 μV, but sometimes the score was quite a bit lower or higher.

We could use the SD of these 10,000 scores as a measure of the consistency of the score across repetitions of the experiment. The SD of the scores in Fig. 22.1d was 1.3 μV, which means that 68% of the single-experiment scores were within 1.3 μV of the mean across the 10,000 scores (because 68% of values will fall within 1 SD of the mean given a normal distribution). The SD of these 10,000 scores seems like a very good way of quantifying the precision of the time-window mean amplitude score from this participant, because it tells us how close to the true value we can expect the score from a single experiment to be (on average). However, this approach is impractical in reality, because it requires repeating the experiment 10,000 times, and it assumes no learning or fatigue.

The magic of Eq. 22.1 is that it allows us to take the single-trial values from a single experiment and use them to estimate the SD of the scores we would obtain from the averaged ERP waveform if we repeated the experiment an infinite number of times (with no learning or fatigue). In other words, the SME we compute by applying Eq. 22.1 to the data from the one experiment shown in Fig. 22.1a, b is an estimate of the SD of the frequency distribution of scores shown in Fig. 22.1d. In technical terms, Fig. 22.1d shows the *sampling distribution* of the time-window mean amplitude score, the standard error of this score is the SD of the sampling distribution, and Eq. 22.1 allows us to estimate this standard error. Equation 22.1 is only an *estimate*, but you can see that the estimated value of 1.28 μV is quite close to the value of 1.3 μV obtained by simulating 10,000 repetitions of the experiment.

To summarize: the SME quantifies the variability (lack of consistency) that we would expect to see in our score if we repeated the experiment for a given participant an infinite number of times (expressed as the SD of the distribution of scores across repetitions). A large SME value indicates low consistency, which means that we cannot be confident that the score we obtain in a single experiment is close to the true value. When our score is a time-window mean amplitude, we can estimate the SME using the data from a single experiment by measuring this score from the single-trial EEG epochs and applying Eq. 22.1 to these single-trial scores. In other words, when we quantify the amplitude of an ERP component by scoring the time-window mean amplitude from the averaged ERP waveform, the SME is the same as the SEM.

22.2.3 SME for Other Types of Scores and for Transformed Waveforms

Unfortunately, Eq. 22.1 is valid only for time-window mean amplitude scores. It is not valid for peak amplitude scores, for peak latency scores, or for any other kind of amplitude or latency score that I have ever seen. However, there is a procedure called *bootstrapping* that you can use to estimate the SME for virtually any amplitude or latency score that can be obtained from an averaged ERP waveform. We call this the *bootstrapped SME* or *bSME*, and it is described in detail in Luck et al. (2021). At present, it cannot be automatically computed by any software package, but it can be computed by means of a relatively simple script in Matlab or another high-level programming language; several example Matlab scripts are available at https://doi.org/10.18115/D58G91.

The bSME also allows you to quantify the data quality for scores that are obtained from difference waves or other transformations of the ERP waveform. For example, imagine that you ran an N170 experiment in which you presented participants with pictures of faces and pictures of cars. You would find a larger N170 for the faces, and you could create a face-minus-car difference wave to isolate the face-specific processing from all of the nonspecific brain activity that is equal in the face-elicited and car-elicited ERP waveforms. You could use the onset time of this difference to quantify the speed at which the brain differentiated between faces and cars, and you could compare the onset times in a patient group and a control group. In other words, your score is the onset time measured from the difference wave. You could obtain a bSME value for this score for each participant to determine whether you are obtaining a good estimate of that participant's true onset time (i.e., the score you would obtain if you had an infinite number of trials).

No matter whether you are using aSME or bSME, it is an estimate of the standard error of the score. That is, it's an estimate of what would happen if you repeated the experiment an infinite number of times in a given participant (assuming no fatigue or learning), obtained the score for each repetition, and took the SD of these scores. It therefore provides a metric of the precision of the score you obtain from a given participant in one experiment.

22.2.4 Combining Scores Across Participants, Channels, and Conditions

You will get a separate SME value for each score you obtain. For example, you will get separate SME values for each participant. If you obtain the score from multiple channels, you will get a separate SME value for each channel for each participant (but if you average across a cluster of channels prior to obtaining the score, you will get a single SME value for the cluster). You will typically get separate values for each experimental condition (separately for each participant and channel), unless you average across conditions or make a difference wave across conditions prior to obtaining the scores. The fact that you get an SME value for each score can be very valuable, because you can determine whether some participants, channels, or conditions are particularly noisy.

Sometimes, however, you want to summarize the data quality for an entire study with a single number. In these cases, you can simply average the SME values across participants. You can also average the SME scores across channels and/or conditions, although you might consider averaging across the channels prior to scoring and/or obtaining scores from difference waves (see Luck & Gaspelin, 2017 for a discussion of the statistical advantages of averaging across channels or making difference waves between conditions prior to scoring). You can also combine the single-participant SME values using the *root mean square* instead of the mean, which has some advantages but is not usually necessary to obtain an overall metric of data quality for a given study (see Luck et al., 2021 for details).

22.2.5 Why the Scoring Method Matters

You might be wondering why I have been putting so much emphasis on the scores we obtain from averaged ERPs rather than the data quality of the EEG itself. One reason for this, as I mentioned earlier, is that the scores are what we put into our statistical analyses and use to test our scientific hypotheses. I don't care how clean or noisy my EEG is except insofar as the noise impacts the scores I will use to test my hypotheses. In this section, we will discuss a second reason for focusing on scores, namely the fact that a given type of noise may make one scoring method very imprecise without having much impact on a different scoring method. In other words, data quality depends on how the scoring method interacts with the noise in the data.

This is illustrated in Fig. 22.2, which shows a noise-free simulated ERP waveform with a P3 component peaking at 400 ms (Fig. 22.2a) and the same waveform with the addition of unrelated high-frequency noise (analogous to muscle noise; Fig. 22.2b). If we score the amplitude of the P3 component as the peak voltage between 300 and 500 ms, the high-frequency contamination has a large impact on the score (i.e., the score is very different when obtained from the original waveform versus the contaminated waveform). However, if we score the P3 amplitude as the time-window mean amplitude between 300 and 500 ms, the high-frequency contamination has only a small impact (i.e., the score is nearly identical for the original and contaminated waveforms). This is because the rapid upward and downward deflections largely cancel out when we average across the 300–500 ms period. For the noisy data in panel B, the SME would be very large for the peak amplitude score but very small for the time-window mean amplitude score.

The waveform in panel B "looks noisy," but that noise doesn't actually have much impact if we are using time-window mean amplitude scores. Perhaps the most important thing that I have learned from my efforts to quantify ERP data quality is that my eyes are not a good judge of how noisy an ERP waveform is. My eyes are drawn to high-frequency variations in the signal, but these high-frequency variations

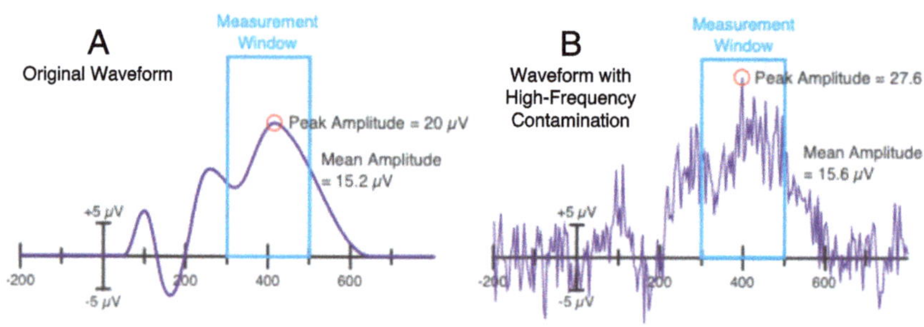

Fig. 22.2 Example of how the effect of noise on a score depends on the scoring method. (**a**) ERP waveform with no noise. (**b**) Same waveform as A, but with high-frequency contamination added. The high-frequency noise distorts the peak amplitude between 300 and 500 ms but has relatively little effect on the mean voltage during this measurement window

have minimal impact in most studies using time-window mean amplitude as the dependent variable. Low-frequency noise is not as visually distinctive but is a bigger problem than high-frequency noise in many ERP studies, especially those examining relatively long-latency components such as P3 and N400.

22.2.6 Example SME Values for 7 Common ERP Components

Now let's see what SME values look like in some actual experiments, using the data from the ERP CORE (Kappenman et al., 2021). This online resource (freely available at https://doi.org/10.18115/D5JW4R) contains data from 40 neurotypical young adults, each of whom was tested in optimized versions of 6 standard ERP paradigms that were designed to isolate 7 common ERP components. Each paradigm took about 10 min. Figure 22.3 shows the SME values, averaged across participants, for each component. Note that these components are described in more detail in Chap. 19.

Difference waves were used to isolate the component of interest from all of the other overlapping components. The P3b component was isolated from an oddball-minus-standard difference wave in an active visual oddball paradigm. The N170 component was isolated from a faces-minus-cars difference wave in a visual categorization paradigm. The mismatch negativity (MMN) was isolated from an oddball-minus-standard difference wave in a passive auditory oddball paradigm. The N400 component was isolated from an unrelated-minus-related difference wave in a visual word priming paradigm. The lateralized readiness potential (LRP) was isolated from a contralateral-minus-ipsilateral difference wave (relative to the response hand) in a flankers paradigm. The error-related negativity (ERN) was isolated from an incorrect-minus-correct difference wave in the same flankers paradigm used for the LRP (but from a different part of the scalp, averaged across contralateral and ipsilateral). The N2pc component was isolated from a contralateral-minus-ipsilateral difference wave (relative to the target side) in a simple visual search paradigm.

The amplitude of each component was scored as either the time-window mean amplitude or the peak amplitude from these difference waves. The latency was scored as either the 50% area latency (the latency that evenly divided the area under the curve into two equal-sized areas) or the peak latency from these same difference waves. A separate, optimized time window and electrode site was chosen for each component.

The results shown in Fig. 22.3 (from Zhang & Luck, 2023) demonstrate that the SME depends on the scoring method. For the amplitude of a component (Fig. 22.3a), the SME was about 20% worse (i.e., 20% larger) for peak amplitude scores than for time-window mean amplitude scores. For the latency of a component (Fig. 22.3b), the SME was up to 400% worse for peak latency scores than for 50% area latency

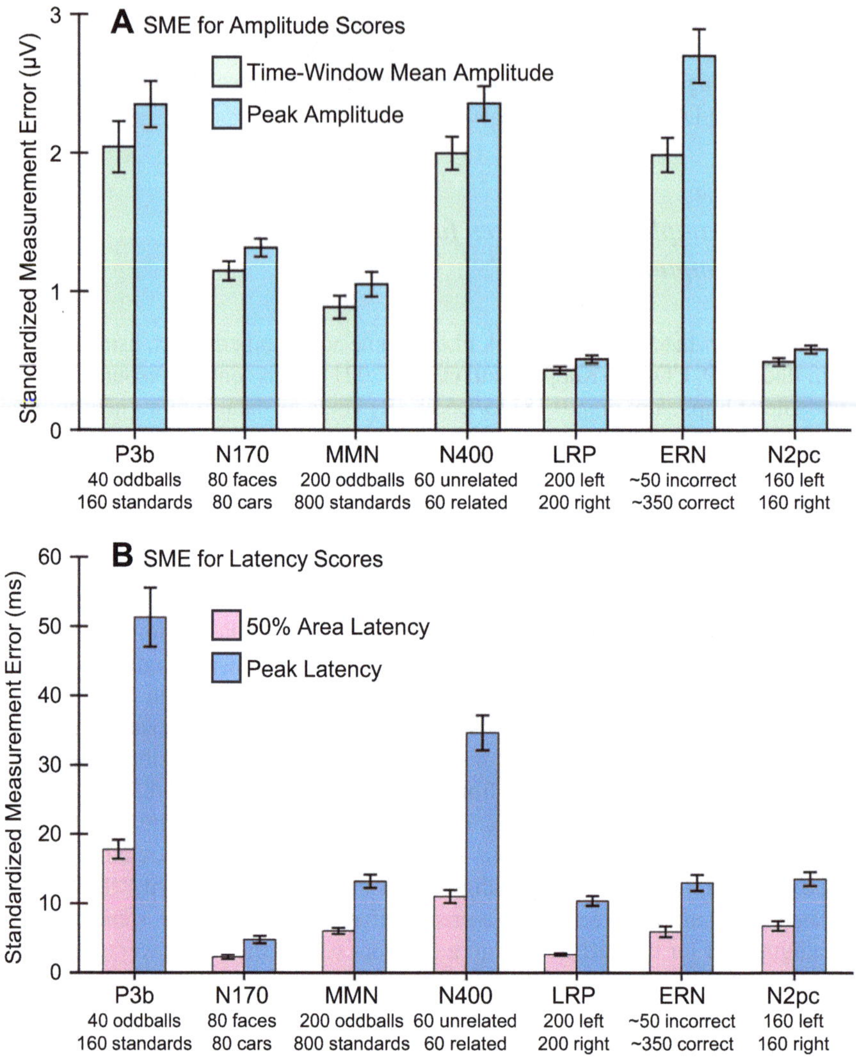

Fig. 22.3 Bootstrapped standardized measurement error (SME) values from the seven ERP components in the ERP CORE dataset (Kappenman et al., 2021), as derived by Zhang and Luck (2023). The number of trials per condition is given for each component. The SME values were obtained via bootstrapping from the difference wave used to isolate a given component (e.g., oddball minus standard for P3b). Separate SME values were obtained from each participant at the optimal electrode site for each component, and these SME values were then averaged across participants, with error bars indicating the standard error of the mean SME value across participants. Separate SME values are shown for two different amplitude scoring methods (**a**: time-window mean amplitude and peak amplitude) and for two different latency scoring methods (**b**: 50% area latency and peak latency)

scores. A lower SME value translates directly into larger effect sizes and greater statistical power (Luck et al., 2021), which means that a real effect in your data is more likely to yield a significant *p* value for a scoring method that produces lower SME values. This is a very direct way in which the SME can help you with your research: It can help you figure out which methods lead to the best data quality and therefore the largest effect sizes and greatest statistical power. Note that time-window mean amplitude and 50% area latency also have many other advantages over peak amplitude and peak latency (see Chapter 9 in Luck, 2014).

Figure 22.3 also shows that the SME values differ greatly among the different ERP components. For the amplitude scores, this can be explained primarily by differences in the number of trials used in each of the paradigms. For the latency scores, however, the SME values are not simply a function of the number of trials. They also vary with the shape of the waveform being measured, with lower SME values for narrow, "sharp" components like N170.

Note that the units of SME are the same as the units of the score. For amplitude scores, the SME quantifies your uncertainty about the true amplitude, and both the score and the SME of that score are in units of microvolts. For latency scores, the SME quantifies your uncertainty about the true latency, and both the score and the SME of that score are in units of milliseconds.

If you have recently set up a new ERP lab, and you would like to know if you are getting reasonable data quality, you can run one or more of the ERP CORE paradigms and compare your SME values to those shown in Fig. 22.3. If you are collecting data from a more challenging population (e.g., children), or in a noisier recording environment (e.g., a hospital room), or with a noisier EEG recording system (e.g., a dry electrode system), you can run one or more of the ERP CORE paradigms and see how much worse your SME values are compared to those in Fig. 22.3 (which come from neurotypical young adults tested in a low-noise environment with a high-quality EEG system). With some simple math, you can figure out how much your effect sizes and statistical power will be reduced by the greater SME values, and you can estimate how many more trials you would need to collect to make up for the noisier data (see Luck et al., 2021). I wish I had been able to compare my data quality with a set of values like these when I first started doing ERP research!

22.2.7 Using SME to Determine Optimal Processing and Analysis Procedures

One of the most valuable aspects of the SME is that it gives you the ability to determine which data processing methods and parameters will yield the best data quality in your own research. That is, you can try several different methods and/or parameters with your own data and see which ones yield the smallest SME values. You can then apply those methods and parameters to future studies. As an example, I will explain how SME can be used to determine the optimal filtering settings for a given study, but must be combined with other factors to avoid problems. But make sure to keep in mind that the SME quantifies only the precision, and you'll need to make sure you don't choose methods that are biased and create bogus differences between groups or conditions.

You might think that the best filter would be the one that produces the best (smallest) SME. However, there are two other important factors to consider. One is that filters reduce the size of the signal as well as the size of the noise. This is illustrated in Fig. 22.4a, which shows a single-participant averaged ERP waveform with and without the application of a 5 Hz low-pass filter. This filter passes low frequencies and attenuates high frequencies, and you can see that it reduces the high-frequency fluctuations and "smooths" the waveform. However, it also dramatically reduces the amplitude of the P1 and N1 peaks. A filter that decreases the signal more than it decreases the noise would do more harm than good.

Figure 22.4b shows what happens when the same filter is applied to an artificial ERP waveform designed to simulate the N170 waveform from the ERP CORE dataset (Kappenman et al., 2021). You can again see that the filter reduces the amplitude of the N170, and you can also see that it "smears out" the waveform, making it appear to onset earlier and offset later.

Figure 22.4c shows the effects of a 2 Hz high-pass filter on an averaged ERP waveform. This filter dramatically reduces the SME for the P3 component in this waveform by reducing slow drifts in the single-trial EEG epochs (which is not easily appreciated when viewing the averaged ERPs). However, it also cuts the P3 amplitude by more than half. Figure 22.4d shows the impact of this filter on the artificial N170 waveform. The filter has only a small impact on the N170 amplitude, but it creates artifactual positive-going peaks right before and right after the N170 wave. If you used this filter, you might write a paper saying that you had discovered new positive-going face-related ERP components, and you might be very embarrassed when someone else publishes a follow-up paper showing that your new components were actually filter artifacts. This has actually happened to researchers, and it was very embarrassing. It almost happened to me when I was in graduate school, but I realized that the effects were filter artifacts before I wrote the paper.

The point here is that, although filters may improve the SME by reducing low-frequency or high-frequency noise, they may also reduce the amplitude of the signal and distort the time course of the waveform. Consequently, the filter that produces the best SME may actually reduce your effect sizes and create bogus effects in your waveforms. My lab has therefore created an approach for determining the optimal filter that takes into account the impact of filtering on the size of the signal, the amount of noise, and the waveform distortion (Zhang et al., 2024a). For amplitude scores, this approach involves estimating the impact of a given filter on the signal-to-noise ratio, defined as the magnitude of the signal (as determined from an artificial waveform) divided by the magnitude of the noise (quantified as the SME in a real dataset). It also involves quantifying the magnitude of the waveform distortion using tools available in ERPLAB Toolbox (see details in Zhang et al., 2024a). We then systematically assess the signal-to-noise ratio and waveform

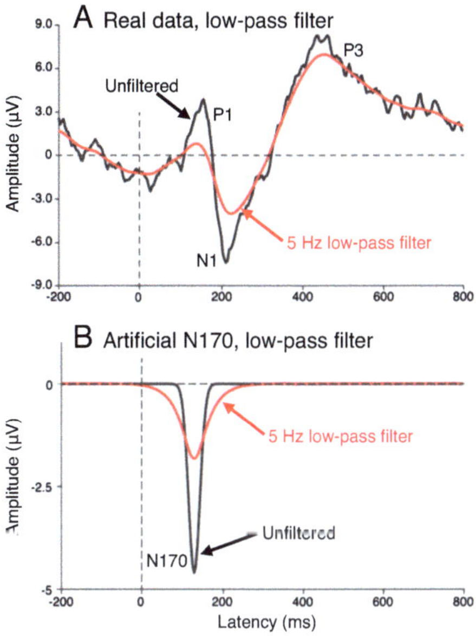

Fig. 22.4 Effects of low-pass and high-pass filters on real and artificial ERP waveforms. (**a**) Averaged ERP waveform from an actual research participant, with and without the application of a noncausal Butterworth low-pass filter with a 5 Hz half-amplitude cutoff and a slope of 48 dB/octave. (**b**) Artificial ERP waveform, with and without the same low-pass filter as in (**a**). (**c**) Averaged ERP waveform, with and without the application of a noncausal Butterworth high-pass filter with a 2 Hz half-amplitude cutoff and a slope of 12 dB/octave. (**d**) Artificial ERP waveform, with and without the same low-pass filter as in panel C. Note that both the low-pass and high-pass filters reduce the size of the signal as well as reducing the noise. Note also that the high-pass filter produces artifactual peaks (highlighted with blue circles) that are more easily observed in the artificial waveforms

distortion across a broad range of filter parameters. From the tested filters, we select the filter that yields the best signal-to-noise ratio without exceeding a threshold for the maximum allowable waveform distortion. The same approach can be used to select the optimal filter for latency scores, except that most filters do not reduce the size of the latency "signal," so the optimal filter is the one that produces the lowest noise (the smallest SME) without exceeding the threshold for waveform distortion.

The optimal filter varies depending on what component you are measuring, how you are measuring it, and the types of noise that are present in your data. As an example, consider low-frequency noise, which mainly arises from skin potentials and movement artifacts. This type of noise causes the signal to drift randomly away from the prestimulus baseline voltage, with more drift at longer latencies, and this can really reduce the precision of amplitude scores obtained from long-latency components such as P3 and N400. This noise has much less impact on shorter-latency components, such as N170 and MMN, because the EEG has not had much time to drift away from the baseline voltage prior to the measurement window.

Table 22.1 Recommended filter settings from Zhang et al. (2024b) for the 7 ERP components in the ERP CORE (in Hz, with a slope of 12 dB/octave), separately for each of four different scoring methods

	Mean amplitude		Peak amplitude		Peak latency		50% area latency	
	High-pass	Low-pass	High-pass	Low-pass	High-pass	Low-pass	High-pass	Low-pass
P3	0.2	≥10 or none	0.2	≥10 (or 10[a])	0.2	10 (or 5[a])	0.2	10 (or 5[a])
N170	0.9	≥30 or none	0.9	≥30 or none (or 30[a])	≤0.9	10–20	≤0.9	10–20
MMN	0.5	≥20 or none	0.5	≥20 or none (or 30[a])	≤0.5	10 (or 5[b])	≤0.5	≥20 or none
N400	0.2	≥10 or none	0.2	≥10 (or 10[a])	0.2	10 (or 5[a])	0.2	10 (or 5[a])
LRP	0.3	≥30 or none	0.3	30 (or 20[a])	0.3	5 (or 10[c])	0.3	5 (or 10[c])
ERN	0.4	≥20 or none	0.4	≥20 or none	0.4	10	0.4	10
N2pc	0.5	≥20 or none	0.5	20–40 (or 20[a])	0.01 or none	10	0.01 or none	5 (or 10[c])

[a]For data with moderate amounts of high-frequency noise
[b]For data with moderate amounts of broadband noise (if distortion of onset and offset times can be tolerated)
[c]To avoid distortion of onset and offset times

If you apply our procedure for determining the optimal filter settings to your own data, it will naturally take into account the component you are measuring, how you are measuring it, and the types of noise that are present in your data. You might not want to take the time to apply our procedure to your own data, so we've determined the optimal settings for the seven ERP components in the ERP CORE (Zhang et al., 2024b). The optimal settings are shown in Table 22.1. If you are measuring these components or similar components, and you have data with a similar noise level, you should be able to adapt these filter settings to your own data. The ERP CORE data were collected from highly cooperative college-student participants using a high-quality gel-based EEG system inside an electrically shielded chamber, so the data were quite clean. However, the table also shows the optimal settings for somewhat noisier data, which we assessed by adding broadband noise to the ERP CORE data prior to determining the optimal settings, so you can still use our results if you have somewhat noisier data. However, if your data have much higher levels of noise, especially high levels of low-frequency noise, you will probably benefit from more aggressive filtering (and you should determine the optimal settings using your own data).

22.2.8 Other Potential Uses of the SME

The SME can also help you determine the optimal approach for other steps of your processing pipeline. For example, we have also used the SME to assess the impact of artifact correction and artifact rejection on data quality, showing that excluding trials with artifacts will improve the overall data quality as long as the number of excluded trials is not too large (Zhang et al., 2024c). In principle, the SME could also be used to determine the optimal artifact rejection threshold for each participant and to decide whether a "bad" channel is so noisy that it should be interpolated. And, as shown in Fig. 22.3, the SME can help you determine which scoring approach will give you the smallest SME.

In principle, you could use the SME to identify participants whose data are so noisy that they should be excluded from the final analyses of a study (see supplementary section S8 in Luck et al., 2021). However, there are three important caveats. First, it is essential to define your criteria for exclusion prior to analyzing the data. Otherwise, you will likely inflate the rate of false positives in your statistical tests. Second, you may end up with a non-representative sample of participants, because the amount of noise may be correlated with other characteristics of your participants. Third, the benefits of excluding participants with noisy data may be outweighed by the reduction of degrees of freedom in your statistical analyses. To avoid this reduction in degrees of freedom, you could instead use the SME as a weighting factor in your statistical analyses, downweighting contributions from participants with noisier data rather than completely excluding those participants.

22.3 Adapting the SME for Time-Frequency Analyses and Other Psychophysiological Signals

It would be conceptually straightforward to adapt the SME for time-frequency analyses and for other psychophysiological signals. The main constraint is that it is valid only for scores obtained after averaging across trials.

For many standard time-frequency analyses, a wavelet transform is applied to convert the single-trial EEG epochs into a time-frequency representation, and then the time-frequency data are averaged across trials. A score is then obtained from the averaged time-frequency representation. When the score is the mean across a range of frequencies and time points (e.g., the mean power between 8 and 12 Hz from 300 to 600 ms), Eq. 22.1 could be used to compute the analytic SME. For more complex scoring algorithms, bootstrapping could be used to obtain the SME.

Similar approaches could be used for other psychophysiological measures, such as the skin conductance response, as long as the data are scored from averaged waveforms. There are many advantages to obtaining ERP, time-frequency, and other psychophysiological scores from the single-trial epochs rather than obtaining the scores from averaged waveforms, especially if the data are analyzed using multilevel models (Bürki et al., 2018; Heise et al., 2022; Volpert-Esmond et al., 2018; Winsler et al., 2018). Some other approach would be needed to quantify data quality in these situations.

22.4 Metrics of Reliability

An alternative approach to quantifying data quality is to compute a measure of *psychometric reliability*. In psychometrics, reliability is defined as the ratio of *true score variance* to *total variance*, where true score variance is variance across the individuals in a sample that reflects their true differences (uncontaminated by measurement error) and total variance is the observed variance that results from a mixture of true differences across individuals and measurement error.

The true score variance is not something you can directly measure, but there are various ways of estimating reliability. Consider, for example, the data shown in Fig. 22.1, in which P3 amplitude was measured from an averaged ERP created by averaging together 20 trials. Imagine that you made one average from the 10 odd-numbered oddball trials and another average from the 10 even-numbered trials. You could then score the P3 amplitude from each of these two averaged ERP waveforms. If these two P3 amplitude scores are quite similar, this would suggest that your data quality is high (i.e., that your P3 amplitude scores are relatively precise). However, you might have simply gotten lucky with the two specific sets of 10 trials used to create each average. The typical approach is therefore to get these two scores (i.e., from the averages of the odd-numbered and even-numbered trials) from each of your participants, and then look at the Pearson *r* correlation between them. If the correlation between the scores from the odd-numbered averages and the scores from the even-numbered averages is high, then the data quality must be high. This is called the *split-half reliability*. The estimate of reliability that you obtain in this manner will underestimate the true reliability, because each averaged ERP waveform has only half the number of trials that will be used in the main analysis, but the Spearman-Brown prophecy formula can be used to adjust for this (Brown, 2018). Also, you can compute Cronbach's alpha instead of split-half reliability, which effectively gives you the reliability averaged across all possible splits of the data (Tavakol & Dennick, 2011).

This approach has two major downsides. One is that it gives you a metric of data quality only for the entire sample of participants, and it tells you nothing about the data quality for individual participants. However, Clayson and his colleagues have developed an approach to reliability that provides estimates of subject-level as well as group-level reliability (Clayson et al., 2021b).

The second and more significant downside is that, because reliability is defined as true score variance divided by total variance, it depends on the amount of real variability across participants and not just the data quality. In general, you will see greater reliability in heterogeneous groups of participants than in homogeneous participants. For example, imagine that you have taken great pains to ensure that all of your participants have gotten a good night's sleep the night before, that they are being tested at their best time of day, that they are not hungry, and they are in a neutral mood. All of these factors will tend to reduce the true score variance (assuming that each participant is tested in a single session) and will therefore reduce the reliability, making it seem as if you have poor data quality. Or imagine that you have used a difference wave to isolate the effects of a new experimental manipulation that produces the same effect in all adult participants. Again, you will have low reliability and might think you have poor data quality. This is sometimes called the *reliability paradox* (Hedge et al., 2018).

However, this does not mean that reliability metrics have no value. If you are studying individual differences—looking at correlations between psychophysiological scores and other measures—reliability is usually the appropriate way to quantify data quality. This is because you must have substantial true score variance to see correlations with other measures, and the ratio of true score variance to total variance will predict the maximum correlation you can expect to obtain given your data quality. Note that you can convert SME values into true score variance and reliability with some simple math, so there is no need to create the averages of odd-numbered and even-numbered trials (see Luck et al., 2021).

A much more advanced and powerful approach to estimating reliability has been developed by Clayson et al. (2021a, b). Whereas the SME, split-half reliability, and Cronbach's alpha are based on classical test theory, this new approach to reliability is based on generalizability theory (Briesch et al., 2014). As a result, it can model factors that impact the consistency of scores across multiple different levels of analysis, including single trials, averages, and differences between conditions. Unlike the SME, it could be used to quantify data quality for single-trial analyses. However, because it is still a measure of reliability, it is still dependent on the amount of true score variance and is therefore best suited for correlational (individual differences) analyses.

22.5 Final Thoughts

Given how noisy the EEG can be, it is rather remarkable that rigorous metrics of ERP data quality have only recently been developed and disseminated. These metrics can be highly valuable for avoiding problematic data and for determining the methods that yield the cleanest data. Here I have focused mainly on the SME, which is now accessible in multiple EEG/ERP software packages. However, the most important point is that you should pay close attention to data quality and quantify it using whatever metrics are most sensible for your research. By optimizing your data quality, you will increase the likelihood that you can publish your research in top-quality journals and you will also increase the likelihood that the conclusions of your research are true. And isn't that what we all want?

References

Briesch, A. M., Swaminathan, H., Welsh, M., & Chafouleas, S. M. (2014). Generalizability theory: A practical guide to study design, implementation, and interpretation. *Journal of School Psychology, 52*(1), 13–35. https://doi.org/10.1016/j.jsp.2013.11.008

Brown, J. D. (2018). Spearman-Brown prophecy formula. In B. B. Frey (Ed.), *The SAGE encyclopedia of educational research, measurement, and evaluation*. SAGE Publications, Inc. https://doi.org/10.4135/9781506326139.n647

Bürki, A., Frossard, J., & Renaud, O. (2018). Accounting for stimulus and participant effects in event-related potential analyses to increase the replicability of studies. *Journal of Neuroscience Methods, 309*, 218–227. https://doi.org/10.1016/j.jneumeth.2018.09.016

Clayson, P. E., Baldwin, S. A., & Larson, M. J. (2021a). Evaluating the internal consistency of subtraction-based and residualized difference scores: Considerations for psychometric reliability analyses of event-related potentials. *Psychophysiology*, e13762. https://doi.org/10.1111/psyp.13762

Clayson, P. E., Brush, C. J., & Hajcak, G. (2021b). Data quality and reliability metrics for event-related potentials (ERPs): The utility of subject-level reliability. *International Journal of Psychophysiology, 165*, 121–136. https://doi.org/10.1016/j.ijpsycho.2021.04.004

Gramfort, A., Luessi, M., Larson, E., Engemann, D. A., Strohmeier, D., Brodbeck, C., Goj, R., Jas, M., Brooks, T., Parkkonen, L., & Hämäläinen, M. (2013). MEG and EEG data analysis with MNE-python. *Frontiers in Neuroscience, 7*. https://doi.org/10.3389/fnins.2013.00267

Hedge, C., Powell, G., & Sumner, P. (2018). The reliability paradox: Why robust cognitive tasks do not produce reliable individual differences. *Behavior Research Methods, 50*(3), 1166–1186. https://doi.org/10.3758/s13428-017-0935-1

Heise, M. J., Mon, S. K., & Bowman, L. C. (2022). Utility of linear mixed effects models for event-related potential research with infants and children. *Developmental Cognitive Neuroscience, 54*, 101070. https://doi.org/10.1016/j.dcn.2022.101070

Kappenman, E. S., & Luck, S. J. (2010). The effects of electrode impedance on data quality and statistical significance in ERP recordings. *Psychophysiology, 47*, 888–904. https://doi.org/10.1111/j.1469-8986.2010.01009.x

Kappenman, E. S., Farrens, J. L., Zhang, W., Stewart, A. X., & Luck, S. J. (2021). ERP CORE: An open resource for human event-related potential research. *NeuroImage, 225*, 117465. https://doi.org/10.1016/j.neuroimage.2020.117465

Lopez-Calderon, J., & Luck, S. J. (2014). ERPLAB: An open-source toolbox for the analysis of event-related potentials. *Frontiers in Human Neuroscience, 8*, 213. https://doi.org/10.3389/fnhum.2014.00213

Luck, S. J. (2014). *An introduction to the event-related potential technique* (2nd ed.). MIT Press.

Luck, S. J., & Gaspelin, N. (2017). How to get statistically significant effects in any ERP experiment (and why you shouldn't). *Psychophysiology, 54*, 146–157. https://doi.org/10.1111/psyp.12639

Luck, S. J., Stewart, A. X., Simmons, A. M., & Rhemtulla, M. (2021). Standardized measurement error: A universal metric of data quality for averaged event-related potentials. *Psychophysiology, 58*, e13793. https://doi.org/10.1111/psyp.13793

Tavakol, M., & Dennick, R. (2011). Making sense of Cronbach's alpha. *International Journal of Medical Education, 2*, 53–55. https://doi.org/10.5116/ijme.4dfb.8dfd

Volpert-Esmond, H. I., Merkle, E. C., Levsen, M. P., Ito, T. A., & Bartholow, B. D. (2018). Using trial-level data and multilevel modeling to investigate within-task change in event-related potentials. *Psychophysiology, 55*(5), e13044. https://doi.org/10.1111/psyp.13044

Winsler, K., Midgley, K. J., Grainger, J., & Holcomb, P. J. (2018). An electrophysiological megastudy of spoken word recognition. *Language, Cognition and Neuroscience, 33*(8), 1063–1082. https://doi.org/10.1080/23273798.2018.1455985

Zhang, G., & Luck, S. J. (2023). Variations in ERP data quality across paradigms, participants, and scoring procedures. *Psychophysiology, 60*, e14264. https://doi.org/10.1111/psyp.14264

Zhang, G., Garrett, D. R., & Luck, S. J. (2024a). Optimal filters for ERP research I: A general approach for selecting filter settings. *Psychophysiology, 61*, e14531. https://doi.org/10.1111/psyp.14531

Zhang, G., Garrett, D. R., & Luck, S. J. (2024b). Optimal filters for ERP research II: Recommended settings for seven common ERP components. *Psychophysiology, 61*, e14530. https://doi.org/10.1111/psyp.14530

Zhang, G., Garrett, D. R., Simmons, A. M., Kiat, J. E., & Luck, S. J. (2024c). Evaluating the effectiveness of artifact correction and rejection in event-related potential research. *Psychophysiology, 61*, e14511. https://doi.org/10.1111/psyp.14511

Part V
What Is EEG Used For?

Chapter 23
EEG in Basic Science and Academic Research

Fernando Cross Villasana

Abstract The EEG possesses qualities that make it a valuable tool in basic scientific research. Thanks to its faster than millisecond temporal precision, EEG can track fast-paced neurological activity and allow us to investigate its relation to cognitive processes. Its portability and non-invasive nature allow large samples to be tested, and facilitate access to populations that may be difficult to test using other techniques. The design of EEG studies and their interpretation largely relies on broader frameworks from experimental psychology and cognitive neuroscience. This enables a systematic approach for relating the brainwave and behavioral observations to neural functions, and for building cognitive models about attention, emotion, language, or memory, among other functions.

This chapter focuses on EEG in the context of the broader frameworks from cognitive neuroscience that are the basis for most academic EEG studies. The emphasis is on the process of how such frameworks can be applied to investigate the links between brain, cognition, and behavior, and how these methods contribute to clarify the meaning of EEG rhythms. The chapter ends with brief examples of contributions from other disciplines to EEG, and how EEG-derived knowledge can contribute to questions in other fields of study.

Keywords Neuropsychology · Mental chronometry · EEG

23.1 Cognitive Neuroscience

Over the decades, cognitive research has enhanced our understanding of faculties like sensation, perception, memory, attention, language, movement, or decision-making, where EEG has played an important role. However, investigating brain function in relation to behavior began long before the development of EEG. For example, systematic observations of the deficits shown by patients with different

F. Cross Villasana (✉)
Brain Products GmbH, Gilching, Germany

T. Warbrick (ed.), *The EEG Handbook*,
https://doi.org/10.1007/978-3-032-20450-9_23

kinds of neurological damage have been used to attribute mental functions to different sections of the brain (Vaidya et al., 2019). This method would become part of the field of neuropsychology. Additionally, in the field of mental chronometry, clever reaction time experiments were devised to make deductions about the cognitive operations involved in a task. The methods and perspectives of neuropsychology and mental chronometry have provided a useful framework for designing EEG experiments and interpreting observations (Linden, 2007). Meanwhile, EEG provides a way to relate theoretical constructs about cognition to activity in the brain, especially with regard to the timing of mental processes (Meyer et al., 1988). These disciplines are deeply interconnected and their methods and perspectives continuously evolve. This chapter presents research examples within the framework of cognitive research, specifically neuropsychology and mental chronometry, to show how EEG integrates into the landscape of cognitive neuroscience.

23.1.1 Neuropsychology and EEG

During the nineteenth century, the idea that particular areas of the brain relate to specific cognitive functions gained traction among the scientific community (Benton & Sivan, 2007). This notion motivated the systematic observation of cognitive deficits produced by lesions in different locations of the brain in patients (Vaidya et al., 2019). Physicians and researchers have used meticulous observations of the patient's behavior, and developed diverse procedures to test their cognitive function. From these works, the concept of *dissociation* appeared (Vaidya et al., 2019), where impairment of a cognitive function would be present after damage to a brain region, without affecting other functions. Such dissociation would suggest that the impaired function is independent of other functions, and can be attributed to a specific brain area. A condition that brings stronger evidence for dissociation is *double dissociation*, where two patients with lesions in different brain regions (A and B) each show a deficit in one cognitive function (X or Y) without effects in the other function (Ramminger et al., 2023; Vaidya et al., 2019). This would imply that these functions are implemented by separate brain sub-systems, and are independent of each other. Aiding such dissociation logic, technologies like EEG can be used to investigate the involved cognitive functions not only in patients, but also in healthy population. In this way, EEG enables studying normal neuro-cognitive function beyond the clinical context. In particular, event-related potentials (ERPs) have been useful in this regard due to their precise timing. More information on ERPs can be found in Chap. 19.

An example of a proposed dissociation is the perception of faces as an independent function from the perception of other objects. This is reflected in the clinical condition of prosopagnosia (Duchaine & Nakayama, 2005; Gerlach & Starrfelt, 2024), where patients are unable to recognize faces, without showing visual or memory impairments, nor impairments in the recognition of other objects (Corrow et al., 2016). Prosopagnosia can occur after a brain injury, mainly to the fusiform face area (FFA) but also to the occipital face area (OFA) or on temporal regions like the posterior superior temporal sulcus. Some patients have developmental prosopagnosia, so they do not acquire face recognition skills despite not suffering adverse neurological events (Corrow et al., 2016). In EEG, a similar dissociation is reflected by the N170 ERP component in the healthy population. The N170 shows enhanced amplitude to the perception of faces (Rossion & Caharel, 2011), and is estimated to originate from the FFA (Gao et al., 2019). In patients with acquired prosopagnosia, N170 modulation is altered when the damage involves both the FFA and OFA, but preserved if the damage is limited only to OFA or to anterior temporal regions (Corrow et al., 2016). This supports the role of FFA and OFA in early face perception, while temporal areas are involved in later stages of face identification (Corrow et al., 2016). It further supports the notion that face perception processes are themselves dissociated from later face memory (Dalrymple et al., 2014; Manippa et al., 2023). In patients with developmental prosopagnosia, the N170 modulation is preserved in many but not all cases, suggesting that some patients preserve perceptual face processing (Towler & Eimer, 2012). However, if the faces are inverted or scrambled, they fail to show additional N170 modulations normally seen in healthy participants. This suggests that patients may develop some compensatory strategies for face perception, but these are not effective under more challenging conditions (Manippa et al., 2023; Towler et al., 2017). In relation to the previous proposal, there is evidence showing that the N170 improves after therapeutic training interventions in patients with developmental prosopagnosia (Manippa et al., 2023). Altogether, these examples from prosopagnosia show how EEG can inform and refine ideas inspired by the neuropsychological concept of dissociation.

Double dissociation is classically portrayed in the distinction between speech production and comprehension reflected respectively in Broca's and Wernicke's aphasia (Ramminger et al., 2023). In Broca's aphasia, patients with damage to a network involving the insula, left inferior frontal gyrus, and adjacent areas have difficulty generating speech but retain the ability to understand language with some limitations (Fridriksson et al., 2015; Pracar et al., 2025; Silva, 2021). In Wernicke's aphasia, patients with damage involving middle-posterior temporal areas show impaired language comprehension, while preserving their ability to generate fluent but incoherent speech, using mixed words and sounds, and neologisms (Matchin et al., 2022; Silva, 2021; Thompson et al., 2015). From a language processing perspective, these deficits reflect impairments in the syntactic and semantic processes for Broca and Wernicke aphasia patients respectively. However, there is still a degree of overlap in the symptoms of these two aphasias that calls for a closer investigation of this proposal. With EEG, the ERP components N400 and P600 related to language comprehension have been used to gain insight into this question. In healthy participants, larger N400 component amplitudes have been observed during sentences with semantic violations, such as the appearance of an unexpected word (Lau et al., 2009). Amplitude modulations of the P600 meanwhile, are proposed to reflect detection of syntactic violations as in incoherent sentences (Meechan et al., 2021). ERP assessments in patients with Broca's aphasia have shown disruptions in both the N400 and P600. While the P600 reflects the expected difficulties with syntax in Broca's aphasia patients, the N400 disruption represents a neural marker that their comprehension of language is still affected (Kotz & Friederici, 2003; Meechan et al., 2021). Moreover, the N400 disruptions show brain-related evidence that semantic language processing is also affected in Broca's aphasia. Wernicke's aphasia is less studied with EEG but it has been observed that these patients generate inconsistent N400 responses (Kotz & Friederici, 2003), as would be expected due to the clearly impaired language comprehension. Using an additional ERP component, the Phonological Mapping Negativity to check auditory processing, Robson et al. (2017) support that Wernicke's aphasia patients already have phonological processing deficits, which likely contribute to comprehension deficits and the related N400 alteration.

Not only can EEG contribute to neuropsychology research, but frameworks developed in neuropsychology can also guide EEG basic research. As an example, Wischnewski et al. (2022) relied on the concept of double dissociation to clarify the relationship between the ongoing EEG beta and mu oscillations to corticospinal excitability. In the experiment, they applied transcranial magnetic stimulation on the motor cortex during the positive or negative phase of the ongoing mu or beta oscillations. They looked at the effect it had on the motor evoked potential as an index of corticospinal excitability. Results showed greater excitability during the peak and the falling phase of beta, and less during the trough and rising phase, suggesting an excitatory role. The opposite was observed for mu: greater excitability during the trough and rising phase, and less during the peak and the falling phase, supporting an inhibitory role. The double dissociation between ongoing beta and mu contributes to the understanding of these rhythms and can help the fine-tuning of transcranial magnetic stimulation to reduce response variability for research and clinical purposes (Wischnewski et al., 2022).

The previous examples show how clinical observations guided by neuropsychology and EEG research inform each other, with particular attention to the logic of dissociation of functions. Though this approach has been fruitful in enhancing our understanding of the brain, the limitations in lesion studies and the logic of dissociation must also be kept in mind (Vaidya et al., 2019). For instance, the lesions in patients rarely affect a single region of the brain. There is also damage to adjacent regions and generalized brain effects related to the injury (e.g., inflammation). Lesions involving a studied region also vary in type and severity between patients. Such conditions make it harder to isolate the effects of the lesion on a single cognitive function and attribute it to a particular brain region. Observations of many patients with overlapping lesions are required to compensate for this disadvantage. Another important aspect is that certain cognitive functions may emerge from distributed networks, rather than singular locations in the brain (Ramminger et al., 2023; Vaidya et al., 2019).

Limitations are not an impediment for research to continue evolving. In fact, it is this type of limitation that fuels the exchange between methodological perspectives, and the use of different techniques in neuroscience. From the prosopagnosia examples above, functional magnetic resonance imaging helped in the detection of a neural network related to face recognition beyond the fusiform gyrus, while the N170 ERP component helps to clarify how damage to the different regions affects the perception or recognition of a face. Furthermore, technical and methodological advances within EEG help to address open questions such as the proposal that distributed networks generate a cognitive function, which can be addressed through connectivity techniques in EEG. As an example, Rutar Gorišek et al. (2016) used connectivity measures in Broca's patients to suggest that Broca's area is part of a broader working memory network that gets disrupted during injury, and this could explain why these patients also show some language comprehension deficits.

23.1.2 Mental Chronometry and EEG

Mental chronometry is a discipline that evolved with the aim of timing mental activity, and disentangling what cognitive processing stages occur behind overt behavior. As such, it has been an important theoretical and methodological source for EEG in cognitive research. At the same time, the EEG's precise time resolution has made it an ideal ally for mental chronometry as it enables the timing of neural markers that happen before or without overt behavior.

Mental chronometry relies on the assumption that separable cognitive processes exist whose final outputs manifest in behavior, and an alteration of one of those cognitive processes will modify behavior (Linden, 2007; Meyer et al., 1988). As a result, it is possible to manipulate task conditions to target a specific cognitive process such as attention, working memory, or decision making. Inferences on the target cognitive process are made based on alterations in reaction time, in accuracy, or speed-accuracy tradeoffs (Meyer et al., 1988). A useful example comes from one of the pioneers in mental chronometry, Franciscus Donders, in the late nineteenth century, who used reaction times to infer cognitive processes (Vidal et al., 2011). In his experiments, he devised a subtraction method to time mental processes by comparing the reaction times between three tasks. In task A "simple reaction time", participants had to press a button as quickly as possible following a stimulus. In task B "choice reaction time", two different stimuli were presented in alternation and participants had to push the correct button that corresponded to the target stimulus. Task B added the processes of stimulus discrimination and response selection. In task C "go-no-go", one of two stimuli could be presented but participants only had to respond to one type of stimulus and refrain the response to the other. Task C then required stimulus discrimination but not response selection. Donders reasoned that by subtracting the reaction times from these tasks, it would be possible to determine the timing of the particular cognitive operations that each task requires (Meyer et al., 1988). As task B required target discrimination and response selection, but task C only needed target discrimination, B–C would eliminate the common process of target discrimination, and reveal the timing needed for response selection. Similarly C–A would eliminate the common process of response execution, and reveal the timing for stimulus discrimination. With EEG, it is possible to check the timing of ERPs during reaction time experiments to complement the subtraction method. In this way, Vidal et al. (2011) observed that Donders' choice

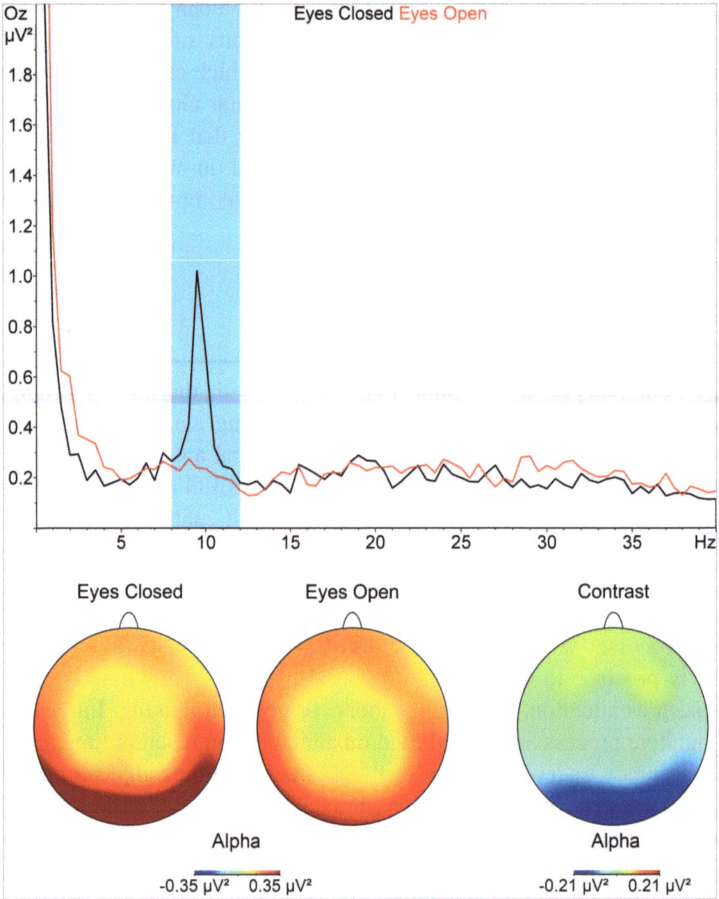

Fig. 23.1 Example of the subtraction method applied to alpha (8–12 Hz) power during an eyes-closed and an eyes-open condition. On top, the spectrum plot shows a clear power reduction effect after eye opening. When mapping alpha over the scalp, anterior alpha levels common to both conditions occlude the alpha reduction localization. The contrast is created by subtracting the spectra of the eyes-closed from the eyes-open condition. This eliminates the common anterior activity and isolates the alpha power reduction effect to the occipital region

and Go/No-go tasks showed differences in brain motor activity and response execution timing in the response process that they have in common. This shows that although Donders' method is helpful, the assumption that additional processes can be "purely inserted" into a task only partially holds (Vidal et al., 2011).

Besides its use in reaction time, the subtraction logic has also been applied in EEG and neuroimaging. The brain activity levels recorded under two conditions are subtracted from each other to eliminate common neural processes, and isolate a specific neural signal (Luck, 2014). A simple example can be observed in Fig. 23.1, where the power levels in different bands are compared between an eyes-closed and an eyes-open condition. The plot of channel Oz shows the classic reduction of power in the alpha band (8–12 Hz) after eye opening. When plotting alpha levels on the scalp, a difference can be spotted in the occipital region, but it is perhaps not as clear, and comes along with common activity in anterior regions. By subtracting eyes-closed from eyes-open activity to create a contrast, the neural activity common to both conditions is largely eliminated, revealing the alpha reduction more clearly in the occipital region.

Like in mental chronometry, the subtraction method has also received criticism in the EEG and imaging domains. It is argued that this method may miss relevant neural modulations produced by the experimental conditions, which are better captured through more advanced methods (Alexander et al., 2015; Vidal et al., 2011). Nonetheless, due to its availability and ease for exploring neural signals, the subtraction method continues to be widely used by researchers.

In mental chronometry, the Donders' subtraction technique requires that the cognitive processes can be inserted or removed by the experimenter, but this is not always possible. Considering this, the "additive factors" method was developed by Saul Sternberg (Linden, 2007). Instead of adding or removing cognitive processes to a task, one identifies factors that may affect the duration of particular processing stages. For example, in a reaction time experiment the modification of the factor "stimulus visual clarity" can be used to affect a stage of stimulus perception; and the modification of the factor "frequency of the target stimulus" to affect a stage of response selection. If altering two factors leads to additive effects on reaction times, it can be proposed that at least two separate processing stages were affected. In this way, by manipulating factors one can make inferences about the kinds of processing stages behind overt behavior (Meyer et al., 1988). Additive factors methodology has been fruitful in the design of EEG experiments, as it allows for disentangling various processing stages within a single experimental task. As an example, in the quest to clarify the meaning of ERP components, Smulders et al. (1995) used a 2-choice reaction time task where either of the factors: clarity of the stimulus or complexity of the response was varied. The task was paired with assessment of the P300 ERP component, known to respond to stimulus evaluation, and the LRP (lateralized readiness potential) component, related to the response preparation stage. Degraded stimulus quality caused delays in both components following the stimulus presentation. Response complexity did not delay the P300, as predicted, however it also did not cause delays in LRP. Response complexity delayed the time between both components and response execution. These results support the idea that the P300 is not involved in response processes. They further suggest that while the LRP precedes response programming, it may not be involved in the previous response selection, since the onset of the LRP was postponed by delays in the previous stimulus evaluation, and the distance from the LRP to the overt response was increased by response complexity. With these results, the authors also support that information transmission in the brain follows consecutive discrete stages during this task (Smulders et al., 1995).

Both Donders' subtraction and Sternberg's additive factors methods rely on the assumption that cognitive processes are discrete and follow each other serially, starting only when the previous stage has concluded (Linden, 2007). These frameworks do not consider the possibility of different processes running in parallel, or that a process could already start with partial information fed forward from the previous stage. Further models have been developed to address these questions. As an example, the cascade model (McClelland, 1979) posits that a processing stage can kick-start after reaching an input threshold from a previous stage, implying that the flow of information is continuous. In the stimulus to response mapping, the Eriksen and Schultz continuous flow model (Coles et al., 1985) proposes that as soon as partially processed visual information is available, multiple competing response processes start in parallel, which are refined with more complete visual information. In EEG, response monitoring processes are likely candidates that reflect continuous and parallel information processing. Going back to choice reaction time, it has been reported that when errors are committed, the ERP component "Error Related Negativity" (ERN; Gehring et al., 2018) becomes prominent following the incorrect response. However, a smaller ERN appears during partial errors where a participant rectifies just before pressing the wrong button, as reflected by hand muscle activity. Moreover, an even smaller ERN is also present during correct responses (Vidal et al., 2000). This suggests that ERN may reflect a response verification process that starts already before the response is emitted (Vidal et al., 2000). With these characteristics, it has been proposed that ERN modulations reflect the model of continuous flow of information (Ullsperger et al., 2014). Although the significance of the ERN is still being refined, it is a useful tool for researching the characteristics of cognitive processes.

The various cognitive models in the field involve questions such as whether information processing between regions of the brain is serial or parallel, discrete or continuous. Here EEG can play a role in clarifying these open questions. At the same time EEG research profits extensively from the framework of mental chronometry. Nonetheless, one must be aware of limitations when combining EEG and mental chronometry. In the case of ERPs, as exemplified above, the putative processes that various ERP components represent do not always coincide with the processing stages proposed in mental chronometric models (Linden, 2007). Components can overlap with each other, so it is not always easy to define their timing (Linden, 2007). Some ERP components seem to result from diverse neural sources from various areas in the brain, possibly representing different cognitive processes. An example is the P300, which is estimated to have multiple brain sources (Huang et al., 2015). P300 is also known to be modulated under different contexts such as sudden unexpected stimuli, or frequent but more complex stimuli (Huang et al., 2015). EEG measurements are also mostly confined to the scalp, while some cognitive processes are related to deeper brain areas. Considering the capabilities and limitations of EEG measurements, conducting a conscientious literature review of the EEG signals and mental chronometric models to use, and a careful experimental design represent the best way to implement a successful study.

23.2 Research on the EEG Signals

Knowing what EEG signals represent is paramount for their use in research. In the examples from the previous section, EEG was mostly implemented to help answer questions about neurocognitive processes. But to achieve a more thorough understanding of the EEG itself, specific studies are required to disentangle the meaning of its various signals. With this aim, multiple complementary approaches are used to provide context for the observed modulations in the EEG: besides behavioral manipulations and mental chronometric models, joint measurements with other techniques like transcranial magnetic stimulation (TMS) or functional magnetic resonance imaging (fMRI) are implemented. A broader electrophysiological perspective in human, animal, and in-vitro studies has also helped to explain the neuronal origins of the EEG, and the physiological states that underlie different EEG rhythms.

A good example is research on the meaning of alpha band oscillations. From its first observation, alpha was seen in contexts that implied down-regulation of cortical activity such as relaxing or closing the eyes. Traditional cognitive experiments show that ongoing alpha decreases transiently during the presentation of stimuli, indicating activation (Klimesch et al., 2007). Meanwhile, withholding a motor response enhances alpha-range rhythms over motor areas, supporting a role of active inhibition (Hummel et al., 2002). Using TMS, the brain's motor area shows a decreased response when preceding alpha levels are higher (Sauseng et al., 2009), moreover, magnetic pulses deployed to visual areas during higher ongoing alpha power generate a smaller response than during the lower alpha power (Taylor & Thut, 2012). In the same way, visual stimuli are less likely to be perceived if they are presented during higher alpha power (Taylor & Thut, 2012). When pairing EEG and fMRI, it can be observed that cortical activation levels are lower when alpha levels are higher (Murta et al., 2015). On a physiological level, one mechanism observed for alpha rhythms in animal models is an excitation-inhibition loop of thalamic neurons that project towards the cortex during periods of inactivity, particularly in visual and motor systems, giving shape to oscillations in the alpha range (Sterman, 1996). Altogether, these studies at different levels paint a more complete picture of alpha, where it plays an inhibitory role over the cortex. It is further proposed that this plays a role in pacing and coordinating cognitive systems (Beste et al., 2023; Jensen & Mazaheri, 2010; Klimesch et al., 2007).

Studying the properties of EEG facilitates its use for answering questions on different aspects of the brain or in other academic areas. For instance, based on the inhibitory properties of alpha rhythms, it has been proposed that the motor and visual systems exert transient inhibition effects on each other, seen as an enhancement in EEG alpha levels over the inactive system when the other is activated (Pfurtscheller & Lopes da Silva, 1999). An example of a practical application is that EEG features can be used as an adjunct measure to monitor the state of consciousness during anesthesia (Hight et al., 2020). For the alpha rhythm in particular, it has been observed that as consciousness fades, the down-regulation of anterior brain regions is reflected by increases in frontally coherent alpha amplitude (e.g., Purdon et al., 2013). Importantly however, this is not a definitive marker and the phenomenon is variable during practice, so it must be complemented with other measures (Hight et al., 2020).

The above examples serve to illustrate how deeper knowledge of particular EEG signals can be used to inform other fields. Besides the alpha example above, similar cases exist for other EEG signals, e.g., ERP components, phase measurements, and raw signal monitoring, among others. This knowledge is applied in multiple fields of research (e.g., social or developmental neuroscience) whose scope is too broad to cover here. The remaining chapters in this section explain practical applications of EEG in greater detail, such as clinical uses, mobile EEG, or brain-computer interfaces.

23.3 Conclusions

EEG is an invaluable tool for basic and academic research of the human brain and cognitive processes. Cognitive EEG research has traditionally relied on concepts and methods from the broader scope of psychology and neuroscience, such as neuropsychology, psychophysics, or mental chronometry. These disciplines provide frames of reference such as the idea of dissociable cognitive functions from neuropsychology, or experimental methods for disentangling covert cognitive processes from mental chronometry. These notions continue to evolve as multidisciplinary research progresses, for example, the idea of discrete and serial processing stages, versus cascade processes. The EEG is also a subject of study in itself, with experiments conducted specifically to improve understanding of the signals it generates. This endeavor is not only helped by cognitive methodologies, but also by interaction with other techniques like TMS or fMRI and relies on knowledge derived from neurophysiology and the physics of the brain.

Besides traditional methodologies, advances in engineering and data processing have enabled more complex analysis of the EEG, as well as its application in novel fields such as using portable devices or brain-computer interfaces outside of laboratories. Together, these new developments and traditional methods complement each other in the pursuit of a deeper comprehension of the brain.

References

Alexander, D. M., Trengove, C., & van Leeuwen, C. (2015). Donders is dead: cortical traveling waves and the limits of mental chronometry in cognitive neuroscience. *Cognitive Processing, 16*(4), 365–375. https://doi.org/10.1007/s10339-015-0662-4

Benton, A. L., & Sivan, A. B. (2007). Clinical neuropsychology: A brief history. *Disease-a-Month, 53*(3), 142–147. https://doi.org/10.1016/j.disamonth.2007.04.003

Beste, C., Munchau, A., & Frings, C. (2023). Towards a systematization of brain oscillatory activity in actions. *Communications Biology, 6*(1), 137. https://doi.org/10.1038/s42003-023-04531-9

Coles, M. G., Gratton, G., Bashore, T. R., Eriksen, C. W., & Donchin, E. (1985). A psychophysiological investigation of the continuous flow model of human information processing. *Journal of Experimental Psychology. Human Perception and Performance, 11*(5), 529–553. https://doi.org/10.1037/0096-1523.11.5.529

Corrow, S. L., Dalrymple, K. A., & Barton, J. J. (2016). Prosopagnosia: Current perspectives. *Eye Brain, 8*, 165–175. https://doi.org/10.2147/eb.S92838

Dalrymple, K. A., Garrido, L., & Duchaine, B. (2014). Dissociation between face perception and face memory in adults, but not children, with developmental prosopagnosia. *Developmental Cognitive Neuroscience, 10*, 10–20. https://doi.org/10.1016/j.dcn.2014.07.003

Duchaine, B., & Nakayama, K. (2005). Dissociations of face and object recognition in developmental prosopagnosia. *Journal of Cognitive Neuroscience, 17*(2), 249–261. https://doi.org/10.1162/0898929053124857

Fridriksson, J., Fillmore, P., Guo, D., & Rorden, C. (2015). Chronic Broca's Aphasia is caused by damage to Broca's and Wernicke's areas. *Cerebral Cortex, 25*(12), 4689–4696. https://doi.org/10.1093/cercor/bhu152

Gao, C., Conte, S., Richards, J. E., Xie, W., & Hanayik, T. (2019). The neural sources of N170: Understanding timing of activation in face-selective areas. *Psychophysiology, 56*(6), e13336. https://doi.org/10.1111/psyp.13336

Gehring, W. J., Goss, B., Coles, M. G. H., Meyer, D. E., & Donchin, E. (2018). The error-related negativity. *Perspectives on Psychological Science, 13*(2), 200–204. https://doi.org/10.1177/1745691617715310

Gerlach, C., & Starrfelt, R. (2024). Evidence for a classical dissociation between face and object recognition in developmental prosopagnosia. *Brain Sciences, 14*(1). https://doi.org/10.3390/brainsci14010107

Hight, D. F., Kaiser, H. A., Sleigh, J. W., & Avidan, M. S. (2020). Continuing professional development module: An updated introduction to electroencephalogram-based brain monitoring during intended general anesthesia. *Canadian Journal of Anaesthesia, 67*(12), 1858–1878. https://doi.org/10.1007/s12630-020-01820-3 (Module de developpement professionnel continu: Une introduction mise a jour du monitorage cerebral par encephalogramme pour une anesthesie generale planifiee).

Huang, W. J., Chen, W. W., & Zhang, X. (2015). The neurophysiology of P 300—An integrated review. *European Review for Medical and Pharmacological Sciences, 19*(8), 1480–1488.

Hummel, F., Andres, F., Altenmuller, E., Dichgans, J., & Gerloff, C. (2002). Inhibitory control of acquired motor programmes in the human brain. *Brain, 125*(Pt 2), 404–420. https://doi.org/10.1093/brain/awf030

Jensen, O., & Mazaheri, A. (2010). Shaping functional architecture by oscillatory alpha activity: Gating by inhibition. *Frontiers in Human Neuroscience, 4*, 186. https://doi.org/10.3389/fnhum.2010.00186

Klimesch, W., Sauseng, P., & Hanslmayr, S. (2007). EEG alpha oscillations: The inhibition-timing hypothesis. *Brain Research Reviews, 53*(1), 63–88. https://doi.org/10.1016/j.brainresrev.2006.06.003

Kotz, S. A., & Friederici, A. D. (2003). Electrophysiology of normal and pathological language processing. *Journal of Neurolinguistics, 16*(1), 43–58. https://doi.org/10.1016/s0911-6044(02)00008-8

Lau, E., Phillips, C., & Poeppel, D. (2009). A cortical network for semantics: (De)constructing the N400. *Nature Reviews. Neuroscience, 9*, 920–933. https://doi.org/10.1038/nrn2532

Linden, D. E. (2007). What, when, where in the brain? Exploring mental chronometry with brain imaging and electrophysiology. *Reviews in the Neurosciences, 18*(2), 159–171. https://doi.org/10.1515/revneuro.2007.18.2.159

Luck, S. J. (2014). *An introduction to the event-related potential technique* (2nd ed.). MIT Press. https://books.google.de/books?id=SzavAwAAQBAJ

Manippa, V., Palmisano, A., Ventura, M., & Rivolta, D. (2023). The neural correlates of developmental prosopagnosia: Twenty-five years on. *Brain Sciences, 13*(10). https://doi.org/10.3390/brainsci13101399

Matchin, W., den Ouden, D. B., Hickok, G., Hillis, A. E., Bonilha, L., & Fridriksson, J. (2022). The Wernicke conundrum revisited: Evidence from connectome-based lesion-symptom mapping. *Brain, 145*(11), 3916–3930. https://doi.org/10.1093/brain/awac219

McClelland, J. L. (1979). On the time relations of mental processes: An examination of systems of processes in cascade. *Psychological Review, 86*, 287–330.

Meechan, R. J. H., McCann, C. M., & Purdy, S. C. (2021). The electrophysiology of aphasia: A scoping review. *Clinical Neurophysiology, 132*(12), 3025–3034. https://doi.org/10.1016/j.clinph.2021.08.023

Meyer, D. E., Osman, A. M., Irwin, D. E., & Yantis, S. (1988). Modern mental chronometry. *Biological Psychology, 26*(1–3), 3–67. https://doi.org/10.1016/0301-0511(88)90013-0

Murta, T., Leite, M., Carmichael, D. W., Figueiredo, P., & Lemieux, L. (2015). Electrophysiological correlates of the BOLD signal for EEG-informed fMRI. *Human Brain Mapping, 36*(1), 391–414. https://doi.org/10.1002/hbm.22623

Pfurtscheller, G., & Lopes da Silva, F. H. (1999). Event-related EEG/MEG synchronization and desynchronization: basic principles. *Clinical Neurophysiology, 110*(11), 1842–1857. https://doi.org/10.1016/s1388-2457(99)00141-8

Pracar, A. L., Biondo, N., Dronkers, N. F., & Ivanova, M. V. (2025). The neuroanatomy of Broca's aphasia [Original Research]. *Frontiers in Language Sciences, 4*. https://doi.org/10.3389/flang.2025.1496209

Purdon, P. L., Pierce, E. T., Mukamel, E. A., Prerau, M. J., Walsh, J. L., Wong, K. F., Salazar-Gomez, A. F., Harrell, P. G., Sampson, A. L., Cimenser, A., Ching, S., Kopell, N. J., Tavares-Stoeckel, C., Habeeb, K., Merhar, R., & Brown, E. N. (2013). Electroencephalogram signatures of loss and recovery of consciousness from propofol. *Proceedings of the National Academy of Sciences of the United States of America, 110*(12), E1142–E1151. https://doi.org/10.1073/pnas.1221180110

Ramminger, J. J., Peper, M., & Wendt, A. N. (2023). Neuropsychological assessment methodology revisited: Metatheoretical reflections. *Frontiers in Psychology, 14*, 1170283. https://doi.org/10.3389/fpsyg.2023.1170283

Robson, H., Pilkington, E., Evans, L., DeLuca, V., & Keidel, J. L. (2017). Phonological and semantic processing during comprehension in Wernicke's aphasia: An N400 and Phonological Mapping Negativity Study. *Neuropsychologia, 100*, 144–154. https://doi.org/10.1016/j.neuropsychologia.2017.04.012

Rossion, B., & Caharel, S. (2011). ERP evidence for the speed of face categorization in the human brain: Disentangling the contribution of low-level visual cues from face perception. *Vision Research, 51*(12), 1297–1311. https://doi.org/10.1016/j.visres.2011.04.003

Rutar Gorišek, V., Zupanc Isoski, V., Belič, A., Manouilidou, C., Koritnik, B., Bon, J., Pečarič Meglič, N., Vrabec, M., Žibert, J., Repovš, G., & Zidar, J. (2016). Beyond aphasia: Altered EEG connectivity in Broca's patients during working memory task. *Brain and Language, 163*, 10–21. https://doi.org/10.1016/j.bandl.2016.08.003

Sauseng, P., Klimesch, W., Gerloff, C., & Hummel, F. C. (2009). Spontaneous locally restricted EEG alpha activity determines cortical excitability in the motor cortex. *Neuropsychologia, 47*(1), 284–288. https://doi.org/10.1016/j.neuropsychologia.2008.07.021

Silva, L. F. L. (2021). Syntactic aspects of impaired speech in interface with prosody and information structure: Data from Broca's and Wernicke's aphasia. *Signótica, 33*(1), 2.

Smulders, F. T., Kok, A., Kenemans, J. L., & Bashore, T. R. (1995). The temporal selectivity of additive factor effects on the reaction process revealed in ERP component latencies. *Acta Psychologica, 90*(1–3), 97–109. https://doi.org/10.1016/0001-6918(95)00032-p

Sterman, M. B. (1996). Physiological origins and functional correlates of EEG rhythmic activities: Implications for self-regulation. *Biofeedback and Self-Regulation, 21*(1), 3–33. https://doi.org/10.1007/bf02214147

Taylor, P. C., & Thut, G. (2012). Brain activity underlying visual perception and attention as inferred from TMS-EEG: A review. *Brain Stimulation, 5*(2), 124–129. https://doi.org/10.1016/j.brs.2012.03.003

Thompson, H. E., Robson, H., Lambon Ralph, M. A., & Jefferies, E. (2015). Varieties of semantic 'access' deficit in Wernicke's aphasia and semantic aphasia. *Brain, 138*(12), 3776–3792. https://doi.org/10.1093/brain/awv281

Towler, J., & Eimer, M. (2012). Electrophysiological studies of face processing in developmental prosopagnosia: Neuropsychological and neurodevelopmental perspectives. *Cognitive Neuropsychology, 29*(5-6), 503–529. https://doi.org/10.1080/02643294.2012.716757

Towler, J., Fisher, K., & Eimer, M. (2017). The cognitive and neural basis of developmental prosopagnosia. *Quarterly Journal of Experimental Psychology (Hove), 70*(2), 316–344. https://doi.org/10.1080/17470218.2016.1165263

Ullsperger, M., Danielmeier, C., & Jocham, G. (2014). Neurophysiology of performance monitoring and adaptive behavior. *Physiological Reviews, 94*(1), 35–79. https://doi.org/10.1152/physrev.00041.2012

Vaidya, A. R., Pujara, M. S., Petrides, M., Murray, E. A., & Fellows, L. K. (2019). Lesion studies in contemporary neuroscience. *Trends in Cognitive Sciences, 23*(8), 653–671. https://doi.org/10.1016/j.tics.2019.05.009

Vidal, F., Hasbroucq, T., Grapperon, J., & Bonnet, M. (2000). Is the 'error negativity' specific to errors? *Biological Psychology, 51*(2–3), 109–128. https://doi.org/10.1016/s0301-0511(99)00032-0

Vidal, F., Burle, B., Grapperon, J., & Hasbroucq, T. (2011). An ERP study of cognitive architecture and the insertion of mental processes: Donders revisited. *Psychophysiology, 48*(9), 1242–1251. https://doi.org/10.1111/j.1469-8986.2011.01186.x

Wischnewski, M., Haigh, Z. J., Shirinpour, S., Alekseichuk, I., & Opitz, A. (2022). The phase of sensorimotor mu and beta oscillations has the opposite effect on corticospinal excitability. *Brain Stimulation, 15*(5), 1093–1100. https://doi.org/10.1016/j.brs.2022.08.005

Chapter 24
Clinical Applications

Cilia Jaeger

Abstract EEG is a valuable tool in clinical research with the potential for broader use in clinical practice. It is commonly used in the early phases of clinical research as an objective tool for studying disease pathophysiology, pharmacodynamic effects on brain activity, and biomarker development. Despite its promise, adoption in routine clinical care remains limited due to the lack of standardized parameters—such as channel number, processing pipelines, analysis techniques, and recording duration—as well as concerns about inter-rater reliability. These challenges have slowed translation, yet EEG continues to play a growing role in exploring biomarkers, disease onset and progression, and therapeutic outcomes. This chapter first defines clinical research and covers why EEG is a cost-effective, beneficial tool for clinical applications. It highlights applications where the use of EEG is well established, such as in epilepsy and sleep research. The chapter also covers more novel applications of EEG as a biomarker in neurodegenerative disease and assessment of drug efficacy in clinical trials. The chapter then ends on the future of using electrophysiology more widely in clinical settings, including using spectral fingerprints and EEG connectivity patterns for evaluating disease onset and progression.

Keywords Biomarkers · Clinical research · Anesthesia · Epilepsy · Sleep disorders · Neurodegenerative disease · Brain-computer interface

24.1 Introduction

EEG has served as an important tool in neuroscience, offering insight into brain function through non-invasive, (near) real-time recordings of neural activity. In academic research, EEG is commonly used to study cognition, sensory processing, and information processing. Translating these findings into clinical practice requires moving beyond controlled laboratory experiments toward clinical research. The

C. Jaeger (✉)
Brain Products GmbH, Gilching, Germany
e-mail: cilia.jaeger@brainproducts.com

T. Warbrick (ed.), *The EEG Handbook*,
https://doi.org/10.1007/978-3-032-20450-9_24

focus shifts to understanding disease processes, identifying measurable indicators of disease progression, and evaluating therapeutic interventions and outcomes. This chapter focuses on the use of EEG in clinical research and clinical applications.

24.2 Definition of Clinical Research and Why EEG Is a Useful Tool

24.2.1 Clinical Research

The NIH defines clinical research as any research conducted within human participants. This includes research that is conducted on materials originating from human participants such as specimens or cognitive phenomena that were directly obtained from the participant by the researcher (NIH, 2024). Epidemiologic and behavioral studies are also classified as clinical research. Also falling under the clinical research umbrella are outcomes research and health service research, which evaluate the effects of healthcare interventions on patient health, quality of life, and care delivery. Many examples of human-oriented research have been described throughout this book, including Chap. 23. This chapter focuses on examples of research for clinical applications. This refers to research conducted in patient populations in clinical trials and studies that test and validate diagnostic tools, treatment interventions, and biomarkers for their use in healthcare.

24.2.2 EEG in Research for Clinical Applications

EEG is a fundamental tool in non-invasive neuroscience research including research for clinical applications. The use of EEG in clinical neurological and psychiatric research is valuable because it is non-invasive, cost-effective, and offers high temporal, (near) real-time whole brain monitoring.

EEG is not as commonly used in clinical applications as other whole-brain imaging techniques, such as MRI, because there is a lack of standardized acquisition and analysis pipelines (Sinha et al., 2016; Jobert et al., 2012). Inter-data variability across EEG data sets can arise from differences in EEG equipment and channel configuration, quality control checks, and data preprocessing and evaluation methods. For example, wet electrodes provide better signal quality than dry electrodes; however, they require more preparation time. Another key challenge in standardizing EEG signal evaluation arises from inter-rater variability of EEG data analysis, particularly related to artifact removal (Tatum et al., 2016; Shirk et al., 2017). Because many EEG artifacts are non-stationary, they cannot fully be removed by automated artifact algorithms. Therefore, most EEG preprocessing pipelines require a subjective manual removal of artifacts from the EEG data. This can result in variability in artifact attenuation or even the removal of the signal of interest (Mumtaz et al., 2021). Some EEG features are more robust to residual artifacts than others. Part of defining standardized EEG methods in clinical research involves finding EEG features that are robust to some variability in inter-rater evaluation, as well as incorporating AI-assisted analysis methods (Rutkowski & Saab, 2025).

Efforts are being made to overcome these challenges by providing standardization recommendations such as those defined by the minimal technical requirements provided by the American Clinical Neurophysiology Society (Sinha et al., 2016). Other standardization guidelines have already been adopted by the EEG community such as standardized electrode layouts defined by the 10–20 system. Further standardization efforts will be discussed at the end of this chapter.

24.2.3 EEG as a Biomarker

An integral part of clinical research is finding new biomarkers that can be used to diagnose a disease and assess the progression and severity of a condition or disease. A biomarker is a measurable biological indicator (Strimbu & Tavel, 2010). Biomarkers can also be used to measure response to drug and therapeutic intervention. In a more classic sense biomarkers are often measurable substances such as identification of a specific genetic mutation or protein levels that are linked to a specific condition. The use of neuroimaging biomarkers to differentiate between normal and pathological processes is gaining increasing interest (Linden, 2012). Biomarkers can further be divided into different types. Prognostic biomarkers are classified as biomarkers that indicate the probability of a disease occurring or progressing. Whereas predictive biomarkers determine the likelihood that a group of individuals will respond to a treatment and are used as an outcome measure for clinical trials (Califf, 2018).

EEG features can be used as biomarkers by providing a measurable electrical pattern of underlying neural activity that correlates with a physiological or pathological state (Fig. 24.1). Aberrant electrical patterns recorded in EEG are used to identify and evaluate disease progression in epilepsy (Britton et al., 2016), Alzheimer's disease (Monllor et al., 2021), sleep disorders (Dai et al., 2021), and psychiatric disorders (Farzan et al., 2017). EEG can also be used to monitor response to treatment, such as antiseizure medication (Porcaro et al., 2024) or to track neurorehabilitation efforts and brain activity after brain injury (Zhang et al., 2022).

The next sections highlight examples in which EEG has been widely accepted as a vital tool in clinical research to assist with establishing techniques for clinical diagnosis and disease monitoring. Examples of more novel applications will also be discussed and finally the chapter will end with a discussion on what is needed to standardize EEG recording and signal analysis methods to make EEG a more commonly used tool in clinical research.

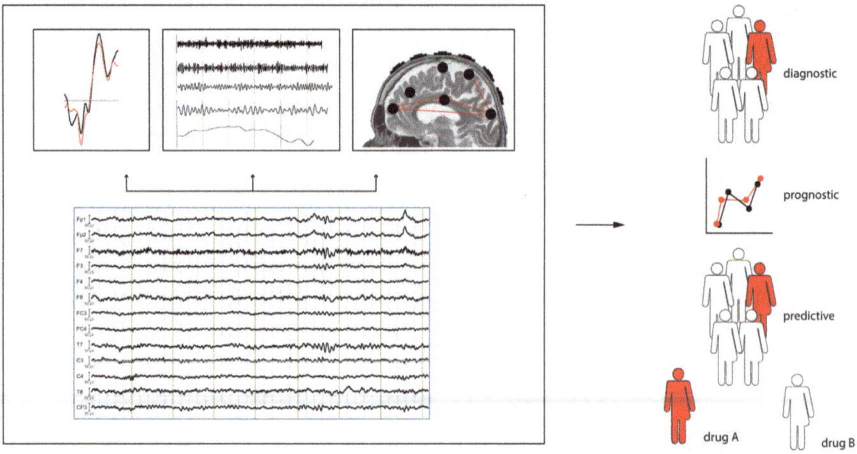

Fig. 24.1 EEG features as biomarkers. Different EEG features such as event-related potentials, frequency patterns of neural oscillations, and EEG connectivity can be used as potential biomarkers. These features can be used as biomarkers. Diagnostic biomarkers are used to identify diseases within a population. Prognostic biomarkers are classified as biomarkers that indicate the probability of a disease occurring or progressing. A predictive biomarker helps identify the probability of a group of individuals that respond to a treatment

24.3 Examples of Clinical Applications of EEG

24.3.1 Epilepsy

The use of EEG as a clinical tool has been well established in the diagnosis and evaluation of epilepsy and sleep disorders. EEG has also been vital for the development of anesthetic agents. Both epilepsy and sleep are discussed in much more detail in their respective dedicated chapters and will only be briefly mentioned here. Yet they provide examples of how EEG has been standardized to be used effectively in a clinical setting.

EEG has become an invaluable tool for epilepsy research because it offers a near real-time, high temporal resolution signal that is time-locked to seizure onset and activity. One of the most useful electrical patterns used to characterize a specific epilepsy syndrome are interictal epileptiform discharges (IEDs). IEDs are large electrophysiological events that can be observed in the EEG intermittently between seizures. The type, location, and frequency of IEDs can be used to localize the onset of seizure activity within a specific brain region and characterize seizure type (Britton et al., 2016).

Combined with video monitoring, long-term EEG recording has thus become the gold standard for clinical epilepsy research (Beniczky & Schomer, 2020). This enables simultaneous assessment of clinical behavior and electrostatic artifacts, facilitating the identification and classification of specific epilepsy syndromes. The presence and frequency of IEDs are also frequently used as predictive biomarkers to evaluate the efficacy of antiseizure medication (Reynolds et al., 2023). To ensure diagnostic efficacy and reliability, guidelines have been developed to define minimum requirements for recording durations, channel counts, and hardware configurations (Beniczky & Schomer, 2020). Standard electrode layouts have also gained widespread acceptance in epilepsy research to ensure EEG is recorded in similar locations across individuals. Additionally, regulatory boards have emerged to establish and assess these guidelines, such as the International League Against Epilepsy or the American Epilepsy Society.

24.3.2 Sleep and Sleep Disorders

Another application in which EEG has been established as a valuable clinical tool is sleep. In Chap. 6, the distinct electrical patterns associated with different sleep stages were introduced. In clinical applications, deviations from these distinct electrical patterns can be used to characterize sleep disorders. During sleep studies, EEG is often combined with other physiological measurements. The combination of these physiological recordings is defined as polysomnography (PSG), which records changes in physiological activity such as eye movements, muscle activity, heart and respiration rate (Jobert et al., 2012).

Polysomnography is commonly used in the diagnosis and evaluation of disease progression in obstructive sleep apnea. Sleep apnea is typically characterized by obstruction of the upper airways, which disrupts respiration and leads to sleep disturbances (Martins & Conde, 2021). In a clinical setting, polysomnography recordings during sleep are used to calculate the number of apnea events, defined as a complete pause of breathing for over 10 s, within a given period. Based on these events, a respiratory disturbance index can be calculated to determine the severity of sleep apnea (O'Connor et al., 2000). PSG also presents itself as a useful and cost-effective tool for at-home sleep recordings. While home-based sleep apnea tests monitor fewer parameters than in-lab recordings for ease of use, they remain a valuable initial screening tool for detecting sleep apnea (Kapur et al., 2017). Additionally, home testing may yield more reliable data, as patients sleep in a familiar environment rather than a hospital setting, where discomfort and unfamiliarity can impact sleep quality and alter results. This approach balances convenience with diagnostic accuracy, demonstrating the value of having portable and cost-effective tools that can be worn by the patient not only within a clinical setting.

Polysomnographic studies have identified gender differences in sleep architecture and the presentation of obstructive sleep apnea. Research suggests that men generally exhibit more apneic events than women. However, women often exhibit more fragmented sleep due to apneic events, as women express a lower arousal threshold that is thought to result from the influence of hormone regulation of arousal and ventilatory control (Martins & Conde, 2021; Bonsignore et al., 2019). Apnea events during REM sleep are also more common in women, whereas apnea events generally only occur during NREM sleep in men (O'Connor et al., 2000). These findings have also been correlated to hormonal differences in men and women and highlight the importance of considering gender-specific factors in the diagnosis and management of sleep disorders. Understanding these differences can lead to more tailored and effective treatment strategies for both men and women.

24.3.3 Anesthesia

EEG is also used to monitor sedation and loss of consciousness in response to anesthetic drugs to evaluate their efficacy, optimize drug administration, and minimize drug-related complications. During the induction of sedation, a decrease in high-frequency activity and an increase in alpha and theta oscillatory activity are observed for most anesthetic drugs. This correlates with drowsiness and loss of awareness of the environment. During the maintenance phase, during which surgery is performed, slow delta waves below 4 Hz and alpha oscillations are predominant in the EEG (Yousefi-Banaem et al., 2020). Intermittent burst suppressions can also be observed, which are characterized by bursts of high-amplitude slow-wave activity followed by flat or isoelectric EEG. For many anesthetic drugs, the presence of burst suppression indicates deep sedation (Shanker et al., 2021). This feature can be used to monitor for oversedation, which can cause complications in patients. Oversedation has been linked to post-operative delirium, especially in older adults (Mei et al., 2020). Studies also show that prolonged exposure to deep sedation may lead to cognitive impairments affecting memory and executive function (Vacas et al., 2021).

In clinical practice, the objective assessment of depth of anesthesia induced by sedative drug is crucial. The bispectral index (BIS) monitor serves as an example for standardizing and objectively analyzing EEG data to monitor sedation levels. The BIS is a number given between a range of 0 and 100, which indicates the level of brain activity. Zero indicates total loss of brain activity, and one hundred indicates wakefulness (Fig. 24.2). The BIS index is calculated using a combination of spectral and bispectral analysis methods. It requires analyzing EEG data from a set of frontal electrodes and removing common artifacts such as muscle activity and eye movement. Time-based measures such as the suppression ratio are calculated from these electrodes. The ratio is defined as the time of isoelectric activity divided by the total time of EEG activity. The EEG signal is also decomposed into different frequency bands to assess the power of the low-frequency bands in comparison to the power of the high-frequency bands. The bispectral analysis then includes measuring phase synchronization between different frequency components in the EEG. Higher degrees of synchronization across the frequency bands indicate a deeper level of anesthesia. A weighting algorithm is used to combine the different EEG-derived parameters to create the bispectral index (Yousefi-Banaem et al., 2020).

BIS monitoring can be applied in near real-time. The raw EEG data is continuously processed and analyzed, providing an update on the bispectral index every 15–30 s. A continuous display of this index can thus help clinicians administer and monitor correct anesthesia levels. However, there are some limitations to the

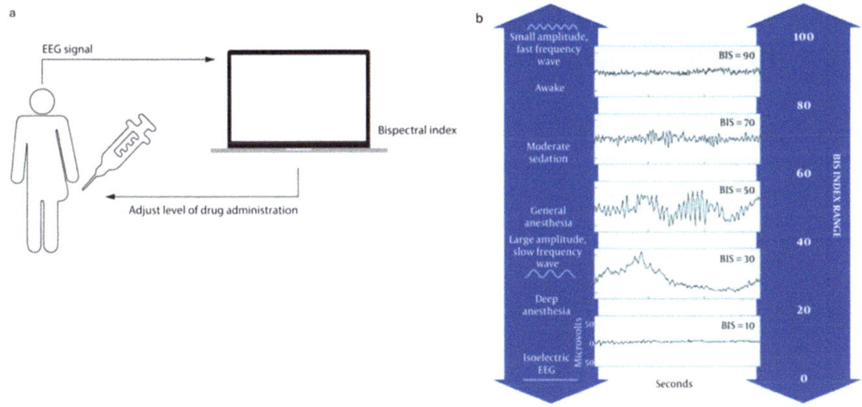

Fig. 24.2 Example demonstrating real-time analysis of the bispectral index. (**a**) The raw EEG data is analyzed in near real time to monitor the level of anesthesia and adjust the level of drug administration. (**b**) The bispectral index is a value ranging from 0 (no brain activity) to 100 (fully awake). The BIS value increases as the brain wave amplitude decreases and the frequency increases. The awake state has a BIS index of 90 and is characterized by small amplitude, fast frequency waves. Deep anesthesia occurs around a score of 30 and is characterized by large amplitude and slow frequency oscillations. Part B of this figure has been reproduced by Yousefi-Banaem et al. (2020) with permission

bispectral index. Since BIS monitoring is proprietary, there is a lack of transparency, difficulty in identifying bias within the algorithm, and a lack of reproducibility of results that were used to verify the algorithm (Connor et al., 2022). Additionally, BIS is sensitive to certain artifacts such as EMG activity (Youefi-Banaem et al., 2020). Certain drugs like ketamine present with distinct EEG patterns leading to an increase in high frequency activity and low frequency power. The BIS score will inaccurately indicate a higher level of wakefulness for ketamine that does not reflect the depth of sedation induced by the drug (Hans et al., 2005; Musizza & Ribaric, 2010). Because of the limitations of the BIS, other clinical tools for assessing EEG and anesthesia depth have been considered, such as the Patient State Index and the Richmond Agitation-Sedation Scale (Han et al., 2023).

This section provides some examples where EEG is well established for assessing neurological disorders and, importantly, where some consensus on recording techniques and analysis has been established. These advances have made EEG indispensable for disease differentiation and for using EEG features as predictive biomarkers to guide treatment decisions, such as assessing drug responsiveness in epilepsy and sedation. The next section will shift focus to EEG's role as a prognostic biomarker.

24.4 Examples of EEG as a Biomarker in Disease Prognostics and Treatment

This subsection will focus on more novel applications using EEG as a biomarker in clinical research focusing on diagnosing and monitoring neurodegenerative diseases and movement disorders.

24.4.1 Alzheimer's Disease and Neurodegeneration

As life expectancy has increased worldwide, the prevalence of neurodegenerative diseases has also strongly increased. Early detection of neurodegenerative diseases can help slow the progression of the disease and improve quality of life. However, finding cost-effective tools to diagnose early signs of neurodegeneration is still a challenge. EEG presents a cost-effective and portable tool that is a potential candidate for detecting Alzheimer's disease (AD) in its presymptomatic stages (Whelan et al., 2022). Along with cognitive impairment and loss of memory, AD is characterized by the presence of clusters of beta-amyloid protein in the brain. While amyloid plaques serve as a good AD biomarker candidate, they can only be detected by using PET imaging, which is costly and invasive (Monllor et al., 2021). Slowing of EEG activity in amyloid-positive compared to amyloid-negative populations has been observed. EEG slowing is characterized by reduction of alpha and beta activity and an increase in delta and theta activity across the brain (Monllor et al., 2021; Hata et al., 2024). Sleep disturbances including reduction in the percentage of slow-wave sleep and REM sleep have been correlated with the progression of AD (Zhang et al., 2022).

Currently, most EEG-related clinical research on AD is focused on validating EEG biomarkers in populations already diagnosed with AD and presenting with cognitive impairment. The EEG abnormalities observed in AD are not exclusive to Alzheimer's and overlap with other psychiatric and neurodegenerative disorders. The complexity of brain activity linked to cognition presents a further challenge. There is distinct individual variability in brain activity associated with cognitive function, and the onset of cortical activity changes linked to dementia is subtle (Horvath et al., 2018; Al-Nuaimi et al., 2021). However, portable EEG headsets and in-ear EEG devices are emerging as a cost-effective solution for collecting long-term and individualized data (Byrom et al., 2018; Musaeus et al., 2022). A study by McWilliams et al. (2021) assessed the performance of a cognitive game with simultaneous EEG recorded from portable headsets that was played repeatedly over a period of 12 weeks within a healthy, aging population. The study characterized cognitive performance in correlation with EEG features that are indicative of cognitive decline with age. Such study designs implementing long-term monitoring can help classify cognitive aging and differentiate between normal and aberrant progression. Together with other diagnostic modalities, portable long-term EEG monitoring has the potential as a non-invasive, cost-effective tool for early diagnosis of AD.

24.4.2 Brain-Computer Interfaces and Movement Disorders

EEG is also a valuable tool in brain-computer interfaces and the treatment of neurological conditions. In short, brain-computer interfaces (BCIs) are systems that create a direct communication pathway between the brain and an external device, allowing for quantification of CNS activity and translation of the CNS activity into an artificial output that replaces or enhances the biological output (Varbu et al., 2022; Wolpaw et al., 2020). One of the most frequent applications of BCI is within the study and treatment of movement disorders (Wolpaw et al., 2002). For example, in studies where the activity from the motor cortex is used to control motor output, such as limb movement in individuals experiencing paralysis or other movement disorders (Zhang et al., 2022; Varbu et al., 2022). Because of its high temporal resolution, EEG can provide near real-time feedback and thus presents a valuable tool for BCI research. Sensorimotor rhythms such as the Mu rhythm and beta rhythm observed in the sensorimotor cortex have been identified and associated with motor planning and movement intention. The desynchronization or decrease in Mu power is observed when preparing a movement or when imagining a movement (Pfurtscheller et al., 2006). This response can be utilized to help initiate a motor response in clinical populations with amyotrophic lateral sclerosis (ALS) who develop severe motor impairments over time. These individuals lose the ability to perform motor control, yet they are still able to perform motor imagery (McFarland, 2020). Desynchronization of the Mu rhythm during motor imagery thus serves as a potential measure that can be used to drive artificially assisted motor commands in ALS populations.

Event-related potentials such as the P300 component also serve as EEG features that can be evaluated and used to facilitate communication between the brain and the artificial system. Individuals with severe progression of ALS lose the ability to speak. A study by Geronimo and Simmons (2020) used speech encoding paradigms in which individuals with ALS were asked to focus on a specific letter to spell a word. The study showed that the P300 response can be used to select the letter the participant is focusing on. This enables the participants to spell words and gives them a method for communicating their thoughts.

While such applications prove beneficial for improving the quality of life for patients affected by movement disorders, BCI applications for initiating behavioral tasks are limited because they require a robust EEG feature that is unique to one task and has little overlap with other cognitive functions (Wolpaw et al., 2020). Furthermore, a major limitation of using scalp EEG in BCI applications is its low spatial resolution. The blurring of signals at the scalp makes it difficult to identify the exact location of a specific neural response, making it difficult to precisely decode an intended action (Mumtaz et al., 2021). For example, within the context of motor movement, distinct brain regions within the motor cortex are responsible for the movement of specific body parts. Due to EEG's low spatial resolution, electrical signals from adjacent brain regions that are responsible for the movement of different parts of the body can overlap, making it challenging to decode the user's intended movement.

24.5 Future of EEG in Clinical Applications

EEG has been widely recognized as an important tool in clinical research. It provides a cost-effective, (near) real-time direct measure of brain activity. With advancements in recording technology and analysis methods, it has the potential to be applied to more clinical applications.

For example, over the last few decades, there has been significant progress in implementing EEG to study various psychiatric disorders. Psychiatric disorders are difficult to diagnose because they can manifest differently in individuals with varying degrees of severity (van den Heuvel et al., 2019; Lenartowicz & Loo, 2014). Many psychiatric disorders also share common symptoms such as cognitive dysfunction or mood disturbances. Differentiating varying degrees of individual differences in brain activity and function within normal and psychiatric populations is crucial yet challenging (Linden, 2012). Recent research has shown that psychiatric disorders often involve disruption of large-scale brain networks that include both slow, sustained changes and fast, transient activity (van den Heuvel et al., 2019; Sadaghiani et al., 2022). While the study of dynamic temporal changes within brain networks or functional connectivity has classically been studied using different imaging modalities such as functional magnetic resonance imaging, EEG-derived connectivity measures are of increasing interest. Due to its high temporal resolution, EEG connectivity analyses can provide important information on fast, transient brain dynamics. Frequency-distinct connectivity changes have also proven important for building a better understanding of how complex information processing and communication across the brain can occur concurrently by having differential coupling and decoupling across and within frequency bands (Sadaghiani et al., 2022).

In addition to common limitations like low spatial resolution and interference from artifacts, the complexity of EEG data associated with brain states makes it hard to use EEG analysis in clinical settings to distinguish normal patterns of behavior and cognition from abnormal patterns (Mumtaz et al., 2021; Babiloni et al., 2021). However, a push for standardizing EEG recording, pre-processing, and analysis methods is important for making EEG a more widely accepted tool in clinical research. The emergence of regulatory boards, such as the American Clinical Neurophysiology Society, to establish best practice guidelines within clinical research is also important.

24.6 Conclusion

EEG is a versatile and accessible tool in clinical neuroscience research. It provides a cost-effective approach for capturing brain dynamics noninvasively in clinical and emerging home-based contexts. EEG has been well established in clinical applications such as sleep and epilepsy, where distinct EEG characteristics have been classified and are used for diagnostic or therapeutic outcome evaluations. Wearable EEG technologies and brain-computer interfaces further provide potential for clinical applications. Yet, despite the advances in EEG technology and research, challenges remain. Particularly, in psychiatric applications, in which heterogeneity and methodological inconsistencies limit clinical translation. Further efforts toward standardized recording and analysis procedures, as well as the integration of EEG with other neuroimaging techniques, may help to address these challenges and make the use of EEG more widespread in clinical applications.

References

Al-Nuaimi, A. H., Bluma, M., Al-Juboori, S. S., Eke, C. S., Jammeh, E., Sun, L., & Ifeachor, E. (2021). Robust EEG based biomarkers to detect Alzheimer's disease. *Brain Sciences, 11*(8). https://doi.org/10.3390/brainsci11081026

Babiloni, C., Arakaki, X., Azami, H., Bennys, K., Blinowska, K., Bonanni, L., Bujan, A., Carrillo, M. C., Cichocki, A., De Frutos-Lucas, J., Del Percio, C., Dubois, B., Edelmayer, R., Egan, G., Epelbaum, S., Escudero, J., Evans, A., Farina, F., Fargo, K., et al. (2021). Measures of resting state EEG rhythms for clinical trials in Alzheimer's disease: Recommendations of an expert panel. *Alzheimers Dement, 17*(9), 1528–1553. https://doi.org/10.1002/alz.12311

Beniczky, S., & Schomer, D. L. (2020). Electroencephalography: Basic biophysical and technological aspects important for clinical applications. *Epileptic Disorders, 22*(6), 697–715. https://doi.org/10.1684/epd.2020.1217

Bonsignore, M. R., Saaresranta, T., & Riha, R. L. (2019). Sex differences in obstructive sleep apnoea. *European Respiratory Review, 28*(154). https://doi.org/10.1183/16000617.0030-2019

Britton, J. W., Frey, L. C., Hopp, J. L., et al. (2016). *Electroencephalography (EEG): An introductory text and atlas of normal and abnormal findings in adults, children, and infants* [Internet]. American Epilepsy Society. Available from: https://www.ncbi.nlm.nih.gov/books/NBK390347/

Byrom, B., Mccarthy, M., Schueler, P., & Muehlhausen, W. (2018). Brain monitoring devices in neuroscience clinical research: The potential of remote monitoring using sensors, wearables, and mobile devices. *Clinical Pharmacology and Therapeutics, 104*(1), 59–71. https://doi.org/10.1002/cpt.1077

Califf, R. M. (2018). Biomarker definitions and their applications. *Experimental Biology and Medicine (Maywood, N.J.), 243*(3), 213–221. https://doi.org/10.1177/1535370217750088

Connor, C.W. (2022) Open reimplementation of the BIS algorithms for depth of anesthesia. *Anesthesia & Analgesia, 135*(4), 855–864. https://doi.org/10.1213/ANE.0000000000006119.

Dai, X. J., Zhang, J., Wang, Y., Ma, Y., & Shi, K. (2021). Editorial: EEG and FMRI for sleep and sleep disorders—Mechanisms and clinical implications. *Frontiers in Neurology, 12*, 749620. https://doi.org/10.3389/fneur.2021.749620

Farzan, F., Atluri, S., Frehlich, M., Dhami, P., Kleffner, K., Price, R., Lam, R. W., Frey, B. N., Milev, R., Ravindran, A., Mcandrews, M. P., Wong, W., Blumberger, D., Daskalakis, Z. J.,

Vila-Rodriguez, F., Alonso, E., Brenner, C. A., Liotti, M., Dharsee, M., et al. (2017). Standardization of electroencephalography for multi-site, multi-platform and multi-investigator studies: Insights from the Canadian biomarker integration network in depression. *Scientific Reports, 7*(1), 7473. https://doi.org/10.1038/s41598-017-07613-x

Geronimo, A., & Simmons, Z. (2020). TeleBCI: Remote user training, monitoring, and communication with an evoked-potential brain-computer interface. *Brain and Computer Interfaces (Abingdon), 7*(3–4), 57–69. https://doi.org/10.1080/2326263X.2020.1848134

Han, L., Drover, D. R., Chen, M. C., Saxena, A. R., Eagleman, S. L., & Nekhendzy, V. (2023). Evaluation of patient state index, bispectral index, and entropy during drug induced sleep endoscopy with dexmedetomidine. *Journal of Clinical Monitoring and Computing, 37*(3), 727–734. https://doi.org/10.1007/s10877-022-00952-9

Hans, P., Dewandre, P.-Y., Brichant, J.F. and Bonhomme, V. (2005) Comparative effects of ketamine on Bispectral Index and spectral entropy of the electroencephalogram under sevoflurane anaesthesia. *British Journal of Anaesthesia, 94*(3), 336–340

Hata, M., Miyazaki, Y., Mori, K., Yoshiyama, K., Akamine, S., Kanemoto, H., Gotoh, S., Omori, H., Hirashima, A., Satake, Y., Suehiro, T., Takahashi, S., & Ikeda, M. (2024). Utilizing portable electroencephalography to screen for pathology of Alzheimer's disease: A methodological advancement in diagnosis of neurodegenerative diseases. *Frontiers in Psychiatry, 15*, 1392158. https://doi.org/10.3389/fpsyt.2024.1392158

Horvath, A., Szucs, A., Csukly, G., Sakovics, A., Stefanics, G., & Kamondi, A. (2018). EEG and ERP biomarkers of Alzheimer's disease: A critical review. *Frontiers in Bioscience (Landmark Ed), 23*(2), 183–220. https://doi.org/10.2741/4587

Jobert, M., Wilson, F. J., Ruigt, G. S., Brunovsky, M., Prichep, L. S., & Drinkenburg, W. H. (2012). Guidelines for the recording and evaluation of pharmaco-EEG data in man: The International Pharmaco-EEG Society (IPEG). *Neuropsychobiology, 66*(4), 201–220. https://doi.org/10.1159/000343478

Kapur, V. K., Auckley, D. H., Chowdhuri, S., Kuhlmann, D. C., Mehra, R., Ramar, K., & Harrod, C. G. (2017). Clinical practice guideline for diagnostic testing for adult obstructive sleep apnea: An American Academy of Sleep Medicine clinical practice guideline. *Journal of Clinical Sleep Medicine, 13*(3), 479–504. https://doi.org/10.5664/jcsm.6506

Lenartowicz, A., & Loo, S. K. (2014). Use of EEG to diagnose ADHD. *Current Psychiatry Reports, 16*(11), 498. https://doi.org/10.1007/s11920-014-0498-0

Linden, D. E. (2012). The challenges and promise of neuroimaging in psychiatry. *Neuron, 73*(1), 8–22. https://doi.org/10.1016/j.neuron.2011.12.014

Martins, F. O., & Conde, S. V. (2021). Gender differences in the context of obstructive sleep apnea and metabolic diseases. *Frontiers in Physiology, 12*, 792633. https://doi.org/10.3389/fphys.2021.792633

Mcfarland, D. J. (2020). Brain-computer interfaces for amyotrophic lateral sclerosis. *Muscle & Nerve, 61*(6), 702–707. https://doi.org/10.1002/mus.26828

Mcwilliams, E. C., Barbey, F. M., Dyer, J. F., Islam, M. N., Mcguinness, B., Murphy, B., Nolan, H., Passmore, P., Rueda-Delagdo, L. M., & Buick, A. R. (2021). Feasibility of repeated assessment of cognitive function in older adults using a wireless, mobile, dry-EEG headset and tablet-based games. *Frontiers in Psychiatry, 12*, 574482. https://doi.org/10.3389/fpsyt.2021.574482

Mei, X., Zheng, H. L., Li, C., Ma, X., Zheng, H., Marcantonio, E., Xie, Z., & Shen, Y. (2020). The effects of propofol and sevoflurane on postoperative delirium in older patients: A randomized clinical trial study. *Journal of Alzheimer's Disease, 76*(4), 1627–1636. https://doi.org/10.3233/JAD-200322

Monllor, P., Cervera-Ferri, A., Lloret, M. A., Esteve, D., Lopez, B., Leon, J. L., & Lloret, A. (2021). Electroencephalography as a non-invasive biomarker of Alzheimer's disease: A forgotten candidate to substitute CSF molecules? *International Journal of Molecular Sciences, 22*(19), 10889. https://doi.org/10.3390/ijms221910889

Mumtaz, W., Rasheed, S., & Irfan, A. (2021). Review of challenges associated with the EEG artifact removal methods. *Biomedical Signal Processing and Control, 68*, 102741. https://doi.org/10.1016/j.bspc.2021.102741

Musaeus, C. S., Waldemar, G., Andersen, B. B., Hogh, P., Kidmose, P., Hemmsen, M. C., Rank, M. L., Kjaer, T. W., & Frederiksen, K. S. (2022). Long-term EEG monitoring in patients with Alzheimer's disease using ear-EEG: A feasibility study. *Journal of Alzheimer's Disease, 90*(4), 1713–1723. https://doi.org/10.3233/Jad-220491

Musizza, B., & Ribaric, S. (2010). Monitoring the depth of anaesthesia. *Sensors (Basel), 10*(12), 10896–10935. https://doi.org/10.3390/s101210896

National Institutes of Health (NIH). (2024). *NIH definition of clinical research.* Office of Research on Women's Health. Available at: https://orwh.od.nih.gov/toolkit/nih-policies-inclusion/definitions

O'Connor, C., Thornley, K. S., & Hanly, P. J. (2000). Gender differences in the polysomnographic features of obstructive sleep apnea. *American Journal of Respiratory and Critical Care Medicine, 161*(5), 1465–1472. https://doi.org/10.1164/ajrccm.161.5.9904121

Pfurtscheller, G., Brunner, C., Schlögl, A., & Lopes Da Silva, F. H. (2006). Mu rhythm (de)-synchronization and EEG single-trial classification of different motor imagery tasks. *NeuroImage, 31*(1), 153–159. https://doi.org/10.1016/j.neuroimage.2005.12.003

Porcaro, C., Seppi, D., Pellegrino, G., Dainese, F., Kassabian, B., Pellegrino, L., De Nardi, G., Grego, A., & Corbetta, M. (2024). Characterization of antiseizure medications effects on the EEG neurodynamic by fractal dimension. *Frontiers in Neuroscience, 18*, 1401068. https://doi.org/10.3389/fnins.2024.1401068

Reynolds, A., Vranic-Peters, M., Lai, A., Grayden, D. B., Cook, M. J., & Peterson, A. (2023). Prognostic interictal electroencephalographic biomarkers and models to assess antiseizure medication efficacy for clinical practice: A scoping review. *Epilepsia, 64*(5), 1125–1174. https://doi.org/10.1111/epi.17548

Rutkowski, J., & Saab, M. (2025). AI-based EEG analysis: New technology and the path to clinical adoption. *Clinical Neurophysiology, 179*, 2110994. https://doi.org/10.1016/j.clinph.2025.2110994

Sadaghiani, S., Brookes, M. J., & Baillet, S. (2022). Connectomics of human electrophysiology. *NeuroImage, 247*, 118788. https://doi.org/10.1016/j.neuroimage.2021.118788

Shanker, A., Abel, J. H., & Schamberg, G. (2021). Etiology of burst suppression EEG patterns. *Frontiers in Psychology, 12*, 673529. https://doi.org/10.3389/fpsyg.2021.673529

Shirk, S. D., Mclaren, D. G., Bloomfield, J. S., Powers, A., Duffy, A., Mitchell, M. B., Ezzati, A., Ally, B. A., & Atri, A. (2017). Inter-rater reliability of preprocessing EEG data: Impact of subjective artifact removal on associative memory task ERP results. *Frontiers in Neuroscience, 11*, 322. https://doi.org/10.3389/fnins.2017.00322

Sinha, S. R., Sullivan, L., Sabau, D., San-Juan, D., Dombrowski, K. E., Halford, J. J., Hani, A. J., Drislane, F. W., & Stecker, M. M. (2016). American Clinical Neurophysiology Society Guideline 1: Minimum technical requirements for performing clinical electroencephalography. *Journal of Clinical Neurophysiology, 33*(4), 303–307. https://doi.org/10.1097/WNP.0000000000000308

Strimbu, K., & Tavel, J. A. (2010). What are biomarkers? *Current Opinion in HIV and AIDS, 5*(6), 463–466. https://doi.org/10.1097/COH.0b013e32833ed177

Tatum, W. O., Olga, S., Ochoa, J. G., Munger Clary, H., Cheek, J., Drislane, F., & Tsuchida, T. N. (2016). American Clinical Neurophysiology Society Guideline 7: Guidelines for EEG reporting. *Journal of Clinical Neurophysiology, 33*(4), 328–332. https://doi.org/10.1097/WNP.0000000000000319

Vacas, S., Cole, D. J., & Cannesson, M. (2021). Cognitive decline associated with anesthesia and surgery in older patients. *JAMA*. https://doi.org/10.1001/jama.2021.4773

van den Heuvel, M.P., Scholtens, L.H. and Kahn, R.S. (2019) Multiscale neuroscience of psychiatric disorders. *Biological Psychiatry, 86*(7), 512–522. https://doi.org/10.1016/j.biopsych.2019.05.015

Varbu, K., Muhammad, N., & Muhammad, Y. (2022). Past, present, and future of EEG-based BCI applications. *Sensors (Basel), 22*(9). https://doi.org/10.3390/s22093331

Whelan, R., Barbey, F. M., Cominetti, M. R., Gillan, C. M., & Rosicka, A. M. (2022). Developments in scalable strategies for detecting early markers of cognitive decline. *Translational Psychiatry, 12*(1), 473. https://doi.org/10.1038/s41398-022-02237-w

Wolpaw, J. R., Birbaumer, N., Heetderks, W. J., & Mcfarland, D. J. (2002). Brain–computer interface technology: A review of the first international meeting. *IEEE Transactions on Rehabilitation Engineering, 8*(2), 164–173. https://doi.org/10.1109/Tr.2002.1012015

Wolpaw, J. R., Millán, J. D. R., & Ramsey, N. F. (2020). Brain-computer interfaces: Definitions and principles. In N. F. Ramsey & J. D. R. Millán (Eds.), *Handbook of clinical neurology* (Vol. 168, pp. 15–23). Elsevier. https://doi.org/10.1016/B978-0-444-63934-9.00002-0

Yousefi-Banaem, H., Goharani, R., Hajiesmaeili, M., Tafrishinejad, A., Zangi, M., et al. (2020). A review of bispectral index utility in neurocritical care patients. *Archives of Neuroscience, 7*(3), e96490. https://doi.org/10.5812/ans.96490

Zhang, Y., Ren, R., Yang, L., Zhang, H., Shi, Y., Okhravi, H. R., Vitiello, M. V., Sanford, L. D., & Tang, X. (2022). Sleep in Alzheimer's disease: A systematic review and meta-analysis of polysomnographic findings. *Translational Psychiatry, 12*(1), 136. https://doi.org/10.1038/s41398-022-01897-y

Chapter 25
From Brain Signals to Neuroadaptive Technology: BCIs for Human-Computer Interaction

Marius Klug, Diana E. Gherman-Nagy, and Thorsten O. Zander

Abstract This chapter introduces brain-computer interfaces (BCIs) with a focus on their role in human-computer interaction and neuroadaptive technology. We first define BCIs and distinguish active, reactive, and passive systems, emphasizing how EEG-based BCIs transform neural activity into real-time system behavior. We then review the main neural signals and sensing modalities used in BCIs, with an emphasis on wearable EEG and emerging form factors for everyday use. The core BCI pipeline is outlined from experimental design and labeling through preprocessing, feature extraction, machine learning, and validation, providing readers with a practical overview of how decoding models are built and assessed. Using motor imagery, P300/SSVEP, and a range of passive paradigms as examples, we illustrate how BCIs can decode intention, workload, error perception, vigilance, and related mental states. Building on this, we describe how passive BCIs enable neuroadaptive systems, from mental state assessment and open-loop feedback to closed-loop and autonomous adaptations. Finally, we discuss key practical challenges, including artifacts, non-stationarity, and cross-user generalization, as well as ethical issues around privacy and neurorights. The chapter concludes by outlining future trajectories for BCIs as a core component of human-centered, adaptive AI systems.

Keywords Brain-computer interface · BCI · Passive BCI · Neuroadaptive systems · Human-computer interaction · Mental state decoding

25.1 Introduction

Imagine you are studying for an important exam, scrolling through an online tutorial late at night. The material is dense, the examples feel rushed, and you catch yourself rereading the same paragraph without taking in a word. Your eyes stay on the page,

M. Klug (✉) · D. E. Gherman-Nagy · T. O. Zander
Neuroadaptive Human-Computer Interaction, Brandenburg University of Technology Cottbus-Senftenberg, Cottbus, Germany
e-mail: marius.klug@b-tu.de

T. Warbrick (ed.), *The EEG Handbook*,
https://doi.org/10.1007/978-3-032-20450-9_25

your mouse hardly moves, yet inside your head, workload and fatigue are rising, and your attention is drifting.

As you continue, the layout on the screen quietly changes. The video pauses and is replaced by a short multiple-choice quiz that checks a single key concept. The pace slows, and a brief recap appears in the margin. A few minutes later, after you answer confidently and your brain activity settles back into a lower workload pattern, the system resumes the normal speed and hides the extra scaffolding. You never told it you were confused. You never clicked a "help" button. The adaptation was triggered by a small EEG headset resting behind your ear that continuously tracks your mental state.

What makes this possible is a passive brain-computer interface: a system that does not wait for explicit commands but instead monitors ongoing brain activity to infer states such as effort, fatigue, and engagement. In this chapter, you will get an overview of how BCIs work, from EEG acquisition through the full processing pipeline, with a special focus on passive BCIs and how they can be integrated into everyday technology to create interfaces that respond not only to what you do, but also to how you think and feel while you are doing it.

25.1.1 Why Connect Brains and Computers?

When humans interact with machines, systems usually infer goals and internal states from behavior: button presses, cursor movements, speech, or gaze patterns. These cues are useful but indirect and often arrive only after critical decisions have been made. Brain-computer interfaces (BCIs) add a complementary information channel by decoding neural activity itself. In clinical settings, BCIs can restore communication and basic control to people with severe motor impairments by translating brain signals into selections or movements (Farwell & Donchin, 1988; Wolpaw et al., 2002). Beyond rehabilitation, BCIs can reveal workload, attention, or error perception during ongoing interaction (Parasuraman & Rizzo, 2007; Zander & Kothe, 2011), enabling adaptive systems that respond not only to what users do, but also to how they currently think and feel.

25.1.2 What Is a BCI?

A brain-computer interface is a system that measures neural activity, extracts informative features, and translates them into outputs that influence an external device or software process (Wolpaw & Wolpaw, 2012). This canonical loop—acquisition, processing, decoding, and system action—distinguishes BCIs from general neuroimaging, where brain signals are recorded for analysis rather than real-time interaction. BCIs can rely on various modalities, but EEG remains the dominant choice because it is safe, portable, inexpensive, and capable of millisecond-level temporal resolution (Mehta & Parasuraman, 2013). In a BCI, the defining property is the real-time transformation of actual brain activity into meaningful system behavior.

25.1.3 Types of BCIs: Active, Reactive, Passive

A widely used taxonomy distinguishes BCIs by the origin and purpose of the neural activity they decode (Zander & Kothe, 2011). *Active BCIs* rely on intentionally generated brain signals that are independent of external stimulation. Users deliberately modulate neural activity to convey commands, as in motor imagery (MI) systems where imagined hand movements control cursors or prostheses (Daly & Wolpaw, 2008). These interfaces emphasize volitional control, often require substantial training, and are most commonly used in clinical settings.

Reactive BCIs decode brain responses that are elicited by external stimuli. Here, the control signal is tied to events in the environment but still shaped by the user's attention. Classic examples include P300-based spellers, where rare or attended items evoke a characteristic potential (Farwell & Donchin, 1988), and steady-state systems that use frequency-tagged flickering targets (Norcia et al., 2015). Users influence the interface by focusing attention on specific stimuli, producing reliable, stimulus-locked responses.

Passive BCIs (pBCIs) interpret neural activity that arises without intentional control. They infer naturally occurring cognitive and affective states such as workload, engagement, fatigue, or error perception (Zander & Kothe, 2011). Rather than replacing explicit input, passive BCIs augment human-computer interaction (HCI) by adding implicit information about user state, enabling systems to adapt to them continuously during complex tasks.

25.1.4 BCIs in Neuroscience and HCI

BCIs occupy a dual position in research and application. In neuroscience, they enable real-time access to cognitive processes during naturalistic behavior, supporting studies that investigate perception, prediction, and action outside controlled laboratory settings (Gramann et al., 2011). Mobile EEG is especially important in this context because many states relevant for interaction—such as workload, attention, or prediction errors—manifest differently in realistic environments. Studying these states during movement, navigation, or immersive tasks increases ecological validity and reveals how neural dynamics behave under conditions that more closely resemble everyday interaction. This also aligns BCI research with the broader shift in HCI toward examining users in real-world contexts rather than exclusively in seated, constrained setups.

Within HCI, BCIs contribute to user modeling by providing direct markers of internal states that are difficult to infer from behavior alone, allowing interfaces to respond to cognitive or affective changes as they occur (Schmidt, 2000; Fischer, 2001). The focus on passive BCIs follows naturally from this intersection. Because passive BCIs decode spontaneously occurring neural activity without requiring intentional control, they add an implicit information channel that complements traditional input modalities (Zander & Kothe, 2011; Zander et al., 2014). Real-time brain state decoding (RBSD) enables systems to detect elevated workload, lapses of attention, or perceived errors and adapt accordingly, supporting applications in driving, aviation, and learning technologies. Passive BCIs thus form a foundation for neuroadaptive systems that adjust their behavior to support performance, maintain safety, or personalize user experience (Krol et al., 2018b). Their ability to supply continuous, fine-grained information about the user's internal state positions them as a central component in future adaptive HCI and human–AI systems.

25.2 Signals and Sensors

25.2.1 Neural Signals for BCIs

BCIs rely on neural signals that differ in temporal resolution, spatial specificity, and practicality for real-time decoding. Electrophysiological methods such as EEG (see also Chap. 2) and MEG measure voltage changes generated by postsynaptic potentials and capture neural dynamics on the millisecond scale (Biasiucci et al., 2019), making them suitable for detecting rapid processes like error responses, attention shifts, or oscillatory markers of workload (Mehta & Parasuraman, 2013). Their spatial resolution is limited, but their speed and portability—especially in the case of EEG—make them the backbone of most BCI systems. Hemodynamic methods such as fNIRS and fMRI instead indirectly measure activity by blood-oxygenation changes and thus offer better spatial localization at the cost of several-second delays. These slower dynamics restrict real-time use to gradual cognitive states rather than fast perceptual or decision-related events. As a result, electrophysiology is favored when BCIs must operate continuously and respond within behaviorally relevant timescales.

Outside of restorative research, such as speech reconstruction (Willett et al., 2023), EEG is particularly prevalent in BCI research because it strikes a practical balance between safety, usability, and information content. EEG systems are relatively inexpensive, portable, and compatible with mobile or immersive setups, enabling studies and applications beyond traditional laboratory environments. EEG also offers the high temporal resolution that is essential for decoding fast neural phenomena in real time. At the same time, EEG data suffer from low signal-to-noise ratio, sensitivity to artifacts, and limited spatial precision due to volume conduction. These constraints require careful preprocessing and feature extraction, and they shape which mental states can be decoded reliably. Despite these limitations, EEG remains the most versatile modality for BCIs because it supports rapid, continuous, and minimally intrusive monitoring of brain activity in real-world contexts.

EEG signals relevant for BCIs fall broadly into two categories: event-related responses and ongoing oscillatory activity. *Event-related potentials (ERPs)* are time-locked amplitude deflections in the EEG that occur in response to discrete sensory, cognitive, or action-related events (see also Chap. 19). Their temporo-spatial structure allows precise identification of components such as the P300, error-related negativity, or fixation-related potentials, each reflecting specific stages of information processing (Farwell & Donchin, 1988; Falkenstein et al., 2000; Donoghue & Voytek, 2022). ERPs form the basis of many reactive BCIs because they offer reliable markers tied to stimulus timing. *Oscillatory rhythms* reflect continuous fluctuations in synchronized neural activity, typically quantified through power in frequency bands such as theta, alpha, beta, or gamma (see also Chap. 18). Changes in these rhythms index sustained processes including workload, attention, motor imagery, or vigilance (Klimesch, 1999; Gevins & Smith, 2003; Gherman et al., 2025). Together, ERPs and oscillations constitute the primary signal families exploited in BCI design.

25.2.2 Wearable EEG and Form Factors

EEG systems vary widely in how electrodes make contact with the scalp, how portable they are, and how well they balance signal quality with user comfort. *Gel-based caps* provide the highest signal quality and remain the standard in research and clinical environments, but they require time-consuming preparation and restrict mobility. *Dry electrodes* reduce setup time and improve comfort, enabling faster deployment in applied settings, though they often trade off signal-to-noise ratio and can be more sensitive to motion artifacts (Zander et al., 2017). *Semi-dry solutions* refer to saline sponge electrodes that have an intermediate signal quality and setup time, but require saline solution to be re-applied over time.

Ear-EEG systems position electrodes in or around the ear canal, offering discrete, stable recordings suitable for long-term use and emerging everyday BCI applications (Debener et al., 2015; Mirkovic et al., 2016). *Consumer headsets,* such as the Interaxon Muse, further emphasize ease of use and accessibility, integrating lightweight materials and wireless transmission. They can still be used to detect oscillations, as well as limited event-related responses (Krigolson et al., 2017), but their lower channel counts and variable fit limit the complexity of signals that can be decoded reliably (Ratti et al., 2017). Finally, *Virtual Reality (VR)-integrated EEG,* such as the OpenBCI Galea, combines neural sensing with immersive head-mounted displays, supporting BCIs in extended reality scenarios while introducing new constraints related to weight, ergonomics, and mechanical coupling during movement. These diverse form factors reflect ongoing efforts to reconcile signal fidelity with practicality and user comfort.

25.3 The BCI Pipeline: From Raw Signals to Decisions

25.3.1 Overview

A BCI transforms neural activity into meaningful system outputs through a series of coordinated stages. The pipeline begins with experimental design, where tasks and labeling strategies determine what neural phenomena can be decoded. Data acquisition and preprocessing follow, addressing artifacts and preparing the signals for reliable analysis. Feature extraction converts continuous EEG into informative representations, which are then passed to machine learning models that learn mappings between neural activity and target states. Finally, trained models are validated and deployed in real time for online interaction.

25.3.2 Experimental Design and Labeling

The structure of a BCI experiment determines the information available for decoding and the quality of the labels used for training. *Event-based paradigms* rely on well-defined, time-locked events such as the presentation of images and sounds, system errors, or fixation onsets. These designs allow the precise stimulus-locked segmentation of EEG and thus enable the subsequent decoding of ERPs. In contrast, *continuous-state elicitation* involves tasks that manipulate cognitive variables over longer periods, such as workload, vigilance, or affect. Here, labels often reflect gradual changes, derived from task difficulty, behavioral markers, or continuous ratings.

Regardless of the paradigm type, having properly labeled data is critical. Noisy labels from ambiguous or imprecise instructions or temporal jitter in the apparatus can erode classifier performance severely and make a proper interpretation of the results impossible. Ensuring a well-defined mental state elicitation, minimizing confounds (e.g., from co-occurring eye movements), and aligning labels precisely with the actual neural events is thus vital in ensuring the validity of the experimental design.

25.3.3 Preprocessing and Artifacts

Raw EEG is susceptible to noise from environmental sources, hardware, and physiological processes (see also Chap. 17). Preprocessing aims to attenuate these disturbances while preserving neural signals relevant for decoding. Standard steps include *band-pass filtering* to remove slow drifts and high-frequency, *notch-filtering* or other tools to remove line noise, the removal of bad channels, and *re-referencing* to the common average. Artifacts from eye blinks and muscle activity, too, can pose significant challenges if they co-occur systematically with labels, especially in mobile or applied settings. Techniques such as *Independent Component Analysis (ICA)* (Bell & Sejnowski, 1995; Hyvärinen et al., 2001) or *Artifact Subspace Reconstruction (ASR)* (Mullen et al., 2015; Chang et al., 2019) can help to suppress these sources but come at the cost of not necessarily being real-time capable, or risking removing brain activity from the data (see also Chap. 15). As in many situations, however, preprocessing can only do so much, and good data begins at the source. Proper EEG setup and data quality monitoring during acquisition are vital to ensure the highest possible data quality for any downstream analysis.

Comparing BCIs with other forms of neuroimaging, one point needs to be emphasized: In real-time BCIs, preprocessing must operate with minimal latency. By this nature, BCI preprocessing, too, is minimal and favors lightweight, causal filters, and reduced-complexity artifact handling.

25.3.4 Feature Extraction

BCI performance depends heavily on how neural activity is represented and which mental states are to be decoded, or elicited in the paradigm. As the neural signals in the EEG were categorized into event-related and oscillatory components, this distinction is relevant for BCI feature extraction, as well. *Time-domain features*, such as ERP windowed mean amplitudes, capture transient components tied to specific events (e.g., error responses). *Frequency-domain features* are used to capture oscillatory activity in frequency bands associated with continuous cognitive processes (e.g., workload). *Spatial filtering methods,* including Common Spatial Patterns (CSPs), enhance class separability in such frequency features by emphasizing spatial differences in activation patterns before extracting the band powers themselves.

A more recent family of representations uses *Riemannian geometry*, modeling covariance matrices on their natural manifold to capture stable spatial relationships in EEG (Congedo et al., 2017; Yger et al., 2017). These approaches have shown strong performance in both event-based and continuous decoding tasks, but require additional processing steps.

25.3.5 Machine Learning and Classifiers

BCIs employ a range of machine learning methods to translate features into pre dictions. *Linear Discriminant Analysis (LDA)* and *logistic regression* remain widely used because they provide strong performance with minimal computational cost and work well with limited training data (Lotte et al., 2018). *Support Vector Machines (SVMs)* can handle nonlinear boundaries, but this is often not necessary in BCI, and linear SVMs are more commonly used. They may exhibit slightly improved performance but require more tuning and computational overhead. Importantly, in BCI, there is often a lack of data points in relation to the number of features—or parameters—the classifier is taking in. For example, in windowed-means approaches, the number of features can exceed one thousand, but the number of samples may be only one hundred per class. Hence, it is particularly important to take care to regularize the data accordingly, independently of the machine learning method employed.

Recently, new approaches employing convolutional neural nets, such as EEGNet (Lawhern et al., 2018), have emerged. These approaches combine the extraction of features and the separation of the classes in one algorithm to learn temporal or spatial structure directly from raw or minimally processed EEG, thus leveraging the power of both data-driven feature extraction and non-linear decoding (Schirrmeister et al., 2017; Roy et al., 2019). While these methods are newer and initially showed promise, deep learning approaches require more substantial computational resources, are prone to overfitting due to their size, and they do not show relevant improvements over classic approaches (Lotte et al., 2018). With larger quantities of data, these methods might become more powerful, however.

25.3.6 Classifier Validation

Validation assesses how well a trained model can *decode neural states*, and it must go beyond reporting the accuracy on the training data, which is inflated because the model has already seen those examples. *Cross-validation* addresses this by partitioning data into train–test folds and estimating performance on unseen data, ideally preserving temporal structure to avoid leakage. A completely *independent test set* provides the most reliable measure of generalization, however. Importantly, it is not enough to assume that a decoding accuracy above chance level is sufficient. Instead, the statistical significance level of "exceeding chance level by chance" needs to be computed for the given number of samples per class, and reported as a baseline that the decoding must exceed (Combrisson & Jerbi, 2015; Mueller-Putz et al., 2008). Here, *data imbalance* can bias both models and accuracy estimates, and although some algorithms, such as LDA, tolerate imbalance reasonably well during training, it is preferable to balance conditions for validation to obtain a proper estimate of the classifier's performance.

Beyond numerical metrics, validation must include inspection of the *neural sources* driving classification to ensure that decoding reflects genuine brain signals rather than artifacts (Haufe et al., 2014; Krol et al., 2018a). Finally, especially pBCIs require assessing *generalizability to new tasks or contexts*, since calibration is often performed in one setting but deployment may occur in others (e.g., Zhang et al., 2018).

25.4 BCI Types Illustrated

25.4.1 Motor Imagery as an Active BCI Paradigm

Motor imagery (MI) BCIs convert imagined movements into commands by exploiting modulations of sensorimotor rhythms. When users imagine left- or right-hand actions, mu (8–13 Hz) and beta (13–30 Hz) power over the sensorimotor cortex shows reliable event-related desynchronization patterns, lateralized to the imagined limb (Pfurtscheller & Lopes da Silva, 1999; Pfurtscheller et al., 2006). A standard MI pipeline uses band-pass filtering around the mu/beta range, spatial filtering such as Common Spatial Patterns to enhance class separability (Ramoser et al., 2000), and a linear classifier such as LDA. MI systems are a canonical example of volitional neural control and are widely used to study intentional BCI communication, neurofeedback, and motor rehabilitation (Daly & Wolpaw, 2008; Blankertz et al., 2008).

25.4.2 P300 and SSVEPs as Reactive BCI Paradigms

Reactive BCIs leverage stimulus-evoked activity that is modulated by attention. In the P300 speller, a matrix of letters is presented, where rows and columns flash at random. Flashes of the attended letter elicit a positive ERP around 300 ms poststimulus. By classifying which stimuli produce strong P300 responses, the system infers the intended selection, and over time, words and sentences can be spelled (Farwell & Donchin, 1988). Decoding typically uses ERP amplitude features with linear classifiers. Steady-state visually evoked potential (SSVEP) BCIs instead use different continuous flicker frequencies on visible objects, where fixations on such objects induce frequency-tagged oscillations in the visual cortex detectable over occipital channels (Vialatte et al., 2010; Norcia et al., 2015). Spectral power in the EEG can then be used to identify the dominant flicker frequency to determine the chosen item. Both paradigms provide high reliability with little user training, making them central for communication BCIs.

25.4.3 Workload, Error, and Other Passive BCI Paradigms

Passive BCIs decode naturally occurring brain activity to estimate cognitive or affective state in real time, without requiring intentional control. Typical targets include workload, vigilance, attention, fatigue, and error perception, operationalized through task manipulations rather than explicit commands (Zander & Kothe, 2011; Arico et al., 2018). Continuous states are commonly inferred from oscillatory markers such as frontal theta or parietal alpha changes for cognitive workload, whereas event-based states rely on ERPs at the level of single trials linked to feedback, surprise, or errors. Processing pipelines mirror those of other BCIs, with continuous states being decoded similarly to MI BCIs, and event-based transient states being decoded similarly to P300 BCIs. These paradigms are not used for communication or control, and do not require conscious attentional shifts or active imagination on the user's side. Instead, they ground neuroadaptive systems or artificial intelligence applications that can adjust to users based on moment-to-moment state detection.

In the literature, workload is one of the most extensively studied targets for passive BCIs. Here, changes in frontal theta and parietal alpha are commonly used to track cognitive demand in working-memory paradigms (Gevins & Smith, 2003; Gerjets et al., 2014; Brouwer et al., 2012). Carefully designed training paradigms have also been shown to be sensitive across tasks (Zhang et al., 2018), while remaining robust across posture and stimulation modality (flat screen/VR) and using the same frontal theta/parietal alpha markers as previous studies (Gherman et al., 2025).

Beyond workload, error and surprise responses provide powerful event-based targets. Error-related negativities and related components can signal when the user detects a mismatch or mistake, whether caused by themselves or by the system (Falkenstein et al., 2000). This has been exploited in passive BCIs that detect interaction errors and trigger automatic "undo" or corrective actions (Chavarriaga et al., 2014; Lopes-Dias et al., 2019; Xavier Fidêncio et al. 2022). Prediction-error responses have also been used to detect unrealistic visuo-haptic interactions and implausible object physics in VR, enabling systems to identify when a simulation violates user expectations and adapt automatically to user preference (Gehrke et al. 2019, 2022, 2025).

Beyond workload and error, passive BCIs have been used to track a range of other mental states. Attention and task engagement are often monitored using spectral indices similar to workload, for example, frontal theta and parietal alpha ratios, or composite "engagement indices" (Pope et al., 1995; Berka et al., 2007). These measures have been applied to vigilance and drowsiness monitoring and safety-critical supervision, sometimes complemented by event-related components such as the P300, which is modulated by stimulus relevance, motivation, and attentional resources (Begleiter et al., 1983; Polich, 2007; Acı et al., 2019; Vortmann et al., 2022; Pawlitzki et al., 2021). A closely related line of work targets relaxation, often operationalized through increased alpha power or related spectral changes, for example, in neurofeedback or stress-reduction settings (Ewing et al., 2016; Krol et al., 2017; Klug, 2022). Another distinct target is the intent to interact: passive BCIs can distinguish spontaneous from goal-directed fixations by analyzing fixation-related potentials, particularly when combined with an eye tracker. Studies have shown that EEG during target fixations carries markers of intentional selection (Protzak et al., 2013), and that gaze fixations used for interaction can be separated from incidental fixations and even smooth pursuit using EEG features (Shishkin et al., 2016; Zhao et al., 2021). Together, these paradigms demonstrate that passive BCIs can access a broad spectrum of cognitive, affective, and intentional states beyond workload and error.

25.5 From Mental State Assessment to Neuroadaptive Systems

25.5.1 Mental State Assessment as a First Stage

The most basic use of passive BCIs is *mental state assessment (MSA)*. Here, brain signals are recorded and decoded to estimate states such as workload or vigilance, but these estimates are used mainly for monitoring or later analysis, not for direct control. A fatigue-monitoring app that records your state while working and lets you review it afterwards is a typical example. Strictly speaking, this does not meet the classical BCI definition, which requires both neural input and a system-level output (Wolpaw & Wolpaw, 2012). The original passive BCI paper instead framed *real-time brain state decoding (RBSD)* as an extension of cognitive monitoring (Zander & Kothe, 2011), in line with *neuroergonomics*, which studies cognition at work and in everyday life (Parasuraman & Rizzo, 2007; Ayaz & Dehais, 2018). Although MSA only uses the "front half" of the BCI loop, it relies on the same decoding methods as interactive systems and is one of the most mature real-world uses of passive BCIs (Arico et al., 2018; Krol et al., 2018b).

25.5.2 Open- and Closed-Loop Adaptation

The next step is to let systems respond to decoded states. *Open-loop adaptation* refers to one-off actions that do not directly change the triggering state, for example, a high-workload estimate could be used to trigger a notification to take a break or highlight a problematic interface element. *Closed-loop adaptation* changes task parameters to influence the state itself, such as increasing automation or suppressing non-essential information when workload rises in a driving or piloting task (Kohlmorgen et al., 2007; Zander & Jatzev, 2012; Andreessen et al., 2021). In both cases, passive BCIs provide implicit input, but in closed-loop systems, the adaptation becomes part of the mechanism that shapes the user's mental state.

25.5.3 Autonomous Adaptation, User Models, and Cognitive Probing

The most powerful use of passive BCIs is to monitor mental states over longer periods and relate them to contextual factors to build *user models* and enable *automated adaptation* that anticipates problematic states (Zander & Jatzev, 2012; Fairclough, 2017; Klaproth et al., 2020). Because decoding alone cannot specify *why* a state occurred, systems can actively manipulate their own settings to elicit diagnostic responses, an approach termed *cognitive probing* (Krol et al., 2020). Over time, such probes refine the user model and support autonomous behavior that increasingly approaches the goal state of the user (Zander et al., 2016). Since these adaptations rely on internal states not explicitly offered as input, they must satisfy the *user-interest criterion*: actions should be justifiable as being in the user's best interest and remain transparent and overridable (Krol & Zander, 2022).

25.6 Practical Challenges

25.6.1 Mobility, Artifacts, and Non-stationarity

Real-world BCIs must operate under movement, changing environments, and varying user states. Mobile EEG introduces additional artifacts from motion, muscle activity, and shifting electrode contact that may result in suboptimal data quality (Gramann et al., 2011; Klug & Gramann, 2021; Klug et al., 2024). While offline analysis approaches, such as ICA, can help reduce this issue for MSA, real-time capable artifact attenuation approaches, such as ASR (Mullen et al., 2015), may be required for fully real-time applications in the wild. Even with good preprocessing, the statistical properties of EEG change over time due to factors such as fatigue, learning, electrode drift, or changing posture, a phenomenon commonly referred to as *non-stationarity* (Blankertz et al., 2008; Samek et al., 2014). Classifiers trained on short, clean calibration sessions can degrade at long-term use or in different contexts. Robust pBCIs therefore require not only better artifact handling but also models that can track slow drifts and rapid changes in data distributions, for example, through adaptive filters, online covariance updates, or periodic recalibration (Lotte et al., 2018).

25.6.2 Cross-User and Cross-Session Generalization

Most current BCIs are still *subject-specific*: they require a dedicated calibration session for each user and often for each new session or task. This calibration burden is problematic for applied and consumer scenarios. Cross-session and cross-user generalization are difficult because of individual differences in anatomy, cognitive strategy, and noise profiles (Lotte et al., 2018). Research on transfer learning and domain adaptation addresses this by aligning feature spaces across users or sessions, re-weighting samples, or learning shared representations that are less sensitive to distributional changes (Jayaram et al., 2016). Riemannian alignment, multitask learning, and few-shot adaptation are promising directions (Yger et al., 2017), but robust zero-calibration BCIs remain rare in practice. For passive BCIs in particular, where the goal is to deploy workload or error decoders across tasks and contexts, improving generalization is one of the central open challenges.

25.6.3 Evaluation in Real Settings

BCI performance is often reported as offline accuracy on well-controlled datasets, but real-world evaluation requires broader metrics. In applied scenarios, *task-relevant outcomes* such as driving performance, error rates, learning gains, or user workload reduction are more informative than raw classification accuracy. Longitudinal and field studies are needed to assess robustness over days or weeks, as well as user acceptance and trust. For pBCIs, *cross-task application* is particularly critical: decoders trained in one paradigm are often applied to different tasks or environments, where signal characteristics and label structure change. Similarly, applying event-based classifiers (e.g., error reaction decoders) continuously to live-streamed data is nontrivial, as the live data may not have event markers at the same temporal accuracy level as the training data, or no markers at all (Pan et al., 2024). Evaluation frameworks must therefore account for streaming performance, latency, false-alarm rates, and the impact of neuroadaptive interventions on actual user behavior.

25.6.4 Ethics, Privacy, and Neurorights

BCIs and pBCIs raise distinctive ethical questions because they access signals that are closely linked to perception, intention, and affect. Even simple workload or error decoders can reveal information that users might prefer to keep private, such as lapses of attention, fatigue, or preference patterns (Krol & Zander, 2022; Ienca & Andorno, 2017). Responsible design requires transparency about which signals are recorded, how long they are stored, who can access them, and how adaptations are triggered. Informed consent should cover not just data acquisition but also the possible inferences and system actions. Recent discussions of *neurorights* have proposed protections for cognitive liberty, mental privacy, and mental integrity in response to emerging neurotechnology (Ligthart et al., 2023; Yuste et al., 2017). For neuroadaptive systems based on passive BCIs, respecting these principles and the user-interest criterion is essential to maintain trust and ensure that powerful decoding methods are used in ways that genuinely benefit users (Krol & Zander, 2022).

25.7 Conclusions and Outlook

25.7.1 Key Takeaways

This chapter has outlined how BCIs transform neural activity into meaningful system behavior. We distinguished active, reactive, and passive BCIs based on whether signals are intentionally produced, stimulus-locked, or arise spontaneously during ongoing cognition (Zander & Kothe, 2011). We then described the core pipeline from experimental design and signal acquisition through preprocessing, feature extraction, machine learning, and validation (Lotte et al., 2018), with ERPs and oscillations providing the main signal families exploited for decoding.

Within this landscape, passive BCIs occupy a special role. They do not replace explicit input but add an implicit channel that tracks workload, attention, error perception, surprise, or intent to interact in real time. We reviewed paradigms and markers for these states and showed how they can be combined with context information to support neuroadaptive systems. Finally, we discussed a spectrum of interactivity—from mental state assessment and open-loop notifications to closed-loop and autonomous adaptations driven by user models and cognitive probing—underpinned by the user-interest criterion as a guardrail for responsible design (Krol et al., 2018b; Krol & Zander, 2022).

25.7.2 Future Trajectories

Several technical and conceptual developments are likely to shape the next generation of BCIs and neuroadaptive systems. On the hardware side, progress in wearable EEG, ear-EEG, and VR-integrated devices will improve comfort and ecological validity, making long-term and mobile applications more realistic. On the algorithmic side, cross-session, cross-subject, and cross-task classifier generalization are the most relevant frontiers at the moment. Both hardware and software improvements are building towards a more user-friendly application of pBCIs, allowing their widespread use.

Perhaps most intriguingly, passive BCIs can provide a rich feedback channel for adaptive AI systems. Real-time estimates of workload, uncertainty, surprise, preference, or agreement could serve as implicit rewards or constraints for interactive learning, helping AI systems to align their behavior with human goals and limits more directly than behavioral logs alone. In such scenarios, pBCIs would not "read thoughts" to replace speech but offer continuous, noisy measurements of how well an AI is supporting the user in context. Combining these capabilities with neurorights-informed safeguards for mental privacy and agency may enable artificial assistants that are not only more capable, but also more genuinely understanding and aligned with the people they serve.

References

Acı, Ç.İ., Kaya, M., & Mishchenko, Y. (2019). Distinguishing mental attention states of humans via an EEG-based passive BCI using machine learning methods. *Expert Systems with Applications*, *134*, 153–166. Available at: https://www.sciencedirect.com/science/article/abs/pii/S0957417419303926

Andreessen, L. M., et al. (2021). Toward neuroadaptive support technologies improving digital reading: Passive BCI-based assessment mental workload imposed text difficulty presentation speed during reading. *User Modeling and User-Adapted Interaction, 31*, 75–104.

Arico, P., et al. (2018). Passive BCI beyond the lab: Current trends and future directions. *Physiological Measurement, 39*(8). https://doi.org/10.1088/1361-6579/aad57e

Ayaz, H., & Dehais, F. (Eds.). (2018). *Neuroergonomics—The brain at work and in everyday life* (1st ed.). Elsevier Academic. https://doi.org/10.1016/C2016-0-02196-4

Begleiter, H., et al. (1983). P3 and stimulus incentive value. *Psychophysiology, 20*(1), 95–101. https://doi.org/10.1111/j.1469-8986.1983.tb00909.x

Bell, A. J., & Sejnowski, T. J. (1995). An information-maximization approach to blind separation and blind deconvolution. *Neural Computation, 7*(6), 1129–1159. https://doi.org/10.1162/neco.1995.7.6.1129

Berka, C., et al. (2007). EEG correlates of task engagement and mental workload in vigilance, learning, and memory tasks. *Aviation, Space, and Environmental Medicine, 78*(5 Suppl), B231–B244. https://www.ncbi.nlm.nih.gov/pubmed/17547324

Biasiucci, A., Franceschiello, B., & Murray, M. M. (2019). Electroencephalography. *Cell*. https://doi.org/10.1016/j.cub.2018.11.052

Blankertz, B., et al. (2008). The Berlin brain-computer interface. In *Lecture notes in computer science* (pp. 79–101). Springer. https://doi.org/10.1007/978-3-540-68860-0_4

Brouwer, A.-M., et al. (2012). Estimating workload using EEG spectral power and ERPs in the n-back task. *Journal of Neural Engineering, 9*(4), 045008. http://stacks.iop.org/1741-2552/9/i=4/a=045008?key=crossref.27952d535478677b655a40d95df34b6d

Chang, C.-Y., et al. (2019). Evaluation of artifact subspace reconstruction for automatic artifact components removal in multi-channel EEG recordings. *IEEE Transactions on Bio-Medical Engineering, 67*(4), 1114–1121. https://doi.org/10.1109/TBME.2019.2930186

Chavarriaga, R., Sobolewski, A., & Millán, J. D. R. (2014). Errare machinale est: The use of error-related potentials in brain-machine interfaces. *Frontiers in Neuroscience, 8*, 208. http://journal.frontiersin.org/article/10.3389/fnins.2014.00208/abstract

Combrisson, E., & Jerbi, K. (2015). Exceeding chance level by chance: The caveat of theoretical chance levels in brain signal classification and statistical assessment of decoding accuracy. *Journal of Neuroscience Methods, 250*, 126–136. https://doi.org/10.1016/j.jneumeth.2015.01.010

Congedo, M., Barachant, A., & Bhatia, R. (2017). Riemannian geometry for EEG-based brain-computer interfaces; a primer and a review. *Brain-Computer Interfaces, 4*(3), 155–174. https://doi.org/10.1080/2326263X.2017.1297192

Daly, J. J., & Wolpaw, J. R. (2008). Brain-computer interfaces in neurological rehabilitation. *Lancet Neurology, 7*(11), 1032–1043. https://doi.org/10.1016/S1474-4422(08)70223-0

Debener, S., et al. (2015). Unobtrusive ambulatory EEG using a smartphone and flexible printed electrodes around the ear. *Scientific Reports, 5*, 16743. https://doi.org/10.1038/srep16743

Donoghue, T., & Voytek, B. (2022). Automated meta-analysis of the event-related potential (ERP) literature. *Scientific Reports, 12*(1), 1867. https://doi.org/10.1038/s41598-022-05939-9

Ewing, K. C., Fairclough, S. H., & Gilleade, K. (2016). Evaluation of an adaptive game that uses EEG measures validated during the design process as inputs to a biocybernetic loop. *Frontiers in Human Neuroscience, 10*, 223. https://doi.org/10.3389/fnhum.2016.00223

Fairclough, S. H. (2017. Physiological computing and intelligent adaptation. In *Emotions and affect in human factors and human-computer interaction* (pp. 539–556). Elsevier. https://doi.org/10.1016/b978-0-12-801851-4.00020-3

Falkenstein, M., et al. (2000). ERP components on reaction errors and their functional significance: A tutorial. *Biological Psychology, 51*(2–3), 87–107. https://pubmed.ncbi.nlm.nih.gov/10686361/

Farwell, L. A., & Donchin, E. (1988). Talking off the top of your head: Toward a mental prosthesis utilizing event-related brain potentials. *Electroencephalography and Clinical Neurophysiology, 70*(6), 510–523. https://doi.org/10.1016/0013-4694(88)90149-6

Fischer, G. (2001). User modeling in human–computer interaction. *User Modeling and User-Adapted Interaction, 11*(1/2), 65–86. https://doi.org/10.1023/a:1011145532042

Gehrke, L., et al. (2019). Detecting visuo-haptic mismatches in virtual reality using the prediction error negativity of event-related brain potentials. In *Proceedings of the 2019 CHI conference on human factors in computing systems*. CHI '19 (pp. 1–11). Association for Computing Machinery. https://doi.org/10.1145/3290605.3300657

Gehrke, L., et al. (2022). Neural sources of prediction errors detect unrealistic VR interactions. *Journal of Neural Engineering, 19*(3), 036002. https://doi.org/10.1088/1741-2552/ac69bc

Gehrke, L., et al. (2025). Neuroadaptive haptics: A proof-of-concept comparing reinforcement learning from explicit ratings and neural signals for adaptive XR systems. *Frontiers in Virtual Reality, 6*, 1616442. https://doi.org/10.3389/frvir.2025.1616442

Gerjets, P., et al. (2014). Cognitive state monitoring and the design of adaptive instruction in digital environments: Lessons learned from cognitive workload assessment using a passive brain-computer interface approach. *Frontiers in Neuroscience, 8*(DEC), 1–21. https://doi.org/10.3389/fnins.2014.00385

Gevins, A., & Smith, M. E. (2003). Neurophysiological measures of cognitive workload during human–computer interaction. *Theoretical Issues in Ergonomics Science, 4*(1–2), 113–131. https://doi.org/10.1080/14639220210159717

Gherman, D. E., et al. (2025). An investigation of a passive BCI's performance for different body postures and presentation modalities. *Biomedical Physics & Engineering Express, 11*(2). https://doi.org/10.1088/2057-1976/adb58b

Gramann, K., et al. (2011). Cognition in action: Imaging brain/body dynamics in mobile humans. *Reviews in the Neurosciences, 22*(6), 593–608. https://doi.org/10.1515/RNS.2011.047

Haufe, S., et al. (2014). On the interpretation of weight vectors of linear models in multivariate neuroimaging. *NeuroImage, 87*, 96–110. https://doi.org/10.1016/j.neuroimage.2013.10.067

Hyvärinen, A., Karhunen, J., & Oja, E. (2001). *Independent component analysis*. Wiley. https://doi.org/10.3156/jsoft.16.150_1

Ienca, M., & Andorno, R. (2017). Towards new human rights in the age of neuroscience and neurotechnology. *Life Sciences, Society and Policy, 13*(1), 5. https://doi.org/10.1186/s40504-017-0050-1

Jayaram, V., et al. (2016). Transfer learning in brain-computer interfaces. *IEEE Computational Intelligence Magazine, 11*(1), 20–31. https://doi.org/10.1109/MCI.2015.2501545

Klaproth, O. W., et al. (2020). A neuroadaptive cognitive model for dealing with uncertainty in tracing pilots' cognitive state. *Topics in Cognitive Science, 12*(3), 1012–1029. https://onlinelibrary.wiley.com/doi/10.1111/tops.12515

Klimesch, W. (1999). EEG alpha and theta oscillations reflect cognitive and memory performance: A review and analysis. *Brain Research. Brain Research Reviews, 29*(2–3), 169–195. https://doi.org/10.1016/s0165-0173(98)00056-3

Klug, M. (2022). Real Virtual Magic–Modifying a VR game with a BCI to enhance immersion. In *Neuroadaptive technologies 2022 conference proceedings*.

Klug, M., & Gramann, K. (2021). Identifying key factors for improving ICA-based decomposition of EEG data in mobile and stationary experiments. *The European Journal of Neuroscience, 54*(12), 8406–8420. https://doi.org/10.1111/ejn.14992

Klug, M., Berg, T., & Gramann, K. (2024). Optimizing EEG ICA decomposition with data cleaning in stationary and mobile experiments. *Scientific Reports, 14*(1), 14119. https://doi.org/10.1038/s41598-024-64919-3

Kohlmorgen, J., et al. (2007). Improving human performance in a real operating environment through real-time mental workload detection. In G. Dornhege, et al. (Eds.), *Toward brain-computer interfacing* (pp. 409–422). MIT Press. https://play.google.com/store/books/details?id=V88swGX83ecC

Krigolson, O. E., et al. (2017). Choosing MUSE: Validation of a low-cost, portable EEG system for ERP research. *Frontiers in Neuroscience, 11*, 109. https://doi.org/10.3389/fnins.2017.00109

Krol, L. R., & Zander, T. O. (2022). Chapter 2—Defining neuroadaptive technology: the trouble with implicit human-computer interaction. In S. H. Fairclough & T. O. Zander (Eds.), *Current research in neuroadaptive technology* (pp. 17–42). Academic. https://www.sciencedirect.com/science/article/pii/B9780128214138000075

Krol, L. R., Freytag, S.-C., & Zander, T. O. (2017). Meyendtris: A hands-free, multimodal tetris clone using eye tracking and passive BCI for intuitive neuroadaptive gaming. In *Proceedings of the 19th ACM international conference on multimodal interaction*. ICMI '17. ICMI '17: INTERNATIONAL CONFERENCE ON MULTIMODAL INTERACTION (pp. 433–437). ACM. https://dl.acm.org/doi/10.1145/3136755.3136805

Krol, L. R., Mousavi, M., et al. (2018a). Towards classifier visualisation in 3D source space. In *2018 IEEE international conference on systems, man, and cybernetics (SMC)* (pp. 71–76). https://ieeexplore.ieee.org/abstract/document/8616017/

Krol, L. R., Andreessen, L. M., & Zander, T. O. (2018b). Passive brain–computer interfaces: A perspective on increased interactivity. In *Brain–computer interfaces handbook* (pp. 69–86). CRC Press. https://www.taylorfrancis.com/chapters/edit/10.1201/9781351231954-3/passive-brain%E2%80%93computer-interfaces-laurens-krol-lena-andreessen-thorsten-zander

Krol, L. R., Haselager, P., & Zander, T. O. (2020). Cognitive and affective probing: A tutorial and review of active learning for neuroadaptive technology. *Journal of Neural Engineering, 17*(1), 012001. https://doi.org/10.1088/1741-2552/ab5bb5

Lawhern, V. J., et al. (2018). EEGNet: A compact convolutional neural network for EEG-based brain-computer interfaces. *Journal of Neural Engineering, 15*(5). https://doi.org/10.1088/1741-2552/aace8c

Ligthart, S., et al. (2023). Minding rights: Mapping ethical and legal foundations of "neurorights." *Cambridge Quarterly of Healthcare Ethics: CQ*, 1–21. https://doi.org/10.1017/S0963180123000245

Lopes-Dias, C., Sburlea, A. I., & Müller-Putz, G. R. (2019). Online asynchronous decoding of error-related potentials during the continuous control of a robot. *Scientific Reports, 9*(1), 17596. https://doi.org/10.1038/s41598-019-54109-x

Lotte, F., et al. (2018). A review of classification algorithms for EEG-based brain-computer interfaces: A 10 year update. *Journal of Neural Engineering, 15*(3). https://doi.org/10.1088/1741-2552/aab2f2

Mehta, R. K., & Parasuraman, R. (2013). Neuroergonomics: A review of applications to physical and cognitive work. *Frontiers in Human Neuroscience, 7*, 889. https://doi.org/10.3389/fnhum.2013.00889

Mirkovic, B., et al. (2016). Target speaker detection with concealed EEG around the ear. *Frontiers in Neuroscience, 10*, 349. https://doi.org/10.3389/fnins.2016.00349

Mueller-Putz, G., et al. (2008). Better than random: A closer look on BCI results. *International Journal of Bioelectromagnetism, 10*(1), 52–55.

Mullen, T. R., et al. (2015). Real-time neuroimaging and cognitive monitoring using wearable dry EEG. *IEEE Transactions on Biomedical Engineering, 62*(11), 2553–2567. https://doi.org/10.1109/TBME.2015.2481482

Norcia, A. M., et al. (2015). The steady-state visual evoked potential in vision research: A review. *Journal of Vision, 15*(6), 4. https://doi.org/10.1167/15.6.4

Pan, Y., Zander, T. O., & Klug, M. (2024). Advancing passive BCIs: A feasibility study of two temporal derivative features and effect size-based feature selection in continuous online EEG-based machine error detection. *Frontiers in Neuroergonomics, 5*, 1346791. https://doi.org/10.3389/fnrgo.2024.1346791

Parasuraman, R., & Rizzo, M. (Eds.). (2007). *Neuroergonomics—The brain at work*. Oxford University Press.

Pawlitzki, J., et al. (2021). *Automation surprise in the neuroadaptive cockpit*. https://lrkrol.com/files/pawlitzki2021neuroergonomics-automationsurprise.pdf. Accessed 8 Oct 2025.

Pfurtscheller, G., & Lopes da Silva, F. H. (1999). Event-related EEG/MEG synchronization and desynchronization: Basic principles. *Clinical Neurophysiology, 110*(11), 1842–1857. https://doi.org/10.1016/s1388-2457(99)00141-8

Pfurtscheller, G., et al. (2006). Mu rhythm (de)synchronization and EEG single-trial classification of different motor imagery tasks. *NeuroImage, 31*(1), 153–159. https://doi.org/10.1016/j.neuroimage.2005.12.003

Polich, J. (2007). Updating P300: An integrative theory of P3a and P3b. *Clinical Neurophysiology, 118*(10), 2128–2148. https://www.ncbi.nlm.nih.gov/pmc/articles/PMC2715154/

Pope, A. T., Bogart, E. H., & Bartolome, D. S. (1995). Biocybernetic system evaluates indices of operator engagement in automated task. *Biological Psychology, 40*(1–2), 187–195. https://doi.org/10.1016/0301-0511(95)05116-3

Protzak, J., Ihme, K., & Zander, T. O. (2013). A passive brain-computer interface for supporting gaze-based human-machine interaction. In *Universal access in human-computer interaction. Design methods, tools, and interaction techniques for eInclusion* (pp. 662–671). Springer. https://link.springer.com/chapter/10.1007/978-3-642-39188-0_71

Ramoser, H., Müller-Gerking, J., & Pfurtscheller, G. (2000). Optimal spatial filtering of single trial EEG during imagined hand movement. *IEEE Transactions on Rehabilitation Engineering, 8*(4), 441–446. https://doi.org/10.1109/86.895946

Ratti, E., et al. (2017). Comparison of medical and consumer wireless EEG systems for use in clinical trials. *Frontiers in Human Neuroscience, 11*. https://www.frontiersin.org/articles/10.3389/fnhum.2017.00398/full

Roy, Y., et al. (2019). Deep learning-based electroencephalography analysis: A systematic review. *Journal of Neural Engineering, 16*(5), 051001. https://doi.org/10.1088/1741-2552/ab260c

Samek, W., Kawanabe, M., & Müller, K.-R. (2014). Divergence-based framework for common spatial patterns algorithms. *IEEE Reviews in Biomedical Engineering, 7*, 50–72. https://doi.org/10.1109/RBME.2013.2290621

Schirrmeister, R. T., et al. (2017). Deep learning with convolutional neural networks for EEG decoding and visualization. *Human Brain Mapping, 38*(11), 5391–5420. https://doi.org/10.1002/hbm.23730

Schmidt, A. (2000). Implicit human computer interaction through context. *Personal Technologies, 4*(2–3), 191–199. https://doi.org/10.1007/bf01324126

Shishkin, S., et al. (2016). Improving eye-brain-computer interface performance by using electroencephalogram frequency components. *Bulletin of Russian State Medical University*, *2*, 36–41. https://cyberleninka.ru/article/n/improving-eye-brain-computer-interface-performance-by-using-electroencephalogram-frequency-components. Accessed 19 June 2022.

Vialatte, F.-B., et al. (2010). Steady-state visually evoked potentials: Focus on essential paradigms and future perspectives. *Progress in Neurobiology*, *90*(4), 418–438. https://doi.org/10.1016/j.pneurobio.2009.11.005. Accessed 20 Nov 2025.

Vortmann, L.-M., Ceh, S., & Putze, F. (2022). Multimodal EEG and eye tracking feature fusion approaches for attention classification in hybrid BCIs. *Frontiers in Computer Science*, *4*. https://www.frontiersin.org/articles/10.3389/fcomp.2022.780580/full

Willett, F. R., et al. (2023). A high-performance speech neuroprosthesis. *Nature, 620*(7976), 1031–1036. https://doi.org/10.1038/s41586-023-06377-x

Wolpaw, J. R., & Wolpaw, E. W. (2012). *Brain–computer interfaces: Something new under the sun*. Oxford University Press. https://academic.oup.com/book/1700/chapter-abstract/141279712?redirectedFrom=fulltext. Accessed 23 June 2023.

Wolpaw, J. R., et al. (2002). Brain–computer interfaces for communication and control. *Clinical Neurophysiology*, *113*(6), 767–791. https://www.sciencedirect.com/science/article/pii/S1388245702000573

Xavier Fidêncio, A., Klaes, C., & Iossifidis, I. (2022). Error-related potentials in reinforcement learning-based brain-machine interfaces. *Frontiers in Human Neuroscience, 16*, 806517. https://doi.org/10.3389/fnhum.2022.806517

Yger, F., Berar, M., & Lotte, F. (2017). Riemannian approaches in brain-computer interfaces: A review. *IEEE Transactions on Neural Systems and Rehabilitation Engineering, 25*(10), 1753–1762. https://doi.org/10.1109/TNSRE.2016.2627016

Yuste, R., et al. (2017). Four ethical priorities for neurotechnologies and AI. *Nature, 551*(7679), 159–163. https://doi.org/10.1038/551159a

Zander, T. O., & Jatzev, S. (2012). Context-aware brain-computer interfaces: Exploring the information space of user, technical system and environment. *Journal of Neural Engineering, 9*(1), 016003. https://doi.org/10.1088/1741-2560/9/1/016003

Zander, T. O., & Kothe, C. (2011). Towards passive brain-computer interfaces: Applying brain-computer interface technology to human-machine systems in general. *Journal of Neural Engineering, 8*(2). https://doi.org/10.1088/1741-2560/8/2/025005

Zander, T. O., et al. (2014). Towards BCI-based implicit control in human–computer interaction. In S. H. Fairclough & K. Gilleade (Eds.), *Advances in physiological computing* (pp. 67–90). Springer. https://doi.org/10.1007/978-1-4471-6392-3_4

Zander, T. O., et al. (2016). Neuroadaptive technology enables implicit cursor control based on medial prefrontal cortex activity. *Proceedings of the National Academy of Sciences, 113*(52), 14898–14903. http://www.pnas.org/lookup/doi/10.1073/pnas.1605155114

Zander, T. O., et al. (2017). Evaluation of a dry EEG system for application of passive brain-computer interfaces in autonomous driving. *Frontiers in Human Neuroscience, 11*, 78. https://doi.org/10.3389/fnhum.2017.00078

Zhang, X., Krol, L. R., & Zander, T. O. (2018). Towards task-independent workload classification: shifting from binary to continuous classification. In *Proceedings—2018 IEEE international conference on systems, man, and cybernetics, SMC 2018* (pp. 556–561). https://doi.org/10.1109/SMC.2018.00104

Zhao, D. G., et al. (2021). A passive BCI for monitoring the intentionality of the gaze-based moving object selection. *Journal of Neural Engineering, 18*(2). https://doi.org/10.1088/1741-2552/abda09

Chapter 26
EEG in Focal Epilepsy and Its Role in the Management of Adult Patients with Drug-Resistant Epilepsy

Boyuan Song, Umair J. Chaudhary, and Louis Lemieux

Abstract The principal role of EEG in focal epilepsy is to assist in diagnosis and classification of epilepsy subtype and identifying and localising epileptic focus. Interictal and ictal EEG patterns both provide very useful information to localise the epileptogenic zone in patients with drug-resistant focal epilepsy, specially those undergoing presurgical evaluation. Scalp EEG, particularly prolonged inpatient video-EEG monitoring, provides very valuable information in patients with drug resistant epilepsy undergoing presurgical evaluation in most clinical situations, whereas intracranial EEG is often needed in a small proportion of patients with drug resistant focal epilepsy undergoing presurgical evaluation to refine the localisation of epileptogenic zone and its relationship with the surrounding eloquent cortex. Advances in AI-driven EEG interpretation hold promise for providing more objective, sensitive, and efficient clinical and research information in the future.

Keywords Focal epilepsy · Drug-resistant epilepsy · Scalp EEG · Intracranial EEG

26.1 Focal Epilepsy

The International League Against Epilepsy (ILAE) defines epilepsy as a brain disorder that meets one of the following conditions: (1) At least two unprovoked seizures >24 h apart; (2) One unprovoked (or reflex) seizure and a probability of further seizures similar to the general recurrence risk (at least 60%) after two unprovoked seizures, occurring over the next 10 years; (3) Diagnosis of an epilepsy

B. Song · L. Lemieux (✉)
UCL Queen Square Institute of Neurology, University College London, London, UK
e-mail: boyuan.song.22@ucl.ac.uk; louis.lemieux@ucl.ac.uk

U. J. Chaudhary
UCL Queen Square Institute of Neurology, University College London, London, UK

National Hospital for Neurology and Neurosurgery, University College London Hospitals NHS Foundation Trust, London, UK
e-mail: umair.chaudhary@ucl.ac.uk

T. Warbrick (ed.), *The EEG Handbook*,
https://doi.org/10.1007/978-3-032-20450-9_26

syndrome (Fisher et al., 2014a). Since 2005, the ILAE uses the following definition of a seizure: "A transient occurrence of signs and/or symptoms due to abnormal excessive or synchronous neuron activity in the brain" (Fisher et al., 2005). Focal seizure activity typically originates within a relatively confined region of the cerebral cortex and subsequently propagates to other brain areas through grey and white matter pathways (Schevon et al., 2012). Focal seizures may originate in subcortical structures (Berg et al., 2010).

Seizures are the hallmark of epilepsy and are characterized by observable clinical signs. Seizures are classified based on their clinical features, as illustrated in Fig. 26.1. The main seizure classes include Focal, Generalized, and Unknown (Beniczky et al., 2025). Focal-onset seizures originate within networks limited to one hemisphere, although they may subsequently propagate to broader cortical or subcortical networks. In contrast, generalized-onset seizures engage bilateral brain networks at seizure onset. The unknown category is used when available clinical or electrophysiological information suggests a seizure onset, but the information is

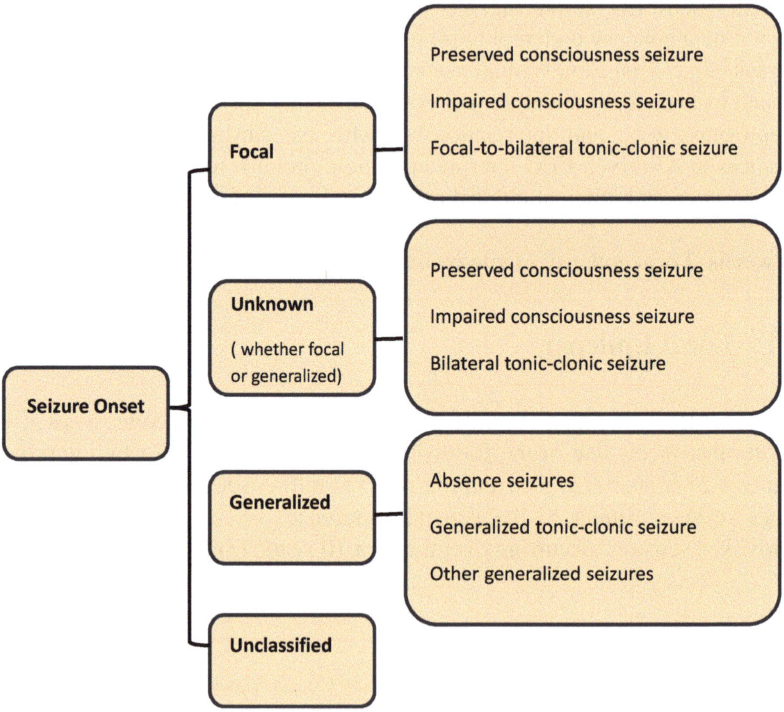

Fig. 26.1 The 2025 updated ILAE classification of seizure types categorise them into further subtypes based on consciousness i.e., preserved consciousness seizure or impaired consciousness seizure. If a seizures starts as focal and then propagates to become a tonic clonic seizure it is classified as focal to bilateral tonic clonic seizure. Under unknown, the subtypes are preserved consciousness seizure, impaired consciousness seizure, and bilateral tonic-clonic seizure. Under generalized, the subtypes are absence seizures, generalized tonic-clonic seizures, and other generalized seizures. (Adapted with permission from Beniczky et al., 2025)

insufficient to determine whether the onset is focal or generalized. A key change from the 2017 ILAE classification is the addition of the "unclassified" category, which is reserved for events confirmed by a clinician to be epileptic but lacking sufficient information for any more specific classification. Electrographic seizures are detected on electrophysiological recordings, typically without accompanying clinical manifestations.

This chapter focuses specifically on adults with focal epilepsy, which the ILAE defines as "originating within networks limited to one hemisphere; they may be discretely localized or more widely distributed" (Berg et al., 2010).

In the early stages of epilepsy diagnosis, structural neuroimaging techniques such as magnetic resonance imaging (MRI) is employed to identify the underlying aetiology (Cendes et al., 2016). Structural imaging can reveal an epileptogenic lesion in approximately 50% of patients with focal-onset seizures (Hakami et al., 2013), although the yield increases with new developments in imaging, e.g. 7T MRI (Klodowski et al., 2025).

Electroencephalography (EEG), a technique with high temporal resolution, is widely used to differentiate paroxysmal electrophysiological discharges, distinguish focal from generalized seizures, and identify syndrome-specific patterns (Smith, 2005). Within the broader diagnostic framework of epilepsy, EEG contributes to multiple aspects of evaluation, including the classification of epilepsy and seizure types, the diagnosis of status epilepticus, the localization of the epileptic focus, and the assessment of seizure-recurrence risk (Misulis et al., 2022).

EEG records the summated electrical activity of populations of excitable neurons. The intrinsic electrical properties of these cells generate local electrical fields that can be recorded using electrodes positioned at varying distances from the source. At shorter distances, recordings capture local field potentials (LFPs), whereas at longer distances such as recorded on the scalp, as in conventional scalp EEG, the recorded activity represents the summation of the activity of extended and/or multiple cortical generators (Lopes da Silva, 2009; Gloor, 1985). To effectively observe and monitor EEG activity, the recorded signals must exhibit sufficient duration and sustained intensity. In standard scalp EEG recordings lasting 10–30 min and, epileptiform discharges are detected in only 30–50% of patients (Pandian et al., 2004). The presence of epileptiform EEG patterns is one of the primary indicators of epilepsy. However, a recent systematic review with meta-analysis concluded that approximately 1.7% of individuals with epileptiform abnormalities do not experience seizures (Aschner et al., 2024).

In clinical practice, long-term video telemetry EEG (VT-EEG) is an important presurgical evaluation in patients with drug-resistant focal epilepsy, typically conducted over several hours or days. By integrating simultaneous video recordings of clinical manifestations, VT-EEG enables clinicians to classify and localize seizure types in people with epilepsy and differentiate between epileptic and non-epileptic/functional (also called 'dissociative'[1]) seizures (Van Patten et al., 2025).

[1]Defined as paroxysmal episodes of behavioural, sensory or motor changes.

26.2 EEG Manifestations of Focal Epilepsy

Epileptiform EEG activity is typically categorized into three states: ictal, postictal, and interictal, corresponding respectively to neural activity occurring during a seizure, after a seizure, and between seizures (Fisher & Engel, 2010). The postictal state is conceptually defined as an abnormal period beginning at the termination of an epileptic seizure and lasting until the brain returns to its baseline functional state. When seizures recur, the interval between them is referred to as the interictal state.

26.2.1 Ictal EEG Patterns

In focal epilepsy, ictal discharges demonstrate characteristic changes in frequency, amplitude, and morphology. The ictal EEG patterns are characterized by spatiotemporal evolution of focal rhythmic activity, which typically demonstrates a progressive increase in amplitude accompanied by a gradual slowing of frequency. This activity may subsequently spread to neighbouring electrodes (Britton et al., 2016). The specific EEG patterns vary depending on the seizure focus. In the absence of any subjective or objective clinical symptoms, such events are classified as subclinical/electrographic seizures.

Six major ictal patterns on scalp EEG (Foldvary et al., 2001), are summarized in Table 26.1. Among the 566 seizures recorded from their cohort of 72 patients with epilepsy, rhythmic delta activity occurred more frequently in seizures arising from the temporal lobe than from extratemporal regions. Theta activity was more commonly associated with mesial and neocortical temporal lobe epilepsy (TLE). There is a higher likelihood of repetitive epileptiform activity in seizures originating from the extratemporal lobe epilepsy. Moreover, generalized ictal onset patterns can be seen in seizures emerging from the mesial frontal and occipital lobes.

Although researchers generally agree that rhythmic theta/delta activity is commonly present in mesial temporal lobe epilepsy, this pattern is not exclusive to temporal lobe pathology. For example, Baumgartner et al. reported that temporal

Table 26.1 Ictal onset patterns on scalp EEG

EEG pattern	Definition
Rhythmic activity	Alpha, theta, or delta frequencies
Paroxysmal fast	Rhythmic activity>=13 Hz
Suppression	Activity <=10 μv in amplitude
Repetitive epileptiform activity	3 or more discharges in sequence
Arrhythmic activity	Irregular, mixed-frequency waveforms
Obscured	Pattern evolving from a period obscured by artefact such that precise time, pattern, and distribution of onset were indiscernible

Text from Foldvary et al. (2001)

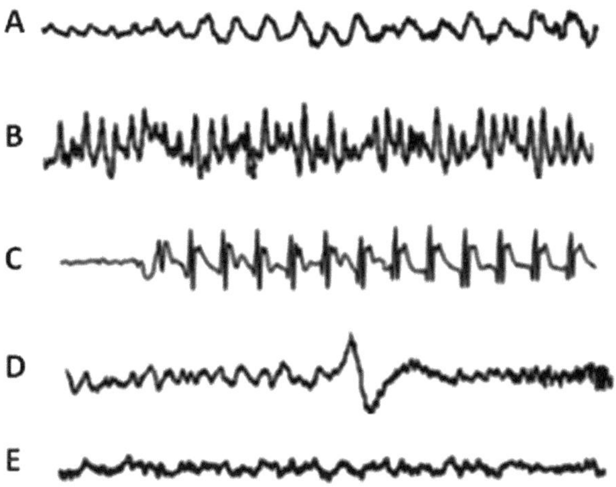

Fig. 26.2 Patterns of seizure onset frequently observed on scalp EEG. (**a**) Rhythmical activity evolving theta, delta, alpha frequencies; (**b**) Rhythmical spiking; (**c**) Spike waves; (**d**) Electro-decremental onset characterized by low-voltage fast activity: (**e**) Clinical seizure without a clear EEG correlate. (Adapted with permission from Fisher et al., 2014a, b)

intermittent rhythmic delta activity (TIRDA) may also be detected in lateral temporal and orbitofrontal irritative zones (Baumgartner et al., 2025).

Examples of seizure onset patterns commonly observed on scalp EEG are illustrated in Fig. 26.2. Importantly, the final conclusion about seizure/epilepsy diagnosis should always be made based on clinical history and seizure semiology and EEG provides additional supportive information (Fisher et al., 2014b).

Temporal lobe seizure onset often presents with rhythmic theta/delta activity localized to the temporal region. Mesial temporal lobe epilepsy (MTLE) is one of the most common forms of localization-related epilepsy, characterized by seizure activity originating from the medial structures of the temporal lobe, including the hippocampus, parahippocampal gyrus, and amygdala (Tatum, 2012). The electrophysiological hallmark of MTLE is the presence of rhythmic temporal theta or alpha activity, typically in the 5–9 Hz range (Pataraia et al., 1998; Tatum, 2012). Extratemporal seizures may exhibit features similar to those arising from the temporal lobe, and characterized by subtle rhythmic changes obscured by muscle artefacts (Britton et al., 2016).

Intracranial EEG (iEEG or icEEG) provides enhanced sensitivity compared with scalp EEG and therefore reveals a wider range of ictal onset patterns. The typical characteristics of iEEG seizure-onset patterns are summarized in Fig. 26.3, with a much greater signal-to-noise ratio and a wider frequency range compared to scalp EEG (Alkawadri et al., 2024; Lagarde et al., 2019; Perucca et al., 2014). In the study by Lagarde et al., (Lagarde et al., 2019) low-voltage fast activity (LVFA) emerged as one of the most common SEEG-defined seizure onset pattern in patients with drug resistant focal epilepsy. LVFA is defined as rhythmic fast oscillations greater than 14 Hz with amplitudes below 30 μV, occurring without an initial overt change in background activity. Although LVFA was observed across all etiological categories, it was overrepresented in malformations of cortical development (MCD) and post-vascular epilepsy. Notably, a burst of polyspikes preceding LVFA was identified exclusively in patients with focal cortical dysplasia (FCD). In addition, LVFA was

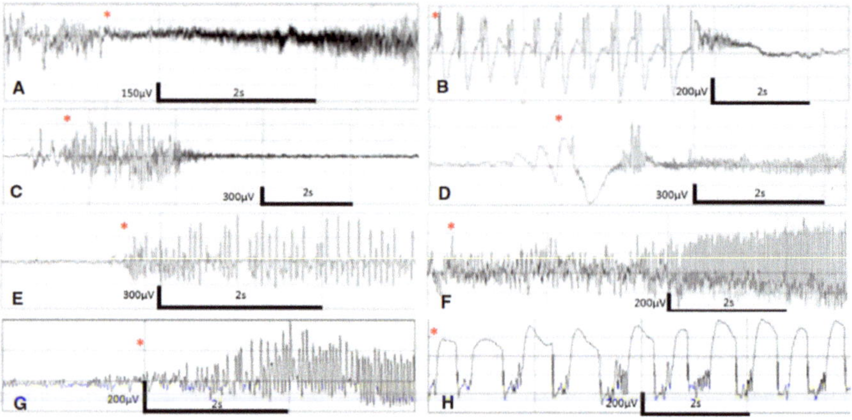

Fig. 26.3 The eight seizure onset patterns from SEEG, with red asterisks marking the seizure onset point. (**a**) Low-voltage fast activity (LVFA). (**b**) Preictal spiking that transitions into LVFA. (**c**) A brief burst of polyspikes—high-frequency (>12 Hz), high-amplitude, short-duration activity lasting less than 5 seconds—followed by LVFA. (**d**) A slow wave or baseline shift (comparable to a DC shift), which precedes LVFA. (**e**) Rhythmic spikes and spike-wave discharges at low frequencies above 6 Hz and consistently below 14 Hz. (**f**) Sharp theta or alpha activity, represented by low-frequency sinusoidal waveforms. (**g**) Sharp beta-frequency activity with beta-band sinusoidal oscillations. (**h**) A delta-brush pattern characterized by bursts of low-amplitude, rapid gamma-frequency activity superimposed on low-frequency delta sinusoidal waves. (Adapted with permission from Lagarde et al., 2019)

often followed by a DC shift or slow wave, particularly in cases involving extensive networks of epileptogenic zones (Lagarde et al., 2019).

26.2.2 *Interictal EEG Patterns*

The interictal EEG plays a crucial role in the diagnosis of epilepsy specifically assisting in establishing the presence of epilepsy, differentiating between focal and generalized seizure disorders, and defining specific epilepsy syndromes (Pillai & Sperling, 2006). Precise characterization and localization of interictal epileptiform abnormalitiesis essential for identifying specific epileptic syndromes, thereby guiding both therapeutic strategies and prognostic evaluation.

The International Federation of Clinical Neurophysiology (IFCN) defines interictal epileptiform discharges (IEDs) as isolated waveforms characterized by a di- or tri-phasic shape with a sharply pointed peak, a duration that differs from the surrounding background activity, and a distinctly asymmetric morphology (Kural et al., 2020). They are typically followed by a slow after-wave and disrupt the ongoing background rhythm at the time of their occurrence. In addition, their voltage topography demonstrates a coherent distribution of negative and positive potentials consistent with a cortical generator oriented radially, obliquely, or tangentially. A spike is defined as a transient with a duration of 20–70 ms, whereas a sharp wave has a duration of 70–200 ms (Kane et al., 2017).

The accurate interpretation of interictal activity requires caution, as certain benign EEG variants may mimic epileptiform discharges and lead to potential misclassification. Notably, some benign EEG variants, normally with unknown aetiology, can appear as patterns resembling epileptiform EEG activity, for example: Small sharp spikes of sleep (SSS), wicket spikes, subclinical rhythmic electrographic discharges in adults (SREDA), rhythmic mid-temporal discharges (RMTD), and Mu rhythm (White et al., 1977). Sometimes a burst of repeating SREDA can evolve to resemble subclinical seizures (Westmoreland & Klass, 1997).

In focal epilepsies, interictal spikes, when interpreted alongside other neuroimaging findings, can provide valuable guidance for localizing the seizure onset zone. The sensitivity of detecting IEDs using scalp EEG must be carefully considered. Goodin et al. reported that 29–55% of patients exhibited positive EEG findings for IEDs (Goodin et al., 1990). When the duration of scalp EEG monitoring is extended and recordings are repeated, sensitivity can increase substantially, reaching 80–90% (Salinsky et al., 1987). Tao et al. demonstrated that at least 10 cm^2 of temporally overlapping cortical activity is required to generate a scalp-recordable EEG spike in humans (Tao et al., 2011).

By contrast, IEDs are highly prevalent in iEEG, with reported sensitivities of 85–95% (Lee et al., 2023; Nayak et al., 2004), compared with the considerably lower and more variable rates observed with scalp EEG (approximately 24–55% across studies) in patients with drug-resistant focal epilepsy (Goodin et al., 1990; Perucca et al., 2014). Because the duration of an iEEG investigation is primarily driven by the need to record a sufficient number of seizures to reliably delineate the seizure-onset zone, the sensitivity for detecting ictal electrophysiological activity is correspondingly high, though still influenced by peri-surgical factors (Bottan et al., 2023; Muthiah et al., 2025).

26.3 Localization of Ictal and Interictal EEG Events

The localization of the epileptogenic zone is a principal objective of EEG recording. The cortical region generating IEDs and the distribution of surface ictal changes are generally broader than the actual cortical area where clinical seizures originate (So et al., 1989). Since the last century, neurologists have emphasized the importance of identifying ictal EEG patterns for accurate localization, as bilateral spikes are frequently recorded in patients with unilateral TLE (So et al., 1989). Ebersole and Pacia reported that nearly all of their cases exhibited a regular 5–9 Hz EEG rhythm, typically lasting for more than five seconds and primarily localized to the subtemporal and temporal electrodes (Ebersole & Pacia, 1996). They also noted that lateralized polymorphic activity in the 2–5 Hz range may indicate seizure onset within the temporal neocortex.

The EEG rhythms associated with seizures arising from extratemporal regions can differ markedly from those originating in the temporal lobe. Rhythmic temporal theta activity is a prominent feature of seizures of temporal origin, with approximately 80% of these cases involving MTLE (Foldvary et al., 2001). Generalized EEG patterns are more common among patients with extratemporal epilepsy, particularly those with mesial frontal or occipital lobe epilepsy.

Some researchers consider that IEDs can provide more precise localization information, as seizure activity detected on scalp EEG often reflects propagation across a wider cortical region (Foldvary et al., 2001). The modulation of interictal activity by local neuronal processes within the seizure onset zone may reflect the effect of distinct cortical vigilance states (Fouad et al., 2022). The accurate localization of IEDs using scalp EEG is a valuable tool in the presurgical evaluation of focal epilepsy. Clusters of lateralized rhythmic sharp waves at 5–10 Hz were detected in 81% of patients whose seizures originated from the mesial temporal lobe (Williamson et al., 1993). IEDs may arise from adjacent cortical regions or propagate from deep-seated generators, allowing them to occur in association with lesions of any depth. Moreover, the waveform morphology of these discharges provides valuable insight into the structural characteristics of the underlying lesion (Cuello-Oderiz et al., 2017).

In addition to ictal EEG rhythm and interictal epileptiform events, certain non-epileptiform EEG patterns, such as interictal slow-wave activity, may also provide evidence of localisation of foci. Previously mentioned TIRDA, typically lasting between 4 and 20 s, has been observed in approximately 25% of patients with TLE and is closely associated with epileptiform discharges(Geyer et al., 1999). On scalp EEG, TIRDA may manifest as 4–7 Hz theta or 1–3 Hz delta activity, either persistent or intermittent, and may appear unilaterally or bilaterally in the temporal regions. Unilateral temporal slowing may be a lateralizing feature (Koutroumanidis et al., 2004).

With advances in EEG technology, researchers have sought to achieve more precise identification of epileptogenic zones through the use of high-density EEG (hdEEG) caps. Use of hdEEG system may provide additional localising information compared with conventional EEG, particularly in cases of frontal lobe epilepsy with seizure onset in midline and parasagittal regions (Feyissa et al., 2017). The hdEEG facilitates the application of electrical source imaging (ESI) analysis for localizing epileptic generators, especially in patients with drug-resistant epilepsy. In a relatively small number of studies ictal ESI have been shown to provided useful information for localization of the epileptogenic focus (Nemtsas et al., 2017).

Note that several factors can influence the accuracy of epileptogenic focus localization using scalp EEG. Because the recorded signals represent the summated activity of neuronal populations in proximity to the electrodes, precise localization of the epileptic focus at a three-dimensional generator level remains challenging. Ictal onset signals can exhibit a significant time delay before propagating to the scalp (Kissani et al., 2001). Due to the inherently low sensitivity of scalp recordings, the spatial and temporal accuracy of the observed activity may be limited. Consequently, iEEG may be needed to provide more precise localization of the epileptogenic zone.

26.4 Intracranial EEG in Presurgical Planning

Generally, iEEG is indicated when non-invasive presurgical evaluations fail to adequately localize the epileptogenic zone (EZ) due to ambiguous scalp EEG findings or non-lesional imaging in patients with drug-resistant focal epilepsy. The main aims for employing iEEG have been summarized by Stjepana Kovac et al. (2017) as follows: (1) to more precisely define the true epileptogenic zone and (2) to localize the EZ and delineate its spatial relationship to adjacent eloquent cortex. In this context, the EZ is defined as the minimal cortical region that must be resected to achieve postoperative seizure freedom (Rosenow & Luders, 2001). By providing high–spatial resolution recordings directly from cortical and subcortical structures, iEEG enables detailed characterization of ictal onset patterns and propagation pathways. This information is critical for determining the boundaries of the epileptogenic zone and for assessing the safety and feasibility of resection or ablation, thereby improving surgical outcomes.

The iEEG encompasses a number of recording techniques that involve the placement of electrode contacts inside the cranium either through surgically created cavities or natural channels such as the foramen ovale. The use of the latter, which has declined since the advent of advanced brain imaging, has been mostly limited to the investigation of mesial temporal lobe epilepsy. The more widely used invasive recording techniques require surgical implantation of electrodes, either in the form of linear or rectangular arrays of disk-shaped contacts ebedded in a plastic membrane (electrocortigography or ECoG) placed on the surface of the cortex through craniotomy or the stereotactic insertion of linear multi-contact ‘stick’ electrodes through burr holes (stereoencephalography or SEEG; also sometimes referred to as ‘depth EEG’*). Examples are shown in Fig. 26.4.

A fundamental characteristic of iEEG is the spatial sensitivity profile of its electrodes, which differs substantially between ECoG and SEEG and has important

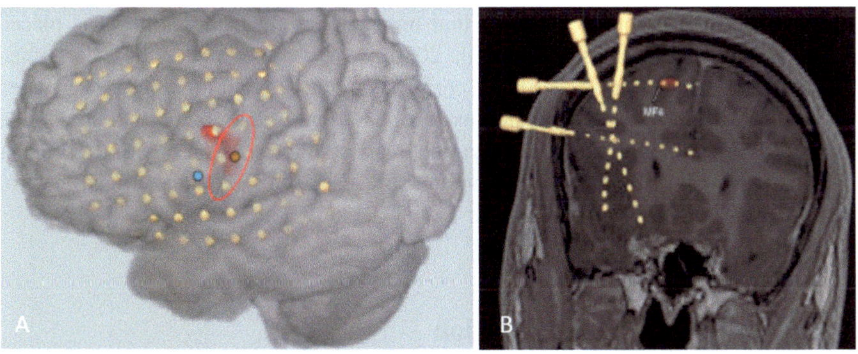

Fig. 26.4 Examples of icEEG implantations. (**a**) A brain model with a mixed subdural grid, where yellow dots represent individual contacts, and orange and blue dots indicate the position of depth electrodes. (**b**) A T1 MRI image showing SEEG implantation targeting the right orbitofrontal and mesial frontal regions and cingulum, marked by yellow dots. (Adapted with permission from Kovac et al., 2017)

implications for electrode placement strategies. ECoG may be conceptualized as a form of "scalp EEG beneath the skull," offering broad two-dimensional coverage of the cortical surface but limited access to the full extent of the neocortex. In contrast, SEEG provides highly focal sampling, with each contact exhibiting a relatively small region of sensitivity confined to the immediate surrounding tissue, including structures located deep within the brain (Lee et al., 2000).

Intracranial EEG is particularly valuable for certain aetiologies. Diehl and Lüders noted that when structural MRI demonstrates mesial temporal sclerosis (MTS) but does not clearly indicate unilateral involvement, iEEG is necessary to determine from which temporal lobe the seizures originate (Diehl & Luders, 2000). For such cases, they recommended depth electrodes, citing the substantial latency of 20–30 s often observed between the detection of ictal activity on depth electrodes and its appearance on subdural electrodes. However, this does not imply that depth electrodes are universally superior. For example, in patients with temporal lobe tumours, Diehl and Lüders emphasized that subdural grid electrodes may be required, particularly when the lesion extends posteriorly or involves eloquent cortical regions. In these cases, subdural electrodes facilitate functional mapping—such as language mapping—which is essential for determining safe resection boundaries.

Given the limits of neurosurgery and the limited number of electrodes that can be implanted, icEEG and SEEG in particular rely on accurate targeting. Given that the objective is the very identification of a putative target for surgical resection, the approach to iEEG implantations can be described as 'hypothesis-driven', based on the results of the non-invasive tests that are therefore a necessary prelude. In other words, icEEG is not an option in cases for which the non-invasive tests do not reveal a possible surgical target (Jobst et al., 2020). The iEEG electrodes are therefore implanted in regions suspected of hosting the seizure onset zone, and conversely situations in which the putative onset zone is deemed surgically unreachable or non-resectable are considered red flags for icEEG. Furthermore, an iEEG investigation which fails to reveal electrophysiological changes that precede or coincide with the onset of ictal clinical signs is likely to be considered an implantation targeting failure because it strongly suggests that none of the electrode contacts is placed within the seizure onset zone (Lee et al., 2000). As a consequence of these combined requirements and constraints, iEEG is performed relatively rarely; nonetheless, when utilized successfully it is considered by some to approximate the gold (some would say 'silver') standard for the localization (and characterization) of the region or regions involved in seizure generation and spread due to its exquisite local electrophysiological sensitivity, in particular SEEG (Blount et al., 2008).

26.5 AI in EEG Interpretation

As we mention in the chapter, EEG is one of the most crucial techniques in clinical diagnosis. However, EEG interpretation is complex, subjective, and based on experience. The clinicians can have different opinions on the same epoch of the EEG, and sometimes may result in misinterpretation. Thus it is important to find a more objective interpretation method. Tveit et al. developed an AI model-based platform, SCORE- AI, aiming to distinguish abnormalities from EEG recordings and provide evidence for clinical decision-making based on classifying the abnormal EEG events (Tveit et al., 2023). They used 30,493 anonymized EEG recordings from multiple centres, a convolutional neural network mode, and demonstrate SCORE AI can reach a high accuracy similar as human experts.

26.6 Conclusion

In summary, scalp and intracranial EEG both serve essential roles in the diagnosis and treatment of epilepsy and are especially critical for surgical planning. By offering complementary information on ictal and interictal activity, EEG enables precise localization of the epileptogenic zone and clear delineation of the functional and pathological networks involved in seizure generation and propagation. Their integration into the presurgical evaluation process is vital for selecting the appropriate and safe surgical strategy, guiding decisions regarding resection. As a result, EEG remains a cornerstone of modern epilepsy care, substantially improving the likelihood of successful outcomes for patients with drug-resistant focal epilepsy.

References

Alkawadri, C. I., Yan, Q., Kocuglu Kinal, A. G., Spencer, D. D., & Alkawadri, R. (2024). Comparison of EEG signal characteristics of subdural and depth electrodes. *Journal of Clinical Neurophysiology, 42*(5), 391–399. https://doi.org/10.1097/WNP.0000000000001139

Aschner, A., Kowal, C., Arski, O., Crispo, J. A. G., Farhat, N., & Donner, E. (2024). Prevalence of epileptiform electroencephalographic abnormalities in people without a history of seizures: A systematic review and meta-analysis. *Epilepsia, 65*(3), 583–599. https://doi.org/10.1111/epi.17864

Baumgartner, C., Baumgartner, J., Duarte, C., Lang, C., Lisy, T., & Koren, J. P. (2025). Role of specific interictal and ictal EEG onset patterns. *Epilepsy & Behavior, 164*, 110298. https://doi.org/10.1016/j.yebeh.2025.110298

Beniczky, S., Trinka, E., Wirrell, E., Abdulla, F., Al Baradie, R., Alonso Vanegas, M., Auvin, S., Singh, M. B., Blumenfeld, H., Bogacz Fressola, A., Caraballo, R., Carreno, M., Cendes, F., Charway, A., Cook, M., Craiu, D., Ezeala-Adikaibe, B., Frauscher, B., French, J., et al. (2025). Updated classification of epileptic seizures: Position paper of the International League Against Epilepsy. *Epilepsia, 66*(6), 1804–1823. https://doi.org/10.1111/epi.18338

Berg, A. T., Berkovic, S. F., Brodie, M. J., Buchhalter, J., Cross, J. H., van Emde Boas, W., Engel, J., French, J., Glauser, T. A., Mathern, G. W., Moshe, S. L., Nordli, D., Plouin, P., & Scheffer, I. E. (2010). Revised terminology and concepts for organization of seizures and epilepsies: Report of the ILAE Commission on Classification and Terminology, 2005–2009. *Epilepsia, 51*(4), 676–685. https://doi.org/10.1111/j.1528-1167.2010.02522.x

Blount, J. P., Cormier, J., Kim, H., Kankirawatana, P., Riley, K. O., & Knowlton, R. C. (2008). Advances in intracranial monitoring. *Neurosurgical Focus, 25*(3), E18. https://doi.org/10.3171/FOC/2008/25/9/E18

Bottan, J. S., Alshahrani, A., Gilmore, G., Steven, D. A., Burneo, J. G., Lau, J. C., McLachlan, R. S., Parrent, A. G., MacDougall, K. W., Diosy, D. C., Mirsattari, S. M., & Suller Marti, A. (2023). Lack of spontaneous typical seizures during intracranial monitoring with stereo-electroencephalography. *Epileptic Disorders, 25*(6), 833–844. https://doi.org/10.1002/epd2.20165

Britton, J. W., Frey, L. C., Hopp, J. L., Korb, P., Koubeissi, M. Z., Lievens, W. E., Pestana-Knight, E. M., & St. Louis, E. K. (2016). *Electroencephalography (EEG): An introductory text and atlas of normal and abnormal findings in adults, children, and infants*. In E. K. St. Louis & L. C. Frey (Eds.). https://www.ncbi.nlm.nih.gov/pubmed/27748095

Cendes, F., Theodore, W. H., Brinkmann, B. H., Sulc, V., & Cascino, G. D. (2016). Neuroimaging of epilepsy. *Handbook of Clinical Neurology, 136*, 985–1014. https://doi.org/10.1016/B978-0-444-53486-6.00051-X

Cuello-Oderiz, C., von Ellenrieder, N., Dubeau, F., & Gotman, J. (2017). Influence of the location and type of epileptogenic lesion on scalp interictal epileptiform discharges and high-frequency oscillations. *Epilepsia, 58*(12), 2153–2163. https://doi.org/10.1111/epi.13922

Diehl, B., & Luders, H. O. (2000). Temporal lobe epilepsy: When are invasive recordings needed? *Epilepsia, 41*(Suppl 3), S61–S74. https://doi.org/10.1111/j.1528-1157.2000.tb01536.x

Ebersole, J. S., & Pacia, S. V. (1996). Localization of temporal lobe foci by ictal EEG patterns. *Epilepsia, 37*(4), 386–399. https://doi.org/10.1111/j.1528-1157.1996.tb00577.x

Feyissa, A. M., Britton, J. W., Van Gompel, J., Lagerlund, T. L., So, E., Wong-Kisiel, L. C., Cascino, G. C., Brinkman, B. H., Nelson, C. L., Watson, R., & Worrell, G. A. (2017). High density scalp EEG in frontal lobe epilepsy. *Epilepsy Research, 129*, 157–161. https://doi.org/10.1016/j.eplepsyres.2016.12.016

Fisher, R. S., & Engel, J. J., Jr. (2010). Definition of the postictal state: When does it start and end? *Epilepsy & Behavior, 19*(2), 100–104. https://doi.org/10.1016/j.yebeh.2010.06.038

Fisher, R. S., van Emde Boas, W., Blume, W., Elger, C., Genton, P., Lee, P., & Engel, J., Jr. (2005). Epileptic seizures and epilepsy: Definitions proposed by the International League Against Epilepsy (ILAE) and the International Bureau for Epilepsy (IBE). *Epilepsia, 46*(4), 470–472. https://doi.org/10.1111/j.0013-9580.2005.66104.x

Fisher, R. S., Acevedo, C., Arzimanoglou, A., Bogacz, A., Cross, J. H., Elger, C. E., Engel, J., Jr., Forsgren, L., French, J. A., Glynn, M., Hesdorffer, D. C., Lee, B. I., Mathern, G. W., Moshe, S. L., Perucca, E., Scheffer, I. E., Tomson, T., Watanabe, M., & Wiebe, S. (2014a). ILAE official report: A practical clinical definition of epilepsy. *Epilepsia, 55*(4), 475–482. https://doi.org/10.1111/epi.12550

Fisher, R. S., Scharfman, H. E., & deCurtis, M. (2014b). How can we identify ictal and interictal abnormal activity? *Advances in Experimental Medicine and Biology, 813*, 3–23. https://doi.org/10.1007/978-94-017-8914-1_1

Foldvary, N., Klem, G., Hammel, J., Bingaman, W., Najm, I., & Luders, H. (2001). The localizing value of ictal EEG in focal epilepsy. *Neurology, 57*(11), 2022–2028. https://doi.org/10.1212/wnl.57.11.2022

Fouad, A., Azizollahi, H., Le Douget, J. E., Lejeune, F. X., Valderrama, M., Mayor, L., Navarro, V., & Le Van Quyen, M. (2022). Interictal epileptiform discharges show distinct spatiotemporal and morphological patterns across wake and sleep. *Brain Communications, 4*(5), fcac183. https://doi.org/10.1093/braincomms/fcac183

Geyer, J. D., Bilir, E., Faught, R. E., Kuzniecky, R., & Gilliam, F. (1999). Significance of interictal temporal lobe delta activity for localization of the primary epileptogenic region. *Neurology, 52*(1), 202–205. https://doi.org/10.1212/wnl.52.1.202

Gloor P. (1985). Neuronal generators and the problem of localization in electroencephalography: application of volume conductor theory to electroencephalography. J Clin Neurophysiol 2(4): 327–54. *Annals of Neurology, 34*(6), 781–787. https://doi.org/10.1097/00004691-198510000-00002. PMID: 4056020.

Goodin, D. S., Aminoff, M. J., & Laxer, K. D. (1990). Detection of epileptiform activity by different noninvasive EEG methods in complex partial epilepsy. *Annals of Neurology, 27*(3), 330–334. https://doi.org/10.1002/ana.410270317

Hakami, T., McIntosh, A., Todaro, M., Lui, E., Yerra, R., Tan, K. M., French, C., Li, S., Desmond, P., Matkovic, Z., & O'Brien, T. J. (2013). MRI-identified pathology in adults with new-onset seizures. *Neurology, 81*(10), 920–927. https://doi.org/10.1212/WNL.0b013e3182a35193

Jobst, B. C., Bartolomei, F., Diehl, B., Frauscher, B., Kahane, P., Minotti, L., Sharan, A., Tardy, N., Worrell, G., & Gotman, J. (2020). Intracranial EEG in the 21st century. *Epilepsy Current, 20*(4), 180–188. https://doi.org/10.1177/1535759720934852

Kane, N., Acharya, J., Beniczky, S., Caboclo, L., Finnigan, S., Kaplan, P. W., Shibasaki, H., Pressler, R., & van Putten, M. J. A. M. (2017). A revised glossary of terms most commonly used by clinical electroencephalographers and updated proposal for the report format of the EEG findings. Revision 2017. *Clinical Neurophysiology Practice, 2*, 170–185. https://doi.org/10.1016/j.cnp.2017.07.002

Kissani, N., Alarcon, G., Dad, M., Binnie, C. D., & Polkey, C. E. (2001). Sensitivity of recordings at sphenoidal electrode site for detecting seizure onset: Evidence from scalp, superficial and deep foramen ovale recordings. *Clinical Neurophysiology, 112*(2), 232–240. https://doi.org/10.1016/s1388-2457(00)00531-9

Klodowski, K., Zhang, M., Jen, J. P., Scoffings, D. J., Morris, R., Lupson, V.,... & Cope, T. E. (2025). Parallel transmit 7 T MRI for adult epilepsy pre-surgical evaluation. *Epilepsia, 66*(7), 2315–2327

Koutroumanidis, M., Martin-Miguel, C., Hennessy, M. J., Akanuma, N., Valentin, A., Alarcon, G., Jarosz, J. M., & Polkey, C. E. (2004). Interictal temporal delta activity in temporal lobe epilepsy: Correlations with pathology and outcome. *Epilepsia, 45*(11), 1351–1367. https://doi.org/10.1111/j.0013-9580.2004.61203.x

Kovac, S., Vakharia, V. N., Scott, C., & Diehl, B. (2017). Invasive epilepsy surgery evaluation. *Seizure, 44*, 125–136. https://doi.org/10.1016/j.seizure.2016.10.016

Kural, M. A., Duez, L., Sejer Hansen, V., Larsson, P. G., Rampp, S., Schulz, R., Tankisi, H., Wennberg, R., Bibby, B. M., Scherg, M., & Beniczky, S. (2020). Criteria for defining interictal epileptiform discharges in EEG: A clinical validation study. *Neurology, 94*(20), e2139–e2147. https://doi.org/10.1212/WNL.0000000000009439

Lagarde, S., Buzori, S., Trebuchon, A., Carron, R., Scavarda, D., Milh, M., McGonigal, A., & Bartolomei, F. (2019). The repertoire of seizure onset patterns in human focal epilepsies: Determinants and prognostic values. *Epilepsia, 60*(1), 85–95. https://doi.org/10.1111/epi.14604

Lee, S. A., Spencer, D. D., & Spencer, S. S. (2000). Intracranial EEG seizure-onset patterns in neocortical epilepsy. *Epilepsia, 41*(3), 297–307. https://doi.org/10.1111/j.1528-1157.2000.tb00159.x

Lee, S., Wu, S., Tao, J. X., Rose, S., Warnke, P. C., Issa, N. P., & van Drongelen, W. (2023). Manifestation of hippocampal interictal discharges on clinical scalp EEG recordings. *Journal of Clinical Neurophysiology, 40*(2), 144–150. https://doi.org/10.1097/WNP.0000000000000867

Misulis, K. E., et al. (2022). Abnormal EEG in epilepsy: Interictal epileptiform and ictal discharges. In *Atlas of EEG, seizure semiology, and management* (3rd ed.). Academic. https://doi.org/10.1093/med/9780197543023.003.0016

Muthiah, N., Reecher, H. M., Maroufi, S. F., Mansouri, A., Harford, E., & Abel, T. J. (2025). Seizure outcomes and complications associated with stereoelectroencephalography versus subdural electrodes for invasive monitoring in epilepsy surgery: A meta-analysis. *Frontiers in Neurology, 16*, 1619288. https://doi.org/10.3389/fneur.2025.1619288

Nayak, D., Valentin, A., Alarcon, G., Garcia Seoane, J. J., Brunnhuber, F., Juler, J., Polkey, C. E., & Binnie, C. D. (2004). Characteristics of scalp electrical fields associated with deep medial temporal epileptiform discharges. *Clinical Neurophysiology, 115*(6), 1423–1435. https://doi.org/10.1016/j.clinph.2004.01.009

Nemtsas, P., Birot, G., Pittau, F., Michel, C. M., Schaller, K., Vulliemoz, S., Kimiskidis, V. K., & Seeck, M. (2017). Source localization of ictal epileptic activity based on high-density scalp EEG data. *Epilepsia, 58*(6), 1027–1036. https://doi.org/10.1111/epi.13749

Pandian, J. D., Cascino, G. D., So, E. L., Manno, E., & Fulgham, J. R. (2004). Digital video-electroencephalographic monitoring in the neurological-neurosurgical intensive care unit: Clinical features and outcome. *Archives of Neurology, 61*(7), 1090–1094. https://doi.org/10.1001/archneur.61.7.1090

Pataraia, E., Lurger, S., Serles, W., Lindinger, G., Aull, S., Leutmezer, F., Bacher, J., Olbrich, A., Czech, T., Novak, K., Deecke, L., & Baumgartner, C. (1998). Ictal scalp EEG in unilateral mesial temporal lobe epilepsy. *Epilepsia, 39*(6), 608–614. https://doi.org/10.1111/j.1528-1157.1998.tb01429.x

Perucca, P., Dubeau, F., & Gotman, J. (2014). Intracranial electroencephalographic seizure-onset patterns: Effect of underlying pathology. *Brain, 137*(Pt 1), 183–196. https://doi.org/10.1093/brain/awt299

Pillai, J., & Sperling, M. R. (2006). Interictal EEG and the diagnosis of epilepsy. *Epilepsia, 47*(Suppl 1), 14–22. https://doi.org/10.1111/j.1528-1167.2006.00654.x

Rosenow, F., & Luders, H. (2001). Presurgical evaluation of epilepsy. *Brain, 124*(Pt 9), 1683–1700. https://doi.org/10.1093/brain/124.9.1683

Salinsky, M., Kanter, R., & Dasheiff, R. M. (1987). Effectiveness of multiple EEGs in supporting the diagnosis of epilepsy: An operational curve. *Epilepsia, 28*(4), 331–334. https://doi.org/10.1111/j.1528-1157.1987.tb03652.x

Schevon, C. A., Weiss, S. A., McKhann, G., Jr., Goodman, R. R., Yuste, R., Emerson, R. G., & Trevelyan, A. J. (2012). Evidence of an inhibitory restraint of seizure activity in humans. *Nature Communications, 3*, 1060. https://doi.org/10.1038/ncomms2056

da Silva, F.L. (2009). EEG: Origin and Measurement. In: Mulert, C., Lemieux, L. (eds) EEG - fMRI. Springer, Berlin, Heidelberg. *Annals of Neurology, 34*(6), 781–787. https://doi.org/10.1007/978-3-540-87919-0_2

Smith, S. J. (2005). EEG in the diagnosis, classification, and management of patients with epilepsy. *Journal of Neurology, Neurosurgery, and Psychiatry, 76*(Suppl 2), ii2–ii7. https://doi.org/10.1136/jnnp.2005.069245

So, N., Gloor, P., Quesney, L. F., Jones-Gotman, M., Olivier, A., & Andermann, F. (1989). Depth electrode investigations in patients with bitemporal epileptiform abnormalities. *Annals of Neurology, 25*(5), 423–431. https://doi.org/10.1002/ana.410250502

Tao, J. X., Chen, X. J., Baldwin, M., Yung, I., Rose, S., Frim, D., Hawes-Ebersole, S., & Ebersole, J. S. (2011). Interictal regional delta slowing is an EEG marker of epileptic network in temporal lobe epilepsy. *Epilepsia, 52*(3), 467–476. https://doi.org/10.1111/j.1528-1167.2010.02918.x

Tatum, W. O. T. (2012). Mesial temporal lobe epilepsy. *Journal of Clinical Neurophysiology, 29*(5), 356–365. https://doi.org/10.1097/WNP.0b013e31826b3ab7

Tveit, J., Aurlien, H., Plis, S., Calhoun, V. D., Tatum, W. O., Schomer, D. L., Arntsen, V., Cox, F., Fahoum, F., Gallentine, W. B., Gardella, E., Hahn, C. D., Husain, A. M., Kessler, S., Kural, M. A., Nascimento, F. A., Tankisi, H., Ulvin, L. B., Wennberg, R., & Beniczky, S. (2023). Automated interpretation of clinical electroencephalograms using artificial intelligence. *JAMA Neurology, 80*(8), 805–812. https://doi.org/10.1001/jamaneurol.2023.1645

Van Patten, R., Richards, J., Blum, A. S., & LaFrance, W. C., Jr. (2025). Evaluation and management of non-epileptic (functional) seizures: Lessons learned. *Epilepsy & Behavior, 167*, 110398. https://doi.org/10.1016/j.yebeh.2025.110398

Westmoreland, B. F., & Klass, D. W. (1997). Unusual variants of subclinical rhythmic electrographic discharge of adults (SREDA). *Electroencephalography and Clinical Neurophysiology, 102*(1), 1–4. https://doi.org/10.1016/s0013-4694(96)96035-6

White, J. C., Langston, J. W., & Pedley, T. A. (1977). Benign epileptiform transients of sleep. Clarification of the small sharp spike controversy. *Neurology, 27*(11), 1061–1068. https://doi.org/10.1212/wnl.27.11.1061

Williamson, P. D., French, J. A., Thadani, V. M., Kim, J. H., Novelly, R. A., Spencer, S. S., Spencer, D. D., & Mattson, R. H. (1993). Characteristics of medial temporal lobe epilepsy: II. Interictal and ictal scalp electroencephalography, neuropsychological testing, neuroimaging, surgical results, and pathology. *Annals of Neurology, 34*(6), 781–787. https://doi.org/10.1002/ana.410340605

Chapter 27
EEG Applications in Neonatal and Paediatric Clinical Neuroscience

Jayvian Mavi and Kimberley Whitehead

Abstract EEG in neonates and children allows for monitoring of the functional dynamics of the developing brain in real time. This is especially important in populations following atypical developmental trajectories, including preterm and acutely ill infants, those with seizures, and children with wider neurodevelopmental conditions. In this chapter, we focus on how EEG can shed light on paediatric sleep, and somatosensory, pain, and interoceptive processing. In particular, we emphasise its value in assessing these parameters *naturalistically*. The chapter is organised into (i) neonatal, and then (ii) paediatric applications. In the first part of the chapter, we illustrate how brain-body interactions during neonatal sleep can be assessed by integrating EEG with other physiological time series. We then describe how neonatal sensory cortical processing of meaningful, realistic inputs—e.g. feeding, self-generated movements—can be evaluated. In the second part, we focus on the use of EEG in children and young adults. We first examine how EEG can be used to assess disrupted sleep architecture. We then appraise its current capabilities in evaluating sensory states, namely chronic pain and interoceptive awareness. Throughout, we highlight the growing potential and feasibility of community-based settings, and more naturalistic paradigms, to improve the ecological validity of EEG assessments. Finally, we draw together the strands of both sections and discuss future directions for developmental EEG research, particularly its translational promise in complex paediatric populations.

Keywords Sleep · Sensory · Motor · Pain · Interoception · Epilepsy · Home

J. Mavi · K. Whitehead (✉)
Digital Health and Applied Technology Assessment (DHATA), King's College London, London, UK
e-mail: jay.mavi.19@ucl.ac.uk; kimberley.whitehead@kcl.ac.uk

T. Warbrick (ed.), *The EEG Handbook*,
https://doi.org/10.1007/978-3-032-20450-9_27

27.1 Introduction

Electrophysiological functions of the human brain develop upon a structural architecture. These structural connections develop rapidly across the third trimester of gestation and through childhood and adolescence. From 23 weeks gestation, sensory thalamic afferents make synaptic connections—first with the subplate, before growing into the cortical plate (Volpe, 2009). Cortico-cortical connections then increase rapidly from 28 weeks gestation (Flower, 1985; Burkhalter et al., 1993), peak in infancy (Huttenlocher & Dabholkar, 1997), then decrease through puberty and early adulthood (Petanjek et al., 2011). Differences in brain structure are associated with variations in sleep amounts (Wang et al., 2024), sensory exposure (Ranger et al., 2013), and intellectual ability (Dierssen & Ramakers, 2006). Crucially, this variance in structural brain development occurs via electrical activity-dependent mechanisms, according to mammalian models (Luhmann et al., 2022) and concordant human data (Benders et al., 2015). This means that EEG can sample the functional dynamics that underlie how sleep, sensory inputs, and neurodevelopmental risk factors both shape and are shaped by emerging neural networks (Whitehead, 2024).

27.2 Neonatal EEG Applications

EEG allows to monitor the developing brain from as early as the limits of viable birth (approximately 23 weeks gestational age (3 months premature)). In populations so young, or otherwise critically ill, EEG is typically only acquired for clinical purposes, especially as these subjects' poor skin integrity means that electrode-related lesions can occur (Pasupuleti et al., 2016). However, in stable preterm and older infants, EEG is well-tolerated enough for research applications (Whitehead et al., 2017). Neonatal EEG recordings have unique technical considerations. For example, artefactual rhythmic patterns can occur according to the side that the infant is lying on, which can be wrongly interpreted as cerebral (Weeke et al., 2017). Nevertheless, a multi-disciplinary approach to overcoming these challenges—ideally incorporating neurophysiology, neonatal nursing and medicine, and signal processing expertise—allows to extract these data's immense value (Lloyd et al., 2015).

27.2.1 Sleep-Wake Monitoring Using Multi-physiological Data in Neonates

In infants, as in the wider population, sleep-wake cycling is a whole-body process, affecting every system. This is even more relevant in the youngest subjects, in whom its assessment relies heavily on retinal-corneal (indexing eye movements), cardio-respiratory, and electromyographic data (indexing muscle tone), as mature EEG sleep biomarkers do not emerge until 2 months of age (Grigg-Damberger, 2016). Such multi-physiological recordings can shed light on sleep architecture. For example, quantifying retinal-corneal potentials shows that EEG power attenuates within sections of rapid eye movement (REM) sleep with more frequent saccades (Whitehead et al., 2019a), indicating organisation *within* sleep states, as well as between them (i.e. REM vs. non-REM sleep). However, it is cardiorespiratory parameters that have been most often studied synchronised to EEG, and these are the focus of this section.

(Throughout this chapter, which spans the neonatal period to young adulthood, we use the terms REM and non-REM sleep for consistency, but note that these are often replaced with active and quiet sleep in neonates, to flag their difference to the mature states.)

EEG-Cardiorespiratory Monitoring

In neonates, across sleep states, there is coherence between EEG power in the delta band and short-term heart rate variability driven by respiratory rate (respiratory sinus arrhythmia) which estimates parasympathetic vagal input to the heart (Mulkey et al., 2021). This indicates co-dependency of cortical and brainstem-vagal activity. Such co-dependency is lower in neonates with seizures, specifically in the direction of EEG → heart rate variability (Frassineti et al., 2022). This implies reduced ability of the disordered cortex to exert top-down control onto the autonomic nervous system. Taking account of heart rate then, alongside EEG, offers a completely new window onto the multidimensional developing nervous system, especially when at its most vulnerable.

Use of respiratory recordings, in addition, allows to detect apnoeas and hypopnoeas—cessation and reduction of breathing respectively—which occur regularly in neonates, especially during REM sleep (Horne, 2014). Apnoeic events have long been reported to depress EEG activity, but often within anecdotal case series of extreme cases, which were briefly visible to the naked eye (Usman et al., 2023). Recent work employing signal processing allows to model subtler interactions, demonstrating how apnoeas usually depress EEG first, and then heart rate and peripheral oxygen saturation levels (Zandvoort et al., 2024) (Fig. 27.1). Apnoea can also occur during electrographic seizures, with a more complex relationship with heart rate: during apnoeic seizures in older infants, heart rate can either increase, decrease, or evolve to do both (e.g. see Fig. 2 in (Maruyama et al., 2022)). Using simple recording modalities then—EEG, cardiac, respiratory, pulse oximetry—which can be acquired at the cot side, opens up rich, nested physiological activity dynamics to investigation.

In sum, studying neonatal sleep-wake EEG alone—without linked physiological data—means that ongoing variance in its characteristics may be treated as 'noise' which can in fact be explained by brain-body interactions. The excellent time resolution of EEG allows to unmask the *direction* of this physiological signal's dependence, including how cortical → autonomic control breaks down during seizures.

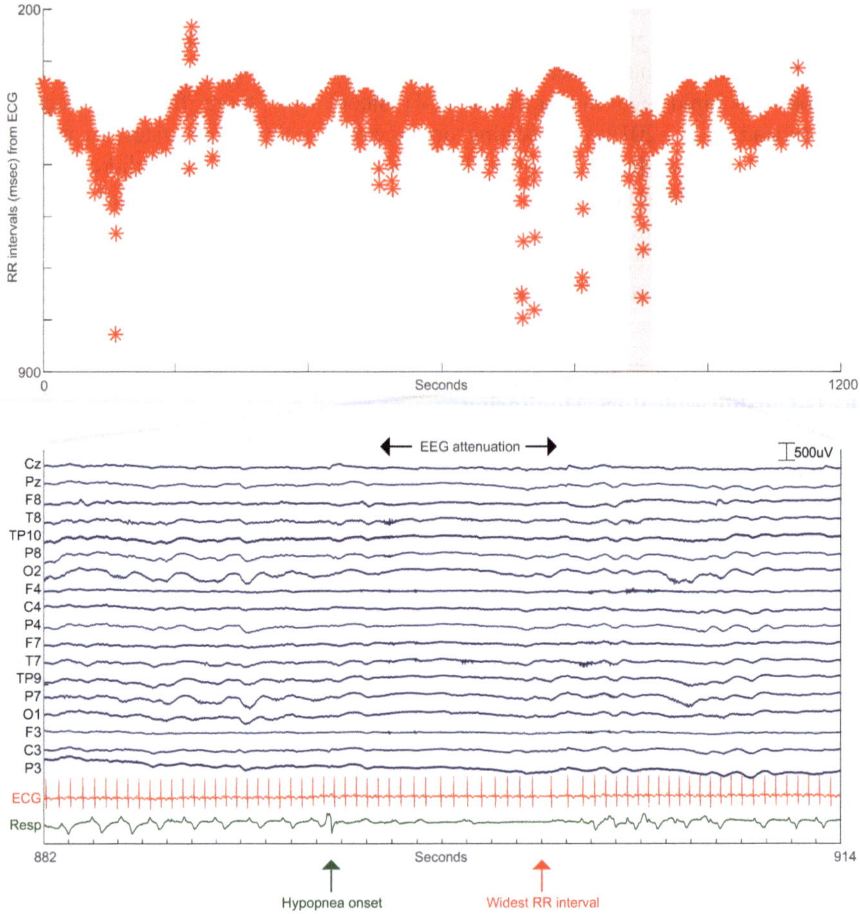

Fig. 27.1 Illustrative data showing how an episode of subtle bradycardia is preceded by a hypopnoea and EEG attenuation. Top panel: Heartbeat (RR) intervals (asterisks) extracted from a 19 min recording segment from an infant of 35 weeks corrected gestational age (gestational age + postnatal age) using the EEG-Beats toolbox (Thanapaisal et al., 2020). Note the RR intervals variability, including intermittent bradycardias with one example shaded pink. Bottom panel: EEG, electrocardiography (ECG) and respiratory movement (Resp) recordings associated with the shaded example bradycardia. The infant is awake, 3 min before a transition into REM sleep. No filters applied except 50 Hz notch filter; DC offset removed. Midline central referential EEG montage

27.2.2 Somatosensory States Monitoring in Neonates

The sleeping neonatal brain can process somatosensory inputs (Hrbek et al., 1973), without awakening (Georgoulas et al., 2021). This has often been researched using experimental stimuli, like taps from a device. These controlled protocols have elucidated many important principles, such as how somatosensory cortical excitability spans hierarchical networks (Whitehead et al., 2019b). However, such protocols lack ecological validity, and are not suitable for the most sensitive parts of the body, like the face. As an example of what is possible using a *naturalistic* paradigm, stimulation gently delivered by the experimenter's forefinger has allowed to map the face within a full cortical body map in preterm infants: with leg, arm, and face representation moving downwards from the central vertex (Donadio et al., 2018).

Given that functional somatosensory representation of the facial area is in place from preterm age, we can also study how it is modulated by sucking—a major primitive reflex observed during REM sleep and wakefulness. EEG recordings show that higher frequency, lower power rhythms are enhanced by sucking in neonates, although some co-linearity between the stimulus and sleep-wake shifts is likely (Lehtonen et al., 2016; Barlow et al., 2014). These effects on the EEG occur especially during feeding, above non-nutritive sucking (pacifier use) (Lehtonen et al., 2016). This could indicate that the 'extra' sensory stimuli when sucking is associated with milk intake (gustatory, olfactory, interoceptive from milk entering the stomach) heightens the engagement of cortical areas. Thus, EEG can capture the fullness of multi-modal sensory experience.

Another naturalistic way to study somatosensory processing is to use the infant's own self-generated movements, for example, by examining coherence between limb electromyographic signals and EEG. Such studies have shown that not only does the proprioceptive and tactile feedback from muscle activation evoke changes in EEG (Milh et al., 2007; Losito et al., 2017; Whitehead et al., 2018; Whitehead et al., 2019c), but causality flows in the other direction too (cortico-muscular coherence) from the first few months after birth (Kanazawa et al., 2014; Ritterband-Rosenbaum et al., 2017). Therefore, EEG can track the early emergence of the mechanisms which will link the somatosensory body map to later voluntary motor control (e.g. reaching for objects) (Kanazawa et al., 2014).

To summarise, as in Sect. 27.2.1, and Sect. 27.2.2, we reiterate how EEG activity does not exist in a vacuum. We show how ongoing naturalistic somatosensory inputs modulate cortical rhythms. Indeed, this may underlie the ability of positive somatosensory inputs—like naturalistic touch and feeding—to counteract the negative impact of nociceptive (painful) stimuli in hospitalised neonates (Maitre et al., 2017). In Sect. 27.3, devoted to children and young adults especially with neurodevelopmental conditions, we expand upon EEG's contribution to tracking complex sensory inputs like pain, after first describing its value in monitoring sleep.

27.3 Paediatric EEG Applications

Unlike preterm and ill neonates, who must be cared for in hospital, the wider paediatric population typically resides at home and participates in everyday community life. For these populations, traditional EEG recordings in a hospital or laboratory pose substantial challenges: the unfamiliar environment, travel-related stress, and demands of prolonged monitoring introduce diverse external factors that may affect the interpretability and reliability of observed activity (Riley et al., 1981; Milne-Ives et al., 2023).

Recent developments in EEG technology have led to advances in both mobile (ambulatory) and home-based systems (Biondi et al., 2021; Rehman et al., 2024). Mobile EEG permits monitoring across diverse and dynamic settings, such as school, but lacks video recording. In contrast, home EEG generally includes video recording, facilitating precise correlation between environmental stimuli, behavioural changes, and EEG activity, but for this reason is typically restricted to a single location. While distinct in their advantages and limitations, both methods offer unique extended windows into brain activity in real-world contexts, e.g. incorporating eating, caregiver contact, and free movement, in addition to increased practicality and inclusiveness (Debener et al., 2015; Lau-Zhu et al., 2019). This enables the capture of neural activity during unstructured behavioural states and ordinary routines, which are often disrupted or unrepresentative in hospital or laboratory settings. This also presents a valuable opportunity for use in younger populations with complex medical needs or neurodevelopmental conditions, such as intellectual disability, whose assessments may be particularly liable to disruption in an unfamiliar setting (Lau-Zhu et al., 2019). Therefore, mobile and home EEG could facilitate accurate monitoring and interpretation of the internal states that are critical to child wellbeing and development, but otherwise remain challenging to capture, particularly for children who are minimally verbal, cognitively impaired, or behaviourally atypical (Cassidy et al., 2023).

While mobile and home EEG are still emerging as research tools in medically complex children, they already see established clinical use in routine paediatric epilepsy diagnostics (Michaeli et al., 2024; Brunnhuber et al., 2014; Carlson et al., 2018), offering an effective alternative to inpatient monitoring for classifying seizures and capturing sleep-related epileptiform activity. These multi-day video-EEG recordings, set within the patients' homes, are annotated with seizure events reported by the patient or carer. Alongside these events, annotations are also made pertaining to the patient's behaviours, activity, and other potentially relevant environmental stimuli, allowing to cross-reference this information to ongoing physiological variability (Fig. 27.2). Thus, this clinical foundation in epilepsy diagnostics highlights an exciting potential for transferability of such home-based assessments to other contexts, like sleep and sensory state monitoring. In Sect. 27.3.1 we first review recent EEG studies investigating sleep, and then in Sect. 27.3.2 highlight the value of assessing sleep in home settings. Following this, in Sect. 27.3.3 we explore EEG studies of sensory states, particularly chronic pain and interoception. While these topics are lesser studied using home EEG when compared to epilepsy or sleep, in Sect. 27.3.4 we discuss relevant emerging findings and consider how naturalistic EEG paradigms may unlock novel avenues for assessing these under-recognised experiences in real-world settings.

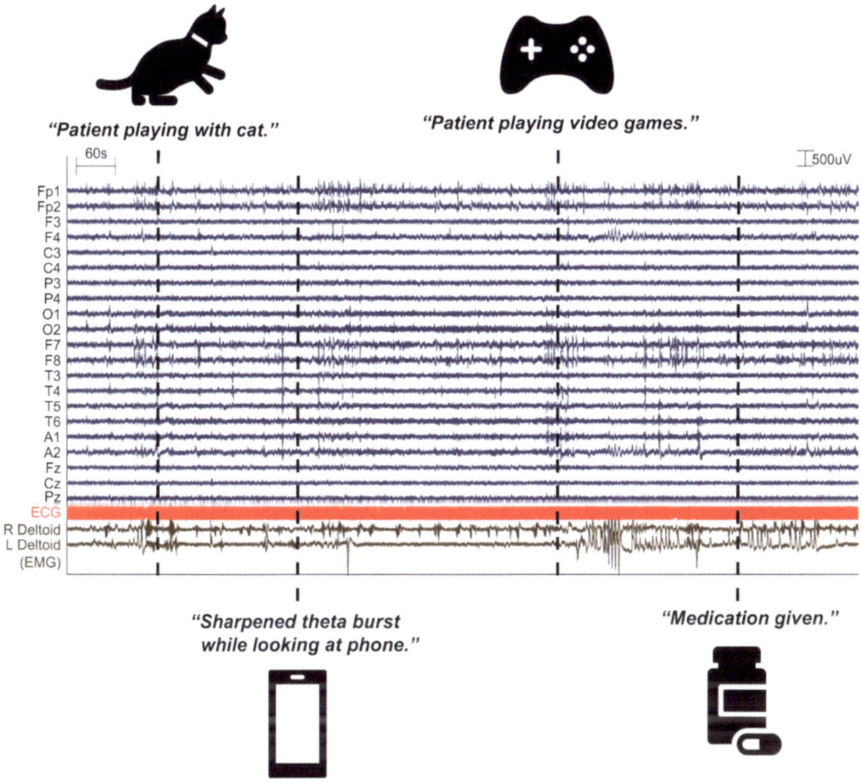

Fig. 27.2 Examples of annotated events occurring during home EEG monitoring. Annotations collated from a deidentified research database of multiple home-video EEG sessions carried out in paediatric populations at King's College Hospital. For illustrative purposes, annotations are superimposed onto a 20 min segment of EEG, electrocardiography (ECG), and electromyography (EMG) activity from a patient of 10 years of age undergoing 24-h home-video EEG monitoring. The child is awake. Midline central referential EEG montage

27.3.1 Sleep Monitoring in Children and Adolescents

From the neonatal period onwards, sleep is a foundational process in brain development. In juvenile mammals, experimental studies demonstrate that sleep supports synaptic plasticity and refinement of cortical circuitry (Li et al., 2017; Yang & Gan, 2012). Consistent with a similar role in human children, sleep quality is associated with learning (Dutil et al., 2018), as well as the regulation of affect and physiological processes. For example, lower sleep quality in otherwise healthy children has been linked to reduced parasympathetic vagal activity and a resultant baseline sympathetic dominance, a physiological profile associated with increased cardiovascular risk (Michels et al., 2013). In children with neurodevelopmental conditions—such as autism, attention deficit hyperactivity disorder (ADHD), developmental epilepsies, and cerebral palsy (CP)—sleep disturbances are common and often deeply interconnected with daytime functioning (Gorgoni et al., 2020; Angriman et al., 2015; Halstead et al., 2021; Hodge et al., 2014).

From 2 months of age, non-REM sleep can be subdivided into stages 1, 2, and 3 (Grigg-Damberger, 2016). In both healthy (Hoedlmoser et al., 2014; Chatburn et al., 2013) and neurodevelopmentally atypical children (Tessier et al., 2015; Bölsterli et al., 2017), sleep stage-specific features captured by EEG, such as sleep spindles and slow-waves, can be used to index sleep quality. In addition to this, their frequency and spatiotemporal characteristics are associated with various aspects of cognitive maturation, memory consolidation, and intellectual ability, making them valuable biomarkers in typical and atypical development (Page et al., 2021). For example, children with Rolandic epilepsy, a condition accounting for up to a quarter of childhood epilepsies (Ross et al., 2020), show reduced sleep spindle density in centrotemporal brain regions during non-REM stage 2 compared to healthy controls, even in the absence of active seizures, and this is associated with impaired sleep-dependent memory consolidation (Kwon et al., 2025).

Children with more severe epilepsies, such as epileptic encephalopathies characterised by Electrical Status Epilepticus in Sleep (ESES), frequently experience acquired regression in language, memory, and behaviour, which even after epileptic seizures resolve, can persist for life (Arican et al., 2021). ESES is generally defined by near-continuous spike-wave discharges during non-REM sleep (Tassinari et al., 2000). These discharges, therefore, severely disrupt normal sleep architecture—particularly slow-wave sleep (non-REM stage 3)—interfering with the neural processes associated with learning and development. One such process is the homeostatic regulation of network synchrony. During wakefulness, neuronal networks become increasingly active and interconnected as we interact with the environment, resulting in greater synaptic strength and higher cortical synchrony. Slow-wave sleep is thought to reflect the recalibration of this activity via downscaling of potentiated synaptic connections to an energetically sustainable state (Tononi & Cirelli, 2006). This recalibration can be indexed with EEG by the slope of slow-waves, calculated by dividing the amplitude change of a wave by the time between its negative peak and subsequent zero crossing. The slope typically decreases over the course of the night as synaptic strength is progressively downregulated (Kurth et al., 2010). In ESES, these slope decreases are impaired, or even completely absent, but recover after remission (Bölsterli et al., 2017). Interestingly, children with the highest degree of typical slope decrease during active ESES were found to have the best cognitive outcomes post-remission of ESES. It has therefore been hypothesised that slope changes in non-REM sleep could be an early prognostic indicator of developmental outcomes.

Collectively, these findings exemplify how EEG can index both subtle and profound changes in cortical activity that persist and evolve throughout sleep, highlighting its value not just as a diagnostic tool, but in tracking and potentially even predicting neurodevelopmental health. However, to fully investigate sleep dynamics and address such hypotheses, it is paramount that the sleep captured is representative and ecologically valid.

27.3.2 Naturalistic Sleep Monitoring in Children and Adolescents

Laboratory-based polysomnography (PSG), which assesses sleep via EEG and other synchronised monitoring, including cardiorespiratory and electromyographic signals, is considered the gold standard for sleep assessment (Rundo & Downey, 2019), but is often poorly tolerated by children with sensory sensitivities, anxiety, or behavioural challenges (Lanzlinger et al., 2023; Coverstone et al., 2014). The laboratory environment, restrictive instrumentation, and rigid procedures frequently provoke arousal and disrupt natural sleep patterns—a phenomenon known as the 'first-night' effect (Agnew Jr. et al., 1966), with meta-analysis demonstrating that this effect is particularly pronounced in children and adolescents compared to young adults (Ding et al., 2022). Consequently, this can render conventional assessments unrepresentative and incomplete.

More naturalistic mobile and home EEG paradigms provide a compelling alternative, allowing for overnight recordings in a familiar environment. Even low-channel density, highly portable systems can reliably detect key features of sleep architecture, including sleep spindles and K-complexes of stage 2, and slow-waves of stage 3 non-REM sleep (Kwon et al., 2021; Mikkelsen et al., 2019). As indicated in Sect. 27.3.1, these features play a crucial role in understanding the role of sleep in both developmentally typical and atypical contexts.

In addition, full PSG is possible in the home, allowing sophisticated integration of EEG with autonomic and behavioural monitoring, and insight into how sleep interacts with broader physiological states in a naturalistic environment. For example, a paediatric home PSG study capturing multiple physiological measures, including EEG, cardiorespiratory metrics, pulse oximetry, and leg electromyographic data, as well as audio recording, demonstrated the feasibility of home sleep assessment. Of 55 PSG sessions, 53 provided sufficient data to support a clinical diagnosis, including cases of obstructive sleep apnoea, central sleep apnoea, and periodic limb movement disorders (Russo et al., 2021). This supplements previous feasibility studies demonstrating clinical utility of home-based assessments for narcolepsy (Blackwell et al., 2017) and epileptic encephalopathy diagnosis (Nagyova et al., 2019; Brunnhuber et al., 2020). In contrast to previous paediatric PSG feasibility studies (Goodwin et al., 2001; Marcus et al., 2014), a high proportion (39%) of children in Russo et al. (2021) also had a prior neurodevelopmental diagnosis, with no significant difference in PSG success rates observed between these children and their typically developing peers. This generalisability is particularly relevant because in children with neurodevelopmental conditions, sleep disturbances may reflect not only primary sleep disorders, but also the effect of sensory states, such as discomfort, pain, or interoceptive imbalance (Onen et al., 2005; Reid et al., 2023; Bynum & Brindle, 2025), which are more likely to be representatively captured at home. The importance of sensory states in this population is addressed in the following section.

27.3.3 Sensory States Monitoring in Children and Adolescents

Children with neurodevelopmental disorders or complex medical conditions often experience states of discomfort that elude not only precise measurement, but even basic classification (Cassidy et al., 2023). Chronic pain and interoceptive difficulties are among the most common and distressing symptoms in this population, yet are frequently under-recognised, undertreated, or poorly characterised. This is largely due to an inability to self-report experiences and the lack of truly sensitive and specific indicators (Barney et al., 2020; Fehlings, 2017; Gomez-Suarez, 2016; Rice et al., 2017). These sensory states often co-occur, fluctuate over time, and manifest in ways that are shaped by the child's underlying neurological profile. Following the same format as we did for sleep in Sects. 27.3.1 and 27.3.2, we first describe sensory states monitoring overall (Sect. 27.3.3), and then focus on naturalistic approaches (Sect. 27.3.4).

Chronic Pain States
EEG has been used to study paediatric pain (Busse et al., 2025), but pain in children with complex needs is often chronic as described above. In such contexts, EEG signatures of pain are harder to study, as it may be difficult to identify the onset and offset, and to what degree any EEG features relate to the underlying aetiology of the pain (e.g., CP) rather than the pain itself. However, carefully controlled juvenile mammalian models of chronic pain have identified a causal relationship between pain and unique electrographic characteristics. These include decreased coherence between thalamus and primary somatosensory cortical activity in the 2–30 Hz range (LeBlanc et al., 2014), and enhanced theta-gamma phase-amplitude coupling (Wang et al., 2016), measured via intracranial electrophysiology.

The pre-clinical literature described above, then, supports the interpretation of EEG signatures of chronic pain states that have been reported in the human paediatric literature. For example, compared to healthy controls, adolescents with chronic musculoskeletal pain showed higher delta and beta power at rest (Ocay et al., 2022). The same study used hot and cold thermal stimuli to investigate EEG activity following acutely painful events. The temperature of such stimuli (i.e., extreme heat or cold) can be calibrated to induce acute thermal pain through the activation of a subclass of peripheral C-fibres known as nociceptive C-fibres. These stimuli showed again unique electrographic profiles between groups, including differences in EEG spectral power, peak frequency, and permutation entropy (a measure of the complexity and unpredictability of ongoing activity). Interestingly, the EEG-derived differences in thermal pain stimulus processing were not reflected in self-reported pain intensity scores between patients and controls. This indicates that EEG may be sensitive to neural processes underlying adolescent chronic musculoskeletal pain-related processing that are inaccessible through verbal report alone (Ocay et al., 2022). Furthermore, machine learning paradigms have been leveraged to successfully distinguish between rest and the onset of cold thermal pain in both adolescent chronic pain patients and controls, with theta-band permutation entropy features being the largest contributor to machine learning model accuracy (Teel et al., 2022). These results, in combination, highlight the potential capability of EEG to not only detect chronic pain states, but also the onset of additional acute pain experience in the absence of self-report.

Interoceptive States

Pain overlaps with interoception, defined as the brain's representation of internal bodily states, particularly related to the gastrointestinal and autonomic systems (Erkin et al., 2010; Mannion et al., 2013). This overlap is highlighted by the shared distributed networks that serve them involving (but not limited to) the insular cortex, anterior cingulate, and somatosensory regions (Wager et al., 2013; Barrett & Simmons, 2015), with functional connectivity between these regions in adults demonstrated with EEG (García-Cordero et al., 2017). Many children with neurodevelopmental conditions experience pain alongside difficulties with interoception. Chronic constipation, gastro-oesophageal reflux, and irregular autonomic arousal are common but difficult to assess, especially in children who have difficulties externally expressing their internal sensations (Barney et al., 2020).

To access/monitor interoceptive processing, the heartbeat-evoked potential (HEP), an EEG-derived marker, has been validated in adolescents (Mai et al., 2018). This, in addition to evidence from other functional imaging modalities (Klabunde et al., 2019), suggests that EEG could quantify interoception and its difficulties in children. Indeed, there is growing evidence that children with autism, ADHD, and other neurodevelopmental conditions exhibit altered resting-state connectivity and atypical cortical responses to visceral cues (Hechler, 2021; Yang et al., 2024). While direct EEG applications in paediatric populations are notably sparse, preliminary evidence in adolescents has shown that, during an emotional face recognition task, there is a positive relationship between the amplitude of the HEP and ADHD symptoms (Rapp et al., 2023). Potential explanations include increased arousal, atypical neural processing of internal bodily processes, and misaligned attentional weighting of task-irrelevant stimuli (i.e., one's own heartbeat). Though challenging to interpret, this early evidence supports the feasibility of EEG-based interoceptive measures in younger age groups and highlights their potential application to more clinically complex and neurodevelopmentally atypical populations.

27.3.4 Naturalistic Sensory States Monitoring in Children and Adolescents

Naturalistic sensory states monitoring can be split into the stimuli themselves, and the environment in which monitoring takes place. Addressing the stimuli first, sensory states in children are generally studied using a mix of artificial (e.g. thermal) and naturalistic endogenous stimuli (e.g. HEP). Overall, though, there is a trend towards greater utilisation of more naturalistic experimental paradigms in research. While still conducted in controlled settings, as with the neonatal research cited above (Maitre et al., 2017), there has been growing importance placed on investigating the effects of meaningful sensory input (Maallo et al., 2022), such as affective touch and parent-child interaction that are so prevalent in children's everyday lives, as opposed to more 'manufactured' stimuli. For instance, neurotypical, CP, and autistic children all display elevated gamma power following affective touch (slow, 'caressing' brush strokes) when compared to non-affective touch (faster, more frequent brush strokes) (Sabater-Gárriz et al., 2025). Slow brush strokes occur at speeds to optimally engage C-tactile afferents—a subclass of C-fibres distinct to those described in the context of pain—which are associated with the social and emotional components of tactile perception (McGlone et al., 2014). Interestingly, affective touch was also correlated with better proprioceptive accuracy in CP patients, compared to non-affective touch. While the underlying mechanisms are incompletely understood, the emotionally meaningful qualities of affective touch may enhance attentional and sensory integration processes (Sacchetti et al., 2021), potentially supporting proprioceptive processes in children with impaired body awareness, such as those with CP. These results highlight the potential utility of naturalistic paradigms to investigate more complex, sensory-affective interactions children may have with carers.

Unlike sleep, even naturalistic sensory paradigms in children are typically recorded exclusively in laboratory rather than community settings. However, the infrastructure for this to extend into the home is building, which has been echoed throughout the preceding sections. Feasibility of mobile and home EEG has been demonstrated across not only healthy paediatric populations (Troller-Renfree et al., 2021), but also in children who are neurodevelopmentally atypical (Milne-Ives et al., 2023; Russo et al., 2021), arguably the populations who benefit most from such technologies' capabilities. Review of existing examples of mobile and home EEG further highlights the potential of naturalistic EEG for sensory states monitoring. The events depicted in Fig. 27.2 are not only environmental, but also complex, multimodal sensory stimuli that may influence and modulate any states of pain and discomfort. By leveraging early evidence of unique neural signatures of sensory states, for example, via utilising automatic detection algorithms which have shown promise in adult EEG (Rockholt et al., 2023), clinicians may be able to continuously monitor and better understand fluctuations of chronic pain or autonomic sensory distress in everyday settings. Furthermore, adoption of home EEG in epilepsy diagnostics demonstrates that such services are not only feasible, but ecologically, economically, and logistically advantageous (Brunnhuber et al., 2023). These factors lay the groundwork for the integration of naturalistic EEG in other domains of paediatric care.

27.4 Future Directions and Conclusion

Looking forward, the ongoing coalescence of numerous factors relevant to naturalistic EEG may be early indicators of an exciting paradigm shift in paediatric EEG. One key driver is the rapid development of portable EEG devices that are less obtrusive and restrictive, and better-tolerated. For example, the aforementioned study distinguishing chronic pain patients from healthy children at rest and following painful stimulation utilised a dry-electrode wearable EEG headset (Teel et al., 2022), which can be recording-ready in under five minutes. Wireless systems have also been used in everyday environments (Wang et al., 2023) and in contexts relevant to medically complex populations, such as assessing preoperative anxiety in children (Xu et al., 2023).

In parallel, the longitudinal integration of neonatal and paediatric EEG may yield further insights. EEG can be used not only for direct observation at a single time point, but also to identify developmental risk and vulnerability across multiple stages of childhood. For instance, reduced sleep spindle activity in infancy has been shown to predict impaired motor function and CP diagnosis at pre-school age (Berja et al., 2024). Similarly, across adolescence, increased cortical arousability during sleep in a paediatric cohort (median age = 9), was found to be a significant predictor of cardiac autonomic imbalance at follow-up seven years later (Rahawi et al., 2025). Together, these findings highlight the utility of EEG-derived biomarkers in tracking developmental trajectories at the intersections of sleep, sensorimotor function, and autonomic activity, and the potential to guide supportive intervention strategies to improve long-term outcomes.

Here we have explored how EEG can be used to examine sleep and sensory processing across early development, beginning in the neonatal period and extending through to childhood and adolescence. In infants, EEG combined with physiological signals such as heart rate, respiration, and muscle activity enable detailed investigation of emerging brain–body interactions. Naturalistic paradigms, including feeding and self-generated movement, offer insight into how meaningful sensory experiences are represented in the cortex from early life. In older children, particularly those with neurodevelopmental conditions or complex medical needs, EEG continues to provide valuable information. When used in everyday settings, EEG allows for the assessment of sleep architecture in ways that are both feasible and ecologically valid. Preliminary findings also suggest that EEG possesses the capability to explore less accessible domains such as chronic pain and interoceptive function, highlighting promising directions for future research.

While often considered as distinctly separate populations in neuroscience research and clinical practice, both neonatal and paediatric EEG research are following common trends of shifts towards ecologically valid approaches. Together, these two strands illustrate a developmental continuum in which EEG serves not only as a tool for observing brain activity crudely/on the whole, but as a method for probing, investigating, and ultimately understanding how neural functions shape, and are shaped by, physiological states, sleep, and sensory experiences over the course of early life.

Acknowledgements KW warmly thanks the families who took part in her neonatal EEG research, from which Fig. 27.1 derives. We gratefully acknowledge Dr Joel Winston, who is a collaborator on the home EEG research underlying the data in Fig. 27.2.

References

Agnew, H. W., Jr., Webb, W. B., & Williams, R. L. (1966). The first night effect: An EEG study of sleep. *Psychophysiology, 2*, 263–266. https://doi.org/10.1111/j.1469-8986.1966.tb02650.x

Angriman, M., Caravale, B., Novelli, L., et al. (2015). Sleep in children with neurodevelopmental disabilities. *Neuropediatrics, 46*, 199–210.

Arican, P., Gencpinar, P., Dundar, N. O., & Tekgul, H. (2021). Electrical status epilepticus during slow-wave sleep (ESES): Current perspectives. *Journal of Pediatric Neurosciences, 16*, 91–96.

Barlow, S. M., Jegatheesan, P., Weiss, S., et al. (2014). Amplitude-integrated EEG and range-EEG modulation associated with pneumatic orocutaneous stimulation in preterm infants. *Journal of Perinatology, 34*, 213–219. https://doi.org/10.1038/jp.2013.150

Barney, C. C., Andersen, R. D., Defrin, R., et al. (2020). Challenges in pain assessment and management among individuals with intellectual and developmental disabilities. *Pain Reports, 5*, e821.

Barrett, L. F., & Simmons, W. K. (2015). Interoceptive predictions in the brain. *Nature Reviews Neuroscience, 16*, 419–429.

Benders, M. J., Palmu, K., Menache, C., et al. (2015). Early brain activity relates to subsequent brain growth in premature infants. *Cerebral Cortex, 25*, 3014–3024. https://doi.org/10.1093/cercor/bhu097

Berja, E. D., Kwon, H., Walsh, K. G., et al. (2024). Infant sleep spindle measures from EEG improve prediction of cerebral palsy. *Clinical Neurophysiology, 167*, 51–60. https://doi.org/10.1016/j.clinph.2024.08.017

Biondi, A., Laiou, P., Bruno, E., et al. (2021). Remote and long-term self-monitoring of electroencephalographic and noninvasive measurable variables at home in patients with epilepsy (EEG@HOME): Protocol for an observational study. *JMIR Research Protocols, 10*, e25309.

Blackwell, J. E., Kingshott, R. N., Weighall, A. R., et al. (2017). *Conducting unattended home sleep studies in children with narcolepsy and healthy matched controls: A feasibility study*. Leeds Proceedings.

Bölsterli, B. K., Gardella, E., Pavlidis, E., et al. (2017). Remission of encephalopathy with status epilepticus (ESES) during sleep renormalizes regulation of slow wave sleep. *Epilepsia, 58*, 1892–1901. https://doi.org/10.1111/epi.13910

Brunnhuber, F., Amin, D., Nguyen, Y., et al. (2014). Development, evaluation and implementation of video-EEG telemetry at home. *Seizure, 23*, 338–343.

Brunnhuber, F., Slater, J., Goyal, S., et al. (2020). Past, present and future of home video-electroencephalographic telemetry: A review of the development of in-home video-electroencephalographic recordings. *Epilepsia, 61*, S3–S10.

Brunnhuber, F., Slater, J. D., Goyal, S., et al. (2023). The unforeseen future: Impacts of the COVID-19 pandemic on home video-EEG telemetry. *Epilepsia, 64*, S12–S22. https://doi.org/10.1111/epi.17473

Burkhalter, A., Bernardo, K. L., & Charles, V. (1993). Development of local circuits in human visual cortex. *The Journal of Neuroscience, 13*, 1916–1931.

Busse, J., Chen, S., Brace, K., & Slater, R. (2025). Use of the electroencephalogram to measure pain in infants and young children. *Journal of Neurosurgical Anesthesiology, 37*, 133–134.

Bynum, L. G., & Brindle, R. C. (2025). The relationship between sleep and interoception. *Journal of Sleep Research, 34*, e14337.

Carlson, S., Kandler, R. H., Moorhouse, D., et al. (2018). Home video telemetry in children: A comparison to inpatient video telemetry. *Seizure, 61*, 209–213.

Cassidy, L., Quirke, M. B., Alexander, D., et al. (2023). Integrated care for children living with complex care needs: An evolutionary concept analysis. *European Journal of Pediatrics, 182*, 1517–1532.

Chatburn, A., Coussens, S., Lushington, K., et al. (2013). Sleep spindle activity and cognitive performance in healthy children. *Sleep, 36*, 237–243. https://doi.org/10.5665/sleep.2380

Coverstone, A. M., Merielle, B., Melissa, S., et al. (2014). Overnight pulse oximetry for evaluation of sleep apnea among children with trisomy 21. *Journal of Clinical Sleep Medicine, 10*, 1309–1315. https://doi.org/10.5664/jcsm.4286

Debener, S., Emkes, R., De Vos, M., & Bleichner, M. (2015). Unobtrusive ambulatory EEG using a smartphone and flexible printed electrodes around the ear. *Scientific Reports, 5*, 16743.

Dierssen, M., & Ramakers, G. J. A. (2006). Dendritic pathology in mental retardation: From molecular genetics to neurobiology. *Genes, Brain and Behavior, 5*, 48–60. https://doi.org/10.1111/j.1601-183X.2006.00224.x

Ding, L., Chen, B., Dai, Y., & Li, Y. (2022). A meta-analysis of the first-night effect in healthy individuals for the full age spectrum. *Sleep Medicine, 89*, 159–165.

Donadio, A., Whitehead, K., Gonzalez, F., et al. (2018). A novel sensor design for accurate measurement of facial somatosensation in pre-term infants. *PLoS ONE, 13*, e0207145. https://doi.org/10.1371/journal.pone.0207145

Dutil, C., Walsh, J. J., Featherstone, R. B., et al. (2018). Influence of sleep on developing brain functions and structures in children and adolescents: A systematic review. *Sleep Medicine Reviews, 42*, 184–201.

Erkin, G., Culha, C., Ozel, S., & Kirbiyik, E. G. (2010). Feeding and gastrointestinal problems in children with cerebral palsy. *International Journal of Rehabilitation Research, 33*, 218–224.

Fehlings, D. (2017). Pain in cerebral palsy: A neglected comorbidity. *Developmental Medicine and Child Neurology, 59*, 782–783.

Flower, M. J. (1985). Neuromaturation of the human fetus. *Journal of Medicine and Philosophy, 10*, 237–252. https://doi.org/10.1093/jmp/10.3.237

Frassineti, L., Manfredi, C., Ermini, D., et al. (2022). Analysis of Brain-Heart Interactions in newborns with and without seizures using the Convergent Cross Mapping approach. In *2022 44th Annual International Conference of the IEEE Engineering in Medicine & Biology Society (EMBC)* (pp. 36–39).

García-Cordero, I., Esteves, S., Mikulan, E. P., et al. (2017). Attention, in and out: Scalp-level and intracranial EEG correlates of interoception and exteroception. *Frontiers in Neuroscience, 11*, 411.

Georgoulas, A., Jones, L., Laudiano-Dray, M. P., et al. (2021). Sleep–wake regulation in preterm and term infants. *Sleep, 44*. https://doi.org/10.1093/sleep/zsaa148

Gomez-Suarez, R. (2016). Difficulties in the diagnosis and management of functional or recurrent abdominal pain in children. *Pediatric Annals, 45*, e388–e393.

Goodwin, J. L., Enright, P. L., Kaemingk, K. L., et al. (2001). Feasibility of using unattended polysomnography in children for research—report of the Tucson Children's Assessment of Sleep Apnea study (TuCASA). *Sleep, 24*, 937–944.

Gorgoni, M., Scarpelli, S., Reda, F., & De Gennaro, L. (2020). Sleep EEG oscillations in neurodevelopmental disorders without intellectual disabilities. *Sleep Medicine Reviews, 49*, 101224.

Grigg-Damberger, M. M. (2016). The visual scoring of sleep in infants 0 to 2 months of age. *Journal of Clinical Sleep Medicine, 12*, 429–445. https://doi.org/10.5664/jcsm.5600

Halstead, E. J., Joyce, A., Sullivan, E., et al. (2021). Sleep disturbances and patterns in children with neurodevelopmental conditions. *Frontiers in Pediatrics, 9*, 637770.

Hechler, T. (2021). Altered interoception and its role for the co-occurrence of chronic primary pain and mental health problems in children. *Pain, 162*, 665–671.

Hodge, D., Carollo, T. M., Lewin, M., et al. (2014). Sleep patterns in children with and without autism spectrum disorders: Developmental comparisons. *Research in Developmental Disabilities, 35*, 1631–1638.

Hoedlmoser, K., Heib, D. P., Roell, J., et al. (2014). Slow sleep spindle activity, declarative memory, and general cognitive abilities in children. *Sleep, 37*, 1501–1512.

Horne, R. S. C. (2014). Cardio-respiratory control during sleep in infancy. *Paediatric Respiratory Reviews, 15*, 163–169. https://doi.org/10.1016/j.prrv.2013.02.012

Hrbek, A., Karlberg, P., & Olsson, T. (1973). Development of visual and somatosensory evoked responses in pre-term newborn infants. *Electroencephalography and Clinical Neurophysiology, 34*, 225–232. https://doi.org/10.1016/0013-4694(73)90249-6

Huttenlocher, P. R., & Dabholkar, A. S. (1997). Regional differences in synaptogenesis in human cerebral cortex. *Journal of Comparative Neurology, 387*, 167–178. https://doi.org/10.1002/(SICI)1096-9861(19971020)387:2<167::AID-CNE1>3.0.CO;2-Z

Kanazawa, H., Kawai, M., Kinai, T., et al. (2014). Cortical muscle control of spontaneous movements in human neonates. *European Journal of Neuroscience, 40*, 2548–2553. https://doi.org/10.1111/ejn.12612

Klabunde, M., Juszczak, H., Jordan, T., et al. (2019). Functional neuroanatomy of interoceptive processing in children and adolescents: A pilot study. *Scientific Reports, 9*, 16184. https://doi.org/10.1038/s41598-019-52776-4

Kurth, S., Jenni, O. G., Riedner, B. A., et al. (2010). Characteristics of sleep slow waves in children and adolescents. *Sleep, 33*, 475–480. https://doi.org/10.1093/sleep/33.4.475

Kwon, S., Kim, H., & Yeo, W.-H. (2021). Recent advances in wearable sensors and portable electronics for sleep monitoring. *iScience*, 24. https://doi.org/10.1016/j.isci.2021.102461

Kwon, H., Chinappen, D. M., Kinard, E. A., et al. (2025). Association of sleep spindle rate with memory consolidation in children with Rolandic epilepsy. *Neurology, 104*, e210232. https://doi.org/10.1212/WNL.0000000000210232

Lanzlinger, D., Kevat, A., Collaro, A., et al. (2023). Tolerance of polysomnography in children with neurodevelopmental disorders compared to neurotypical peers. *Journal of Clinical Sleep Medicine, 19*, 1625–1631.

Lau-Zhu, A., Lau, M. P. H., & McLoughlin, G. (2019). Mobile EEG in research on neurodevelopmental disorders: Opportunities and challenges. *Developmental Cognitive Neuroscience, 36*, 100635. https://doi.org/10.1016/j.dcn.2019.100635

LeBlanc, B. W., Lii, T. R., Silverman, A. E., et al. (2014). Cortical theta is increased while thalamocortical coherence is decreased in rat models of acute and chronic pain. *Pain, 155*, 773–782. https://doi.org/10.1016/j.pain.2014.01.013

Lehtonen, J., Valkonen-Korhonen, M., Georgiadis, S., et al. (2016). Nutritive sucking induces age-specific EEG-changes in 0–24 week-old infants. *Infant Behavior and Development, 45*(Part A), 98–108. https://doi.org/10.1016/j.infbeh.2016.10.005

Li, W., Ma, L., Yang, G., & Gan, W.-B. (2017). REM sleep selectively prunes and maintains new synapses in development and learning. *Nature Neuroscience, 20*, 427–437.

Lloyd, R., Goulding, R., Filan, P., & Boylan, G. (2015). Overcoming the practical challenges of electroencephalography for very preterm infants in the neonatal intensive care unit. *Acta Paediatrica, 104*, 152–157. https://doi.org/10.1111/apa.12869

Losito, E., Eisermann, M., Vignolo, P., et al. (2017). Benign neonatal sleep myoclonus evokes somatosensory responses. *Journal of Clinical Neurophysiology, 34*, 484. https://doi.org/10.1097/WNP.0000000000000412

Luhmann, H. J., Kanold, P. O., Molnár, Z., & Vanhatalo, S. (2022). Early brain activity: Translations between bedside and laboratory. *Progress in Neurobiology, 213*, 102268. https://doi.org/10.1016/j.pneurobio.2022.102268

Maallo, A. M. S., Duvernoy, B., Olausson, H., & McIntyre, S. (2022). Naturalistic stimuli in touch research. *Current Opinion in Neurobiology, 75*, 102570.

Mai, S., Wong, C. K., Georgiou, E., & Pollatos, O. (2018). Interoception is associated with heartbeat-evoked brain potentials (HEPs) in adolescents. *Biological Psychology, 137*, 24–33. https://doi.org/10.1016/j.biopsycho.2018.06.007
Maitre, N. L., Key, A. P., Chorna, O. D., et al. (2017). The dual nature of early-life experience on somatosensory processing in the human infant brain. *Current Biology, 27*, 1048–1054. https://doi.org/10.1016/j.cub.2017.02.036
Mannion, A., Leader, G., & Healy, O. (2013). An investigation of comorbid psychological disorders, sleep problems, gastrointestinal symptoms and epilepsy in children and adolescents with autism spectrum disorder. *Research in Autism Spectrum Disorders, 7*, 35–42.
Marcus, C. L., Traylor, J., Biggs, S. N., et al. (2014). Feasibility of comprehensive, unattended ambulatory polysomnography in school-aged children. *Journal of Clinical Sleep Medicine, 10*, 913–918.
Maruyama, S., Jain, P., Parbhoo, K., et al. (2022). Prolonged video-EEG and heart rate variability can elucidate autonomic dysregulation in infantile apneic seizures. *Pediatric Neurology, 127*, 48–55. https://doi.org/10.1016/j.pediatrneurol.2021.11.007
McGlone, F., Wessberg, J., & Olausson, H. (2014). Discriminative and affective touch: Sensing and feeling. *Neuron, 82*, 737–755.
Michaeli, Y., Blumkin, L., Medvedovsky, M., et al. (2024). Home-video EEG monitoring in a pediatric setting. *Helivon, 10*. https://doi.org/10.1016/j.heliyon.2024.e35108
Michels, N., Clays, E., Buyzere, M. D., et al. (2013). Children's sleep and autonomic function: Low sleep quality has an impact on heart rate variability. *Sleep, 36*, 1939–1946. https://doi.org/10.5665/sleep.3234
Mikkelsen, K. B., Kappel, S. L., Hemmsen, M. C., et al. (2019). *Discrimination of sleep spindles in Ear-EEG* (pp. 6697–6700). IEEE.
Milh, M., Kaminska, A., Huon, C., et al. (2007). Rapid cortical oscillations and early motor activity in premature human neonate. *Cerebral Cortex, 17*, 1582–1594. https://doi.org/10.1093/cercor/bhl069
Milne-Ives, M., Duun-Henriksen, J., Blaabjerg, L., et al. (2023). At home EEG monitoring technologies for people with epilepsy and intellectual disabilities: A scoping review. *Seizure: European Journal of Epilepsy, 110*, 11–20.
Mulkey, S. B., Hitchings, L., Persaud, R., et al. (2021). Cerebral cortical autonomic connectivity in low-risk term newborns. *Clinical Autonomic Research, 31*, 415–424. https://doi.org/10.1007/s10286-021-00793-7
Nagyova, R., Horsburgh, G., Robertson, A., & Zuberi, S. M. (2019). The clinical utility of ambulatory EEG in childhood. *Seizure, 64*, 45–49. https://doi.org/10.1016/j.seizure.2018.11.020
Ocay, D. D., Teel, E. F., Luo, O. D., et al. (2022). Electroencephalographic characteristics of children and adolescents with chronic musculoskeletal pain. *PAIN Reports, 7*.
Onen, S. H., Onen, F., Courpron, P., & Dubray, C. (2005). How pain and analgesics disturb sleep. *The Clinical Journal of Pain, 21*, 422–431.
Page, J. M., Wakschlag, L. S., & Norton, E. S. (2021). Nonrapid eye movement sleep characteristics and relations with motor, memory, and cognitive ability from infancy to preadolescence. *Developmental Psychobiology, 63*, e22202.
Pasupuleti, A., Amling, J., Chang, T., et al. (2016). Skin integrity during prolonged EEG recording in hospitalized neonatal and pediatric patients (P3.247). *Neurology, 86*, P3.247.
Petanjek, Z., Judaš, M., Šimić, G., et al. (2011). Extraordinary neoteny of synaptic spines in the human prefrontal cortex. *PNAS, 108*, 13281–13286. https://doi.org/10.1073/pnas.1105108108
Rahawi, A. H., He, F., Fang, J., et al. (2025). Association of novel EEG biomarkers of sleep depth and cortical arousability with cardiac autonomic modulation in adolescents. *Sleep*, zsaf018. https://doi.org/10.1093/sleep/zsaf018
Ranger, M., Chau, C. M. Y., Garg, A., et al. (2013). Neonatal pain-related stress predicts cortical thickness at age 7 years in children born very preterm. *PLoS ONE, 8*, e76702. https://doi.org/10.1371/journal.pone.0076702

Rapp, L., Mai-Lippold, S. A., Georgiou, E., & Pollatos, O. (2023). Elevated EEG heartbeat-evoked potentials in adolescents with more ADHD symptoms. *Biological Psychology, 184*, 108698.
Rehman, M., Higdon, L. M., & Sperling, M. R. (2024). Long-term home EEG recording: Wearable and implantable devices. *Journal of Clinical Neurophysiology, 41*, 200–206.
Reid, M. J., Quigg, M., & Finan, P. H. (2023). Sleep-EEG in comorbid pain and insomnia: Implications for the treatment of pain disorders. *Pain Reports, 8*, e1101.
Rice, J., Skuza, P., Baker, F., et al. (2017). Identification and measurement of dystonia in cerebral palsy. *Developmental Medicine and Child Neurology, 59*, 1249–1255.
Riley, T., Porter, R., White, B., & Penry, J. (1981). The hospital experience and seizure control. *Neurology, 31*, 912–912.
Ritterband-Rosenbaum, A., Herskind, A., Li, X., et al. (2017). A critical period of corticomuscular and EMG–EMG coherence detection in healthy infants aged 9–25 weeks. *The Journal of Physiology, 595*, 2699–2713. https://doi.org/10.1113/JP273090
Rockholt, M. M., Kenefati, G., Doan, L. V., et al. (2023). In search of a composite biomarker for chronic pain by way of EEG and machine learning: Where do we currently stand? *Frontiers in Neuroscience, 17*, 1186418.
Ross, E. E., Stoyell, S. M., Kramer, M. A., et al. (2020). The natural history of seizures and neuropsychiatric symptoms in childhood epilepsy with centrotemporal spikes (CECTS). *Epilepsy & Behavior, 103*, 106437.
Rundo, J. V., & Downey, R. (2019). Chapter 25—Polysomnography. In K. H. Levin & P. Chauvel (Eds.), *Handbook of clinical neurology* (pp. 381–392). Elsevier.
Russo, K., Greenhill, J., & Burgess, S. (2021). Home (Level 2) polysomnography is feasible in children with suspected sleep disorders. *Sleep Medicine, 88*, 157–161. https://doi.org/10.1016/j.sleep.2021.10.024
Sabater-Gárriz, Á., Mingorance, J. A., Mestre-Sansó, F., et al. (2025). Affective touch enhances low gamma activity during hand proprioceptive perception in children with different neurodevelopmental conditions. *Frontiers in Human Neuroscience, 19*, 1538428.
Sacchetti, S., McGlone, F., Cazzato, V., & Mirams, L. (2021). The off-line effect of affective touch on multisensory integration and tactile perceptual accuracy during the somatic signal detection task. *PLoS One, 16*, e0261060.
Tassinari, C. A., Rubboli, G., Volpi, L., et al. (2000). Encephalopathy with electrical status epilepticus during slow sleep or ESES syndrome including the acquired aphasia. *Clinical Neurophysiology, 111*, S94–S102.
Teel, E. F., Ocay, D. D., Blain-Moraes, S., & Ferland, C. E. (2022). Accurate classification of pain experiences using wearable electroencephalography in adolescents with and without chronic musculoskeletal pain. *Frontiers in Pain Research, 3*, 991793.
Tessier, S., Lambert, A., Chicoine, M., et al. (2015). Intelligence measures and stage 2 sleep in typically-developing and autistic children. *International Journal of Psychophysiology, 97*, 58–65.
Thanapaisal, S., Mosher, S., Trejo, B., & Robbins, K. (2020). EEG-Beats: Automated analysis of heart rate variability (HVR) from EEG-EKG. 2020.07.21.211862.
Tononi, G., & Cirelli, C. (2006). Sleep function and synaptic homeostasis. *Sleep Medicine Reviews, 10*, 49–62.
Troller-Renfree, S. V., Morales, S., Leach, S. C., et al. (2021). Feasibility of assessing brain activity using mobile, in-home collection of electroencephalography: Methods and analysis. *Developmental Psychobiology, 63*, e22128. https://doi.org/10.1002/dev.22128
Usman, F., Marchant, S., Baxter, L., et al. (2023). The effect of acute respiratory events and respiratory stimulants on EEG-recorded brain activity in neonates: A systematic review. *Clinical Neurophysiology Practice, 8*, 203–225. https://doi.org/10.1016/j.cnp.2023.11.002
Volpe, J. J. (2009). Brain injury in premature infants: A complex amalgam of destructive and developmental disturbances. *The Lancet Neurology, 8*, 110–124. https://doi.org/10.1016/S1474-4422(08)70294-1

Wager, T. D., Atlas, L. Y., Lindquist, M. A., et al. (2013). An fMRI-based neurologic signature of physical pain. *New England Journal of Medicine, 368*, 1388–1397.

Wang, J., Wang, J., Xing, G.-G., et al. (2016). Enhanced gamma oscillatory activity in rats with chronic inflammatory pain. *Frontiers in Neuroscience, 10*, 489.

Wang, J.-W., Zhang, D.-W., & Johnstone, S. J. (2023). Portable EEG for assessing attention in educational settings: A scoping review. *Acta Psychologica, 255*, 104933.

Wang, X., Groot, E. R. D., Tataranno, M. L., et al. (2024). Machine learning-derived active sleep as an early predictor of white matter development in preterm infants. *The Journal of Neuroscience, 44*. https://doi.org/10.1523/JNEUROSCI.1024-23.2023

Weeke, L. C., Ooijen, I. M. V., Groenendaal, F., et al. (2017). Rhythmic EEG patterns in extremely preterm infants: Classification and association with brain injury and outcome. *Clinical Neurophysiology, 128*, 2428–2435. https://doi.org/10.1016/j.clinph.2017.08.035

Whitehead, K. (2024). Co-developing sleep-wake and sensory foundations for cognition in the human fetus and newborn. *Developmental Cognitive Neuroscience*, 101487. https://doi.org/10.1016/j.dcn.2024.101487

Whitehead, K., Jones, L., Laudiano Dray, P., et al. (2017). Full 10-20 EEG application in hospitalised neonates is not associated with an increase in stress hormone levels. *Clinical Neurophysiology Practice, 3*, 20–21. https://doi.org/10.1016/j.cnp.2017.12.002

Whitehead, K., Meek, J., & Fabrizi, L. (2018). Developmental trajectory of movement-related cortical oscillations during active sleep in a cross-sectional cohort of pre-term and full-term human infants. *Scientific Reports, 8*, 17516. https://doi.org/10.1038/s41598-018-35850-1

Whitehead, K., Slobodina, M., Meek, J., & Fabrizi, L. (2019a). Fronto-central slow cortical activity is attenuated during phasic events in rapid eye movement sleep at full-term birth. *Early Human Development, 136*, 45–48. https://doi.org/10.1016/j.earlhumdev.2019.07.007

Whitehead, K., Papadelis, C., Laudiano-Dray, M. P., et al. (2019b). The emergence of hierarchical somatosensory processing in late prematurity. *Cerebral Cortex, 29*, 2245–2260. https://doi.org/10.1093/cercor/bhz030

Whitehead, K., Jones, L., Laudiano-Dray, M. P., et al. (2019c). Event-related potentials following contraction of respiratory muscles in pre-term and full-term infants. *Clinical Neurophysiology, 130*, 2216–2221. https://doi.org/10.1016/j.clinph.2019.09.008

Xu, R. Y., Chow, C. H., Poole, K. L., et al. (2023). Mobile quantitative EEG assessment and children's preoperative anxiety: A psychometric and feasibility study. *Journal of Technology in Behavioral Science, 8*, 47–56.

Yang, G., & Gan, W. (2012). Sleep contributes to dendritic spine formation and elimination in the developing mouse somatosensory cortex. *Developmental Neurobiology, 72*, 1391–1398.

Yang, H., Zhang, Y., Hu, H., et al. (2024). Relationship between interoception and autistic traits: A resting-state functional connectivity study. *Journal of Autism and Developmental Disorders, 54*, 3290–3300.

Zandvoort, C. S., Dereymaeker, A., Baxter, L., et al. (2024). Apnoea suppresses brain activity in infants. *Imaging Neuroscience, 2*, 1–14. https://doi.org/10.1162/imag_a_00236

Chapter 28
Sleep

Masako Tamaki

Abstract Why do we sleep? Although we spend about one-third of our lives asleep, the roles of sleep remain largely unclear. Studies using animal models have shown that sleep contributes to homeostasis and plasticity, which are critical for adapting to changing environments. In humans, advances in neuroimaging techniques have enabled researchers to measure brain activity during sleep with increasingly high spatiotemporal resolution. The brain undergoes dynamic changes that contribute to distinct aspects of learning during both non-rapid eye movement (NREM) sleep and rapid eye movement (REM) sleep. When sleeping in a new and unfamiliar environment, asymmetric sleep patterns emerge. In this chapter, we first review basic characteristics of sleep, measured by polysomnography including EEG, which will be followed by recent findings on how sleep contributes to memory and learning, and their link to metabolite waste clearance, using EEG concurrently with various neuroimaging methods.

Keywords Sleep · Polysomnography · Multimodal neuroimaging · Learning · Memory

28.1 What Is Sleep?

Sleep is not merely a time when the mind shuts down. Some may experience feeling as though they "must have been sleeping", yet also aware of wake-like moments, while for others, the time between getting into bed at night and waking up in the morning may feel as if it passes in an instant. Some may recall entering a vivid and bizarre world of dreams.

Beyond subjective experiences, objectively measured sleep also indicates it is not simply a period of rest (Tamaki et al., 2020a, b; Uji et al., 2025). Some brain regions remain highly active (Tamaki et al., 2020a; Maquet et al., 2000; Yotsumoto et al.,

M. Tamaki (✉)
RIKEN Center for Brain Science, Wako, Japan
e-mail: masako.tamaki@riken.jp

T. Warbrick (ed.), *The EEG Handbook*,
https://doi.org/10.1007/978-3-032-20450-9_28

2009), and brain networks are reconfigured during sleep (Albouy et al., 2013; Bastian et al., 2022). These processes during sleep enable us to stabilize episodic memory, improve skills, and make effective decisions in the days that follow (Tamaki et al., 2007, 2008, 2009, 2013, 2020b; Yotsumoto et al., 2009; Killgore et al., 2006; Tamaki & Sasaki, 2022; Diekelmann & Born, 2010). These suggest that sleep is not a single state, but can take various forms, and it plays a critical role in higher cognitive functions.

28.1.1 Stages of Sleep

Let's take a look at the physiological changes that occur as sleep progresses (Fig. 28.1). We have a gold standard for classifying human sleep into different stages (Berry et al., 2017; Uji & Tamaki, 2023). While these are just labels determined by experts in sleep research, they are nevertheless powerful tools for understanding macroscopic sleep structures. Sleep is largely divided into non-rapid eye movement (NREM) sleep and rapid eye movement (REM) sleep (Fig. 28.1a) (Berry et al., 2017; Uji & Tamaki, 2023). NREM sleep is further divided into three stages, N1–N3, by the depth of sleep. So, there are wakefulness, stages N1–N3, and REM sleep, five types in total.

When EEG patterns are carefully monitored, distinct features can be observed that characterize the various stages of sleep (Fig. 28.1b; see also Sect. 28.2). During NREM sleep, alpha waves, which occur continuously during wakefulness, disappear, and instead, theta waves associated with subjective visual experiences occur. The eyes move slowly, known as slow eye movements (SEMs). These are the characteristics of sleep onset period (drowsy state). As the sleep onset period passes, a spindle-shaped EEG rhythm known as sleep spindles begin to appear. Sleep spindles are considered an objective marker for "stable sleep". As sleep deepens further, large-amplitude EEG waves known as slow waves (0.5–4 Hz) emerge, often accompanied by sleep spindles interspersed among them. These are the hallmarks of the deepest stage of NREM sleep, known as slow-wave sleep. At this point, the arousal threshold is at its highest, meaning it is most difficult to wake a person. During slow-wave sleep, EEG shows a state of synchrony, with abundant slow waves and sleep spindles. This is because large populations of neurons fire together at the same time. Communication between the thalamus and the cortex is thought to be essential for generating these oscillatory activities during NREM sleep (Steriade, 2000, 2005).

During REM sleep, the eyes dart rapidly back and forth. These movements appear surprisingly fast and random. By contrast to NREM sleep, EEG during REM sleep shift to low-amplitude, irregular patterns. At the same time, heart rate and breathing become variable. In humans, REM sleep is short at the beginning of the night, lengthening progressively across the night, and reaching its longest duration toward early morning when the core body temperature reaches the lowest, showing a clear circadian rhythm. REM sleep is also known as the stage in which vivid, emotional,

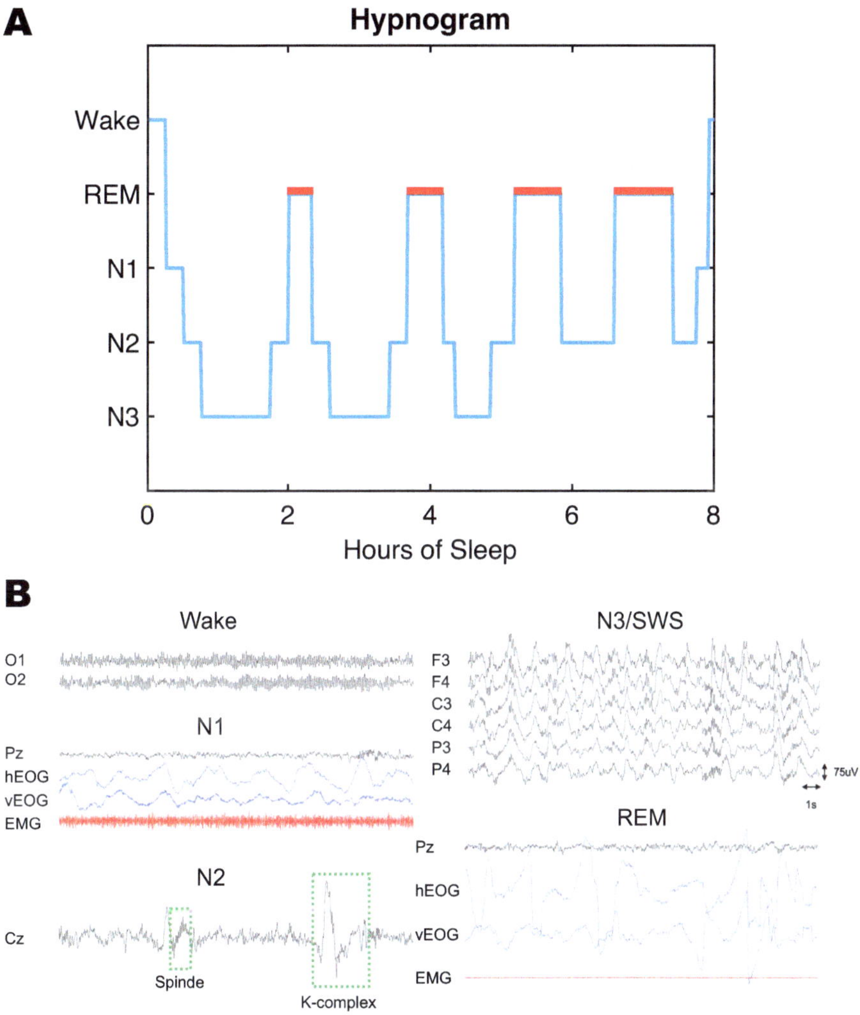

Fig. 28.1 Examples of a hypnogram and physiological recordings during sleep. (**a**) An example hypnogram. The hypnogram depicts the cyclic alternation between NREM and REM sleep roughly every 90 min. N3 predominates early in the night, whereas the proportions of N2 and REM sleep increase in the latter half of the night. (**b**) Representative recordings of EEG, horizontal and vertical EOG, and EMG across different sleep stages, including wakefulness, N1, N2, N3, and REM. During wakefulness, alpha waves (8–13 Hz) appear continuously. N1 is a drowsy state in which alpha activity diminishes and theta waves (4–8 Hz) emerge in EEG signals, while EOG shows slow eye movements (SEMs). N2 is characterized by sleep spindles (11–16 Hz) and K-complexes. N3 shows large, synchronous slow waves (0.5–4 Hz). During REM sleep, EEG signals exhibit mixed high-frequency, low-amplitude activity resembling that of N1 or wakefulness. EOG displays periodic bursts of rapid eye movements, whereas EMG recordings show markedly reduced muscle tone. Negative polarity is up in the scale. *hEOG* horizontal EOG, *vEOG* vertical EOG. (Cited from Uji & Tamaki, 2023; Fig. 1)

and often strange dreams occur. According to the gold standard, there's only one stage for REM sleep, but some studies have shown at least two substates exist, the phasic and tonic periods. The phasic period is when eyes are moving rapidly with twitches occur intermittently, and the tonic period is when eyes are seemingly still. Previous studies have suggested these two substates serve different roles in vigilance (Wehrle et al., 2007; Simor et al., 2020; Ermis et al., 2010; Takahara et al., 2002), but their functional roles in plasticity are unclear.

Additionally, it remains controversial whether the eye movements during REM sleep have any roles. Some say these are due to scanning of objects (Aserinsky & Kleitman, 1953; Dement & Kleitman, 1957; Senzai & Scanziani, 2022) (the Scanning Hypothesis), while these are purely random (Hobson, 2009) (the AIM model). Or these could relate to genetically programmed neural functions (Jouvet, 1998) (the Programming Hypothesis). Thus, dream research remains full of mysteries and continues to fascinate scientists and nonscientists. Of note, dreaming and REM sleep are not mutually exclusive; people can also dream outside of REM sleep, for example, during the sleep-onset period, at the transition from wakefulness to stable sleep (Suzuki et al., 2004; Horikawa et al., 2013). Researchers have succeeded in reading out visual objects that appeared in sleepers' sleep-onset dreams using a method called decoding (Horikawa et al., 2013). The sleep-onset period seems to be associated with insight (Lacaux et al., 2021)! However, the neural mechanisms are still unclear.

28.1.2 How Sleep Changes with Age

Sleep also changes across the lifespan (Carskadon, 1986). In newborns, sleep and wake alternate every 2–4 h. At this stage, total sleep time is very long, about 14–17 h per day, and REM sleep makes up roughly half of total sleep. Over the first 2 years of life, however, the proportion of REM sleep decreases sharply, stabilizing at around 25%. NREM sleep develops somewhat later than REM sleep. It is not clearly observed immediately after birth, but gradually emerges between 2 and 6 months of age. Slow-wave sleep decreases by about 40% between around age 10 and adolescence. Thereafter, the amount of slow-wave sleep continues to decline gradually with aging. Total sleep time also decreases with age. Before adolescence, children typically need 10–13 h. During adolescence, 8–10 h, then in adulthood, 7–9 h. And in older adults, 7–8 h or shorter. Unfortunately, sleep efficiency, which is the proportion of sleep while in bed, also declines significantly with age. This is related to the increased time it takes to fall asleep and the higher frequency of nighttime awakenings in older adults. These sleep structure changes can be captured by polysomnography, including EEG, as indicated below.

28.2 Measuring Human Sleep

How do we measure sleep objectively? A method that detects sleep by integrating multiple physiological measurements is known as polysomnography (PSG) (Carskadon & Dement, 2011). PSG typically consists of three measures: EEG, the electrooculogram (EOG), and the electromyogram (EMG). EEG measures brain activity using electrodes placed on the scalp, and EOG measures eye movements using electrodes placed around the eyes. EMG measures muscle tone using electrodes attached to the chin. Thus, by combining these measures, one can determine the patterns of EEG, whether the eyes are moving, and if so, whether they move slowly or rapidly, and whether muscle tone is strong enough to support posture or is relaxed.

These indices allow researchers and clinicians to assess the depth and type of sleep. For example, if rapid eye movements (REMs) are observed on the EOG recording, while EMG activity is low and EEG patterns are desynchronized, the participant is likely to be in REM sleep rather than wakefulness. If the eyes move rapidly, while alpha waves are continuously present on the EEG and EMG activity is high, the participant may be awake. In sleep clinics, PSG is used for the diagnosis of sleep disorders.

In addition to EEG, EOG, and EMG, respiration, heart rate, and leg movements may be monitored to evaluate sleep disorders, for example, sleep apnea syndrome, restless leg syndrome, or periodic limb movement disorder.

28.2.1 The Various Forms of Sleep

As mentioned at the beginning of this chapter, the brain does not go to sleep or stay awake as a whole. By contrast, sleep or wakefulness can occur regionally, like in gradation. Below I'll begin by introducing an interesting form of sleep in animals, the unihemispheric sleep (Rattenborg et al., 2000; Lyamin et al., 2008), then some findings in human sleep, showing asymmetric patterns in the brain in association with the first-night effect, and brain usage.

28.2.2 Unihemispheric Sleep

The unihemispheric sleep, or unihemispheric slow-wave sleep, is a form of sleep where only one hemisphere of the brain sleeps while the other remains awake (Rattenborg et al., 2000; Lyamin et al., 2008). Certain marine mammals such as dolphins, whales, seals, and mallard ducks, have been found to be capable of unihemispheric sleep. Unihemispheric sleep is measured by the amount of slow waves (delta power) or the state of eyes (closed or open) for each brain hemisphere.

Bihemispheric sleep indicates delta power being higher than during wakefulness in both sides of the brain, while unihemispheric sleep denotes a state where one of the brain hemispheres showing greater delta power while the other showing somewhat comparable to wakefulness. Typically, the eye contralateral to the brain hemisphere showing greater delta power is closed, and the other remains open, when an animal is in unihemispheric sleep.

Several hypotheses have been proposed as to why certain animals have unihemispheric sleep. One possibility is that unihemispheric sleep serves a protective function essential for survival, such as allowing migratory birds to keep flying, dolphins to continue swimming, or enabling the detection of predators.

For example, mallards rest in groups, and those positioned at the edge of the group are at higher risk of predation (Rattenborg et al., 1999a, b). Could unihemispheric sleep in mallards be associated with a survival strategy? If so, ducks on the outside of the group may be more likely to exhibit unihemispheric sleep. A study has found that this is exactly what has been observed (Rattenborg et al., 1999b). Mallards tend to keep the eye facing outward open, while the contralateral (corresponding) hemisphere of the brain remains awake. At the same time, they close the inward-facing eye, allowing the corresponding hemisphere to sleep deeply. In this way, mallards can rest their brain while simultaneously monitoring for potential predators. Sleep deprivation studies, in which animals are experimentally deprived of sleep, have shown that unihemispheric sleep involves sleep homeostasis. For example, when bottlenose dolphins are sleep-deprived for each hemisphere, the sleep-deprived hemisphere shows increased slow-wave activity the following ‘recovery’ day, indicating homeostatic rebound (Oleksenko et al., 1992).

28.2.3 The First-Night Effect, Asymmetric Sleep, and Local Sleep

What about humans? Many people have trouble sleeping in a new hotel room. Taking longer to fall asleep, waking up during the night, and not feeling fully rested the next morning are common experiences when traveling or staying over at a friend’s place. This phenomenon is well-known and called the first-night effect (FNE) in sleep research (Agnew et al., 1966; Tamaki et al., 2005, 2014, 2024). The term “first-night effect” is used because it is most pronounced on the first night (day) of an experiment. In sleep studies, participants are asked to sleep in a laboratory, but because the environments of the lab are so different from their usual sleep environment, it is often difficult to sleep well, affecting the quality of sleep. For this reason, an adaptation session, a practice session to help participants get used to sleeping in a new environment, is usually conducted before the main experimental session.

How the brain activity during sleep is altered in new environments remained a mystery for a long time. However, a previous study showed that a safety-monitoring system is involved (Tamaki et al., 2016). Based on findings of unihemispheric sleep in animals, Tamaki et al. (2016) hypothesized that humans also have an intrinsic system for environmental monitoring which could be asymmetric in association with the first-night effect.

They examined sleep depth (the strength of delta activity) across various brain regions and analyzed the "networks" of the brain, which are the functional groupings of regions, such as the attention network, the sensorimotor network, the default mode network (DMN), and the visual network (Raichle, 2015). The DMN is particularly interesting because it shows increased activity during mind-wandering, a state in which one is not focused on a task but instead lets thoughts drift automatically (Raichle, 2015; Buckner et al., 2008; Andrews-Hanna et al., 2010; Mason et al., 2007). For instance, DMN activity decreases during focused task performance but rises when the mind is not engaged in specific external demands.

In the first experiment, magnetoencephalography was used concurrently with EEG to measure slow-wave activity during slow-wave sleep, then using individual structural brain information from MRI, the strength of sleep was measured in several brain networks, including the DMN. They found that in the DMN, the depth of sleep differed between the brain hemispheres on the first night of the sleep experiment (Fig. 28.2a). Specifically, on Day 1 when the first-night effect was present, the left hemisphere of the DMN showed lighter sleep. Once the FNE subsided on Day 2, this interhemispheric difference was mitigated. The amount of the FNE measured by sleep-onset latency was significantly related to the amount of asymmetry (asymmetry index) on Day 1 (Fig. 28.2b, c).

They next tested whether the FNE is related to the hemispheric asymmetry in vigilance. To examine this, the amplitude of the N3 component, an event-related potential that correlates with the strength of vigilance during sleep, was measured from EEG measured during slow-wave sleep. This may be surprising: the brain is not

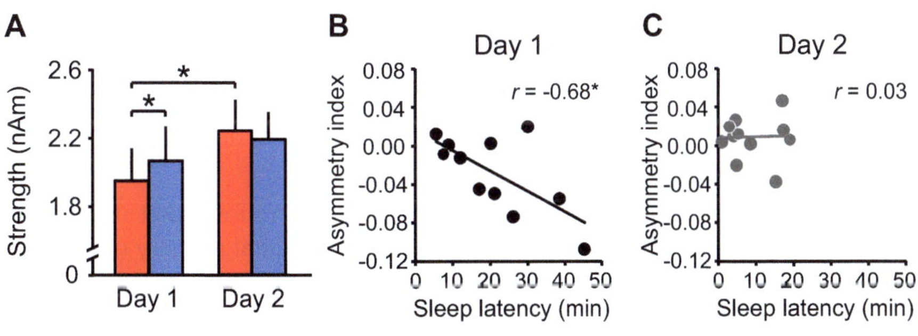

Fig. 28.2 Asymmetric sleep related to the first-night effect in humans. (**a**) Interhemispheric differences in slow-wave activity in the default mode network. Red, left hemisphere; blue, right hemisphere. (**b**) Relationship between the sleep onset latency and the asymmetry index on Day 1. (**c**) Relationship between the sleep onset latency and the asymmetry index on Day 2. (Cited from Tamaki et al., 2016; Fig. 1)

completely shut off from the environment during sleep. Instead, it continues to monitor the environment, even during deep sleep. The researchers wondered whether the vigilance specific to sleep was somehow altered by FNE. They found that on Day 1, during slow-wave sleep, deviant (rare, unexpected) sounds elicited larger N3 responses in the left hemisphere, showing asymmetry in vigilance. This asymmetry in vigilance was attenuated on Day 2 when the FNE subsided.

Finally, they tested whether the FNE is related to a protective function during sleep. If so, the presentation of deviant sounds during sleep should result in a rapid awakening. Indeed, on Day 1, when the left hemisphere detects deviant sounds, participants were able to wake up and produce quicker behavioral responses. These findings suggest that when humans sleep in a new environment, parts of the left hemisphere remain in a lighter sleep state to monitor the surroundings.

These are just a few examples of how EEG can be applied in human sleep research.

Other forms of regional sleep are also known. These include use-dependent regional sleep where a part of the brain shows stronger slow wave activity in response to brain usage or stimulation that occurred prior to sleep (Kattler et al., 1994; Seitz & Roland, 1993).

28.3 Offline Learning Process During Sleep

Have you ever experienced being unable to master a task no matter how many times you practiced it during the day, only to find yourself able to perform it effortlessly the next morning? Research has shown that brain plasticity, the ability of the brain to change in response to stimuli and the environment, undergoes major fluctuations not only during practice, but also offline during subsequent sleep (Tamaki et al., 2020a, b; Tamaki & Sasaki, 2022). These changes strongly influence the degree to which skills improve and become stabilized after sleep. In other words, sleep is not merely a state of rest; it is also a period in which the brain actively changes.

28.3.1 NREM Sleep and Learning

Memory is broadly classified into declarative memory (information that can be explicitly described, such as facts or episodes) and procedural memory (knowledge of "how" to do things, such as riding a bicycle, which is difficult to verbalize) (Squire, 2004).

Brain activities during NREM sleep play important roles in strengthening and stabilizing learning and memory (for reviews see Diekelmann & Born, 2010; Tononi & Cirelli, 2014; Yamada et al., 2023). For example, sleep spindles are thought to be associated with the replay and reactivation of neural activity that occurred during wakeful training. Experiments have shown that presenting auditory stimulation during NREM sleep can enhance spindle activity and improve later memory recall, whereas disrupting spindle activity through auditory stimulation can impair performance. It is thought that during NREM sleep, reactivation of task-related neural activity linked to spindles enhances neural plasticity and supports memory consolidation.

Another important brain activity during NREM sleep is slow waves (Steriade, 2006; Bernardi et al., 2018; Staresina, 2024). Slow waves are classified into slow oscillations (<1 Hz) and delta waves (0.5–4 Hz). Slow oscillations alternate between depolarized "up states," when neuronal activity is high, and hyperpolarized "down states," when neuronal activity is largely silent (Steriade, 2005). Spindles and hippocampal sharp-wave ripples tend to cluster during up states, and these events are thought to be critical for memory processes. Animal studies have suggested that delta waves may weaken memory traces, whereas slow oscillations may strengthen them (Kim et al., 2019). Whether the same functional distinction exists in humans remains an important question for future research.

These are further examples of how EEG allows us to investigate spontaneously occurring neural activity during sleep in a noninvasive manner.

28.3.2 REM Sleep and Learning

Compared with NREM sleep, the role of REM sleep in learning and memory is less well understood. Theta rhythms are characteristic of REM sleep, and have been reported to be involved in memory processing (Popa et al., 2010; Boyce et al., 2016), but compared with spindles and slow waves, relatively little is known about their functional significance.

In humans, Tamaki et al. (2020a) reported that REM sleep contributes to the stabilization of visual learning (a form of perceptual learning), making newly acquired skills more resistant to disruption. Visual learning refers to improvements in perceptual skills (e.g., detecting motion direction or line orientation) that persist over time following visual experience. The degree of stabilization by sleep can be assessed by testing whether learning effects observed before sleep are maintained after sleep without being disrupted. The study investigated the role of REM sleep in stabilizing visual learning by using retrograde interference. For example, if participants train on Task A and then immediately train on a similar but different Task B, performance on Task A typically does not improve, because training on Task B disrupts consolidation of Task A. This is known as retrograde interference, the negative effect of later learning on earlier learning. Conversely, when earlier learning disrupts later learning, it is referred to as anterograde interference. The study found that when both NREM and REM sleep occurred between training on Tasks A and B, training on Task B no longer interfered with learning on Task A. However, if only NREM sleep occurred and REM sleep was absent during the interval, Task B training disrupted Task A learning. These findings suggest that REM sleep plays a crucial role in protecting learning from interference and thereby supports memory stabilization.

28.4 Combining EEG with Neuroimaging Methods for Investigating Sleep

Over the past few decades, non-invasive brain imaging techniques for measuring human brain activity have improved dramatically. When MRI and EEG are combined, they complement each other's limitations, allowing the acquisition of data with high spatiotemporal resolution (Uji & Tamaki, 2023; Warbrick, 2022). This method allows us to investigate the brain network changes in high spatial resolution, specific oscillations, or evoked potentials, measured by EEG. Adding EOG and EMG, comprising PSG, with MRI, the method has evolved into unique 'simultaneous recording techniques' specifically for human sleep research.

28.4.1 Active Brain Networks During Sleep

Using fMRI with EEG, studies have found that different brain networks are activated depending on the depth of sleep or existence of brain oscillations. For example, increased activation is found in the thalamus to tones during NREM sleep when sleep spindles are absent (Schabus et al., 2024). Slow waves are associated with increases in the sensorimotor network, frontal, precuneus, and hippocampal areas (Uji et al., 2025; Dang-Vu et al., 2008) in addition to the thalamus. The brain regions recruited during light NREM sleep involve an arousal-related circuit, while those during deep NREM sleep involve synaptic plasticity and homeostatic regulation (Uji et al., 2025). There are only a small number of studies investigated REM sleep using fMRI (Uji et al., 2025; Wehrle et al., 2005, 2007; Miyauchi et al., 2009). These studies have some key brain regions activating during REMs, including the thalamus and visual cortices. Recently, using a simultaneous fMRI and PSG method, one study has successfully characterized brain regions associated with sawtooth waves during REM sleep (Uji et al., 2025). While the brain regions activated for sawtooth waves were largely the same as those activated for REMs, the inferior frontal cortex was specific to sawtooth waves. The inferior frontal cortex may serve as a potential source of sawtooth wave generation (Frauscher et al., 2020). How brain activation related to sawtooth waves is involved in learning and memory, or in dream generation, may be another interesting question to pursue.

28.4.2 Measuring the Balance of Excitation and Inhibition in the Human Brain

At present, the only neuroimaging technique that allows non-invasive estimation of neurotransmitter concentrations correlated with brain plasticity is magnetic resonance spectroscopy (MRS). Using MRS, it is possible to estimate the concentrations of excitatory neurotransmitter glutamate and inhibitory neurotransmitter gamma-aminobutyric acid (GABA) in specific brain regions (Edden & Barker, 2007; Mescher et al., 1998; Muthukumaraswamy et al., 2009). From the ratio of these concentrations (glutamate/GABA), one can derive the excitation-inhibition (E/I) balance. Previous studies have reported that the E/I balance in the early visual cortex correlates with plasticity in visual learning (Shibata et al., 2017).

How, then, can we measure the E/I balance during sleep? A group has developed a technique for simultaneous acquisition of MRS and polysomnography within the MRI environment (Tamaki et al., 2020a, 2021, 2024) (Fig. 28.3). Using this combined method, they examined the E/I balance in the early visual cortex during NREM and REM sleep. Results showed that the E/I balance increased beyond wake levels during NREM sleep (i.e., became more excitatory), but decreased below wake levels during REM sleep (i.e., became more inhibitory; Fig. 28.3a) (Tamaki et al., 2020a).

Is sleep-related E/I balance linked to learning? Correlations between visual learning task performance and sleep E/I balance revealed that the E/I balance during NREM sleep was associated with the rate of post-sleep performance improvement (the so-called "offline gain," referring to skill enhancement without additional training; Fig. 28.3b). By contrast, the E/I balance during REM sleep correlated with the stabilization of learning (Fig. 28.3c). These findings indicate that brain plasticity fluctuates dynamically during sleep, and suggest that NREM and REM sleep contribute to different aspects of learning through opposing neurochemical processes.

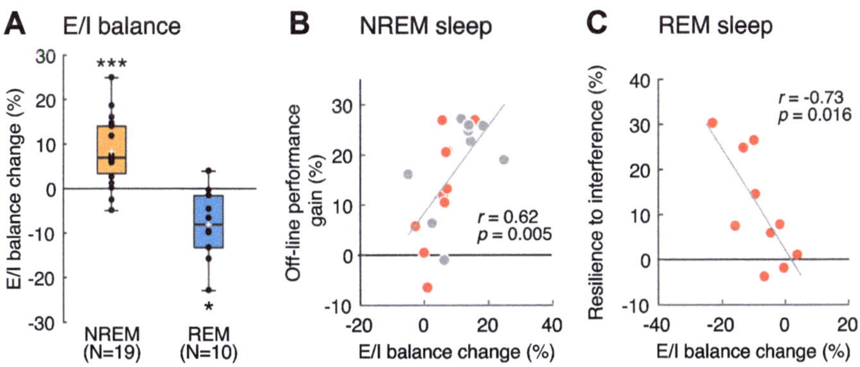

Fig. 28.3 E/I balance during sleep. (**a**) The E/I balance changes during NREM (orange) and REM (blue) sleep. During NREM sleep, the E/I balance is significantly higher than during the wake baseline, whereas during REM sleep, the E/I balance is significantly lower than the wake baseline. (**b**) The correlation between E/I balance changes during NREM sleep and offline performance gains. Red, NREM + REM group. Gray, NREM-only group. (**c**) The correlation between E/I balance changes during REM sleep and stabilization of learning. (Cited from Tamaki et al., 2020a; Fig. 1c, d, and f, respectively)

28.4.3 The Cerebrospinal Fluid Dynamics in Human Sleep

How does sleep maintain healthy brain functions and are alterations in sleep associated with diseases? Cerebrospinal fluid (CSF) during sleep has recently been proposed as a crucial mechanism for brain function. Accumulating evidence indicates that the brain metabolite waste is reduced specifically during deep sleep when CSF flows are increased. Animal studies have shown that the amount of CSF tracer influx increases during sleep compared to during wakefulness (Xie et al., 2013). Furthermore, sleep was associated with βA clearance (Xie et al., 2013; Cankar et al., 2024). How CSF dynamics are driven in the healthy human brain during deep sleep had remained unclear, while several studies have reported findings during light sleep (Fultz et al., 2019).

One of the reasons why deep sleep in humans had long been unreported is that fMRI produces substantial acoustic noise. If you've ever been inside an MRI scanner, you know how loud it can be! As a result, recording brain activity during deep sleep using fMRI is extremely difficult.

A group of scientists set out to tackle this challenge. As described above, a study had previously succeeded in measuring brain activity during deep sleep using MRS, an MRI technique distinct from conventional fMRI (Tamaki et al., 2020a, 2021, 2024). In this study, they took advantage of the fact that MRS produces intermittent rather than continuous scanner noise, and accordingly adopted a sparse fMRI approach that acquires brain signals intermittently and slowly, instead of the conventional continuous fMRI method that generates persistent high-pitched noise. Furthermore, by simultaneously recording physiological signals such as EEG, electrooculography (EOG), and electromyography (EMG), they succeeded in obtaining brain activity data during slow-wave sleep and REM sleep, the states that have been notoriously difficult to measure with conventional MRI techniques (Uji et al., 2025).

As a result, the study has revealed that CSF dynamics are facilitated, especially during deep sleep in healthy young human participants (Uji et al., 2025). The CSF signals are tightly linked to spontaneous brain oscillations. Slow waves and sleep spindles during slow-wave sleep, the deepest NREM sleep, are followed by short-cycle, frequent, yet moderate changes in the CSF signals, the fMRI signals measured from the lateral ventricles (Fig. 28.4b). Interestingly, slow waves during light sleep and arousals (brief awakening from sleep) are also followed by CSF signal changes, but the changes are slow, infrequent, and steep (Fig. 28.4a). During REM sleep as well, CSF signals are time-locked to neural events. Rapid eye movements and sawtooth waves are linked to CSF signal changes.

They also found that these brain oscillations and neural events recruit different brain networks depending on the depth of sleep, with deep sleep involving memory and homeostatic circuits, while lighter sleep involves sensory and motor networks (Fig. 28.5). Thus, human deep sleep may have a specific way of facilitating CSF dynamics in tight link to learning and memory. However, how the CSF dynamics during deep sleep are involved in metabolic clearance in healthy humans remains unclear. Investigating how CSF alterations are involved in various clinical

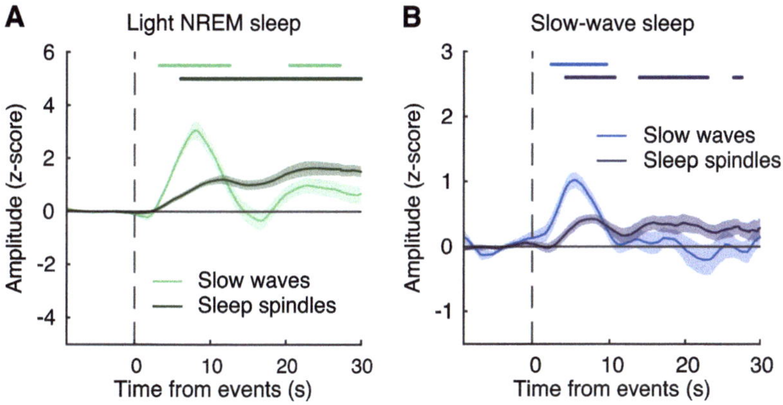

Fig. 28.4 CSF dynamics during sleep. (**a**) CSF signal changes to slow waves and sleep spindles during light NREM sleep. During light NREM sleep, sleep spindles were correlated with significant CSF signal changes at 6 s after onset, whereas slow waves preceded a large CSF signal peak at 8 s after onset. (**b**) CSF signals changes to slow waves and sleep spindles during deep NREM sleep or slow-wave sleep. During slow-wave sleep, slow waves triggered CSF signal changes peaking at 5.5 s, in which latency was significantly shorter and smaller in amplitude. Sleep spindles preceded CSF signal changes at 4 s after their onset, which lasted for approximately 10 s. (Cited from Uji et al., 2025; Fig. 1d, e)

conditions in humans is also needed (Elabasy et al., 2025). To this end, advanced noninvasive along with invasive methods will be useful (Elabasy et al., 2025; Hirschler et al., 2025).

28.5 Conclusions

In this chapter, I've introduced how EEG has been utilized in human sleep research. These studies demonstrate that EEG can be used not only to detect macroscopic sleep structure but also to assess the depth of sleep within brain networks, to probe the plasticity processes underlying it, and potentially to capture the brain's cleaning processes.

Strikingly, EEG becomes even more powerful when combined with other modalities, including structural MRI, functional MRI, MRS, and MEG, providing rich information about various offline brain processes.

The studies presented here focus on healthy young adults; however, it is well established that sleep problems become more prevalent with aging, and individuals with neuropsychiatric disorders almost invariably experience sleep disturbances. Moving forward, it will be important to investigate how sleep is altered in clinical populations and older adults using EEG and multimodal approaches.

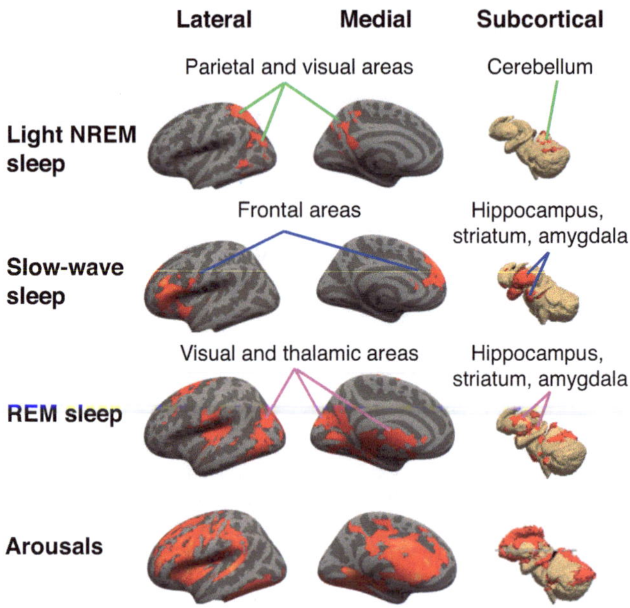

Fig. 28.5 Brain regions recruited during various stages of sleep and for arousals. The red-highlighted parts indicate brain areas activated to slow waves during light NREM (top row) and slow-wave sleep (the second row), rapid eye movements during REM sleep (the third row), and arousals (the bottom row). Light NREM sleep activates sensory and motor regions including the parietal and visual cortices and cerebellum. Slow-wave sleep engages regions involved in plasticity and homeostatic processing including the prefrontal and hippocampal regions, striatum, and amygdala. REM sleep recruits visual and plasticity circuits including the visual areas, the thalamus, the hippocampus, the striatum, and the amygdala. Arousals accompanied widespread brain activation not localized to specific brain regions. (Citated from Uji et al., 2025; Fig. 5. More details can be found in SI Appendix, Tables S4-10 in Uji et al., 2025)

References

Agnew, H. W., Webb, W. B., & Williams, R. L. (1966). The first night effect: An EEG study of sleep. *Psychophysiology, 2*(3), 263–266.

Albouy, G., Sterpenich, V., Vandewalle, G., Darsaud, A., Gais, S., Rauchs, G., et al. (2013). Interaction between hippocampal and striatal systems predicts subsequent consolidation of motor sequence memory. Robertson E, editor. *PLoS ONE, 8*(3), e59490.

Andrews-Hanna, J. R., Reidler, J. S., Sepulcre, J., Poulin, R., & Buckner, R. L. (2010). Functional-anatomic fractionation of the brain's default network. *Neuron, 65*(4), 550–562.

Aserinsky, E., & Kleitman, N. (1953). Regularly occurring periods of eye motility, and concomitant phenomena, during sleep. *Science, 118*, 273–274.

Bastian, L., Samanta, A., Ribeiro De Paula, D., Weber, F. D., Schoenfeld, R., Dresler, M., et al. (2022). Spindle–slow oscillation coupling correlates with memory performance and connectivity changes in a hippocampal network after sleep. *Human Brain Mapping, 43*(13), 3923–3943.

Bernardi, G., Siclari, F., Handjaras, G., Riedner, B. A., & Tononi, G. (2018). Local and widespread slow waves in stable NREM sleep: Evidence for distinct regulation mechanisms. *Frontiers in Human Neuroscience, 12*, 248.

Berry, R., Brooks, R., Gamaldo, C., Harding, S., Lloyd, R., Quan, S., et al. (2017). AASM scoring manual updates for 2017 (Version 2.4). *Journal of Clinical Sleep Medicine, 13*(5), 665–666.
Boyce, R., Glasgow, S. D., Williams, S., & Adamantidis, A. (2016). Causal evidence for the role of REM sleep theta rhythm in contextual memory consolidation. *Science, 352*(6287), 812–816.
Buckner, R. L., Andrews-Hanna, J. R., & Schacter, D. L. (2008). The brain's default network: Anatomy, function, and relevance to disease. *Annals of the New York Academy of Sciences, 1124*(1), 1–38.
Cankar, N., Beschorner, N., Tsopanidou, A., Qvist, F. L., Colaço, A. R., Andersen, M., et al. (2024). Sleep deprivation leads to non-adaptive alterations in sleep microarchitecture and amyloid-β accumulation in a murine Alzheimer model. *Cell Reports, 43*(11), 114977.
Carskadon, M. A. (1986). Guidelines for the Multiple Sleep Latency Test (MSLT): A standard measure of sleepiness. *Sleep, 9*(4), 519–524.
Carskadon, M. A., & Dement, W. C. (2011). Chapter 2—Normal human sleep: An overview (p. 21).
Dang-Vu, T. T., Schabus, M., Desseilles, M., Albouy, G., Boly, M., Darsaud, A., et al. (2008). Spontaneous neural activity during human slow wave sleep. *National Academy of Sciences of the United States of America, 105*(39), 15160–15165.
Dement, W., & Kleitman, N. (1957). Cyclic variations in EEG during sleep and their relation to eye movements, body motility, and dreaming. *Electroencephalography and Clinical Neurophysiology, 9*(4), 673–690.
Diekelmann, S., & Born, J. (2010). The memory function of sleep. *Nature Reviews. Neuroscience, 11*(2), 114–126.
Edden, R. A. E., & Barker, P. B. (2007). Spatial effects in the detection of γ-aminobutyric acid: Improved sensitivity at high fields using inner volume saturation. *Magnetic Resonance in Medicine, 58*(6), 1276–1282.
Elabasy, A., Helakari, H., Väyrynen, T., Rajna, Z., Huotari, N., Raitamaa, L., et al. (2025). Sleep increases propagation speed of physiological brain pulsations [Internet]. [cited 2025 May 9]. Available from: http://biorxiv.org/lookup/doi/10.1101/2025.01.09.632145
Ermis, U., Krakow, K., & Voss, U. (2010). Arousal thresholds during human tonic and phasic REM sleep: Arousal threshold in tonic and phasic REM. *Journal of Sleep Research, 19*(3), 400–406.
Frauscher, B., Von Ellenrieder, N., Dolezalova, I., Bouhadoun, S., Gotman, J., & Peter-Derex, L. (2020). Rapid eye movement sleep sawtooth waves are associated with widespread cortical activations. *The Journal of Neuroscience, 40*(46), 8900–8912.
Fultz, N. E., Bonmassar, G., Setsompop, K., Stickgold, R. A., Rosen, B. R., Polimeni, J. R., et al. (2019). Coupled electrophysiological, hemodynamic, and cerebrospinal fluid oscillations in human sleep. *Science, 366*(6465), 628–631.
Hirschler, L., Runderkamp, B. A., Decker, A., Van Harten, T. W., Scheyhing, P., Ehses, P., et al. (2025). Region-specific drivers of CSF mobility measured with MRI in humans. *Nature Neuroscience* [Internet]. [cited 2025 Oct 16]. Available from: https://www.nature.com/articles/s41593-025-02073-3
Hobson, J. A. (2009). REM sleep and dreaming: Towards a theory of protoconsciousness. *Nature Reviews. Neuroscience, 10*(11), 803–813.
Horikawa, T., Tamaki, M., Miyawaki, Y., & Kamitani, Y. (2013). Neural decoding of visual imagery during sleep. *Science, 340*(6132), 639–642.
Jouvet, M. (1998). Paradoxical sleep as a programming system. *Journal of Sleep Research, 7*(S1), 1–5.
Kattler, H., Dijk, D. J., & Borbely, A. A. (1994). Effect of unilateral somatosensory stimulation prior to sleep on the sleep EEG in humans. *Journal of Sleep Research, 3*(3), 159–164.
Killgore, W. D. S., Balkin, T. J., & Wesensten, N. J. (2006). Impaired decision making following 49 h of sleep deprivation. *Journal of Sleep Research, 15*(1), 7–13.
Kim, J., Gulati, T., & Ganguly, K. (2019). Competing roles of slow oscillations and delta waves in memory consolidation versus forgetting. *Cell, 179*(2), 514–526.e13.

Lacaux, C., Andrillon, T., Bastoul, C., Idir, Y., Fonteix-Galet, A., Arnulf, I., et al. (2021). Sleep onset is a creative sweet spot. *Science Advances, 7*(50), eabj5866.

Lyamin, O., Manger, P., Ridgway, S., Mukhametov, L., & Siegel, J. (2008). Cetacean sleep: An unusual form of mammalian sleep. *Neuroscience and Biobehavioral Reviews, 32*(8), 1451–1484.

Maquet, P., Laureys, S., Peigneux, P., Fuchs, S., Petiau, C., Phillips, C., et al. (2000). Experience-dependent changes in cerebral activation during human REM sleep. *Nature Neuroscience, 3*(8), 831–836.

Mason, M. F., Norton, M. I., Van Horn, J. D., Wegner, D. M., Grafton, S. T., & Macrae, C. N. (2007). Wandering minds: The default network and stimulus-independent thought. *Science, 315*(5810), 393–395.

Mescher, M., Merkle, H., Kirsch, J., Garwood, M., & Gruetter, R. (1998). Simultaneous in vivo spectral editing and water suppression. *NMR in Biomedicine, 11*, 7.

Miyauchi, S., Misaki, M., Kan, S., Fukunaga, T., & Koike, T. (2009). Human brain activity time-locked to rapid eye movements during REM sleep. *Experimental Brain Research, 192*(4), 657–667.

Muthukumaraswamy, S. D., Edden, R. A. E., Jones, D. K., Swettenham, J. B., & Singh, K. D. (2009). Resting GABA concentration predicts peak gamma frequency and fMRI amplitude in response to visual stimulation in humans. *National Academy of Sciences of the United States of America, 106*(20), 8356–8361.

Oleksenko, A. I., Mukhametov, L. M., Polyakova, I. G., Supin, A. Y., & Kovalzon, V. M. (1992). Unihemispheric sleep deprivation in bottlenose dolphins. *Journal of Sleep Research, 1*(1), 40–44.

Popa, D., Duvarci, S., Popescu, A. T., Lena, C., & Pare, D. (2010). Coherent amygdalocortical theta promotes fear memory consolidation during paradoxical sleep. *National Academy of Sciences of the United States of America, 107*(14), 6516–6519.

Raichle, M. E. (2015). The brain's default mode network. *Annual Review of Neuroscience, 38*(1), 433–447.

Rattenborg, N. C., Lima, S. L., & Amlaner, C. J. (1999a). Facultative control of avian unihemispheric sleep under the risk of predation. *Behavioural Brain Research, 105*(2), 163–172.

Rattenborg, N. C., Lima, S. L., & Amlaner, C. J. (1999b). Half-awake to the risk of predation. *Nature, 397*(6718), 397–398.

Rattenborg, N. C., Amlaner, C. J., & Lima, S. L. (2000). Behavioral, neurophysiological and evolutionary perspectives on unihemispheric sleep. *Neuroscience and Biobehavioral Reviews, 24*(8), 817–842.

Schabus, M., Dang-Vu, T. T., Heib, D. P. J., Boly, M., Desseilles, M., Vandewalle, G., et al. (2024). The fate of incoming stimuli during NREM sleep is determined by spindles and the phase of the slow oscillation. *Frontiers in Neurology* [Internet]. [cited 2024 Aug 12], *3*. Available from: http://journal.frontiersin.org/article/10.3389/fneur.2012.00040/abstract

Seitz, R. J., & Roland, P. E. (1993). Activity changes in the human brain due to vibratory stimulation of the hand studied with positron emission tomography. In *Imaging of the brain in psychiatry and related fields* (pp. 177–182). Springer.

Senzai, Y., & Scanziani, M. (2022). A cognitive process occurring during sleep is revealed by rapid eye movements. *Science, 377*(6609), 999–1004.

Shibata, K., Sasaki, Y., Bang, J. W., Walsh, E. G., Machizawa, M. G., Tamaki, M., et al. (2017). Overlearning hyperstabilizes a skill by rapidly making neurochemical processing inhibitory-dominant. *Nature Neuroscience, 20*(3), 470–475.

Simor, P., van der Wijk, G., Nobili, L., & Peigneux, P. (2020). The microstructure of REM sleep: Why phasic and tonic? *Sleep Medicine Reviews, 52*, 101305.

Squire, L. R. (2004). Memory systems of the brain: A brief history and current perspective. *Neurobiology of Learning and Memory, 82*(3), 171–177.

Staresina, B. P. (2024). Coupled sleep rhythms for memory consolidation. *Trends in Cognitive Sciences, 28*(4), 339–351.
Steriade, M. (2000). Corticothalamic resonance, states of vigilance and mentation. *Neuroscience, 101*(2), 243–276.
Steriade, M. (2005). Sleep, epilepsy and thalamic reticular inhibitory neurons. *Trends in Neurosciences, 28*(6), 317–324.
Steriade, M. (2006). Grouping of brain rhythms in corticothalamic systems. *Neuroscience, 137*(4), 1087–1106.
Suzuki, H., Uchiyama, M., Tagaya, H., Ozaki, A., Kuriyama, K., Aritake, S., et al. (2004). Dreaming during non-rapid eye movement sleep in the absence of prior rapid eye movement sleep. *Sleep, 27*(8), 1486–1490.
Takahara, M., Nittono, H., & Hori, T. (2002). Comparison of the event-related potentials between tonic and phasic periods of rapid eye movement sleep. *Psychiatry and Clinical Neurosciences, 56*(3), 257–258.
Tamaki, M., & Sasaki, Y. (2022). Sleep-dependent facilitation of visual perceptual learning is consistent with a learning-dependent model. *Journal of Neuroscience, 42*(9), 1777–1790.
Tamaki, M., Nittono, H., Hayashi, M., & Hori, T. (2005). Examination of the first-night effect during the sleep-onset period. *Sleep, 28*(2), 195–202.
Tamaki, M., Nittono, H., & Hori, T. (2007). Efficacy of overnight sleep for a newly acquired visuomotor skill: Efficacy of sleep for visuomotor skill. *Sleep and Biological Rhythms, 5*(2), 111–116.
Tamaki, M., Matsuoka, T., Nittono, H., & Hori, T. (2008). Fast sleep spindle (13–15 Hz) activity correlates with sleep-dependent improvement in visuomotor performance. *Sleep, 31*(2), 204–211.
Tamaki, M., Matsuoka, T., Nittono, H., & Hori, T. (2009). Activation of fast sleep spindles at the premotor cortex and parietal areas contributes to motor learning: A study using sLORETA. *Clinical Neurophysiology, 120*(5), 878–886.
Tamaki, M., Huang, T. R., Yotsumoto, Y., Hamalainen, M., Lin, F. H., Nanez, J. E., et al. (2013). Enhanced spontaneous oscillations in the supplementary motor area are associated with sleep-dependent offline learning of finger-tapping motor-sequence task. *Journal of Neuroscience, 33*(34), 13894–13902.
Tamaki, M., Bang, J. W., Watanabe, T., & Sasaki, Y. (2014). The first-night effect suppresses the strength of slow-wave activity originating in the visual areas during sleep. *Vision Research, 99*, 154–161.
Tamaki, M., Bang, J. W., Watanabe, T., & Sasaki, Y. (2016). Night watch in one brain hemisphere during sleep associated with the first-night effect in humans. *Current Biology, 26*(9), 1190–1194.
Tamaki, M., Wang, Z., Barnes-Diana, T., Guo, D., Berard, A. V., Walsh, E., et al. (2020a). Complementary contributions of non-REM and REM sleep to visual learning. *Nature Neuroscience, 23*(9), 1150–1156.
Tamaki, M., Berard, A. V., Barnes-Diana, T., Siegel, J., Watanabe, T., & Sasaki, Y. (2020b). Reward does not facilitate visual perceptual learning until sleep occurs. *Proceedings of the National Academy of Science USA, 117*(2), 959–968.
Tamaki, M., Watanabe, T., & Sasaki, Y. (2021). Coregistration of magnetic resonance spectroscopy and polysomnography for sleep analysis in human subjects. *STAR Protocols, 2*(4), 100974.
Tamaki, M., Yamada, T., Barnes-Diana, T., Wang, Z., Watanabe, T., & Sasaki, Y. (2024). First-night effect reduces the beneficial effects of sleep on visual plasticity and modifies the underlying neurochemical processes. *Scientific Reports, 14*(1), 14388.
Tononi, G., & Cirelli, C. (2014). Sleep and the price of plasticity: From synaptic and cellular homeostasis to memory consolidation and integration. *Neuron, 81*(1), 12–34.
Uji, M., & Tamaki, M. (2023). Sleep, learning, and memory in human research using noninvasive neuroimaging techniques. *Neuroscience Research, 189*, 66–74.

Uji, M., Li, X., Saotome, A., Katsumata, R., Waggoner, R. A., Suzuki, C., et al. (2025). Human deep sleep facilitates cerebrospinal fluid dynamics linked to spontaneous brain oscillations and neural events. *Proceedings of the National Academy of Science USA, 122*(41), e2509626122.

Warbrick, T. (2022). Simultaneous EEG-fMRI: What have we learned and what does the future hold? *Sensors, 22*(6), 2262.

Wehrle, R., Czisch, M., Kaufmann, C., Wetter, T. C., Holsboer, F., Auer, D. P., et al. (2005). Rapid eye movement-related brain activation in human sleep: a functional magnetic resonance imaging study. *NeuroReport, 16*(8), 853–857.

Wehrle, R., Kaufmann, C., Wetter, T. C., Holsboer, F., Auer, D. P., Pollmächer, T., et al. (2007). Functional microstates within human REM sleep: First evidence from fMRI of a thalamocortical network specific for phasic REM periods: Thalamocortical network in phasic REM sleep. *European Journal of Neuroscience, 25*(3), 863–871.

Xie, L., Kang, H., Xu, Q., Chen, M. J., Liao, Y., Thiyagarajan, M., et al. (2013). Sleep drives metabolite clearance from the adult brain. *Science, 342*(6156), 373–377.

Yamada, T., Watanabe, T., & Sasaki, Y. (2023). Plasticity–stability dynamics during post-training processing of learning. *Trends in Cognitive Sciences* [Internet]. [cited 2023 Nov 11]. Available from: https://www.cell.com/trends/cognitive-sciences/abstract/S1364-6613(23)00249-8

Yotsumoto, Y., Sasaki, Y., Chan, P., Vasios, C. E., Bonmassar, G., Ito, N., et al. (2009). Location-specific cortical activation changes during sleep after training for perceptual learning. *Current Biology, 19*(15), 1278–1282.

Chapter 29
Mobile Electroencephalography

Olave E. Krigolson, Mathew Rocha Hammerstrom, and Katherine Boere

Abstract Electroencephalography (EEG) research has traditionally been associated with a participant being seated at a table, with many electrodes affixed to a "swimming cap", which in turn is connected to an amplifier connected to one or more computers. While this type of experimental setup affords the ability to ask an array of research questions, it also greatly restricts ecological validity among other factors. The advent of mobile EEG—EEG systems that are lightweight, easy to put on, and wireless—allows a researcher to take EEG research out of the laboratory and into the "real world". In this chapter, we define what mobile EEG is and provide a brief historical context of the development of mobile EEG. We then look at the range of mobile EEG systems available to researchers, the technical considerations that need to be considered when using these systems, and provide thoughts on the validation of said systems. Finally, we end with an application section in which we describe a range of mobile EEG "use cases" to give the reader examples of how mobile EEG can be used.

Keywords EEG · ERP · Mobile technology · Ecological validity · Neuroscience · Cognitive neuroscience

29.1 What Is Mobile EEG?

Traditionally, the collection of electroencephalographic (EEG) data has been associated with desk-mounted, wired, large array systems (32+ channels) that are inherently "not mobile" (Fig. 29.1). While this sort of experimental setup was

O. E. Krigolson (✉)
Theoretical and Applied Neuroscience Laboratory, School of Exercise Science, University of Victoria, Victoria, BC, Canada
e-mail: krigolson@uvic.ca

M. R. Hammerstrom · K. Boere
Department of Exercise Science, Physical and Health Education, The University of Victoria, Victoria, BC, Canada
e-mail: mathewhammerstrom@uvic.ca; katherineboere@uvic.ca

T. Warbrick (ed.), *The EEG Handbook*,
https://doi.org/10.1007/978-3-032-20450-9_29

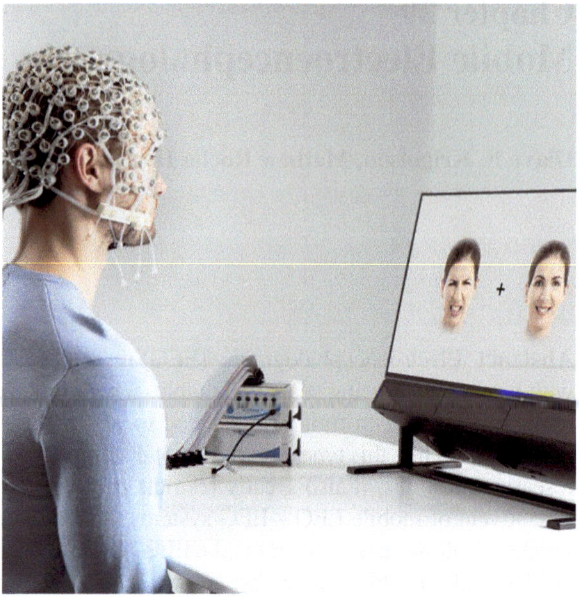

Fig. 29.1 A "traditional" EEG system, the Brain Products actiChamp (Brain Products GmbH, Gilching, Germany)

deemed necessary to collect EEG data, there are obvious problems that stem from the methodology in this format. First, the technical requirements of lab-based EEG systems restrict ecological validity—only a small portion of the daily tasks that we engage in take place sitting in front of a computer in isolation. And we would assume, even fewer wearing a swimming-style cap with many electrodes streaming from said cap connected to an amplifier and a computer. As such, it is fair to question whether the cognitive, motor, and perceptual phenomena that are studied in this manner with EEG parallel the neural processes that subserve us in the real world. Second, another obvious and related problem that stems from the use of traditional EEG systems is that the equipment itself restricts the brain processes that can be studied. For example, we know that being outside impacts neural processing (e.g., Boere et al., 2023a, b), yet, it is very hard to be in nature when one is wired to a computer. In a similar manner, what about movement? A vast portion of the human brain is dedicated to moving our bodies from one location to another, yet traditional EEG data collection by its very nature makes movement problematic. With these two issues in mind—ecological validity and the restrictions in what can be studied due to the nature of the methodology - over the past 15 years or so researchers have begun to look for "mobile EEG" solutions (Fig. 29.2).

First, it is probably important to define what is meant by mobile EEG (mEEG). One has to consider whether the cap is wired to an amplifier, whether the movement of the participant is restricted by proximity to the amplifier, and other factors such as these. Indeed, the concept of mEEG is a true spectrum that on one end has a traditional EEG amplifier with wired electrodes and a cap carried with or by the participant and on the other end a Bluetooth headband-style system running off of a mobile device such as a phone or a tablet. Here, we will attempt to be inclusive of all

Fig. 29.2 A "mobile" EEG system, the Brain Products X.on (Brain Products GmbH, Gilching, Germany)

mEEG solutions. What truly matters, in a sense, is that it is feasible given the research question and capacity to conduct mEEG research.

It is hard to pinpoint the exact start of "mobile EEG" (mEEG) research. In 1993, Sirevaag and colleagues collected EEG data while pilots were flying rotary-wing aircraft. Going back further, one would even have to consider the seminal work done by Berger in 1924 to be, in a sense, "mobile"—at least in the sense that his system could be moved between rooms (MRI scanner, for example, cannot do this). With that said, credit must be given to the work by Debener et al. (2012)[1], who had participants perform an auditory oddball task while seated indoors but importantly also while walking outside. mEEG data was collected from a modified Emotiv EEG headset, a standard EEG cap connected to a laptop running OpenVibe software that was carried within a backpack, which in the walking outside condition made this study truly mobile. The data from this study clearly demonstrated evoked auditory components of the human event-related brain potential (ERPs) thus highlighting that it was possible to make these kinds of measurements outside of the laboratory. In a similar manner, Scalon and colleagues in a series of studies (e.g., 2017) demonstrated that auditory ERPs could be collected using a similar setup while participants

[1] It is important to note that other researchers and several commercial companies had collected EEG data prior to the Debener study. For example, Emotiv Inc. launched the EPOC EEG headset commercially and Lin et al. (2009) were arguably the first researchers to demonstrate mobile EEG technology.

Fig. 29.3 EEG being recorded while riding a bike

were riding a bicycle (Fig. 29.3). In their work, EEG data was collected from a cap with electrodes using a Brain Products V-Amp system (Brain Products GmbH, Gilching, Germany) connected to a Windows laptop carried in a backpack. Similar to the work done by Debener and colleagues while walking, Scalon et al. were able to capture auditory ERP components while participants were riding a bicycle. In other work using the Muse Mobile EEG headband, Kovacevic et al. (2015) collected neurofeedback data from 523 participants in a single night, and Hashemi et al. (2016) presented an EEG study with 6029 participants performing a meditation exercise, research that demonstrates the ability of mEEG to capture larger data sets than possible in a traditional laboratory environment.

Perhaps the most important question that has to be asked with regard to mEEG is why use it? While we have hinted at the reasons above, we will now clearly state them. First and foremost is of course the opportunity to get outside of the laboratory and study cognitive phenomena in the real world. Indeed, the Scanlon et al. (2017) study highlights this as we think it is quite obvious that riding a bicycle outside is considerably different from riding a stationary bike inside from a perceptual-motor perspective. As such, perhaps the strongest reason to use mEEG in research is to increase ecological validity, a construct that is paid homage to in Research Methods classes across the globe, but is rarely addressed adequately in EEG studies in Psychology and Neuroscience. The second key advantage to mEEG is ease of use/setup time. In work done by our research group using the Muse mEEG system, we had an average setup time of under 60 s (Krigolson et al., 2021). Of course, the setup time for mEEG systems is wholly dependent on the mEEG system in question, which is noted below when various systems are reviewed. But overall, there is little to no doubt that ease of use/setup time is a massive advantage when using mEEG relative to a traditional computer-based cap—amplifier EEG system. Finally, mEEG is low cost by its very nature, especially when one considers the available commercial devices. Lowering the cost of EEG data capture leads to a natural increase in sample size which is of growing importance in a research field that has been criticized for underpowered research leading to non-replicable findings (Ioannidis, 2005).

29.2 Range of mEEG Systems

It is impossible to list here all of the mEEG systems that are currently available, especially within the commercial space. In an attempt to classify current systems, we will broadly group them into what we will call commercial and research systems. Before this however, it is worth noting four key features that differ between systems. One, the number and placement of electrodes—for mobile systems this spans from a single recording electrode to 32+ recording electrodes. Two, the data quality of the individual systems. Three, the manner in which data is transmitted (Bluetooth, Wireless, Wired). Four, ease of setup of the system. When considering the use of a mEEG system, it is important to examine each of these four features.

Commercial Systems Without question, the Muse EEG headband is the most popular and the most distributed mEEG system in the world with over 500,000 units currently in the market. The Muse system has four electrodes located at AF7, AF8, TP9, TP10, and also comes with an accelerometer, gyroscope, and a pulse oximeter (www.choosemuse.com; Fig. 29.4).

Further, the latest version of Muse, Athena, also comes with functional near-infrared spectroscopy sensors. Importantly, work by our research group and others has demonstrated that the EEG data quality of the Muse is sufficient for conducting peer-reviewed research studies (Fickling et al., 2020; Krigolson et al., 2017, 2021: see below for more detail). It is important to note that there are numerous other mEEG systems capable of collecting research-grade data—for instance, other research groups have demonstrated similar findings with systems such as the OpenBCI Cyton (Qiu et al., 2019), the Neurosky Mindflex (Katona et al., 2014),

Fig. 29.4 The Muse S Athena. A combined EEG and fNIRS system

and the Emotiv Epoc+ (Kotowski et al., 2018; Mercado-Aguirre et al., 2019). A full review of all of the commercially available mEEG systems is beyond this chapter, but others such as Bateson et al. (2017) and Sawangjai et al. (2020) have attempted to do so in review papers. With that said, it is worth noting that new commercial mEEG systems are released on a regular basis making it almost impossible to stay current with the latest technology.

Research Systems As with commercial mEEG systems, there are non-commercial "research" mEEG systems available for research as well. Again, there are too many systems to list here so we will endeavor to highlight some prominent systems to highlight key differences in mEEG technology. Some research-grade mEEG systems provide full head coverage and a large number of electrodes (20+). For instance, the CGX Quick 20 and Quick 32 have 20 or 32 electrodes respectively and transmit data via Bluetooth to a recording computer or tablet. Other research-grade mEEG systems that try to achieve full head coverage rely on traditional EEG caps connected to an amplifier, but the amplifier is not directly wired to a computer (e.g., the Brain Products LiveAmp). Other research-grade mEEG systems use a reduced electrode array but still provide high-quality EEG data from the available electrodes. An excellent example of this is the CGX Patch system that only has two electrodes and is worn on the forehead only (see Boere et al., 2023a, b). Recently, the release of the Brain Products X.on (Brain Products GmbH, Gilching, Germany) has provided researchers with a research-grade mEEG system with seven electrodes that provides semi-full head coverage. While the X.on has yet to be fully validated—specifically there are no published papers with the device—the evidence that is available suggests that the X.on is an excellent compromise between data quality, head coverage, and mobility (see https://www.peereeg.com/results.html for preliminary results). Other companies in the mEEG research space with excellent headsets include the APEX system by TMSI, the Bittium BrainStatus system, the Smarting systems by mBraintrain, the Enobio system by Neuroelectrics, and a series of systems with varying configurations by Bitbrain, among others.

29.3 Technical Considerations

When considering the use of a mEEG for research, there are several factors that differentiate these systems from more traditional lab-based EEG systems that need to be considered and evaluated.

Wired Versus Wireless First and foremost, one must consider whether the EEG electrodes will be wired to the amplifier or not. The advantage of a wired system is that data quality is typically better with a wired system, and one does not have to worry about data loss due to Bluetooth/wireless dropout. However, the obvious constraint is that the participant is wired to the amplifier. Some researchers have created a compromise wherein the participant carries the amplifier with them in a pack or in some form of mounting device. Again, the restraint here is obvious—the participant needs to carry the amplifier with them and the wires to the EEG cap could potentially interfere with head or other movements. There is no correct answer here, it is simply worth noting that when choosing a wired or a wireless solution the pros and cons of the system must be evaluated with regard to the proposed research methodology.

Data Lag/Loss If one chooses to use a wireless mEEG system, then the issue of data lag/loss must be considered. Bluetooth technology comes with a guarantee of data packet order, but there is an inherent lag between the "start" of data collection and the actual arrival of data on the recording device. This lag is especially a potential problem if one is considering mEEG studies of event-related brain potentials (see Event Markers and Timing Below). With wireless technology, the lag is in principle less of a problem and thus may not be as much of a factor. However, with both Bluetooth and wireless data transmission, signal interference is a real issue that can lead to data loss or corruption and as such has to be factored into research design (i.e., Tiparaju et al., 2021).

The Number of Channels and Channel Placement Typically, mEEG systems have fewer electrodes than more traditional EEG systems, although this may not be the case if one uses a wired mEEG solution. Perhaps of more importance is the location of the electrodes in mEEG systems. Typically, the key debate is whether or not the system attempts to get through hair. In other words, some mEEG systems only attempt to position electrodes directly on the skin; the Muse for instance only has electrodes on the forehead and directly above the ears. Other mEEG systems like the CGX Quick 20 attempt to make contact with the scalp through the hair. The systems that attempt full head/through the hair coverage typically have specially designed electrodes to penetrate through the hair. The position of electrodes is important as some EEG signals of interest might not be visible at the electrodes that a given mEEG has. For instance, in the work in our laboratory we have been able to measure the N200 and P300 components of the human event-related brain potential (ERP) but have had little success measuring the P100 and N100 ERP components which are typically measured over the occipital cortex. It is worth noting here that EEG and ERP signatures tend to look different when measured at non-traditional electrode sites. In our "Choosing Muse" paper (Krigolson et al., 2017), we did measure a P300 ERP component at electrodes TP9 and TP10, but the morphology of the component was significantly different than the morphology of the component when measured at the more typical location of Pz. Again, when choosing a mEEG system, the number and placement of electrodes is a key factor that must be considered in the research design process.

Wet Versus Dry Another consideration when selecting a mEEG system for research is whether to use a wet or dry system. Traditional systems rely on conductive gel or a saline solution to reduce the impedance between an electrode and the scalp, thus yielding better overall signal quality. For example, the Brain Products X. on uses sponge electrodes that are soaked in saline prior to data collection. However, to promote ease of use and mobility, some mEEG systems utilize dry electrodes wherein the electrode itself is made of a special material to promote conductivity. Typically, these systems operate with higher impedances which can have an impact on signal quality. Further, these systems typically rely on compression to maintain good contact between the electrodes and the scalp, and as such, the issue of participant comfort must be considered. It is beyond the scope of this chapter to dive into this issue further, but our recommendation when considering the use of a dry electrode system is to: one, assess data quality, and two, try on the system to ensure the level of comfort is sufficient for the participants that you will be working with.

Data Quality When utilizing mEEG systems at some level one must accept that the data quality will not be the same as that of a more traditional lab-based research systems such as those made by Brain Products or BioSemi. Again, the purpose of this chapter is not to provide table summaries that directly compare all available mEEG systems, their characteristics, and their data quality. With that said, when considering the use of a mEEG system our recommendation would be to see if there are comparison review papers which include the system being considered (e.g., Bateson et al., 2017; Sawangjai et al., 2020) or whether there are published research papers using the potential mEEG system. For example, in our own work (Krigolson et al., 2017) we did demonstrate that the data quality of the Muse EEG system was not the same as that of the Brain Products actiChamp system that we were using for comparison, but we did demonstrate that the Muse EEG system was capable of measuring event-related brain potentials. Other systems (e.g., Emotiv Epoc, etc.) have also been shown to be able to collect publication-quality EEG data.

Event Markers and Timing Specific to ERP studies, when working with mEEG systems, the issue of event-markers must be considered. In brief, in an ERP study, one typically wants to mark the data at the exact point in time associated with stimulus onset (visual, auditory, etc.) to be able to accurately generate ERP waveforms (see Luck, 2014 for considerable detail with regard to the issue of event-marker timing jitter). With mEEG systems, the wired connection between the stimulus generation computer and the amplifier is typically not present—but it could be if one uses a wired mEEG solution. As such, in the majority of mEEG studies without a wired connection, the researcher has to live with the issue of marker timing jitter. In systems such as the Muse EEG system, this could be as much as 20 ms (see Krigolson et al., 2017) but with mEEG systems with higher sampling rate that transmit data one sample at a time, this jitter drops considerably (i.e., the CGX series of headsets). Again, when working with a mEEG system that has not been previously validated for the collection of ERP data, one would need to conduct pilot research to assess the feasibility of doing so with said system (see Chap. 10 for more detail).

Data Availability When working with mEEG systems, the issue of data availability has to be considered. In general, this is a non-issue with research mEEG systems as the expectation is that the researcher will want to access the raw EEG data for analysis purposes. However, with commercial mEEG systems, access to data is not assured. For example, when working with the Muse EEG system, it is possible to access raw EEG data through Muse's own research platform, but this is something that the researcher needs to sign up for and it is not the same as direct access to the raw data in real time. Alternatively, there are numerous ways to "hack" the Muse and stream the raw data to a mobile device or a laptop, but typically these methods require knowledge of Python and/or MATLAB and require a non-trivial knowledge of programming and device interface skills. Or, one can purchase software—Mind Monitor for instance is an application on the Apple and Google stores that connects and records raw data from a Muse EEG system. However, it does not easily afford the ability to perform ERP experiments. Other systems, such as the Emotiv systems require researchers to pay for access to the raw data. Other commercial systems are encrypted or the communication protocols are not publicly available. In short, when working with mEEG technology one has to first check whether or not the raw EEG data will be readily available with a given system.

Software One of the challenges of working with mEEG systems is the software that will be used for experimental data collection, and potentially even to run the paradigms themselves. All research mEEG systems come with core software that typically allows for the recording of raw EEG data. Further, there are a limited number of commercially available solutions that also allow a researcher to easily run ERP paradigms (e.g., PEER: www.peereeg.com). With that said, when working with commercial mEEG systems the issue of software becomes more problematic. Systems such as the Emotiv and Muse do come with research software to allow the collection of certain types of EEG data. More commonly however, the researcher needs to use MATLAB or Python and either program their own acquisition software or find existing software on the internet. Easily, the most popular choice here is the Lab Streaming Layer series of applications that allow the collection of data from not only mEEG systems, but also a wide range of other research devices.

29.4 Validation

As noted previously, it is hard to pinpoint the exact "start" of mEEG research, which makes it hard to qualify when mEEG methodology was "validated". Indeed, as noted above, Debener et al. (2012) were able to collect reliable EEG data while someone was walking outside while wearing an EEG cap with an amplifier carried with them. As such, one could argue that this study in itself is a validation of the efficacy of mEEG. More extensive validation of mEEG can be seen in work done by our research group wherein we compared the EEG data we collected from a research-grade system (a Brain Products actiChamp) with the Muse EEG headband. In our study, *Choosing Muse* (Krigolson et al., 2017), we had participants perform two standard ERP tasks while consecutively having EEG recorded with an actiChamp and a Muse EEG system. While we did not statistically compare the results of the two data sets, visual inspection of the ERPs clearly demonstrates that the waveform morphology was similar between the two systems (Fig. 29.5).

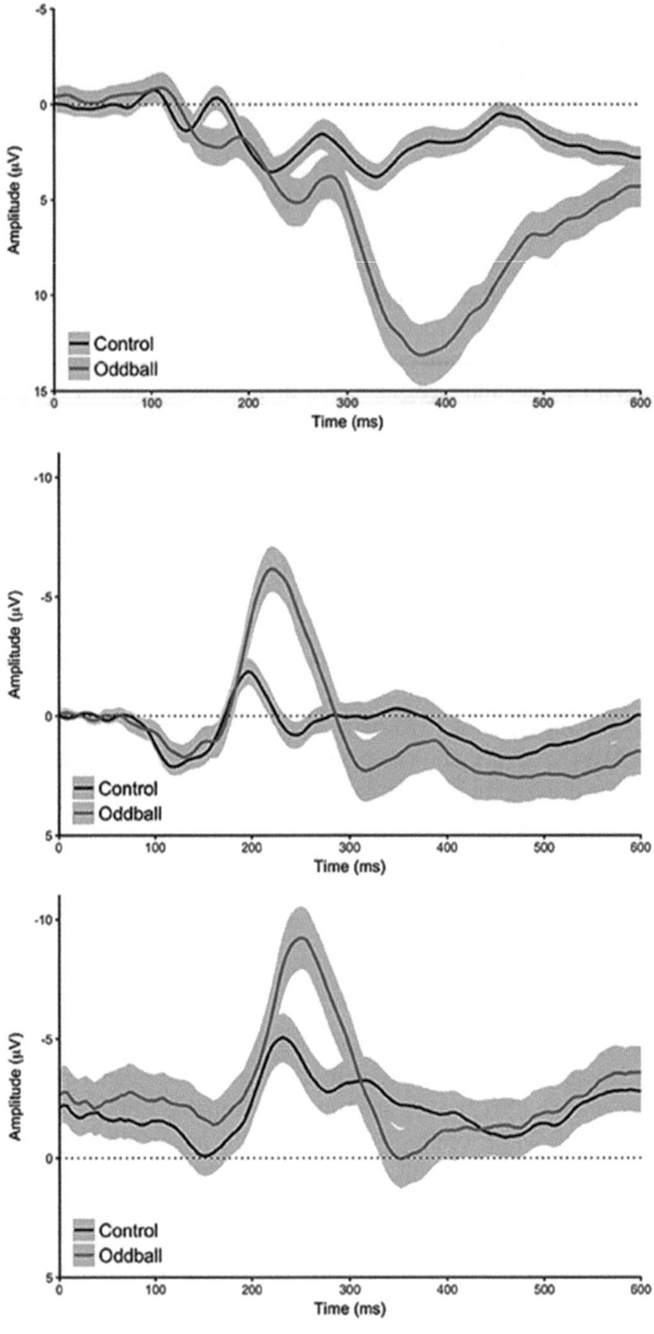

Fig. 29.5 The P300 measured at Pz (top panel) TP9 and TP10 (middle panel) with a Brain Products actiChamp and an InterAxon Muse (bottom panel). (From Choosing Muse, Krigolson et al., 2017)

To be fair, the vast majority of mEEG systems have not been validated in this manner—in a direct head-to-head comparison with a research-grade system. So, when one considers the use of a mEEG system it is warranted to search for publications with the system of choice, and ideally to request access to data from said system for evaluation. Only by examining data recorded from a mEEG system can one truly ascertain the validation of the system for research purposes.

29.5 Application

In this last section, we will review work done by our laboratory group and others to highlight some of the use cases of mEEG in research. Specifically, we will briefly review some use cases for mEEG with a focus on cognitive fatigue, clinical uses, health and exercise, and sports.

Cognitive Fatigue

Cognitive fatigue is a mental state in which the brain is "tired" resulting in increased errors and accidents while driving (Fletcher et al., 2005), flying (Goode, 2003), operating heavy machinery (Tran et al., 2020), making medical decisions (Cammu and Haentjens, 2012), and a wide range of other areas impossible to list here. Importantly, it is well established that cognitive (or mental) fatigue has a negative impact on brain performance (Borghini et al., 2014; Dinges et al., 1997; Hopstaken et al., 2015; Lal & Craig, 2002; Trejo et al., 2015).

In recent work, we demonstrated that we were able to use mEEG to measure cognitive fatigue in a hospital environment (see Krigolson et al., 2025)—a problem that is known to result in an increased incidence of medical errors (Cammu & Haentjens, 2012). More specifically, we studied cognitive fatigue in medical students who were taking part in a simulated night on call. The purpose of the simulated night on call was to provide clinical practice for the students and also to expose them to a long-duration shift without a potential negative impact on actual patients. Before each student started their shift, we used a Muse EEG headband to measure the amplitude of the P300 ERP component, an event-related potential response that has been demonstrated to be sensitive to cognitive fatigue (Kathner et al., 2014; Kato et al., 2009; Lamti et al., 2016; Schmidt et al., 2009). Our results demonstrated that P300 amplitude was reduced following the 12 h simulated night on call in 15 of the 16 medical students, a result that paralleled self-reported increases in cognitive fatigue (Fig. 29.6).

Importantly, our work here suggests that mEEG technology could be deployed in medical environments as a means to measure cognitive fatigue. This is of particular importance because it is well accepted that the current "gold standard" method of self-report by doctors and nurses is not reliable (Schmidt et al., 2009; Moller et al., 2006; Philip et al., 1997; 2003; Belz et al., 2004; Lenne et al., 1997; Baranski, 2007; Horne & Baulk, 2004; Lisper et al., 1986; Nordbakke & Sagberg, 2007).

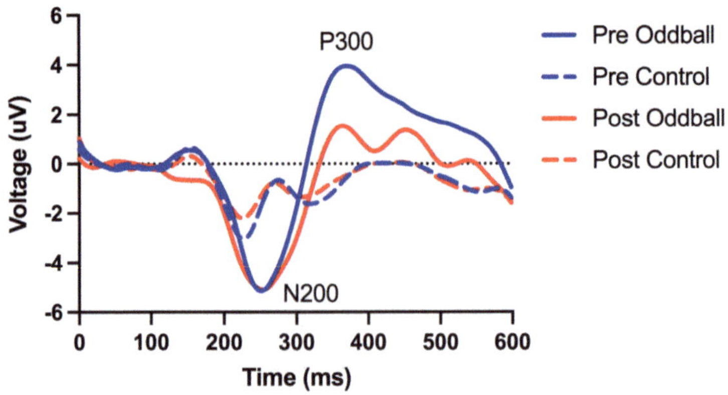

Fig. 29.6 P300 amplitude is reduced following a simulated 12 h night on call. (Figure is from Krigolson et al., 2025)

In follow-up work, we extended this initial work by assessing cognitive fatigue and lack of sleep in the general public. Here, we asked people to self-report their self-perceived level of cognitive fatigue in addition to the number of hours they slept the night before and the number of hours they had been awake. Subsequent to this, we had people perform a standard visual oddball task on an Apple iPad running PEER software while we recorded EEG data with a Muse EEG headband. Importantly, our study highlighted two key advantages of using mEEG. One, data collection (including experimental setup and take-down) took less than 7 min on average per person. Two, we were able to test a large sample size ($n = 1000$) in ecologically valid environments (e.g., shopping malls, hospitals, work environments). In line with our previous work, we were able to demonstrate that our EEG data predicted cognitive fatigue, although in this instance we utilized a multiple regression model that included numerous predictor variables. Together, the two aforementioned studies, in addition to numerous studies not included here, demonstrate mEEG as a viable tool for the assessment of cognitive fatigue.

29.6 Mild Cognitive Impairment

A potential patient-oriented clinical use for mEEG is the detection of mild cognitive impairment (MCI). MCI is defined as an objectively measured deterioration of cognitive function that does not notably impact one's ability to perform the regular, independent activities associated with daily life (Langa & Levine, 2014). Importantly, individuals with MCI have a greater risk of developing dementia (Gauthier et al., 2006; Gillis et al., 2019; Langa & Levine, 2014; Owens et al., 2020), and in many persons MCI represents the prodromal stage of Alzheimer's disease. However, the ability to accurately diagnose MCI has not kept up with the rising prevalence rate (Patnode et al., 2020). Prevalence estimates of MCI vary wildly due to different definitions and availability of different diagnostic test methods (Owens et al., 2020). For example, current prevalence estimates of MCI in the US are 22.3% for people aged 65–74, 28.5% for people aged 75–84, and 38.0% for people aged 85 and older (Rajan et al., 2021).

Our research group recently collected mEEG data using a CGX Dev Kit with four channels (https://www.cgxsystems.com/dev-kit) from 200 people who were also scored for MCI using both the Montreal Cognitive Assessment (MoCA) and the Repeatable Battery for the Assessment of Neuropsychological Status (RBANS). During mEEG data collection, we had participants play both a 1-back and 2-back nBack task. Interestingly, our results revealed that we were able to predict both MoCA and RBANS scores from the mEEG data recorded during performance of the nBack tasks using multiple regression models where the EEG features (power for each electrode at 1 Hz increments) were the predictor variables. Importantly, as with our work with cognitive fatigue, our finding here (Krigolson et al., submitted) suggests that mEEG can be deployed clinically as a screening tool—in this case for the detection of MCI.

29.7 mEEG and Health and Exercise

Physical activity confers wide-ranging benefits for mood, fatigue, and cognition, yet most neurophysiological evidence comes from tightly controlled laboratory tasks (Chang et al., 2025). Such constraints make real-world translation difficult, because context, motivation, and sensory input vary markedly outside the lab. mEEG bridges this gap by allowing brain activity to be measured before, during, and after exercise in authentic settings without tethering participants to bulky equipment.

We first tested whether a short outdoor walk would restore attention more effectively than an equivalent indoor walk, using behavioral performance and P300 amplitude as converging indices of attentional resource allocation (Boere et al., 2023a, b). Thirty healthy young adults completed two self-paced, ~15 min walks—one indoors along a quiet campus corridor, the other outdoors on a shaded wooded trail—in a counterbalanced, within-subject design. A visual oddball task was administered on an iPad while participants wore a Muse headband immediately before and after each walk. Although duration and pace were matched, only the outdoor walk improved measures of attentional capacity: reaction times to oddball targets quickened and P300 amplitude rose (Fig. 29.7), whereas both measures remained stable after the indoor walk. These results indicate that brief exposure to natural stimuli can replenish attentional reserves beyond the benefits of light aerobic movement alone. Importantly, this work also underscores mEEG's promise for aiding the development of real world interventions to promote brain health through simple lifestyle practices.

Building on this work to examine the effects of sustained exertion, we next assessed the impact of a 50 km mountain ultra marathon on executive function and attentional capacity (Boere et al., 2025a, b, c, d). Seventy-six experienced

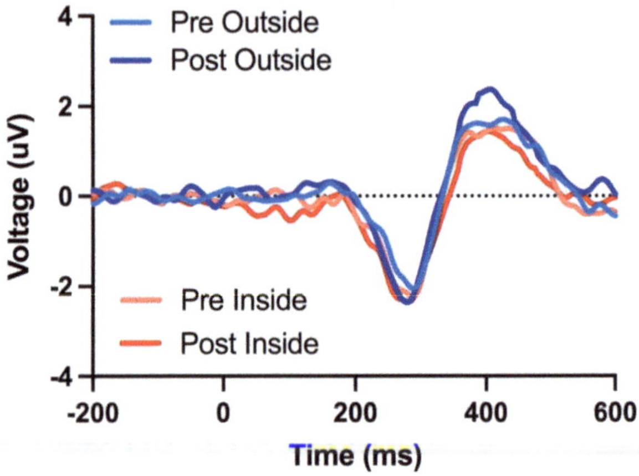

Fig. 29.7 Exercising outside enhances P300 amplitude after a 15 min walk. (From Boere et al., 2023a, b)

runners completed the same oddball task while wearing the X.on immediately pre- and postrace. Runners took an average of 7 h and 12 min to complete the race, with lower N200 and P300 amplitudes upon finishing indicating reduced inhibitory control and attentional capacity. Moreover, the magnitude of P300 attenuation correlated with stress scores on the DASS, suggesting that individual psychological state modulates vulnerability to exertional cognitive fatigue. mEEG remains sensitive and feasible even under extreme physical demand, extending its utility well beyond brief laboratory-style exercise.

29.8 mEEG in Sports

Athletes have several tools to work on their physical fitness, but less is available for improving their mental acuity. The primary explanation for this is that neuroimaging methods are typically stationary and subject to movement artifacts, confining them to laboratory environments. As a result, studies including athletes are often limited to "closed sports" such as golf, where performance is internally generated, movement is limited, and the environment can be heavily controlled (Park et al., 2015). The emergence of mobile EEG (mEEG) provides the means to measure cognition on the field or in the gym, with several avenues to improve athletic performance.

One fruitful approach to employing mEEG in athletics is the prediction of performance. For example, Pluta et al. (2018) had baseball players complete a baseline mEEG recording before batting practice, and their performance was judged by three expert coaches based on their pitch recognition and swing form, power, and contact. They found that all participants with higher pre-performance frontal beta power showed worse performance in all metrics (Fig. 29.8). Indeed, this is in line with evidence from several other performance applications, which all show that increased beta power both predicts and follows the commission of motor skill errors (Kilavik et al., 2013; Palucci Vieira et al., 2022; Ruiz et al., 2011). Given that frontal

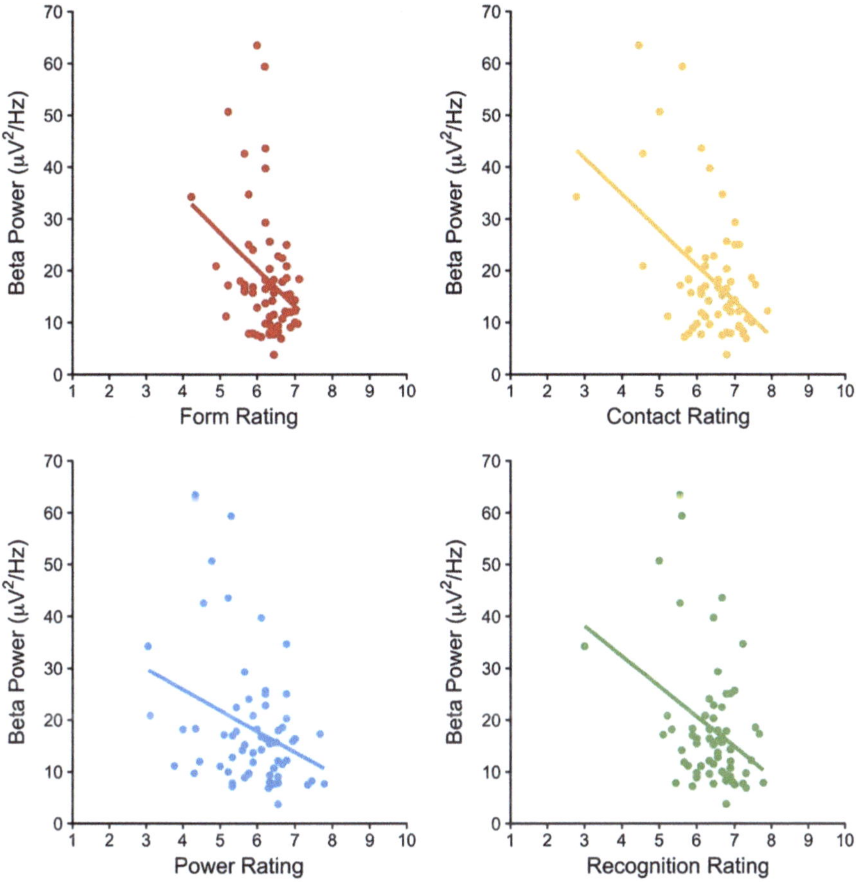

Fig. 29.8 Decreased EEG beta power predicts baseball batting performance. (From Pluta et al., 2018)

beta is thought to closely reflect our motor systems implementing external error information (Engel & Fries, 2010; Mustile et al., 2021), it is an enticing target for a better understanding of athlete performance. However, this work is largely based on pre-motor preparation and cannot reflect an athlete's dynamic motor activity.

A complete understanding of naturalistic sports requires mEEG measurement *during* performance. In one innovative example, Carey et al. (2024) had golfers putt from different positions while mEEG data were recorded and synchronized to the onset of the putter contacting the ball. They found that golfers exhibited increased theta power and decreased readiness potentials in the 1–2 s before unsuccessful putts, suggesting a disruption in motor plan implementation and updating during performance. Recording mEEG during movement still resulted in considerable data loss, not only due to limb movements but eye movements as well. However, it is now possible to reduce the impact of these artifacts with eye tracking or kinematics, which can be used to model movement and improve mEEG data quality (Grasso-Cladera et al., 2024; Ladouce et al., 2022).

Alternatively, mEEG can be used to study an athlete's overall cognitive state to incorporate non-physical performance metrics. Indeed, it's becoming increasingly common for athletes to experience "burnout"—severe mental exhaustion in response to frequent, fatiguing stress (Perlman & Hartman, 2016), which in turn causes decreased sports performance as well as general decreases in cognitive ability (Deligkaris et al., 2014; Dubuc-Charbonneau & Durand-Bush, 2015; Gustafsson et al., 2017). Using mEEG, athletes and their coaches can reliably measure the impact of these deficits, as athletes experiencing burnout exhibit reduced theta activity (Hammerstrom et al., 2023).

The advent of mEEG and its widespread use clearly has important applications for sports, and by extension athletes. One can imagine regular mEEG recordings, both before and after sports execution, to examine cognitive factors affecting motor performance. Indeed, targeting EEG components with neurofeedback can improve performance (Cheng et al., 2015; Wu et al., 2024). Extending this, mEEG data provides a means to understand how other cognitive factors, such as burnout, are affecting athletes over time, and use this information to intervene.

29.9 Conclusions

mEEG provides a valid methodology for researchers who want to collect EEG data in the "real world". Indeed, the myriad of research studies that utilize mEEG has grown exponentially in the past ten years. As more mEEG systems become available, one can speculate that this trend will continue because at the end of the day, researchers ideally want to study cognitive phenomena where they occur as opposed to in a laboratory.

mEEG Tips

We will end by providing ten tips to consider when conducting mEEG research.

1. Always use a research-validated mEEG system (unless the point of your research is to validate the system).
2. When choosing a system, pick a system that has electrodes where you want them (or as close as possible).
3. When choosing a system, consider the pros and cons of wet versus dry systems relative to your use case.
4. When choosing a system, pick a system that has published research measuring the EEG/ERP phenomena that you are interested in.
5. When choosing a system, ensure that there is a software solution in place that is easy to use to conduct the research you want to conduct.
6. Ensure that you are prepared to analyze the output data, most mEEG software does not give you the final result you want—you will still have to preprocess the data.

7. Be aware that the preprocessing of mEEG data comes with limitations, some of the more complex analysis techniques are not available to you.
8. Be aware that most mEEG systems are "disposable"—they will wear out considerably faster than more expensive lab-based systems.
9. Think through the practical considerations of doing research "in the field". It is vastly different than doing research in a lab.
10. mEEG systems are not a replacement for lab-based systems. If you are doing basic scientific research, start with a lab-based system before you go "mobile".

References

Baranski, J. V. (2007). Fatigue, sleep loss, and confidence in judgment. Journal of Experimental Psychology: *Applied, 13*(4), 182–196. https://doi.org/10.1037/1076-898X.13.4.182

Bateson, A. D., Baseler, H. A., Paulson, K. S., Ahmed, F., & Asghar, A. U. (2017). Categorisation of mobile EEG: A researcher's perspective. *Biomed research international, 2017*(1), 5496196. https://doi.org/10.1155/2017/5496196

Belz, S. M., Robinson, G. S., & Casali, J. G. (2004). Temporal separation and self-rating of alertness as indicators of driver fatigue in commercial motor vehicle operators. *Human Factors, 46*(1), 154–169. https://doi.org/10.1518/hfes.46.1.154.30393

Boere, K., Lloyd, K., Binstead, G., & Krigolson, O. E. (2023a). Exercising is good for the brain but exercising outside is potentially better. *Scientific Reports, 13*(1), 877. https://doi.org/10.1038/s41598-022-26093-2

Boere, K., Parsons, E., Binsted, G., & Krigolson, O. E. (2023b). How low can you go? Measuring human event-related brain potentials from a two-channel EEG system. *International Journal of Psychophysiology, 187*, 20–26. https://doi.org/10.1016/j.ijpsycho.2023.02.005

Boere, K., Copithorne, F., & Krigolson, O. E. (2025a). The impact of a two-hour endurance run on brain activity monitored over 24 h. *Exp Brain Res, 243*, 101. https://doi.org/10.1007/s00221-025-07056-1

Boere, K., Kremple, R., Walsh, E., Li, H., McLaughlin, L., Krigolson, O., & Blomkvist, A. (2025b). Task-evoked EEG reveals neural processing differences in aphantasia. *Scientific Reports*. https://doi.org/10.21203/rs.3.rs-6675918/v1

Boere, K., Young, N., Dauphinee, R., Copithorne, F., Kirby, B.S., Heath, M., & Krigolson, O.E. (2025c). Ultramarathon racing modulates task-related neural activity and alters behavioral indices of executive function.

Boere, K., Young, N., Dauphinee, R., Copithorne, F., Kirby, B.S., Krigolson, O. E., & Stellingwerff, T. (2025d). Working memory is impaired following a marathon race and associated with low energy availability in females.

Borghini, G., Astolfi, L., Vecchiato, G., Mattia, D., & Babiloni, F. (2014). Measuring neurophysiological signals in aircraft pilots and car drivers for the assessment of mental workload, fatigue and drowsiness. *Neuroscience & Biobehavioral Reviews, 44*, 58–75. https://doi.org/10.1016/j.neubiorev.2012.10.003

Cammu, H., & Haentjens, P. (2012). Perceptions of fatigue–and perceived consequences–among Flemish obstetricians-gynaecologists: A survey. *The European Journal of Contraception & Reproductive Health Care, 17*(4), 314–320. https://doi.org/10.3109/13625187.2012.672664

Carey, L. M., Alexandrou, G., Ladouce, S., Kourtis, D., Berchicci, M., Hunter, A. M., & Donaldson, D. I. (2024). Commit to your putting stroke: Exploring the impact of quiet eye duration and neural activity on golf putting performance. *Frontiers in Psychology, 15*, 1424242. https://doi.org/10.3389/FPSYG.2024.1424242/BIBTEX

Chang, Y. K., Ren, F. F., Li, R. H., Ai, J. Y., Kao, S. C., & Etnier, J. L. (2025). Effects of acute exercise on cognitive function: A meta-review of 30 systematic reviews with meta-analyses. *Psychological bulletin, 151*(2), 240–259. https://doi.org/10.1037/bul0000460

Cheng, M. Y., Huang, C. J., Chang, Y. K., Koester, D., Schack, T., & Hung, T. M. (2015). Sensorimotor rhythm neurofeedback enhances golf putting performance. *Journal of Sport and Exercise Psychology, 37*(6), 626–636. https://doi.org/10.1123/JSEP.2015-0166

Debener, S., Minow, F., Emkes, R., Gandras, K., & de Vos, M. (2012). How about taking a low-cost, small, and wireless EEG for a walk? *Psychophysiology, 49*(11), 1617–1621. https://doi.org/10.1111/j.1469-8986.2012.01471.x

Deligkaris, P., Panagopoulou, E., Montgomery, A. J., & Masoura, E. (2014). Job burnout and cognitive functioning: A systematic review. *Work & Stress, 28*(2), 107–123.

Dinges, D. F., Pack, F., Williams, K., Gillen, K. A., Powell, J. W., Ott, G. E., et al. (1997). Cumulative sleepiness, mood disturbance, and psychomotor vigilance performance decrements during a week of sleep restricted to 4–5 hours per night. *Sleep, 20*(4), 267–277.

Dubuc-Charbonneau, N., & Durand-Bush, N. (2015). Exploring burnout in junior elite athletes: A qualitative investigation. *The Sport Psychologist, 29*(3), 266–278. https://doi.org/10.1123/JCSP.2014-0036

Engel, A. K., & Fries, P. (2010). Beta-band oscillations—Signalling the status quo? *Current Opinion in Neurobiology, 20*(2), 156–165. https://doi.org/10.1016/j.conb.2010.02.015

Fickling, S. D., Bollinger, F. H., Gurm, S., Pawlowski, G., Liu, C. C., Hajra, S. G., Song, X., & D'Arcy, R. C. N. (2020). Distant sensor prediction of event-related potentials. *IEEE Transactions on Biomedical Engineering, 67*(10), 2916–2924. https://doi.org/10.1109/TBME.2020.2973617

Fletcher, A., McCulloch, K., Baulk, S. D., & Dawson, D. (2005). Countermeasures to driver fatigue: A review of public awareness campaigns and legal approaches. *Australian and New Zealand Journal of Public Health, 29*(5), 471–476. https://doi.org/10.1111/j.1467-842X.2005.tb00229.x

Gauthier, S., Reisberg, B., Zaudig, M., Petersen, R. C., Ritchie, K., Broich, K., et al. (2006). Mild cognitive impairment. *The Lancet, 367*(9518), 1262–1270. https://doi.org/10.1016/S0140-6736(06)68542-5

Gillis, C., Mirzaei, F., Potashman, M., Ikram, M. A., & Maserejian, N. (2019). The incidence of mild cognitive impairment: A systematic review and data synthesis. *Alzheimer's & dementia: diagnosis, assessment & disease monitoring, 11*, 248–256. https://doi.org/10.1016/j.dadm.2019.01.004

Goode, J. H. (2003). Are pilots at risk of accidents due to fatigue? *Journal of safety research, 34*(3), 309–313. https://doi.org/10.1016/s0022-4375(03)00033-1

Grasso-Cladera, A., Bremer, M., Ladouce, S., & Parada, F. (2024). A systematic review of mobile brain/body imaging studies using the P300 event-related potentials to investigate cognition beyond the laboratory. *Cognitive, Affective and Behavioral Neuroscience, 24*(4), 631–659. https://doi.org/10.3758/S13415-024-01190-Z/METRICS

Gustafsson, H., DeFreese, J. D., & Madigan, D. J. (2017). Burnout in athletes: A systematic review and critique. *Sports Medicine, 47*(9), 1737–1750. https://doi.org/10.1007/s40279-017-0717-8

Hammerstrom, M. R., Ferguson, T., Pepler, H., Pluta, A., Binsted, G., & Krigolson, O. (2023). Using neural signals to investigate athlete burnout. *Journal for Sports Neuroscience, 1*(2). https://nsuworks.nova.edu/neurosports/vol1/iss2/10

Hashemi, A., Pino, L. J., Moffat, G., Mathewson, K. E., Aimone, C., Bennett, P. J., Sekuler, A. B., & Chau, T. (2016). Characterizing population EEG dynamics throughout adulthood. *eNeuro, 3*(6), ENEURO.0275-16.2016. https://doi.org/10.1523/ENEURO.0275-16.2016

Hopstaken, J. F., Van Der Linden, D., Bakker, A. B., & Kompier, M. A. (2015). A multifaceted investigation of the link between mental fatigue and task disengagement. *Psychophysiology, 52*(3), 305–315. https://doi.org/10.1111/psyp.12339

Horne, J. A., Baulk, S. D. (2004). Awareness of sleepiness when driving. *Psychophysiology, 41*(1), 161–165. https://doi.org/10.1046/j.1469-8986.2003.00130.x

Ioannidis, J. P. A. (2005). Why most published research findings are false. *PLoS Med, 2*(8), e124. https://doi.org/10.1371/journal.pmed.0020124

Käthner, I., Wriessnegger, S. C., Müller-Putz, G. R., Kübler, A., & Halder, S. (2014). Effects of mental workload and fatigue on the P300, alpha and theta band power during operation of an ERP (P300) brain–computer interface. *Biological psychology, 102*, 118–129. https://doi.org/10.1016/j.biopsycho.2014.07.014

Kato, Y., Endo, H., & Kizuka, T. (2009). Mental fatigue and impaired response processes: event-related brain potentials in a Go/NoGo task. *International Journal of Psychophysiology, 72*(2), 204–211. https://doi.org/10.1016/j.ijpsycho.2008.12.008

Katona, J., Farkas, I., Ujbányi, T., Dukan, P., & Kovári, A. (2014). Evaluation of the NeuroSky MindFlex EEG headset brain waves data. In *2014 IEEE 12th international symposium on Applied Machine Intelligence and Informatics (SAMI)* (pp. 91–94). IEEE. https://doi.org/10.1109/SAMI.2014.6822382

Kilavik, B. E., Zaepffel, M., Brovelli, A., MacKay, W. A., & Riehle, A. (2013). The ups and downs of beta oscillations in sensorimotor cortex. *Experimental Neurology, 245*, 15–26. https://doi.org/10.1016/j.expneurol.2012.09.014

Kotowski, K., Stapor, K., Leski, J., & Kotas, M. (2018). Validation of Emotiv EPOC+ for extracting ERP correlates of emotional face processing. *Biocybernetics and Biomedical Engineering, 38*(4), 773–781. https://doi.org/10.1016/j.bbe.2018.06.006

Kovacevic, N., Ritter, P., Tays, W., Moreno, S., & McIntosh, A. R. (2015). 'My Virtual Dream': Collective neurofeedback in an immersive art environment. *PLOS ONE, 10*(7), e0130129. https://doi.org/10.1371/journal.pone.0130129

Krigolson, O. E., Williams, C. C., Norton, A., Hassall, C. D., & Colino, F. L. (2017). Choosing MUSE: Validation of a low-cost, portable EEG system for ERP research. *Frontiers in Neuroscience, 11*, 243179. https://doi.org/10.3389/fnins.2017.00109

Krigolson, O. E., Hammerstrom, M. R., Abimbola, W., Trska, R., Wright, B. W., Hecker, K. G., & Binsted, G. (2021). Using muse: Rapid mobile assessment of brain performance. *Frontiers in Neuroscience, 15*, 634147. https://doi.org/10.3389/fnins.2021.634147

Krigolson, O. E., Howse, H., Hammerstrom, M. R., Walzak, A., Wright, B., & Hecker, K. G. (2025). Using EEG to assess cognitive fatigue in real time: A medical simulation study. *Medical Science Educator*, 1–9. https://doi.org/10.1007/s40670-025-02421-9

Ladouce, S., Mustile, M., Ietswaart, M., & Dehais, F. (2022). Capturing cognitive events embedded in the real world using mobile electroencephalography and eye-tracking. *Journal of Cognitive Neuroscience, 34*(12), 2237–2255. https://doi.org/10.1162/JOCN_A_01903

Lal, S. K., & Craig, A. (2002). Driver fatigue: Electroencephalography and psychological assessment. *Psychophysiology, 39*(3), 313–321. https://doi.org/10.1017/s0048577201393095

Lamti, H. A., Gorce, P., Ben Khelifa, M. M., & Alimi, A. M. (2016). When mental fatigue maybe characterized by Event Related Potential (P300) during virtual wheelchair navigation. *Computer methods in biomechanics and biomedical engineering, 19*(16), 1749–1759. https://doi.org/10.1080/10255842.2016.1183198

Langa, K. M., & Levine, D. A. (2014). The diagnosis and management of mild cognitive impairment: A clinical review. *JAMA, 312*(23), 2551–2561. https://doi.org/10.1001/jama.2014.13806

Lenne, M. G., Triggs, T. J., Redman, J. R. (1997). Time of day variations in driving performance. *Accident Analysis & Prevention, 29*(4), 431–437. https://doi.org/10.1016/S0001-4575(97)00022-5

Li, J., Li, H., Wang, H., Umer, W., Fu, H., & Xing, X. (2019). Evaluating the impact of mental fatigue on construction equipment operators' ability to detect hazards using wearable eye-tracking technology. *Automation in construction, 105*, 102835. https://doi.org/10.1016/j.autcon.2019.102835

Lin, C. T., Ko, L. W., Chang, C. J., Chiou, J. C. (2009). Wearable and wireless brain-computer interface and its applications. In D. D. Schmorrow, I. V. Estabrooke, & M. Grootjen (Eds.), Foundations of Augmented Cognition. Neuroergonomics and Operational Neuroscience (LNAI 5638, pp. 741–748). Springer. https://doi.org/10.1007/978-3-642-02812-0_84 [KB1]

Lisper, H. O., Laurell, H., & van Loon, J. (1986). Relation between time to falling asleep behind the wheel on a closed track and changes in subsidiary reaction time during prolonged driving on a motorway. *Ergonomics, 29*(3), 445–453. https://doi.org/10.1080/00140138608968278

Luck, S. J. (2014). *An introduction to the event-related potential technique* (2nd ed.). MIT Press.

Mercado-Aguirre, M., Gutiérrez-Ruiz, K., & Contreras-Ortiz, S. H. (2019). Acquisition and analysis of cognitive evoked potentials using an Emotiv headset for ADHD evaluation in children. In *2019 XXII Symposium on Image, Signal Processing and Artificial Vision (STSIVA)* (pp. 1–5). IEEE. https://doi.org/10.1109/STSIVA.2019.8730225

Moller, H. J., Kayumov, L., Bulmash, E. L., Nhan, J., & Shapiro, C. M. (2006). Simulator performance, microsleep episodes, and subjective sleepiness: Normative data using convergent methodologies to assess driver drowsiness. *Journal of Psychosomatic Research, 61*(3), 335–342. https://doi.org/10.1016/j.jpsychores.2006.04.007

Mustile, M., Kourtis, D., Ladouce, S., Learmonth, G., Edwards, M. G., Donaldson, D. I., & Ietswaart, M. (2021). Mobile EEG reveals functionally dissociable dynamic processes supporting real-world ambulatory obstacle avoidance: Evidence for early proactive control. *European Journal of Neuroscience, 54*(12), 8106–8119. https://doi.org/10.1111/ejn.15120

Nordbakke, S., Sagberg, F. (2007). Sleepy at the wheel: Knowledge, symptoms and behaviour among car drivers. Transportation Research Part F: *Traffic Psychology and Behaviour, 10*(1), 1–10. https://doi.org/10.1016/j.trf.2006.03.003 [KB2].

Owens, D. K., Davidson, K. W., Krist, A. H., Barry, M. J., Cabana, M., et al. (2020). Screening for cognitive impairment in older adults: US Preventive Services Task Force recommendation statement. *JAMA, 323*(8), 757–763. https://doi.org/10.1001/jama.2020.0435

Palucci Vieira, L. H., Lüttke, C. S., & Frazer, A. K. (2022). Beta oscillations and motor performance: A systematic review and meta-analysis. *Neuroscience & Biobehavioral Reviews, 134*, 104570. https://doi.org/10.1016/j.neubiorev.2021.12.034

Park, J. L., Fairweather, M. M., & Donaldson, D. I. (2015). Making the case for mobile cognition: EEG and sports performance. *Neuroscience & Biobehavioral Reviews, 52*, 117–130. https://doi.org/10.1016/J.NEUBIOREV.2015.02.014

Patnode, C. D., Perdue, L. A., Rossom, R. C., Rushkin, M. C., Redmond, N., Thomas, R. G., & Lin, J. S. (2020). Screening for cognitive impairment in older adults: Updated evidence report and systematic review for the US Preventive Services Task Force. *JAMA, 323*(8), 764–785. https://doi.org/10.1001/jama.2019.22258

Perlman, D., & Hartman, M. E. (2016). Burnout in elite athletes: A systematic review. *International Journal of Sports Physiology and Performance, 11*(7), 904–907. https://doi.org/10.1123/ijspp.2015-0698

Philip, P., Ghorayeb, I., Leger, D., Menny, J. C., Bioulac, B., Dabadie, P., Guilleminault, C. (1997). Objective measurement of sleepiness in summer vacation long-distance drivers. *Electroencephalography and Clinical Neurophysiology, 102*(5), 383–389. https://doi.org/10.1016/S0921-884X(96)96511-X

Philip, P., Sagaspe, P., Taillard, J., Moore, N., Guilleminault, C., Sanchez-Ortuno, M., Åkerstedt, T., Bioulac, B. (2003). Fatigue, sleep restriction, and performance in automobile drivers: A controlled study in a natural environment. *Sleep, 26*(3), 277–284. https://doi.org/10.1093/sleep/26.3.277 [KB3].

Pluta, A., Williams, C. C., Binsted, G., Hecker, K. G., & Krigolson, O. E. (2018). Chasing the zone: Reduced beta power predicts baseball batting performance. *Neuroscience Letters, 686*, 150–154. https://doi.org/10.1016/j.neulet.2018.09.004

Qiu, J. M., Ramon, C., Schimpf, P. H., Haueisen, J., & McEwan, A. (2019). Assessing feedback response with a wearable, consumer-grade EEG system. *Frontiers in Human Neuroscience, 13*, 258. https://doi.org/10.3389/fnhum.2019.00258

Rajan, K. B., Weuve, J., Barnes, L. L., McAninch, E. A., Wilson, R. S., & Evans, D. A. (2021). Population estimate of people with clinical AD and mild cognitive impairment in the United

States (2020–2060). *Alzheimer's & Dementia, 17*(12), 1966–1975. https://doi.org/10.1002/alz.12362

Ruiz, M. H., Strübing, F., Hummel, F. C., & Gerloff, C. (2011). Functional role of interhemispheric synchronization in movement control. *Current Biology, 21*(22), 1973–1977. https://doi.org/10.1016/j.cub.2011.10.036

Sawangjai, P., Hompoonsup, S., Leelaarporn, P., Kongwudhikunakorn, S., & Wilaiprasitporn, T. (2020). Consumer grade EEG measuring sensors as research tools: A review. *IEEE Sensors Journal, 20*(8), 3996–4024. https://doi.org/10.1109/JSEN.2019.2962874

Scanlon, J. E. M., Sieben, A. J., Holyk, K. R., & Mathewson, K. E. (2017). Your brain on bikes: P3, MMN/N2b, and baseline noise while pedaling a stationary bike. *Psychophysiology, 54*(6), 927–937. https://doi.org/10.1111/psyp.12850

Schmidt, E. A., Schrauf, M., Simon, M., Fritzsche, M., Buchner, A., & Kincses, W. E. (2009). Drivers' misjudgement of vigilance state during prolonged monotonous daytime driving. *Accident Analysis & Prevention, 41*(5), 1087–1093.

Tipparaju, V. V., Mallires, K. R., Wang, D., Tsow, F., & Xian, X. (2021). Mitigation of data packet loss in bluetooth low energy-based wearable healthcare ecosystem. *Biosensors, 11*(10), 350. https://doi.org/10.3390/bios11100350

Tran, Y., Craig, A., Craig, R., Chai, R., Nguyen, H. (2020). The influence of mental fatigue on brain activity: Evidence from a systematic review with meta-analyses. *Psychophysiology, 57*(5), e13554. https://doi.org/10.1111/psyp.13554

Trejo, L. J., Kubitz, K., Rosipal, R., Kochavi, R. L., Montgomery, L. D. (2015). EEG-based estimation and classification of mental fatigue. *Psychology, 6*(5), 572–589. https://doi.org/10.4236/psych.2015.65055

Wu, J. H., Chueh, T. Y., Yu, C. L., Wang, K. P., Kao, S. C., Gentili, R. J., Hatfield, B. D., & Hung, T. M. (2024). Effect of a single session of sensorimotor rhythm neurofeedback training on the putting performance of professional golfers. *Scandinavian Journal of Medicine and Science in Sports, 34*(1), e14540. https://doi.org/10.1111/SMS.14540;PAGE:STRING:ARTICLE/CHAPTER

Chapter 30
Understanding the Creative Brain in Action

Aime J. Aguilar-Herrera, Maxine Annel Pacheco-Ramírez, Yoshua E. Lima-Carmona, Lianne Sanchez-Rodriguez, and Jose L. Contreras-Vidal

Abstract Understanding the social and creative brain in real-world contexts is a key challenge in neuroscience and crucial for revealing how art shapes cognition, emotion, and human connection. Artistic environments provide ideal conditions for studying neuroaesthetics—how the brain responds to and is transformed by art. Mobile electroencephalography (EEG) and mobile brain-body imaging (MoBI) technologies enable researchers to collect neural, motion, and physiological data during natural artistic activities such as dance, music, acting, and art appreciation. This approach supports the development of arts-based interventions that promote brain health and wellbeing while offering insights into neural mechanisms of creativity and social interaction.

This chapter presents a structured MoBI framework for studying the creative brain in action, emphasizing multidisciplinary collaboration, cultural awareness, and integration of technologies like brain-computer interfaces (BCI), adaptive noise canceling, and generative AI (ANC). A multimodal pipeline for data denoising and analysis is introduced for hyperscanning and interactive public environments such as theaters and museums. The chapter discusses methodological and ethical challenges—data interoperability, logistics, and balancing scientific rigor with artistic expression—while providing a roadmap for applying MoBI in art-science collaborations that expand neuroscience beyond the lab.

Keywords Mobile EEG · Mobile brain-body imaging (MoBI) · Neuroaesthetics · Creativity · Brain-computer interface (BCI) · Generative AI · Hyperscanning · Data processing · Team science

---

A. J. Aguilar-Herrera · M. A. Pacheco-Ramírez · Y. E. Lima-Carmona · L. Sanchez-Rodriguez · J. L. Contreras-Vidal (✉)
Department of Electrical and Computer Engineering, University of Houston, Houston, TX, USA
e-mail: jlcontreras-vidal@uh.edu

T. Warbrick (ed.), *The EEG Handbook*,
https://doi.org/10.1007/978-3-032-20450-9_30

30.1 Introduction

For much of its history, neuroscience has pursued the study of the brain in controlled isolation—stripped of the body, divorced from the environment, and severed from the sensorial, emotional, and social textures of everyday life. In the name of experimental rigor, participants have been instructed to lie motionless, press buttons, and suppress spontaneous behavior, lest neural signals be contaminated by the noise of movement. These methodological constraints, while invaluable for foundational insights into perception, memory, and attention, have produced a science of cognition largely constrained to the artificial stillness of the laboratory completely devoid of the environmental and social context. The result is a body of knowledge rich in precision but often impoverished in ecological relevance—our understanding of the brain has been built, paradoxically, on its disconnection from the world it was shaped to navigate (Sanctis et al., 2025).

In recent years, virtual reality (VR) and other immersive simulation environments have been proposed as partial remedies to this ecological gap. While VR enables the recreation of sensory-rich environments under experimental control, it often fails to replicate the full sensorimotor, emotional, and existential dimensions of real-world experience. For example, VR prototypes can evoke engagement and positive affect but are frequently reported to lack ecological validity and sensory fidelity compared to their physical counterparts (Pizzolante et al., 2024). Likewise, studies directly comparing VR and real environments with identical spatial designs demonstrate that VR induces heightened arousal and altered physiological signatures rather than the comfort and preference typically observed in real spaces (Kobayashi et al., 2025). Similarly, multisensory simulations reveal that mismatched cues—particularly olfactory stimuli—can disrupt immersion and elicit negative affect such as disgust, underscoring the fragility of emotional authenticity in VR (Alshaer, 2025).

By contrast, Mobile Brain/Body Imaging (MoBI) technologies offer a complementary and, in many ways, paradigm-shifting alternative. By placing EEG systems onto free-behaving individuals, like musicians in performance, children wandering through an exhibit, and actors mid-monologue, we enter a new era of real-world brain research that is not only methodologically novel but also socially resonant (King, 2021; Jungnickel et al., 2019).

We offer a roadmap for employing mobile EEG to investigate cognition in both artistic and everyday environments. Our exploration begins with a survey of the rapidly expanding landscape of MoBI applications in the arts—where real-time neuroimaging has been used to explore the neural signatures of creativity, attention, and flow during live performance and collaborative artistic processes (Cruz-Garza et al., 2019; Contreras-Vidal et al., 2019a). These studies highlight how mobile neuroimaging provides access to cognitive states that are difficult, if not impossible, to evoke in laboratory settings—such as creative improvisation or real-time audience engagement.

We then introduce the MoBI+ framework, a structured, interdisciplinary approach to conducting research in naturalistic settings. This framework is intentionally designed to support collaboration among artists, scientists, engineers, and artificial intelligence systems. It provides guidance across all phases of the research process: from experimental design and technical implementation to data analysis and public engagement. It is a system that values rigor without rigidity and complexity without chaos, balancing scientific precision with adaptability to real-world contexts.

The creative arts, in particular, offer a uniquely fertile ground for this work. From dancers and musicians to interactive installations and virtual reality experiences, MoBI enables the study of aesthetic and social processes as they occur in real time. These contexts reveal how the brain supports complex behaviors like improvisation, empathy, collaboration, and sensory integration (Cruz-Garza et al., 2019).

Alongside technical innovation, MoBI invites philosophical reconsideration: What is cognition when freed from the laboratory? How do the arts scaffold mental processes? Can science be participatory, collaborative, and culturally situated? These questions lie at the heart of a growing movement toward a neuroscience that is both ecologically valid and socially resonant.

Through the case studies and frameworks that follow, we demonstrate how MoBI enables the study of the brain not as a static organ, but as a dynamic system, living, moving, sensing, and creating within the world.

30.2 General Framework and System Overview

At first glance, MoBI may seem like a straightforward extension of cognitive neuroscience into the wild: equip participants with a wireless EEG cap, press record, and watch as the brain "performs" under naturalistic conditions. Yet beneath this surface-level simplicity lies a deeper conceptual and methodological evolution. Rather than replacing traditional neuroimaging approaches, MoBI builds upon them by expanding the range of environments and behaviors accessible to scientific inquiry. It offers a means of studying cognition not as a disembodied, isolated phenomenon, but as one that is dynamically embedded in the lived experience of movement, social interaction, and sensory context. In doing so, MoBI invites researchers to reconsider foundational assumptions about where and how cognition unfolds, pushing toward models that are not only neurally precise, but behaviorally and culturally situated.

Figure 30.1 illustrates the breadth and flexibility of MoBI implementations across diverse artistic and cultural contexts. A compelling example is *Meeting of Minds* (Fig. 30.1a), an interdisciplinary performance project uniting dancers, neuroengineers, and musicians. This collaboration follows a more classical MoBI approach, employing hyperscanning (simultaneous EEG recording from two performers) to examine the neural correlates of interaction. The choreography was designed to embed distinct experimental conditions within the dance, allowing us to examine how different modes of interaction affect brain activity. These conditions included eye contact, synchronized movement, physical touch, and combinations of

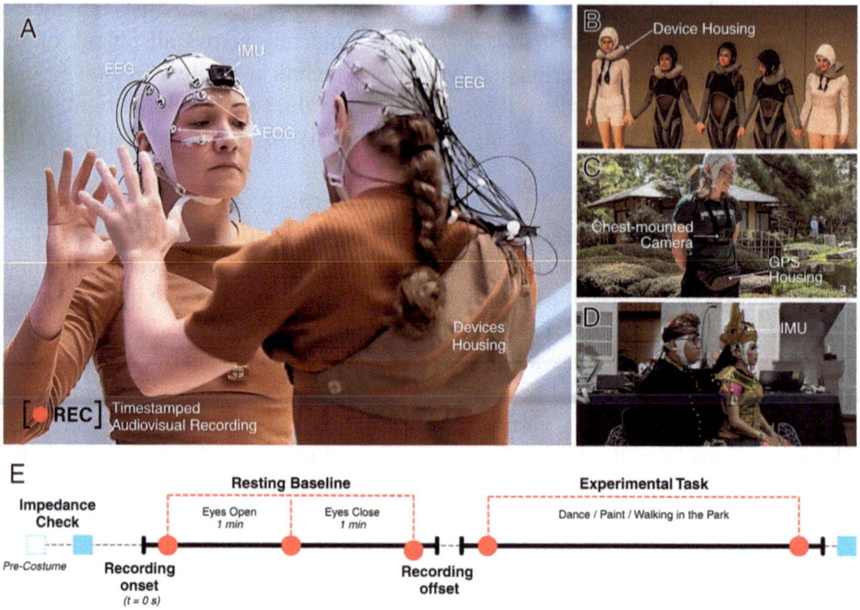

Fig. 30.1 Mobile Brain/Body Imaging (MoBI) system setup, experimental workflow in representative art-science performances. (**a**) Standard MoBI configuration with 32-ch EEG (BrainAmp DC, 28 scalp + 4 EOG, 1 kHz) and IMUs (APDM Opal, 9-axis, 128 Hz), housed in a wearable pack adapted to each performance. (**b**) The Slowest Wave performance with. (**c**) Brain On Nature project. (**d**) Balinese Gamelan, Brain, Mind and Body performance in collaboration with Udayana University and Institut Seni Indonesia (ISI) Denpasar. (**e**) Typical experimental timeline illustrating impedance check, resting baseline (eyes open/eyes closed), and the experimental task phase

all three, offering a naturalistic yet controlled framework for probing social and sensorimotor processes.

To ensure the technology is integrated seamlessly with the performance, the team collaborated with a local tailor to create a custom fanny pack that discreetly housed the EEG wireless transmitters. This allowed performers to move freely without compromising comfort, aesthetics, or data quality.

At its core, the system integrates a coordinated suite of mobile, wearable technologies, each serving a distinct role in capturing neural and behavioral activity. This general design can be adapted to meet the specific needs of a given experiment, research question, or artistic context. It also provides a baseline that we recommend for newcomers to the field, while remaining open to the integration of new technologies or additional physiological sensors as they become available.

- Wireless EEG system: A multichannel EEG cap (typically 28 channels or more) connected to wireless transmitters enables high-resolution data acquisition during full-body movement.

- Electrooculography (EOG): A 4-channel configuration around the eyes records horizontal and vertical movements, supporting effective detection and removal of ocular artifacts from EEG data.
- Inertial Measurement Units (IMUs): Head-mounted IMUs capture motion dynamics including acceleration, rotation, and magnetic orientation. Data are resampled to match EEG sampling rates (e.g., 1000 Hz) to ensure precise temporal alignment. While primarily used for head motion tracking, IMUs can also be extended to other body sites depending on study demands.
- Manual trigger device: A custom-built synchronization unit (e.g., SyncBox) provides time-stamped event markers across all recording modalities (EEG, EOG, IMU), ensuring reliable segmentation of experimental phases and behavioral events.
- Audiovisual recordings: All sessions—whether rehearsals, live performances, or public demonstrations—are captured using timestamped video. These recordings serve a dual purpose: enabling behavioral annotation and synchronization across modalities, and supporting communication with wider audiences through public engagement and storytelling.

Taken together, these components form a generalizable MoBI infrastructure: a mobile, modular laboratory that can be scaled or reconfigured to suit diverse research settings. Extensions might include additional physiological signals (e.g., EMG, ECG, respiration), integration with motion capture or eye-tracking systems, or embedding new adaptive sensors as technologies evolve. Importantly, such adaptability is not incidental; it is the product of iterative, context-sensitive design developed in close collaboration with artists, engineers, and cultural practitioners.

Figure 30.2 illustrates one specific implementation of this framework, adapted for our experimental work. The participant setup integrates a wireless EEG cap (actiCAP, Brain Products GmbH, Gilching, Germany) connected to a MOVE transmitter via IDC ribbon cable. The transmitter sends data wirelessly to a MOVE receiver, which connects to the amplifier and recording system via fiber optics. Simultaneously, IMUs (APDM Opal sensors) transmit motion data to an access point and are logged via *Motion Studio* software. A central SyncBox delivers triggers to all devices, enabling temporal alignment across modalities. Timestamped audiovisual recordings capture task-relevant cues and behavioral contexts, which are crucial for annotation and interpretation.

The system is reconfigured prior to each study to meet the demands of the task and setting. Design modifications account not only for technical requirements but also for aesthetic, cultural, and comfort considerations. In *The Slowest Wave* (Fig. 30.1b), for instance, the standard hardware was embedded in a soft neck pillow worn by dancers, allowing freedom of movement while concealing transmitters and amplifiers. In the *Brain on Nature* project (Fig. 30.1c), participants walked through a public park while equipped with a chest-mounted camera and GPS tracker, enabling synchronization of neural data with first-person video and precise geolocation. In the *Balinese Gamelan, Brain, Mind, and Body* performance (Fig. 30.1d), cultural sensitivity required spatial reconfiguration of the hardware: the traditional *galungan* crown necessitated moving the IMU from the forehead to the top of the head to avoid

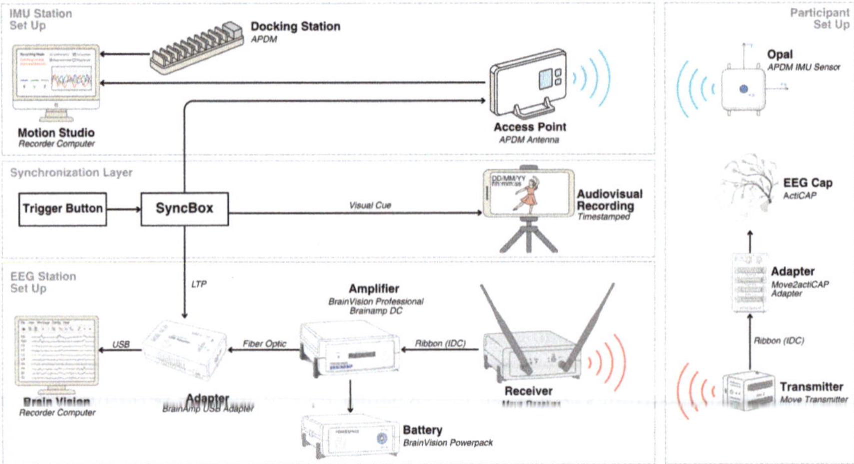

Fig. 30.2 Mobile Brain/Body Imaging (MoBI) system architecture. A typical setup integrating EEG (BrainAmp DC, Brain Products GmbH, Gilching, Germany), IMUs (APDM Opal), audiovisual recording, and a synchronization layer (SyncBox) to enable multimodal, time-aligned data collection in naturalistic environments. The system is adapted per study to align with task demands, movement constraints, and aesthetic requirements

interference. These examples highlight how the general MoBI framework adapts to different environments, whether theatrical, natural, or ritual, through context-sensitive, collaborative design.

Figure 30.1e illustrates the general timeline used in our MoBI studies. Each session begins with an impedance check to verify signal quality; in certain cases where costumes might interfere with electrode contact, an additional pre-costume impedance check is performed to establish a baseline before attire is added. Following this, the recording starts with a standardized resting baseline consisting of one minute of eyes-open and one minute of eyes-closed conditions. Participants then proceed to the experimental task, which may involve rehearsal, live performance, or another naturalistic activity. Finally, a post-performance impedance check is conducted to ensure that signal integrity was maintained throughout the session.

These examples highlight that MoBI research in naturalistic settings is not plug-and-play. It demands flexibility—technically, logistically, and culturally. Each deployment is a negotiation between form and function, between empirical rigor and embodied expression. It is a methodological choreography where tools, traditions, and technologies must move in sync. The result is a system not only capable of studying the brain in motion but also of engaging with the world in which that brain is embedded.

Challenges Along the Way

MoBI research into real-world contexts is not simply a matter of relocating laboratory tools—it is a process of reimagining how science is conceived, practiced, and shared. Unlike traditional experiments, MoBI studies are often grounded in transdisciplinary ideation, where artists, engineers, and neuroscientists must first learn to co-inhabit each other's conceptual and technical languages. The question "Where do we start?" is not rhetorical; it marks the beginning of a collaborative design process that may span weeks of co-creation, iterative rehearsals, and negotiation between aesthetic intention and scientific rigor. In this liminal phase, storyboards, improvisation, and even philosophical reflection can become as essential as technical schematics.

Logistical and financial limitations often serve as the gravitational force pulling even the most inspired MoBI visions back to earth. Rehearsal spaces—frequently shared or overbooked—may offer limited availability and time to troubleshoot, while participating artists juggle touring, teaching, or other creative commitments. Coordinating across these moving parts demands foresight and flexibility. Legal and ethical considerations add further complexity: informed consent must cover not only neural recordings but also high-resolution video, sound, and imagery that may feature performers or public figures. This requires carefully balancing visibility with privacy—ensuring that participants can proudly share their work while maintaining agency over how their likeness and identity are used in research dissemination.

Funding, likewise, must stretch far beyond equipment costs. Travel, housing, venue rental, and liability insurance can accumulate quickly. MoBI projects often span institutions and time zones, requiring stipends not only for scientists and engineers but also for dancers, musicians, designers, and technical staff. When the event is public-facing (as many MoBI studies aim to be), additional resources are needed for media outreach, community engagement, and audience facilitation. Each component carries its own timeline, budget, and risk profile.

On the technical front, limitations arise not from a lack of creativity, but from the boundaries of current tools and infrastructure. Hyperscanning studies, for instance, depend on having access to several compatible EEG systems. These systems are often in limited supply, and some may no longer be in production, making repair and replacement difficult. Wireless signals can interfere with stage electronics; ambient noise from lighting rigs or mobile devices can degrade signal quality. In these contexts, real-world setups are inherently unstable. A scalable MoBI methodology must be built not for perfection, but for resilience, through redundancy, modular design, repairable systems, and field-tested workflows.

And then, environments change. What functions in a controlled theater may break down in the unpredictability of an open-air festival. Lighting conditions may shift unexpectedly, a security guard might block a camera mid-performance, or a photographer may enter the frame during a crucial moment. Rather than resisting such variability, successful MoBI research embraces environmental fluidity. Modular rigs, portable backups, and adaptable data structures become essential. A pre-visit or dry run can reduce uncertainty, but improvisation remains a vital tool in the field.

In this context, reproducibility takes on a different shape. It is less about rigid control and more about establishing clear baselines—standardized conditions such as eyes-open/eyes-closed segments, impedance measurements, and time-synchronized multimodal data streams—and maintaining consistent file structures and metadata schemas to support traceability. The goal is not to replicate by reduction, but to identify meaningful patterns that emerge across the complex textures of real-world behavior.

Finally, there is the matter of the audience. MoBI research often unfolds in public, on stages, in galleries, or among museum visitors. In such contexts, the audience becomes part of the system: not only watching, but engaging, questioning, and interpreting. A thoughtful setup anticipates this role, incorporating signage, opportunities for dialogue, and channels for public feedback. Providing accessible explanations of the system—along with real-time visualizations of the data streams or brain activity—further enhances audience engagement, transforming the technical apparatus into a tool for storytelling and connection. The EEG cap thus becomes more than a measurement device; it becomes a symbolic interface, a point of curiosity, conversation, and shared discovery.

MoBI research demands flexibility, patience, and a willingness to collaborate across domains and vocabularies. Yet when executed well, it offers something rare: a neuroscience that moves with the world, listens to its rhythms, and speaks in more than one language. The question, then, is *how this vision translates into practice*.

30.3 Spectrum of Studies

MoBI + Arts research began in laboratory and university settings, where early studies focused on testing signal quality during natural movement and collaborative artistic exploration. As the methods and wearable designs improved, these experiments expanded into rehearsal studios, galleries, and small public demonstrations. Over time, as we became increasingly able to visualize brain activity in real time, audiences began to respond not only to the artistic performance but also to the live unfolding of cognitive and emotional dynamics. This sparked conversation, curiosity, and a sense of shared discovery—people began asking questions, reflecting on their own experiences, and requesting more opportunities to see and participate in these works.

These exchanges encouraged new collaborations with professional artists, cultural institutions, and eventually international performance companies. Each project taught us how to adapt MoBI to different environments—lighting, acoustics, costumes, movement styles, and even outdoor conditions—allowing the work to expand across theaters, museums, hospitals, nature trails, and community spaces. What began as feasibility testing gradually evolved into a reciprocal ecosystem in which MoBI both supports existing artistic practices and inspires entirely new creative forms.

Figure 30.3 summarizes this development over time. The letter markers (A–T) indicate when each project occurred, while the color coding identifies the artistic domain (e.g., dance, music, visual art, museum engagement, or nature-based experiences). The figure is not meant to categorize or rank the projects, but to illustrate

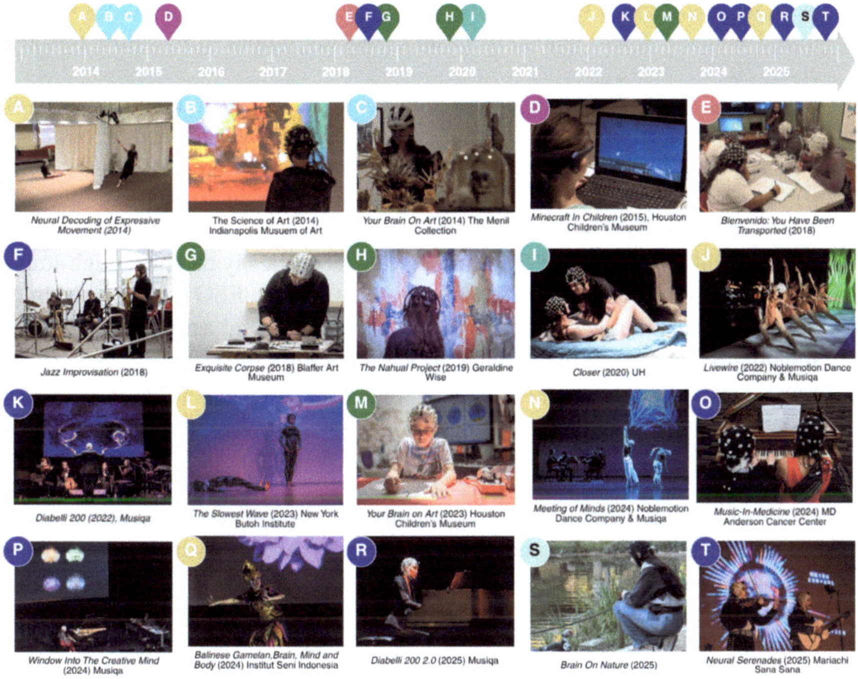

Fig. 30.3 A visual timeline illustrating the Brain + Arts collaborative projects from 2014 to 2025

how MoBI moved from academic spaces into full-scale productions and public settings, eventually reaching international venues and outdoor environments. The timeline highlights the growing role of public participation, dialogue, and co-creation between science and the arts.

The overview that follows provides additional detail about these projects. For clarity and ease of reference, the list is organized alphabetically, rather than by timeline position or color grouping. This allows the reader to explore the range of MoBI applications across contexts, audiences, and artistic modes, independent of when each project took place.

(a) *Acting:* In a study of "neuro-acting," we examined interpersonal brain synchronization among student actors of varying levels of theatrical experience across three staged performances. Drawing upon scenes selected by theater professionals for their emotional and narrative depth, the study employed hyperscanning (simultaneous EEG recordings from multiple participants) to measure neural coherence during live dramatic interaction. Shared gazes—moments of direct eye contact between performers—were associated with increased inter-brain connectivity, suggesting a neural signature for co-regulated attention and affective resonance. These findings not only offer a window into actor-actor dynamics but also suggest pathways for exploring actor-audience coupling and the cognitive infrastructure of performance itself. This work underscores the value of MoBI in extending performance studies into the neural domain, where embodiment and empathy unfold in real time (Hendry et al., 2025).

(b) *Art Appreciation:* This study explored how individuals cognitively engage with artworks in the natural setting of real-world museums. Conducted across exhibitions in the United States and Mexico, participants wore mobile EEG systems—both dry and gel-based—while freely viewing curated pieces. The aim was twofold: to capture spontaneous neural responses to visual art and to assess the feasibility of various EEG technologies in acoustically complex, high-traffic public environments. Findings confirmed that mobile EEG can reliably track patterns of attention and engagement during unstructured aesthetic experience. Importantly, the study highlighted critical trade-offs in equipment design: gel caps yielded higher data quality but required longer setup, while dry systems enhanced participant comfort and public adaptability. These design tensions are particularly salient for neuroaesthetic research seeking both ecological validity and technical rigor (Kontson et al., 2015; Herrera-Arcos et al., 2017).

(c) *Creative Writing:* In educational and outdoor environments, we investigated the neural correlates of the creative writing process—particularly how sensory input and autobiographical memory are translated into narrative expression. Participants engaged in structured writing tasks while equipped with mobile EEG systems, allowing researchers to track shifts in cognitive state during key stages such as ideation, revision, and free composition. Preliminary analyses indicate that different writing stages elicit distinct neural signatures, with increased frontal theta during generative flow and parietal desynchronization during sensory recall and emotional introspection. By studying writing as a lived, embodied practice rather than a laboratory abstraction, this work helps bridge the gap between neuroscience and literary creativity, and lays the groundwork for new pedagogical applications in cognitive education and expressive therapies (Cruz Garza et al., 2020).

(d) *Dance:* Here, we deployed our MoBI approach to investigate the neural basis of expressive, spontaneous, and choreographed movements. Early explorations in controlled settings demonstrated that expressive movement qualities could be decoded from brain activity, laying the foundation for larger-scale artistic collaborations (Cruz-Garza et al., 2014). LiveWire expanded this work into a full-length choreography, where hyperscanning revealed intra- and inter-brain dynamics as dancers performed in rehearsal and on stage (Pacheco-Ramírez et al., 2024). In Meeting of Minds, we examined how brain-to-brain coupling evolved as dancers transitioned from states of discord to collaboration, highlighting the neural basis of social interaction through movement. The Slowest Wave further explored dance as a medium for probing states of prolonged attention and altered consciousness, blending choreography with neurotechnology in an experimental performance context (Theofanopoulou et al., 2024). Most recently, our Bali project integrated Gamelan dance and music traditions, offering a non-Western lens on collective creativity and neural synchrony. Across these studies, real-time brain visualizations during performances unveiled enhanced neural connections as dancers engaged in interactive movements. Together, these findings inspire ongoing artistic collaborations, expanding the horizons of creativity while pushing the boundaries of social neuroscience and brain–computer interfaces.

(e) *Drawing:* Here, we used MoBI to capture brain activity as participants engaged in drawing tasks in naturalistic conditions. In Exquisite Corpse, artists collaborated on an improvisational drawing game, enabling us to study the neural dynamics of co-creation as their sketches unfolded collectively (Cruz-Garza et al., 2017). Across these studies, we observed how creative visual expression recruits brain networks for attention, hand–eye coordination, and imagination, offering new perspectives on the embodied nature of artistic thought.

(f) *Gustatory Experiences:* In this study, we extended MoBI methodologies to the domain of taste, investigating the neural processes underlying gustatory perception and multisensory integration during wine tasting (González-España et al., 2023). In structured tasting sessions, participants were equipped with mobile EEG systems while flavor, aroma, and contextual variables were systematically varied. These experiments revealed how sensory and contextual cues shape brain activity in real time, highlighting the complex interplay between perception, expectation, and experience. By bringing neurotechnology into the traditionally aesthetic practice of wine tasting, this work demonstrates the potential of MoBI to capture the richness of multisensory creativity in ecological settings, expanding the scope of artistic neuroscience beyond visual and performing arts.

(g) *Interactive Art:* This project highlights mobile EEG as a powerful creative medium, where real-time brain signals are transformed into dynamic forms that animate architecture and reshape perception (Todd et al., 2019). Brain activity is directly linked to the shifting movement and color of ceiling panels, allowing thought and sensation to alter the very structure of the space. In doing so, the work externalizes inner cognitive and motor states, turning them into visible and tangible transformations. The result is an immersive spatial experience where architecture becomes a living reflection of the mind, blurring the line between neural activity and artistic expression.

(h) *Music:* In the study of music, mobile EEG was implemented among jazz musicians, revealing synchronized neural activity during improvisation (Ramírez-Moreno et al., 2023). These findings served as the foundation for subsequent projects including "Diabelli 200", which integrated real-time visualizations of performers' brain activity to enrich our comprehension of the intersection of music and neuroscience. More recent projects include "A Window into the Creative Mind", which provides unique insights into the neural signatures of artistic creativity during the improvisation of variations on musical pieces, and "Music in Medicine" (Contreras-Vidal, 2025), which explores how the power of music can be harnessed to heal by influencing brain dynamics in therapeutic contexts.

(i) *Nature Appreciation:* MoBI technologies were implemented to understand the neural, physiological, and psychological mechanisms underlying human appreciation of natural environments. Participants are fitted with a lightweight mobile

EEG system with an integrated accelerometer, synchronized with a GPS sensor while walking through a natural park. Additionally, self-report assessments are implemented to track psychological factors such as mood, anxiety, and attentional states. This multimodal framework allows the understanding of how moment-to-moment environmental exposures shape cognitive-emotional states and support well-being, while demonstrating the potential of MoBI to study complex human experiences in natural environments.

(j) *Painting:* Using MoBI technology, the first painting study takes inspiration from the surrealist game Exquisite Corpse, providing new insights into the cognitive–motor processes underlying collaborative art improvisation (Nijholt, 2019). The second study, the Nahual Project, examines AI's role in creativity and well-being by capturing EEG data as an artist creates a work of art in real time. Together, these projects highlight the intersection of neuroscience, technology, and artistic expression.

(k) *Videogames:* In the following study, MoBI technology is used to capture simultaneous EEG and head-movement data from children engaging in a Minecraft video-game task at the Children's Museum of Houston. The experiment offers rich insight into how age, gender, and gaming skill level influence temporal and spectral neural responses in naturalistic, free-behavior contexts (Sujatha Ravindran et al., 2019).

(l) *Into the Artist's Mind (An 18-Month Longitudinal Study):* Using context-aware MoBI technology, we conducted an 18-month longitudinal study to investigate the creative brain in naturalistic settings. The research captured neural and behavioral data across the distinct phases of ideation, planning, prototyping, and production of an artistic installation. This approach offers a rare, continuous perspective on the dynamic interplay between cognitive processes and artistic creation over extended time scales (Contreras-Vidal et al., 2019b).

In addition, we provide a continuously updated project table that offers further detail for each study, including artistic domain, venue and location, performance and rehearsal history, audience engagement scale, research tasks, study aims, participant enrollment, hyperscanning configurations, equipment used, data-sharing status, funding sources, and media documentation. Because several projects remain ongoing or continue to tour and develop, the table is maintained as a living document and updated as new performances and datasets become available:

Access the project table: BOA Projects Outreach—Studies Review

Together, these projects demonstrate the adaptability of MoBI across artistic genres, cultural contexts, and research environments. They reflect not only the versatility of mobile neuroimaging tools but also the range of questions that emerge when neuroscience steps outside the lab and into public life.

As these studies accumulated, a clearer picture began to form: despite their diversity, successful MoBI projects shared key methodological patterns and design principles. These were not prescribed protocols, but adaptable practices developed through interdisciplinary collaboration, trial and error, and reflection in the field.

30.4 MoBI+ Framework

While the catalog above demonstrates the diversity of contexts in which MoBI can be applied, it also revealed recurring challenges—and, more importantly, consistent strategies that led to successful outcomes. Through iterative experimentation across disciplines and venues, a flexible framework has gradually emerged to support mobile EEG research in naturalistic, creative, and culturally embedded environments.

This section outlines that framework. It is not a rigid protocol, but a set of working principles designed to guide researchers, artists, and technologists through the complex process of designing and conducting MoBI studies outside the laboratory. These principles are grounded in lived experience, from negotiating the technical constraints of live performance, to navigating ethical questions in community-based settings.

Early, Interdisciplinary Collaboration

Effective MoBI studies begin with cross-disciplinary collaboration. Artists, scientists, engineers, and cultural practitioners must work together from the outset, ideally before research questions or protocols are finalized. Co-development often includes brainstorming sessions, movement rehearsals, and site visits, allowing teams to identify not only opportunities but also disciplinary constraints, production workflows, and audience expectations. For example, designing an attention-monitoring task for an actor cannot rely on inserting discrete trials into a script. It requires aligning experimental design with the narrative arc and emotional rhythm of the performance. These early interactions shape both the scientific approach and the integrity of the artistic work.

Technical Adaptation and Rehearsal

In naturalistic settings, equipment must adapt to the realities of live performance. EEG hardware is modified to accommodate costumes, headpieces, and movement, without compromising signal quality. We have implemented strategies such as internal padding for stability, sensor-safe zones on the face, and full technical rehearsals under lighting and physical conditions that mirror actual performance environments. These rehearsals are essential. They serve as pilot tests for equipment durability, signal integrity, and system synchronization, helping to identify and resolve vulnerabilities prior to data collection.

Cultural Literacy and Ethical Protocols

Because many studies occur in culturally specific or public settings, cultural awareness and ethics must be built into every stage of the research. Working with participants whose identities, traditions, or social roles carry significance requires more than institutional approval. It requires sustained dialogue, cultural literacy, and clear communication. In studies involving traditional performing arts, we have encountered rituals and garments with deep symbolic meaning. Sensor placement on the head or face, for instance, may not be appropriate without cultural consultation or adaptation. In such cases, we work closely with advisors, community leaders, or participants themselves to co-design respectful protocols.

We also adjust our visual outputs accordingly. Neurofeedback displays projected during a performance should enhance, not distract from, the artist's intent. Informed consent must be tailored to the setting, especially when working with children, public figures, or multilingual communities. We include language that clarifies how EEG data, photos, or video footage may be used, and when live data will be visualized. Participants are briefed on whether audiences can see individual signals or only group-level trends. These decisions directly impact participant comfort and public trust and help foster long-term collaborations rooted in transparency and mutual respect.

Integrating Scientific Rigor with Environmental Variability
Balancing public engagement with experimental rigor requires deliberate design. Unlike controlled laboratory settings, public performances introduce a host of uncontrolled variables: ambient noise, shifting lighting, spontaneous performer gestures, and dynamic audience interactions. Rather than treating these factors as disruptions, we incorporate them into the research framework. Behavioral markers—such as motion tracking, audio cues, or video annotation—are aligned with EEG data, while multimodal data streams provide contextual layers that help interpret neural signals. In this way, ecological validity is preserved without sacrificing data quality.

Public Engagement as Scientific Practice
Audience interaction is not incidental but central to MoBI+Arts research. Performances and demonstrations often take place in settings where the audience is physically proximate and cognitively engaged. Before a performance, audiences are introduced to the EEG system through brief explanations and accessible visualizations, tailored to the specific group, whether K–12 students, museum visitors, or general audiences. After the performance, Q&A sessions invite dialogue across diverse experience levels. These conversations frequently spark new perspectives, raise unanticipated questions, and even suggest future research directions. In these contexts, the audience becomes an active participant in the scientific process, broadening both the scope and the societal relevance of the research.

Open Data and Reproducibility
A critical but often overlooked part of the framework is data sharing and open practices. From the start, we organize our studies for reproducibility, using formats like BIDS (Brain Imaging Data Structure) and logging all sensor configurations, annotations, and metadata. In addition to sharing raw EEG data, we publish detailed guides—such as electrode placement maps tailored to specific costumes or performance types (Hendry et al., 2025; Pacheco-Ramírez et al., 2024; Theofanopoulou et al., 2024)—to reduce the entry barrier for other research teams and practitioners interested in replicating the work.

Integrating Emerging Technologies
Finally, the framework leaves room for emerging technologies such as brain-computer interfaces (BCIs), generative AI, and real-time feedback systems. These tools enable new forms of interactivity and expression, allowing performers to influence

sound, lighting, or visualizations with their neural signals. For instance, we have used real-time EEG to drive generative visuals based on attention or emotional arousal. While these integrations require significant technical coordination, they enhance both artistic and scientific outcomes when implemented thoughtfully.

Future Directions and Framework Evolution
As the field of MoBI continues to expand, this framework must evolve to support more complex, distributed, and scalable research environments. One key direction involves enabling group-based and multisite studies. Many creative and community-based activities involve collective expression—such as ensemble performance, collaborative storytelling, or large-scale participatory installations. Scaling hyperscanning beyond two or three participants requires synchronized timing across mobile systems, consistent event annotation, and hardware/software reliability under diverse conditions. Developing lightweight, replicable setups that can be deployed across different venues and cultural contexts will be essential for broader adoption.

Another area for refinement involves multimodal data integration and preprocessing. MoBI studies often involve EEG data recorded alongside motion sensors, audio, video, and physiological measures. Each of these data streams is subject to its own noise and variability in real-world conditions. Streamlining preprocessing pipelines to manage movement artifacts, synchronize streams, and support contextual labeling remains an ongoing challenge. Advances in semi-automated artifact detection, machine learning-assisted annotation, and modular toolkits will help interdisciplinary teams process and interpret data more efficiently—even those without deep EEG expertise.

Ultimately, improving this framework is not just a technical goal—it's about maintaining an adaptable, ethical, and collaborative model for research in public life. With each new study, we gain a deeper understanding of how culture, context, and logistics shape the possibilities and limits of mobile neuroimaging. By continuing to document what works, revise what doesn't, and stay responsive to diverse partners, we can build a more inclusive and sustainable foundation for MoBI research. The next section outlines the ongoing challenges and limitations we have encountered in this work—insights we hope will help future teams anticipate complexity and design more resilient and impactful studies.

30.5 Processing Multimodal MoBI Data

Preprocessing EEG data in Mobile Brain/Body Imaging (MoBI) contexts presents a distinct set of challenges compared to traditional, laboratory-based recordings. In MoBI studies, participants are not stationary but instead engaged in natural behaviors within complex and often unpredictable environments. These real-world conditions introduce a wide range of physiological and mechanical artifacts, including those from eye movements, muscle activity, posture shifts, and sensor displacement, all of which can obscure the underlying neural signals of interest.

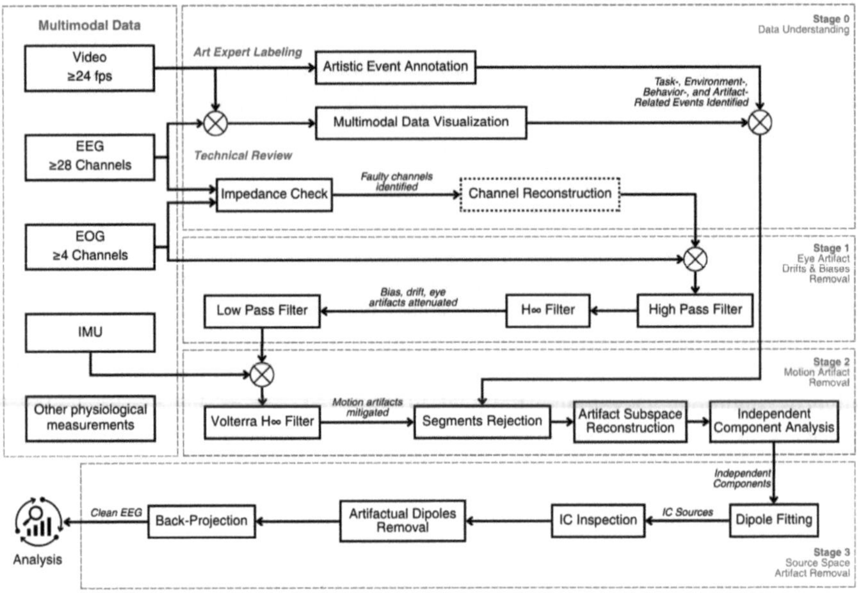

Fig. 30.4 Multimodal data processing pipeline for MoBI studies in naturalistic settings. The framework supports synchronized preprocessing and artifact removal across video, EEG, EOG, and IMU signals. It consists of four stages: Stage 0 (Data Understanding and Annotation), Stage 1 (Eye Artifact Removal), Stage 2 (Motion Artifact Removal), and Stage 3 (Source-Space Artifact Removal). This full pipeline is used for offline data cleaning and analysis. A simplified real-time implementation using H-infinity ANC is currently employed during live performances for visualization, but is not yet configured for closed-loop neurofeedback or BCI control

Figure 30.4 illustrates our multimodal processing pipeline designed to manage this complexity. The framework accommodates real-world noise, supports adaptive artifact removal, and preserves ecologically meaningful signal features. It is structured into four major stages: Stage 0: Data Understanding, Stage 1: Eye Artifact Removal, Stage 2: Motion Artifact Removal, and Stage 3: Source-Space Artifact Removal.

At Stage 0, raw data streams—including EEG ($\geq$28 channels), EOG ($\geq$4 channels), IMU data, and timestamped video ($\geq$24 fps)—are synchronized and aligned to a common sampling rate. An initial impedance check flags noisy or disconnected EEG channels, which may be reconstructed if necessary. Simultaneously, art domain experts (e.g., choreographers, composers) annotate artistic or behaviorally significant events, such as shifts in rhythm, posture, eye contact, or audience response. These annotations help to define task-relevant segments and establish ground truth for interpretation. A multimodal data visualization layer allows for real-time inspection and cross-modal verification of artifacts and cognitive events.

Stage 1 focuses on ocular artifact removal. Traditional EEG filtering pipelines are insufficient in naturalistic settings where eye movements, blinks, and gaze shifts occur continuously and unpredictably. To address this, we apply a robust adaptive noise cancellation method that treats each EEG channel as an independent subsystem (Kilicarslan et al., 2016). This technique selectively filters eye-related artifacts—such as blinks, lateral saccades, and amplitude drifts—without distorting spatial or temporal dynamics of nearby channels.

Stage 2 addresses motion artifacts, which are especially prevalent in MoBI studies involving full-body movement, expressive gestures, or walking. Unlike traditional artifact rejection techniques (e.g., Independent Component Analysis or Artifact Subspace Reconstruction), which often underperform in movement-rich contexts, we employ a Volterra-based nonlinear adaptive filtering method (Kilicarslan & Contreras Vidal, 2019). This approach models complex motion artifacts—such as those generated by gait harmonics or upper-limb gestures—and suppresses them without removing task-relevant brain activity. It is robust, real-time compatible, and generalizable to various ecological contexts, including unstructured, improvisational performances.

After filtering, data undergo segment rejection, artifact subspace reconstruction, and independent component analysis (ICA) to isolate and remove residual noise. The pipeline is designed to minimize unnecessary data loss while retaining neural signals relevant to embodied cognition and creative behavior.

Stage 3 focuses on source-space artifact removal. Cleaned EEG data are subjected to back-projection, followed by dipole fitting and independent component inspection. Components are classified and spatially localized using a combination of automated heuristics and manual validation. Artifactual dipoles—those associated with muscle noise or residual motion—are removed prior to analysis. The resulting dataset represents a high-fidelity, source-level reconstruction suitable for event-related or network-based analyses.

Parallel to these technical procedures is the equally essential task of event labeling and segmentation. In contrast to lab-based paradigms, MoBI studies rarely rely on time-locked stimuli. Instead, they require post-hoc annotation of complex, unscripted behaviors that may be culturally or artistically significant. In these cases, expert annotation—sometimes down to the frame—is necessary to identify meaningful segments for comparison (e.g., shifts in tone, role transitions, audience interaction). This work often involves dramaturgs, composers, and other domain experts, and requires the development of ontology-aware coding schemas that bridge scientific and artistic frameworks.

The time cost of this process should not be underestimated. Creating a single analysis-ready dataset may involve weeks of iterative work—filtering, synchronizing, aligning metadata, and segmenting events across modalities. The pipeline must support both algorithmic scalability and human interpretability, ensuring that datasets are robust enough for analysis yet transparent enough to support cross-disciplinary review.

Together, these strategies form a flexible, real-world-ready data processing framework. By balancing automation with expert annotation, and artifact removal with ecological validity, the system supports the scientific goals of MoBI while honoring the creative and embodied realities in which it operates.

30.6 Challenges and Limitations

In controlled laboratory environments, experimental rigor is prioritized through standardization and isolation. In contrast, MoBI studies conducted in naturalistic settings must contend with the complexities of real-world contexts. While these contexts introduce invaluable ecological validity, enabling the study of human behavior as embedded in social, emotional, and physical environments. They also present multifactorial challenges that impact data acquisition, system design, and long-term project sustainability.

Technological Barriers
While consumer-grade EEG devices have improved, many MoBI applications—particularly those involving hyperscanning—still require research-grade equipment that is not easily scalable. Synchronizing neural recordings across multiple mobile units is prone to latency jitter, sampling drift, and connection loss, especially in WiFi-congested environments such as galleries or performance venues. Achieving temporal alignment across multiple EEG systems, audio-video feeds, and movement data remains an open technical frontier.

Hardware also presents limitations in form factor and wearability. For instance, electrodes must maintain consistent contact during movement, yet be discreet enough to blend into costumes or avoid occlusion in video capture. In live performance, even small issues—like a forehead electrode detaching due to sweat or facial makeup—can compromise entire sessions. Solutions such as customized caps, adhesives, or electrode holders must be tailored per study, often requiring engineering intervention.

Moreover, interoperability between devices is rarely seamless. Real-time systems must juggle streaming protocols (e.g., LSL, OSC), signal buffering, and computational loads across platforms, often with minimal documentation or community support. Signal dropout or latency spikes can interrupt feedback loops, making closed-loop BCI applications especially fragile in public-facing demonstrations.

Logistical and Environmental Constraints
The logistical footprint of a MoBI study is often larger than anticipated. Participants—whether dancers, actors, musicians, or visitors—must arrive early for headset fitting, signal quality testing, and calibration. These procedures can be time-consuming and may interfere with rehearsal schedules, particularly in artistic settings where time is limited and performance preparation is paramount.

In fieldwork settings, venue limitations often dictate what is technically feasible. Power outlets, lighting conditions, background noise, and WiFi bandwidth can all influence data quality. Performing in open-air settings, for example, introduces humidity and wind artifacts that affect both sensors and behavior. Coordinating with venue staff, obtaining permissions, and ensuring safety protocols (e.g., wireless interference testing) add further overhead.

Aesthetic integration is another important, though less technical, concern. The visual presence of EEG headsets, cables, or amplifiers may clash with the artistic intent or disrupt the experience of the audience. In interactive installations, participants may feel self-conscious wearing equipment that appears medical or foreign. Designing aesthetically mindful and culturally sensitive setups requires input from artists and often bespoke fabrication.

Ethical and Intellectual Property Considerations

The intersection of neuroscience, the arts, and public engagement raises pressing ethical and legal questions that remain insufficiently addressed in current MoBI research practices. One such concern involves music and multimedia stimuli, which are frequently integral to MoBI-art collaborations. While these elements may be used under educational or scientific pretexts, their incorporation in publicly shared datasets implicates copyright law. At present, no clear, standardized mechanism exists for the sharing of copyrighted media in neuroimaging studies, even when the stimuli are critical to the interpretation of results. This lack of clarity hampers data sharing and undermines reproducibility.

There is a similar need to protect the rights of artists and public figures who participate in MoBI experiments, particularly in performance settings. These individuals may be identifiable in video recordings, event annotations, or biometric traces, raising concerns about consent, publicity rights, and long-term data governance. As MoBI research increasingly engages with high-visibility participants and public platforms, guidelines must be developed to ensure ethical representation, protect creative ownership, and respect participant privacy in both research outputs and dissemination activities.

Sustainability and Institutional Support

The operationalization of MoBI research often relies on unstable, short-term funding structures. Many collaborative projects between neuroscience and the arts are funded through time-limited grants that do not support long-term maintenance of equipment, personnel, or data infrastructure. The creation of open-source pipelines and shared resources is a promising step, but such efforts require sustained institutional commitment and cross-sector coordination.

Finally, researchers are frequently expected to balance the dual demands of experimental rigor and public impact. Designing participatory, context-rich experiences while maintaining scientific reproducibility remains an unsolved challenge. Furthermore, existing evaluation metrics fail to capture the full impact of MoBI projects, particularly those situated at the interface of research, performance, and public education. Novel frameworks are needed to assess community engagement, cross-cultural resonance, and non-quantifiable outcomes in a manner that respects the hybrid nature of this emerging domain.

30.7 Conclusion

Across theaters, classrooms, and galleries, mobile EEG has shown us that brain activity in real-world settings is not only measurable—it is meaningfully shaped by the surrounding environment, movement, and social interaction. From actors sharing a scene to dancers navigating improvised choreography, our studies reaffirm a fundamental insight: cognition is embodied, relational, and responsive to context. MoBI enables us to study these phenomena as they unfold, not in isolation, but within the complex rhythms of everyday and artistic life.

What have we learned? That rigorous neuroscience can coexist with artistic expression. With thoughtful planning, EEG equipment can be embedded into costumes and performances. That movement and emotion are not confounds to be minimized, but essential components of the experiences we aim to study. And most importantly, that public-facing research can deepen both scientific insight and community engagement—when it is conducted with care, cultural awareness, and collaboration.

We encourage researchers to build upon and adapt the MoBI framework presented in this chapter. It is not a one-size-fits-all protocol, but a flexible guide designed for real-world complexity. It supports diverse teams in crafting research that is technically sound, ethically grounded, and relevant to the communities it involves.

Looking forward, the potential of interdisciplinary EEG research lies in its ability to scale—across sites, cultures, and time. This will require more robust tools for multimodal data integration, more accessible workflows, and continued collaboration between artists, engineers, neuroscientists, and educators. As new technologies emerge—such as generative AI, real-time feedback systems, and portable BCI applications—they must be integrated with attention to both scientific validity and human experience.

This work is already being advanced by initiatives such as the NSF BRAIN AccelNet: Movement, Music, and BrainHealth, and the NIH U24 Music and Dementia Research Network. These networks create new pathways for transdisciplinary research, workforce training, and STEAM outreach—ensuring that discoveries are not only made, but shared and translated across domains.

Ultimately, MoBI research challenges us to rethink where and how neuroscience happens. By stepping outside the lab, we open new doors—not just for data collection, but for inclusion, creativity, and meaningful collaboration. It is in these spaces—shared, dynamic, and alive—that we may come to better understand the brain as it truly functions: in motion, in connection, and in context.

Acknowledgments We extend our deepest gratitude to the many contributors who made this work possible. This chapter is the result of sustained collaboration among artists, scientists, students, and institutions who shared a commitment to exploring the brain in action. We thank our interdisciplinary collaborators and research participants for their creativity, time, and trust. Support from funding agencies—including the National Science Foundation (NSF BRAIN #2137255, AccelNet #2412731, and NCS programs), the National Institutes of Health (NIH), and the University of

Houston—was instrumental in developing and refining the MoBI framework. We also acknowledge the vital contributions of our partner institutions, including Rice University, the Menil Collection, Sam Houston State University, the Children's Museum of Houston, the Midtown Arts and Theater Center Houston (MATCH), and the Museo de Arte Contemporáneo de Monterrey (MARCO), for providing platforms to integrate neuroscience into public, artistic, and educational spaces.

Appendix

List: Traveling with MoBI Equipment

(a) Typical items needed for a mobile EEG setup.
A standard MoBI travel kit includes the core EEG recording system (often a wireless 28-channel cap with amplifiers and transmitters) along with complementary sensing equipment such as EOG electrodes, inertial measurement units (IMUs), and manual trigger devices. We also pack laptops with the acquisition and synchronization software pre-installed, spare electrodes, conductive gels, batteries, chargers, and appropriate power adapters for the destination country. Rugged transport cases with custom foam inserts protect sensitive gear from shocks and humidity. Backup cables, connectors, and video cameras are also part of the kit. Finally, we carry printed and digital copies of technical documentation, electrode placement diagrams, and setup schematics.

(b) Packing tips for international travel.
We carry the duplicates from the most irreplaceable components such as amplifiers, transmitters, and dongles to reduce the risk of loss or damage. Items are packed in lockable, impact-resistant cases, with each component labeled with its name, model, and serial number. A photographic inventory accompanies every component, serving as both an internal checklist and a reference for customs inspections. Power compatibility is addressed in advance, with voltage converters and plug adapters included for all destinations. Scheduling extra time before and after flights is essential to accommodate inspection procedures, especially when traveling with multiple cases of sensitive electronics.

(c) Customs and logistical considerations.
International transportation of MoBI equipment often involves complex customs procedures, particularly when equipment is temporarily exported and re-imported for research purposes. One highly effective solution is the *ATA Carnet*, an internationally recognized customs document that functions as a "passport for goods." The Carnet permits the temporary, duty-free importation of professional equipment, thereby streamlining border crossings and minimizing the risk of delays or unexpected fees.

Our team adopted the ATA Carnet as a standard practice for all international equipment transportation. The process for obtaining one is relatively straightforward:

1. Prepare a detailed inventory listing each item's description, serial number, approximate value, and country of origin.
2. Submit this information via the official ATA Carnet portal (https://www.atacarnet.com/).
3. Pay an issuance fee, which is determined by the value of the equipment and the number of countries to be visited.

References

Alshaer, A. (2025). The role of sensory experiences in evoking emotional responses in virtual reality. *Journal of Umm Al-Qura University for Engineering and Architecture, 16*(1), 172–184.

Pacheco-Ramírez, M. A., Klatka, A., Aguilar-Herrera, A. J., Lima Carmona, Y. E., Sánchez-Rodríguez, L., Das, D., Lang, F., Beus, S., Kaplan, D., Belcher, R., Rui, M., & Contreras-Vidal, J. L. (2025) Music-in-medicine: Mobile brain-body imaging of professional musicians during live in-hospital concerts [modified 2026 Feb 09; cited 2026 Feb 24]; Available from: https://doi.org/10.6084/m9.figshare.c.7982714

Contreras-Vidal, J. L., Cruz-Garza, J. G., Robleto, D., Azorín, J. M., & Nam, C. S. (2019a). Introduction: The confluence of art, neuroscience, and creativity through mobile brain–body imaging. In J. L. Contreras-Vidal, D. Robleto, J. G. Cruz-Garza, J. M. Azorín, & C. S. Nam CS (Eds.), *Mobile brain-body imaging and the neuroscience of art, innovation and creativity* [Internet]. Springer [cited 2025 Sept 30] (pp. 1–3). Available from: https://doi.org/10.1007/978-3-030-24326-5_1

Contreras-Vidal, J. L., Robleto, D., Cruz-Garza, J. G., Azorín, J. M., & Nam, C. S. (Eds.). (2019b). *Mobile brain-body imaging and the neuroscience of art, innovation and creativity* [Internet] (Springer series on bio- and neurosystems, Vol. 10). Springer [cited 2025 Sept 30]. Available from: http://link.springer.com/10.1007/978-3-030-24326-5

Cruz Garza, J. G., Sujatha Ravindran, A., Kopteva, A. E., Rivera Garza, C., & Contreras-Vidal, J. L. (2020). Characterization of the stages of creative writing with mobile EEG using generalized partial directed coherence. *Frontiers in Human Neuroscience* [Internet] [cited 2025 Oct 2], *14*. Available from: https://www.frontiersin.org/journals/human-neuroscience/articles/10.3389/fnhum.2020.577651/full

Cruz-Garza, J. G., Hernandez, Z. R., Nepaul, S., Bradley, K. K., & Contreras-Vidal, J. L. (2014). Neural decoding of expressive human movement from scalp electroencephalography (EEG). *Frontiers in Human Neuroscience* [Internet] [cited 2025 Sept 30], *8*. Available from: https://www.frontiersin.org/journals/human-neuroscience/articles/10.3389/fnhum.2014.00188/full

Cruz-Garza, J. G., Brantley, J. A., Nakagome, S., Kontson, K., Megjhani, M., Robleto, D., et al. (2017). Deployment of mobile EEG technology in an art museum setting: Evaluation of signal quality and usability. *Frontiers in Human Neuroscience* [Internet] [cited 2025 Sept 30], *11*. Available from: https://www.frontiersin.org/journals/human-neuroscience/articles/10.3389/fnhum.2017.00527/full

Cruz-Garza, J. G., Chatufale, G., Robleto, D., & Contreras-Vidal, J. L. (2019). Your brain on art: A new paradigm to study artistic creativity based on the 'exquisite corpse' using mobile brain-body imaging. In A. Nijholt (Ed.), *Brain art: Brain-computer interfaces for artistic expression* [Internet]. Springer [cited 2025 Sept 30] (pp. 283–308). Available from: https://doi.org/10.1007/978-3-030-14323-7_10

González-España, J. J., Back, K. J., Reynolds, D., & Contreras-Vidal, J. L. (2023). Decoding taste from EEG: Gustatory evoked potentials during wine tasting. In *2023 IEEE international conference on Systems, Man, and Cybernetics (SMC)* [Internet]. 2023 [cited 2025 Sept 30] (pp. 4253–4258). Available from: https://ieeexplore.ieee.org/abstract/document/10394408

Hendry, M. F., Cruz-Garza, J. G., Delgado-Jiménez, E. A., Lima-Carmona, Y. E., Aguilar-Herrera, A. J., Ramírez-Moreno, M. A., et al. (2025). Mobile brain-body imaging and visual data of theatrical actors during rehearsal and performance. *Scientific Data, 12*(1), 1421.

Herrera-Arcos, G., Tamez-Duque, J., Acosta-De-Anda, E. Y., Kwan-Loo, K., de-Alba, M., Tamez-Duque, U., et al. (2017). Modulation of neural activity during guided viewing of visual art. *Frontiers in Human Neuroscience* [Internet] [cited 2025 Oct 2], *11*. Available from: https://www.frontiersin.org/journals/human-neuroscience/articles/10.3389/fnhum.2017.00581/full

Jungnickel, E., Gehrke, L., Klug, M., & Gramann, K. (2019). Chapter 10—MoBI—Mobile brain/body imaging. In H. Ayaz & F. Dehais (Eds.), *Neuroergonomics* [Internet]. Academic [cited 2025 Sept 30] (pp. 59–63). Available from: https://www.sciencedirect.com/science/article/pii/B9780128119266000105

Kilicarslan, A., & Contreras Vidal, J. L. (2019). Characterization and real-time removal of motion artifacts from EEG signals. *Journal of Neural Engineering, 16*(5), 056027.

Kilicarslan, A., Grossman, R. G., & Contreras-Vidal, J. L. (2016). A robust adaptive denoising framework for real-time artifact removal in scalp EEG measurements. *Journal of Neural Engineering, 13*(2), 026013.

King. (2021). Using mobile brain/body imaging to advance research in arts, health, and related therapeutics. *European Journal of Neuroscience—Wiley Online Library* [Internet]. [cited 2025 Sept 30]. Available from: https://onlinelibrary.wiley.com/doi/10.1111/ejn.15313

Kobayashi, T., Jadram, N., Ninomiya, S., Suzuki, K., & Sugaya, M. (2025). A pilot study on emotional equivalence between VR and real spaces using EEG and heart rate variability. *Sensors, 25*(13), 4097.

Kontson, K., Megjhani, M., Brantley, J. A., Cruz-Garza, J. G., Nakagome, S., Robleto, D., et al. (2015). 'Your Brain on Art': Emergent cortical dynamics during aesthetic experiences. *Frontiers in Human Neuroscience* [Internet] [cited 2025 Oct 2], *9*. Available from: https://www.frontiersin.org/journals/human-neuroscience/articles/10.3389/fnhum.2015.00626/full

Nijholt, A. (Ed.). (2019). *Brain art: Brain-computer interfaces for artistic expression* [Internet]. Springer [cited 2025 Sept 30]. Available from: http://link.springer.com/10.1007/978-3-030-14323-7

Pacheco-Ramírez, M. A., Ramírez-Moreno, M. A., Kukkar, K., Rao, N., Huber, D., Brandt, A. K., et al. (2024). Neural dynamics of creative movements during the rehearsal and performance of "LiveWire". *Scientific Data, 11*(1), 1208.

Pizzolante, M., Bartolotta, S., Sarcinella, E. D., Chirico, A., & Gaggioli, A. (2024). Virtual vs. real: Exploring perceptual, cognitive and affective dimensions in design product experiences. *BMC Psychology, 12*(1), 10.

Ramírez-Moreno, M. A., Cruz-Garza, J. G., Acharya, A., Chatufale, G., Witt, W., Gelok, D., et al. (2023). Brain-to-brain communication during musical improvisation: a performance case study [Internet]. *F1000Research* [cited 2025 Sept 30]. Available from: https://f1000research.com/articles/11-989

Sanctis, P. D., Solis-Escalante, T., Seeber, M., Wagner, J., Ferris, D. P., & Gramann, K. (2025) Time to move: Brain dynamics underlying natural action and cognition [cited 2025 Sept 30]; Available from: https://onlinelibrary.wiley.com/doi/10.1111/ejn.15562

Sujatha Ravindran, A., Mobiny, A., Cruz-Garza, J. G., Paek, A., Kopteva, A., & Contreras Vidal, J. L. (2019). Assaying neural activity of children during video game play in public spaces: A deep learning approach. *Journal of Neural Engineering, 16*(3), 036028.

Theofanopoulou, C., Paez, S., Huber, D., Todd, E., Ramírez-Moreno, M. A., Khaleghian, B., et al. (2024). Mobile brain imaging in butoh dancers: From rehearsals to public performance. *BMC Neuroscience, 25*(1), 62.

Todd, E., Cruz-Garza, J. G., Moreau, A., Templeton, J., & Contreras-Vidal, J. L. (2019). Self-conscience/physical memory: An immersive, kinetic art installation driven by real-time and archival EEG signals. In A. Nijholt (Ed.), *Brain art: Brain-computer interfaces for artistic expression* [Internet]. Springer [cited 2025 Sept 30] (pp. 309–223). Available from: https://doi.org/10.1007/978-3-030-14323-7_11

Part VI
Multimodal EEG Applications

Chapter 31
Combining EEG and Transcranial Electric Brain Stimulation

Heiko I. Stecher, Sreekari Vogeti, Daniel Strüber, and Christoph S. Herrmann

Abstract Neuromodulation techniques, such as transcranial electric stimulation, offer the possibility of modulating brain activity and, in turn, behavior. Electroencephalography (EEG) is the ideal tool to demonstrate such alterations of brain activity and prove that neuromodulation actually achieves a modulation of brain activity. However, care must be taken to use the right approaches and technical equipment. This book chapter reviews recent developments and provides recommendations for recording EEG before, during, and after brain stimulation sessions.

Keywords Aperiodic brain activity · Artifact removal · Closed-loop stimulation · EEG · Neuromodulation · Periodic brain activity · Transcranial alternating current stimulation · Transcranial direct current stimulation · Transcranial random noise stimulation

H. I. Stecher · D. Strüber
Experimental Psychology Lab, Department of Psychology, Carl-von-Ossietzky Universität, Oldenburg, Germany

S. Vogeti
Experimental Psychology Lab, Department of Psychology, Carl-von-Ossietzky Universität, Oldenburg, Germany

Cluster for Excellence "Hearing for All", Carl-von-Ossietzky Universität, Oldenburg, Germany

C. S. Herrmann (✉)
Experimental Psychology Lab, Department of Psychology, Carl-von-Ossietzky Universität, Oldenburg, Germany

Cluster for Excellence "Hearing for All", Carl-von-Ossietzky Universität, Oldenburg, Germany

Research Center Neurosensory Science, Carl von Ossietzky Universitßt, Oldenburg, Germany
e-mail: christoph.herrmann@uol.de; christoph.herrmann@uni-oldenburg.de

T. Warbrick (ed.), *The EEG Handbook*,
https://doi.org/10.1007/978-3-032-20450-9_31

31.1 Introduction

Over the last two decades, transcranial electrical stimulation (tES) has been rediscovered as a valuable tool for both fundamental neuroscience and medical therapy (Guleyupoglu et al., 2013). Directly influencing the brain activity via electric fields provides a critical approach to intervention that shifts our understanding of the relationship between measurable brain activity and observable brain function from a purely correlative perspective to a causal framework (Bergmann et al., 2016; Miniussi et al., 2012; Herrmann et al., 2016; Herrmann & Strüber, 2017). Whereas early transcranial electrical stimulation studies by Merton and Morton (1980) relied on painful, supra-threshold high-voltage stimulation with limited focality, today's protocols use weak, voltage-limited currents that are informed by neuroimaging-based anatomical insights. These currents are generally considered subthreshold, because they do not directly trigger neuronal firing. They are also well tolerated, producing only mild sensations during stimulation (Turi et al., 2013; Poreisz et al., 2007). These weak currents (usually in the range of 1–2 mA at the scalp) subtly alter the membrane potentials of neurons in the stimulated cortical regions. According to the momentary polarity of the applied electrical field, they alter the neuron's natural probability to fire (firing is more likely during anodal stimulation and less likely during cathodal stimulation). This approach allows for the suppression or excitation of a targeted region when a constant electrical field is applied, as in transcranial direct current stimulation (tDCS), or to rhythmically alternate between excitation and inhibition, as in the family of alternating current protocols such as tACS. Rhythmic stimulation can then be used to alter endogenous brain oscillations via synchronization or desynchronization.

However, applying stimulation and observing subsequent changes in brain functions, such as alterations in perception or cognition, does not alone establish a causal link between neural activity and function. Because the brain is a highly complex system, stimulation can have many unforeseen effects on brain activity. Therefore, it is important to confirm that the stimulation has achieved the intended modulatory effect on brain activity and, in turn, on perception or cognition. This chapter will explore how electrophysiological measures can be used to demonstrate the effects of tES

31.2 Types of Electric Brain Stimulation

Electric brain stimulation dates back to Roman times when electric fish were used to treat headache (Wexler, 2017b). Between the 1740s and 1930s, many experiments on electric stimulation of animal and human tissue were carried out with precursors of the modern battery, the Leyden jar and the voltaic pile (Cohen Kadosh & Elliott, 2013). These experiments were carried out by Luigi Galvani, inventor of the frog leg experiment, and Alessandro Volta, inventor of the voltaic pile. These experiments ultimately led to the development of medical devices for therapy referred to as medical batteries. These batteries were used to stimulate various tissues, including the human brain (Wexler, 2017a). In 2000, electric brain stimulation experienced a renaissance with the development of transcranial direct current stimulation or tDCS (Nitsche & Paulus, 2000).

In its simplest form, tDCS delivers a direct current to the scalp via two electrodes: the anode (+) and the cathode (−). This current passes through the skull, altering the membrane voltage of neurons. Depolarization increases the likelihood of neurons firing in response to incoming postsynaptic currents, while hyperpolarization reduces this likelihood.

In addition to direct current, tES also employs alternating current, as seen in tACS, which can modulate ongoing brain oscillations through entrainment (Herrmann et al., 2013). Another form of tES, transcranial random noise stimulation (tRNS), uses band-limited white noise for stimulation, likely producing neural effects via stochastic resonance (Antal & Herrmann, 2016). Additionally, amplitude-modulated alternating currents have been utilized in transcranial brain stimulation, where the amplitude of a high-frequency carrier wave, such as 80 Hz, is modulated by the phase of a lower frequency, such as 10 Hz (Witkowski et al., 2016).

Generally, tES methods face the challenge that deeper brain regions cannot be stimulated without more intensely affecting superficial brain regions. However, it has been demonstrated that applying two sinusoidal currents with different frequencies, f_1 and f_2, through separate electrode pairs can produce a beat frequency at the point where the current pathways overlap. This beat frequency manifests as an amplitude-modulated oscillation, with a carrier frequency of $f_c = (f_1 + f_2)/2$ and a modulation frequency of $f_m = |f_1 - f_2|$. Known as temporal interference stimulation (tTIS), this technique was first demonstrated in animals (Grossman et al., 2017) and has since been applied also in humans (von Conta et al., 2022).

For all of the aforementioned brain stimulation methods, verifying their ability to modulate brain activity is crucial, with EEG recordings serving as the most effective method for evaluation.

31.3 Online Effects

Noninvasive brain stimulation aims to modulate neural activity, and the millisecond temporal resolution of EEG makes it ideal for assessing neural activity during stimulation (i.e., *online effects*). However, electrical stimulation using signals in the range of several volts generates significant artifacts, which are orders of magnitude larger than the endogenous EEG signal, which is measured in microvolts. This section outlines the challenges of conventional artifact rejection methods in tDCS, tRNS, and tACS. It then examines phantom-based evaluations, which provide a controlled approach to characterizing stimulation artifacts and validating artifact removal techniques. Finally, it discusses innovative strategies for obtaining reliable, analyzable EEG data during stimulation.

31.3.1 *Conventional Artifact Removal Strategies*

31.3.1.1 Transcranial Direct Current Stimulation (tDCS)

Since tDCS typically does not involve frequency-specific stimulation, artifacts can theoretically be removed using a high-pass filter, provided that the EEG signal remains unsaturated. While artifacts during the ramp-up and ramp-down periods have been removed successfully (e.g., Roy et al., 2014), persistent low frequency artifacts remain in the data throughout the stimulation period. These artifacts vary significantly across channels and over time, making them challenging to characterize or eliminate without prior knowledge of the stimulation artifact's properties (Gebodh et al., 2019; Mancini et al., 2015; Roy et al., 2014). Furthermore Gebodh et al. (2019) identified several complex, nonstationary physiological artifacts—such as cardiac, ocular, motion, and myogenic distortions—that are difficult to address using conventional artifact removal techniques. For a detailed discussion on employing spatial and temporal filters to remove stimulation artifacts, see Gebodh et al. (2019).

31.3.1.2 Transcranial Random Noise Stimulation (tRNS)

TRNS is believed to modulate neural activity through stochastic resonance rather than frequency-specific entrainment. Most research on tRNS effects has focused on motor cortex excitability (for reviews see Potok et al., 2022; Van der Groen et al., 2022). To our knowledge, no systematic studies have characterized EEG artifacts resulting from low-frequency (LF) or full-spectrum tRNS. Several studies suggest that high-frequency (HF) tRNS may be more effective than LF-tRNS, based primarily on findings from motor cortex research (for discussion see Potok et al., 2022). As HF-tRNS employs stimulation frequencies beyond the endogenous EEG range, the high-frequency component is typically attenuated using low-pass filters (e.g., Rufener et al., 2017; Van Doren et al., 2014). Using this approach, Rufener et al. (2017) observed reduced peak latencies of the early auditory event-related potentials (P50 and N1), which was associated with improved detection of near-threshold stimuli, while stimuli below and above the individual perception threshold were unaffected.

31.3.1.3 Transcranial Alternating Current Stimulation (tACS)

In tACS, the challenge is amplified because stimulation frequencies overlap with the endogenous EEG frequencies targeted for modulation, resulting in artifacts within those frequency ranges. Therefore, a key difficulty lies in disentangling true neural responses from stimulation-induced artifacts.

Multiple artifact rejection methods have been proposed for this purpose: temporal filtering, template subtraction, principle component analysis (PCA) (Helfrich et al., 2014b,a), and spatial filtering (Guarnieri et al., 2020; Haslacher et al., 2021).

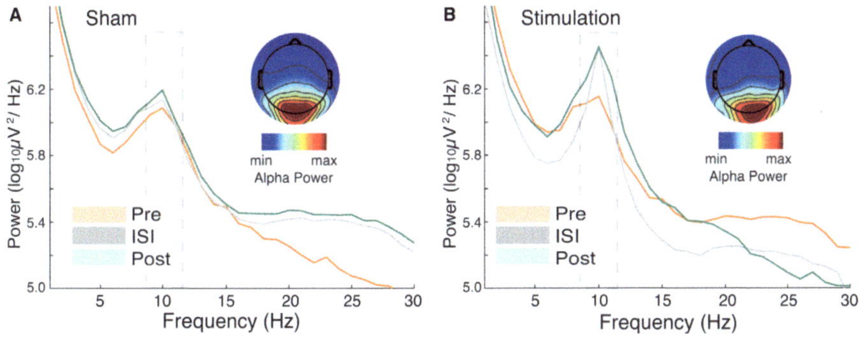

Fig. 31.1 Grand average power at electrode POz before, during, and after 20 min of 10 Hz tACS (B) or sham (A) during a visual oddball task. The orange lines depict the average power before stimulation, the gray lines depict recovered power during stimulation, and the green lines depict the power after stimulation. While there are no significant differences during sham stimulation around 10 Hz between conditions during sham stimulation, during tACS an increase is visible during (ISI) and after tACS stimulation (adapted from Helfrich et al. (2014b) and published with permission by Elsevier)

Figure 31.1 illustrates how a combination of template subtraction and PCA was used to recover power increases during and after alpha-stimulation. While the abovementioned methods are suitable to reduce linear effects of the stimulation, they are insufficient to remove nonlinear effects. These include artifacts arising from interactions between stimulation and other physiological processes (Noury et al., 2016; Noury & Siegel, 2017).

At present, there is no consensus on the effectiveness of artifact rejection methods in fully eliminating stimulation artifacts in tACS studies (for discussions, see Neuling et al., 2017; Noury et al., 2016; Noury and Siegel, 2018). Incomplete artifact removal can result in systematic noise that may be mistaken for evidence of neural entrainment. Therefore, establishing standardized criteria to evaluate the success of stimulus artifact removal would significantly advance the assessment of "online" neural EEG effects of entrainment in humans.

31.3.2 Validating Artifact Removal Methods: Phantom-Based Evaluations

One strategy for addressing this challenge is the use of a "phantom" head model, which replicates the conductive properties of the head but lacks neural activity. Phantom models can be utilized in two complementary ways: (1) to systematically characterize the stimulation artifact and (2) to compare and validate artifact removal methods in controlled conditions. A range of phantom models have been proposed, from simple setups such as water melons (Guidetti et al., 2025) and gelatin head models (Kohli & Casson, 2019) to more sophisticated versions (e.g., Hunold et al., 2018, 2020) that better approximate the conductivity of the head.

To characterize the stimulation artifact, Roy et al. (2014) recorded EEG from a melon model during anodal, cathodal, and sham tDCS. In both types of active stimulation, they observed polarity specific artifacts across the frequency spectrum in electrodes near the stimulation sites. These artifacts were identified via independent component analysis (ICA) and subsequently compared with ICA decompositions of human EEG to facilitate artifact removal. Additionally, they detected low-frequency drift artifacts throughout the stimulation period, prompting them to filter out frequencies below 2 Hz in further analyses.

A limitation of using ICA alone is the need for visual inspection and matching of participants' ICs to reference ICs derived from the phantom, a method particularly effective for anodal stimulation (Mancini et al., 2015; Roy et al., 2014). To address this issue, Mancini et al. (2015) proposed an automated filtering approach that avoids manual IC selection and is adaptable to various stimulation types. Similar to Roy et al. (2014), Mancini et al. (2015) begin by capturing tDCS artifacts using a phantom model. These ICs are used to identify corresponding artifactual ICs in human EEG data, generating a reference signal correlated with the tDCS artifact. An adaptive filter then employs this reference signal to minimize the artifact in each EEG channel. Subsequently, ICs from the cleaned data are inspected to confirm artifact removal. Unlike the ICA-only approach, this method eliminates the need for visual IC inspection for each participant once a reference signal is established. However, validation still relied on visual inspection of the ICs obtained after adaptive filtering. Although this approach advances artifact rejection, the authors note the need for more robust validation methods and assessment across a wider range of anode-cathode configurations to fully validate the technique.

To validate artifact removal methods, phantoms with a "ground truth" EEG signal free of stimulation artifacts can be used (e.g., Kohli and Casson, 2019; Vosskuhl et al., 2020). For example, Vosskuhl et al. (2020) constructed a "phantom head" using multiple concentric plastic spheres filled with a saline solution to mimic the conductivity of the human brain and scalp. An artificial dipolar current source within the phantom head played prerecorded, artifact-free EEG signals, while tACS was applied. This setup allowed a direct comparison of multiple artifact removal techniques, including simple sine wave subtraction, template subtraction, and single space projection, with the latter proving most effective.

This work illustrates how phantom models provide a controlled environment for identifying, classifying, developing, and evaluating strategies to remove stimulation artifacts from human EEG data.

31.3.3 Innovative Approaches to Minimize Artifacts

Beyond developing sophisticated artifact rejection and assessment methods, another strategy for obtaining analyzable EEG data during stimulation involves using alternative types of stimulation that differ from endogenous neural activity. These approaches typically focus on generating stimulation waveforms that are distinct from EEG signals or delivering stimulation at frequencies outside of the physiological EEG range, allowing them to be filtered out effectively.

31.3.3.1 Non-Sinusoidal Waveforms

One such method involves using non-sinusoidal stimulation shapes that are distinct from endogenous oscillations. For instance, Dowsett and Herrmann (2016) employed a sawtooth-shaped current as the stimulation waveform. Sawtooth waveforms generate harmonics in the frequency spectrum, which can be reduced with artifact removal algorithms while preserving the peak at the fundamental frequency of interest. This approach leverages harmonics that are distinct from neural EEG signals, making the artifact predictable and easier to remove.

31.3.3.2 Amplitude-Modulated tACS (AM-tACS)

In AM-tACS low-frequency envelopes of the high-frequency carrier signals are used to modulate the endogenous activity. Since the high-frequency carrier frequencies exceed the endogenous EEG frequency range, the artifact can be removed from the EEG using low-pass filters (Witkowski et al., 2016) or spatial filters (Haslacher et al., 2021). However, Kasten et al. (2018) demonstrated that there can be nonlinear artifacts that transfer from stimulation devices and contaminate the data. Therefore, further development of the stimulation devices is necessary to minimize these effects and enhance the effectiveness of this approach.

31.3.3.3 Transcranial Temporal Interference Stimulation (tTIS)

TTIS avoids nonlinear artifacts associated with stimulation devices which are common in AM-tACS, by using pure sine waves that interact within the brain to produce a frequency that aligns with the endogenous target frequency (von Conta et al., 2022). Although these artifacts can be mitigated with a low-pass filters, von Conta et al. (2022) emphasized that selecting an EEG system with a larger dynamic range, such as a 24-bit amplifier capable of sampling a wide range of amplitudes crucial for accurately recovering endogenous EEG signals.

31.3.3.4 Summary of Advantages and Limitations

These innovative stimulation techniques aim to either distinguish the artifact from neural EEG or shift it outside the physiological frequency range enabling the recovery of neural EEG using conventional artifact rejection methods. Alternative waveforms introduce complex harmonics, while high-frequency approaches require EEG amplifiers with a large dynamic range to prevent hardware-related artifacts. Despite these challenges, such methods provide promising avenues for acquiring reliable neural EEG signals during tES.

31.4 Offline Effects

EEG can also be used to evaluate effects of transcranial brain stimulation that outlast stimulation offset (also known as *offline* effects or *aftereffects*). Studies employing this method typically compare EEG measures, such as power or phase locking value, across various time periods and conditions.

Common comparisons include post- versus pre-stimulation, post-stimulation versus post-sham, or, ideally the difference between post- and pre-stimulation compared to a corresponding difference in a control condition (e.g., sham, shunt, or alternative-frequency stimulation). In addition, a within-participants design is preferable as compared to a between-participants design due to the decreased variance and thus increased statistical power. This ideal comparison is crucial because certain EEG measures vary systematically regardless of electrical stimulation. For example, alpha power increases and peak alpha frequency decreases over time (Benwell et al., 2019). A systematic examination of individual differences is still needed to determine whether between-participant comparisons are essential.

In stimulation methods involving oscillatory components, spectral changes may be expected in target frequencies, as well as adjacent frequencies and harmonics. Consequentially, studies employing these methods typically examine modulations of spectral power, phase alignment, or coherence. In contrast, non-oscillatory stimulation forms, which are not expected to produce frequency-specific effects, are commonly assessed using measures like event-related potentials (ERPs) to detect changes in cortical excitability that may not directly correspond to oscillatory activity.

This section reviews EEG measures that are typically used to evaluate offline effects.

31.4.1 Spectral Power

In tACS, Zaehle et al. (2010) were the first to use EEG to examine changes in cortical excitability induced by stimulation. They applied stimulation at the individual alpha frequency (IAF) over the occipital cortex and observed an increase in alpha power increased post-stimulation compared to pre-stimulation in the tACS group relative to the sham group. Similar increases in spectral power within the stimulated frequency band have been reported in numerous studies across various tasks (e.g., Aktürk et al., 2022; D'Atri et al., 2019; Helfrich et al., 2014b; Kasten et al., 2016; Neuling et al., 2013; Stecher et al., 2021; Vossen et al., 2015). For an overview see Veniero et al. (2015). Figure 31.2 depicts an example of alpha power increase following IAF tACS.

However, several studies failed to observe such effects (e.g., Harada et al., 2020; Lafleur et al., 2021; Neuling et al., 2013; Stecher and Herrmann, 2018; Strüber et al., 2015; Wischnewski et al., 2019). These inconsistencies likely arise from variations

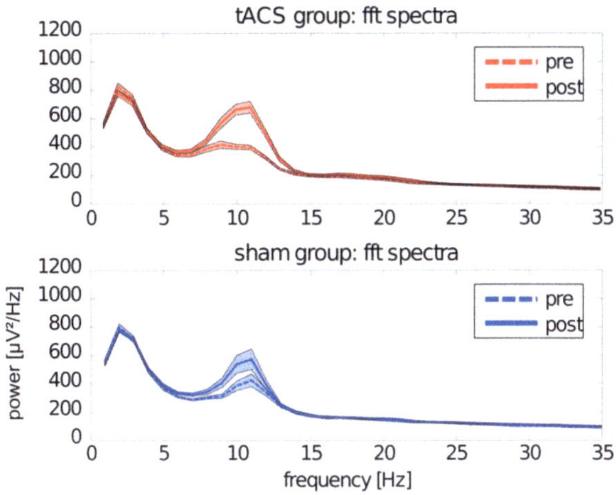

Fig. 31.2 Average power at electrode Pz before and after 20 min of IAF tACS (top) and sham (bottom) during an eyes-open auditory detection task. The dashed lines indicate the pre-stimulation power, solid lines indicate post-stimulation power, and the shaded areas represent the standard error of the mean. Even after sham, alpha power increases over time from pre- to post-intervention (blue spectra). After tACS, this increase is significantly stronger (red spectra). This nicely illustrates why a sham group is necessary for tACS (adapted from Neuling et al. (2013) under the Creative Commons Attribution License (CC BY 3.0))

in experimental design, including differences in tasks, stimulation parameters, and statistical comparisons between pre- and post-stimulation. For example, Stecher and Herrmann (2018) conducted an exploratory analysis suggesting that a mismatch between the stimulation frequency and the participants' IAF could account for the absence of aftereffects in their study. Similarly, Kasten and Herrmann (2022) and Neuling et al. (2013) propose that the state of the stimulated network may influence whether or not aftereffects occur.

Spectral power modulations have also been observed in other types of stimulation. For instance, HF-tRNS has been shown to enhance individual gamma activity in the auditory cortex following an auditory task (Rufener et al., 2017) and during a visual task (Ghin et al., 2021). In contrast, studies investigating resting state EEG have yielded mixed results. While Ke et al. (2024) observed no frequency specific effects of tRNS on resting EEG, Van Doren et al. (2014) reported power increase in all frequencies, and (Mohsen et al., 2019) observed power increases following both LF- and HF-tRNS applied over multiple sites. These findings suggest that gamma modulations may be task-dependent, primarily occurring when tRNS is paired with sensory or cognitive tasks. However, not all task related studies support this conclusion. For example, Schoisswohl et al. (2021) found no effects of LF- or HF-tRNS on the power of any frequency band in resting state EEG. These mixed results underscore the need to clarify the effects of tRNS on spectral power and whether they depend primarily on task engagement and stimulation parameters.

Polarity-specific modulations of spectral power have also been reported with tDCS. For example, Boonstra et al. (2016) observed increases in spectral power below 15 Hz and decreases above 15 Hz following anodal tDCS. Similarly, Spitoni et al. (2013) applied cathodal and anodal tDCS to the posterior parietal areas and found that only anodal tDCS increased alpha power post-stimulation. This finding was replicated by Mangia et al. (2014), who also noted increases in beta power. Together, these studies suggest that tDCS may modulate oscillatory activity in a polarity and frequency-dependent manner. However, further systematic research is needed to determine the conditions under which these effects are consistent and reliable.

In summary, although modulations of spectral power persisting beyond the stimulation period are frequently reported, the inconsistent findings in the literature highlight the need for systematic research to explore moderating factors, including frequency matching, stimulation parameters, state of the network, and task dependency.

31.4.2 Phase Locking/Phase Coherence

Matsumoto et al. (2010) reported polarity-specific effects of tDCS on mu-rhythm desynchronization, with anodal stimulation leading to increased desynchronization and cathodal stimulation resulting in decreased desynchronization. Beyond local effects, interhemispheric connectivity has also been modulated using tACS. Helfrich et al. (2014a) applied bilateral 40 Hz tACS over the parieto-occipital cortex, either in-phase or antiphase. Their findings indicated an increase of interhemispheric gamma coherence for in-phase stimulation, whereas antiphase stimulation impaired functional coupling.

Although the number of studies examining phase locking and coherence using EEG is limited, the results suggest that stimulation can affect network synchrony. Similar to the effects on spectral power, methodological factors such as frequency matching, electrode montage, and network state are likely to impact phase locking and coherence.

31.4.3 ERPs

Studies combining EEG and tDCS have reported polarity-specific stimulation effects on visual, auditory, and sensory ERPs. An early study by Antal et al. (2004) analyzed the effects of tDCS over the occipital cortex with EEG recordings before and after electrical stimulation. They found that occipital stimulation increased the amplitude of visual ERP components N70 and P1, while their latencies remained unaffected. Zaehle et al. (2011) applied tDCS to the left temporal and temporoparietal cortex, observing that anodal tDCS significantly increased P50 amplitudes, whereas cathodal tDCS increased N1 amplitudes. Similarly, a study assessing medial frontal potentials associated with task performance monitoring, such as the error-related negativity (ERN) and the feedback-related negativity (FRN), found that cathodal tDCS reduced the ERN and FRN amplitudes to levels comparable to the control condition, while anodal tDCS elicited significantly larger ERNs and FRNs compared to sham stimulation. However, other studies reported no changes in the ERP amplitudes or latencies following anodal tDCS (Conley et al., 2016; Kunzelmann et al., 2018). These discrepancies likely reflect methodological differences, including variations in stimulation parameters, electrode placement, and experimental tasks. To achieve reliable ERP modulations, these factors must be systematically investigated and optimized.

HF-tRNS has also been shown to modulate the latency and amplitudes of ERPs. For example, Rufener et al. (2017) observed reduced peak latencies of P50 and N1 components following HF-tRNS applied to the auditory cortex. While many stimulation studies apply stimulation during a task to evaluate behavioral measures online and then assess neural measures offline, fewer have explored the effects of stimulation delivered prior to a task on subsequent behavior and neural activity. The mechanisms underlying tRNS offline effects remain poorly understood, whereas online effects are mostly attributed to stochastic resonance. Studying pre-task stimulation effects could clarify whether the same mechanisms operate when stimulation is applied before versus during the task. For instance, Ghin et al. (2021) applied HF-tRNS before a visual motion detection discrimination task, and recorded EEG, finding no effects of HF-tRNS on visual evoked potentials (P1, N1, and P2). Only a limited number of studies have directly examined the tRNS effects on ERPs, with mixed results. More work is needed to determine whether and how tRNS alters ERPs.

31.4.4 Further Measures

Less commonly assessed measures include peak alpha frequency shifts (see Millard et al., 2024, for a meta-analysis), as well as functional connectivity (e.g., Schwab et al., 2019; Polanía et al., 2011; Mohsen et al., 2019), and entropy measures (e.g., Nascimento et al., 2019) providing insights into network-level dynamics.

31.4.5 Online Versus Offline Measures

Offline measures can provide valuable insights into the persistence of stimulation effects beyond the simulation period, but they may not reflect the same underlying mechanisms as those engaged during the stimulation. For instance, in tACS online effects are likely driven by entrainment, whereas offline effects involve plasticity-related processes (Vogeti et al., 2022). This distinction is crucial, as relying solely on offline measures to infer how stimulation affects neural dynamics during stimulation may result in inaccurate conclusions.

31.5 Differentiation Between Aperiodic and Periodic Activity

31.5.1 Rationale

Researchers familiar with EEG recognize that the spectral representation of brain activity does not only display periodic activity (such as the classical brain rhythms alpha, beta, gamma, delta, and theta) as distinct peaks in the spectrum but also consistently includes additional activity whose power follows a 1/f distribution (Buzsaki, 2006; Pritchard, 1992).

This "1/f-noise floor" (also referred to as "pink noise," "scale-invariant activity," or "fractal component") represents aperiodic (or "arrhythmic") brain activity. Endogenous brain rhythms, such as alpha, beta, gamma, delta, and theta, are considered truly periodic, characterized by a sinusoidal pattern with a continuous sequence of positive and negative deflections of equal amplitude and duration. In contrast, aperiodic activity encompasses all other types of neuronal activity that do not follow a continuous sine-wave pattern, such as unipolar voltage deflections. However, spectral analysis methods, like, for example, the Fourier transformation, primarily measure how closely a given signal resembles cosines of various frequencies. But even non-cosine-shaped signals (e.g., a Gaussian peak) partially resemble cosines across multiple frequencies, resulting in their representation in the spectral domain as 1/f-activity.

Recent evidence suggests that regarding 1/f-activity as mere "noise" is inadequate (He et al., 2010; He, 2014). Multiple studies have demonstrated that the level of 1/f-activity is functionally relevant for various cognitive processes (Miller et al., 2014; Ouyang et al., 2020). Moreover, failing to account for changes in 1/f-activity can lead to distorted effects (Gyurkovics et al., 2021). Additional research suggests that the specific exponent of the power spectrum is closely related to the current balance between excitatory and inhibitory neuronal activity (Gao et al., 2017).

In the context of neuromodulation research, this means that solely analyzing the total spectral activity without accounting for changes in aperiodic activity may mask, attenuate, or exaggerate intervention effects. Moreover, the intervention itself could modify the shape of aperiodic activity, potentially indicating changes in the overall ratio between excitation and inhibition. Therefore, it is essential to decompose the spectral outcomes of neuromodulation studies into periodic and aperiodic components and asses each separately for intervention effects (Keil et al., 2022; Donoghue et al., 2022), as demonstrated in several recent studies (Kasten et al., 2024; Masina et al., 2025; Venugopal et al., 2025; Davis et al., 2023).

31.5.2 Procedure

To separate aperiodic and periodic activity from the total spectrum, the aperiodic activity must first be estimated as a fractal component. This component is then

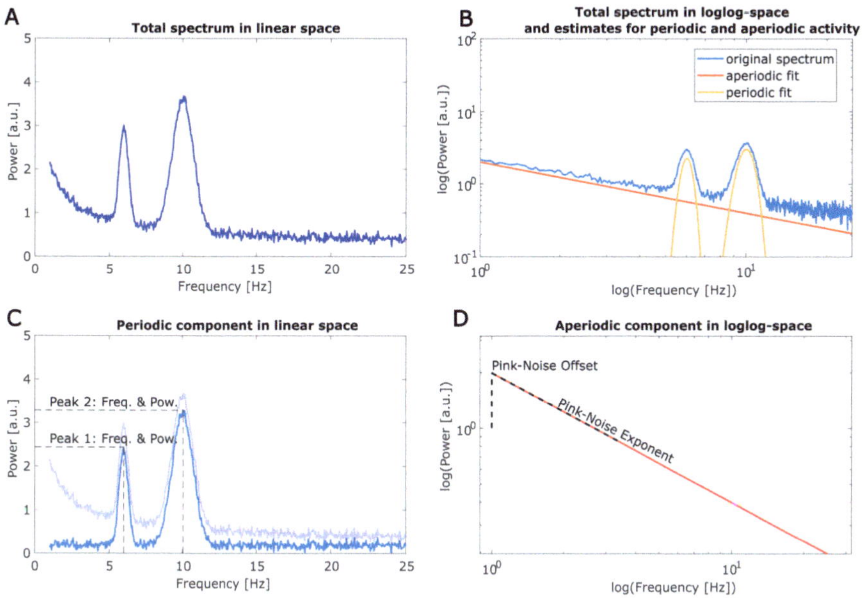

Fig. 31.3 Basic principle of dividing spectral data into periodic and aperiodic components. (**a**) Original total spectrum composed of aperiodic activity (1/f-floor) and two oscillatory signals (visible as peaks in the spectrum). The magnitude of the peaks represents the magnitude of the periodic activity plus the underlying noise floor. (**b**) The same spectrum as viewed on a double-logarithmic scale (blue line). On this scale the aperiodic noise floor resembles a straight line. Through iterative fitting procedures, the total spectrum is divided into an aperiodic fit (red) and an oscillatory fit (yellow). (**c**) Subtracting the aperiodic fit from the total spectrum (transparent blue curve) yields the corrected purely periodic spectrum (solid blue), from which we can now extract the corrected peak-frequency and peak-power values for further analysis of periodic activity. (**d**) The aperiodic fit (red) can be used for the characterization of the pink-noise activity: We can extract the offset ($\log_{10}$ of the intercept in log-log space) and the exponent (-slope in log-log space) of the pink-noise activity

"subtracted" from the full spectrum, isolating the "truly" periodic activity. Note that a subtraction in logarithmic space is mathematically a division in linear space, rendering the term ambivalent. Some studies (e.g., Kasten et al., 2024) perform the correction as a subtraction in linear space, while others perform it in log space (Donoghue et al., 2020). Which approach is the "correct" one is currently debated (Gyurkovics et al., 2021) and depends on the specific research question. Figure 31.3 depicts the basic thought process behind the general approach.

Seven distinct approaches have been proposed, each addressing the problem from a different perspective:

1. **SPECPARAM** (Spectral Parameterization; Donoghue et al., 2020) works by iteratively modeling periodic activity as Gaussian peaks and the fractal component as a line in log-log space. Note that this approach was formerly called FOOOF (Fitting Oscillations and One-Over-F).

2. **IRASA** (Irregular Resampling Auto-Spectral Analysis; Wen and Liu, 2016) separates oscillatory and fractal (1/f) components by resampling the signal at non-integer rates and computing the geometric mean of the power spectra at these rates to isolate the fractal component. Subtracting this fractal component from the total spectrum isolates the oscillatory signals.
3. **eBOSC** (extended Better Oscillation Detection; Kosciessa et al., 2020) identifies oscillations by setting thresholds based on a statistical model of the background (1/f) noise, enabling differentiation of oscillations from noise. It refines the traditional BOSC approach (Whitten et al., 2011; Caplan et al., 2001) by incorporating longer time windows and enhancements to improve the detection of oscillations in the presence of 1/f noise.
4. **PAWNextra** (Pink and White Noise Extraction; Barry and De Blasio, 2021) employs a traditional Welch periodogram to estimate the power spectrum, followed by techniques to distinguish the oscillatory peaks from the background 1/f slope and white noise. It utilizes model fitting to extract the power of oscillations relative to the 1/f baseline.
5. **MODAL** (Multiple Oscillations Detection Algorithm; Watrous et al., 2018) detects oscillatory frequency bands by fitting an aperiodic component using robust linear regression. Frequencies that deviate by more than one standard deviation from this fit are classified as periodic activity.
6. **BSD** (Bayesian Spectral Decomposition; Medrano et al., 2025) models neural power spectra as a composite of Gaussian peaks and aperiodic activity with added noise. The parameters of these components are estimated by employing a Bayesian framework that accounts for uncertainty.
7. **SPRiNT** (Spectral Parameterization Resolved in Time; Wilson et al., 2022) is an extension to the SPECPARAM approach that works on time-frequency data.

SPECPARAM is built into the Brainstorm (Tadel et al., 2011) and Fieldtrip (Oostenveld et al., 2011) toolboxes and also available as Python code. IRASA is implemented in Fieldtrip as well and also available for Python. EBOSC and SPRiNT have MATLAB code available on GITHUB, and BSD is stated to be made available as a toolbox as part of the SPM software package.

SPECPARAM and IRASA are currently the most widely used methods (SPECPARAM: 1223 citations, IRASA: 307 citations). For both methods, Gerster et al. (2022) conducted a detailed analysis using artificial spectra with known ground truth to identify common challenges. SPECPARAM performs well, when there are no plateaus (disrupting the power-law distribution) within the fitting range. Oscillatory peaks at the border of the fitting range cannot be accurately modeled, and oscillations lacking distinct peaks are indistinguishable from the noise floor. For IRASA, the resampling procedure extends the evaluated frequency range beyond the fitted range, which can introduce biases toward lower slopes if the evaluated range encounters high-pass filter edges or high-frequency noise. To mitigate this, small resampling factors should be used. However, broad oscillatory peaks require high resampling factors for accurate detection by the algorithm, necessitating a trade-off between these challenges. Additionally, as with SPECPARAM, IRASA struggles to detect oscillations that do not manifest as a clear peak.

For all methods, there is no universal set of parameters that fits every recording, and relying solely on goodness-of-fit metrics might not prevent overfitting as was pointed out for SPECPARAM by Ostlund et al. (2022). Recently, Wilson et al. (2024) proposed an algorithm in a preprint to assist with model selection for SPECPARAM and SPRiNT, with code available for MATLAB (ms-SPECPARAM and ms-SPRinT). Generally, the most effective approach for all methods involves starting with the recommended default parameters, followed by visual inspection before adjusting the parameters to better align with the specific experimental data.

31.6 Closed-Loop Systems

Most stimulation studies employ a predetermined set of parameters to elicit their effects. Ideally, individual anatomy and individual brain activity features like individual peak-frequencies are taken into account, but once the stimulation has started, it usually follows a fixed waveform.

It has been established in several studies, however, that the effects of stimulation are dependent on the brain's current state (Neuling et al., 2013; Ruhnau et al., 2016; Feurra et al., 2013; Wei et al., 2024), with only certain states actually being affected (Kasten & Herrmann, 2022; Luo et al., 2025). Moreover, endogenous brain oscillations can change over time, for instance, the peak alpha frequency tends to decrease with time on task (Benwell et al., 2019). Consequently, parameters that were fitting at the onset of stimulation might become ineffective as an experiment progresses. Concurrent EEG recording during stimulation provides an opportunity to adapt to these changes in brain activity by applying stimulation selectively during certain states or adjusting the phase and frequency to align with a targeted endogenous oscillation. Such setups are commonly referred to as "closed-loop system" or "brain state-dependent NTBS" (Bergmann et al., 2016; Bergmann, 2018; Thut et al., 2017). Figure 31.4 illustrates the general schematic of such a system. A control computer is connected to both a neuroimaging device (e.g., EEG, MEG, and fMRI) and a stimulator (e.g., TMS, tDCS, tACS, and transcranial ultrasound stimulator), which are both connected to a participant. The control computer receives neuro-imaging data in (near) real time, extracts relevant brain activity features, and adjusts or triggers stimulation based on these features. Due to its relatively low cost and practicality, EEG is a prime-candidate for the neuroimaging component of such a system.

Numerous studies have highlighted the need for and outlined the principles of closed-loop brain stimulation systems (Soleimani et al., 2023; Frohlich & Townsend, 2021; Zarubin et al., 2018; Iturrate et al., 2018; Kohli et al., 2017; Shirinpour et al., 2025; Schestatsky et al., 2013; Xiong et al., 2025). However,

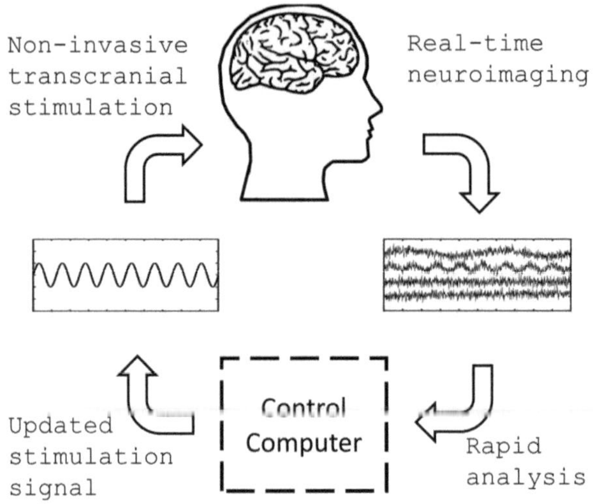

Fig. 31.4 Schematic of a closed-loop brain stimulation system: A participant or patient receives ongoing brain stimulation. Neuroimaging data (e.g., EEG) is concurrently recorded and fed to a control computer via remote data access, Lab Streaming Layer (LSL), or a similar protocol. The control computer performs a rapid analysis of brain activity, like current state, or power, frequency or phase of a brain rhythm. Based on the extracted features, the control computer either triggers the application of a predefined waveform or updates a waveform and feeds it to the attached stimulator via a digital-to-analog converter

only a limited number of studies have implemented this concept. This scarcity may stem from two primary technical challenges in EEG/TES closed-loop designs. The first challenge is the requirement for rapid computational processing to acquire EEG data, performing data-processing routines, feature extraction, and the generation of waveforms. All system latencies must be minimal or at least predictable to ensure stimulation aligns precisely with the specific brain activity features. In practice, this demands dedicated real-time systems, such as the Real-Time eXperiment Interface (RTXI) (Patel et al., 2017) or field programmable gate array (FPGA) (Wilde et al., 2015). The second challenge in EEG/TES closed-loop systems is the artifact induced by electrical stimulation in the EEG signal (see Sect. 31.3). To address this, many studies have adopted an intermittent closed-loop approach rather than true real-time processing. In such a system stimulation is delivered in short blocks of limited length, with feature extraction performed on EEG segments recorded between these blocks. For example, in a proof-of-principle study, Leite et al. (2017) developed a protocol that triggered a continuous block of tDCS when the EEG exhibited a certain alpha-to-beta ratio. Similarly, in tACS studies, stimulation is delivered in short blocks, interleaved with stimulation-free EEG intervals, during which new stimulation features are extracted. Schwippel et al. (2024) implemented 120-s of tACS at individual alpha frequency when the alpha power within a 10-s observation window exceeded a predefined threshold as a treatment for Major Depressive Disorder. To selectively enhance sleep spindles, Lustenberger et al. (2016)

employed an RTXI system to deliver 1-s bursts of 12 Hz tACS followed by a 6.5-s timeout, triggered when ongoing band-pass-filtered brain activity exhibited five rectified sigma peaks exceeding an individual threshold. Stecher et al. (2021) implemented a design using 8-s tACS blocks, alternating with 8-s epochs of EEG recordings, continuously adjusting the stimulation frequency to match the current prevalent alpha peak-frequency. Zarubin et al. (2020) employed an intermittent design with 1-s stimulation blocks, predicting the alpha rhythm phase for the upcoming block using a Hilbert-transformed 250-ms window before each stimulation block, allowing tACS to be applied either in-phase or antiphase with the endogenous alpha rhythm. Similarly, Ketz et al. (2018) and Jones et al. (2018) utilized a setup with a running 5-s buffer to detect slow-waves in the EEG, fitting a sine-wave matched in frequency and phase, and applying it for five cycles. Both setups rely on accurate estimates of system latencies, including the time required to receive EEG data, process it, and initiate stimulation.

Three studies have implemented true real-time closed-loop approaches, with feature extraction occurring simultaneously with ongoing stimulation, each employing distinct methods to obtain usable EEG data despite stimulation artifacts. For tDCS, as long as the EEG data does not clip, the artifact primarily causes a spectral offset at 0 Hz without affecting the spectrum of typical brain oscillations. Employing a combination of standard low-pass and high-pass filters along with basic artifact rejection, Caravati et al. (2024) adjusted ongoing tDCS amplitude—either increasing or decreasing it—based on statistical metrics derived from a sliding 3-s window of EEG data to improve performance in a continuous performance task. In tACS studies, when the stimulated frequency band and the evaluated band of interest do not overlap, a narrow band-pass filter around the band of interest can yield artifact free EEG recordings. Boyle and Fröhlich (2013) utilized this approach with an RTXI system, continuously applying 40 Hz tACS while monitoring alpha activity in the EEG. To obtain stimulation artifact-free EEG data, they applied a narrow band-pass filter in the alpha range and delivered 2-s bursts of 1 mA tACS when alpha power exceeded a predefined threshold. Otherwise, the system constantly applied 0.1 mA tACS. Lastly, tTIS and AM-tACS theoretically produce artifact-free data in the band of interest, even during stimulation, as only the carrier signal should appear in the spectrum. In practice, however, residual artifacts may still occur at the envelope frequency (see Sect. 31.3.3.2). By combining AM-tACS and stimulation artifact source separation in a real-time computer, Haslacher et al. (2021) successfully restored phase-estimates of the ongoing alpha activity, enabling stimulation at six different phases by continuously adjusting the envelope-phase and frequency of the AM-tACS.

31.7 Technical Requirements

The technical requirements for combining EEG and tES depend primarily on the research objective: Whether the focus is on offline or online effects. For offline effects, the stimulation component imposes minimal constrains on the EEG setup.

Most EEG systems feature high input impedance in the range of Mega-Ωs which safeguards the amplifier from the applied current. However, the placement of stimulation electrodes might overlap with standard EEG electrode sites, limiting the spatial resolution of the recordings. Using a "high definition" stimulation setup (hd-tDCS or hd-tACS), composed of multiple smaller electrodes in favor of few large electrodes can not only increase focality of stimulation but can also allow the placement of more EEG electrodes (Roy et al., 2014). Care must also be taken to prevent bridging between closely positioned electrodes. Additional requirements are determined by the frequency and amplitude of the brain activity under study. The EEG sampling rate should adhere to the Nyquist-Shannon sampling theorem to accurately capture the oscillation of interest, with the general recommendation to over-sample by at least 3–5 times the target frequency (Luck, 2014). For example, to study high gamma activity up to 150 Hz, a minimum sampling rate of at least 450 Hz is required. Given that brain oscillations follow a power-law distribution, high frequency bands like gamma typically produce small amplitudes of less than 1 μV in the EEG (Busch et al., 2006). Therefore, a system with a sufficient voltage resolution, typically 0.1 μV per bit, is essential for accurate measurement (see Chap. 12: Hardware for Recording EEG and Peripheral Physiology).

When the focus is on physiological online effects in the EEG during active stimulation, two primary challenges must be addressed to accurately recover true brain activity: 1. EEG saturation and 2. distortion caused by the stimulation artifact. Saturation occurs when the EEG amplitude exceeds the dynamic range of the amplifier's analog-to-digital (A/D) converter. As illustrated in Fig. 31.5, the tACS-artifact can reach amplitudes in the millivolt range, depending on factors such as impedance, the proximity of recording electrode to stimulation site, and the stimulation amplitude. This huge artifact can drive the EEG signal into saturation if the amplifier's A/D range is insufficient. Saturation is characterized by blocking (or clipping) of the signal, where the amplifier continuously records the maximum or minimum value upon reaching the A/D limit, effectively obliterating any recoverable information about brain activity (see Fig. 31.6). To mitigate saturation, selecting an EEG system with a wide A/D range is critical. For example, the 24-bit actiCHamp device from Brain Products offers an A/D range of ±400 mV, significantly broader than their 16-bit BrainAmp system, which is limited to ±16 mV. Additionally, maintaining low impedances for both stimulation and recording electrodes is essential to minimize the artifact and ensure accurate EEG recordings.

The distortions of waveforms in EEG recordings during brain stimulation stem from various sources, primarily rooted in the nonlinear response characteristics of the involved systems. Nonlinear response functions, inherent to all electrical equipment, result in a nonproportional mapping of the ideal input signal (the digital stimulation signal) to the output (the stimulation waveform in the EEG, see Fig. 31.7). Kasten et al. (2018) explored these nonlinear properties in an AM-tACS setup, demonstrating that nonlinearities are introduced by the stimulation device, the EEG hardware, and, to a lesser extent, the digital-to-analog converter driving the stimulation, all of which impact the accuracy of the recorded stimulation signal. Nonlinearities pose significant challenges when analyzing EEG data during

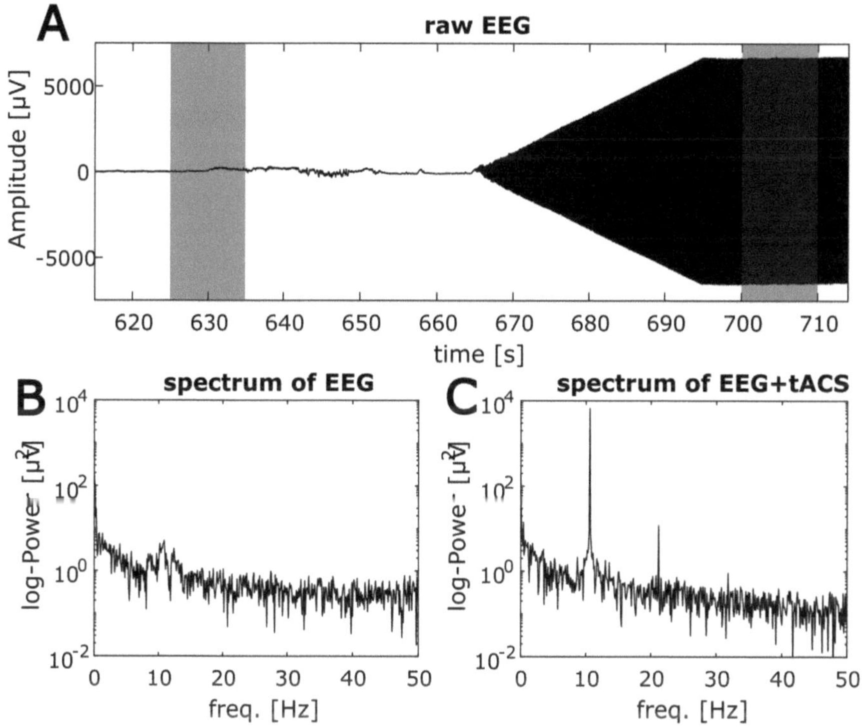

Fig. 31.5 Example of a tACS-artifact size in an EEG recoding. (**a**) A 110 seconds EEG recording during onset of tACS at 2 mA and 10 Hz, recorded from electrode Pz, with stimulation applied over Cz and Oz. The normal EEG signal is situated in the μV range, while the tACS-artifact reaches a peak-to-peak amplitude of 10mV. The gray regions depict 10 second windows for the FFT-spectra below. (**b**) Semi-logarithmic plot of a normal EEG spectrum with a visible alpha peak at 10.9 Hz. (**c**) Semi-logarithmic plot of an EEG containing tACS. Notice the peak at the stimulation frequency that is 3 orders of magnitude above any physiological activity. Harmonics are visible at 20 and 30 Hz

concurrent stimulation, affecting both artifact removal strategies (such as template subtraction or spatial filtering) and approaches that focus on uncontaminated frequency bands (e.g., tTIS, AM-tACS). Distortions of the ideal waveform by nonlinear transfer effects can introduce (sub)harmonic components or the envelope frequency of tTIS or AM-tACS stimulation into the EEG spectrum, complicating accurate analysis of brain activity. To minimize nonlinearities, in transcranial electrical stimulation, the choice of electrode type is critical. Typically, three different types of electrodes are used: carbonized rubber electrodes enveloped in saline-soaked sponges, bar rubber electrodes affixed with adhesive conductive paste, or Ag/AgCl electrodes applied to the scalp with an appropriate electrolyte. Carbonized rubber electrodes, however, are polarizable (Agrebi et al., 2017; Zhu & Deegan, 2024; Stecher et al., 2025), leading to charge accumulation on their own that risks amplifier saturation and inherently causes nonlinear response functions (Fig. 31.7). Therefore, it might be preferable to use Ag/AgCl electrodes for stimulation, which are

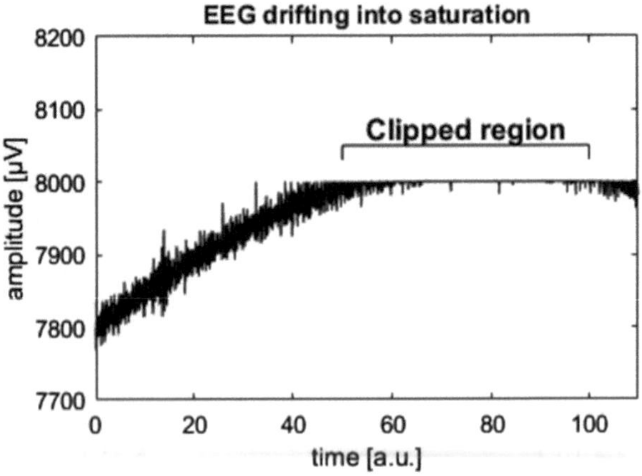

Fig. 31.6 Example of EEG signal drifting (naturally or by applied stimulation into the limit of the amplifiers A/D range). In the clipped region, all recorded samples have the same value. Any physiological information is lost from this time period

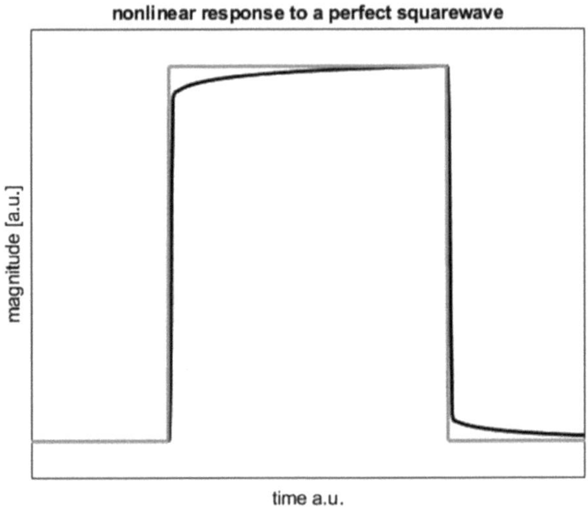

Fig. 31.7 Example of nonlinear component response in a stimulation + EEG setup. The gray line shows the perfect digital square-wave, before passing through the digital/analog converter, the stimulation device, rubber electrodes, EEG electrodes, and the EEG amplifier. The black line shows how this square was recorded in the EEG. Note the distortions at both rising and falling edges of the square

nonpolarizable. However, it should be considered that Ag/AgCl electrodes have a comparatively higher price and limited lifespan (Langenbach et al., 2020).

On the stimulator side, the sampling rate of the generated signal must be sufficient to avoid aliasing of the waveform in the EEG. If the stimulator's sampling rate is lower than that of the EEG, the recorded waveform will appear as a staircase, introducing subharmonic components into the EEG spectrum. For optimal results, especially when employing template subtraction for artifact correction (Vosskuhl et al., 2020), the recording and stimulation devices should be synchronized using a shared clock-channel. Otherwise, even minor clock drifts between the systems can cause phase divergence and distortions of the recorded waveform.

31.8 Conclusion

In conclusion, we strongly advocate for verifying that brain stimulation methods effectively modulate brain activity by demonstrating changes in EEG recordings, whether online or offline. As discussed, this necessitates specific recording hardware and careful measures to eliminate or minimize artifacts.

As an EEG amplifier that can definitely cope with the voltages of tES without going into saturation, we can recommend the 24-bit actiCHamp by Brain Products. We hope that readers find our suggestions helpful.

The website *Combining EEG and tACS* that is also linked from the Brain Products website lists some of our key papers on the topic:

https://uol.de/allgemeine-psychologie/combining-eeg-and-tacs.

Conflicts of Interest
CSH holds a patent on brain stimulation. All other authors declare no conflicts of interest.

Funding
CSH was funded by the Deutsche Forschungsgemeinschaft (DFG, German Research Foundation) under Germany's Excellence Strategy—EXC 2177/1—Project ID 390895286. CSH was also funded by the research training group on Neuromodulation of Motor and Cognitive Function in Brain Health and Disease (RTG 2783).

References

Agrebi, F., Ghorbel, N., Ladhar, A., Bresson, S., & Kallel, A. (2017). Enhanced dielectric properties induced by loading cellulosic nanowhiskers in natural rubber: Modeling and analysis of electrode polarization. *Materials Chemistry and Physics, 200*, 155–163.

Aktürk, T., de Graaf, T. A., Güntekin, B., Hanoğlu, L., & Sack, A. T. (2022). Enhancing memory capacity by experimentally slowing theta frequency oscillations using combined EEG-tACS. *Scientific Reports, 12*(1), 14199.

Antal, A., & Herrmann, C. S. (2016). Transcranial alternating current and random noise stimulation: Possible mechanisms. *Neural Plasticity, 2016*(1), 3616807.

Antal, A., Kincses, T. Z., Nitsche, M. A., Bartfai, O., & Paulus, W. (2004). Excitability changes induced in the human primary visual cortex by transcranial direct current stimulation: Direct electrophysiological evidence. *Investigative Ophthalmology & Visual Science, 45*(2), 702–707.

Barry, R. J., & De Blasio, F. M. (2021). Characterizing pink and white noise in the human electroencephalogram. *Journal of Neural Engineering, 18*(3), 034001.

Benwell, C. S., London, R. E., Tagliabue, C. F., Veniero, D., Gross, J., Keitel, C., & Thut, G. (2019). Frequency and power of human alpha oscillations drift systematically with time-on-task. *Neuroimage, 192*, 101–114.

Bergmann, T. O. (2018). Brain state-dependent brain stimulation. *Frontiers in Psychology, 9*, 2108.

Bergmann, T. O., Karabanov, A., Hartwigsen, G., Thielscher, A., & Siebner, H. R. (2016). Combining non-invasive transcranial brain stimulation with neuroimaging and electrophysiology: Current approaches and future perspectives. *Neuroimage, 140*, 4–19.

Boonstra, T. W., Nikolin, S., Meisener, A.-C., Martin, D. M., & Loo, C. K. (2016). Change in mean frequency of resting-state electroencephalography after transcranial direct current stimulation. *Frontiers in Human Neuroscience, 10*, 270.

Boyle, M. R., & Fröhlich, F. (2013). EEG feedback-controlled transcranial alternating current stimulation. In *2013 6th International IEEE/EMBS Conference on Neural Engineering (NER)* (pp. 140–143). IEEE.

Busch, N. A., Herrmann, C. S., Müller, M. M., Lenz, D., & Gruber, T. (2006). A cross-laboratory study of event-related gamma activity in a standard object recognition paradigm. *Neuroimage, 33*(4), 1169–1177.

Buzsaki, G. (2006). *Rhythms of the Brain.* Oxford University Press.

Caplan, J. B., Madsen, J. R., Raghavachari, S., & Kahana, M. J. (2001). Distinct patterns of brain oscillations underlie two basic parameters of human maze learning. *Journal of Neurophysiology, 86*(1), 368–380.

Caravati, E., Barbeni, F., Chiarion, G., Raggi, M., & Mesin, L. (2024). Closed-loop transcranial electrical neurostimulation for sustained attention enhancement: A pilot study towards personalized intervention strategies. *Bioengineering, 11*(5), 467.

Cohen Kadosh, R., & Elliott, P. (2013). Neuroscience: Brain stimulation has a long history. *Nature, 500*(7464), 529.

Conley, A. C., Fulham, W., Marquez, J. L., Parsons, M. W., & Karayanidis, F. (2016). No effect of anodal transcranial direct current stimulation over the motor cortex on response-related ERPs during a conflict task. *Frontiers in Human Neuroscience, 10*, 384.

D'Atri, A., Scarpelli, S., Gorgoni, M., Alfonsi, V., Annarumma, L., Giannini, A. M., Ferrara, M., Ferlazzo, F., Rossini, P. M., & De Gennaro, L. (2019). Bilateral theta transcranial alternating current stimulation (tACS) modulates EEG activity: When tACS works awake it also works asleep. *Nature and Science of Sleep*, 343–356.

Davis, M.-C., Fitzgerald, P. B., Bailey, N. W., Sullivan, C., Stout, J. C., Hill, A. T., & Hoy, K. E. (2023). Effects of medial prefrontal transcranial alternating current stimulation on neural activity and connectivity in people with Huntington's disease and neurotypical controls. *Brain Research, 1811*, 148379.

Donoghue, T., Haller, M., Peterson, E. J., Varma, P., Sebastian, P., Gao, R., Noto, T., Lara, A. H., Wallis, J. D., Knight, R. T., et al. (2020). Parameterizing neural power spectra into periodic and aperiodic components. *Nature Neuroscience, 23*(12), 1655–1665.

Donoghue, T., Schaworonkow, N., & Voytek, B. (2022). Methodological considerations for studying neural oscillations. *European Journal of Neuroscience, 55*(11–12), 3502–3527.

Dowsett, J., & Herrmann, C. S. (2016). Transcranial alternating current stimulation with sawtooth waves: Simultaneous stimulation and EEG recording. *Frontiers in Human Neuroscience, 10*, 135.

Feurra, M., Pasqualetti, P., Bianco, G., Santarnecchi, E., Rossi, A., & Rossi, S. (2013). State-dependent effects of transcranial oscillatory currents on the motor system: what you think matters. *Journal of Neuroscience, 33*(44), 17483–17489.

Frohlich, F., & Townsend, L. (2021). Closed-loop transcranial alternating current stimulation: Towards personalized non-invasive brain stimulation for the treatment of psychiatric illnesses. *Current Behavioral Neuroscience Reports, 8*(2), 51–57.

Gao, R., Peterson, E. J., & Voytek, B. (2017). Inferring synaptic excitation/inhibition balance from field potentials. *Neuroimage, 158*, 70–78.

Gebodh, N., Esmaeilpour, Z., Adair, D., Chelette, K., Dmochowski, J., Woods, A. J., Kappenman, E. S., Parra, L. C., & Bikson, M. (2019). Inherent physiological artifacts in EEG during tDCS. *Neuroimage, 185*, 408–424.

Gerster, M., Waterstraat, G., Litvak, V., Lehnertz, K., Schnitzler, A., Florin, E., Curio, G., & Nikulin, V. (2022). Separating neural oscillations from aperiodic 1/f activity: Challenges and recommendations. *Neuroinformatics, 20*(4), 991–1012.

Ghin, F., O'Hare, L., & Pavan, A. (2021). Electrophysiological aftereffects of high-frequency transcranial random noise stimulation (HF-tRNS): An EEG investigation. *Experimental Brain Research, 239*(8), 2399–2418.

Grossman, N., Bono, D., Dedic, N., Kodandaramaiah, S. B., Rudenko, A., Suk, H.-J., Cassara, A. M., Neufeld, E., Kuster, N., Tsai, L.-H., Pascual-Leone, A., & Boyden, E. S. (2017). Noninvasive deep brain stimulation via temporally interfering electric fields. *Cell, 169*(6), 1029.e16–1041.e16.

Guarnieri, R., Brancucci, A., D'Anselmo, A., Manippa, V., Swinnen, S. P., Tecchio, F., & Mantini, D. (2020). A computationally efficient method for the attenuation of alternating current stimulation artifacts in electroencephalographic recordings. *Journal of Neural Engineering, 17*(4), 046038.

Guidetti, M., Ferrara, R., Montemagno, K., Maiorana, N. V., Bocci, T., Marceglia, S., Oliveri, S., Bianchi, A. M., & Priori, A. (2025). The "Cocombola Study": A Physical Phantom Model for tDCS-Induced Electric Field Distribution. *Bioengineering, 12*(4), 346.

Guleyupoglu, B., Schestatsky, P., Edwards, D., Fregni, F., & Bikson, M. (2013). Classification of methods in transcranial electrical stimulation (tES) and evolving strategy from historical approaches to contemporary innovations. *Journal of Neuroscience Methods, 219*(2), 297–311.

Gyurkovics, M., Clements, G. M., Low, K. A., Fabiani, M., & Gratton, G. (2021). The impact of 1/f activity and baseline correction on the results and interpretation of time-frequency analyses of EEG/MEG data: A cautionary tale. *Neuroimage, 237*, 118192.

Harada, T., Hara, M., Matsushita, K., Kawakami, K., Kawakami, K., Anan, M., & Sugata, H. (2020). Off-line effects of alpha-frequency transcranial alternating current stimulation on a visuomotor learning task. *Brain and Behavior, 10*(9), e01754.

Haslacher, D., Nasr, K., Robinson, S. E., Braun, C., & Soekadar, S. R. (2021). Stimulation artifact source separation (SASS) for assessing electric brain oscillations during transcranial alternating current stimulation (tACS). *Neuroimage, 228*, 117571.

He, B. J. (2014). Scale-free brain activity: past, present, and future. *Trends in Cognitive Sciences, 18*(9), 480–487.

He, B. J., Zempel, J. M., Snyder, A. Z., & Raichle, M. E. (2010). The temporal structures and functional significance of scale-free brain activity. *Neuron, 66*(3), 353–369.

Helfrich, R. F., Knepper, H., Nolte, G., Strüber, D., Rach, S., Herrmann, C. S., Schneider, T. R., & Engel, A. K. (2014a). Selective modulation of interhemispheric functional connectivity by HD-tACS shapes perception. *PLoS biology, 12*(12), e1002031.

Helfrich, R. F., Schneider, T. R., Rach, S., Trautmann-Lengsfeld, S. A., Engel, A. K., & Herrmann, C. S. (2014b). Entrainment of brain oscillations by transcranial alternating current stimulation. *Current Biology, 24*(3), 333–339.

Herrmann, C. S., Rach, S., Neuling, T., & Strüber, D. (2013). Transcranial alternating current stimulation: A review of the underlying mechanisms and modulation of cognitive processes. *Frontiers in Human Neuroscience, 7*, 279.

Herrmann, C. S., & Strüber, D. (2017). What can transcranial alternating current stimulation tell us about brain oscillations? *Current Behavioral Neuroscience Reports, 4*(2), 128–137.

Herrmann, C. S., Strüber, D., Helfrich, R. F., & Engel, A. K. (2016). EEG oscillations: from correlation to causality. *International Journal of Psychophysiology, 103*, 12–21.

Hunold, A., Machts, R., & Haueisen, J. (2020). Head phantoms for bioelectromagnetic applications: a material study. *Biomedical engineering online, 19*(1), 87.

Hunold, A., Strohmeier, D., Fiedler, P., & Haueisen, J. (2018). Head phantoms for electroencephalography and transcranial electric stimulation: A skull material study. *Biomedizinische Technik, 63*(6), 683–689.

Iturrate, I., Pereira, M., & Millán, J. d. R. (2018). Closed-loop electrical neurostimulation: Challenges and opportunities. *Current Opinion in Biomedical Engineering, 8*, 28–37.

Jones, A. P., Choe, J., Bryant, N. B., Robinson, C. S., Ketz, N. A., Skorheim, S. W., Combs, A., Lamphere, M. L., Robert, B., Gill, H. A., et al. (2018). Dose-dependent effects of closed-loop tACS delivered during slow-wave oscillations on memory consolidation. *Frontiers in Neuroscience, 12*, 867.

Kasten, F. H., Dowsett, J., & Herrmann, C. S. (2016). Sustained aftereffect of α-tACS lasts up to 70 min after stimulation. *Frontiers in Human Neuroscience, 10*, 245.

Kasten, F. H., & Herrmann, C. S. (2022). The hidden brain-state dynamics of tACS aftereffects. *NeuroImage, 264*, 119713.

Kasten, F. H., Lattmann, R., Strüber, D., & Herrmann, C. S. (2024). Decomposing the effects of α-tACS on brain oscillations and aperiodic 1/f activity. *Brain Stimulation: Basic, Translational, and Clinical Research in Neuromodulation, 17*(3), 721–723.

Kasten, F. H., Negahbani, E., Fröhlich, F., & Herrmann, C. S. (2018). Non-linear transfer characteristics of stimulation and recording hardware account for spurious low-frequency artifacts during amplitude modulated transcranial alternating current stimulation (AM-tACS). *Neuroimage, 179*, 134–143.

Ke, S.-C., Lo, Y.-H., & Tseng, P. (2024). No frequency-specific effect of transcranial random noise stimulation on resting EEG. *Journal of Integrative Neuroscience, 23*(3), 59.

Keil, A., Bernat, E. M., Cohen, M. X., Ding, M., Fabiani, M., Gratton, G., Kappenman, E. S., Maris, E., Mathewson, K. E., Ward, R. T., et al. (2022). Recommendations and publication guidelines for studies using frequency domain and time-frequency domain analyses of neural time series. *Psychophysiology, 59*(5), e14052.

Ketz, N., Jones, A. P., Bryant, N. B., Clark, V. P., & Pilly, P. K. (2018). Closed-loop slow-wave tACS improves sleep-dependent long-term memory generalization by modulating endogenous oscillations. *Journal of Neuroscience, 38*(33), 7314–7326.

Kohli, S., & Casson, A. J. (2019). Removal of gross artifacts of transcranial alternating current stimulation in simultaneous EEG monitoring. *Sensors, 19*(1).

Kohli, S., Krachunov, S., & Casson, A. (2017). Towards closed-loop transcranial electrical stimulation: a comparison of methods for real time tES-EEG artefact removal using a phantom head model. In *2nd International Brain Stimulation Conference*.

Kosciessa, J. Q., Grandy, T. H., Garrett, D. D., & Werkle-Bergner, M. (2020). Single-trial characterization of neural rhythms: Potential and challenges. *NeuroImage, 206*, 116331.

Kunzelmann, K., Meier, L., Grieder, M., Morishima, Y., & Dierks, T. (2018). No effect of transcranial direct current stimulation of the auditory cortex on auditory-evoked potentials. *Frontiers in neuroscience, 12*, 880.

Lafleur, L.-P., Murray, A., Desforges, M., Pacheco-Barrios, K., Fregni, F., Tremblay, S., Saint-Amour, D., Lepage, J.-F., & Théoret, H. (2021). No aftereffects of high current density 10 Hz and 20 Hz tACS on sensorimotor alpha and beta oscillations. *Scientific Reports, 11*(1), 21416.

Langenbach, B. P., Savic, B., Baumgartner, T., & Knoch, D. (2020). Repeated anodal HD-tDCS stimulation might render silver chloride electrodes unreliable. *Brain Stimulation, 13*(3), 525–526.

Leite, J., Morales-Quezada, L., Carvalho, S., Thibaut, A., Doruk, D., Chen, C.-F., Schachter, S. C., Rotenberg, A., & Fregni, F. (2017). Surface EEG-transcranial direct current stimulation (tDCS) closed-loop system. *International Journal of Neural Systems, 27*(06), 1750026.

Luck, S. J. (2014). *An introduction to the event-related potential technique*. MIT Press.

Luo, H., Ye, X., Cai, H.-T., Wang, M., Wang, Y., Liu, Q., Xu, Y., Mao, Z., Cai, Y., Hong, J., et al. (2025). Frequency-specific and state-dependent neural responses to brain stimulation. *Molecular Psychiatry*, 1–11.

Lustenberger, C., Boyle, M. R., Alagapan, S., Mellin, J. M., Vaughn, B. V., & Fröhlich, F. (2016). Feedback-controlled transcranial alternating current stimulation reveals a functional role of sleep spindles in motor memory consolidation. *Current Biology, 26*(16), 2127–2136.

Mancini, M., Pellicciari, M. C., Brignani, D., Mauri, P., De Marchis, C., Miniussi, C., & Conforto, S. (2015). Automatic artifact suppression in simultaneous tDCS-EEG using adaptive filtering. In *2015 37th Annual International Conference of the IEEE Engineering in Medicine and Biology Society (EMBC)* (pp. 2729–2732). IEEE.

Mangia, A. L., Pirini, M., & Cappello, A. (2014). Transcranial direct current stimulation and power spectral parameters: A tDCS/EEG co-registration study. *Frontiers in Human Neuroscience, 8*, 601.
Masina, F., Napoli, E., Santacesaria, P., Giustiniani, A., Zago, S., Marino, M., Pezzetta, R., Montemurro, S., Manzo, N., Cona, G., et al. (2025). Transcranial alternating current stimulation selectively modulates aperiodic EEG component: Unveiling alternative mechanisms of modulation. *Clinical Neurophysiology*, 2110929.
Matsumoto, J., Fujiwara, T., Takahashi, O., Liu, M., Kimura, A., & Ushiba, J. (2010). Modulation of mu rhythm desynchronization during motor imagery by transcranial direct current stimulation. *Journal of NeuroEngineering and Rehabilitation, 7*(1), 27.
Medrano, J., Alexander, N. A., Seymour, R. A., & Zeidman, P. (2025). BSD: A Bayesian framework for parametric models of neural spectra. *European Journal of Neuroscience, 61*(10), e70149.
Merton, P., & Morton, H. (1980). Stimulation of the cerebral cortex in the intact human subject. *Nature, 285*(5762), 227–227.
Millard, S. K., Speis, D. B., Skippen, P., Chiang, A. K., Chang, W.-J., Lin, A. J., Furman, A. J., Mazaheri, A., Seminowicz, D. A., & Schabrun, S. M. (2024). Can non-invasive brain stimulation modulate peak alpha frequency in the human brain? A systematic review and meta-analysis. *European Journal of Neuroscience, 60*(3), 4182–4200.
Miller, K. J., Honey, C. J., Hermes, D., Rao, R. P., Ojemann, J. G., et al. (2014). Broadband changes in the cortical surface potential track activation of functionally diverse neuronal populations. *NeuroImage, 85*, 711–720.
Miniussi, C., Brignani, D., & Pellicciari, M. C. (2012). Combining transcranial electrical stimulation with electroencephalography: A multimodal approach. *Clinical EEG and Neuroscience, 43*(3), 184–191.
Mohsen, S., Mahmoudian, S., Talebian, S., & Pourbakht, A. (2019). Multisite transcranial Random Noise Stimulation (tRNS) modulates the distress network activity and oscillatory powers in subjects with chronic tinnitus. *Journal of Clinical Neuroscience, 67*, 178–184.
Nascimento, D. C., Depetri, G., Stefano, L. H., Anacleto, O., Leite, J. P., Edwards, D. J., Santos, T. E., & Louzada Neto, F. (2019). Entropy analysis of high-definition transcranial electric stimulation effects on EEG dynamics. *Brain Sciences, 9*(8), 208.
Neuling, T., Rach, S., & Herrmann, C. S. (2013). Orchestrating neuronal networks: Sustained after-effects of transcranial alternating current stimulation depend upon brain states. *Frontiers in Human Neuroscience, 7*, 161.
Neuling, T., Ruhnau, P., Weisz, N., Herrmann, C. S., & Demarchi, G. (2017). Faith and oscillations recovered: On analyzing EEG/MEG signals during tACS. *Neuroimage, 147*, 960–963.
Nitsche, M. A., & Paulus, W. (2000). Excitability changes induced in the human motor cortex by weak transcranial direct current stimulation. *The Journal of Physiology, 527*(Pt 3), 633–639.
Noury, N., Hipp, J. F., & Siegel, M. (2016). Physiological processes non-linearly affect electrophysiological recordings during transcranial electric stimulation. *Neuroimage, 140*, 99–109.
Noury, N., & Siegel, M. (2017). Phase properties of transcranial electrical stimulation artifacts in electrophysiological recordings. *Neuroimage, 158*, 406–416.
Noury, N., & Siegel, M. (2018). Analyzing EEG and MEG signals recorded during tES, a reply. *Neuroimage, 167*, 53–61.
Oostenveld, R., Fries, P., Maris, E., & Schoffelen, J.-M. (2011). FieldTrip: open source software for advanced analysis of MEG, EEG, and invasive electrophysiological data. *Computational Intelligence and Neuroscience, 2011*(1), 156869.
Ostlund, B., Donoghue, T., Anaya, B., Gunther, K. E., Karalunas, S. L., Voytek, B., & Pérez-Edgar, K. E. (2022). Spectral parameterization for studying neurodevelopment: How and why. *Developmental Cognitive Neuroscience, 54*, 101073.
Ouyang, G., Hildebrandt, A., Schmitz, F., & Herrmann, C. S. (2020). Decomposing alpha and 1/f brain activities reveals their differential associations with cognitive processing speed. *Neuroimage, 205*, 116304.

Patel, Y. A., George, A., Dorval, A. D., White, J. A., Christini, D. J., & Butera, R. J. (2017). Hard real-time closed-loop electrophysiology with the real-time experiment interface (RTXI). *PLoS Computational Biology, 13*(5), e1005430.

Polanía, R., Nitsche, M. A., & Paulus, W. (2011). Modulating functional connectivity patterns and topological functional organization of the human brain with transcranial direct current stimulation. *Human Brain Mapping, 32*(8), 1236–1249.

Poreisz, C., Boros, K., Antal, A., & Paulus, W. (2007). Safety aspects of transcranial direct current stimulation concerning healthy subjects and patients. *Brain Research Bulletin, 72*(4–6), 208–214.

Potok, W., Van der Groen, O., Bächinger, M., Edwards, D., & Wenderoth, N. (2022). Transcranial random noise stimulation modulates neural processing of sensory and motor circuits, from potential cellular mechanisms to behavior: A scoping review. *Eneuro, 9*(1).

Pritchard, W. S. (1992). The brain in fractal time: 1/f-like power spectrum scaling of the human electroencephalogram. *International Journal of Neuroscience, 66*(1–2), 119–129.

Roy, A., Baxter, B., & He, B. (2014). High-definition transcranial direct current stimulation induces both acute and persistent changes in broadband cortical synchronization: A simultaneous tDCS–EEG study. *IEEE Transactions on Biomedical Engineering, 61*(7), 1967–1978.

Rufener, K. S., Ruhnau, P., Heinze, H.-J., & Zaehle, T. (2017). Transcranial random noise stimulation (tRNS) shapes the processing of rapidly changing auditory information. *Frontiers in Cellular Neuroscience, 11*, 162.

Ruhnau, P., Neuling, T., Fuscà, M., Herrmann, C. S., Demarchi, G., & Weisz, N. (2016). Eyes wide shut: Transcranial alternating current stimulation drives alpha rhythm in a state dependent manner. *Scientific Reports, 6*.

Schestatsky, P., Morales-Quezada, L., & Fregni, F. (2013). Simultaneous EEG monitoring during transcranial direct current stimulation. *Journal of Visualized Experiments, 76*, 50426.

Schoisswohl, S., Langguth, B., Gebel, N., Poeppl, T. B., Kreuzer, P. M., & Schecklmann, M. (2021). Electrophysiological evaluation of high and low-frequency transcranial random noise stimulation over the auditory cortex. *Progress in Brain Research, 263*, 95–108.

Schwab, B. C., Misselhorn, J., & Engel, A. K. (2019). Modulation of large-scale cortical coupling by transcranial alternating current stimulation. *Brain stimulation, 12*(5), 1187–1196.

Schwippel, T., Pupillo, F., Feldman, Z., Walker, C., Townsend, L., Rubinow, D., & Frohlich, F. (2024). Closed-loop transcranial alternating current stimulation for the treatment of major depressive disorder: An open-label pilot study. *American Journal of Psychiatry, 181*(9), 842–845.

Shirinpour, S., Alekseichuk, I., Güth, M. R., Haigh, Z., Wischnewski, M., & Opitz, A. (2025). Bayesian temporal prediction: A robust algorithm for real-time EEG phase-dependent brain stimulation. *IEEE Transactions on Biomedical Engineering*.

Soleimani, G., Nitsche, M. A., Bergmann, T. O., Towhidkhah, F., Violante, I. R., Lorenz, R., Kuplicki, R., Tsuchiyagaito, A., Mulyana, B., Mayeli, A., et al. (2023). Closing the loop between brain and electrical stimulation: towards precision neuromodulation treatments. *Translational psychiatry, 13*(1), 279.

Spitoni, G. F., Cimmino, R. L., Bozzacchi, C., Pizzamiglio, L., & Di Russo, F. (2013). Modulation of spontaneous alpha brain rhythms using low-intensity transcranial direct-current stimulation. *Frontiers in Human Neuroscience, 7*, 529.

Stecher, H. I., & Herrmann, C. S. (2018). Absence of alpha-tACS aftereffects in darkness reveals importance of taking derivations of stimulation frequency and individual alpha variability into account. *Frontiers in Psychology, 9*, 984.

Stecher, H. I., Notbohm, A., Kasten, F. H., & Herrmann, C. S. (2021). A comparison of closed loop vs. fixed frequency tACS on modulating brain oscillations and visual detection. *Frontiers in Human Neuroscience, 15*, 661432.

Stecher, H. I., Schulz, P., Meiser, A., Hunold, A., Schellhorn, K., & Herrmann, C. S. (2025). Carbonized rubber electrodes can cause a DC-offset in transcranial alternating current stimulation. bioRxiv (pp. 2025–09).

Strüber, D., Rach, S., Neuling, T., & Herrmann, C. S. (2015). On the possible role of stimulation duration for after-effects of transcranial alternating current stimulation. *Frontiers in Cellular Neuroscience, 9*, 311.

Tadel, F., Baillet, S., Mosher, J. C., Pantazis, D., & Leahy, R. M. (2011). Brainstorm: A user-friendly application for MEG/EEG analysis. *Computational Intelligence and Neuroscience, 2011*(1), 879716.

Thut, G., Bergmann, T. O., Fröhlich, F., Soekadar, S. R., Brittain, J.-S., Valero-Cabré, A., Sack, A. T., Miniussi, C., Antal, A., Siebner, H. R., et al. (2017). Guiding transcranial brain stimulation by EEG/MEG to interact with ongoing brain activity and associated functions: A position paper. *Clinical Neurophysiology, 128*(5), 843–857.

Turi, Z., Ambrus, G. G., Janacsek, K., Emmert, K., Hahn, L., Paulus, W., & Antal, A. (2013). Both the cutaneous sensation and phosphene perception are modulated in a frequency-specific manner during transcranial alternating current stimulation. *Restorative Neurology and Neuroscience, 31*(3), 275–285.

Van der Groen, O., Potok, W., Wenderoth, N., Edwards, G., Mattingley, J. B., & Edwards, D. (2022). Using noise for the better: The effects of transcranial random noise stimulation on the brain and behavior. *Neuroscience & Biobehavioral Reviews, 138*, 104702.

Van Doren, J., Langguth, B., & Schecklmann, M. (2014). Electroencephalographic effects of transcranial random noise stimulation in the auditory cortex. *Brain stimulation, 7*(6), 807–812.

Veniero, D., Vossen, A., Gross, J., & Thut, G. (2015). Lasting EEG/MEG aftereffects of rhythmic transcranial brain stimulation: Level of control over oscillatory network activity. *Frontiers in Cellular Neuroscience, 9*, 477.

Venugopal, R., Sasidharan, A., Bhowmick, K., Nagaraj, N., Udupa, K., John, J. P., & Kutty, B. M. (2025). Personalized theta transcranial alternating current stimulation and gamma transcranial alternating current stimulation bring differential neuromodulatory effects on the resting electroencephalogram: Characterizing the temporal, spatial, and spectral dimensions of transcranial alternating current stimulation. *Neuromodulation: Technology at the Neural Interface, 28*(3), 425–433.

Vogeti, S., Boetzel, C., & Herrmann, C. S. (2022). Entrainment and spike-timing dependent plasticity–a review of proposed mechanisms of transcranial alternating current stimulation. *Frontiers in Systems Neuroscience, 16*, 827353.

von Conta, J., Kasten, F. H., Ćurčić-Blake, B., Schellhorn, K., & Herrmann, C. S. (2022). Characterizing low-frequency artifacts during transcranial temporal interference stimulation (tTIS). *Neuroimage: Reports, 2*(3), 100113.

Vossen, A., Gross, J., & Thut, G. (2015). Alpha power increase after transcranial alternating current stimulation at alpha frequency (α-tACS) reflects plastic changes rather than entrainment. *Brain Stimulation, 8*(3), 499–508.

Vosskuhl, J., Mutanen, T. P., Neuling, T., Ilmoniemi, R. J., & Herrmann, C. S. (2020). Signal-space projection suppresses the tACS artifact in EEG recordings. *Frontiers in Human Neuroscience, 14*, 536070.

Watrous, A. J., Miller, J., Qasim, S. E., Fried, I., & Jacobs, J. (2018). Phase-tuned neuronal firing encodes human contextual representations for navigational goals. *Elife, 7*, e32554.

Wei, J., Alamia, A., Yao, Z., Huang, G., Li, L., Liang, Z., Zhang, L., Zhou, C., Song, Z., & Zhang, Z. (2024). State-dependent tACS effects reveal the potential causal role of prestimulus alpha traveling waves in visual contrast detection. *Journal of Neuroscience, 44*(27).

Wen, H.-H., & Liu, Z. (2016). Separating fractal and oscillatory components in the power spectrum of neurophysiological signal. *Nature Communications, 7*, 12356.

Wexler, A. (2017a). The medical battery in the united states (1870–1920): Electrotherapy at home and in the clinic. *Journal of the History of Medicine and Allied Sciences, 72*(2), 166–192.

Wexler, A. (2017b). Recurrent themes in the history of the home use of electrical stimulation: Transcranial direct current stimulation (tDCS) and the medical battery (1870–1920). *Brain Stimulation, 10*(2), 187–195.

Whitten, T. A., Hughes, A. M., Dickson, C. T., & Caplan, J. B. (2011). A better oscillation detection method robustly extracts EEG rhythms across brain state changes: The human alpha rhythm as a test case. *Neuroimage, 54*(2), 860–874.

Wilde, C., Bruder, R., Binder, S., Marshall, L., & Schweikard, A. (2015). Closed-loop transcranial alternating current stimulation of slow oscillations. *Current Directions in Biomedical Engineering, 1*(1), 85–88.

Wilson, L. E., Castanheira, J. d. S., Kinder, B. L., & Baillet, S. (2024). Model selection for spectral parameterization. *bioRxiv*.

Wilson, L. E., da Silva Castanheira, J., & Baillet, S. (2022). Time-resolved parameterization of aperiodic and periodic brain activity. *eLife, 11*, e77348.

Wischnewski, M., Engelhardt, M., Salehinejad, M., Schutter, D., Kuo, M.-F., & Nitsche, M. (2019). NMDA receptor-mediated motor cortex plasticity after 20 Hz transcranial alternating current stimulation. *Cerebral Cortex, 29*(7), 2924–2931.

Witkowski, M., Garcia-Cossio, E., Chander, B. S., Braun, C., Birbaumer, N., Robinson, S. E., & Soekadar, S. R. (2016). Mapping entrained brain oscillations during transcranial alternating current stimulation (tACS). *Neuroimage, 140*, 89–98.

Xiong, G., Li, F., Liu, J., Zhang, T., Zhang, H., & Yang, C. (2025). A tentative closed-loop system for rehabilitation by using tDCS and EEG. In *2025 9th International Conference on Biomedical Engineering and Applications (ICBEA)* (pp. 36–40). IEEE.

Zaehle, T., Beretta, M., Jäncke, L., Herrmann, C. S., & Sandmann, P. (2011). Excitability changes induced in the human auditory cortex by transcranial direct current stimulation: direct electrophysiological evidence. *Experimental Brain Research, 215*(2), 135–140.

Zaehle, T., Rach, S., & Herrmann, C. S. (2010). Transcranial alternating current stimulation enhances individual alpha activity in human EEG. *PloS One, 5*(11), e13766.

Zarubin, G., Gundlach, C., Nikulin, V., & Bogdan, M. (2018). Real-time phase detection for EEG-based tACS closed-loop system. In *Proceedings of the 6th International Congress on Neurotechnology, Electronics and Informatics* (Vol. 13, p. 20). SciTePress Seville.

Zarubin, G., Gundlach, C., Nikulin, V., Villringer, A., & Bogdan, M. (2020). Transient amplitude modulation of alpha-band oscillations by short-time intermittent closed-loop tACS. *Frontiers in Human Neuroscience, 14*, 366.

Zhu, X., & Deegan, T. (2024). Origin and mitigation of nonlinear artefacts in concurrent LFP and EEG recording. Presented at the *International Meeting on Temporal Interference Brain Stimulation*. https://www.imperial.ac.uk/dementia-research-institute/events/international-meeting-on-temporal-interference-brain-stimulation/talk-recordings. Video, accessed June 23, 2026

Chapter 32
TMS–EEG: A Tool to Probe Key Features of Human Thalamocortical Circuits

Silvia Casarotto and Mario Rosanova

Abstract Transcranial magnetic stimulation combined with electroencephalography (TMS–EEG) offers a unique opportunity to investigate human thalamocortical circuits non-invasively from a causal perspective. Delivering a controlled perturbation—the TMS pulse—to a targeted cortical region and recording the ensuing TMS-evoked potentials (TEPs) reveals physiological features that complement those accessible through observation alone, such as through the resting EEG. These features include cortical excitability, effective connectivity, oscillatory dynamics, and recurrent interactions across cortical and thalamocortical circuits. Methodological advances—ranging from TMS-compatible EEG equipment to biological artefact-control procedures—now allow researchers to isolate genuine neural responses from stimulation-related confounds with increasing reliability. These insights are turning TMS–EEG into a versatile tool for basic and clinical neuroscience, enabling the characterization of circuit-level mechanisms across behavioral states, from wakefulness and sleep to anesthesia and disorders of consciousness, as well as fostering the development of biomarkers relevant for stroke, psychiatric disorders, neurodegeneration, and recovery of function. This chapter provides an overview of the instrumentation, experimental procedures, physiological principles, and representative applications of TMS–EEG, highlighting both its strengths and limitations. We argue that TMS–EEG holds substantial promise for advancing mechanistic models of thalamocortical function in humans and for developing clinically meaningful biomarkers.

Keywords TMS–EEG · Thalamocortical circuits · TMS-evoked potentials (TEPs) · Effective connectivity · Neurophysiological biomarkers

S. Casarotto
Department of Biomedical and Clinical Sciences, University of Milan, Milan, Italy

IRCCS Fondazione Don Carlo Gnocchi ONLUS, Milan, Italy

M. Rosanova (✉)
Department of Biomedical and Clinical Sciences, University of Milan, Milan, Italy
e-mail: mario.rosanova@unimi.it

T. Warbrick (ed.), *The EEG Handbook*,
https://doi.org/10.1007/978-3-032-20450-9_32

32.1 Introduction

The thalamocortical system plays a central role in human perception, motor behavior, cognition, and consciousness. These functions arise from interactions between local cortical circuits and long-range cortico-cortical and thalamocortical loops that regulate excitability, synchronize oscillations, and sustain recurrent processing. Dysregulation of these mechanisms characterizes many neurological and psychiatric disorders, in which altered excitability, connectivity, or oscillatory dynamics impair information processing and behavior. Understanding how these processes fail—and how they may be restored—requires tools capable of probing thalamocortical function directly and non-invasively. Transcranial magnetic stimulation combined with electroencephalography (TMS-EEG) is uniquely suited to this goal, as it captures how neural activity elicited by a focal perturbation propagates through thalamocortical circuits.

Early TMS-evoked potentials (TEPs) components (10–50 ms) primarily reflect local cortical excitability and the recruitment of intracortical and thalamocortical circuits, indexing changes in membrane excitability, synaptic efficacy, or neuromodulatory state. At longer latencies (>50 ms), TEPs reveal the propagation of activity across connected regions, providing a direct measure of effective connectivity that is highly sensitive to brain state, despite preserved structural connectivity—for example, during NREM sleep, anesthesia, or disorders of consciousness. In addition, TMS can elicit sustained oscillatory responses at frequencies characteristic of specific cortical targets, reflecting the natural frequencies of corticothalamic modules and conveying information about excitation–inhibition balance and network organization. State-dependent modulation of these responses distinguishes conditions with preserved causal interactions, such as wakefulness or the minimally conscious state, from those in which large-scale dynamics collapse, as in deep sleep, anesthesia, or unresponsive wakefulness syndrome. The degree of such complex causal interactions can be quantified by TMS–EEG–derived measures such as the Perturbational Complexity Index (PCI), which discriminates between clinically similar conditions with high sensitivity.

This chapter provides a focused overview of the methodological and physiological foundations of TMS–EEG, with particular emphasis on its application to thalamocortical circuits. We review the instrumentation and experimental procedures required for reliable data acquisition, summarize the physiological interpretation of TEPs, oscillatory responses, connectivity and complexity metrics, and present representative applications in basic, clinical, and translational neuroscience.

32.2 TMS–EEG: A Brief Overview

TMS is a non-invasive brain stimulation technique based on Faraday's law of electromagnetic induction. A brief, high-intensity current pulse—typically several kiloamperes and lasting a few hundred microseconds—is delivered through a stimulation coil (Ilmoniemi et al., 1999), generating a rapidly time-varying magnetic field that penetrates the scalp and skull with minimal attenuation and induces an electric field in a focal region of the cerebral cortex (Ilmoniemi et al., 1997). Peak magnetic field strengths usually range between 1 and 3 tesla and rise within ~50–100 microseconds, providing TMS with submillisecond temporal precision. This temporal accuracy allows controlled perturbation of neuronal membrane potentials and modulation of ongoing and oscillatory brain activity, enabling causal interrogation of human brain circuits that are otherwise inaccessible to purely observational methods (Hernandez-Pavon et al., 2023; Koponen et al., 2018b).

Electroencephalography (EEG) records the electrical activity generated by large populations of cortical neurons via scalp electrodes, offering millisecond temporal resolution and spatial resolution that depends on electrode density and head conductivity (Ilmoniemi & Sarvas, 2019). While EEG alone provides limited causal inference, it yields a direct measure of brain dynamics as they evolve over time.

Because of the intense electric field generated by the TMS pulse, combining TMS with EEG required the development of dedicated TMS-compatible amplifiers. Early attempts in 1989 demonstrated the feasibility of recording transcallosal and cerebellar–cortical responses (Cracco et al., 1989; Amassian et al., 1992), but the use of conventional EEG amplifiers resulted in large pulse-related artifacts that prevented reliable recording under the coil. Fully TMS-compatible amplifiers introduced about twenty years ago (Ilmoniemi et al., 1997; Virtanen et al., 1999) suppressed these electromagnetic artifacts and enabled the reliable recording of artifact-free TMS-evoked potentials (TEPs) within a few milliseconds (Ilmoniemi et al., 1997; Paus et al. 2001; Massimini et al., 2005). More recently, DC amplifiers with wide dynamic range and high sampling rates ($\geq$5 kHz) have further improved the ability to capture TEPs without long-lasting pulse artifacts (Bonato et al., 2006).

The concurrent application of TMS and EEG enables the recording of neural responses to a precisely timed external perturbation. This combined approach allows investigators to (i) evoke and measure cortical responses, (ii) probe how ongoing brain activity is perturbed by stimulation, and (iii) implement state-dependent or closed-loop stimulation paradigms in which TMS is adapted to the current brain state (Bergmann, 2018; Hernandez-Pavon et al., 2023; Zrenner & Ziemann, 2024).

Together, these features make TMS–EEG a powerful framework for studying the causal organization, state dependence, and dynamical properties of human cortical and thalamocortical circuits in health and disesase (Ziemann et al., 2026 10.1016/j.clinph.2025.2111487).

32.2.1 *Equipment*

The acquisition of high-quality TMS–EEG data depends on the careful selection and integration of dedicated hardware and software components (Fig. 32.1, Panel A). A typical system includes (i) the TMS stimulator and coil, (ii) a TMS-compatible EEG amplifier, and (iii) a neuronavigation system for accurate, reproducible targeting. Additional elements—such as a graphical user interface for real-time monitoring and

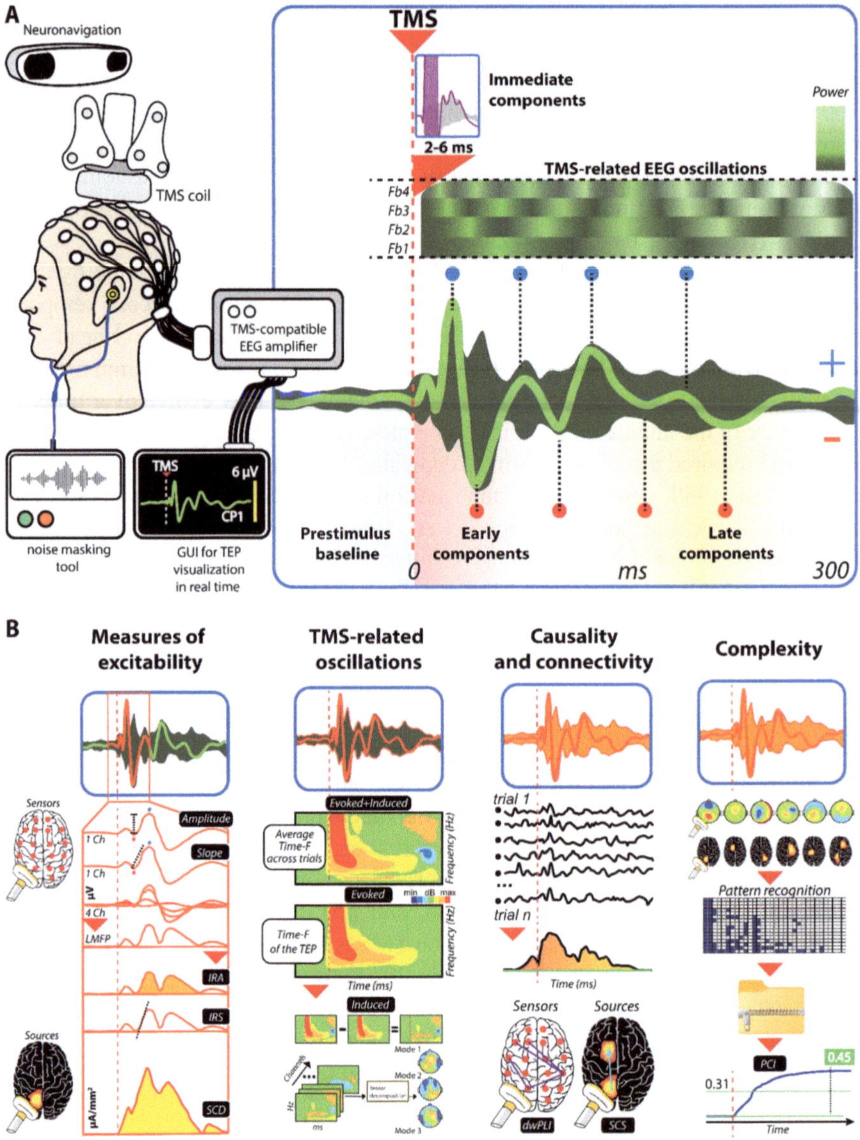

Fig. 32.1 *Panel A (left; modified from Ziemann et al., 2026 10.1016/j.clinph.2025.2111487)* shows a schematic representation of the main elements used to acquire TMS–EEG signals: a neuronavigated TMS coil, a TMS-compatible EEG amplifier receiving signals from an EEG cap with TMS-compatible electrodes designed to reduce TMS-induced eddy currents, a Graphic User Interface to visualize TEPs in real time, and TMS-compatible ergonomic in-ear earphones playing noise masking generated by a specifically devised tool *Panel A (right; modified from Ziemann et al., 2026 10.1016/j.clinph.2025.2111487)* displays an example of TEPs recorded from the posterior parietal cortex (modified from Rosanova et al., 2009), shown for one channel (thick light-green trace) and across all channels as a butterfly plot (dark-green shaded area). Positive (blue circles) and negative (red circles) peaks corresponding to TEP components are marked. Background colors indicate the prestimulus baseline (light blue), early components (light red), and late components (light yellow). A symbolic representation of oscillatory activity in four frequency bands (Fb1–Fb4,

an auditory noise-masking system—are increasingly used to enhance data quality and interpretability (Casarotto et al., 2022; Hernandez-Pavon et al., 2023; Ilmoniemi & Kicic, 2010; Russo et al., 2022). Overall, instrumentation choice plays a decisive role in the reliability and clinical utility of TMS–EEG recordings.

The spatial extent and focality of cortical activation depend on TMS coil geometry, stimulus intensity (SI), target site, and coil–cortex distance (Deng et al., 2013; Ilmoniemi et al., 1999). Because the magnetic field decays rapidly with distance and the induced electric field approaches zero toward the head center, TMS preferentially activates superficial cortical regions (Heller & van Hulsteyn, 1992; Thielscher et al., 2011). Responses are further shaped by target location, E-field orientation, SI, and ongoing brain state (Casarotto et al., 2010; Fecchio et al., 2017).

TMS-evoked potentials (TEPs) are the best-characterized TMS–EEG signals. They represent EEG responses time-locked to the TMS pulse, reflecting direct neural activation and subsequent synaptic and network activity, and appear as reproducible waveform deflections that can be quantitatively analyzed across trials and subjects (Casarotto et al., 2010; Lioumis et al., 2009).

Integration of TMS and EEG requires careful attention to connectivity and electromagnetic interference. Connectivity may be one-way or bidirectional, ensuring precise synchronization and timing control between systems. Inevitably, TMS introduces artifacts—mainly from the magnetic pulse and secondarily from stimulator recharge (Hernandez-Pavon et al., 2023; Ilmoniemi et al., 2015). Mitigation strategies include electromagnetic shielding, careful cable management, and appropriate device configuration. Addressing these issues is essential for successful TMS–EEG integration and reliable signal acquisition.

Fig. 32.1 (continued) increasing in frequency) is shown as rectangles with alternating light and dark green shading. The inset illustrates immediate TEP components recorded over M1. (Modified from Beck et al., 2024)

Panel B (left to right; modified from Ziemann et al., 2026 10.1016/j.clinph.2025.2111487) illustrates:

Excitability measures—examples include the peak-to-peak amplitude and slope of the first TEP component at the single-channel level. Similar measures can be derived from the Local Mean Field Power (LMFP) within the first 50 ms, such as the Immediate Response Area (IRA) and Immediate Response Slope (IRS) (modified from Casarotto et al., 2013; 10.1007/s10548-012-0256-8). Cortical excitability can also be estimated at the source level using indices such as the Significant Current Density (SCD; modified from Casali et al., 2010)

Oscillatory activity—schematic representation of methods used to extract evoked and induced TMS-related oscillations (time–frequency plots modified from Rosanova et al., 2009; pipeline modified from Harquel et al., 2024)

Causality and connectivity—examples of local causality measures (Phase Locking Factor, PLF) and connectivity indices computed at the sensor level (directed Weighted Phase Lag Index, dWPLI) and at the source level. (Significant Current Scattering, SCS; modified from Casali et al., 2010)

Perturbational Complexity Index (PCI)—schematic representation of the main computational steps. (Modified from Casali et al., 2013; and Rosanova et al., 2018)

32.2.2 TMS Electronics and Coils

TMS stimulators are commonly classified according to pulse waveform, most notably monophasic and biphasic. Depending on waveform and hardware capabilities, they can deliver single-pulse, paired-pulse, repetitive TMS (rTMS), and patterned protocols such as theta-burst stimulation (TBS), offering flexibility for probing cortical excitability and network dynamics, but also posing methodological challenges for concurrent EEG recordings.

A key requirement for TMS–EEG acquisition is control over capacitor recharge delay, as the recharge process can generate artifacts in EEG signals, appearing as slow drifts, spikes, or abrupt shifts that may overlap with physiological responses (Veniero et al., 2009). Many modern stimulators therefore allow delaying recharge to shift such artifacts outside the relevant time window.

Monophasic and biphasic pulses differ in amplitude and temporal structure of the induced electric field. Although technical details are beyond the scope of this chapter (see Hernandez-Pavon et al., 2023), monophasic pulses generally require higher stimulation thresholds than biphasic ones (Säisänen et al., 2008), and the optimal choice depends on study aims and signal quality considerations.

Evidence on waveform effects on TEPs remains limited and often confounded by current direction. Some studies indicate that monophasic pulses can increase the amplitude of certain components while being less reliable in eliciting very early responses (Casula et al., 2018; Guidali et al., 2023). Coil orientation and induced current direction can modulate polarity and amplitude of components between ~20 and 45 ms (Bonato et al., 2006; Casula et al., 2018), with some reports of greater global mean field power for monophasic stimulation, though interpretation is complicated by differences in pulse duration (Casula et al., 2018).

More recent studies suggest stronger sensitivity of early components to pulse waveform and current direction. Guidali et al. (2023) showed that during M1 stimulation, amplitude and latency of a ~15 ms component vary across posterior–anterior (PA), anterior–posterior (AP), and lateral–medial (LM) currents, with monophasic responses critically dependent on orientation and AP yielding the largest responses, consistent with Casula et al. (2022), and with AP direction also delaying responses.

Supporting the central role of electric field orientation, Granö et al. (2025) demonstrated direction-dependent modulation of early (~20–40 ms) components in the pre-SMA using 36 E-field orientations, reinforcing that waveform and coil orientation critically shape TEP amplitude and latency.

Coil selection depends on the stimulation protocol and target region, as geometry, size, and winding determine focality and depth of stimulation. The figure-of-eight coil remains the most widely used in TMS–EEG due to its improved focality compared to circular coils (Ueno et al., 1988). Although coil type does not strongly affect electromagnetic artifact duration (Veniero et al., 2009), it can influence electrode exposure and peripheral muscle activation, potentially contaminating EEG signals (Hernandez-Pavon et al., 2023). Muscle artifacts are particularly challenging to remove (Hernandez-Pavon et al., 2012, 2022), so stimulation over scalp, facial, and neck muscles should be carefully considered; double-cone coils, for example, can evoke strong contractions (Fernandez et al., 2021).

Finally, multi-locus TMS (mTMS) enables electronically controlled stimulation of multiple cortical sites without moving the coil (Koponen et al., 2018a; Nieminen et al., 2022). By supporting rapid, precise adjustments and EEG-feedback control, mTMS may reduce artifacts and allow more accurate targeting of specific brain networks (Lioumis et al., 2025; Sinisalo et al., 2024).

32.2.3 TMS-Compatible EEG Amplifiers

One of the main methodological challenges in simultaneous TMS–EEG is the intense electric field produced by the TMS pulse, which can transiently saturate conventional EEG amplifiers and compromise signal recovery for milliseconds to seconds (Wagner et al., 2007 10.1146/annurev.bioeng.9.061206.133100). Early solutions employed a sample-and-hold circuit to stabilize the EEG signal during and immediately after stimulation (Virtanen et al., 1999). By briefly freezing the EEG signal following the pulse, this approach prevented amplifier saturation and enabled reliable recording once the hold period ended (Iramina et al., 2003; Virtanen et al., 1999).

More recently, sample-and-hold systems have been replaced by TMS-compatible EEG amplifiers designed to function in rapidly varying magnetic fields. These devices combine high dynamic range with very high sampling rates, minimizing saturation and allowing continuous EEG acquisition. At ~20 kHz sampling, the pulse artifact may last little more than 2 ms, whereas the more commonly used ~5 kHz rate typically limits it to ~5 ms, balancing temporal resolution and data volume (Jamil et al., 2024).

DC-coupled amplifiers are generally preferred over AC-coupled systems because high-pass filters in AC devices can interact with the TMS artifact, introducing artificial drifts around the pulse. DC systems better preserve slow components and facilitate offline artifact correction.

Finally, high-quality TMS–EEG still depends on careful EEG preparation. Elevated impedances and suboptimal wiring can exacerbate stimulation artifacts, and electrode-lead geometry can critically influence recorded signals in the presence of strong induced fields.

32.2.4 Neuronavigated TMS

Neuronavigation is strongly recommended in TMS–EEG experiments to ensure stable and reproducible control of coil position, orientation, and angulation within and across sessions, particularly in longitudinal designs (Lioumis & Rosanova, 2022; Sack et al., 2009). Extensive methodological work shows that navigated

TMS (nTMS) markedly improves both accuracy and precision of coil placement relative to the intended cortical target (Caulfield et al., 2022; Hannula & Ilmoniemi, 2017), an especially important advantage in TMS–EEG, where small deviations can alter the amplitude, latency, or topography of TMS-evoked responses.

Neuronavigation is essential in studies involving structural brain lesions, where anatomical distortions or severely damaged cortex may not generate reliable EEG responses (Gosseries et al., 2015). Accurate, anatomy-based targeting ensures stimulation of viable tissue and interpretable signals. More generally, obtaining robust TEPs or TMS-induced oscillations requires delivering many stimuli, making stable coil positioning a prerequisite for adequate signal-to-noise ratio.

In nTMS systems, coil position and orientation are tracked in real time relative to the head and brain anatomy, maintaining alignment with the target throughout the session and across visits (Danner et al., 2008; Ilmoniemi & Kicic, 2010; Ruohonen & Karhu, 2010). This minimizes inter-trial variability and prevents unintended stimulation of adjacent regions (Bashir et al., 2011).

Neuronavigation systems can also store coil and stimulation parameters, enabling reliable retargeting across sessions and enhancing reproducibility in longitudinal and multi-session designs (Casarotto et al., 2010; Lioumis et al., 2009). Overall, neuronavigation represents a critical component of modern TMS–EEG instrumentation when reliable and physiologically meaningful measures are required.

32.2.5 Graphical User Interface (GUI) for Visualizing TEPs in Real Time

A dedicated graphical user interface (GUI) is an effective means of improving TMS–EEG data quality by enabling real-time evaluation of recorded signals during acquisition (Fig. 32.1, Panel A). Such tools allow the experimenter to assess, online, whether stimulation parameters—such as target location, stimulus intensity (SI), and coil orientation—are producing clear local EEG deflections beneath the coil.

Real-time visualization also supports the early detection and correction of stimulation-related artifacts that might otherwise go unnoticed and require extensive offline processing. These include artifacts from scalp and cranial muscle activity, auditory-evoked responses, slow decay components, and recharge-related disturbances. Addressing these issues during acquisition helps avoid aggressive post hoc corrections that may distort genuine neural signals.

Freely available GUIs now provide customizable visualization of both single-trial and averaged TEPs, enabling interactive adjustment of stimulation parameters on a subject-specific basis. Their use has been shown to substantially improve TMS–EEG data quality by reducing artifact burden and enhancing the reliability and interpretability of evoked responses (Casarotto et al., 2022).

32.2.6 Noise Masking Tools

During TMS delivery, a loud click is produced by rapid mechanical vibration of the coil windings. This sound is precisely time-locked to the TMS pulse and reliably elicits auditory-evoked potentials that overlap with genuine cortical responses (Nikouline et al., 1999; ter Braack et al., 2015), representing a major biological confound, especially at early and mid-latencies.

A widely used strategy to reduce this contamination is continuous presentation of masking noise through in-ear earphones during stimulation. By diminishing the perceptual salience of the coil click, masking substantially attenuates auditory-evoked responses and improves the specificity of TMS-evoked potentials.

Effective masking does not necessarily require high sound levels. More efficient and comfortable masking can be achieved by tailoring noise spectra to match the acoustic profile of the specific coil click. Freely available tools enable generation of such customized masking sounds, allowing effective suppression of auditory responses at lower volumes (Russo et al., 2022). Optimized noise masking is therefore a key methodological element in modern TMS–EEG protocols.

32.3 TEPs and Their Components

32.3.1 Definition and Measurement

TEPs are sequences of positive and negative voltage deflections observed in the EEG signal and time-locked to the delivery of a TMS pulse (Fig. 32.1, Panel A and B). These deflections—commonly referred to as TEP waves or components—can extend over several hundred milliseconds and are conventionally labeled according to their polarity (negative, N; positive, P) and their latency in milliseconds. TEPs represent a specific class of EEG-evoked potentials and, similar to classical sensory evoked responses, typically require the delivery of tens to hundreds of TMS pulses to achieve adequate signal-to-noise ratio through trial averaging.

In standard TMS–EEG acquisition, the continuous EEG is segmented into epochs time-locked to each TMS pulse and subsequently averaged across trials. This procedure suppresses ongoing, non-phase-locked EEG activity while enhancing stimulus-locked responses, allowing TEP components to emerge as statistically robust deviations from baseline and background noise (Ilmoniemi et al., 2021). However, unlike sensory evoked potentials, TMS directly activates cortical tissue and affords a high degree of flexibility in stimulation parameters, including coil position, orientation, and the magnitude and direction of the induced electric field. This versatility represents both a major strength and a methodological challenge for TMS–EEG.

When targeting the primary motor cortex (M1), stimulation efficacy can be readily verified by the presence of motor-evoked potentials (MEPs) recorded from a target muscle, allowing stimulation parameters to be optimized on a physiological basis. In contrast, when stimulating cortical regions outside M1, such immediate peripheral readouts are unavailable and anatomical landmarks may be less well defined. As a consequence, confirming effective engagement of the targeted cortical circuits becomes more challenging.

To address this limitation, several freely available software tools have been developed to optimize stimulation parameters—most notably intensity and coil orientation—in real time by maximizing the signal-to-noise ratio of early TEP components (typically <50 ms) recorded directly beneath the stimulation site (Casarotto et al., 2022; Parmigiani et al., 2025; Tervo et al., 2022). Empirical and modeling studies indicate that robust local cortical activation requires an induced electric field exceeding approximately 100 V/m, a threshold necessary to reliably elicit early TEP components with high signal-to-noise ratio (Casarotto et al., 2022; Raffin et al., 2020). The use of neuronavigation further enhances targeting accuracy and promotes stability of stimulation parameters across trials and sessions, thereby improving reproducibility (Lioumis & Rosanova, 2022).

TEP components elicited by stimulation of M1 are among the most thoroughly characterized and include the N15, P30, N45, P60, N100, and P180 components (Beck et al., 2024; Farzan & Bortoletto, 2022). Another frequently investigated target is the dorsolateral prefrontal cortex (DLPFC), where single-pulse TMS typically evokes components labeled P25, N40, P60, N100, and P180 (Farzan & Bortoletto, 2022). Notably, early components differ in latency and morphology between stimulation sites, suggesting that early TEPs are sensitive to region-specific cortical dynamics.

Although late TEP components are more susceptible to biological confounds, there is increasing evidence that, when such confounds are carefully controlled, these later responses can also retain a degree of target specificity (Belardinelli et al., 2019; D'Ambrosio et al., 2022; De Martino et al., 2023; Fecchio et al., 2017, 2025; Ferrarelli et al., 2012; Rosanova et al., 2009).

Recent evidence suggests that TEPs from M1 stimulation can also capture the immediate cortical response to TMS (Beck et al., 2024 10.1016/j.brs.2024.06.008; Stango et al., 2026 10.1016/j.neuroimage.2025.121666). When biphasic pulses are combined with TMS-compatible EEG amplifiers operating at very high sampling rates, and when cranio-facial muscle activation is minimized through real-time visualization of TEPs prior to recording (Casarotto et al., 2022), it becomes possible to detect so-called immediate TEPs (i-TEPs) as early as ~2 ms after stimulation. These responses closely resemble the I-waves observed in epidural recordings following TMS, characterized by a series of peaks spaced approximately 1.2–1.5 ms apart (600–800 Hz). i-TEPs are largest when stimulating over the motor hotspot and rapidly diminish as the coil is displaced along the rostro-caudal axis (Nuyts et al., 2025), suggesting a high degree of spatial specificity. Whether similarly short-latency responses can be reliably elicited from non-motor cortical regions remains an open question (Fig. 32.1, Panel A).

32.3.2 Neural Underpinnings of TEPs

Despite extensive investigation, the precise neural mechanisms underlying individual TEP components are not yet fully understood. Nevertheless, converging evidence from animal models, intracranial recordings, and computational studies has begun to elucidate their physiological basis.

I-TEPs are strongly influenced by the direction of the induced biphasic current, mirroring the dependence of I-waves on current direction during TMS (Ziemann, 2020). This sensitivity suggests that i-TEPs primarily reflect the local activation of intracortical circuits directly beneath the stimulation site. Consistent with this interpretation, intracellular recordings in the primary somatosensory cortex of rodents, as well as single- and multi-unit recordings in M1 of non-human primates and rodents, have demonstrated neuronal firing within the first 5 ms following a TMS pulse (Boyer et al., 2022; Li et al., 2017; Mueller et al., 2014), providing a plausible substrate for i-TEPs and associated synaptic activity.

Early TEP components are also thought to reflect activity within local cortical circuits directly engaged by the TMS pulse. In rodents, extracellular recordings have shown that single-pulse TMS applied over motor-equivalent cortex evokes neuronal firing during the first 50 ms after stimulation, with a peak around 20 ms (Li et al., 2017). Comparable findings have been reported in non-human primates, where TMS of M1 (Mueller et al., 2014) and posterior parietal cortex (Romero et al., 2019) induced sustained neuronal firing between 20 and 40 ms. Earlier single-unit recordings in anesthetized cats further demonstrated increased firing rates within the first 100 ms following TMS of primary visual cortex (Moliadze et al., 2003). At the mesoscale, recent work combining TMS with intracranial recordings in non-human primates revealed that early local field potential components are highly target-specific and systematically vary with changes in stimulation site (Perera et al., 2023).

Collectively, these observations support the view that early TEP components are driven by the firing of local neuronal populations and the synchronous summation of excitatory and inhibitory postsynaptic potentials within the stimulated cortex. It is conceivable that dendritic processes, including suprathreshold events such as Ca^{2+} spikes, may contribute to early TEPs and be detectable at the scalp level (Suzuki & Larkum, 2017). TMS may also indirectly recruit inhibitory interneurons targeting apical dendrites, thereby modulating dendritic Ca^{2+} signaling (Murphy et al., 2016).

Neuronal facilitation induced by TMS can persist for several hundred milliseconds (up to ~500 ms) following stimulation (Moliadze et al., 2003), suggesting that late TEP components are also shaped by local circuit dynamics. However, recent multi-scale evidence indicates that these later components increasingly reflect the engagement of distributed and subcortical networks. In particular, a study integrating TMS–EEG, intracranial recordings in humans and rodents, and computational modeling demonstrated that TEP components around 200 ms following premotor TMS are strongly influenced by thalamic rebound activity and are modulated by task-dependent changes in the thalamocortical state (Russo et al., 2025).

32.4 TMS-Related EEG Oscillations

32.4.1 Definition and Measurement

TMS–EEG responses can be characterized not only in the time domain—yielding TEPs—but also in the frequency and time–frequency domains, enabling the investigation of oscillatory activity elicited by the TMS pulse (Wang et al., 2024). Time–frequency analyses provide complementary information by capturing transient and sustained changes in spectral power that may not be apparent in the averaged time-domain signal.

Within this framework, a fundamental distinction is made between phase-locked (evoked) and non–phase-locked (induced) oscillatory activity. Phase-locked activity is consistently aligned in phase with the onset of the TMS pulse and therefore survives trial averaging. As such, it contributes to the TEP waveform and can also be detected by band-pass filtering the averaged EEG signal or through time–frequency decomposition. In contrast, induced activity is time-locked to the TMS event but varies in phase across trials; consequently, it is canceled out by averaging and becomes detectable only through time–frequency analysis.

The most widely used approach for time–frequency decomposition of TMS–EEG data is the event-related spectral perturbation (ERSP). This wavelet-based method quantifies trial-averaged changes in spectral power as a function of time and frequency relative to a prestimulus baseline, yielding a two-dimensional representation of power modulation (in decibels) following the TMS pulse (Ferrarelli et al., 2012; Grandchamp & Delorme, 2011; Makeig et al., 2004; Rosanova et al., 2009). Alternative implementations with higher temporal resolution have been proposed, such as the use of the Hilbert–Huang transform instead of wavelets (Pigorini et al., 2011).

From the ERSP, it is also possible to derive the natural frequency of a cortical region, defined as the frequency showing maximal power in the early post-stimulus time window (typically within the first 200–300 ms). This natural frequency corresponds to the dominant oscillatory mode elicited by a brief perturbation of the targeted cortical circuit and can be conceptualized, in analogy with physical systems, as the intrinsic frequency at which the system tends to oscillate following a transient input (Ferrarelli et al., 2012; Rosanova et al., 2009).

When applied to TMS–EEG data, ERSP reflects total spectral power and therefore captures a mixture of evoked and induced oscillatory activity. Several methodological approaches have been developed to disentangle these two components. A commonly used strategy involves subtracting the averaged evoked response from each single trial prior to time–frequency decomposition, thereby isolating induced oscillations (Premoli et al., 2017). More recently, data-driven methods such as parallel factor analysis (PARAFAC) have been introduced to address the high dimensionality of TMS-induced oscillatory datasets, which may span multiple electrodes, time points, frequency bins, subjects, and experimental conditions (Belardinelli et al., 2021; Harquel & Hummel, 2023; Tangwiriyasakul et al., 2019). PARAFAC decomposes the data into a limited set of distinct and interpretable oscillatory modes, facilitating physiological interpretation.

32.4.2 *Neural Underpinnings*

Evidence from invasive animal studies provides important insights into the neural mechanisms underlying TMS-related oscillatory activity. In anesthetized cats, TMS applied over the primary visual cortex was shown to induce sustained neuronal firing for up to 500 ms following stimulation, with recurrent peaks of facilitation occurring at approximately 100 ms intervals—corresponding to oscillations in the alpha frequency range (~10 Hz) once a sufficient stimulation intensity was exceeded (Moliadze et al., 2003).

Similar observations have been reported in non-human primates. When targeting the parietal cortex, TMS did not reliably evoke spiking activity in neurons located more than 2 mm from the stimulation center; however, it induced robust oscillations in neuronal firing rates, predominantly at frequencies below 10 Hz, which are likely to contribute to the TMS–EEG signal (Romero et al., 2019). Complementing these findings, intracranial recordings in humans and non-human primates have demonstrated widespread increases in spectral power lasting several hundred milliseconds after TMS, reflecting a combination of evoked and induced oscillatory responses that appear to be region-specific (Solomon et al., 2024).

Together, these studies indicate that TMS can engage intrinsic oscillatory dynamics of cortical circuits, both locally and across distributed networks, giving rise to sustained rhythmic activity detectable at the scalp level.

32.5 Physiological Interpretation and Emerging Clinical Applications of TMS–EEG

Taken together, the available evidence indicates that early TEP components—typically within 10–50 ms—primarily reflect local reverberatory activity within the targeted cortical region and can therefore serve as reliable readouts of cortical excitability, quantified through measures such as amplitude, slope, or local power (Bliss & Lømo, 1973; Vyazovskiy et al., 2008) (Fig. 32.1, Panel B). Early TEPs increase with prolonged wakefulness and are modulated by circadian rhythms in healthy individuals (Huber et al., 2013; Ly et al., 2016), are sensitive to neuromodulatory interventions including rTMS, tDCS, and tSMS (Esser et al., 2006; Pellicciari et al., 2013; Casula et al., 2014; Romero Lauro et al., 2014; Shibata et al., 2024), differentiate the effects of single- versus paired-pulse TMS (Ferreri et al., 2011), reveal pharmacological modulation of GABAergic and glutamatergic neurotransmission (Belardinelli et al., 2021; Darmani & Ziemann, 2019), and capture neuromodulatory effects of dopaminergic treatment in Parkinson's disease (Casarotto et al., 2019; Leodori et al., 2022; Turco et al., 2018). Later TEP components, especially after M1 stimulation, provide complementary physiological insight: the N100 shows a broad scalp distribution consistent with multiple generators, correlates with the cortical silent period (Farzan et al., 2013;

Nikulin et al., 2003; Rogasch et al., 2013), and is sensitive to GABAB receptor activity (Darmani & Ziemann, 2019), while even later components such as the P180 and N280 likely reflect large-scale network propagation. Beyond time-domain measures, TMS-related oscillations offer important insight into cortical and thalamocortical dynamics. Early clinical observations demonstrated that unilateral thalamectomy markedly reduces beta-band oscillations following stimulation of the ipsilesional M1 (Van der Werf et al., 2006), implicating cortico–thalamo–cortical reverberations. Extensive work shows that different cortical areas express distinct natural frequencies, with anterior regions tending toward beta–gamma oscillations and posterior regions toward alpha–beta rhythms (D'Ambrosio et al., 2022; De Martino et al., 2024; Ferrarelli et al., 2012; Garcia et al., 2011; Rosanova et al., 2009; Tscherpel et al., 2020; Vallesi et al., 2021), a gradient confirmed by robotic mapping (Harquel et al., 2016) and supported by modeling (Cona et al., 2011). Even adjacent cortical sites can display distinct dynamics (Lioumis et al., 2025; Passera et al., 2022). Natural frequencies are reduced in schizophrenia (Donati et al., 2025; Ferrarelli et al., 2012) and delirium (Bai et al., 2023) and replaced by stereotypical slow waves in perilesional cortex after stroke (Bai et al., 2023; Sarasso et al., 2020; Tscherpel et al., 2020); regional differentiation collapses during NREM sleep (Massimini et al., 2007) and in the vegetative state (Rosanova et al., 2018). Although the precise separation between evoked and induced oscillations remains debated (David et al., 2006), ongoing rhythms likely emerge from interactions between local and distributed oscillators (Wang et al., 2024), and TMS elicits both phase-locked and non-phase-locked activity (Pellicciari et al., 2017; Solomon et al., 2024). Pharmaco-TMS–EEG studies show that induced oscillations—such as early alpha/beta synchronization followed by bilateral desynchronization—are strongly modulated by GABAergic mechanisms (Premoli et al., 2017). Advanced approaches such as PARAFAC reveal reproducible long-lasting oscillatory modes, including late alpha and sensorimotor beta patterns, differentially shaped by glutamatergic and GABAergic neurotransmission (Belardinelli et al., 2021; Harquel & Hummel, 2023; Tangwiriyasakul et al., 2019). Overall, both TEP components and TMS-induced oscillations provide sensitive markers of cortical excitability, excitation–inhibition balance, and large-scale network dynamics, with significant relevance for understanding neurological and psychiatric disorders (Ghatak et al., 2021).

32.6 Global Measures of the Brain Response to TMS and State Dependency

In wakefulness, a single TMS pulse initiates a cascade of neural events that begins with the rapid recruitment of neuronal populations in the targeted cortical region. The resulting volleys of action potentials propagate along cortico–cortical and cortico–subcortical pathways, engaging distributed brain regions that in turn send feedback activity back toward the initially stimulated cortex. This recurrent, large-scale activation is likely expressed in the sequential components of TMS-evoked potentials (TEPs) and in the accompanying TMS-related oscillatory activity, as supported by computational modeling studies (Momi et al., 2023). Crucially, these system-level dynamics depend on the capacity of the brain to sustain effective connectivity, that is, the ability of one area to causally influence the activity of another connected region (Friston, 2002) (Fig. 32.1, Panel B).

Nearly two decades ago, Massimini and colleagues provided a seminal demonstration that the spatiotemporal complexity of TMS-evoked responses is strongly state-dependent (Massimini et al., 2005). They showed that the fast recurrent waves and long-range activations typically observed during wakefulness vanish rapidly at sleep onset. During light sleep (N1/N2), early TEP components (often within the first 50 ms) become relatively more prominent, whereas later waves are markedly attenuated. During deep sleep (N3), the initial positive deflection increases substantially in amplitude and becomes delayed, followed by a large negative slow wave peaking at approximately 150 ms. Importantly, this positive–negative sequence remains spatially localized near the stimulation site (Massimini et al., 2005, 2007), illustrating how complex network interactions sustained by effective connectivity can collapse within minutes despite an intact structural connectome. Building on these observations, several indices have subsequently been proposed to quantify effective connectivity and the global properties of TEP dynamics (Casali et al., 2010; De Martino et al., 2025).

The enhanced early positive TEP component observed during sleep should not be interpreted as a straightforward marker of increased cortical excitability, as it might be in wakefulness. Instead, it likely reflects stronger postsynaptic depolarization in neurons that are globally hyperpolarized during sleep (Sachdev et al., 2004), together with increased local synchrony of neuronal firing (Wörgötter et al., 1998). This initial response is followed by a prominent negative wave and a pronounced suppression of high-frequency (>20 Hz) activity, consistent with sustained hyperpolarization and reduced neuronal firing. These dynamics closely resemble cortical OFF-periods—transient phases of neuronal silence that underlie slow waves and delta activity—and are thought to be driven by intrinsic hyperpolarizing mechanisms (Massimini et al., 2024; Rosanova et al., 2018).

Critically, OFF-periods disrupt not only ongoing cortical activity but also effective connectivity. After a down-state, neuronal firing tends to resume in a relatively desynchronized manner, breaking the causal and temporally coherent relationship with the initial TMS input (Rosanova et al., 2018). This collapse of input–output coherence undermines the brain's capacity for integrated processing and is captured by a reduction in the perturbational complexity index (PCI), a global measure of the spatiotemporal complexity of the TMS-evoked response (Casali et al., 2013) (Fig. 32.1, Panel B).

A similar cascade—characterized by OFF-periods, impaired long-range interactions, and reduced PCI—has been documented during general anesthesia (Sarasso et al., 2015), in patients with disorders of consciousness following severe brain injury (Casarotto et al., 2016), and locally within perilesional tissue after stroke (Sarasso et al., 2020; Tscherpel et al., 2020). Importantly, these alterations can be reversible. The re-emergence of complex, causally structured TMS-evoked dynamics has been shown to accompany recovery of consciousness (Rosanova et al., 2018; Rosanova et al., 2012) and to parallel functional improvement in the course of post-stroke recovery (Sarasso et al., 2025). Understanding how global TMS–EEG responses vary across brain states therefore has direct implications for the diagnostic interpretation of TEPs and may contribute to the development of individualized prognostic and rehabilitation strategies.

References

Amassian, V. E., Cracco, R. Q., Maccabee, P. J., & Cracco, J. B. (1992). Cerebello-frontal cortical projections in humans studied with the magnetic coil. *Electroencephalography and Clinical Neurophysiology/Evoked Potentials Section, 85*, 265–272. https://doi.org/10.1016/0168-5597(92)90115-R

Bai, Y., Belardinelli, P., Thoennes, C., Blum, C., Baur, D., Laichinger, K., Lindig, T., Ziemann, U., & Mengel, A. (2023). Cortical reactivity to transcranial magnetic stimulation predicts risk of post-stroke delirium. *Clinical Neurophysiology, 148*, 97–108. https://doi.org/10.1016/j.clinph.2022.11.017

Bashir, S., Edwards, D., & Pascual-Leone, A. (2011). Neuronavigation increases the physiologic and behavioral effects of low-frequency rTMS of primary motor cortex in healthy subjects. *Brain Topography, 24*, 54–64. https://doi.org/10.1007/s10548-010-0165-7

Beck, M. M., Heyl, M., Mejer, L., Vinding, M. C., Christiansen, L., Tomasevic, L., & Siebner, H. R. (2024). Methodological choices matter: A systematic comparison of TMS—EEG studies targeting the primary motor cortex. *Human Brain Mapping, 45*, e70048. https://doi.org/10.1002/hbm.70048

Belardinelli, P., Biabani, M., Blumberger, D. M., Bortoletto, M., Casarotto, S., David, O., Desideri, D., Etkin, A., Ferrarelli, F., Fitzgerald, P. B., Fornito, A., Gordon, P. C., Gosseries, O., Harquel, S., Julkunen, P., Keller, C. J., Kimiskidis, V. K., Lioumis, P., Miniussi, C., Rosanova, M., Rossi, S., Sarasso, S., Wu, W., Zrenner, C., Daskalakis, Z. J., Rogasch, N. C., Massimini, M., Ziemann, U., & Ilmoniemi, R. J. (2019). Reproducibility in TMS–EEG studies: A call for data sharing, standard procedures and effective experimental control. *Brain Stimulation, 12*, 787–790. https://doi.org/10.1016/j.brs.2019.01.010

Belardinelli, P., König, F., Liang, C., Premoli, I., Desideri, D., Müller-Dahlhaus, F., Gordon, P. C., Zipser, C., Zrenner, C., & Ziemann, U. (2021). TMS-EEG signatures of glutamatergic neurotransmission in human cortex. *Scientific Reports, 11*, 8159. https://doi.org/10.1038/s41598-021-87533-z

Bergmann, T. O. (2018). Brain state-dependent brain stimulation. *Frontiers in Psychology, 9*, 2108. https://doi.org/10.3389/fpsyg.2018.02108

Bliss, T. V., & Lomo, T. (1973). Long-lasting potentiation of synaptic transmission in the dentate area of the anaesthetized rabbit following stimulation of the perforant path. *The Journal of Physiology, 232*, 331–356. https://doi.org/10.1113/jphysiol.1973.sp010273

Bonato, C., Miniussi, C., & Rossini, P. M. (2006). Transcranial magnetic stimulation and cortical evoked potentials: A TMS/EEG co-registration study. *Clinical Neurophysiology, 117*, 1699–1707. https://doi.org/10.1016/j.clinph.2006.05.006

Boyer, M., Baudin, P., Stengel, C., Valero-Cabré, A., Lohof, A. M., & Charpier, S., et al. (2022). In vivo low-intensity magnetic pulses durably alter neocortical neuron excitability and spontaneous activity. *The Journal of Physiology, 600*, 4019–37. https://doi.org/10.1113/JP283244

Casali, A. G., Casarotto, S., Rosanova, M., Mariotti, M., & Massimini, M. (2010). General indices to characterize the electrical response of the cerebral cortex to TMS. *NeuroImage, 49*, 1459–1468. https://doi.org/10.1016/j.neuroimage.2009.09.026

Casali, A. G., Gosseries, O., Rosanova, M., Boly, M., Sarasso, S., Casali, K. R., Casarotto, S., Bruno, M.-A., Laureys, S., Tononi, G., & Massimini, M. (2013). A theoretically based index of consciousness independent of sensory processing and behavior. *Science Translational Medicine, 5*, 198ra105. https://doi.org/10.1126/scitranslmed.3006294

Casarotto, S., Romero Lauro, L. J., Bellina, V., Casali, A. G., Rosanova, M., Pigorini, A., Defendi, S., Mariotti, M., & Massimini, M. (2010). EEG responses to TMS are sensitive to changes in the perturbation parameters and repeatable over time. *PLoS ONE, 5*, e10281. https://doi.org/10.1371/journal.pone.0010281

Casarotto, S., Comanducci, A., Rosanova, M., Sarasso, S., Fecchio, M., Napolitani, M., Pigorini, A., Casali, A. G., Trimarchi, P. D., Boly, M., Gosseries, O., Bodart, O., Curto, F., Landi, C., Mariotti, M., Devalle, G., Laureys, S., Tononi, G., & Massimini, M. (2016). Stratification of unresponsive patients by an independently validated index of brain complexity. *Annals of Neurology, 80*, 718–729. https://doi.org/10.1002/ana.24779

Casarotto, S., Turco, F., Comanducci, A., Perretti, A., Marotta, G., Pezzoli, G., Rosanova, M., & Isaias, I. U. (2019). Excitability of the supplementary motor area in Parkinson's disease depends on subcortical damage. *Brain Stimulation, 12*, 152–160. https://doi.org/10.1016/j.brs.2018.10.011

Casarotto, S., Fecchio, M., Rosanova, M., Varone, G., D'Ambrosio, S., Sarasso, S., Pigorini, A., Russo, S., Comanducci, A., Ilmoniemi, R. J., & Massimini, M. (2022). The RT-TEP tool: Real-time visualization of TMS-evoked potentials to maximize cortical activation and minimize artifacts. *Journal of Neuroscience Methods, 370*, 109486. https://doi.org/10.1016/j.jneumeth.2022.109486

Casula, E. P., Tarantino, V., Basso, D., Arcara, G., Marino, G., Toffolo, G. M., Rothwell, J. C., & Bisiacchi, P. S. (2014). Low-frequency rTMS inhibitory effects in the primary motor cortex: Insights from TMS-evoked potentials. *NeuroImage, 98*, 225–232. https://doi.org/10.1016/j.neuroimage.2014.04.065

Casula, E. P., Rocchi, L., Hannah, R., & Rothwell, J. C. (2018). Effects of pulse width, waveform and current direction in the cortex: A combined cTMS-EEG study. *Brain Stimulation, 11*, 1063–1070. https://doi.org/10.1016/j.brs.2018.04.015

Casula, E. P., Leodori, G., Ibáñez, J., Benussi, A., Rawji, V., Tremblay, S., Latorre, A., Rothwell, J. C., & Rocchi, L. (2022). The effect of coil orientation on the stimulation of the pre-supplementary motor area: A combined TMS and EEG study. *Brain Sciences, 12*. https://doi.org/10.3390/brainsci12101358

Caulfield, K. A., Fleischmann, H. H., Cox, C. E., Wolf, J. P., George, M. S., & McTeague, L. M. (2022). Neuronavigation maximizes accuracy and precision in TMS positioning: Evidence from 11,230 distance, angle, and electric field modeling measurements. *Brain Stimulation, 15*, 1192–1205. https://doi.org/10.1016/j.brs.2022.08.013

Cona, F., Zavaglia, M., Massimini, M., Rosanova, M., & Ursino, M. (2011). A neural mass model of interconnected regions simulates rhythm propagation observed via TMS-EEG. *NeuroImage, 57*, 1045–1058. https://doi.org/10.1016/j.neuroimage.2011.05.007

Cracco, R. Q., Amassian, V. E., Maccabee, P. J., & Cracco, J. B. (1989). Comparison of human transcallosal responses evoked by magnetic coil and electrical stimulation. *Electroencephalography and Clinical Neurophysiology/Evoked Potentials Section, 74*, 417–424. https://doi.org/10.1016/0168 5597(89)90030 0

D'Ambrosio, S., Jiménez-Jiménez, D., Silvennoinen, K., Zagaglia, S., Perulli, M., Poole, J., Comolatti, R., Fecchio, M., Sisodiya, S. M., & Balestrini, S. (2022). Physiological symmetry of transcranial magnetic stimulation-evoked EEG spectral features. *Human Brain Mapping, 43*, 5465–5477. https://doi.org/10.1002/hbm.26022

Danner, N., Julkunen, P., Könönen, M., Säisänen, L., Nurkkala, J., & Karhu, J. (2008). Navigated transcranial magnetic stimulation and computed electric field strength reduce stimulator-dependent differences in the motor threshold. *Journal of Neuroscience Methods, 174*, 116–122. https://doi.org/10.1016/j.jneumeth.2008.06.032

Darmani, G., & Ziemann, U. (2019). Pharmacophysiology of TMS-evoked EEG potentials: A mini-review. *Brain Stimulation, 12*, 829–831. https://doi.org/10.1016/j.brs.2019.02.021

David, O., Kilner, J. M., & Friston, K. J. (2006). Mechanisms of evoked and induced responses in MEG/EEG. *NeuroImage, 31*, 1580–1591. https://doi.org/10.1016/j.neuroimage.2006.02.034

De Martino, E., Casali, A., Casarotto, S., Hassan, G., Rosanova, M., Graven-Nielsen, T., & Ciampi De Andrade, D. (2023). Acute pain drives different effects on local and global cortical excitability in motor and prefrontal areas: Insights into interregional and interpersonal differences in pain processing. *Cerebral Cortex, 33*, 9986–9996. https://doi.org/10.1093/cercor/bhad259

De Martino, E., Casali, A., Casarotto, S., Hassan, G., Couto, B. A., Rosanova, M., Graven-Nielsen, T., & De Andrade, D. C. (2024). Evoked oscillatory cortical activity during acute pain: Probing brain in pain by transcranial magnetic stimulation combined with electroencephalogram. *Human Brain Mapping, 45*, e26679. https://doi.org/10.1002/hbm.26679

De Martino, E., Nascimento Couto, B. A., Jakobsen, A., Casali, A. G., Bonde-Heriksen, P. D., Graven-Nielsen, T., & de Andrade, D. C. (2025). Pretreatment TMS-EEG connectivity assessment as a potential predictor of rTMS effectiveness in chronic pain: A feasibility pilot study. *Clinical Neurophysiology, 176*, 2010681. https://doi.org/10.1016/j.clinph.2025.02.268

Deng, Z.-D., Lisanby, S. H., & Peterchev, A. V. (2013). Electric field depth–focality tradeoff in transcranial magnetic stimulation: Simulation comparison of 50 coil designs. *Brain Stimulation, 6*, 1–13. https://doi.org/10.1016/j.brs.2012.02.005

Donati, F. L., Mayeli, A., Nascimento Couto, B. A., Sharma, K., Janssen, S., Krafty, R. J., Casali, A. G., & Ferrarelli, F. (2025). Prefrontal oscillatory slowing in early-course schizophrenia is associated with worse cognitive performance and negative symptoms: A transcranial magnetic stimulation–electroencephalography study. *Biological Psychiatry: Cognitive Neuroscience and Neuroimaging, 10*, 158–166. https://doi.org/10.1016/j.bpsc.2024.07.013

Esser, S. K., Huber, R., Massimini, M., Peterson, M. J., Ferrarelli, F., & Tononi, G., (2006). A direct demonstration of cortical LTP in humans: a combined TMS/EEG study. *Brain Res Bull, 69*, 86–94. https://doi.org/10.1016/j.brainresbull.2005.11.003

Farzan, F., & Bortoletto, M. (2022). Identification and verification of a "true" TMS evoked potential in TMS-EEG. *Journal of Neuroscience Methods, 378*, 109651. https://doi.org/10.1016/j.jneumeth.2022.109651

Farzan, F., Barr, M. S., Hoppenbrouwers, S. S., Fitzgerald, P. B., Chen, R., Pascual-Leone, A., & Daskalakis, Z. J. (2013). The EEG correlates of the TMS-induced EMG silent period in humans. *NeuroImage, 83*, 120–134. https://doi.org/10.1016/j.neuroimage.2013.06.059

Fecchio, M., Pigorini, A., Comanducci, A., Sarasso, S., Casarotto, S., Premoli, I., Derchi, C.-C., Mazza, A., Russo, S., Resta, F., Ferrarelli, F., Mariotti, M., Ziemann, U., Massimini, M., & Rosanova, M. (2017). The spectral features of EEG responses to transcranial magnetic stimulation of the primary motor cortex depend on the amplitude of the motor evoked potentials. *PLoS ONE, 12*, e0184910. https://doi.org/10.1371/journal.pone.0184910

Fecchio, M., Russo, S., Couto, B. A., Mikulan, E., Pigorini, A., Furregoni, G., Hassan, G., D'Ambrosio, S., Solbiati, M., Viganò, A., Parmigiani, S., Sarasso, S., Casarotto, S., Massimini, M., Casali, A. G., & Rosanova, M. (2025). The specific spatiotemporal evolution of TMS-evoked potentials reflects the engagement of cortical circuits. *bioRxiv*. https://doi.org/10.1101/2025.06.25.661535

Fernandez, L., Biabani, M., Do, M., Opie, G. M., Hill, A. T., Barham, M. P., Teo, W.-P., Byrne, L. K., Rogasch, N. C., & Enticott, P. G. (2021). Assessing cerebellar-cortical connectivity using concurrent TMS-EEG: A feasibility study. *Journal of Neurophysiology, 125*, 1768–1787. https://doi.org/10.1152/jn.00617.2020

Ferrarelli, F., Sarasso, S., Guller, Y., Riedner, B. A., Peterson, M. J., Bellesi, M., Massimini, M., Postle, B. R., & Tononi, G. (2012). Reduced natural oscillatory frequency of frontal thalamocortical circuits in schizophrenia. *Archives of General Psychiatry, 69*, 766–774. https://doi.org/10.1001/archgenpsychiatry.2012.147

Ferreri, F., Pasqualetti, P., Määttä, S., Ponzo, D., Ferrarelli, F., Tononi, G., Mervaala, E., Miniussi, C., & Rossini, P. M. (2011). Human brain connectivity during single and paired pulse transcranial magnetic stimulation. *NeuroImage, 54*, 90–102. https://doi.org/10.1016/j.neuroimage.2010.07.056

Friston, K. (2002). Beyond phrenology: What can neuroimaging tell us about distributed circuitry? *Annual Review of Neuroscience, 25*, 221–250. https://doi.org/10.1146/annurev.neuro.25.112701.142846

Garcia, J. O., Grossman, E. D., & Srinivasan, R. (2011). Evoked potentials in large-scale cortical networks elicited by TMS of the visual cortex. *Journal of Neurophysiology, 106*, 1734–1746. https://doi.org/10.1152/jn.00739.2010

Ghatak, S., Talantova, M., McKercher, S. R., & Lipton, S. A. (2021). Novel therapeutic approach for excitatory/inhibitory imbalance in neurodevelopmental and neurodegenerative diseases. *Annual Review of Pharmacology and Toxicology, 61*, 701–721. https://doi.org/10.1146/annurev-pharmtox-032320-015420

Gosseries, O., Sarasso, S., Casarotto, S., Boly, M., Schnakers, C., Napolitani, M., Bruno, M.-A., Ledoux, D., Tshibanda, J.-F., Massimini, M., Laureys, S., & Rosanova, M. (2015). On the cerebral origin of EEG responses to TMS: Insights from severe cortical lesions. *Brain Stimulation, 8*, 142–149. https://doi.org/10.1016/j.brs.2014.10.008

Grandchamp, R., & Delorme, A. (2011). Single-trial normalization for event-related spectral decomposition reduces sensitivity to noisy trials. *Frontiers in Psychology, 2*. https://doi.org/10.3389/fpsyg.2011.00236

Granö, I., Mutanen, T. P., Nieminen, A. E., Nieminen, J. O., Souza, V. H., Ilmoniemi, R. J., & Lioumis, P. (2025). The dependency of TMS-evoked potentials on electric-field orientation in the cortex. *Brain Stimulation, 18*, 990–992. https://doi.org/10.1016/j.brs.2025.04.021

Guidali, G., Zazio, A., Lucarelli, D., Marcantoni, E., Stango, A., Barchiesi, G., & Bortoletto, M. (2023). Effects of transcranial magnetic stimulation (TMS) current direction and pulse waveform on cortico-cortical connectivity: A registered report TMS-EEG study. *The European Journal of Neuroscience, 58*, 3785–3809. https://doi.org/10.1111/ejn.16127

Hannula, H., & Ilmoniemi, R. J. (2017). Basic principles of navigated TMS. In M. Krieg (Ed.), *Navigated transcranial magnetic stimulation in neurosurgery* (pp. 3–29). Springer. https://doi.org/10.1007/978-3-319-54918-7_1

Harquel, S., Bacle, T., Beynel, L., Marendaz, C., Chauvin, A., & David, O. (2016). Mapping dynamical properties of cortical microcircuits using robotized TMS and EEG: Towards functional cytoarchitectonics. *NeuroImage, 135*, 115–124. https://doi.org/10.1016/j.neuroimage.2016.05.009

Harquel. S., Hummel, F. (2023). Brain oscillatory modes as a proxy of stroke recovery. https://doi.org/10.1101/2023.02.01.23285324

Harquel, S., Cadic-Melchior, A., Morishita, T., Fleury, L., Ceroni, M., Menoud, P., Brügger, J., Beanato, E., Meyer, N. H., Evangelista, G. G., Egger, P., Van de Ville, D., Blanke, O., Micera, S., Léger, B., Adolphsen, J., Jagella, C., Mühl, A., Constantin, C., Alvarez, V., Vuadens, P., Turlan, J.-L., Bonvin, C., Koch, P. J., Wessel, M. J., & Hummel, F. C. (2024). Brain oscillatory modes as a proxy of stroke recovery. *medRxiv*. https://doi.org/10.1101/2023.02.01.23285324

Heller, L., & van Hulsteyn, D. B. (1992). Brain stimulation using electromagnetic sources: Theoretical aspects. *Biophysical Journal, 63*, 129–138. https://doi.org/10.1016/S0006-3495(92)81587-4

Hernandez-Pavon, J. C., Metsomaa, J., Mutanen, T., Stenroos, M., Mäki, H., Ilmoniemi, R. J., & Sarvas, J. (2012). Uncovering neural independent components from highly artifactual TMS-evoked EEG data. *Journal of Neuroscience Methods, 209*, 144–157. https://doi.org/10.1016/j.jneumeth.2012.05.029

Hernandez-Pavon, J. C., Kugiumtzis, D., Zrenner, C., Kimiskidis, V. K., & Metsomaa, J. (2022). Removing artifacts from TMS-evoked EEG: A methods review and a unifying theoretical framework. *Journal of Neuroscience Methods, 376*, 109591. https://doi.org/10.1016/j.jneumeth.2022.109591

Hernandez-Pavon, J. C., Veniero, D., Bergmann, T. O., Belardinelli, P., Bortoletto, M., Casarotto, S., Casula, E. P., Farzan, F., Fecchio, M., Julkunen, P., Kallioniemi, E., Lioumis, P., Metsomaa, J., Miniussi, C., Mutanen, T. P., Rocchi, L., Rogasch, N. C., Shafi, M. M., Siebner, H. R., Thut, G., Zrenner, C., Ziemann, U., & Ilmoniemi, R. J. (2023). TMS combined with EEG: Recommendations and open issues for data collection and analysis. *Brain Stimulation, 16*, 567–593. https://doi.org/10.1016/j.brs.2023.02.009

Huber, R., Mäki, H., Rosanova, M., Casarotto, S., Canali, P., Casali, A. G., Tononi, G., & Massimini, M. (2013). Human cortical excitability increases with time awake. *Cerebral Cortex, 23*, 332–338. https://doi.org/10.1093/cercor/bhs014

Ilmoniemi, R. J., & Kičić, D. (2010). Methodology for combined TMS and EEG. *Brain Topography, 22*, 233–248. https://doi.org/10.1007/s10548-009-0123-4

Ilmoniemi, R. J., & Sarvas, J. (2019). *Brain signals: Physics and mathematics of MEG and EEG*. MIT Press.

Ilmoniemi, R. J., Virtanen, J., Ruohonen, J., Karhu, J., Aronen, H. J., Näätänen, R., & Katila, T. (1997). Neuronal responses to magnetic stimulation reveal cortical reactivity and connectivity. *NeuroReport, 8*, 3537–3540. https://doi.org/10.1097/00001756-199711100-00024

Ilmoniemi, R. J., Ruohonen, J., & Karhu, J. (1999). Transcranial magnetic stimulation—a new tool for functional imaging of the brain. *Critical Reviews in Biomedical Engineering, 27*, 241–284.

Ilmoniemi, R. J., Hernandez-Pavon, J. C., Makela, N. N., Metsomaa, J., Mutanen, T. P., Stenroos, M., & Sarvas, J. (2015). Dealing with artifacts in TMS-evoked EEG. *Conference Proceedings: Annual International Conference of the IEEE Engineering in Medicine and Biology Society, 2015*, 230–233. https://doi.org/10.1109/EMBC.2015.7318342

Ilmoniemi, R. J., Rogasch, N. C., & Casarotto, S. (2021). TMS measures explored by EEG recordings. In *The Oxford handbook of transcranial stimulation* (pp. 508–538). Oxford University Press. https://doi.org/10.1093/oxfordhb/9780198832256.013.17

Iramina, K., Maeno, T., Nonaka, Y., & Ueno, S. (2003). Measurement of evoked electroencephalography induced by transcranial magnetic stimulation. *Journal of Applied Physics, 93*, 6718–6720. https://doi.org/10.1063/1.1558635

Jamil, Z., Saisanen, L., Demjan, M., Reijonen, J., & Julkunen, P. (2024). The effect of stimulation intensity, sampling frequency, and sample synchronization in TMS-EEG on the TMS pulse artifact amplitude and duration. *IEEE Transactions on Neural Systems and Rehabilitation Engineering, 32*, 2612–2620. https://doi.org/10.1109/TNSRE.2024.3429176

Koponen, L. M., Nieminen, J. O., & Ilmoniemi, R. J. (2018a). Multi-locus transcranial magnetic stimulation-theory and implementation. *Brain Stimulation, 11*, 849–855. https://doi.org/10.1016/j.brs.2018.03.014

Koponen, L. M., Nieminen, J. O., Mutanen, T. P., & Ilmoniemi, R. J. (2018b). Noninvasive extraction of microsecond-scale dynamics from human motor cortex. *Human Brain Mapping, 39*, 2405–2411. https://doi.org/10.1002/hbm.24010

Leodori, G., De Bartolo, M. I., Guerra, A., Fabbrini, A., Rocchi, L., Latorre, A., Paparella, G., Belvisi, D., Conte, A., Bhatia, K. P., Rothwell, J. C., & Berardelli, A. (2022). Motor cortical network excitability in Parkinson's disease. *Movement Disorders, 37*, 734–744. https://doi.org/10.1002/mds.28914

Li, B., Virtanen, J. P., Oeltermann, A., Schwarz, C., Giese, M. A., Ziemann, U., & Benali, A. (2017). Lifting the veil on the dynamics of neuronal activities evoked by transcranial magnetic stimulation. *eLife, 6*, e30552. https://doi.org/10.7554/eLife.30552

Lioumis, P., & Rosanova, M. (2022). The role of neuronavigation in TMS–EEG studies: Current applications and future perspectives. *Journal of Neuroscience Methods, 380*, 109677. https://doi.org/10.1016/j.jneumeth.2022.109677

Lioumis, P., Kičić, D., Savolainen, P., Mäkelä, J. P., & Kähkönen, S. (2009). Reproducibility of TMS-evoked EEG responses. *Human Brain Mapping, 30*, 1387–1396. https://doi.org/10.1002/hbm.20608

Lioumis, P., Roine, T., Granö, I., Aydogan, D. B., Ukharova, E., Souza, V. H., Kičić, D., Ilmoniemi, R. J., & Makris, N. (2025). Optimization of TMS target engagement: current state and future perspectives. *Frontiers in Neuroscience, 19*, 1517228. https://doi.org/10.3389/fnins.2025.1517228

Ly, J. Q. M., Gaggioni, G., Chellappa, S. L., Papachilleos, S., Brzozowski, A., Borsu, C., Rosanova, M., Sarasso, S., Middleton, B., Luxen, A., Archer, S. N., Phillips, C., Dijk, D.-J., Maquet, P., Massimini, M., & Vandewalle, G. (2016). Circadian regulation of human cortical excitability. *Nature Communications, 7*, 11828. https://doi.org/10.1038/ncomms11828

Makeig, S., Debener, S., Onton, J., & Delorme, A. (2004). Mining event-related brain dynamics. *Trends in Cognitive Sciences, 8*, 204–210. https://doi.org/10.1016/j.tics.2004.03.008

Massimini, M., Ferrarelli, F., Huber, R., Esser, S. K., Singh, H., & Tononi, G. (2005). Breakdown of cortical effective connectivity during sleep. *Science, 309*, 2228–2232. https://doi.org/10.1126/science.1117256

Massimini, M., Ferrarelli, F., Esser, S. K., Riedner, B. A., Huber, R., Murphy, M., Peterson, M. J., & Tononi, G. (2007). Triggering sleep slow waves by transcranial magnetic stimulation. *Proceedings of the National Academy of Sciences USA, 104*, 8496–8501. https://doi.org/10.1073/pnas.0702495104

Massimini, M., Corbetta, M., Sanchez-Vives, M. V., Andrillon, T., Deco, G., Rosanova, M., & Sarasso, S. (2024). Sleep-like cortical dynamics during wakefulness and their network effects following brain injury. *Nature Communications, 15*, 7207. https://doi.org/10.1038/s41467-024-51586-1

Moliadze, V., Zhao, Y., Eysel, U., & Funke, K. (2003). Effect of transcranial magnetic stimulation on single-unit activity in the cat primary visual cortex. *The Journal of Physiology, 553*, 665–679. https://doi.org/10.1113/jphysiol.2003.050153

Momi, D., Wang, Z., & Griffiths, J. D. (2023). TMS-evoked responses are driven by recurrent large-scale network dynamics. *eLife, 12*. https://doi.org/10.7554/eLife.83232

Mueller, J. K., Grigsby, E. M., Prevosto, V., Petraglia, F. W., Rao, H., Deng, Z.-D., Peterchev, A.V., Sommer, M. A., Egner, T., Platt, M. L., Grill, W. M., (2014). Simultaneous transcranial magnetic stimulation and single-neuron recording in alert non-human primates. *Nature Neuroscience 17*, 1130–1136. https://doi.org/10.1038/nn.3751

Murphy, S. C., Palmer, L. M., Nyffeler, T., Müri, R. M., & Larkum, M. E. (2016). Transcranial magnetic stimulation (TMS) inhibits cortical dendrites. *eLife, 5*, e13598. https://doi.org/10.7554/eLife.13598

Nieminen, J. O., Sinisalo, H., Souza, V. H., Malmi, M., Yuryev, M., Tervo, A. E., Stenroos, M., Milardovich, D., Korhonen, J. T., Koponen, L. M., & Ilmoniemi, R. J. (2022). Multi-locus transcranial magnetic stimulation system for electronically targeted brain stimulation. *Brain Stimulation, 15*, 116–124. https://doi.org/10.1016/j.brs.2021.11.014

Nikouline, V., Ruohonen, J., & Ilmoniemi, R. J. (1999). The role of the coil click in TMS assessed with simultaneous EEG. *Clinical Neurophysiology, 110*, 1325–1328. https://doi.org/10.1016/S1388-2457(99)00070-X

Nikulin, V. V., Kičić, D., Kähkönen, S., & Ilmoniemi, R. J. (2003). Modulation of electroencephalographic responses to transcranial magnetic stimulation: evidence for changes in cortical excitability related to movement. *The European Journal of Neuroscience, 18*, 1206–1212. https://doi.org/10.1046/j.1460-9568.2003.02858.x

Nuyts, M., Beck, M. M., Banach, A., Thielscher, A., Meesen, R., Tomasevic, L., Siebner, H. R., & Christiansen, L. (2025). Rostro-caudal TMS mapping of immediate transcranial evoked potentials reveals a pericentral crescendo-decrescendo pattern. *NeuroImage, 319*, 121446. https://doi.org/10.1016/j.neuroimage.2025.121446

Parmigiani, S., Cline, C. C., Sarkar, M., Forman, L., Truong, J., Ross, J. M., Gogulski, J., & Keller, C. J. (2025). Real-time optimization to enhance noninvasive cortical excitability assessment in the human dorsolateral prefrontal cortex. *Clinical Neurophysiology*, S1388245725003049. https://doi.org/10.1016/j.clinph.2025.02.261

Passera, B., Chauvin, A., Raffin, E., Bougerol, T., David, O., Harquel, S., (2022). Exploring the spatial resolution of TMS-EEG coupling on the sensorimotor region. *NeuroImage 259*, 119419. https://doi.org/10.1016/j.neuroimage.2022.119419

Paus, T., Sipila, P. K., & Strafella, A. P. (2001). Synchronization of neuronal activity in the human primary motor cortex by transcranial magnetic stimulation: An EEG study. *Journal of Neurophysiology, 86*, 1983–1990. https://doi.org/10.1152/jn.2001.86.4.1983

Pellicciari, M. C., Brignani, D., & Miniussi, C. (2013). Excitability modulation of the motor system induced by transcranial direct current stimulation: A multimodal approach. *NeuroImage, 83*, 569–580. https://doi.org/10.1016/j.neuroimage.2013.06.076

Pellicciari, M. C., Ponzo, V., Caltagirone, C., & Koch, G. (2017). Restored asymmetry of prefrontal cortical oscillatory activity after bilateral theta burst stimulation treatment in a patient with major depressive disorder: A TMS-EEG study. *Brain Stimulation, 10*, 147–149. https://doi.org/10.1016/j.brs.2016.09.006

Perera, N. D., Alekseichuk, I., Shirinpour, S., Wischnewski, M., Linn, G., Masiello, K., Butler, B., Russ, B. E., Schroeder, C. E., Falchier, A., & Opitz, A. (2023). Dissociation of centrally and peripherally induced transcranial magnetic stimulation effects in nonhuman primates. *The Journal of Neuroscience, 43*, 8649–8662. https://doi.org/10.1523/JNEUROSCI.1016-23.2023

Pigorini, A., Casali, A. G., Casarotto, S., Ferrarelli, F., Baselli, G., Mariotti, M., Massimini, M., & Rosanova, M. (2011). Time-frequency spectral analysis of TMS-evoked EEG oscillations by means of Hilbert-Huang transform. *Journal of Neuroscience Methods, 198*, 236–245. https://doi.org/10.1016/j.jneumeth.2011.04.013

Premoli, I., Bergmann, T. O., Fecchio, M., Rosanova, M., Biondi, A., Belardinelli, P., & Ziemann, U. (2017). The impact of GABAergic drugs on TMS-induced brain oscillations in human motor cortex. *NeuroImage, 163*, 1–12. https://doi.org/10.1016/j.neuroimage.2017.09.023

Raffin, E., Harquel, S., Passera, B., Chauvin, A., Bougerol, T., & David, O. (2020). Probing regional cortical excitability via input-output properties using transcranial magnetic stimulation and electroencephalography coupling. *Human Brain Mapping, 41*, 2741–2761. https://doi.org/10.1002/hbm.24975

Rogasch, N. C., Thomson, R. H., Daskalakis, Z. J., & Fitzgerald, P. B. (2013). Short-latency artifacts associated with concurrent TMS-EEG. *Brain Stimulation, 6*, 868–876. https://doi.org/10.1016/j.brs.2013.04.004

Romero Lauro, L. J., Rosanova, M., Mattavelli, G., Convento, S., Pisoni, A., Opitz, A., Bolognini, N., & Vallar, G. (2014). TDCS increases cortical excitability: Direct evidence from TMS–EEG. *Cortex, 58*, 99–111. https://doi.org/10.1016/j.cortex.2014.05.003

Romero, M. C., Davare, M., Armendariz, M., & Janssen, P. (2019). Neural effects of transcranial magnetic stimulation at the single-cell level. *Nature Communications, 10*. https://doi.org/10.1038/s41467-019-10638-7

Rosanova, M., Casali, A. G., Bellina, V., Resta, F., Mariotti, M., & Massimini, M. (2009). Natural frequencies of human corticothalamic circuits. *The Journal of Neuroscience, 29*, 7679–7685. https://doi.org/10.1523/JNEUROSCI.0445-09.2009

Rosanova, M., Gosseries, O., Casarotto, S., Boly, M., Casali, A. G., Bruno, M.-A., Mariotti, M., Boveroux, P., Tononi, G., Laureys, S., & Massimini, M. (2012). Recovery of cortical effective connectivity and recovery of consciousness in vegetative patients. *Brain, 135*, 1308–1320. https://doi.org/10.1093/brain/awr340

Rosanova, M., Fecchio, M., Casarotto, S., Sarasso, S., Casali, A. G., Pigorini, A., Comanducci, A., Seregni, F., Devalle, G., Citerio, G., Bodart, O., Boly, M., Gosseries, O., Laureys, S., & Massimini, M. (2018). Sleep-like cortical OFF-periods disrupt causality and complexity in the brain of unresponsive wakefulness syndrome patients. *Nature Communications, 9*, 4427. https://doi.org/10.1038/s41467-018-06871-1

Ruohonen, J., & Karhu, J. (2010). Navigated transcranial magnetic stimulation. *Neurophysiologie Clinique/Clinical Neurophysiology, 40*, 7–17. https://doi.org/10.1016/j.neucli.2010.01.006
Russo, S., Sarasso, S., Puglisi, G. E., Dal Palù, D., Pigorini, A., Casarotto, S., D'Ambrosio, S., Astolfi, A., Massimini, M., Rosanova, M., & Fecchio, M. (2022). TAAC—TMS Adaptable Auditory Control: A universal tool to mask TMS clicks. *Journal of Neuroscience Methods, 370*, 109491. https://doi.org/10.1016/j.jneumeth.2022.109491
Russo, S., Claar, L. D., Furregoni, G., Marks, L. C., Krishnan, G., Zauli, F. M., Hassan, G., Solbiati, M., d'Orio, P., Mikulan, E., Sarasso, S., Rosanova, M., Sartori, I., Bazhenov, M., Pigorini, A., Massimini, M., Koch, C., & Rembado, I. (2025). Thalamic feedback shapes brain responses evoked by cortical stimulation in mice and humans. *Nature Communications, 16*, 3627. https://doi.org/10.1038/s41467-025-58717-2
Sachdev, R. N. S., Ebner, F. F., & Wilson, C. J. (2004). Effect of subthreshold up and down states on the whisker-evoked response in somatosensory cortex. *Journal of Neurophysiology, 92*, 3511–3521. https://doi.org/10.1152/jn.00347.2004
Sack, A. T., Cohen Kadosh, R., Schuhmann, T., Moerel, M., Walsh, V., & Goebel, R. (2009). Optimizing functional accuracy of TMS in cognitive studies: A comparison of methods. *Journal of Cognitive Neuroscience, 21*, 207–221. https://doi.org/10.1162/jocn.2009.21126
Säisänen, L., Julkunen, P., Niskanen, E., Danner, N., Hukkanen, T., Lohioja, T., Nurkkala, J., Mervaala, E., Karhu, J., & Könönen, M. (2008). Motor potentials evoked by navigated transcranial magnetic stimulation in healthy subjects. *Journal of Clinical Neurophysiology, 25*, 367–372. https://doi.org/10.1097/WNP.0b013e31818e7944
Sarasso, S., Boly, M., Napolitani, M., Gosseries, O., Charland-Verville, V., Casarotto, S., Rosanova, M., Casali, A. G., Brichant, J.-F., Boveroux, P., Rex, S., Tononi, G., Laureys, S., & Massimini, M. (2015). Consciousness and complexity during unresponsiveness induced by propofol, xenon, and ketamine. *Current Biology, 25*, 3099–3105. https://doi.org/10.1016/j.cub.2015.10.014
Sarasso, S., D'Ambrosio, S., Fecchio, M., Casarotto, S., Viganò, A., Landi, C., Mattavelli, G., Gosseries, O., Quarenghi, M., Laureys, S., Devalle, G., Rosanova, M., & Massimini, M. (2020). Local sleep-like cortical reactivity in the awake brain after focal injury. *Brain, 143*, 3672–3684. https://doi.org/10.1093/brain/awaa338
Sarasso, S., D'Ambrosio, S., Russo, S., Bernardelli, L., Hassan, G., Comanducci, A., De Giampaulis, P., Vecchia, L. D., Lanzone, J., & Massimini, M. (2025). Reduction of sleep-like perilesional slow waves and clinical evolution after stroke: A TMS-EEG study. *Clinical Neurophysiology, 18*, 593–594. https://doi.org/10.1016/j.clinph.2025.2110746
Shibata, S., Onishi, H., & Mima, T. (2024). TMS-EEG signatures of the effects of transcranial static magnetic field stimulation (tSMS) on cortical excitability. *Scientific Reports, 14*, 22394. https://doi.org/10.1038/s41598-024-72875-1
Sinisalo, H., Rissanen, I., Kahilakoski, O.-P., Souza, V. H., Tommila, T., Laine, M., Nyrhinen, M., Ukharova, E., Granö, I., Soto, A. M., Matsuda, R. H., Rantala, R., Guidotti, R., Kičić, D., Lioumis, P., Mutanen, T., Pizzella, V., Marzetti, L., Roine, T., Stenroos, M., Ziemann, U., Romani, G. L., & Ilmoniemi, R. J. (2024). Modulating brain networks in space and time: Multi-locus transcranial magnetic stimulation. *Clinical Neurophysiology, 158*, 218–224. https://doi.org/10.1016/j.clinph.2023.12.007
Solomon, E. A., Wang, J. B., Oya, H., Howard, M. A., Trapp, N. T., Uitermarkt, B. D., Boes, A. D., & Keller, C. J. (2024). TMS provokes target-dependent intracranial rhythms across human cortical and subcortical sites. *Brain Stimulation*, S1935861X24000962. https://doi.org/10.1016/j.brs.2024.05.014
Suzuki, M., Larkum, M. E., (2017). Dendritic calcium spikes are clearly detectable at the cortical surface. *Nat Commun 8*, 276. https://doi.org/10.1038/s41467-017-00282-4
Tangwiriyasakul, C., Premoli, I., Spyrou, L., Chin, R. F., Escudero, J., & Richardson, M. P. (2019). Tensor decomposition of TMS-induced EEG oscillations reveals data-driven profiles of antiepileptic drug effects. *Scientific Reports, 9*, 17057. https://doi.org/10.1038/s41598-019-53565-9

ter Braack, E. M., de Vos, C. C., & van Putten, M. J. A. M. (2015). Masking the auditory evoked potential in TMS–EEG: A comparison of various methods. *Brain Topography, 28*, 520–528. https://doi.org/10.1007/s10548-013-0312-z

Tervo, A. E., Nieminen, J. O., Lioumis, P., Metsomaa, J., Souza, V. H., Sinisalo, H., Stenroos, M., Sarvas, J., & Ilmoniemi, R. J. (2022). Closed-loop optimization of transcranial magnetic stimulation with electroencephalography feedback. *Brain Stimulation, 15*, 523–531. https://doi.org/10.1016/j.brs.2022.01.016

Thielscher, A., Opitz, A., & Windhoff, M. (2011). Impact of the gyral geometry on the electric field induced by transcranial magnetic stimulation. *NeuroImage, 54*, 234–243. https://doi.org/10.1016/j.neuroimage.2010.07.061

Tscherpel, C., Dern, S., Hensel, L., Ziemann, U., Fink, G. R., & Grefkes, C. (2020). Brain responsivity provides an individual readout for motor recovery after stroke. *Brain, 143*, 1873–1888. https://doi.org/10.1093/brain/awaa127

Turco, F., Canessa, A., Olivieri, C., Pozzi, N. G., Palmisano, C., Arnulfo, G., Marotta, G., Volkmann, J., Pezzoli, G., & Isaias, I. U. (2018). Cortical response to levodopa in Parkinson's disease patients with dyskinesias. *European Journal of Neuroscience, 48*, 2362–2373. https://doi.org/10.1111/ejn.14114

Ueno, S., Tashiro, T., & Harada, K. (1988). Localized stimulation of neural tissues in the brain by means of a paired configuration of time-varying magnetic fields. *Journal of Applied Physics, 64*, 5862–5864. https://doi.org/10.1063/1.342181

Vallesi, A., Del Felice, A., Capizzi, M., Tafuro, A., Formaggio, E., Bisiacchi, P., Masiero, S., & Ambrosini, E. (2021). Natural oscillation frequencies in the two lateral prefrontal cortices induced by Transcranial Magnetic Stimulation. *NeuroImage, 227*, 117655. https://doi.org/10.1016/j.neuroimage.2020.117655

Van Der Werf, Y. D., Sadikot, A. F., Strafella, A. P., & Paus, T. (2006). The neural response to transcranial magnetic stimulation of the human motor cortex. II. Thalamocortical contributions. *Experimental Brain Research, 175*, 246–255. https://doi.org/10.1007/s00221-006-0548-x

Veniero, D., Bortoletto, M., & Miniussi, C. (2009). TMS-EEG co-registration: On TMS-induced artifact. *Clinical Neurophysiology, 120*, 1392–1399. https://doi.org/10.1016/j.clinph.2009.04.023

Virtanen, J., Ruohonen, J., Näätänen, R., & Ilmoniemi, R. J. (1999). Instrumentation for the measurement of electric brain responses to transcranial magnetic stimulation. *Medical & Biological Engineering & Computing, 37*, 322–326. https://doi.org/10.1007/BF02513307

Vyazovskiy, V. V., Cirelli, C., Pfister-Genskow, M., Faraguna, U., & Tononi, G. (2008). Molecular and electrophysiological evidence for net synaptic potentiation in wake and depression in sleep. *Nature Neuroscience, 11*, 200–208. https://doi.org/10.1038/nn2035

Wang, Q., Gong, A., Feng, Z., Bai, Y., & Ziemann, U. (2024). Interactions of transcranial magnetic stimulation with brain oscillations: A narrative review. *Frontiers in Systems Neuroscience, 18*, 1489949. https://doi.org/10.3389/fnsys.2024.1489949

Wörgötter, F., Suder, K., Zhao, Y., Kerscher, N., Eysel, U. T., & Funke, K. (1998). State-dependent receptive-field restructuring in the visual cortex. *Nature, 396*, 165–168. https://doi.org/10.1038/24157

Ziemann, U. (2020). I-waves in motor cortex revisited. *Experimental Brain Research, 238*, 1601–1610. https://doi.org/10.1007/s00221-020-05764-4

Zrenner, C., & Ziemann, U. (2024). Closed-loop brain stimulation. *Biological Psychiatry, 95*, 545–552. https://doi.org/10.1016/j.biopsych.2023.09.014

Chapter 33
Combining EEG and fMRI

David W. Carmichael, Rebecca Meagher, Anna Sadilova, Tracy Warbrick, and Cilia Jaeger

Abstract In this chapter we will explore the motivation for simultaneous measurements of electroencephalography (EEG) together with functional magnetic resonance imaging (fMRI), termed EEG-fMRI. When measured together, these modalities provide a unique opportunity to measure human brain activity with a level of precision in space and time that is not possible with each modality alone. However, recording both signals together is challenging. To circumvent these challenges, specialised hardware and a range of signal processing methodologies are used to enable low-risk and high-fidelity recordings. EEG-fMRI is particularly valuable for applications where spontaneous brain activity, either physiological or pathological, is of interest. Spontaneous rhythmic activity has been studied since the start of EEG recordings, but EEG-fMRI has allowed us to view it in a new light. Importantly, in clinical neuroscience, combined EEG-fMRI has provided insights into epilepsy and sleep. Furthermore, it has also advanced our understanding of how EEG and fMRI signals relate to each other at a local and brain network scale. We summarise the current state of the art, how technical challenges have been met, and provide some examples of valuable insights made with EEG-fMRI in applications such as sleep and epilepsy.

Keywords Multimodal brain imaging · EEG-fMRI · Sleep · Epilepsy

D. W. Carmichael (✉) · R. Meagher · A. Sadilova
School of Biomedical Engineering and Imaging Sciences, King's College London, London, UK
e-mail: david.carmichael@kcl.ac.uk; rebecca.meagher@kcl.ac.uk; anna.sadilova@kcl.ac.uk

T. Warbrick · C. Jaeger
Brain Products GmbH, Gilching, Germany
e-mail: tracy.warbrick@brainproducts.com; cilia.jaeger@brainproducts.com

T. Warbrick (ed.), *The EEG Handbook*,
https://doi.org/10.1007/978-3-032-20450-9_33

33.1 Motivation for Combined EEG and fMRI

In recent years, neuroscience has moved toward multimodal approaches that combine different methods of measuring brain activity. This stems from the recognition that no single imaging technique can capture the brain's full complexity. Some methods excel in tracking when things happen in the brain, while others reveal where these processes take place. Each method often captures a certain biophysical process related, more or less directly, to underlying neural activity. Integrating these complementary views allows researchers to build a more comprehensive understanding of neural mechanisms.

The integration of electroencephalography (EEG) and functional magnetic resonance imaging (fMRI) has become particularly influential for recordings in human subjects. EEG records electrical signals directly from the scalp, reflecting neuronal activity, whereas fMRI detects slower, blood-based (haemodynamic) responses indirectly linked with neural metabolism (Murta et al., 2015). When acquired simultaneously, these two techniques allow researchers to link the sub-millisecond dynamics of neural activity derived from EEG with the millimetre precision of structural brain localisation derived from fMRI, offering an excellent spatiotemporal view of brain function non-invasively.

Simultaneous EEG-fMRI is especially useful for studying spontaneous or unpredictable brain activity, such as resting-state networks, epileptic discharges, or sleep. Under these conditions, it is difficult to derive associations between activity measured during recordings of each modality made in separate session. Acquiring both datasets in a single experiment ensures that all multimodal measurements are obtained in the same brain state and cognitive or physiological condition, minimising variability across time. Moreover, simultaneous recordings allow variability in alertness, attention, stimulus habituation, or even subtle head position differences to be captured in both modalities. The result is a more complete approach to exploring how patterns of electrical activity unfold across the brain's structural and functional landscape.

33.1.1 Spatiotemporal Precision

EEG and fMRI each offer distinct advantages and limitations, making their combination particularly powerful. EEG measures voltage changes generated by large populations of synchronously active neurons in the cortex via post-synaptic potentials. These signals reflect the brain's electrical activity with very high temporal precision (sub-millisecond), which is optimal for tracking the timing of neural events. However, because the electrical potentials are intrinsically weak, and affected by the skull and other tissues before reaching scalp electrodes, determining the exact locations of their sources is challenging. Deep or tangentially oriented brain signals are especially difficult to detect, resulting in limited spatial resolution (Nunez & Srinivasan, 2006).

In contrast, fMRI detects changes in blood oxygenation, flow, and volume that accompany neural activation. This blood oxygen level–dependent (BOLD) signal provides a spatially precise but temporally slow view of brain function. The underlying mechanism, known as neurovascular coupling, links increased neural activity to local increases in blood flow and oxygen (Kim & Ogawa, 2012). While this offers excellent localisation of active brain regions, the BOLD signal lags neural events by several seconds, making it unsuitable for capturing rapid fluctuations in brain activity that can be hard to disambiguate from regional variability in haemodynamic responses (Glover, 2011).

By combining EEG and fMRI, researchers can overcome these limitations. The fast temporal dynamics recorded by EEG can be mapped onto the spatially detailed brain images from fMRI, providing a more complete picture of when and where neural processes occur. This integrated approach has the potential to unlock new avenues in the study of cognition, perception, and neurological disorders, enabling insights that neither technique could achieve alone.

33.1.2 Capturing States and Spontaneous Brain Activity

EEG and fMRI can be used to assess brain activity that either occurs spontaneously or is time-locked to a stimulus or task, yet they differ fundamentally in the type of brain activity they measure. EEG can be used to study stimulus-induced activations through event-related potentials (ERPs), which are precisely time-locked to the presentation of a stimulus or performance of a task (more information in Chap. 19). EEG can also capture spontaneous changes in brain activity via neural oscillations across different frequency bands (see Chap. 18), reflecting ongoing brain states and network interactions (Chap. 6)

In fMRI, stimulus-induced brain activation is measured through the BOLD response: neural activity in a given region increases local blood flow and oxygenation, altering the ratio of deoxygenated to oxygenated haemoglobin and producing an increase in pixel intensity (Ogawa et al., 1990). This produces a characteristic time course, the hemodynamic response, with peak activation occurring approximately 5–6 s after the underlying neural event (Buxton et al., 2004). To quantify where in the brain this response occurs, the hemodynamic response is often used within a general linear model (GLM) framework. The known timing of the stimulus is convolved with the hemodynamic response to create a predicted BOLD signal, which is then compared to the observed signal voxel by voxel. The resulting statistical parameter estimates indicate which voxels show activity reliably coupled to the stimulus (Friston et al., 1994), thereby generating spatial maps of activation.

In addition to task-based paradigms, fMRI can also be used to study resting state activity, which examines spontaneous low-frequency fluctuations (0.01–0.1 Hz) in the BOLD signal that are spatially coherent. Regions whose signals fluctuate together in time are interpreted as being functionally connected (Biswal et al., 1997; Lowe et al., 2000). Analysis of these intrinsic correlations has revealed large-scale networks such as the default mode, attention, sensory, and motor networks (Fox et al., 2005; van den Heuvel & Hulshoff Pol, 2010). These networks can be used to study brain states and dynamics by analysing changes within and between networks associated with development, aging, disease, or altered states of consciousness.

While both EEG and fMRI can assess event-related and spontaneous neural activity, they differ in temporal and physiological characteristics. EEG captures neural signals on a millisecond timescale and provides a fine-grained temporal view of neural synchronisation and oscillatory activity. Oscillations serve as a tool to study the rapid dynamics of neural coordination (Uhlhaas et al., 2008). In contrast, fMRI reflects a slower metabolic response to altered neural activity levels due to neurovascular coupling. Through resting-state analyses, fMRI provides information about the spatial architecture of brain activity and the organisation of large-scale networks. Combining these modalities allows researchers to link the fast, oscillatory dynamics observed in EEG with the slower reconfiguration of resting-state networks measured by fMRI (Phadikar et al., 2025; Sadaghiani et al., 2022). However, it is important to account for the distinct temporal and spatial properties of each signal, as these can influence interpretation when integrating EEG and fMRI data. The next section will describe approaches for combining these modalities in joint analyses.

33.2 Data Integration

To fully exploit simultaneously recorded EEG and fMRI, the two data sets must be integrated. As described in Sect. 33.1.2, the two modalities have different temporal and spatial scales. As such, several data integration strategies have been developed to leverage the strengths of each while also taking the different measurement scales into account (see Fig. 33.1). The choice of data integration approach depends on the research or clinical question. Here we provide an overview of the available methods, for detailed reviews see (Abreu et al., 2018; Huster et al., 2012; Jorge et al., 2014).

EEG-fMRI data integration methods can be broadly divided into symmetrical and asymmetrical. Symmetrical approaches integrate both modalities into a joint analysis and can be model-driven or data-driven. Model-driven approaches use complex neurovascular coupling dynamics to predict activation, but they require the explicit definition of the common neuronal process underlying both the EEG and fMRI signals (Daunizeau et al., 2007; Valdes-Sosa et al., 2009). Data-driven approaches avoid the need to model complex neurovascular coupling dynamics, therefore avoiding potential bias, and include joint ICA (Calhoun et al., 2006, 2009), using an information theoretic framework (Ostwald et al., 2010, 2011), or multivariate machine learning (De Martino et al., 2010).

In asymmetrical approaches, information from one modality is used to predict or constrain the other modality: EEG-informed fMRI or fMRI-informed EEG. Asymmetrical approaches have been used more frequently in the literature than

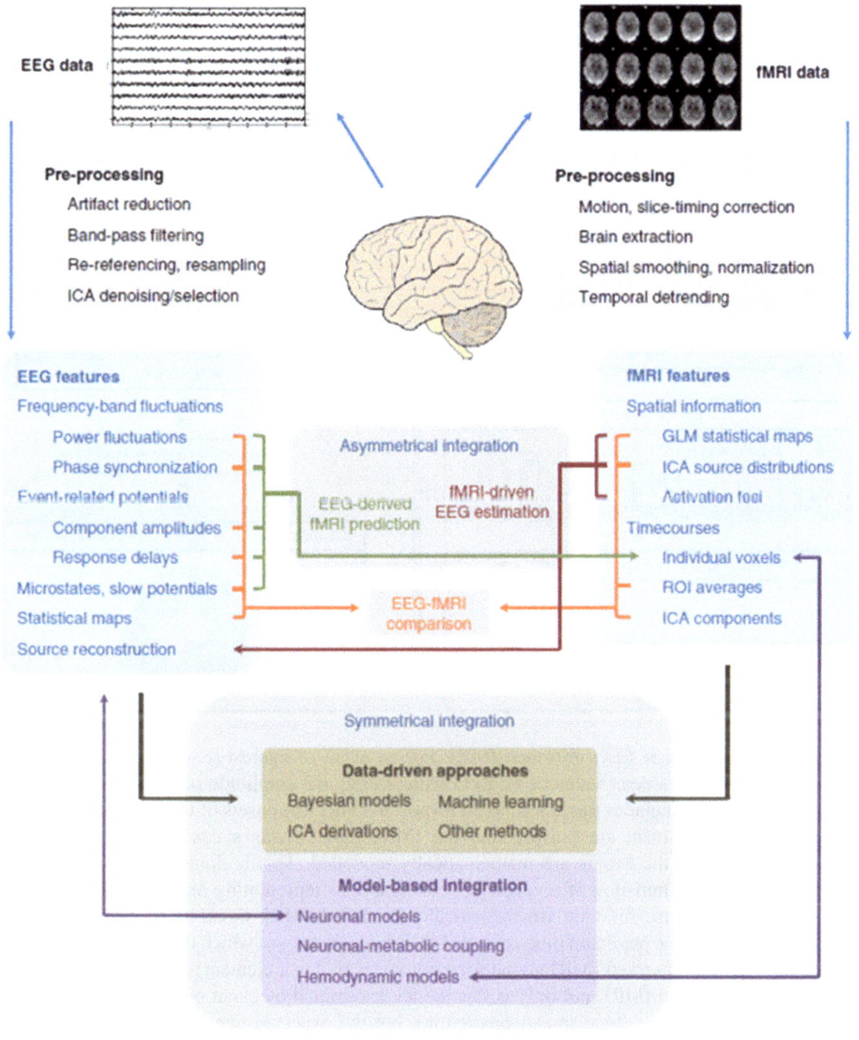

Fig. 33.1 A general scheme of the main EEG-fMRI data integration approaches proposed in the literature, including purely comparative, asymmetrical, and symmetrical techniques. (Reproduced with permission from Jorge et al., 2014)

symmetrical approaches and of the asymmetrical approaches EEG-informed fMRI is by far the most common. This is likely due to the rich temporal dynamics of the EEG data that can be readily integrated into fMRI analysis as a model of expected changes in fMRI signals. EEG-informed fMRI uses the properties of events measured in the EEG such as onset, duration, or power to identify associated fMRI signal related to these events (Debener et al., 2005; Laufs et al., 2003) (Fig. 33.2). These methods can

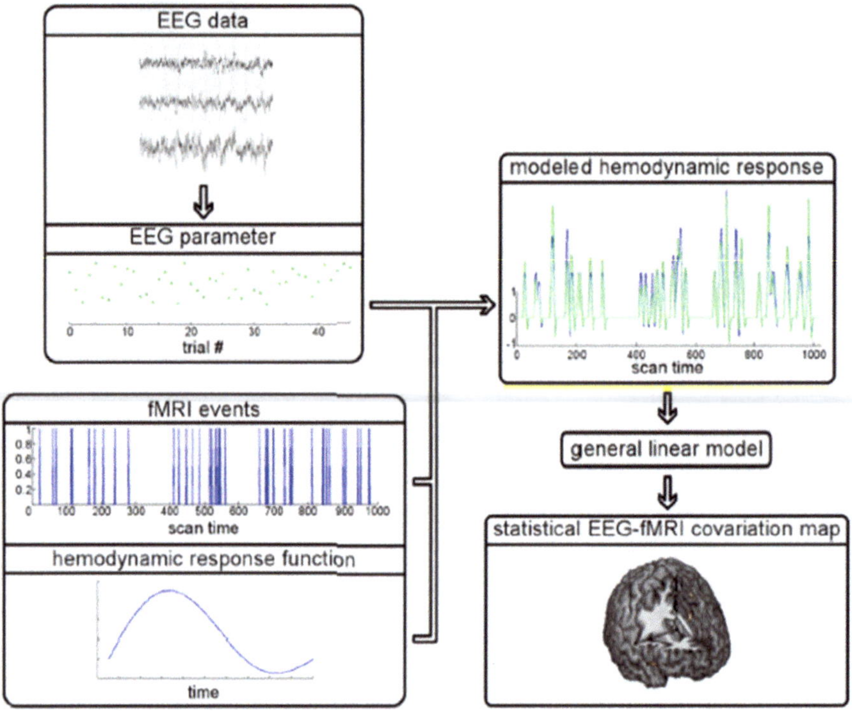

Fig. 33.2 Schematic of EEG-informed fMRI. For an event of interest (e.g., the occurrence of a response error), a parameter value of an EEG feature (e.g., the amplitude of an ERP) is extracted from every trial that includes this event. With respect to fMRI, the onsets of these events during the course of the experiment are known, and the fMRI signal changes caused by hemodynamic responses following the events are mathematically modelled. This is done via convolution: the multiplication and summation of a vector of zeros and ones representing the event onsets and the hemodynamic response function (mathematically modelled fMRI signal changes following an event). The result is the predicted time course of fMRI signal changes which can then be statistically compared with the observed fMRI signal for each voxel (volume element) of a brain scan. In the case of EEG-informed fMRI, not only is this model determined by event onsets and the hemodynamic response function (blue model prediction), but the expected hemodynamic responses are additionally parameterised using the extracted EEG feature: signal changes following events on trials with a large EEG response are scaled up as compared with trials with smaller EEG responses (green model prediction). Although beyond the scope of this chapter, for this analysis knowledge of the hemodynamic response function is required and maybe less certain in some applications such as epilepsy or in neonates. (Reproduced under open access license agreement CC BY-NC-SA 3.0 from: Huster et al., 2012)

be further subdivided into univariate and multivariate approaches. A time-course representative of the phenomenon of interest can be obtained from a single (or a limited number) of channels and then used to predict the BOLD signal, this is considered a univariate approach. Whereas multivariate techniques use multiple EEG channels to capture spatial information, for example, functional connectivity measures across different channels or spatial correlation measures.

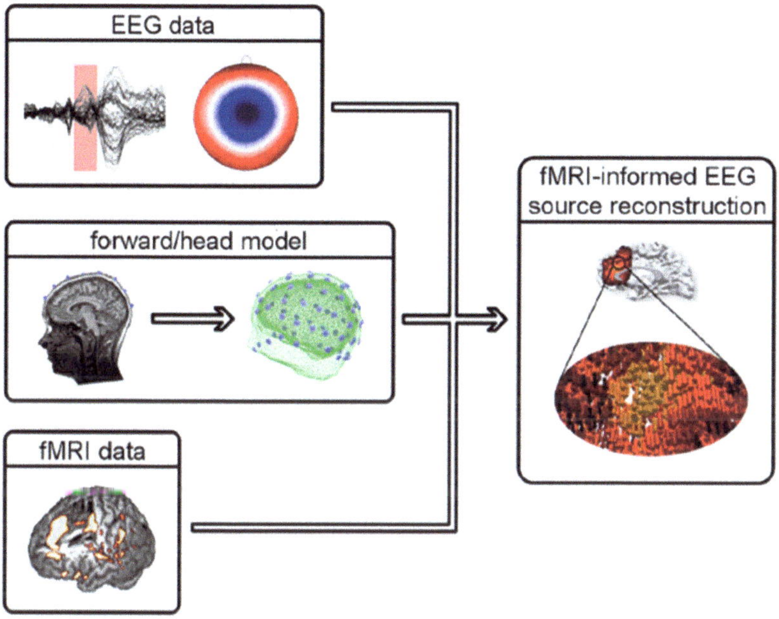

Fig. 33.3 Illustration of fMRI-informed EEG source reconstruction. To estimate the location and activity of active cortical patches in the brain that lead to measurable EEG signal changes on the scalp, forward or head models are constructed from individual MR images. Here, volumes representing skin, skull, and brain tissue have been extracted. Based on such a model and the EEG time courses as well as corresponding scalp topographies (the pattern of EEG activity as recorded on a participant's head), EEG sources can be inferred (inverse modelling). Statistical maps from a standard fMRI analysis are used to further constrain possible source constellations. The procedure used here for inverse modelling computes a high number of dipolar sources distributed across the brain, each of which is characterised by its position, orientation (pointing direction of an arrow), and strength (as indicated by colouring). (Reproduced under open access license agreement CC BY-NC-SA 3.0 from: Huster et al., 2012)

A frequent goal of fMRI-informed EEG (Fig. 33.3) is to solve the spatial EEG inverse problem by guiding source analysis using fMRI results (see Chap. 20 for further information on the inverse problem). A potential reason for fewer studies on fMRI-informed EEG is that simultaneous measurements are not always necessary or recommended in the context of source reconstruction (Huster et al., 2012; Liu et al., 2009). However, the fMRI-informed EEG can achieve better neuronal source reconstructions than using EEG alone (Babiloni et al., 2000, 2002). fMRI-informed EEG is also used to aid the interpretation of fMRI changes. For example, spontaneous changes in brain state such as those in resting-state fMRI can be associated with certain spectral responses aiding their interpretation (Laufs, 2008; Mantini et al., 2007).

33.3 Challenges of Simultaneous Recordings

Acquiring EEG and fMRI data simultaneously poses significant technical and methodological challenges. The two techniques rely on very different physical principles: EEG measures electrical potentials at the scalp, while fMRI detects magnetic resonance signals from protons (hydrogen nuclei). Their coexistence within the same environment creates complex interactions that can affect both safety and data quality. These challenges can be broadly divided into three categories: safety considerations, EEG artifacts introduced by the MRI environment, and MRI signal degradation caused by the EEG equipment, we consider each below.

33.3.1 Safety

Recording EEG inside an MRI scanner introduces complex interactions due to its exposure to the magnetic fields essential to MRI operation. Despite equipment being available that poses a very low risk to participant safety (if used within the instructions for use), it is still crucial to understand these interactions, and consequently how to mitigate them. MRI uses several distinct magnetic fields, each of which interacts differently with EEG equipment:

1. The *static magnetic field* (B_o) is always present and very strong (up to 7 Tesla in ultra-high-field systems). All EEG hardware used in the scanner must be MR-conditional, meaning it is constructed from non-ferromagnetic materials (Ives et al., 1993). This prevents dangerous projectile effects or mechanical forces on the equipment. Introducing standard EEG hardware into the scanner bore would pose severe safety risks because it would become a projectile with lethal capability.
2. The *gradient fields* are switching spatially variable magnetic fields used to encode spatial information in the image in the range of Hz-kHz. These fields can induce electrical currents in the conductive EEG wires and in the scalp itself, potentially leading to stimulation or even tissue damage if uncontrolled.
3. The *RF Field* (B_1^+) is a Radiofrequency magnetic field applied to excite tissue magnetised (by the static field B_o) so that it produces a measurable signal. The RF transmit field can couple strongly with any elongated conductor such as the conductive leads in the EEG cap. Wires at certain B_o field-dependent lengths (owing to its relation to wavelength) can become resonant structures. This will mean that they strongly couple with the RF field and will have significant current flow within them. These cause local "hot spots" of E-field and therefore specific absorption rate (SAR), a measure of energy deposited in tissue. At worst this can cause localised patient burns from local induced temperature increases.

To counter these challenges, modern MR conditional EEG cap designs have evolved considerably. Early systems introduced resistors close to the electrodes mainly to limit currents in tissue (Lemieux et al., 1997), while more recent designs, such as those proposed by Phong Le and colleagues, segment long leads into shorter non-resonant lengths to minimise RF coupling and heating (Lê et al., 2022). Extensive testing has shown that, when proper precautions are taken, EEG–fMRI can be performed safely. Commercially available MR-compatible EEG systems are considered conditional for use at field strengths up to 3 T, and ongoing research is extending this capability to 7 T. Recent studies have demonstrated that even at 7 T, with modern multiband echo-planar imaging (EPI) protocols, local heating remains well within acceptable limits (Martinez et al., 2025). Importantly, no documented cases of injury related to EEG–fMRI have been reported in regulatory or scientific databases to date (Hawsawi et al., 2017), a testament to the community's rigorous safety standards, and the importance of careful EEG cap design.

33.3.1.1 EEG Artifacts Induced by the MRI Environment

Recording EEG inside an MRI scanner generates several sources of interference that are typically orders of magnitude larger than the neural signals of interest. These artifacts can obscure the underlying EEG signal and make it difficult to gain meaningful insights from EEG. The EEG signal also remains susceptible to artifacts seen under laboratory conditions, such as muscle artifact or ocular artifact (see Chaps. 15 and 17 for further details). Therefore, mitigating the effects of scanner-related artifacts through hardware design and sophisticated signal processing is crucial to the final signal quality. The most prominent artifacts include:

1. *The Gradient Artifact (GA):* occurring due to large voltage fluctuations induced by the rapidly switching magnetic field gradient fields used for the spatial encoding of fMRI.
2. *Ballistocardiogram (BCG) artifacts:* distortions linked to cardiac pulsation and heartbeat-related movement of the head in the static field B_0.
3. *Motion and vibration artifacts:* caused by participant movement or vibrations in the scanner environment such as the mechanical helium pump.

Gradient Artifact (GA)

Gradient artifacts arise from electromagnetic induction during fast gradient switching, particularly in EPI sequences used for fMRI. The resulting voltages can exceed true neural signals by several hundredfold, rendering the raw EEG unreadable (Allen et al., 2000; Bullock et al., 2021).

Early solutions involved interleaving EEG acquisition with MRI sequences ("stepping stone sampling" (Bullock et al., 2021)), but this limited the temporal continuity of recordings. Modern approaches rely on post-processing techniques, the most common being Average Artifact Subtraction (AAS) (Allen et al., 2000). AAS exploits the high periodicity of MRI gradient waveforms by creating an average artifact template from several neighbouring epochs and subtracts it from the EEG data (Mulert & Lemieux, 2022). This approach is computationally efficient but sensitive to motion, which can alter the artifact shape between repetitions (Zhang et al., 2019). To improve robustness, variants of AAS use weighted sliding windows. These methods can also be extended to incorporate motion information within the correction making it adaptable but this relies on knowledge of participant motion, for example, using a tracking device (Laufs, 2008; Maziero et al., 2016, 2021).

Additional approaches include model-based correction using known gradient waveforms, and blind source separation (BSS) techniques such as Independent Component Analysis (ICA) or Canonical Correlation Analysis (CCA) (Abreu et al., 2018). These methods can adapt to nonstationary artifacts but must be applied cautiously to avoid removing genuine brain activity (Bullock et al., 2021).

Ballistocardiogram (BCG) Artifact

The BCG artifact, or pulse artifact, is produced by subtle head movements and scalp deformations caused by the cardiac cycle in the static magnetic field (Debener et al., 2008). The pulsatile motion of blood and small rotations of the head generate electromagnetic induction in the EEG leads, while conductive blood flow itself can induce voltages through the Hall effect (Müri et al., 1998).

The amplitude of BCG artifacts typically lies in the same range as EEG signals (tens of microvolts), and their variability makes removal difficult (Debener et al., 2008). At higher magnetic fields, both the amplitude and spatial variability of the BCG increase (Debener et al., 2008). Template-based subtraction (Allen et al., 2000), similar to AAS, is often inadequate because the BCG waveform changes dynamically over time. More adaptive methods, such as Optimal Basis Sets (OBS) and PCA-based approaches (Niazy et al., 2005), can better capture this variability. Many pipelines also incorporate concurrent ECG or photoplethysmography (PPG) recordings to provide reference signals for identifying pulse-related components (Bullock et al., 2021).

Beyond signal processing, EEG hardware modifications can help capture and suppress artifacts at their source. Techniques such as reference-layer caps introduce dedicated sensors that record environmental and motion-related noise independently of true EEG activity. These reference signals can then be subtracted in real time before digitisation, preserving the dynamic range of the amplifier and improving data quality (Chowdhury et al., 2014). Carbon wire loops (CWL) can also be used for direct artifact recording and may offer a more flexible alternative to reference-layer caps (Abbott et al., 2015). CWLs consist of small conductive loops placed near the EEG electrodes or other points of potential artifact induction. As they can be easily added to standard MR-compatible EEG caps, they do not require specialised hardware or cap redesign. These loops detect the same electromagnetic disturbances that affect the EEG electrodes, leads, and wires, such as those produced by head motion, BCG-related movement, or environmental vibrations (e.g., from the MR helium pump or ventilation systems), without recording any neural activity. The CWL signals are then used in post-processing as regressors to remove these artifacts from the EEG data, ideally resulting in uncontaminated neural signal only. Looking forward, CWL-based recordings also hold potential for real-time artifact correction, which could be especially valuable for neurofeedback or brain–computer interface (BCI) applications conducted during fMRI.

More generally, the increased availability and use of AI methods is likely to result in even further improvements in the reliability and simplicity and speed of EEG artefact correction within the scanner environment and early work towards this has already shown promising results (Shahriar et al., 2025).

33.3.1.2 MRI Signal Degradation Caused by EEG Equipment

EEG caps, electrodes, and leads can distort both the static magnetic field (B_o effects) and the RF transmit field (B_1^+ effects), degrading image homogeneity and reducing signal-to-noise ratio (SNR).

1. *B_o effects:* Differences in magnetic susceptibility between EEG components (electrode wires, resistors, adhesives, etc.) and biological tissue create local static magnetic field inhomogeneities, and these issues scale with field strength (Carmichael, 2022). These magnetic field perturbations can lead to geometric distortions or signal dropouts in images, particularly in rapid EPI sequences used in fMRI (Jezzard & Clare, 1999) owing to their long readout durations. Although these image artefacts generally do not strongly affect brain regions in adults, with electrode artefact typically localised mainly to the scalp and skull (Debener et al., 2008; Mullinger et al., 2008), they require greater attention in neonatal imaging due to smaller distance from the electrode to the brain.
2. *B_1^+ effects:* Conductive materials in the EEG cap can perturb the RF excitation field, leading to uneven excitation and a reduction in its efficiency across the brain and a further loss of SNR (Carmichael, 2022). Receive efficiency will also be similarly affected. These effects can be minimised using high-resistance or segmented leads that are then below the dimensions required to strongly interact with the wavelength of the RF fields. This in turn reduces surface currents and associated RF shielding (Lê et al., 2022).

Although both effects can compromise image quality, most studies have found that they do not preclude reliable fMRI acquisition (Bonmassar et al., 2001; Neuner et al., 2014). With careful design and calibration, EEG–fMRI data can achieve sufficient quality for both research and clinical use. The balance between artifact suppression and preservation of genuine signals remains a key consideration, as overly aggressive correction can distort neural information. This is true for any artifact attenuation technique not only those for correcting scanner-related artifacts (see Chap. 17)

In summary, simultaneous EEG–fMRI is a technically demanding but highly rewarding multimodal approach. Achieving safe measurements and high-quality data requires careful attention to hardware compatibility, electromagnetic interactions, and sophisticated signal correction methods.

33.4 Clinical Neuroscience Applications

EEG-fMRI has opened new avenues across both clinical and basic neuroscience domains. Clinically, it has become an important tool for conditions characterised by transient, spontaneous neural events, most notably epilepsy, where it aids in localising the epileptogenic zone and mapping epileptic networks in children (Centeno et al., 2016) and adults (Aghakhani et al., 2004; Bénar et al., 2006; Krakow et al., 1999). From a research perspective, it enables exploration of ongoing brain dynamics in contexts such as sleep, resting-state fluctuations, and sensory or cognitive processing, revealing how electrical rhythms and hemodynamic responses interact across brain states.

The following sections outline several representative applications that exemplify the unique advantages of EEG–fMRI. Rather than providing an exhaustive survey, the aim is to illustrate how this multimodal approach has advanced understanding of both pathological and physiological brain processes, while also highlighting emerging directions made possible by recent technical developments.

33.4.1 Sleep

Sleep is an active state of unconsciousness characterised by cyclical transitions between rapid eye movement (REM) and non-rapid eye movement (NREM) stages, each marked by distinct physiological and electrophysiological signatures (Brinkman et al., 2025). As such, it is an important example of a spontaneous brain state change. Typically, cycling through all sleep stages occurs in approximately 90-min intervals each night, demonstrating structured neural dynamics fundamentally different from wakefulness (Patel et al., 2025). NREM sleep progresses through stages N1–N3, or light to deep sleep, with the latter featuring high-amplitude delta activity thought to underlie restorative and memory-related processes (Besedovsky et al., 2017; Kim et al., 2019). Conversely, REM sleep presents a wake-like electrophysiological pattern with vivid dreaming, muscle atonia, and rapid eye movements (Blumberg et al., 2020). Despite extensive research efforts, the precise functions of sleep and its distinct stages remain unexplained.

EEG has long been a cornerstone of sleep research, being the basis on which sleep stages were first identified (Campbell, 2009) (see also Chap. 28 for further information). However, its limited spatial resolution limits the possible research outcomes and interpretations (Yun, 2024). To address this, EEG-fMRI has emerged as a powerful methodological approach, combining the temporal precision of EEG with the spatial specificity of fMRI, enabling the simultaneous investigation of local oscillatory events and large-scale network dynamics (Pereira et al., 2025).

Current EEG-fMRI sleep research focuses on several major areas, including sleep regulation of cerebrospinal fluid (CSF) movement as a mechanism for metabolic clearance during sleep (Miao et al., 2024; Uji et al., 2025; Xie et al., 2013), which is increasingly investigated as a potentially critical function of sleep. EEG-fMRI also allows for the identification of neurophysiological biomarkers in sleep disorders such as insomnia and narcolepsy (Engström et al., 2014; Li et al., 2022). In these contexts, EEG provides stage-specific signatures, e.g., spindles and slow waves, while fMRI reveals how functional connectivity reorganises across sleep-wake cycles. Together, these measures offer complementary windows onto sleep-related pathology.

Utilising the complementary strengths of EEG and fMRI, Li et al. (2022) have demonstrated alterations in connectivity during wakefulness and across all NREM stages within the salience network in individuals with insomnia. Given the salience network's central role in arousal regulation and the coordination of state transitions between internally and externally oriented attention, these disruptions may provide mechanistic insight into insomnia pathophysiology. Notably, the study found further discrepancies during awakenings that interrupt sleep and during rest stages preceding sleep onset. These findings suggest that rest and transitional periods may represent a valuable target for further investigation, particularly concerning connectivity between regions such as the anterior insula and the thalamus/putamen, or the amygdala and the rostral anterior cingulate cortex (Li et al., 2022).

Beyond connectivity-based analyses, recent efforts have also focused on enhancing sleep stage classification through multimodal data integration, moving beyond conventional EEG-based scoring approaches (Yang et al., 2024). Recent advances in fMRI-based sleep classification have provided important insights into the dynamics of sleep state transitions. Notably, work applying data-driven modelling approaches (e.g. Hidden Markov and recurrent neural network models) has revealed that transitions between wakefulness and successive sleep stages can be predicted with remarkably high accuracy (Tagliazucchi & Laufs, 2014; Yang et al., 2024; Yu et al., 2023). These analyses identify a set of finely resolved brain states which can be mapped onto standard sleep stages as well as intermediate transitional states (Yang et al., 2024). This high-resolution characterisation demonstrates that transitional periods are not merely brief or unstable fluctuations but instead reflect structured, stage-specific patterns of large-scale brain dynamics.

In parallel, EEG–fMRI is informing interventional approaches during sleep, including neurofeedback and non-invasive brain stimulation techniques such as transcranial alternating current stimulation (tACS) and transcranial magnetic stimulation (TMS). These approaches provide new avenues for modulating sleep oscillations, probing causal mechanisms, and developing therapeutic interventions for sleep-related and neuropsychiatric conditions (Lambert-Beaudet et al., 2021; Massimini et al., 2007; Motamedi et al., 2023).

Substantial progress has been made in leveraging the complementary strengths of each modality to better understand sleep physiology. As mentioned earlier, recent work has identified robust fMRI signatures of distinct sleep stages, enabling the prediction of EEG-defined sleep architecture from fMRI alone with high consistency (Yang et al., 2024; Yu et al., 2023). This cross-modal decoding has not only validated the tight coupling between electrophysiological and haemodynamic markers of brain states but has also provided a powerful framework for examining how macroscopic brain dynamics evolve across transitions in sleep and consciousness. Such findings suggest that multimodal data can reveal network properties that are inaccessible to either modality in isolation, offering deeper insight into the neural mechanisms governing state changes.

33.4.2 Epilepsy

Epilepsy was the first main driver of simultaneous EEG–fMRI development, largely because of the technique's ability to capture both the spatial distribution of epileptic activity with fMRI using the timing from EEG. For approximately 1/3 of patients with focal epilepsy that are pharmaco-resistant, surgery remains the only curative treatment available with major associated improvements in quality of life (Galan et al., 2021). However, its success is reliant on the identification of the epileptogenic tissue that is capable of producing seizures. In this context, techniques utilising EEG, the key diagnostic tool to identify epileptiform activity, are crucial.

Non-invasive scalp EEG is safe and widely used, and if recorded with a dense electrode array, can be used to localise activity. However, EEG source localisation is spatially blurred by volume conduction and the ill-posed inverse problem, limiting localisation accuracy (see Chap. 2) (Michel et al., 2004) particularly for deeper sources. Simultaneous EEG–fMRI is therefore considered as a non-invasive, spatially informed complement to EEG for presurgical mapping. Here, the sensitivity to brain activity across the whole brain at a scale of mm-cm is a potential advantage. The most common analytic strategy for identifying epilepsy-related BOLD changes involves event-related modelling of the EEG signal.

In epilepsy EEG–fMRI protocols, experienced clinical neurophysiologists mark epileptiform events in the recorded EEG, typically interictal epileptiform discharges (IEDs) and, when they occur, ictal events. These events are convolved with a haemodynamic response model to generate regressors used in mass univariate fMRI general linear model analysis. Peaks or clusters of BOLD signal change that covary with the timing of EEG events are then interpreted as haemodynamic correlates of epileptic activity and can point toward candidate epileptogenic regions (Bagshaw et al., 2005; Jacobs et al., 2009; Lemieux et al., 2001; LeVan & Gotman, 2009). This event-based approach is straightforward and widely used, but it can only capture activity that is both temporally locked to the marked EEG events and sufficiently frequent during the scan to produce detectable BOLD effects (Abreu et al., 2018). It is therefore limited to the epileptic activity that is captured in the relatively short time window typically achievable and the sensitivity of scalp EEG.

Nevertheless, over the past two decades, numerous studies have demonstrated the clinical utility of EEG–fMRI in localising the EZ, especially in patients with inconclusive conventional findings. For instance, Thornton et al. (2010) reported that the degree of spatial overlap between BOLD signal maxima and resected tissue correlated strongly with postoperative seizure freedom, suggesting EEG–fMRI can predict surgical outcomes (Thornton et al., 2010). Similarly, Pittau et al. (2012)

showed that EEG–fMRI enhances localisation of the epileptic focus by detecting BOLD changes temporally coupled to IEDs, providing valuable spatial context for scalp EEG findings (Pittau et al., 2012). Additionally, Zijlmans et al. (2007) demonstrated that EEG–fMRI could reveal epileptogenic foci in patients previously considered unsuitable for surgery due to an unclear localisation based on other modalities (Zijlmans et al., 2007).

Traditional tools for localising the EZ include scalp EEG and intracranial EEG (icEEG). IcEEG is often regarded as the clinical "gold standard" because it records local field potentials directly from the brain (Ebrahimzadeh et al., 2021; Rosenow & Lüders, 2001), but it is invasive, samples only limited brain regions, and is practical only when there is a good non-invasive hypothesis to guide electrode placement for subsequent patient monitoring (Rosenow & Lüders, 2001). Intracranial EEG shows that there is extensive ongoing epileptic activity occurring that is not scalp-visible. To this end, there have been studies attempting to mitigate the coverage of intracranial EEG and the sensitivity of scalp EEG by performing combined intracranial EEG and fMRI recordings. Owing to the much greater risk when imaging a patient with a brain implant, careful safety considerations are needed (Carmichael et al., 2010; Vulliemoz et al., 2009). While limited in scope and practically hard to achieve, the feasibility of these recordings has been demonstrated (Chaudhary et al., 2021; Federico et al., 2005; Vulliemoz et al., 2010).

In summary, precise localisation of the epileptogenic zone remains a critical challenge in the management of pharmaco-resistant epilepsy. Traditional methods such as icEEG and scalp EEG each have limitations, being invasive or spatially coarse, respectively. Simultaneous EEG–fMRI offers a non-invasive means to bridge this gap, capturing both the rapid electrical discharges and their spatially distributed haemodynamic correlates. By combining interictal and ictal information within a single framework, EEG–fMRI has improved our understanding of epileptogenic networks, enhanced presurgical mapping, and shown potential to guide more successful surgical interventions.

Several EEG–fMRI studies have defined the networks involved in the generation of generalised spike wave (GSW) epileptic activity, the hallmark of GGE/IGE syndromes. Common findings across studies show activation of a cortico-subcortical network composed of mid-frontal regions, thalami, caudate, and cerebellum during the occurrence of generalised spike-waves. That there was an extensive but not all-encompassing involvement of the neo-cortex in human GSW was an important finding. Around this time there was also substantial progress made in understanding the pathophysiology of GSW in animal models that indicated a focal cortical generation of discharges (Lüttjohann & van Luijtelaar, 2015), in parallel came the identification of key genetic abnormalities that were shared between the various clinical syndromes (JME, JAE, CAE, GTCS) suggesting a related pathological substrate. A network neuroscience perspective can identify alterations in network topography that might be relevant, but it does not obviously allow access to these key questions; are GSW predictable or have a predisposing state and is this a localised phenomenon? Addressing this question requires knowledge of the timing of GSW occurrence as the marker of the state change, which requires EEG recording. Studies that have done this have found network states and patterns that appear to increase the probability of GSW events (Tangwiriyasakul et al., 2018).

In summary, EEG-fMRI has important potential in clinical localisation of epilepsy although practical limitations related to the complexity of data acquisition and analysis limit its routine application. The ability to localise and study brain dynamics related to epileptic activity has improved our understanding of its pathophysiology.

33.5 Neuroscience Applications

33.5.1 Spontaneous Rhythms

EEG-fMRI has enabled a better understanding of the functional role of EEG oscillations in both task-related and spontaneous measurements. Measuring spontaneous neuronal events and their haemodynamic correlates cannot be done in separate sessions but EEG-fMRI allows us to investigate relationships between EEG oscillations and the BOLD response, for example, in resting state networks, functional connectivity, and pathological spontaneous activity, specifically epilepsy. Early research on simultaneous EEG–fMRI focused primarily on the relationship between alpha-band power and hemodynamic changes in the brain. However, findings regarding the association between alpha power and the BOLD response were not always aligned (Goldman et al., 2002; Laufs et al., 2003). Subsequent work has confirmed that oscillations in different EEG frequency bands show distinct relationships with BOLD changes, both at rest and during task performance.

For example, Mayhew et al. (2013) reported a significant interaction between spontaneous alpha-power amplitude and the magnitude of both positive and negative BOLD responses to visual stimulation (Mayhew et al., 2013). They proposed that spontaneous "baseline" fluctuations in alpha activity can modulate the amplitude and shape of visual BOLD responses suggesting a functional role for resting state network dynamics. Additionally, using an attentional monitoring task, Scheeringa et al. (2011) showed that neural processes reflected in high-gamma power, and those underlying alpha and beta power make independent contributions to explaining BOLD variance. This supports previous work on BOLD–gamma coupling in animals. This is noteworthy because intracranial animal studies cannot easily be replicated in humans, and probing these mechanisms non-invasively with EEG–fMRI represents a major advance in understanding the electrophysiological basis of cognition-related hemodynamic responses in humans.

Advances in ultra-high-field (UHF, 7T) fMRI now enable high-resolution, laminar-level measurements, allowing researchers to investigate brain responses across cortical depths. When combined with simultaneous EEG, a unique opportunity to examine the mechanisms for generating human oscillatory activity arises. For example, Scheeringa et al. (2016) demonstrated that distinct neural dynamics contribute to the BOLD signal across cortical laminae (Scheeringa et al., 2016). During a visual attention task, variations in alpha-band power were associated with BOLD changes in both deep and superficial cortical layers, whereas gamma-band activity related only to BOLD responses in middle and superficial layers. More recently, Marsh et al. (2025) investigated alpha-BOLD depth profiles during an eyes-open/eyes-closed paradigm, finding that the alpha signal is generated in superficial and deep cortical layers suggesting a contribution from top-down processes (Marsh et al., 2025).

These findings align with work on BOLD–EEG coupling in animals. Combining high-resolution laminar fMRI with EEG offers a promising approach for translational research linking human EEG–fMRI findings with systems-level neuroscience in animals. One domain where such an approach could be particularly valuable is the research and treatment of drug-resistant epilepsy. Mapping the neural effects of antiepileptic drugs onto seizure-generating cortical microcircuits in humans remains challenging because of their sub-mm scale. Laminar fMRI, which can resolve activity across cortical depths, offers a potential solution to this mapping problem. A non-invasive measure of laminar responses in humans could complement the findings from animal models and electrophysiological studies, advancing our understanding of the layer-specific mechanisms of seizure generation and propagation in humans (Aitken et al., 2025). Furthermore, given the prominent role of simultaneous EEG-fMRI in epilepsy research (see Sect. 33.4.2) integrating EEG with laminar-level UHF fMRI represents a powerful methodological development for future clinical and translational studies.

33.6 Future Perspectives

33.6.1 Ultra-High Field EEG-fMRI

The growing interest in ultra-high-field (UHF) MRI, defined as imaging at static magnetic field strengths of 7 T or greater, reflects a broader effort to obtain richer and more detailed insights into human brain organisation and function. Among neuroimaging modalities, fMRI has been a major driver of UHF innovation because higher field strengths promise exceptional spatial resolution, sensitivity, and specificity to subtle haemodynamic changes that may be undetectable at conventional field strengths (Uğurbil, 2014). When this enhanced spatial precision is combined with the millisecond-scale temporal resolution of EEG, the result is a uniquely powerful multimodal approach capable of mapping brain activity with unprecedented spatiotemporal detail, potentially advancing our understanding of the human brain connectome (Wirsich et al., 2021).

One of the most prominent benefits of UHF MRI is the approximately linear increase in signal-to-noise ratio (SNR) with the static magnetic field (B_o) (Coffey et al., 2013). These gains enable finer spatial resolution and improved detection of small or subtle signal changes by reducing partial-volume effects. In fMRI, this higher SNR translates into enhanced sensitivity to BOLD signal changes together with enhanced contrast. Because of the differing magnetic susceptibility of blood, the local changes in BOLD signal are increased with field strength (Olman et al., 2012; Opheim et al., 2021; Yacoub et al., 2003). At UHF, the contribution of large draining veins to the BOLD signal is also reduced due to shorter venous T_2, increasing the relative weight of extravascular signals originating from microvasculature (Uğurbil, 2021). This shift brings the haemodynamic signal spatially closer to the underlying neuronal generators, aligning particularly well with EEG's electrophysiological nature.

While fMRI benefits substantially from increased field strength, the simultaneous acquisition of EEG at UHF introduces unique engineering and methodological challenges. Shorter RF wavelengths at 7 T can lead to standing-wave or resonant-antenna effects along EEG cables, necessitating careful design of lead paths and cable geometries (Jorge et al., 2015). Consequently, simultaneous EEG–fMRI at UHF demands stringent safety evaluation and impedance control in conductive wires to limit RF heating (Angelone et al., 2006; Mullinger et al., 2008). Beyond safety, the quality of EEG data itself can degrade with increasing field strength. Head motion and BCG within the static field induce larger voltages which scale approximately linearly with B_o (Debener et al., 2008; Mullinger et al., 2008). Furthermore, gradient artefacts may become more pronounced because of the stronger and faster-switching gradients often used at UHF. These effects complicate post-processing and challenge conventional template-based correction approaches.

Despite these challenges, careful system design and safety protocols have demonstrated that simultaneous EEG–fMRI at UHF can be performed safely, with minimal risk of heating or injury when appropriate precautions are followed (Angelone et al., 2006; Jorge et al., 2015; Mullinger et al., 2008). Continued hardware optimisation has focused on improving EEG system design to minimise its interaction with the MRI's electromagnetic environment and to maintain data quality (Lê et al., 2022; Meyer et al., 2020). These efforts have enabled successful applications of UHF EEG–fMRI in research exploring large-scale functional connectomes (Wirsich et al., 2021) and epileptic networks (Rosenow & Lüders, 2001). As development continues, the technique is likely to expand into domains already established at lower field strengths, such as sleep research (Horovitz et al., 2008) and assessment of resting-state connectivity in unresponsive patients (Vitali et al., 2015). It may eventually support clinical applications, including presurgical evaluation in focal epilepsy, where the increased sensitivity can potentially mitigate the difficulty in detecting changes when EEG-defined events are rare (Vitali et al., 2015).

33.6.2 Neuromodulation

Neuromodulation is an interesting prospect from both research and clinical perspectives. In a research context, manipulating brain activity as an independent variable makes it possible to investigate causal dependencies. From a clinical perspective, it offers treatment opportunities that target specific neural circuitries.

Neuromodulation methods include deep brain stimulation (DBS), transcranial magnetic stimulation (TMS), and transcranial electrical stimulation using direct (tDCS) or alternating (tACS) current. Chapters 31 and 32 cover transcranial electric and magnetic stimulation, respectively. The reader is referred to those chapters for a description of the methods and how they are used with EEG. Here, we cover some applications where simultaneous EEG-fMRI has been used alongside neuromodulation methods.

We start with a neuroscience application: investigating how brain stimulation using TMS propagates through the brain. This is crucial to our understanding of neuromodulation methods and how they could be used in research or treatment. EEG-fMRI lends itself well to investigating brain stimulation. TMS allows the controlled stimulation of one network node, EEG has the temporal resolution to capture fast neuronal fluctuations, and fMRI allows TMS-evoked propagation patterns to be monitored with high spatial resolution including deep lying brain areas (Peters et al., 2020).

Predictably, there are technical challenges associated with combined TMS-EEG-fMRI measurements, nevertheless, their feasibility has been demonstrated in terms of hardware and MR signal quality (Peters et al., 2013). However, artifact from the TMS stimulator can influence the EEG data and interact with other artifacts specific to EEG-fMRI, such as gradient artifact. Peters et al. (2020) showed that usable EEG signals were only possible in pre-TMS intervals and 300 ms or later than the TMS pulse. Despite these limitations, they were able to map whole-brain TMS signal propagation using fMRI as a function of the pre-TMS oscillatory state measured with simultaneous EEG (Peters et al., 2020). EEG-fMRI combined with TMS is a powerful tool for investigating causal dependencies within brain networks.

Transcranial alternating current stimulation (tACS) can be used to entrain endogenous brain oscillations. Used in combination with EEG-fMRI, the temporal and spatial dynamics of oscillatory activity can be investigated. Clancy et al. (2022) showed that tACS of the alpha source in the occipitoparietal cortex augmented alpha oscillations and strengthened BOLD and alpha-frequency oscillatory connectivity (Clancy et al., 2022). This supports the idea that the intrinsic connectivity networks have an electrophysiological origin underpinning them and clearly shows the benefits of neuronal and hemodynamic measurements in response to the same stimulation.

Temporal interference stimulation (TIS) is a prospective neuromodulation method that could benefit from combination with simultaneous EEG-fMRI. By applying two high-frequency electrical alternating currents, TIS can non-invasively target deep brain regions by generating low-frequency amplitude-modulated envelope in the targeted region (Violante et al., 2023). There are a few major areas of research interest when applying TIS with EEG or fMRI. As an example, a recent study by Schaeffer et al. (2025) used simultaneous TIS-EEG to target slow wave generation in the left ventromedial prefrontal cortex using 1Hz amplitude modulation. In this study, they have shown that while stimulating in NREM sleep, they were able to enhance slow wave-activity, thereby enhancing sleep and restorative function (Schaeffer et al., 2025). Other studies, combining TIS and fMRI, are emerging. Zheng et al. (2025) have shown an improved task-modulated connectivity between MFG and IPL after TIS, which correlated with better memory performance (Zheng et al., 2025).

With a surging interest in applying TIS for modulation during sleep (for sleep disorders, epilepsy, and other disorders presenting with sleep dysfunctions), the potential advantages of simultaneous TIS-EEG-fMRI are becoming clear. Not unlike previous examples, there are many challenges to be addressed in terms of safety, experimental design, and effective data analysis. However, current studies utilising simultaneous TIS-EEG (Missey et al., 2025; Schaeffer et al., 2025), TIS-fMRI (Zheng et al., 2025), as well as EEG-fMRI (Li et al., 2022), are providing valuable context and information for future research optimisation.

Neuromodulation can also be achieved non-invasively using neurofeedback. Neurofeedback uses real-time measures of brain activity to encourage self-regulation of a targeted brain activity with the end goal being behaviour modification. While both EEG and fMRI have been used separately in neurofeedback (Micoulaud-Franchi et al., 2015; Thibault et al., 2018), their combination in this context is less well investigated.

Again, there are technical challenges associated with this application but here the difficulties lie in data processing speed. The EEG and fMRI data need to be processed quickly to provide meaningful, timely feedback to the participant. Strategies for setting up real-time EEG and fMRI neurofeedback are described by Zotev et al. (2014) and Mano et al. (2017). Given that multimodal neurofeedback offers some potential benefits, it is worth the time and effort required to do it effectively. For example, targeting disorder-specific brain activity identified by two very different imaging modalities might have stronger effects than the standalone application of either method.

So far, EEG-fMRI neurofeedback has been used in several clinical applications. Patients with major depressive disorder (MDD) were able to increase frontal alpha and beta asymmetry and increase functional connectivity in target fMRI measures. These modulations were associated with reductions in state depression, anxiety, confusion, and total mood disturbance, and increase in state happiness (Zotev & Bodurka, 2020). Stroke rehabilitation patients can also benefit from EEG-fMRI neurofeedback. Lioi et al. (2020) used motor imagery neurofeedback for upper limb recovery in stroke patients; the patients were able to upregulate activity in the target motor areas (Lioi et al., 2020). Finally, PTSD patients were able to upregulate amygdala (fMRI) activity during a happy emotion inducing task. A reduction in PTSD severity and comorbid depression severity was seen after the training (Zotev et al., 2018). Neurofeedback was only provided using the fMRI data, but the simultaneously recorded EEG allowed the electrophysiological correlates of the fMRI-based neurofeedback training to be investigated.

Together, these studies in basic research and across different patient populations suggest a role for EEG-fMRI in neuromodulation research and treatment.

33.7 Conclusion

Simultaneous EEG-fMRI is one of the most powerful tools available for the non-invasive study of human brain function. It is a well-established neuroimaging technique with widespread applications in basic research and clinical studies. This has been enabled by considerable safety investigations, hardware developments, and significant progress in data integration strategies. The result is not only a better understanding of the neural dynamics of brain function but also the potential to facilitate clinical intervention development.

References

Abbott, D. F., Masterton, R. A. J., Archer, J. S., Fleming, S. W., Warren, A. E. L., & Jackson, G. D. (2015). Constructing carbon fiber motion-detection loops for simultaneous EEG–fMRI. *Frontiers in Neurology, 5*. https://doi.org/10.3389/fneur.2014.00260

Abreu, R., Leal, A., & Figueiredo, P. (2018). EEG-Informed fMRI: A review of data analysis methods. *Frontiers in Human Neuroscience, 12*. https://doi.org/10.3389/fnhum.2018.00029

Aghakhani, Y., Bagshaw, A. P., Bénar, C. G., Hawco, C., Andermann, F., Dubeau, F., & Gotman, J. (2004). fMRI activation during spike and wave discharges in idiopathic generalized epilepsy. *Brain, 127*, 1127–1144. https://doi.org/10.1093/brain/awh136

Aitken, F., Winston, J. S., O'Muircheartaigh, J., & Carmichael, D. W. (2025). The potential of laminar functional MRI in refining the understanding of epilepsy in humans. *Brain*, awaf320. https://doi.org/10.1093/brain/awaf320

Allen, P. J., Josephs, O., & Turner, R. (2000). A method for removing imaging artifact from continuous EEG recorded during functional MRI. *NeuroImage, 12*, 230–239. https://doi.org/10.1006/nimg.2000.0599

Angelone, L. M., Vasios, C. E., Wiggins, G., Purdon, P. L., & Bonmassar, G. (2006). On the effect of resistive EEG electrodes and leads during 7 T MRI: Simulation and temperature measurement studies. *Magnetic Resonance Imaging, 24*, 801–812. https://doi.org/10.1016/j.mri.2006.01.006

Babiloni, F., Carducci, F., Cincotti, F., Del Gratta, C., Roberti, G. M., Romani, G. L., Rossini, P. M., & Babiloni, C. (2000). Integration of high resolution EEG and functional magnetic resonance in the study of human movement-related potentials. *Methods of Information in Medicine, 39*, 179–182.

Babiloni, F., Babiloni, C., Carducci, F., Del Gratta, C., Romani, G. L., Rossini, P. M., & Cincotti, F. (2002). Cortical source estimate of combined high resolution EEG and fMRI data related to voluntary movements. *Methods of Information in Medicine, 41*, 443–450.

Bagshaw, A. P., Hawco, C., Bénar, C.-G., Kobayashi, E., Aghakhani, Y., Dubeau, F., Pike, G. B., & Gotman, J. (2005). Analysis of the EEG–fMRI response to prolonged bursts of interictal epileptiform activity. *NeuroImage, 24*, 1099–1112. https://doi.org/10.1016/j.neuroimage.2004.10.010

Bénar, C.-G., Grova, C., Kobayashi, E., Bagshaw, A. P., Aghakhani, Y., Dubeau, F., & Gotman, J. (2006). EEG–fMRI of epileptic spikes: Concordance with EEG source localization and intracranial EEG. *NeuroImage, 30*, 1161–1170. https://doi.org/10.1016/j.neuroimage.2005.11.008

Besedovsky, L., Ngo, H.-V. V., Dimitrov, S., Gassenmaier, C., Lehmann, R., & Born, J. (2017). Auditory closed-loop stimulation of EEG slow oscillations strengthens sleep and signs of its immune-supportive function. *Nature Communications, 8*, 1984. https://doi.org/10.1038/s41467-017-02170-3

Biswal, B. B., Van Kylen, J., & Hyde, J. S. (1997). Simultaneous assessment of flow and BOLD signals in resting-state functional connectivity maps. *NMR in Biomedicine, 10*, 165–170. https://doi.org/10.1002/(sici)1099-1492(199706/08)10:4/5<165::aid-nbm454>3.0.co;2-7

Blumberg, M. S., Lesku, J. A., Libourel, P.-A., Schmidt, M. H., & Rattenborg, N. C. (2020). What is REM sleep? *Current Biology, 30*, R38–R49. https://doi.org/10.1016/j.cub.2019.11.045

Bonmassar, G., Hadjikhani, N., Ives, J. R., Hinton, D., & Belliveau, J. W. (2001). Influence of EEG electrodes on the BOLD fMRI signal. *Human Brain Mapping, 14*, 108–115. https://doi.org/10.1002/hbm.1045

Brinkman, J. E., Reddy, V., & Sharma, S. (2025). Physiology of sleep. In *StatPearls*. StatPearls Publishing.

Bullock, M., Jackson, G. D., & Abbott, D. F. (2021). Artifact reduction in simultaneous EEG-fMRI: A systematic review of methods and contemporary usage. *Frontiers in Neurology, 12*. https://doi.org/10.3389/fneur.2021.622719

Buxton, R. B., Uludağ, K., Dubowitz, D. J., & Liu, T. T. (2004). Modeling the hemodynamic response to brain activation. *NeuroImage, 23*(Suppl 1), S220–S233. https://doi.org/10.1016/j.neuroimage.2004.07.013

Calhoun, V. D., Adali, T., Pearlson, G. D., & Kiehl, K. A. (2006). Neuronal chronometry of target detection: Fusion of hemodynamic and event-related potential data. *NeuroImage, 30*, 544–553. https://doi.org/10.1016/j.neuroimage.2005.08.060

Calhoun, V. D., Liu, J., & Adali, T. (2009). A review of group ICA for fMRI data and ICA for joint inference of imaging, genetic, and ERP data. *NeuroImage, 45*, S163–S172. https://doi.org/10.1016/j.neuroimage.2008.10.057

Campbell, I. G. (2009). EEG recording and analysis for sleep research. *Current Protocols in Neuroscience* Chapter 10, Unit10.2. https://doi.org/10.1002/0471142301.ns1002s49

Carmichael, D. (2022). Image quality issues. In C. Mulert & L. Lemieux (Eds.), *EEG—fMRI: Physiological basis, technique, and applications* (pp. 213–246). Springer. https://doi.org/10.1007/978-3-031-07121-8_10

Carmichael, D. W., Thornton, J. S., Rodionov, R., Thornton, R., McEvoy, A. W., Ordidge, R. J., Allen, P. J., & Lemieux, L. (2010). Feasibility of simultaneous intracranial EEG-fMRI in humans: A safety study. *NeuroImage, 49*, 379–390. https://doi.org/10.1016/j.neuroimage.2009.07.062

Centeno, M., Tierney, T. M., Perani, S., Shamshiri, E. A., StPier, K., Wilkinson, C., Konn, D., Banks, T., Vulliemoz, S., Lemieux, L., Pressler, R. M., Clark, C. A., Cross, J. H., & Carmichael, D. W. (2016). Optimising EEG-fMRI for localisation of focal epilepsy in children. *PLoS ONE, 11*, e0149048. https://doi.org/10.1371/journal.pone.0149048

Chaudhary, U. J., Centeno, M., Carmichael, D. W., Diehl, B., Walker, M. C., Duncan, J. S., & Lemieux, L. (2021). Mapping epileptic networks using simultaneous intracranial EEG-fMRI. *Frontiers in Neurology, 12*. https://doi.org/10.3389/fneur.2021.693504

Chowdhury, M. E. H., Mullinger, K. J., Glover, P., & Bowtell, R. (2014). Reference layer artefact subtraction (RLAS): A novel method of minimizing EEG artefacts during simultaneous fMRI. *NeuroImage, 84*, 307–319. https://doi.org/10.1016/j.neuroimage.2013.08.039

Clancy, K. J., Andrzejewski, J. A., You, Y., Rosenberg, J. T., Ding, M., & Li, W. (2022). Transcranial stimulation of alpha oscillations up-regulates the default mode network. *Proceedings of the National Academy of Sciences of the United States of America, 119*. https://doi.org/10.1073/pnas.2110868119

Coffey, A. M., Truong, M. L., & Chekmenev, E. Y. (2013). Low-field MRI can be more sensitive than high-field MRI. *Journal of Magnetic Resonance, 237*, 169–174. https://doi.org/10.1016/j.jmr.2013.10.013

Daunizeau, J., Grova, C., Marrelec, G., Mattout, J., Jbabdi, S., Pélégrini-Issac, M., Lina, J. M., & Benali, H. (2007). Symmetrical event-related EEG/fMRI information fusion in a variational Bayesian framework. *NeuroImage, 36*, 69–87. https://doi.org/10.1016/j.neuroimage.2007.01.044

De Martino, F., Valente, G., de Borst, A. W., Esposito, F., Roebroeck, A., Goebel, R., & Formisano, E. (2010). Multimodal imaging: An evaluation of univariate and multivariate methods for simultaneous EEG/fMRI. *Magnetic Resonance Imaging, 28*, 1104–1112. https://doi.org/10.1016/j.mri.2009.12.026

Debener, S., Ullsperger, M., Siegel, M., Fiehler, K., von Cramon, D. Y., & Engel, A. K. (2005). Trial-by-trial coupling of concurrent electroencephalogram and functional magnetic resonance imaging identifies the dynamics of performance monitoring. *The Journal of Neuroscience, 25*, 11730–11737. https://doi.org/10.1523/jneurosci.3286-05.2005

Debener, S., Mullinger, K. J., Niazy, R. K., & Bowtell, R. W. (2008). Properties of the ballistocardiogram artefact as revealed by EEG recordings at 1.5, 3 and 7 T static magnetic field strength. *International Journal of Psychophysiology, Integration of EEG and fMRI, 67*, 189–199. https://doi.org/10.1016/j.ijpsycho.2007.05.015

Ebrahimzadeh, E., Shams, M., Seraji, M., Sadjadi, S. M., Rajabion, L., & Soltanian-Zadeh, H. (2021). Localizing epileptic foci using simultaneous EEG-fMRI recording: Template component cross-correlation. *Frontiers in Neurology, 12*. https://doi.org/10.3389/fneur.2021.695997

Engström, M., Hallböök, T., Szakacs, A., Karlsson, T., & Landtblom, A.-M. (2014). Functional magnetic resonance imaging in narcolepsy and the Kleine–Levin syndrome. *Frontiers in Neurology, 5*, 105. https://doi.org/10.3389/fneur.2014.00105

Federico, P., Archer, J. S., Abbott, D. F., & Jackson, G. D. (2005). Cortical/subcortical BOLD changes associated with epileptic discharges. *Neurology, 64*, 1125–1130. https://doi.org/10.1212/01.WNL.0000156358.72670.AD

Fox, M. D., Snyder, A. Z., Vincent, J. L., Corbetta, M., Van Essen, D. C., & Raichle, M. E. (2005). The human brain is intrinsically organized into dynamic, anticorrelated functional networks. *Proceedings of the National Academy of Sciences of the United States of America, 102*, 9673–9678. https://doi.org/10.1073/pnas.0504136102

Friston, K. J., Holmes, A. P., Worsley, K. J., Poline, J.-P., Frith, C. D., & Frackowiak, R. S. J. (1994). Statistical parametric maps in functional imaging: A general linear approach. *Human Brain Mapping, 2*, 189–210. https://doi.org/10.1002/hbm.460020402

Galan, F. N., Beier, A. D., & Sheth, R. D. (2021). Advances in epilepsy surgery. *Pediatric Neurology, 122*, 89–97. https://doi.org/10.1016/j.pediatrneurol.2021.06.004

Glover, G. H. (2011). Overview of functional magnetic resonance imaging. *Neurosurgery Clinics of North America, Functional Imaging, 22*, 133–139. https://doi.org/10.1016/j.nec.2010.11.001

Goldman, R. I., Stern, J. M., Engel, J., Jr., & Cohen, M. S. (2002). Simultaneous EEG and fMRI of the alpha rhythm. *Neuroreport, 13*, 2487–2492. https://doi.org/10.1097/01.wnr.0000047685.08940.d0

Hawsawi, H. B., Carmichael, D. W., & Lemieux, L. (2017). Safety of simultaneous scalp or intracranial EEG during MRI: A review. *Frontiers of Physics, 5*. https://doi.org/10.3389/fphy.2017.00042

Horovitz, S. G., Fukunaga, M., de Zwart, J. A., van Gelderen, P., Fulton, S. C., Balkin, T. J., & Duyn, J. H. (2008). Low frequency BOLD fluctuations during resting wakefulness and light sleep: A simultaneous EEG-fMRI study. *Human Brain Mapping, 29*, 671–682. https://doi.org/10.1002/hbm.20428

Huster, R. J., Debener, S., Eichele, T., & Herrmann, C. S. (2012). Methods for simultaneous EEG-fMRI: An introductory review. *The Journal of Neuroscience, 32*, 6053–6060. https://doi.org/10.1523/jneurosci.0447-12.2012

Ives, J. R., Warach, S., Schmitt, F., Edelman, R. R., & Schomer, D. L. (1993). Monitoring the patient's EEG during echo planar MRI. *Electroencephalography and Clinical Neurophysiology, 87*, 417–420. https://doi.org/10.1016/0013-4694(93)90156-P

Jacobs, J., LeVan, P., Moeller, F., Boor, R., Stephani, U., Gotman, J., & Siniatchkin, M. (2009). Hemodynamic changes preceding the interictal EEG spike in patients with focal epilepsy investigated using simultaneous EEG-fMRI. *NeuroImage, 45*, 1220–1231. https://doi.org/10.1016/j.neuroimage.2009.01.014

Jezzard, P., & Clare, S. (1999). Sources of distortion in functional MRI data. *Human Brain Mapping, 8*, 80–85. https://doi.org/10.1002/(SICI)1097-0193(1999)8:2/3%253C80::AID-HBM2%253E3.0.CO;2-C

Jorge, J., van der Zwaag, W., & Figueiredo, P. (2014). EEG-fMRI integration for the study of human brain function. *NeuroImage, 102*(Pt 1), 24–34. https://doi.org/10.1016/j.neuroimage.2013.05.114

Jorge, J., Grouiller, F., Ipek, Ö., Stoermer, R., Michel, C. M., Figueiredo, P., van der Zwaag, W., & Gruetter, R. (2015). Simultaneous EEG–fMRI at ultra-high field: Artifact prevention and safety assessment. *NeuroImage, 105*, 132–144. https://doi.org/10.1016/j.neuroimage.2014.10.055

Kim, S.-G., & Ogawa, S. (2012). Biophysical and physiological origins of blood oxygenation level-dependent fMRI signals. *Journal of Cerebral Blood Flow and Metabolism, 32*, 1188–1206. https://doi.org/10.1038/jcbfm.2012.23

Kim, J., Gulati, T., & Ganguly, K. (2019). Competing roles of slow oscillations and delta waves in memory consolidation versus forgetting. *Cell, 179*, 514–526.e13. https://doi.org/10.1016/j.cell.2019.08.040

Krakow, K., Woermann, F. G., Symms, M. R., Allen, P. J., Lemieux, L., Barker, G. J., Duncan, J. S., & Fish, D. R. (1999). EEG-triggered functional MRI of interictal epileptiform activity in patients with partial seizures. *Brain, 122*, 1679–1688. https://doi.org/10.1093/brain/122.9.1679

Lambert-Beaudet, F., Journault, W.-G., Rudziavic Provençal, A., & Bastien, C. H. (2021). Neurofeedback for insomnia: Current state of research. *World Journal of Psychiatry, 11*, 897–914. https://doi.org/10.5498/wjp.v11.i10.897

Laufs, H. (2008). Endogenous brain oscillations and related networks detected by surface EEG-combined fMRI. *Human Brain Mapping, 29*, 762–769. https://doi.org/10.1002/hbm.20600

Laufs, H., Kleinschmidt, A., Beyerle, A., Eger, E., Salek-Haddadi, A., Preibisch, C., & Krakow, K. (2003). EEG-correlated fMRI of human alpha activity. *NeuroImage, 19*, 1463–1476. https://doi.org/10.1016/s1053-8119(03)00286-6

Lê, T. P., Gruetter, R., Jorge, J., & Ipek, Ö. (2022). Segmenting electroencephalography wires reduces radiofrequency shielding artifacts in simultaneous electroencephalography and functional magnetic resonance imaging at 7 T. *Magnetic Resonance in Medicine, 88*, 1450–1464. https://doi.org/10.1002/mrm.29298

Lemieux, L., Allen, P. J., Franconi, F., Symms, M. R., & Fish, D. K. (1997). Recording of EEG during fMRI experiments: Patient safety. *Magnetic Resonance in Medicine, 38*, 943–952. https://doi.org/10.1002/mrm.1910380614

Lemieux, L., Salek-Haddadi, A., Josephs, O., Allen, P., Toms, N., Scott, C., Krakow, K., Turner, R., & Fish, D. R. (2001). Event-related fMRI with simultaneous and continuous EEG: description of the method and initial case report. *NeuroImage, 14*, 780–787. https://doi.org/10.1006/nimg.2001.0853

LeVan, P., & Gotman, J. (2009). Independent component analysis as a model-free approach for the detection of BOLD changes related to epileptic spikes: a simulation study. *Human Brain Mapping, 30*, 2021–2031. https://doi.org/10.1002/hbm.20647

Li, Y., Zou, G., Shao, Y., Yao, P., Liu, J., Zhou, S., Hu, S., Xu, J., Guo, Y., Gao, J.-H., Zou, Q., & Sun, H. (2022). Sleep discrepancy is associated with alterations in the salience network in patients with insomnia disorder: An EEG-fMRI study. *NeuroImage: Clinical, 35*, 103111. https://doi.org/10.1016/j.nicl.2022.103111

Lioi, G., Butet, S., Fleury, M., Bannier, E., Lécuyer, A., Bonan, I., & Barillot, C. (2020). A multi-target motor imagery training using bimodal EEG-fMRI neurofeedback: A pilot study in chronic stroke patients. *Frontiers in Human Neuroscience, 14*, 37. https://doi.org/10.3389/fnhum.2020.00037

Liu, Z., Zhang, N., Chen, W., & He, B. (2009). Mapping the bilateral visual integration by EEG and fMRI. *NeuroImage, 46*, 989–997. https://doi.org/10.1016/j.neuroimage.2009.03.028

Lowe, M. J., Dzemidzic, M., Lurito, J. T., Mathews, V. P., & Phillips, M. D. (2000). Correlations in low-frequency BOLD fluctuations reflect cortico-cortical connections. *NeuroImage, 12*, 582–587. https://doi.org/10.1006/nimg.2000.0654

Lüttjohann, A., & van Luijtelaar, G. (2015). Dynamics of networks during absence seizure's on- and offset in rodents and man. *Frontiers in Physiology, 6*. https://doi.org/10.3389/fphys.2015.00016

Mano, M., Lécuyer, A., Bannier, E., Perronnet, L., Noorzadeh, S., & Barillot, C. (2017). How to build a hybrid neurofeedback platform combining EEG and fMRI. *Frontiers in Neuroscience, 11*, 140. https://doi.org/10.3389/fnins.2017.00140

Mantini, D., Perrucci, M. G., Del Gratta, C., Romani, G. L., & Corbetta, M. (2007). Electrophysiological signatures of resting state networks in the human brain. *Proceedings of the National Academy of Sciences, 104*, 13170–13175. https://doi.org/10.1073/pnas.0700668104

Marsh, D. C., Sokoliuk, R., Aquino, K. M., Pakenham, D. O., Wilson, R., Panchuelo, R. S., Brookes, M. J., Hanslmayr, S., Mayhew, S. D., Francis, S. T., & Mullinger, K. J. (2025). Cortical-layer EEG-fMRI at 7T: Experimental setup and analysis pipeline to elucidate generating mechanisms of alpha oscillations. *bioRxiv*, 2025.09.09.674189. https://doi.org/10.1101/2025.09.09.674189

Martinez, C. S., Wirsich, J., Jäger, C., Warbrick, T., Vulliémoz, S., Lemay, M., Bastiaansen, J., Wiest, R., & Jorge, J. (2025). An optimized framework for simultaneous EEG-fMRI at 7T enabling safe, high-quality human brain imaging with millisecond temporal resolution and sub-millimeter spatial resolution. Imaging. *Neuroscience, 3*, IMAG.a.983. https://doi.org/10.1162/IMAG.a.983

Massimini, M., Ferrarelli, F., Esser, S. K., Riedner, B. A., Huber, R., Murphy, M., Peterson, M. J., & Tononi, G. (2007). Triggering sleep slow waves by transcranial magnetic stimulation. *Proceedings of the National Academy of Sciences, 104*, 8496–8501. https://doi.org/10.1073/pnas.0702495104

Mayhew, S. D., Ostwald, D., Porcaro, C., & Bagshaw, A. P. (2013). Spontaneous EEG alpha oscillation interacts with positive and negative BOLD responses in the visual-auditory cortices and default-mode network. *NeuroImage, 76*, 362–372. https://doi.org/10.1016/j.neuroimage.2013.02.070

Maziero, D., Velasco, T. R., Hunt, N., Payne, E., Lemieux, L., Salmon, C. E. G., & Carmichael, D. W. (2016). Towards motion insensitive EEG-fMRI: Correcting motion-induced voltages and gradient artefact instability in EEG using an fMRI prospective motion correction (PMC) system. *NeuroImage, 138*, 13–27. https://doi.org/10.1016/j.neuroimage.2016.05.003

Maziero, D., Stenger, V. A., & Carmichael, D. W. (2021). Unified retrospective EEG motion educated artefact suppression for EEG-fMRI to suppress magnetic field gradient artefacts during motion. *Brain Topography, 34*, 745–761. https://doi.org/10.1007/s10548-021-00870-0

Meyer, M. C., Scheeringa, R., Webb, A. G., Petridou, N., Kraff, O., & Norris, D. G. (2020). Adapted cabling of an EEG cap improves simultaneous measurement of EEG and fMRI at 7T. *Journal of Neuroscience Methods, 331*, 108518. https://doi.org/10.1016/j.jneumeth.2019.108518

Miao, A., Luo, T., Hsieh, B., Edge, C. J., Gridley, M., Wong, R. T. C., Constandinou, T. G., Wisden, W., & Franks, N. P. (2024). Brain clearance is reduced during sleep and anesthesia. *Nature Neuroscience, 27*, 1046–1050. https://doi.org/10.1038/s41593-024-01638-y

Michel, C. M., Murray, M. M., Lantz, G., Gonzalez, S., Spinelli, L., & Grave de Peralta, R. (2004). EEG source imaging. *Clinical Neurophysiology, 115*, 2195–2222. https://doi.org/10.1016/j.clinph.2004.06.001

Micoulaud-Franchi, J. A., McGonigal, A., Lopez, R., Daudet, C., Kotwas, I., & Bartolomei, F. (2015). Electroencephalographic neurofeedback: Level of evidence in mental and brain disorders and suggestions for good clinical practice. *Neurophysiologie Clinique, 45*, 423–433. https://doi.org/10.1016/j.neucli.2015.10.077

Missey, F., Acerbo, E., Dickey, A. S., Trajlinek, J., Studnička, O., Lubrano, C., Silva, M. D. A, Brady, E., Všianský, V., Szabo, J., Dolezalova, I., Fabo, D., Pail, M., Gutekunst, C.-A., Migliore, R., Migliore, M., Lagarde, S., Carron, R., Karimi, F., et al. (2025). Non-invasive temporal interference stimulation of the hippocampus suppresses epileptic biomarkers in patients with epilepsy: Biophysical differences between kilohertz and amplitude modulated stimulation. https://doi.org/10.1101/2024.12.05.24303799

Motamedi, G. K., Jeliazkov, P. G., Oyegbile-Chidi, T. O., Song, S. S., Sharma, V. D., Mete, M., Nawar, S., Turkeltaub, P. E., Won Cho, Y., & Wu, J.-Y. (2023). Transcranial alternating current stimulation (tACS) as a treatment for insomnia. *The Canadian Journal of Neurological Sciences, 50*, 446–449. https://doi.org/10.1017/cjn.2022.33

Mulert, C., & Lemieux, L. (Eds.). (2022). *EEG - fMRI: Physiological basis, technique, and applications*. Springer. https://doi.org/10.1007/978-3-031-07121-8

Mullinger, K., Brookes, M., Stevenson, C., Morgan, P., & Bowtell, R. (2008). Exploring the feasibility of simultaneous electroencephalography/functional magnetic resonance imaging at 7 T. *Magnetic Resonance Imaging, Proceedings of the International School on Magnetic Resonance and Brain Function, 26*, 968–977. https://doi.org/10.1016/j.mri.2008.02.014

Müri, R. M., Felblinger, J., Rösier, K. M., Jung, B., Hess, C. W., & Boesch, C. (1998). Recording of electrical brain activity in a magnetic resonance environment: Distorting effects of the static magnetic field. *Magnetic Resonance in Medicine, 39*, 18–22. https://doi.org/10.1002/mrm.1910390105

Murta, T., Leite, M., Carmichael, D. W., Figueiredo, P., & Lemieux, L. (2015). Electrophysiological correlates of the BOLD signal for EEG-informed fMRI. *Human Brain Mapping, 36*, 391–414. https://doi.org/10.1002/hbm.22623

Neuner, I., Arrubla, J., Felder, J., & Shah, N. J. (2014). Simultaneous EEG–fMRI acquisition at low, high and ultra-high magnetic fields up to 9.4 T: Perspectives and challenges. *NeuroImage, Multimodal Data Fusion, 102*, 71–79. https://doi.org/10.1016/j.neuroimage.2013.06.048

Niazy, R. K., Beckmann, C. F., Iannetti, G. D., Brady, J. M., & Smith, S. M. (2005). Removal of FMRI environment artifacts from EEG data using optimal basis sets. *NeuroImage, 28*, 720–737. https://doi.org/10.1016/j.neuroimage.2005.06.067

Nunez, P. L., & Srinivasan, R. (2006). *Electric fields of the brain: The neurophysics of EEG*. Oxford University Press. https://doi.org/10.1093/acprof:oso/9780195050387.001.0001

Ogawa, S., Lee, T. M., Kay, A. R., & Tank, D. W. (1990). Brain magnetic resonance imaging with contrast dependent on blood oxygenation. *Proceedings of the National Academy of Sciences of the United States of America, 87*, 9868–9872. https://doi.org/10.1073/pnas.87.24.9868

Olman, C. A., Harel, N., Feinberg, D. A., He, S., Zhang, P., Ugurbil, K., & Yacoub, E. (2012). Layer-specific fMRI reflects different neuronal computations at different depths in human V1. *PLoS ONE, 7*, e32536. https://doi.org/10.1371/journal.pone.0032536

Opheim, G., van der Kolk, A., Markenroth Bloch, K., Colon, A. J., Davis, K. A., Henry, T. R., Jansen, J. F. A., Jones, S. E., Pan, J. W., Rössler, K., Stein, J. M., Strandberg, M. C., Trattnig, S., Van de Moortele, P.-F., Vargas, M. I., Wang, I., Bartolomei, F., Bernasconi, N., Bernasconi, A., et al. (2021). 7T epilepsy task force consensus recommendations on the use of 7T MRI in clinical practice. *Neurology, 96*, 327–341. https://doi.org/10.1212/WNL.0000000000011413

Ostwald, D., Porcaro, C., & Bagshaw, A. P. (2010). An information theoretic approach to EEG-fMRI integration of visually evoked responses. *NeuroImage, 49*, 498–516. https://doi.org/10.1016/j.neuroimage.2009.07.038

Ostwald, D., Porcaro, C., & Bagshaw, A. P. (2011). Voxel-wise information theoretic EEG-fMRI feature integration. *NeuroImage, 55*, 1270–1286. https://doi.org/10.1016/j.neuroimage.2010.12.029

Patel, A. K., Reddy, V., Shumway, K. R., & Araujo, J. F. (2025). Physiology, sleep stages. In *StatPearls*. StatPearls Publishing.

Pereira, M., Chen, X., Paltarzhytskaya, A., Pacheco, Y., Muller, N., Bovy, L., Lei, X., Chen, W., Ren, H., Song, C., Lewis, L. D., Dang-Vu, T. T., Czisch, M., Picchioni, D., Duyn, J., Peigneux, P., Tagliazucchi, E., & Dresler, M. (2025). Sleep neuroimaging: Review and future directions. *Journal of Sleep Research, 34*, e14462. https://doi.org/10.1111/jsr.14462

Peters, J. C., Reithler, J., Schuhmann, T., de Graaf, T., Uludag, K., Goebel, R., & Sack, A. T. (2013). On the feasibility of concurrent human TMS-EEG-fMRI measurements. *Journal of Neurophysiology, 109*, 1214–1227. https://doi.org/10.1152/jn.00071.2012

Peters, J. C., Reithler, J., Graaf, T. A., Schuhmann, T., Goebel, R., & Sack, A. T. (2020). Concurrent human TMS-EEG-fMRI enables monitoring of oscillatory brain state-dependent gating of cortico-subcortical network activity. *Communications Biology, 3*, 40. https://doi.org/10.1038/s42003-020-0764-0

Phadikar, S., Pusuluri, K., Iraji, A., & Calhoun, V. D. (2025). Integrating fMRI spatial network dynamics and EEG spectral power: insights into resting state connectivity. *Frontiers in Neuroscience, 19*. https://doi.org/10.3389/fnins.2025.1484954

Pittau, F., Dubeau, F., & Gotman, J. (2012). Contribution of EEG/fMRI to the definition of the epileptic focus. *Neurology, 78*, 1479–1487. https://doi.org/10.1212/WNL.0b013e3182553bf7

Rosenow, F., & Lüders, H. (2001). Presurgical evaluation of epilepsy. *Brain, 124*, 1683–1700. https://doi.org/10.1093/brain/124.9.1683

Sadaghiani, S., Brookes, M. J., & Baillet, S. (2022). Connectomics of human electrophysiology. *NeuroImage, 247*, 118788. https://doi.org/10.1016/j.neuroimage.2021.118788

Schaeffer, E. L., Haber, I., Fan, Z., Bruno, S., Mat, B., Alauddin, T., Vigueras, G., Neumann, L., Smith, R., Missey, F., Williamson, A., Achermann, P., Beerli, S., Capstick, M., Neufeld, E., Kuster, N., Goldman, R. I., Davidson, R. J., Albantakis, L., et al. (2025). Enhancement of sleep slow wave activity using transcranial electrical stimulation with temporal interference. https://doi.org/10.1101/2025.08.11.25333452

Scheeringa, R., Fries, P., Petersson, K. M., Oostenveld, R., Grothe, I., Norris, D. G., Hagoort, P., & Bastiaansen, M. C. (2011). Neuronal dynamics underlying high- and low-frequency EEG oscillations contribute independently to the human BOLD signal. *Neuron, 69*, 572–583. https://doi.org/10.1016/j.neuron.2010.11.044

Scheeringa, R., Koopmans, P. J., van Mourik, T., Jensen, O., & Norris, D. G. (2016). The relationship between oscillatory EEG activity and the laminar-specific BOLD signal. *Proceedings of the National Academy of Sciences of the United States of America, 113*, 6761–6766. https://doi.org/10.1073/pnas.1522577113

Shahriar, K. A., Bhuiyan, E. H., Luo, Q., Chowdhury, M. E. H., & Zhou, X. J. (2025). Deep learning for gradient and BCG artifacts removal in EEG during simultaneous fMRI. https://doi.org/10.48550/arXiv.2507.22263

Tagliazucchi, E., & Laufs, H. (2014). Decoding wakefulness levels from typical fMRI resting-state data reveals reliable drifts between wakefulness and sleep. *Neuron, 82*, 695–708. https://doi.org/10.1016/j.neuron.2014.03.020

Tangwiriyasakul, C., Perani, S., Centeno, M., Yaakub, S. N., Abela, E., Carmichael, D. W., & Richardson, M. P. (2018). Dynamic brain network states in human generalized spike-wave discharges. *Brain, 141*, 2981–2994. https://doi.org/10.1093/brain/awy223

Thibault, R. T., MacPherson, A., Lifshitz, M., Roth, R. R., & Raz, A. (2018). Neurofeedback with fMRI: A critical systematic review. *NeuroImage, 172*, 786–807. https://doi.org/10.1016/j.neuroimage.2017.12.071

Thornton, R., Laufs, H., Rodionov, R., Cannadathu, S., Carmichael, D. W., Vulliemoz, S., Salek-Haddadi, A., McEvoy, A. W., Smith, S. M., Lhatoo, S., Elwes, R. D. C., Guye, M., Walker, M. C., Lemieux, L., & Duncan, J. S. (2010). EEG correlated functional MRI and postoperative outcome in focal epilepsy. *Journal of Neurology, Neurosurgery & Psychiatry, 81*, 922–927. https://doi.org/10.1136/jnnp.2009.196253

Uğurbil, K. (2014). Magnetic resonance imaging at ultrahigh fields. *IEEE Transactions on Biomedical Engineering, 61*, 1364–1379. https://doi.org/10.1109/TBME.2014.2313619

Uğurbil, K. (2021). Ultrahigh field and ultrahigh resolution fMRI. *Current Opinion in Biomedical Engineering, 18*, 100288. https://doi.org/10.1016/j.cobme.2021.100288

Uhlhaas, P. J., Haenschel, C., Nikolić, D., & Singer, W. (2008). The role of oscillations and synchrony in cortical networks and their putative relevance for the pathophysiology of schizophrenia. *Schizophrenia Bulletin, 34*, 927–943. https://doi.org/10.1093/schbul/sbn062

Uji, M., Li, X., Saotome, A., Katsumata, R., Waggoner, R. A., Suzuki, C., Ueno, K., Aritake, S., & Tamaki, M. (2025). Human deep sleep facilitates cerebrospinal fluid dynamics linked to spontaneous brain oscillations and neural events. *Proceedings of the National Academy of Sciences, 122*, e2509626122. https://doi.org/10.1073/pnas.2509626122

Valdes-Sosa, P. A., Sanchez-Bornot, J. M., Sotero, R. C., Iturria-Medina, Y., Aleman-Gomez, Y., Bosch-Bayard, J., Carbonell, F., & Ozaki, T. (2009). Model driven EEG/fMRI fusion of brain oscillations. *Human Brain Mapping, 30*, 2701–2721. https://doi.org/10.1002/hbm.20704

van den Heuvel, M. P., & Hulshoff Pol, H. E. (2010). Exploring the brain network: A review on resting-state fMRI functional connectivity. *European Neuropsychopharmacology, 20*, 519–534. https://doi.org/10.1016/j.euroneuro.2010.03.008

Violante, I. R., Alania, K., Cassarà, A. M., Neufeld, E., Acerbo, E., Carron, R., Williamson, A., Kurtin, D. L., Rhodes, E., Hampshire, A., Kuster, N., Boyden, E. S., Pascual-Leone, A., & Grossman, N. (2023). Non-invasive temporal interference electrical stimulation of the human hippocampus. *Nature Neuroscience, 26*, 1994–2004. https://doi.org/10.1038/s41593-023-01456-8

Vitali, P., Di Perri, C., Vaudano, A. E., Meletti, S., & Villani, F. (2015). Integration of multimodal neuroimaging methods: A rationale for clinical applications of simultaneous EEG-fMRI. *Functional Neurology, 30*, 9–20.

Vulliemoz, S., Thornton, R., Rodionov, R., Carmichael, D. W., Guye, M., Lhatoo, S., McEvoy, A. W., Spinelli, L., Michel, C. M., Duncan, J. S., & Lemieux, L. (2009). The spatio-temporal mapping of epileptic networks: Combination of EEG–fMRI and EEG source imaging. *NeuroImage, 46*, 834–843. https://doi.org/10.1016/j.neuroimage.2009.01.070

Vulliemoz, S., Rodionov, R., Carmichael, D. W., Thornton, R., Guye, M., Lhatoo, S. D., Michel, C. M., Duncan, J. S., & Lemieux, L. (2010). Continuous EEG source imaging enhances analysis of EEG-fMRI in focal epilepsy. *NeuroImage, 49*, 3219–3229. https://doi.org/10.1016/j.neuroimage.2009.11.055

Wirsich, J., Jorge, J., Iannotti, G. R., Shamshiri, E. A., Grouiller, F., Abreu, R., Lazeyras, F., Giraud, A.-L., Gruetter, R., Sadaghiani, S., & Vulliémoz, S. (2021). The relationship between EEG and fMRI connectomes is reproducible across simultaneous EEG-fMRI studies from 1.5T to 7T. *NeuroImage, 231*, 117864. https://doi.org/10.1016/j.neuroimage.2021.117864

Xie, L., Kang, H., Xu, Q., Chen, M. J., Liao, Y., Thiyagarajan, M., O'Donnell, J., Christensen, D. J., Nicholson, C., Iliff, J. J., Takano, T., Deane, R., & Nedergaard, M. (2013). Sleep drives metabolite clearance from the adult brain. *Science, 342*. https://doi.org/10.1126/science.1241224

Yacoub, E., Duong, T. Q., Van De Moortele, P.-F., Lindquist, M., Adriany, G., Kim, S.-G., Uğurbil, K., & Hu, X. (2003). Spin-echo fMRI in humans using high spatial resolutions and high magnetic fields. *Magnetic Resonance in Medicine, 49*, 655–664. https://doi.org/10.1002/mrm.10433

Yang, F. N., Picchioni, D., de Zwart, J. A., Wang, Y., van Gelderen, P., & Duyn, J. H. (2024). Reproducible, data-driven characterization of sleep based on brain dynamics and transitions from whole-night fMRI. *eLife, 13*. https://doi.org/10.7554/eLife.98739.1

Yu, Y., Qiu, Y., Li, G., Zhang, K., Bo, B., Pei, M., Ye, J., Thompson, G. J., Cang, J., Fang, F., Feng, Y., Duan, X., Tong, C., & Liang, Z. (2023). Sleep fMRI with simultaneous electrophysiology at 9.4 T in male mice. *Nature Communications, 14*, 1651. https://doi.org/10.1038/s41467-023-37352-9

Yun, S. (2024). Advances, challenges, and prospects of electroencephalography-based biomarkers for psychiatric disorders: A narrative review. *Journal of Yeungnam Medical Science, 41*, 261–268. https://doi.org/10.12701/jyms.2024.00668

Zhang, S., Hennig, J., & LeVan, P. (2019). Direct modelling of gradient artifacts for EEG-fMRI denoising and motion tracking. *Journal of Neural Engineering, 16*, 056010. https://doi.org/10.1088/1741-2552/ab2b21

Zheng, S., Zhang, Y., Huang, K., Zhuang, J., Lü, J., & Liu, Y. (2025). Temporal interference stimulation boosts working memory performance in the frontoparietal network. *Human Brain Mapping, 46*, e70160. https://doi.org/10.1002/hbm.70160

Zijlmans, M., Huiskamp, G., Hersevoort, M., Seppenwoolde, J.-H., van Huffelen, A. C., & Leijten, F. S. S. (2007). EEG-fMRI in the preoperative work-up for epilepsy surgery. *Brain, 130*, 2343–2353. https://doi.org/10.1093/brain/awm141

Zotev, V., & Bodurka, J. (2020). Effects of simultaneous real-time fMRI and EEG neurofeedback in major depressive disorder evaluated with brain electromagnetic tomography. *Neuroimage Clinical, 28*, 102459. https://doi.org/10.1016/j.nicl.2020.102459

Zotev, V., Phillips, R., Yuan, H., Misaki, M., & Bodurka, J. (2014). Self-regulation of human brain activity using simultaneous real-time fMRI and EEG neurofeedback. *NeuroImage, 85*(Pt 3), 985–995. https://doi.org/10.1016/j.neuroimage.2013.04.126

Zotev, V., Phillips, R., Misaki, M., Wong, C. K., Wurfel, B. E., Krueger, F., Feldner, M., & Bodurka, J. (2018). Real-time fMRI neurofeedback training of the amygdala activity with simultaneous EEG in veterans with combat-related PTSD. *Neuroimage Clinical, 19*, 106–121. https://doi.org/10.1016/j.nicl.2018.04.010

Chapter 34
Bridging Brain and Body: Complementing EEG with Peripheral Physiological Signals

Ignacio Rebollo, Daniel S. Kluger, and Marie Loescher

Abstract Cognitive neuroscience is increasingly moving beyond a brain-centric perspective to study the brain as part of an integrated organism. This chapter advocates for this paradigm shift by detailing how neural measures can be complemented with non-invasive recordings of peripheral physiological signals. We focus on three vital visceral rhythms of the heart (~1 Hz), the lungs (~0.2–0.3 Hz), and the stomach (~0.05 Hz)—that provide tractable access to the continuous, bidirectional communication between the central and peripheral nervous systems. For each rhythm, we outline the underlying physiology, practical guidance for recording (e.g., electrocardiography, respiratory belts, electrogastrography), and key derived metrics (e.g., heart rate variability, respiratory phase, gastric amplitude). We then illustrate how these signals are used to investigate functional brain-body coupling, from heartbeat-evoked potentials and respiration-locked neural oscillations to stomach-brain synchronization. By integrating these bodily rhythms with neural measures, researchers can gain a more holistic understanding of how peripheral physiology shapes affective, cognitive, and perceptual functions.

Keywords Interoception · Brain-Body · Heart · Stomach · Respiration · Visceral rhythms

---

Authors Daniel S. Kluger and Marie Loescher have contributed equally to this chapter.

I. Rebollo (✉)
Department of Gut Brain Interactions, German Institute of Human Nutrition, Nuthetal, Germany
e-mail: ignacio.rebollo@dife.de

D. S. Kluger
Institute for Biomagnetism and Biosignal Analysis, University of Münster, Münster, Germany

M. Loescher
Laboratoire de Neurosciences Cognitives et Computationnelles, Département d'Etudes Cognitives, École Normale Supérieure, Université PSL, INSERM U960, Paris, France

T. Warbrick (ed.), *The EEG Handbook*,
https://doi.org/10.1007/978-3-032-20450-9_34

34.1 Introduction: The Brain-in-a-Body

Cognitive neuroscience has moved from viewing the brain in isolation to studying it as part of an integrated brain–body system. Central nervous system (CNS) activity is continuously shaped by signals from the peripheral nervous system (PNS) and visceral organs (Engelen et al., 2023), and in turn modulates their function (Cechetto, 2014; Quadt et al., 2022; Saper, 2002). This bidirectional exchange is not only essential to maintain the organism's homeostasis, but is increasingly recognized to shape affective, cognitive, and perceptual functions across the brain (Berntson & Khalsa, 2021; Azzalini et al., 2019; Engelen et al., 2023).

This chapter introduces three key visceral rhythms—cardiac, respiratory, and gastric—that offer tractable access to brain–body interactions. Each of these continuously generated rhythms can be recorded non-invasively and has been shown to shape neural dynamics and cognition. We proceed from the faster rhythm of the heart (~1 Hz), to the intermediate rhythm of respiration (~0.2–0.3 Hz), and finally to the slow gastric rhythm (~0.05 Hz). In each section, we begin with the physiology of the organ and its characteristic rhythm, turn to the practicalities of recording (e.g., the electrocardiogram, respiratory belts, or electrogastrogram), and explain key derived metrics such as heart rate variability, respiration-phase measures, or gastric amplitude. Finally, we illustrate how these signals are used to investigate functional coupling between body and brain, from heartbeat-evoked potentials to respiration-locked neural oscillations and stomach–brain synchronization. By proposing to integrate neural measures with these bodily rhythms, this chapter moves beyond a brain-centric perspective towards a holistic understanding of the neural dynamics and cognition.

34.2 Cardiac Signals

34.2.1 *Physiology of the Heart*

The heart's primary function is to generate the pressure necessary for systemic blood circulation. It is a muscular organ with four chambers—two atria through which blood enters, and two ventricles which pump it out—separated by valves that ensure one-way blood flow. Contractions are triggered within the heart itself: pacemaker cells in the sinoatrial node spontaneously depolarize, initiating a wave of electrical activity followed by mechanical contraction that spreads through the atria (atrial systole) and then the ventricles (ventricular systole). Contraction is followed by repolarization and relaxation (diastole), allowing the chambers to refill. Together, *systole* and *diastole* form the two alternating phases of each cardiac cycle (Shaffer et al., 2014).

Heart rate and contractions are regulated locally, but also by the central nervous system (CNS). At rest, parasympathetic (mainly vagal) influence predominates, maintaining a normal heart rate of one contraction every ~900 ms in healthy adults. When needed, sympathetic fibres accelerate and increase contraction force. These dynamic influences, regulated by an extensive central network (Valenza et al., 2019), allow the heart to rapidly adapt to environmental and physiological demands—and high HRV is generally a sign of health (Shaffer et al., 2014).

The heart also continuously sends ascending signals to the brain. The best-studied pathway are *baroreceptors* in the carotid sinuses and aortic arch, which fire with each heartbeat—with various latencies (Bolter et al., 2011)—in response to arterial stretch, and reach the brain through the vagus and glossopharyngeal nerves (Benarroch, 2008). Additional mechanoreceptors in the heart, pulmonary vessels, and peripheral tissues detect cardiac contractions or vascular pulsatility, for instance (Birznieks et al., 2012; Bishop et al., 2011). Recent evidence suggests that vascular pulsations may even be transduced directly in the brain through mechanosensitive astrocytes and neurons (Jammal Salameh et al., 2024; Kim et al., 2016). Cardiac signals then not only reach brainstem and core interoceptive regions—the posterior insula and somatosensory cortices—but are represented across cingulate, medial prefrontal, parietal, and occipital cortices (Engelen et al., 2023).

34.2.2 Recording Cardiac Activity

Two main methods are commonly used to record cardiac signals in neuroscientific research: electrocardiography (ECG) and photoplethysmography (PPG).

34.2.2.1 Electrocardiography (ECG)

The ECG records the heart's electrical activity through electrodes on the skin, detecting the voltage differences generated by cardiac depolarization and repolarization (Sattar & Chhabra, 2025). The summed currents can be approximated as a moving *cardiac dipole*: a net vector whose orientation shifts during each heartbeat and produces different waveforms depending on electrode placement.

In Practice
Electrical activity is recorded as a voltage difference, or *lead*, either between two electrodes (bipolar), or between one electrode and a common average (unipolar). Einthoven's original three bipolar lead system formed a triangle between the limbs: I (left arm—right arm), II (left leg—right arm), and III (left leg—left arm). Although later expanded to the twelve-lead clinical standard (Mirvis & Goldberger, 2012), simpler setups are sufficient for neuroscience. A common minimal setup uses a modified lead II, with electrodes under the right clavicle and on the left lower abdomen, and a ground shared with EEG (Fig. 34.1a). Chest placements, although unsuited for clinical diagnosis (Jowett et al., 2005), reduce motion artefacts (Mason & Likar, 1966) and increase signal amplitude. We further recommend adding a third

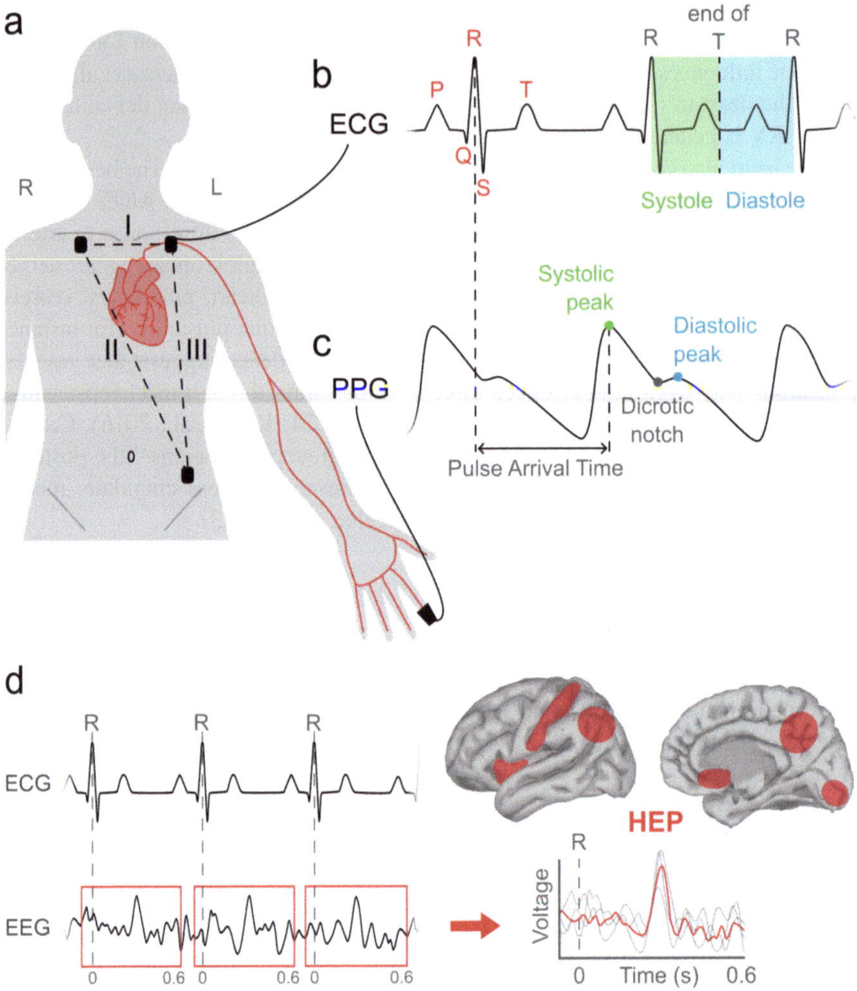

Fig. 34.1 Recording of cardiac activity and heartbeat-evoked potentials (HEPs). (**a**) Modified Einthoven's triangle electrode placement with the resulting bipolar leads I-III for the recording of the ECG, and pulse oximeter finger clip for the recording of the PPG. (**b**) Schematic ECG waveform as recorded by lead II, with PQRST components marked. Each cardiac cycle can be divided into two phases, systole (green) and diastole (blue). We show the recommended ventricular definition, with phases lasting from R-peak to end of T and from end of T to next R-peak (Aufan et al., 2023; Caparco et al., 2025). (**c**) Schematic PPG waveform and main components. The systolic peak appears delayed compared to the ventricular depolarization marked by the ECG's R-peak due to the pulse arrival time. (**d**) Computation of the neural HEP: neural (e.g., EEG) recordings are epoched into short time-windows time-locked to an event within the ECG such as the R-peak (left). Right bottom: The average over epochs yields the HEP. Right top: Locations within the cortex in which HEPs have been reported (Engelen et al., 2023). Abbreviations: *ECG* electrocardiogram, *PPG* photoplethysmography, *EEG* electroencephalography, *HEP* heartbeat-evoked potential

electrode under the left clavicle to also obtain leads I and III (Fig. 34.1a). Ag/AgCl electrodes with conductive gel are standard. To reduce impedance, skin is cleaned with alcohol and lightly abraded if necessary. Sampling rates of min. 500 Hz are recommended to capture sharp waveform components.

The ECG Waveform

Lead II yields the familiar textbook waveform (Fig. 34.1b): as ventricular depolarization runs roughly parallel to this axis, it records a characteristic, clear signal. The trace shows the main cardiac electrical events: the P-wave (atrial depolarization), QRS complex (ventricular depolarization), and T-wave (ventricular repolarization). The corresponding mechanical events can be inferred with near millisecond precision (Aufan et al., 2023), allowing to define ventricular systole from R-peak to the end of the T-wave, and ventricular diastole from the end of the T-wave to the next R-peak. For consistency, we recommend this physiologically grounded and widely used definition of cardiac phases, though alternatives exist (Caparco et al., 2025).

34.2.2.2 Photoplethysmography (PPG)

PPG is an optical technique that measures blood volume changes within the microvasculature (Kyriacou & Allen, 2021). A light source illuminates the tissue and a photodetector records variations in transmitted or reflected light. In transmission mode, used in pulse oximeter clips, light passes through thin tissue such as the finger. Wearable devices use reflectance mode, with source and detector on the same side. Recent camera-based approaches extract PPG remotely from subtle skin colour changes. Fundamentally, PPG is an indirect measure of cardiac activity compared to ECG; but its low cost, ease of setup, and flexible formats make it a valuable alternative when limitations are kept in mind.

In Practice

Signal quality depends on recording site. Pulse oximeters provide the most robust signals, with earlobe probes less sensitive to vascular tone than finger probes (Clayton et al., 1991). Devices typically sample at 20–100 Hz, but 100–200 Hz is recommended to obtain pulse rate variability metrics comparable to ECG (Burma et al., 2024). Because PPG is sensitive to vasoconstriction and motion artefacts, participants should be kept still in a temperature-controlled environment.

The PPG Waveform

The PPG waveform consists of a steady baseline (*DC*), reflecting tissue and venous absorption, and a pulsatile component (*AC*), corresponding to transient arterial blood volume changes. The sharp (systolic) upstroke marks the arrival of the pulse wave, followed by a peak and a slower (diastolic) decline that may feature a dicrotic notch from aortic valve closure (Fig. 34.1c). Unlike the ECG, wave morphology is not determined by a single electrical event but depends on stroke volume, arterial stiffness, vascular tone, and respiration, and its precise physiological underpinnings remain debated (Kyriacou & Allen, 2021). The PPG pulse also lags the true heartbeat due to the *pulse arrival* or *transit time*, which varies across sites and individuals. In practice, the relative timing of successive pulses provides a reliable heart rate estimate (Burma et al., 2024); but for precise temporal alignment with neural signals, ECG remains the method of choice.

34.2.3 Key Derived Metrics and Applications

34.2.3.1 Cardiac Activity as a Marker of Autonomic State

Even before integrating it with neural measures, cardiac activity—whether as a control or the target of your research—reveals much about physiological state.

Inter-Beat-Intervals and Heart Rate
The simplest metric is the Inter-Beat-Interval (IBI), the time difference between successive R-peaks in the ECG or pulse peaks in the PPG. Across experimental paradigms, IBIs are systematically and instantaneously affected by stimuli and behaviour, from conscious perception (e.g., Grund et al., 2022) to erroneous responses (e.g., Danev & de Winter, 1971). Averaged over time, IBIs yield heart rate, which as-is already provides a coarse index of arousal, for instance (e.g., as used in Bolt et al., 2025; Wang et al., 2018).

Heart Rate Variability
The variability of IBIs over time, or heart rate variability (HRV), reflects autonomic regulation and CNS integrity. HRV is notably associated with executive functions, decision-making, and emotional regulation, and abnormal patterns in HRV have been robustly linked to many neurological and psychological conditions, including dementia, epilepsy, anxiety, and schizophrenia (for a review, see Arakaki et al., 2023). We briefly present three main methods to quantify HRV (Electrophysiology, 1996; Shaffer et al., 2014).

In the time domain, successive IBIs are compared directly. Measures such as RMSSD (root mean square of successive differences) and pNN50 (percentage of intervals differing by >50 ms) capture rapid fluctuations and can be computed from short recordings. In the geometric domain, variability is visualized, often with Poincaré plots where each IBI is plotted against the next, providing a subject- or condition-level "fingerprint". In the frequency domain, the IBI series is converted into a spectrum. High-frequency HRV (HF-HRV, ~0.15–0.4 Hz) is considered a key marker of parasympathetic (vagal) tone and is tied to respiration (see also 34.3.1 Physiology of Respiration). Low-frequency HRV (LF-HRV, ~0.04–0.15 Hz) is thought to reflect both sympathetic and parasympathetic influences. Frequency-domain analyses require at least five minutes of continuous data. Across all methods, artifact-free recordings are essential, as spurious or ectopic beats can quickly distort HRV estimates.

34.2.3.2 Investigating Cardiac-Brain Coupling

Beyond reflecting autonomic state, the heart and brain continuously influence each other. Studying this coupling has opened a rich line of research on how cardiac and neural activity dynamically interact.

Bi-directional Heart-Brain Connectivity

Cardiac-brain coupling can be quantified by relating HRV to cortical dynamics, for instance through phase-amplitude coupling (Sargent et al., 2024) or similarity indices between HRV and neural timeseries (Candia-Rivera et al., 2024). Such approaches have revealed altered coupling in conditions such as Parkinson's disease (Candia-Rivera et al., 2024) and schizophrenia (Sargent et al., 2025), as well as across physiological states like stress (Chalmers et al., 2020). However, conclusively determining the direction of influence (heart-to-brain or vice versa) remains a challenge for correlational methods.

Cardiac Phase

Another approach examines how perception and behaviour vary across the cardiac cycle. For example, a study might present a sensory stimulus at specific moments within the cardiac cycle, and investigate the degree to which perceptual behaviour differs between systole, when baroreceptor firing peaks, and diastole. In this example, cardiac phase is treated as a simple biphasic contrast between systole and diastole; a more fine-grained approach consists in treating it as a continuous circular variable, by interpolating phase angles across the cycle. Cardiac phase effects have been reported in very diverse domains, from perception (Grund et al., 2022) to emotional processing (Garfinkel et al., 2014). They are often interpreted within predictive coding accounts, which propose that the brain suppresses predicted cardiac afferences to reduce internal noise (e.g., Skora et al., 2022). Yet, results can also show unexpectedly complex patterns, and are sometimes inconsistent (see Caparco et al., 2025 for a recent discussion). Importantly, systole can produce strong artefacts in neural recordings (the *cardiac field artefact*), so behavioural measures are generally more reliable for phase-related analyses.

Heartbeat-Evoked Potentials

Finally, cardiac–brain coupling can be assessed by directly measuring the neural responses elicited at each heartbeat. These heartbeat-evoked potentials (HEPs) are obtained by averaging EEG or MEG activity time-locked to the R- or T-peak of the ECG, and have been observed across many cortical regions (Fig. 34.1d; for a review, Engelen et al., 2023). Compared to stimulus-evoked potentials, however, recording and analysis require extra caution because of the cyclical nature of the signal and the cardiac field artefact (Azzalini et al., 2019; Steinfath et al., in press).

HEP amplitude scales with interoceptive attention to the heart (Petzschner et al., 2019), but has also been linked to a remarkably broad spectrum of other domains—from exteroceptive perception and emotion to consciousness, clinical disorders, and beyond (e.g., Park et al., 2014; Candia-Rivera et al., 2021; for a review, see Engelen et al., 2023). Perhaps most striking is their consistent role in self-related processing (e.g., Babo-Rebelo et al., 2016; Loescher et al., 2025), motivating the hypothesis that the heartbeat provides an internal reference for coordinating distributed neural activity, and thereby contributing to the emergence of a unified sense of self (Park & Tallon-Baudry, 2014; Loescher et al., in press). More generally, HEPs illustrate how cardiac measures can move from being simple controls to central variables in neuroscientific research.

34.3 Respiratory Signals

34.3.1 Physiology of Respiration and Neural Regulation

At rest, human respiration seems effortless and comprises the two phases of active (but automatic) inspiration and passive expiration (Fleming et al., 2011). At a respiratory frequency of ~0.25 Hz (or about 15 breaths per minute), the rhythmicity of each breath is initiated and coordinated by coupled oscillators periodically driving respiration, most prominently the preBötzinger complex located in the medulla oblongata of the brain stem (Moore et al., 2013). Although this microcircuit typically controls respiration autonomously, respiration is also under top-down control to allow adaptive breathing, for example, during speaking, laughing, and crying (McKay et al., 2003). Hence, there is a bidirectional interplay of the cortex and rhythmic pattern generators of respiration: Efferent respiratory signals from the preBötzinger complex project to suprapontine nuclei (via locus coeruleus and olfactory nuclei) as well as to the central medial thalamus, which is directly connected to limbic and sensorimotor cortical areas (Del Negro et al., 2018; Fig. 34.2).

What makes respiration unique among the physiological signals we consider here is the fact that its rhythmicity is generated within the brain. Compared to cardiac and gastric signals which directly originate within the heart and stomach, respectively, the central pattern generator of respiration has critical implications for neuroscientific research: Just to name one example, there is a unilateral anatomical projection from respiratory microcircuits in the preBötzinger complex to the locus coeruleus, a key structure for norepinephrine release located in the more dorsal brainstem (Aston-Jones & Cohen, 2005). Based on this functional connection, both animal and human studies have provided evidence that respiratory activity is involved in norepinephrine-modulated arousal states in the brain (Yackle et al., 2017; Kluger et al., 2024).

While the mechanisms linking respiration to brain state changes are not yet fully understood, cross-species evidence so far suggests that one key contributing factor are entrained oscillations within the early olfactory cortex: With every breath, airflow within the respiratory tract triggers mechanoreceptors connected to the olfactory bulb, the brain's first olfactory relay station (Lundström et al., 2011). Here, these mechanical signals are translated into infraslow neural rhythms

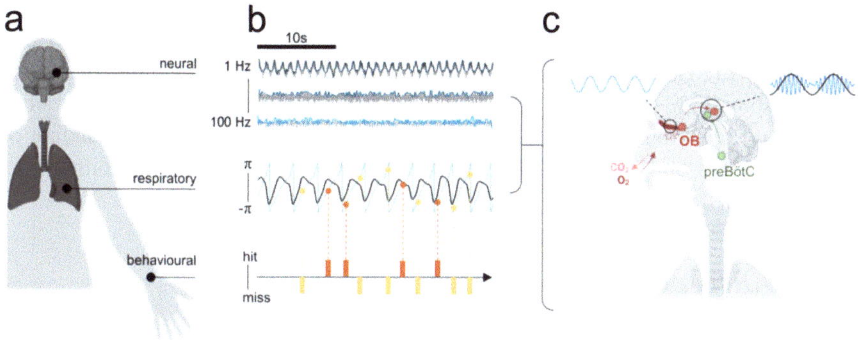

Fig. 34.2 Principles of respiration-brain coupling. (**a**) In studies of respiration-brain coupling, time-locked neural, respiratory, and behavioural signals can be recorded simultaneously. (**b**) Neural time series are oftentimes transformed into frequency space to investigate neural oscillations (top). Respiratory phase is extracted from the raw breathing signal (green and black lines in the middle panel, respectively). Both neural and behavioural measures can then be analysed according to the respiratory phase at which they occurred (bottom). (**c**) Green arrows show the feedforward initiation of respiratory dynamics originating in the preBötzinger complex of the brainstem. Red arrows illustrate the feedback connection by which the circulating airstream in the nasal cavity triggers mechanosensory neurons. This phase-phase coupling (top inset) is then propagated through higher-order brain areas so that the phase of respiration drives the amplitude of neural responses throughout the cortex

(or oscillations) whose phase closely follows that of the respiratory rhythm (*phase-phase coupling*). Research in mice has conclusively demonstrated that these slow respiration-induced oscillations are transmitted through adjacent piriform cortex and subsequent cortico-limbic circuits to modulate the amplitude of faster neural oscillations in upstream cortical areas (Ito et al., 2014). This link is commonly termed *phase-amplitude coupling*, a well-established mechanism of neural information transfer across spatiotemporal scales (Canolty & Knight, 2010). In the respiratory domain, this coupling cascade provides an intuitive explanation of how the link between (nasal) respiratory phase and neural oscillations could be implemented and why this link dissipates when we switch from nasal to oral breathing. However, not all brain activity is oscillatory, and respiratory modulations of so-called aperiodic (i.e. non-rhythmic) neural signals have been shown to remain intact for oral breathing (Kluger et al., 2023). Together, these findings suggest that there are multiple ways in which neural activity could be coupled to the act of breathing, most likely involving changes in CO_2 (Feldman et al., 2003) and pH-dependent modulation of adenosine and ATP levels (Ito et al., 2003; Gourine et al., 2005).

Besides its close relationship to changes in brain states like arousal, the respiratory rhythm is immediately coupled to other physiological parameters as well: Most prominently, there is a tight link between respiratory and cardiac activity by means of the so-called respiratory sinus arrhythmia (or respiratory heart rate variability; Menuet et al., 2025). This coupling is primarily driven by rhythmic changes in parasympathetic activity which inhibits cardiac signals and is strongest during expiration, resulting in a decrease in heart rate. Conversely, during inspiration, this inhibitory influence decreases and leads to an increase in heart rate (Goldberger et al., 2001). Consequently, absolute heart rate changes between inspiration and expiration can again be voluntarily controlled through changes in the frequency and depth of breathing (Sroufe, 1971). Hence, simultaneous recordings of respiratory and electrophysiological activity hold great potential for a deeper understanding of functional brain-body interactions, particularly in the cardiorespiratory domain.

34.3.2 Recording Methods

Respiratory signals can be recorded with a variety of methods ranging from respiratory belts over thermistors and direct measures of nasal airflow with a nasal cannula. Oftentimes, the experimental setup and research question will determine which recording method (or combination of methods) is best suited. If, for example, nasal and oral breathing are compared within the same experimental session, it may be advisable to combine a nasal thermistor and a respiration belt on the chest.

34.3.2.1 Respiratory Belts

Respiratory belts are the most commonly used method to record respiratory activity in neuroscientific applications: They are affordable, easy to use, and well tolerated by participants, particularly when worn over clothing. While there are different sensor types available for respiratory belts, they all record the circumference of the thorax—it increases during inhalation and decreases during exhalation. Flexible placement of the respiratory belt or the use of multiple belts allows for experimental control when participants are asked to breathe in a certain way, e.g. during abdominal vs chest breathing. In any application, some care has to be taken while attaching the respiratory belt: Live signal monitoring is ideal to ensure that the belt is neither too loose, which results in no meaningful data being recorded, nor too tight, which would constrict the participant's breathing and could result in clipping artefacts in case the sensor is, e.g. stretched beyond its capacity.

34.3.2.2 Nasal Thermistors and Pressure Sensors

Other popular acquisition methods record nasal airflow directly, either with a thermistor placed below the nostrils or a nasal cannula attached to a flow meter. Using a small temperature sensor, thermistor time series reflects the respiratory signal as changes between warmer air temperature during exhalation and colder air temperature during inhalation. A nasal cannula is placed inside the nostrils and allows for precise measures of air flow (e.g., volume and speed). Depending on the research question, the experimenter should consider that the placement of sensors in the participant's face might induce slight discomfort or other biases to the recording. One example would be a study in which participants should ideally not be aware that their respiratory activity is of critical interest, e.g. to avoid changes in top-down control of their breathing pattern. However, both methods yield precise and convenient-to-use respiratory time series and allow for the extraction of all key metrics for further analyses.

34.3.2.3 Key Derived Metrics and Applications in Cognitive Neuroscience

Respiratory Rate

Respiratory rate quantifies how many breaths a participant takes in a certain amount of time, e.g. in one minute. Here, one full breathing cycle can be defined as the peak-to-peak distance in the recorded respiratory time series. In neuroscientific research, respiratory rate is commonly used as a measure of physiological arousal (Shea, 1996), physical exercise (Kayser, 2003), and breathlessness (Guz, 1997), as well as a clinical marker for certain patient groups (Cretikos et al., 2008). At rest, the variability of respiratory rate over longer time scales can be quantified as respiratory variability, which has been linked to anxiety (Van Diest et al., 2006). Moreover, respiratory rate is a prime behavioural measure during breathing interventions like breathholds, slow breathing, or box breathing, which in turn can be related to neural changes measured with electrophysiology. For example, a systematic review on slow breathing practices (referring to fewer than 10 breaths per minute) revealed a systematic increase in alpha power and a decrease in theta power across EEG studies (Zaccaro et al., 2018). Note that, in order to avoid hyperventilation, voluntary changes in respiratory rate will always influence another key metric, namely respiratory volume.

Respiratory Volume

The depth of each breath, quantified as the thoracic circumference, is arguably the most direct measure derived from respiration belt recordings. Here, the amplitude of respiratory time series can be used as a measure for interventions like voluntary breathholds or deep vs regular breathing. One hypothesis is that the inhalation phase of voluntary deep breathing drives slowly adapting stretch receptors in the lungs, which in turn modulates autonomic nervous system activity (Noble & Hochmann, 2019). On the neural level, such behavioural manipulations have previously been associated not only with changes in brain activity (Karjalainen et al., 2025), but also in brain-body connectivity such as cortico-muscular communication (Kluger & Gross, 2020).

Respiratory Phase

In general, respiratory phase refers to angular information ($-\pi$ to π) across one full breathing cycle. Continuous vectors of respiratory phase can be extracted from respiratory time series recorded with any of the methods described above, e.g. using a Hilbert transform (which is common, but not recommended) or interpolation approaches. For phase interpolation, one would identify peaks and troughs in the respiratory signal—corresponding to the end of inhalation and exhalation in respiration belt data, respectively—and interpolate between each peak and trough. Recent studies in the field of brain-body neuroscience have exploited this fine-grained phase information to relate respiratory signals to systematic changes in brain activity (Zelano et al., 2016; Kluger & Gross, 2021), perceptual processing (Braendholt et al., 2025; Kluger et al., 2021), cognition (Arshamian et al., 2018; Perl et al., 2019), and voluntary action (Park et al., 2020). In these applications, one would not only consider brain activity or behavioural responses at critical time points in an EEG experiment, but also corresponding respiration phase. Moreover, phase binning—the grouping of time points corresponding to similar respiration phases—allows for analyses of continuous respiration-related changes in brain states even in the absence of triggered events, e.g. during resting state.

34.4 Gastric Signals

34.4.1 Anatomy and Physiology of the Stomach

The stomach's primary functions are the temporary storage and mechanical breakdown of food. Through coordinated muscle contractions and the secretion of acids and enzymes, it reduces ingested material to a semi-liquid state before its controlled release into the intestine for absorption. The pace of this mechanical activity is set by an intrinsic electrical rhythm generated within the stomach wall. This rhythm, known as the *gastric slow wave* or *gastric rhythm*, originates from specialized pacemaker cells called Interstitial Cells of Cajal (ICCs, Sanders et al., 2006). In humans, the dominant frequency of the gastric rhythm is approximately 3 cycles per minute (0.05 Hz), a range clinically defined as *normogastria* (typically 2–4 cpm (Koch & Stern, 2004).

Importantly, the gastric slow wave does not directly cause contractions. Instead, it creates a periodic electrical excitability, or "window", during which smooth muscle contractions can be triggered. The actual contractions are initiated by enteric motor neurons, whose cell bodies lie within the gastric wall (Sanders et al., 2006). These neurons in turn stimulate the smooth muscle to contract. Remarkably, the entire system exhibits significant autonomy—the gastric rhythm persists even when the stomach is surgically decoupled from the central nervous system (Suzuki et al., 1986). However, the ultimate decision to contract is subject to top-down regulation by the autonomic nervous system, which can either increase gastric motility (parasympathetic) or decrease it (sympathetic, Feinle-Bisset et al., 2004).

The different anatomical regions of the stomach have specialized functions that influence electrical activity patterns (Fig. 34.3a). The corpus (middle stomach) contains the highest density of ICCs, from which the gastric rhythm propagates caudally (O'Grady et al., 2010). The antrum, located in the most distal portion of the stomach, serves as the primary site for gastric contractions and grinding activity (Kim et al., 2022). This makes the antrum region the optimal target for

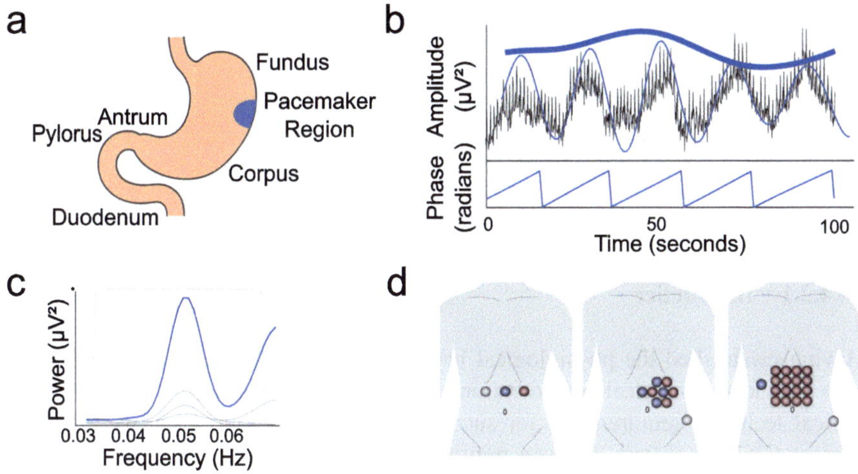

Fig. 34.3 Gastric anatomy, signal characteristics, and EGG recording approaches. (**a**) Schematic of the human stomach showing the location of the pacemaker region in the corpus, where interstitial cells of Cajal initiate the gastric rhythm that propagates towards the antrum and pylorus. (**b**) Example of raw electrogastrography (EGG) recording (black trace) and extracted gastric rhythm (blue trace). The corresponding amplitude envelope (upper blue trace) and instantaneous phase (bottom panel) are derived using the Hilbert transform. (**c**) Power spectral density of EGG signals across electrodes. The dominant peak at ~0.05 Hz reflects the gastric slow wave, within the normogastric range (0.033–0.066 Hz). (**d**) Electrode configurations used in EGG research, ranging from classic single bipolar montages (left), to multi-channel arrangements (middle), to high-density electrode arrays (right) for mapping gastric activity with higher spatial resolution

electrogastrographic recordings when contractile activity is of primary interest, while the corpus is preferred when the pacesetter potential is of greatest interest.

34.4.2 Electrogastrography

Electrogastrography (EGG) is the primary non-invasive technique for recording gastric myoelectrical (i.e. muscular) activity, and consists of electrodes placed on the skin of the abdomen (Koch & Stern, 2004). The raw EGG signal represents a composite signal containing faster oscillations from cardiac activity (~1 Hz) and respiratory activity (~0.3 Hz) superimposed on the gastric rhythm (Fig. 34.3b, black). However, due to the large frequency difference between these physiological signals and the gastric rhythm (~0.05 Hz), the gastric component can be cleanly isolated through frequency analysis as a distinct peak in the normogastric range (Fig. 34.3c).

Once isolated, the EGG signal at 0.05 Hz reflects both the pacesetter potential of ICCs and associated smooth muscle contractions, though these components cannot be disentangled using standard recording techniques. The relative contribution of these components varies dramatically based on digestive state and electrode positioning: During fasted conditions, when gastric motility is minimal, the EGG primarily captures continuous pacemaker activity from a relatively quiescent stomach. Conversely, in the fed state, gastric contractions contribute substantially to the recorded signal, often producing several orders of magnitude increases in amplitude. This state-dependency has important implications for experimental design and data interpretation.

34.4.2.1 In Practice

Having established the physiological basis and signal characteristics of EGG, we now turn to the practical considerations for successful recording. One of the most critical technical requirements for successful EGG recording is the use of direct current (DC) amplifiers, as even minimal high-pass filtering can distort the very low-frequency gastric signals. The use of Ag/AgCl electrodes with conductive gel is standard. To reduce impedance and noise, the skin should be cleaned with alcohol and abraded to achieve electrode impedances below 20 kΩ.

34.4.2.2 Electrode Placement

With respect to electrode location, classic EGG studies (Oczka et al., 2024) relied on a single bipolar channel, three electrode setup. This layout consists of a ground electrode on the left costal margin, a positive recording electrode on the mid-abdomen between the xiphoid process and umbilicus, and a negative recording electrode on the left upper quadrant of the abdomen, yielding a single bipolar channel (Fig. 34.3d right). However, this single-channel approach has significant limitations: the stomach's position can vary considerably between individuals, often placing it far from the recording electrode, and artifacts can easily contaminate the single channel, making signal recovery difficult (Gharibans et al., 2017).

Therefore, modern approaches employ multi-electrode configurations to more robustly capture gastric signals (Simonian et al., 2004). By using multiple recording sites, researchers can identify the optimal electrode location for each individual and maintain signal quality even when artifacts affect individual channels. Recent clinical gastroenterology applications increasingly use high-density arrays (16–40 electrodes, Fig. 34.3c, middle) to map gastric slow wave propagation comprehensively. However, such elaborate setups are often impractical for psychophysiological research, particularly when combining EGG with other techniques like EEG or fMRI. An effective compromise involves multi-channel montages of 4–8 channels, with the specific configuration determined by study design and time constraints (Wolpert et al., 2020). For electromagnetically challenging environments like MRI suites, a 4-channel bipolar setup provides enhanced signal quality (Rebollo et al., 2018, Fig. 34.3d middle).

Electrodes should be arranged in a grid spanning from the xiphoid process to below the umbilicus, extending laterally to the left mid-clavicular line (see Wolpert et al., 2020 for detailed placement instructions). This spatial arrangement ensures comprehensive coverage of the abdomen and maximizes the chances of obtaining at least one electrode with good quality signal. When gastric contractile activity is the primary focus, electrode placement should be guided using ultrasound to identify the position of the antrum in each individual. Electrodes can then be placed in a grid over the antrum (Gharibans et al., 2017). When ultrasound guidance is not possible, or when the focus is on the pacemaker potential rather than contractions, external anatomical landmarks provide reliable placement guidance.

34.4.2.3 Other Practical Considerations

Given the slow cycle of the gastric rhythm, longer recordings are necessary compared to other physiological signals. Recording duration should span at least 10–15 min to capture adequate number gastric cycles (approximately 30–45 cycles). Successful EGG recording also requires careful participant screening. Optimal results are achieved with individuals having a Body Mass Index (BMI) between 18 and 25, as adipose tissue can significantly attenuate the gastric signal. Participants should be screened for medications affecting gastric motility, including proton pump inhibitors or prokinetic agents. Finally, shaving or trimming of abdominal hair is necessary for successful recordings, and participants should be informed of this beforehand.

34.4.3 Signal Analysis Pipeline

EGG analysis involves several sequential steps to extract clean gastric signals from the raw recordings. A typical analysis pipeline begins with examining the frequency spectrum of all channels to identify clear peaks at 0.05 Hz. The channel exhibiting the largest peak in the 0.033–0.066 Hz range is selected for subsequent analysis, while the others are discarded (Fig. 34.3c). The next step is to isolate the gastric rhythm from activity unrelated to the stomach. For this, a bandpass filter is applied around the participant's individual peak frequency. A typical filter configuration spans ±0.015 Hz around the peak frequency (e.g. 0.035–0.065 Hz for a 0.05 Hz peak), capturing physiological frequency variations while excluding respiratory artifacts (Fig. 34.3b, blue line). Finally, application of the Hilbert transform to filtered data yields instantaneous phase and amplitude envelopes of the gastric rhythm. Here, the phase component is more related to the gastric pacesetter potential bottom), while the amplitude is proportional to the degree of stomach contractions (Fig 3b, thick blue line). Detailed code and instructions for implementing this pipeline can be found in Wolpert et al. (2020).

34.4.4 Key Derived Metrics and Applications

34.4.4.1 EGG Signals as an Index of Autonomic Modulation

The processed EGG signal yields two primary metrics that serve as indices of gastric myoelectrical activity and its modulation by the autonomic and central nervous systems. The first one is the dominant frequency of the EGG. The peak frequency in the normogastric range (2–4 cpm, or 0.033–0.066 Hz) reflects the underlying pacemaker rhythm set by the interstitial cells of Cajal. Deviations from this range, such as tachygastria (>4 cpm) or bradygastria (<2 cpm), are considered gastric dysrhythmias. The second metric is the spectral power at the dominant EGG peak frequency. This metric is thought to reflect the strength of gastric contractile activity.

Empirical studies linking these metrics to psychological states reveal a complex and often inconsistent picture (Wolpert et al., 2020). For example, stressors have been reported to increase tachygastric power (Gianaros et al., 2001; Muth et al., 1999), though not consistently (Riezzo et al., 2009), and findings on disgust (Harrison et al., 2010; Meissner et al., 2011) are similarly mixed. Emotional induction studies typically show no frequency change, with inconsistent amplitude effects (Baldaro et al., 1996; Chen et al., 2005). These discrepancies are likely due to methodological differences in tasks, participant states (e.g., fed/fasted), and a lack of standardization. Thus, while promising, EGG requires careful experimental control.

34.4.4.2 Stomach-Brain Interactions

Beyond measuring descending cognitive influences on the stomach, a major emerging application of EGG is the study of how ascending gastric signals can impact brain and behaviour. This research has revealed that the gastric rhythm contributes to the large-scale organization of resting-state brain activity. The first study to demonstrate this link used magnetoencephalography (MEG), and found that the phase of the gastric rhythm modulates the amplitude of neural oscillations in the alpha band (8–12 Hz) over parieto-occipital cortex (Richter et al., 2017). Follow-up studies using EEG have shown that this coupling is higher during stress than during relaxation (Jeanne et al., 2023), and higher during resting state than during working memory (Balasubramani et al., 2022). Using fMRI, it has been shown that the gastric rhythm is coupled with activity in a widespread cortical network, including interoceptive and exteroceptive regions from all sensory modalities (Choe et al., 2020; Levakov et al., 2023; Müller et al., 2021; Rebollo et al., 2018; Rebollo & Tallon-Baudry, 2022). Recently, the strength of this coupling in fronto-parietal regions has been shown to correlate with negative mental health (Banellis et al., 2025), and the gastric phase has been shown to influence motor evoked potentials, independently of respiratory and cardiac rhythms (Engelen et al., 2025). Together, these findings establish the stomach as a source of clinically and cognitively relevant rhythmic input to the brain, positioning EGG as an essential tool for uncovering the physiological basis of gut-brain communication.

34.5 Conclusion and Outlook

Across this chapter we have moved from the rapid rhythm of the heart, through the intermediate pace of respiration, to the slow cycles of the stomach. Each system shows how bodily oscillations can be recorded, quantified, and related to neural dynamics, demonstrating that peripheral physiology is not background noise but a constitutive element of brain function.

Several lessons stand out. Recording cardiac, respiratory, and gastric activity alongside EEG requires careful coordination: the most reliable strategy is to route all signals through the same amplifier or synchronize them with precise trigger pulses. Proper labelling, cable management, and documentation of trigger alignment are essential, as are sampling rates and recording durations that match the timescale of each rhythm. Just as important is interpreting autonomic–neural coupling within the physiological context of each organ rather than treating it as a generic signal.

Looking ahead, challenges remain in standardizing acquisition protocols, developing robust analytic pipelines, and moving from descriptive correlations towards mechanistic models. Yet the opportunities are substantial. Autonomic rhythms provide unique entry points into interoception, arousal, self-related processes, and mental health. By embedding the brain in its bodily context, multimodal recordings allow cognitive neuroscience to move beyond a brain-centric perspective and towards a richer account of how physiology shapes behaviour and experience.

References

Arakaki, X., Arechavala, R. J., Choy, E. H., Bautista, J., Bliss, B., Molloy, C., Wu, D.-A., Shimojo, S., Jiang, Y., Kleinman, M. T., & Kloner, R. A. (2023). The connection between heart rate variability (HRV), neurological health, and cognition: A literature review. *Frontiers in Neuroscience, 17*. https://doi.org/10.3389/fnins.2023.1055445

Arshamian, A., Iravani, B., Majid, A., & Lundström, J. N. (2018). Respiration modulates olfactory memory consolidation in humans. *The Journal of Neuroscience, 38*, 10286–10294.

Aston-Jones, G., & Cohen, J. D. (2005). An integrative theory of locus coeruleus-norepinephrine function: Adaptive gain and optimal performance. *Annual Review of Neuroscience, 28*, 403–450.

Aufan, M. R., Jost, Z. T., Miller, N. J., Sharifov, O. F., Gupta, H., Perry, G. J., Wells, J. M., Denney, T. S., & Lloyd, S. G. (2023). Electrocardiogram to determine mitral and aortic valve opening and closure. *Cardiovascular Engineering and Technology, 14*, 447–456. https://doi.org/10.1007/s13239-023-00664-4

Azzalini, D., Rebollo, I., & Tallon-Baudry, C. (2019). Visceral signals shape brain dynamics and cognition. *Trends in Cognitive Sciences, 23*(6), 488–509. https://doi.org/10.1016/j.tics.2019.03.007

Babo-Rebelo, M., Richter, C. G., & Tallon-Baudry, C. (2016). Neural responses to heartbeats in the default network encode the self in spontaneous thoughts. *The Journal of Neuroscience, 36*, 7829–7840. https://doi.org/10.1523/JNEUROSCI.0262-16.2016

Balasubramani, P. P., Walke, A., Grennan, G., Perley, A., Purpura, S., Ramanathan, D., Coleman, T. P., & Mishra, J. (2022). Simultaneous gut-brain electrophysiology shows cognition and satiety specific coupling. *Sensors, 22*(23), 9242. https://doi.org/10.3390/s22239242

Baldaro, B., Battacchi, M. W., Codispoti, M., Tuozzi, G., Trombini, G., Bolzani, R., & Palomba, D. (1996). Modifications of electrogastrographic activity during the viewing of brief film sequences. *Perceptual and Motor Skills, 82*(3_suppl), 1243–1250.

Banellis, L., Rebollo, I., Nikolova, N., & Allen, M. (2025). Stomach–brain coupling indexes a dimensional signature of mental health. *Nature Mental Health, 3*(8), 899–908. https://doi.org/10.1038/s44220-025-00468-6

Benarroch, E. E. (2008). The arterial baroreflex. *Neurology, 71*, 1733–1738. https://doi.org/10.1212/01.wnl.0000335246.93495.92

Berntson, G. G., & Khalsa, S. S. (2021). Neural circuits of interoception. *Trends in Neurosciences, 44*(1), 17–28. https://doi.org/10.1016/j.tins.2020.09.011

Birznieks, I., Boonstra, T. W., & Macefield, V. G. (2012). Modulation of human muscle spindle discharge by arterial pulsations—Functional effects and consequences. *PLoS ONE, 7*, e35091. https://doi.org/10.1371/journal.pone.0035091

Bishop, V. S., Malliani, A., & Thorén, P. (2011). Cardiac mechanoreceptors. In *Comprehensive physiology* (pp. 497–555). Wiley. https://doi.org/10.1002/cphy.cp020315

Bolt, T., Wang, S., Nomi, J. S., Setton, R., Gold, B. P. D. B., Frederick, B., Yeo, B. T. T., Chen, J. J., Picchioni, D., Duyn, J. H., Spreng, R. N., Keilholz, S. D., Uddin, L. Q., & Chang, C. (2025). Autonomic physiological coupling of the global fMRI signal. *Nature Neuroscience*, 1–9. https://doi.org/10.1038/s41593-025-01945-y

Bolter, C. P., Turner, M. J., & Barrett, C. J. (2011). The relationship between aortic baroreceptor activity and arterial pressure is not monotonic. *The Journal of Physiological Sciences, 61*, 151–160. https://doi.org/10.1007/s12576-011-0132-4

Brændholt, M., et al. (2025). The respiratory cycle modulates distinct dynamics of affective and perceptual decision-making. *PLoS Computational Biology, 21*, e1013086.

Burma, J. S., Griffiths, J. K., Lapointe, A. P., Oni, I. K., Soroush, A., Carere, J., Smirl, J. D., & Dunn, J. F. (2024). Heart rate variability and pulse rate variability: Do anatomical location and sampling rate matter? *Sensors (Basel), 24*, 2048. https://doi.org/10.3390/s24072048

Candia-Rivera, D., Annen, J., Gosseries, O., Martial, C., Thibaut, A., Laureys, S., & Tallon-Baudry, C. (2021). Neural responses to heartbeats detect residual signs of consciousness during resting state in postcomatose patients. *The Journal of Neuroscience, 41*, 5251–5262. https://doi.org/10.1523/JNEUROSCI.1740-20.2021

Candia-Rivera, D., Vidailhet, M., Chavez, M., & De Vico Fallani, F. (2024). A framework for quantifying the coupling between brain connectivity and heartbeat dynamics: Insights into the disrupted network physiology in Parkinson's disease. *Human Brain Mapping, 45*, e26668. https://doi.org/10.1002/hbm.26668

Canolty, R. T., & Knight, R. T. (2010). The functional role of cross-frequency coupling. *Trends in Cognitive Science (Regular Edition), 14*, 506–515.

Caparco, A., Lopez-Martin, G., & Galvez-Pol, A. (2025). Phase confusion: How inconsistent cardiac labeling obscures interoception research. *Biological Psychology, 199*, 109078. https://doi.org/10.1016/j.biopsycho.2025.109078

Cechetto, D. F. (2014). Cortical control of the autonomic nervous system. *Experimental Physiology, 99*(2), 326–331. https://doi.org/10.1113/expphysiol.2013.075192

Chalmers, T., Maharaj, S., Lees, T., Lin, C. T., Newton, P., Clifton-Bligh, R., McLachlan, C. S., Gustin, S. M., & Lal, S. (2020). Impact of acute stress on cortical electrical activity and cardiac autonomic coupling. *Journal of Integrative Neuroscience, 19*, 239–248. https://doi.org/10.31083/j.jin.2020.02.74

Chen, D. D., Xu, X., Wang, Z., & Chen, J. D. Z. (2005). Alteration of gastric myoelectrical and autonomic activities with audio stimulation in healthy humans. *Scandinavian Journal of Gastroenterology, 40*(7), 814–821. https://doi.org/10.1080/00365520510015656

Choe, A. S., Tang, B., Smith, K. R., Honari, H., Lindquist, M. A., Caffo, B. S., & Pekar, J. J. (2020). Phase-locking of resting-state brain networks with the gastric basal electrical rhythm. *bioRxiv*, 2020.10.06.328054. https://doi.org/10.1101/2020.10.06.328054

Clayton, D. G., Webb, R. K., Ralston, A. C., Duthie, D., & Runciman, W. B. (1991). Pulse oximeter probes A comparison between finger, nose, ear and forehead probes under conditions of poor perfusion. *Anaesthesia, 46*, 260–265. https://doi.org/10.1111/j.1365-2044.1991.tb11492.x

Cretikos, M. A., Bellomo, R., Hillman, K., Chen, J., Finfer, S., & Flabouris, A. (2008). Respiratory rate: The neglected vital sign. *Medical Journal of Australia, 188*(11), 657–659.

Danev, S. G., & de Winter, C. R. (1971). Heart rate deceleration after erroneous responses. *Psychologische Forschung, 35*, 27–34. https://doi.org/10.1007/BF00424472

Del Negro, C. A., Funk, G. D., & Feldman, J. L. (2018). Breathing matters. *Nature Reviews. Neuroscience, 19*, 351–367.

Electrophysiology, T.F. of the E.S. of C. the N.A.S. of P. (1996). Heart rate variability. *Circulation, 93*, 1043–1065. https://doi.org/10.1161/01.CIR.93.5.1043

Engelen, T., Solcà, M., & Tallon-Baudry, C. (2023). Interoceptive rhythms in the brain. *Nature Neuroscience, 26*(10), 1670–1684. https://doi.org/10.1038/s41593-023-01425-1

Engelen, T., Schuhmann, T., Sack, A. T., & Tallon-Baudry, C. (2025). The cardiac, respiratory, and gastric rhythms independently modulate motor corticospinal excitability in humans (p. 2024.09.10.612221). *bioRxiv*. https://doi.org/10.1101/2024.09.10.612221

Feinle-Bisset, C., O'Donovan, D., & Horowitz, M. (2004). Gastric motility. In L. R. Johnson (Ed.), *Encyclopedia of gastroenterology* (pp. 132–138). Elsevier. https://doi.org/10.1016/B0-12-386860-2/00301-X

Feldman, J. L., Mitchell, G. S., & Nattie, E. E. (2003). Breathing: Rhythmicity, plasticity, chemosensitivity. *Annual Review of Neuroscience, 26*, 239–266.

Fleming, S., et al. (2011). Normal ranges of heart rate and respiratory rate in children from birth to 18 years of age: A systematic review of observational studies. *Lancet, 377*, 1011–1018.

Garfinkel, S. N., Minati, L., Gray, M. A., Seth, A. K., Dolan, R. J., & Critchley, H. D. (2014). Fear from the heart: Sensitivity to fear stimuli depends on individual heartbeats. *The Journal of Neuroscience, 34*, 6573–6582. https://doi.org/10.1523/JNEUROSCI.3507-13.2014

Gharibans, A. A., Kim, S., Kunkel, D. C., & Coleman, T. P. (2017). High-resolution electrogastrogram: A novel, noninvasive method for determining gastric slow-wave direction and speed. *IEEE Transactions on Biomedical Engineering, 64*(4), 807–815. https://doi.org/10.1109/TBME.2016.2579310

Gianaros, P. J., Quigley, K. S., Mordkoff, J. T., & Stern, R. M. (2001). Gastric myoelectrical and autonomic cardiac reactivity to laboratory stressors. *Psychophysiology, 38*(4), 642–652.

Goldberger, J. J., Challapalli, S., Tung, R., Parker, M. A., & Kadish, A. H. (2001). Relationship of heart rate variability to parasympathetic effect. *Circulation, 103*, 1977–1983.

Gourine, A. V., Llaudet, E., Dale, N., & Spyer, K. M. (2005). ATP is a mediator of chemosensory transduction in the central nervous system. *Nature, 436*, 108–111.

Grund, M., Al, E., Pabst, M., Dabbagh, A., Stephani, T., Nierhaus, T., Gaebler, M., & Villringer, A. (2022). Respiration, heartbeat, and conscious tactile perception. *The Journal of Neuroscience, 42*, 643–656. https://doi.org/10.1523/JNEUROSCI.0592-21.2021

Guz, A. (1997). Brain, breathing and breathlessness. *Respiration Physiology, 109*(3), 197–204.

Harrison, N. A., Gray, M. A., Gianaros, P. J., & Critchley, H. D. (2010). The embodiment of emotional feelings in the brain. *Journal of Neuroscience, 30*(38), 12878–12884. https://doi.org/10.1523/JNEUROSCI.1725-10.2010

Ito, H., Kanno, I., Ibaraki, M., Hatazawa, J., & Miura, S. (2003). Changes in human cerebral blood flow and cerebral blood volume during hypercapnia and hypocapnia measured by positron emission tomography. *Journal of Cerebral Blood Flow and Metabolism, 23*, 665–670.

Ito, J., et al. (2014). Whisker barrel cortex delta oscillations and gamma power in the awake mouse are linked to respiration. *Nature Communications, 5*, 3572.

Jammal Salameh, L., Bitzenhofer, S. H., Hanganu-Opatz, I. L., Dutschmann, M., & Egger, V. (2024). Blood pressure pulsations modulate central neuronal activity via mechanosensitive ion channels. *Science, 383*, eadk8511. https://doi.org/10.1126/science.adk8511

Jeanne, R., Piton, T., Minjoz, S., Bassan, N., Le Chenechal, M., Semblat, A., Hot, P., Kibleur, A., & Pellissier, S. (2023). Gut-brain coupling and multilevel physiological response to biofeedback relaxation after a stressful task under virtual reality immersion: A pilot study. *Applied Psychophysiology and Biofeedback, 48*(1), 109–125. https://doi.org/10.1007/s10484-022-09566-y

Jowett, N., Turner, A., Cole, A., & Jones, P. (2005). Modified electrode placement must be recorded when performing 12-lead electrocardiograms. *Postgraduate Medical Journal, 81*, 122–125. https://doi.org/10.1136/pgmj.2004.021204

Karjalainen, S., Kujala, J., & Parviainen, T. (2025). Neural activity is modulated by spontaneous and volitionally controlled breathing. *Biological Psychology, 197*, 109026.

Kayser, B. (2003). Exercise starts and ends in the brain. *European Journal of Applied Physiology, 90*(3), 411–419.

Kim, K. J., Ramiro Diaz, J., Iddings, J. A., & Filosa, J. A. (2016). Vasculo-neuronal coupling: Retrograde vascular communication to brain neurons. *The Journal of Neuroscience, 36*, 12624–12639. https://doi.org/10.1523/JNEUROSCI.1300-16.2016

Kim, M., Heo, G., & Kim, S.-Y. (2022). Neural signalling of gut mechanosensation in ingestive and digestive processes. *Nature Reviews Neuroscience, 23*(3), 135–156. https://doi.org/10.1038/s41583-021-00544-7

Kluger, D. S., & Gross, J. (2020). Depth and phase of respiration modulate cortico-muscular communication. *NeuroImage, 222*, 117272.

Kluger, D. S., & Gross, J. (2021). Respiration modulates oscillatory neural network activity at rest. *PLoS Biology, 19*, e3001457.

Kluger, D. S., Balestrieri, E., Busch, N. A., & Gross, J. (2021). Respiration aligns perception with neural excitability. *eLife, 10.*

Kluger, D. S., et al. (2023). Modulatory dynamics of periodic and aperiodic activity in respiration-brain coupling. *Nature Communications, 14*, 4699.

Kluger, D. S., Gross, J., & Keitel, C. (2024). A dynamic link between respiration and arousal. *The Journal of Neuroscience, 44.*

Koch, K. L., & Stern, R. M. (2004). *Handbook of electrogastrography*. Oxford University Press.

Kyriacou, P. A., & Allen, J. (Eds.). (2021). *Photoplethysmography: Technology, signal analysis and applications*. Academic.

Levakov, G., Ganor, S., & Avidan, G. (2023). Reliability and validity of brain-gastric phase synchronization. *Human Brain Mapping, 44*(14), 4956–4966. https://doi.org/10.1002/hbm.26436

Loescher, M., Clément, A., & Tallon-Baudry, C. (in press). The neurophysiology of subjectivity. In L. Melloni & U. Olcese (Eds.), *The scientific study of consciousness–experimental and theoretical approaches*. Springer Nature.

Loescher, M., Haggard, P., & Tallon-Baudry, C. (2025). Interoception vs. Exteroception: Cardiac interoception competes with tactile perception, yet also facilitates self-relevance encoding. Proceedings of the National Academy of Sciences, 122(49), e2516229122. https://doi.org/10.1073/pnas.2516229122

Lundström, J. N., Boesveldt, S., & Albrecht, J. (2011). Central processing of the chemical senses: An overview. *ACS Chemical Neuroscience, 2*, 5–16.

Mason, R. E., & Likar, I. (1966). A new system of multiple-lead exercise electrocardiography. *American Heart Journal, 71*, 196–205. https://doi.org/10.1016/0002-8703(66)90182-7

McKay, L. C., Evans, K. C., Frackowiak, R. S. J., & Corfield, D. R. (2003). Neural correlates of voluntary breathing in humans. *Journal of Applied Physiology, 95*, 1170–1178.

Meissner, K., Muth, E. R., & Herbert, B. M. (2011). Bradygastric activity of the stomach predicts disgust sensitivity and perceived disgust intensity. *Biological Psychology, 86*(1), 9–16. https://doi.org/10.1016/j.biopsycho.2010.09.014

Menuet, C., et al. (2025). Redefining respiratory sinus arrhythmia as respiratory heart rate variability: An international Expert Recommendation for terminological clarity. *Nature Reviews. Cardiology*. https://doi.org/10.1038/s41569-025-01160-z

Mirvis, D. M., & Goldberger, A. L. (2012). Electrocardiography. In R. O. Bonow, D. L. Mann, G. F. Tomaselli, D. L. Bhatt, S. D. Solomon, & E. Braunwald (Eds.), *Braunwald's heart disease: A textbook of cardiovascular medicine* (pp. 126–167). Elsevier.

Moore, J. D., et al. (2013). Hierarchy of orofacial rhythms revealed through whisking and breathing. *Nature, 497*, 205–210.

Müller, S., Teckentrup, V., Rebollo, I., Hallschmid, M., & Kroemer, N. B. (2021). Vagus nerve stimulation increases stomach-brain coupling via a vagal afferent pathway. 2021.10.07.463517. https://doi.org/10.1101/2021.10.07.463517

Muth, E. R., Koch, K. L., Stern, R. M., & Thayer, J. F. (1999). Effect of autonomic nervous system manipulations on gastric myoelectrical activity and emotional responses in healthy human subjects. *Psychosomatic Medicine, 61*(3), 297.

Noble, D. J., & Hochman, S. (2019). Hypothesis: Pulmonary afferent activity patterns during slow, deep breathing contribute to the neural induction of physiological relaxation. *Frontiers in Physiology, 10*, 1176.

O'Grady, G., Du, P., Cheng, L. K., Egbuji, J. U., Lammers, W. J. E. P., Windsor, J. A., & Pullan, A. J. (2010). Origin and propagation of human gastric slow-wave activity defined by high-resolution mapping. *American Journal of Physiology. Gastrointestinal and Liver Physiology, 299*(3), G585–G592. https://doi.org/10.1152/ajpgi.00125.2010

Oczka, D., Augustynek, M., Penhaker, M., & Kubicek, J. (2024). Electrogastrography measurement systems and analysis methods used in clinical practice and research: Comprehensive review. *Frontiers in Medicine, 11*, 1369753. https://doi.org/10.3389/fmed.2024.1369753

Park, H.-D., & Tallon-Baudry, C. (2014). The neural subjective frame: From bodily signals to perceptual consciousness. *Philosophical Transactions of the Royal Society, B: Biological Sciences, 369*, 20130208. https://doi.org/10.1098/rstb.2013.0208

Park, H.-D., Correia, S., Ducorps, A., & Tallon-Baudry, C. (2014). Spontaneous fluctuations in neural responses to heartbeats predict visual detection. *Nature Neuroscience, 17*, 612–618. https://doi.org/10.1038/nn.3671

Park, H.-D., et al. (2020). Breathing is coupled with voluntary action and the cortical readiness potential. *Nature Communications, 11*, 289.

Perl, O., et al. (2019). Human non-olfactory cognition phase-locked with inhalation. *Nature Human Behaviour, 3*, 501–512.

Petzschner, F. H., Weber, L. A., Wellstein, K. V., Paolini, G., Do, C. T., & Stephan, K. E. (2019). Focus of attention modulates the heartbeat evoked potential. *NeuroImage, 186*, 595–606. https://doi.org/10.1016/j.neuroimage.2018.11.037

Quadt, L., Critchley, H., & Nagai, Y. (2022). Cognition, emotion, and the central autonomic network. *Autonomic Neuroscience, 238*, 102948. https://doi.org/10.1016/j.autneu.2022.102948

Rebollo, I., & Tallon-Baudry, C. (2022). The sensory and motor components of the cortical hierarchy are coupled to the rhythm of the stomach during rest. *Journal of Neuroscience*. https://doi.org/10.1523/JNEUROSCI.1285-21.2021

Rebollo, I., Devauchelle, A.-D., Béranger, B., & Tallon-Baudry, C. (2018). Stomach-brain synchrony reveals a novel, delayed-connectivity resting-state network in humans. *eLife, 7*, e33321. https://doi.org/10.7554/eLife.33321

Richter, C. G., Babo-Rebelo, M., Schwartz, D., & Tallon-Baudry, C. (2017). Phase-amplitude coupling at the organism level: The amplitude of spontaneous alpha rhythm fluctuations varies with the phase of the infra-slow gastric basal rhythm. *NeuroImage, 146*, 951–958. https://doi.org/10.1016/j.neuroimage.2016.08.043

Riezzo, G., Indrio, F., Raimondi, F., Montagna, O., Salvia, G., Massimo, B., Polimeno, L., Cavallo, L., & Francavilla, R. (2009). Maturation of gastric electrical activity, gastric emptying and intestinal permeability in preterm newborns during the first month of life. *Italian Journal of Pediatrics, 35*, 6. https://doi.org/10.1186/1824-7288-35-6

Sanders, K. M., Koh, S. D., & Ward, S. M. (2006). Interstitial cells of cajal as pacemakers in the gastrointestinal tract. *Annual Review of Physiology, 68*(1), 307–343. https://doi.org/10.1146/annurev.physiol.68.040504.094718

Saper, C. B. (2002). The central autonomic nervous system: Conscious visceral perception and autonomic pattern generation. *Annual Review of Neuroscience, 25*, 433–469. https://doi.org/10.1146/annurev.neuro.25.032502.111311

Sargent, K. S., Martinez, E. L., Reed, A. C., Guha, A., Bartholomew, M. E., Diehl, C. K., Chang, C. S., Salama, S., Popov, T., Thayer, J. F., Miller, G. A., & Yee, C. M. (2024). Oscillatory coupling between neural and cardiac rhythms. *Psychological Science, 35*, 517–528. https://doi.org/10.1177/09567976241235932

Sargent, K., Martinez, E., Reed, A., Guha, A., Bartholomew, M., Diehl, C., Chang, C., Salama, S., Subotnik, K., Ventura, J., Nuechterlein, K., Miller, G., & Yee, C. (2025). Brain-body dysconnectivity: Deficient autonomic regulation of cortical function in first-episode schizophrenia. *Psychological Medicine, 55*, e1. https://doi.org/10.1017/S0033291724003428

Sattar, Y., & Chhabra, L. (2025). Electrocardiogram. In *StatPearls*. StatPearls Publishing.

Shaffer, F., McCraty, R., & Zerr, C. L. (2014). A healthy heart is not a metronome: An integrative review of the heart's anatomy and heart rate variability. *Frontiers in Psychology, 5*. https://doi.org/10.3389/fpsyg.2014.01040

Shea, S. A. (1996). Behavioural and arousal-related influences on breathing in humans. *Experimental Physiology: Translation and Integration, 81*(1), 1–26.

Simonian, H. P., Panganamamula, K., Parkman, H. P., Xu, X., Chen, J. Z., Lindberg, G., Xu, H., Shao, C., Ke, M.-Y., Lykke, M., Hansen, P., Barner, B., & Buhl, H. (2004). Multichannel electrogastrography (EGG) in normal subjects: A multicenter study. *Digestive Diseases and Sciences, 49*(4), 594–601. https://doi.org/10.1023/B:DDAS.0000026304.83214.50

Skora, L. I., Livermore, J. J. A., & Roelofs, K. (2022). The functional role of cardiac activity in perception and action. *Neuroscience & Biobehavioral Reviews, 137*, 104655. https://doi.org/10.1016/j.neubiorev.2022.104655

Sroufe, L. A. (1971). Effects of depth and rate of breathing on heart rate and heart rate variability. *Psychophysiology, 8*(5), 648–655.

Steinfath, P., Herzog, N., Fourcade, A., Sander, C., Nikulin, V., & Villringer, A. (in press). Validating genuine changes in Heartbeat evoked potentials using pseudotrials and surrogate procedures. https://doi.org/10.1101/2024.08.24.609348

Suzuki, N., Prosser, C. L., & Dahms, V. (1986). Boundary cells between longitudinal and circular layers: Essential for electrical slow waves in cat intestine. *American Journal of Physiology. Gastrointestinal and Liver Physiology, 250*(3), G287–G294.

Valenza, G., Sclocco, R., Duggento, A., Passamonti, L., Napadow, V., Barbieri, R., & Toschi, N. (2019). The central autonomic network at rest: Uncovering functional MRI correlates of time-varying autonomic outflow. *NeuroImage, 197*, 383–390. https://doi.org/10.1016/j.neuroimage.2019.04.075

Van Diest, I., Thayer, J. F., Vandeputte, B., Van de Woestijne, K. P., & Van den Bergh, O. (2006). Anxiety and respiratory variability. *Physiology & Behavior, 89*(2), 189–195.

Wang, C.-A., Baird, T., Huang, J., Coutinho, J. D., Brien, D. C., & Munoz, D. P. (2018). Arousal effects on pupil size, heart rate, and skin conductance in an emotional face task. *Frontiers in Neurology, 9*. https://doi.org/10.3389/fneur.2018.01029

Wolpert, N., Rebollo, I., & Tallon-Baudry, C. (2020). Electrogastrography for psychophysiological research: Practical considerations, analysis pipeline, and normative data in a large sample. *Psychophysiology*, e13599. https://doi.org/10.1111/psyp.13599

Yackle, K., et al. (2017). Breathing control center neurons that promote arousal in mice. *Science, 355*, 1411–1415.

Zaccaro, A., Piarulli, A., Laurino, M., Garbella, E., Menicucci, D., Neri, B., & Gemignani, A. (2018). How breath-control can change your life: A systematic review on psycho-physiological correlates of slow breathing. *Frontiers in Human Neuroscience, 12*, 409421.

Zelano, C., et al. (2016). Nasal respiration entrains human limbic oscillations and modulates cognitive function. *The Journal of Neuroscience, 36*, 12448–12467.

Part VII
Writing and Reading EEG Research Papers

Chapter 35
Writing Up Your EEG Research

Shivakumar Viswanathan and Tracy Warbrick

Abstract An important end goal for many EEG research studies is publication in a peer-reviewed scientific article. Therefore, learning to write up your research for publication is an important skill. In this chapter, we provide advice on what to include in your manuscript and why. Deciding what to include is a particular concern for EEG studies, which can include a large amount of technical information. Here, we consider an article as a document with diverse readers with differing expectations and priorities. Based on this view, we provide strategies and resources by considering the key readers of an article and their expectations related to structuring, scientific content, and reproducibility.

Keywords EEG · Scientific writing · Scientific communication · Journal publication · Reproducibility · Peer review

35.1 Introduction

The scientific article is a major outcome of a research study. It is a formal document that scientists use to describe why and how a study was conducted and, importantly, the study's findings. To make it a reliable source of information, this document is published only after review by other scientists (i.e., peers) to ensure that it meets specific standards related to content, accuracy, and ethics. The published article is the official channel to make a study's findings available to other scientists and the general public. It is also how scientific findings enter into the record of scientific knowledge (or the "literature") and become a citable resource for other studies. Due to its importance, writing up research findings for publication is an integral part of doing scientific research and a key professional demand for researchers.

In this chapter, our focus is on researchers starting out on their publication journey to help them answer a common question: *What do I need to report in my*

S. Viswanathan (✉) · T. Warbrick
Brain Products GmbH, Gilching, Germany
e-mail: shivakumar.viswanathan@brainproducts.com

T. Warbrick (ed.), *The EEG Handbook*,
https://doi.org/10.1007/978-3-032-20450-9_35

EEG paper? Our emphasis will be on the Methods and Results sections. Writing to communicate your findings also needs you to place your work in the context of related work in the literature. Therefore, reading and evaluating scientific papers is an important complementary skill, and is covered in Chap. 36. We recommend that you read both chapters.

35.2 Structure Your Article to Manage Complexity

Writing up an EEG study involves reporting a large volume of technical information. A useful way to manage this complexity is by appropriately structuring your manuscript.

The modern scientific article is a highly structured document with a predictable organization (see Table 35.1). This structured format is relatively recent. Previously, articles were published in scientific journals with the format of a letter or essay. For example, Shepard and Metzler's classic article (Shepard & Metzler, 1971), titled *Mental rotation of three-dimensional objects*, was published in the journal *Science* as a continuous 2.5-page article without sections or references. This "letter" format is intended to inform a general scientific reader about high-profile findings that might be outside the reader's specialty. Therefore, the detailed specifics of how the scientific study was conducted are kept to a minimum (or reported as appendices to the main article). This letter format continues to be used by general journals like *Nature* and *Science*, where a journal issue might have articles from a wide variety of fields, from neuroscience to astrophysics to molecular biology.

Unlike general journals, discipline-specific journals have a narrower scope that is often evident from the journal's title (e.g., *Journal of Cognitive Neuroscience*, *Clinical Neurophysiology*). Articles in these journals also aim to inform readers about new findings but these readers are assumed to have discipline-specific interests and expertise. Therefore, there is a higher emphasis on reporting the specifics of how the research findings were obtained. To accommodate the increased detail, these journals require articles to have defined sections and sub-sections with specific expectations about the information content and presentation style for each section. The usual sections and their expected content are listed in Table 35.1.

The use of predictable sections and sub-sections helps the reader to find relevant information easily and reliably. When information is available as expected, reading can be perceived as being easy (Gopen & Swan, 1990). This is especially relevant for a scientific article, which typically reports complex technical context.

Structuring is not just helpful for the reader. You can leverage it to manage complexity when writing. Planning your writing in terms of sections and sub-sections can allow you to categorize information by (sub)section and decide what needs to be said about this information in each section. Therefore, it can be valuable to familiarize yourself with how articles are structured in your field.

While reading articles in your field, pay close attention to how they structure the presented information (also see Chap. 36). Furthermore, journal websites provide

Table 35.1 Typical sections of a scientific article

Section	Description
Title	Sentence fragment describing the content of the article. It is typically read first by a reader and can influence whether they choose to read the full article. This first encounter might be in the results of a web-search, in the reference section of another article, or table of contents of a journal. The form of the title is often discipline-specific but continues to evolve (Haggan, 2004; Pottier et al., 2024; Ball, 2009; Soler, 2007)
Abstract	Short paragraph (~150–500 words) that follows the title and summarizes the article's content. Like the title, the abstract is intended to help readers decide whether to read the article. The structuring and information content of an abstract can be discipline-specific (Haynes, 2017; Nakayama et al., 2005)
Introduction	Article's first section describing the scientific question, its background context, and the aims of the study. Its length and focus can differ between disciplines. The Introduction (where the question is asked) and Discussion section (where the question is answered) are the start and end points respectively of an article's narrative or "story"
Materials & Methods	Dedicated section that describes all technical procedures and details involved in conducting the study. It can have dedicated sub-sections to describe the population and participants, the apparatus/hardware/software used, experiment design and protocol, EEG acquisition and analysis details, statistical procedures, and so on. This section usually follows the introduction. However, general journals can place it at the end of the article to deemphasize its relevance for the non-expert scientific reader
Results	Dedicated section to report the quantitative findings of the study (i.e., based on procedures described in the Materials & Methods). The findings usually take the form of observations and deductions supported by numerical data summaries, statistical tests, and data visualizations
Discussion	Closing section that links the findings from the Results to the scientific hypothesis and aims from the Introduction section. It can include evaluations of how the findings relate to current state of knowledge in the field, and propose explanations/models. This section is often followed by a brief Conclusion section
References	A formatted full listing of all articles cited in the main article as well as information about materials (equipment, software, etc.) used to conduct the study

instructions for authors on the expected structure of manuscripts for submission. Consult these author instructions for the main journals in your field as they can guide your writing even before you decide where to submit your manuscript. A useful starting point is the widely used style guide for journals published by the American Psychological Association (2020), which provides detailed structuring instructions for psychology studies.

35.3 Anticipate the Concerns of Critical Expert Readers

Articles in general journals and discipline-specific journals are structured on differing assumptions about the target reader, for example, whether this reader is an expert from the same discipline or not. Apart from this target reader, there are two types of readers of high relevance to an article: (1) journal editors and (2) reviewers. These two readers are involved in the *peer review* process (discussed below). They see the submitted manuscript and influence whether it will be published in a particular journal. Anticipating and addressing the expectations of these two readers can be a valuable guide for your writing, even before submitting your manuscript for review.

Before publication, a manuscript submitted to a journal undergoes a review process by other scientists (or peers) (Drozdz & Ladomery, 2024; Benos et al., 2007). Although there are different models for peer review, a schematic of a typical model is shown in Fig. 35.1. This peer review process is intended to ensure that a published article meets the specified community standards related to content, accuracy, and ethics.

Editors and reviewers have different roles in the peer review process.

Editor The *editor* serves as a representative of the journal and is responsible for all decisions related to the manuscript, for example, should it be sent to review or not, should it be accepted for publication or not. These decisions are based on scientific

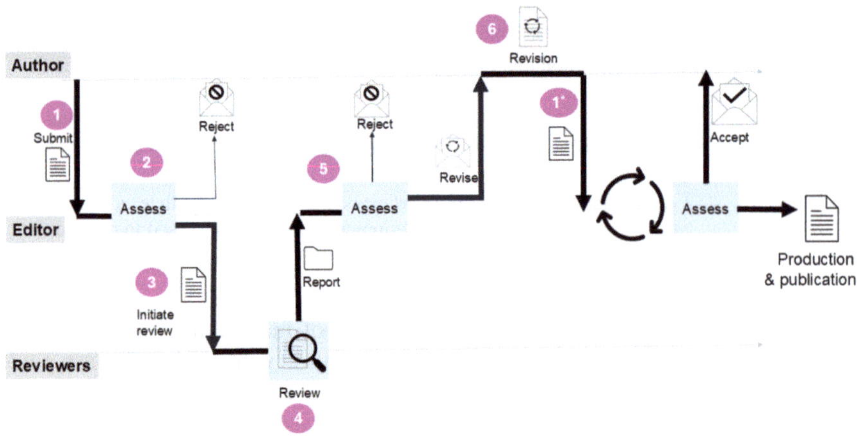

Fig. 35.1 Timeline of a typical peer review process. Authors (upper timeline) submit the manuscript (1). The editor (middle timeline) assesses the manuscript to determine suitability for the journal (2). The manuscript can be rejected at this stage or sent onward to reviewers (3). The reviewers (bottom timeline) evaluate the manuscript (4) and send their recommendations to the editor. Based on the recommendations, the editor can reject, accept (not shown), or return to the authors for revisions (thick line) (5). Authors revise the manuscript based on reviewer suggestions (6) and resubmit it (1*). The process of evaluation/review/revision is iterative and proceeds until the manuscript is either accepted or rejected. The accepted manuscript is sent for production and publication

merits as well as other journal-specific considerations (e.g., suitability for the journal's target audience, expected impact on the field, etc.). To help with this decision, editors invite expert reviewers (usually 2 or more) to evaluate a manuscript's suitability for publication.

Reviewer The *reviewer* is ideally a scientist with expertise in the topic of the manuscript. They act as representatives of the target scientific reader (rather than the journal). Their task is to evaluate a manuscript's scientific content and relevance based on different criteria provided by the editor (see below). Reviewers provide a written assessment of the manuscript that usually includes the reviewer's concerns, requests for clarification from the authors, and suggested revisions to improve the manuscript.

The editor uses the assessments from all reviewers to decide if manuscript should be accepted, rejected, or returned to the authors for revision. Irrespective of the editor's final decision, the reviewer assessments are shared with the authors as it provides them with important critical feedback on the study's scientific content. This feedback and revisions to address them can influence the content, and readability of the published article.

Even though the above steps happen after submission, anticipating potential reviewer concerns can be a valuable guide for writing even before submission. Reviewers and editors consider a variety of criteria related to an article's logic, accuracy and completeness of technical details, relevance to current knowledge, writing clarity, and so on. In recent years, these criteria have been coded into convenient checklists for reviewers to help reduce bias and arbitrariness. Song et al. (2021) identified over 1800 criteria from reviewer checklists across multiple publications and journal websites and found that a large percentage of these criteria focused on three sections related to an article's core scientific content, namely, Methods (30.3%), Results (12.8%), and Discussion (10.5%).

Although these checklists are intended for reviewers, you can apply them to your own draft to anticipate how a critical reader might evaluate your article. This can be a valuable guide for your writing. You can find examples of reviewer checklists on the websites of the main journal publishers in your field, for example, Springer Nature (https://www.springernature.com/gp/reviewers) and Wiley (https://authorservices.wiley.com/Reviewers/journal-reviewers/index.html).

35.4 Deciding What to Include in the Methods Section

A scientific article describes research findings, as well as *how* they were obtained. Although the specifics can differ from study to study, when writing your Methods section include sufficient information to (1) demonstrate scientific validity and (2) ensure reproducibility.

To illustrate the difference, consider how you would describe the procedure to bake a cake. One aspect of the how-to procedure has to do with the material and the actions involved, for example, the ingredients used (e.g., flour, sugar), the equipment needed (e.g., oven), and the sequence of actions to transform the ingredients into a cake. This procedure would be *reproducible* if another person could follow it to bake the cake successfully. However, another aspect of this how-to is the rationale of the procedure, namely, why certain ingredients were used (rather than others), why the steps of the procedure are performed in a certain sequence (rather than another), and so on. The knowledge and considerations behind these reasons help to demonstrate the *validity* of the how-to procedure as a way to bake a cake.

35.4.1 Demonstrate Scientific Validity

The article should provide sufficient information to demonstrate the *scientific validity* of the methods, findings, and statistical inferences (Abelson & McBurney, 2001).

Experimental EEG studies often involve extended chains of theoretical and numerical reasoning to connect together the scientific hypotheses, the experimental details, and the statistical findings. This reasoning depends on multiple interdependent numerical procedures, parameter choices, and assumptions. This complexity carries the unavoidable possibility of errors. Validity here refers to the accuracy of the scientific reasoning in a study and is not about the findings themselves.

Readers have to assess a study's validity from what is presented in the article. This carries the risk of misunderstanding by readers (Gopen & Swan, 1990). Misunderstandings can have many general consequences. A misunderstanding related to one part of the article can cause confusion about the key findings. Furthermore, readers might over/underestimate the importance of the reported findings. However, an important concern is that misunderstandings can affect the perceived validity of the study. For example, readers might mistakenly perceive reasoning errors or false assumptions. The cost of misunderstanding about scientific validity can be high for authors especially during peer review (e.g., rejection, demands for tedious revisions).

Striving to "show your work" can help reduce misunderstandings about validity.

One way to demonstrate scientific validity is by providing principled justifications of method and parameter choices, namely, *why* certain settings were used rather than others (Picton et al., 2000; Keil et al., 2014; Luck & Gaspelin, 2017). These justifications might rely on analytical considerations specific to the experiment. Common conventions and prior studies in your field can be another source of justification. However, beware of blindly using defaults from the literature as studies can differ in many ways, and prior studies can share the same errors.

Crucially, avoid circular justifications where a parameter setting is justified based on the eventual results, that is, setting X was chosen because it produced "good results" (Kriegeskorte et al., 2009; Luck & Gaspelin, 2017; Stefan & Schonbrodt, 2023). To illustrate this circularity, consider an experiment conducted to test if a certain EEG measure is greater in experimental condition A than in B (i.e., A > B).

Suppose A > B is supported (i.e., low p-value) for one parameter setting but not for another equally plausible setting. Here, it would be circular reasoning to use the p-value to decide which outcome (A > B versus not) to report as being "true". This form of circular justification severely effects the scientific validity and inferential reproducibility (Goodman et al., 2016) no matter how compelling the findings might be.

Recommended Resources The following consensus articles provide detailed guidelines on reporting validity-relevant information in general and for EEG studies, with useful checklists.

- Picton et al. (2000): Guidelines for using human event-related potentials to study cognition: Recording standards and publication criteria
- Keil et al. (2014): Publication guidelines and recommendations for studies using electroencephalography and magnetoencephalography
- Keil et al. (2022): Recommendations and publication guidelines for studies using frequency domain and time-frequency domain analyses of neural time series
- Babiloni et al. (2020): Recommendations on frequency and topographic analysis of Resting State EEG rhythms
- Wilkinson (1999): Statistical methods in psychology journals: Guidelines and explanations

35.4.2 Ensure Reproducibility

The article should provide sufficient information for the study to be *reproducible*. Goodman et al. (2016) identify three levels of reproducibility of increasing importance (Goodman et al., 2016; Plesser, 2017):

- *Methods reproducibility:* There is sufficient information for the described methods to be exactly repeated in an independent study (e.g., by researchers in a different lab).
- *Results reproducibility:* The same (or highly similar) results can be obtained by closely reproducing the methods in an independent study.
- *Inferential reproducibility:* The methods and results reproduction lead to qualitatively similar conclusions as the original study (i.e., the results when reproduced do not allow diverse interpretations).

Methods reproducibility is a minimum requirement for the other forms of reproducibility.

As described earlier, remember that scientific validity and reproducibility are related but distinct concepts. A study can be scientifically *invalid* but still be reproducible. For example, a manuscript might report that EEG data acquisition used a high-pass filter to exclude frequencies below 2 Hz. However, this filter setting might mistakenly exclude crucial frequencies required to test the hypothesis of interest. Nonetheless, the methods of the study might be reported with sufficient detail to be independently reproduced (and thus potentially help detect the error). Conversely, even if a scientific study were valid, its validity cannot be independently verified without reproducible methods.

Recommended Resources The following articles provide detailed parameter-by-parameter guides on reproducible reporting in general and for EEG studies, also with useful checklists.

- Pernet et al. (2020): Issues and recommendations from the OHBM COBIDAS MEEG committee for reproducible EEG and MEG research
- Lang and Altman (2015): Statistical Analyses and Methods in the Published Literature (SAMPL guidelines)

35.5 Advice on Presenting Results

Unlike the standardized structure of the Methods section, the Results section does not have a typical format since every study has its own unique findings. Nevertheless, structuring the Results section is important especially since it reports novel information obtained from the study. Here are some useful elements to consider when reporting your findings.

Are the Experimental Assumptions Satisfied?
EEG experiments usually study participants when they are in well-defined states (e.g., rest but not asleep). These states can be produced by direct instruction (e.g., rest) or when participants perform particular tasks. Therefore, demonstrating that these experimental manipulations of the participant's state worked as intended is a crucial pre-condition for the EEG results.

For example, in a Go/NoGo task, participants are presented with a sensory stimulus on each trial with instructions to execute a behavioral response (e.g., press a button) only for certain stimuli (i.e., Go) but not for other stimuli (i.e., NoGo). This task is designed to induce a preference to respond so that NoGo stimuli require an effortful inhibition of the response (Verbruggen et al., 2019). Confirming that these target cognitive states were achieved is critical *before* comparing EEG across the Go and NoGo conditions. This confirmation can, for example, be readily obtained from the behavioral data.

To avoid circularity, where possible report independent measures to confirm that the key experimental assumptions were satisfied.

Does the Data Have the Required Quality and Characteristics for the Study?
A key next step is to establish whether the EEG data has adequate quality to answer the question of interest (this is sometimes referred to as face validity). For example, in experiments using visual stimuli, a basic test of face validity might involve

demonstrating that the EEG data contains expected visual-evoked sensory responses. This might not be critical for the key conditions of interest but provides a useful quality check. If the classical visual-evoked responses were *not* detectable, it would raise concerns about the suitability of the data for the tests of interest. Reporting these validity tests increases confidence in the experimental study and results.

What Are the Results from Pre-planned Tests?
A planned study is often focused on answering specific questions. However, it is often necessary to report other additional results, e.g., from post-hoc tests to clarify unexpected outcomes. Distinguishing between the results of planned versus unplanned tests and their relationship can be valuable to help the reader distinguish between primary and secondary findings.

One approach to ensure that the Results maintain clarity is to use an "assertion-evidence" strategy. With this strategy, each sub-section heading describes/asserts a result (e.g., "Larger lateralization of beta-power following Go trials than NoGo trials"). The text of the sub-section then provides the results/evidence to support the assertion in the heading as well as any relevant post-hoc tests. With this strategy, a reader can follow the logic of the results by reading the sub-section headings with clear expectations of what the text would contain. By contrast, when the results are organized by topic headings (e.g., "Behavioral results", "ERP results", etc.), these content expectations are weaker and carry the risk of being less focused.

The Role of Narrative
The Results section is expected to report the quantitative findings (what I found) but not to discuss or interpret these findings (what does it mean). However, this expectation is often understood as a rule. However, this is not a rule against *describing* the data with narrative.

The narrative in the Results section can have multiple roles. For example, when presenting a complex data plot, it can be valuable to describe the key features of relevance in the plot, for example, that a particular variable X is increasing. These "qualitative" aspects might not be obvious from the outputs of a statistical test. Such narrative descriptions enable the statistical findings to be concretely understood and help to link individual results to other results that came immediately before and that will follow. Also see Chap. 36 for details on what information to include in plots. Another common misunderstanding is to describe parameters from a statistical test (e.g., p-values) without describing the associated features of the data and the inferences (what was different from what and by how much). Additionally, it can be valuable in the Results to remind readers of why something is being tested in relation to details presented in the Methods.

35.6 Conclusion

Despite the complexities along the way, seeing your research eventually published can be a motivating achievement. It is also how your research contributes to advancing scientific knowledge. Therefore, it is important to continuously develop your writing skills. The continuing efforts to demonstrate validity and ensure reproducibility in your scientific reporting can help you to improve how you design and conduct your research studies. These efforts, one article at a time, contribute to a valid and reproducible scientific literature that serves the community.

References

American Psychological Association. (2020). *Publication manual of the American Psychological Association: The official guide to APA style* (7th Ed.).

Abelson, R. P., & Mcburney, P. (2001). Statistics as principled. *Argument, 21*, 275–278.

Babiloni, C., Barry, R. J., Basar, E., Blinowska, K. J., Cichocki, A., Drinkenburg, W., Klimesch, W., Knight, R. T., Lopes Da Silva, F., Nunez, P., Oostenveld, R., Jeong, J., Pascual-Marqui, R., Valdes-Sosa, P., & Hallett, M. (2020). International Federation of Clinical Neurophysiology (IFCN)—EEG research workgroup: Recommendations on frequency and topographic analysis of resting state EEG rhythms. Part 1: Applications in clinical research studies. *Clinical Neurophysiology, 131*, 285–307.

Ball, R. (2009). Scholarly communication in transition: The use of question marks in the titles of scientific articles in medicine, life sciences and physics 1966–2005. *Scientometrics, 79*, 667–679.

Benos, D. J., Bashari, E., Chaves, J. M., Gaggar, A., Kapoor, N., Lafrance, M., Mans, R., Mayhew, D., Mcgowan, S., Polter, A., Qadri, Y., Sarfare, S., Schultz, K., Splittgerber, R., Stephenson, J., Tower, C., Walton, R. G., & Zotov, A. (2007). The ups and downs of peer review. *Advances in Physiology Education, 31*, 145–152.

Drozdz, J. A., & Ladomery, M. R. (2024). The peer review process: Past, present, and future. *British Journal of Biomedical Science, 81*, 12054.

Goodman, S. N., Fanelli, D., & Ioannidis, J. P. (2016). What does research reproducibility mean? *Science Translational Medicine, 8*, 341ps12.

Gopen, G., & Swan, J. (1990). The science of scientific writing. *American Scientist, 78*, 550–558.

Haggan, M. (2004). Research paper titles in literature, linguistics and science: Dimensions of attraction. *Journal of Pragmatics, 36*, 293–317.

Haynes, R. B. (2017). Improving reports of research by more informative abstracts: A personal reflection. *Journal of the Royal Society of Medicine, 110*, 249–254.

Keil, A., Debener, S., Gratton, G., Junghofer, M., Kappenman, E. S., Luck, S. J., Luu, P., Miller, G. A., & Yee, C. M. (2014). Committee report: Publication guidelines and recommendations for studies using electroencephalography and magnetoencephalography. *Psychophysiology, 51*, 1–21.

Keil, A., Bernat, E. M., Cohen, M. X., Ding, M., Fabiani, M., Gratton, G., Kappenman, E. S., Maris, E., Mathewson, K. E., Ward, R. T., & Weisz, N. (2022). Recommendations and publication guidelines for studies using frequency domain and time-frequency domain analyses of neural time series. *Psychophysiology, 59*, e14052.

Kriegeskorte, N., Simmons, W. K., Bellgowan, P. S., & Baker, C. I. (2009). Circular analysis in systems neuroscience: The dangers of double dipping. *Nature Neuroscience, 12*, 535–540.

Lang, T. A., & Altman, D. G. (2015). Basic statistical reporting for articles published in biomedical journals: The "Statistical Analyses and Methods in the Published Literature" or the SAMPL Guidelines. *International Journal of Nursing Studies, 52*, 5–9.

Luck, S. J., & Gaspelin, N. (2017). How to get statistically significant effects in any ERP experiment (and why you shouldn't). *Psychophysiology, 54*, 146–157.

Nakayama, T., Hirai, N., Yamazaki, S., & Naito, M. (2005). Adoption of structured abstracts by general medical journals and format for a structured abstract. *Journal of the Medical Library Association, 93*, 237–242.

Pernet, C., Garrido, M. I., Gramfort, A., Maurits, N., Michel, C. M., Pang, E., Salmelin, R., Schoffelen, J. M., Valdes-Sosa, P. A., & Puce, A. (2020). Issues and recommendations from the OHBM COBIDAS MEEG committee for reproducible EEG and MEG research. *Nature Neuroscience, 23*, 1473–1483.

Picton, T. W., Bentin, S., Berg, P., Donchin, E., Hillyard, S. A., Johnson, R., Miller, G. A., Ritter, W., Ruchkin, D. S., Rugg, M. D., & Taylor, M. J. (2000). Guidelines for using human event-related potentials to study cognition: Recording standards and publication criteria. *Psychophysiology, 37*, 127–152.

Plesser, H. E. (2017). Reproducibility vs. Replicability: A brief history of a confused terminology. *Frontiers in Neuroinformatics, 11*, 76.

Pottier, P., Lagisz, M., Burke, S., Drobniak, S. M., Downing, P. A., Macartney, E. L., Martinig, A. R., Mizuno, A., Morrison, K., Pollo, P., Ricolfi, L., Tam, J., Williams, C., Yang, Y., & Nakagawa, S. (2024). Title, abstract and keywords: A practical guide to maximize the visibility and impact of academic papers. *Proceedings of the Biological Sciences, 291*, 20241222.

Shepard, R. N., & Metzler, J. (1971). Mental rotation of three-dimensional objects. *Science, 171*, 701–703.

Soler, V. (2007). Writing titles in science: An exploratory study. *English for Specific Purposes, 26*, 90–102.

Song, E., Ang, L., Park, J. Y., Jun, E. Y., Kim, K. H., Jun, J., Park, S., & Lee, M. S. (2021). A scoping review on biomedical journal peer review guides for reviewers. *PLoS One, 16*, e0251440.

Stefan, A. M., & Schonbrodt, F. D. (2023). Big little lies: A compendium and simulation of p-hacking strategies. *Royal Society Open Science, 10*, 220346.

Verbruggen, F., Aron, A. R., Band, G. P., Beste, C., Bissett, P. G., Brockett, A. T., Brown, J. W., Chamberlain, S. R., Chambers, C. D., Colonius, H., Colzato, L. S., Corneil, B. D., Coxon, J. P., Dupuis, A., Eagle, D. M., Garavan, H., Greenhouse, I., Heathcote, A., Huster, R. J., et al. (2019). A consensus guide to capturing the ability to inhibit actions and impulsive behaviors in the stop-signal task. *eLife, 8*.

Wilkinson, L. (1999). Statistical methods in psychology journals: Guidelines and explanations. *American Psychologist, 54*, 594–604.

Chapter 36
How to Evaluate an EEG Research Paper

Tracy Warbrick and Michael Hoppstädter

Abstract A good scientist must acquire a comprehensive overview of their research field. Consequently, evaluating research papers is an essential skill that is part of our daily work as scientists. In this chapter, we provide guidance on how to read research papers critically with a focus on the methods and the results sections. For the results section we pay particular attention to figures because not only do they contain valuable information, but they can also be difficult to interpret. Our aim is to encourage readers to reflect on why they are reading a paper, identify what they would like to learn from it, and to read its content critically.

Keywords Critical reading · Scientific paper · EEG figures · Plot · Waveform · Topography

36.1 Introduction

The scientific paper is how we communicate our findings to the scientific community. Learning to read papers effectively will help you build your knowledge and frame your own work in the context of the literature. It will also help you develop your writing skills. Guidance on writing a paper is covered in Chap. 35. These chapters are complementary, and we recommend reading both.

36.2 What Are You Reading and Why?

Reading a scientific paper is different to reading say, a newspaper or novel. It requires active engagement with the intent to understand (Carey et al., 2020). Scientific writing is characterised by precise use of language and specialised

---

T. Warbrick (✉) · M. Hoppstädter
Brain Products GmbH, Gilching, Germany
e-mail: tracy.warbrick@brainproducts.com; michael.hoppstaedter@brainproducts.com

T. Warbrick (ed.), *The EEG Handbook*,
https://doi.org/10.1007/978-3-032-20450-9_36

vocabulary, and it can be challenging to fully understand a paper. At first it can take time to read and understand a paper thoroughly, don't be discouraged, it's a skill you can develop with practice.

36.2.1 *Why Do We Read Papers*

Your reasons for reading a paper can differ depending on the stage of your project or your current interests. For example, when learning about a new topic or trying to define your research question, you might be more focused on the introduction or discussion sections. If you are planning your study, the methods section is likely to be most relevant to you. If you are trying to interpret your findings you might search for other studies with similar or contrasting results.

It's important to be aware of your own intentions and biases when reading papers, they might affect your judgement of a paper's value or usefulness. Consider whether you view a particular paper positively because it supports your own findings. In contrast, do you dismiss papers that don't support your own hypotheses and findings? Self-reflection as a reader can help you critically assess the value of each paper and get the most out of your reading.

36.2.2 *What Are You Reading?*

Evaluating the quality, reliability, and accuracy of a paper begins with considering its source and purpose. The following questions can help guide your critical evaluation:

What is the source? Is it a peer-reviewed journal paper, a thesis, or a popular science article? While peer-reviewed work is considered the gold standard in scientific literature, you should still read these papers critically, being published doesn't guarantee quality or accuracy (van Noorden, 2023). Conversely, non-peer-reviewed sources, such as PhD theses, can also be valuable and reliable. Just be aware that while a thesis has been reviewed by the examiners it hasn't gone through the peer review process (see Chap. 35, for further details on the peer review process). Likewise, popular science publications can contain accurate and reliable information, but keep in mind they reflect the author's own perspective on the topic rather than primary research.

What type of paper is it? Is it a primary source such as original research or a case study, or a secondary source such as a review paper or a meta-analysis? Many of the papers you read will be original research papers and we will focus on these in this chapter. When reading review papers, you should differentiate between narrative and systematic reviews. A systematic review must follow a strict protocol for inclusion and exclusion of papers (Page et al., 2021a, b). A narrative review is open to the biases and choices of the author and the story they want to tell. These can still be useful; you just need to be aware of the author's goal.

When was the paper published? If it is essential that you read work done with current technology, then newer papers will be more relevant. However, not all new research is good and old research is bad. There are many older papers that are frequently cited because they are the original work and are still relevant. Consider whether the publication date is relevant to your reasons for reading the paper. In some cases, it can be useful to read papers pre- and post- the date of what you're reading for comparison.

Are there any conflicts of interest? Journals require authors to declare any conflicts of interest such as commercial relationships to products or companies mentioned in their paper, patent-licensing agreements, and funding sources for the research. These relationships should not influence the integrity of the study, but they must be declared so readers can evaluate the paper in context.

36.3 How to Read Critically

Whatever the source, type, or date of the paper, reading critically and assessing the reliability and quality of what is reported is essential. A paper should not be considered definitive just because it's published, is written by a prominent scientist, or features in a high-impact journal. Your goal as a critical reader is to understand what the authors did and why they did it that way, what they found, and why they interpreted the findings that way. You might have tested a hypothesis in a different way or have interpreted the results differently. Being aware of these possibilities and acknowledging your own questions and interpretations is important.

36.3.1 Some Tips to Develop Your Critical Reading Skills

The best advice is to read often and read actively. Active reading involves being engaged with the material and reading with questions in mind (Sun, 2020). For example, why is the research question important, did they use the correct method to test their hypothesis, how do the conclusions fit with what I currently understand on this topic? A structured, questioning approach can also develop consistency in how you read papers.

There are different approaches to reading a paper, and it can take time to find a strategy that works for you. One useful approach:

- First Read: Read the paper once fully from beginning to end to get an overview. This can be a quick read through without focusing on details.

- Second Read: Focus on the questions you have in mind and which parts of the paper can provide answers. You don't need to read the paper in order from beginning to end. Focus on details and take notes.

Context and comparison can help you get a feel for the paper and how to evaluate it. Reading papers cited in the paper you're reading can provide context. Also consider whether the paper is a part of a larger body of work from the author or research group.

Finally, discussing papers with others can enhance your understanding and develop your evaluation skills. Your peers or advisors are a good place to start and journal clubs have been shown to improve scientific literacy and confidence (Kozeracki et al., 2006).

36.4 Content of a Scientific Paper

There is an accepted convention for the content and structure of scientific papers (see Table 35.1 in Chap. 35). Although the specific order might differ depending on the journal and field of study, the main body of an original research paper will typically contain an introduction, a methods section, a results section, and a discussion. The components are described in Chap. 35. In this chapter we will focus on the methods and results sections, specifically in relation to EEG data.

While the methods and results can be considered the most objective parts of the paper (their purpose is to describe what was done and what was found), they are still influenced by the author (Carey et al., 2020). These sections have also been identified as the most challenging for students and early career researchers (in the biological sciences) to understand (Hubbard & Dunbar 2017).

36.4.1 The Methods Section

A methods section is often quite dense and difficult to read, and it can be tempting to skip over the technical details. But critically evaluating the methods is crucial to how well you can judge the conclusions of the paper. A method is a systematic way of doing a piece of research, and a methods section should describe the experiment in a way that it is reproducible, complete, and transparent. A clear method allows readers to assess whether the experiment was done appropriately, comparison across studies, and the replication of studies (Baker, 2016). It is beyond the scope of this chapter to cover all methodological details of an EEG study. For a comprehensive description of what should be included in EEG publications, we direct readers to some published guidelines (Keil et al., 2014, Picton et al., 2000, Pernet et al., 2020). Here, we will use an example experiment to show how ambiguous, missing, or incorrect methodological information can influence interpretation of the findings.

Our example experiment investigates the effect of stress on the P300 component of the ERP in patients with schizophrenia compared to a control group. The P300 component is described in more detail in Chap. 19. The important points to note here are that the P300 is elicited when an infrequent stimulus is presented (oddball effect), it is known to differ in patient groups (such as patients with schizophrenia compared to controls), and is affected by attention and emotional factors (such as stress).

36.4.1.1 Non-EEG-Specific Factors

Participants Who are being studied, for example, adults, adolescents, patients? Are there different groups of participants, for example, patients and controls. If so, are the control group well-matched to the experimental group on variables that might affect the outcome measures, for example, age, gender, or educational level?

In our example, we have a patient group and a control group. In a well-designed study, the groups would be matched on factors that could affect the results. If our patient group was drawn from the community and the control group from university students, the groups are likely to differ in age. Given that ERP amplitudes and latencies are affected by age, a mismatch in age would affect the interpretation of the results. Would you be confident about assigning any effects to the schizophrenia diagnosis if the groups were not age-matched? Factors that require matching will differ across experiments; your goal is to be aware of what is relevant to your field of study.

Variables and Experimental Design Think about what is being measured and what is manipulated. In our example, we are investigating the effects of stress. Ideally, both groups would be measured under stress and in a stress-free state. The order of these conditions should also be counterbalanced across participants. For example, if all controls did no-stress first and all patients did stress first, could you say that any differences were a result of stress? Or could it be that the order of the conditions that had an effect? Missing information about the design of the study limits how confident we can be in the findings.

Statistical Approach Your main question should be whether the statistical approach is appropriate for the experimental design and hypotheses. In our example, we would expect planned analyses to test the differences between groups, the difference between conditions, and any interaction between group and condition. If the paper doesn't clearly state the planned analysis for all variables, it should be cause for concern.

36.4.1.2 EEG-Specific Factors

Paradigm and Associated Timings and Trial Numbers In EEG research, there are several well-known paradigms for investigating specific perceptual and cognitive phenomena. When reading a paper, you should consider whether the appropriate paradigm was used for testing the hypothesis. A schematic of the paradigm can be very helpful when done well. It should include the types of stimuli, duration of presentation, inter-trial interval, and number of trials.

For our example P300 experiment, we know that target (or oddball) stimuli should be no more than 20% of trials. If our example experiment reported using 50% target stimuli, it deviates from the established design, and we cannot be confident that they are measuring a P300 oddball effect.

Hardware and Software Used Earlier we mentioned some guideline papers that include a comprehensive list of information that should be included for EEG studies (Keil et al., 2014, Picton et al., 2000, Pernet et al., 2020) you should consult these for specific details. But as a rule, a methods section should include what equipment and software was used and what parameters were used for the recording.

In our example, the P300, it is common to use a linked mastoid reference. However, this is not possible with a referential EEG amplifier where one reference electrode is used during recording. If our example experiment doesn't state the recording reference and whether mastoids were also recorded, it becomes difficult to compare it to other P300 research.

Data Analysis Parameters Processing parameters should also be accurately and completely reported. It's important to show how the author got from the raw data to the results presented in the paper.

An important point for our example P300 experiment is whether the data were re-referenced. As stated above, re-referencing to mastoids would be expected for a P300 study. If this is not reported for the recording and data processing steps, we can't confidently compare the findings to other papers that use the standard, established approach.

36.4.2 The Results Section

Reading a results section can be a daunting task, especially for those new to scientific literature. In fact, Hubbard and Dunbar (2017) found that inexperienced readers tend to undervalue the results section and are less likely to engage in critical evaluation compared to experienced readers. Here we provide guidance to help you evaluate and interpret results more confidently. We continue using our example P300 experiment from the previous section.

The results should be presented in a systematic way that states each finding clearly and relates these findings to the hypotheses tested. To support these statements, researchers can use statistical findings, tables, and figures. If the writer does this well, understanding the results and extracting relevant information should be easy. A useful question is 'can I clearly state what the findings are after reading this results section?'. If the answer is no, you should look a little closer as to why that is:

- Start by going back to the hypotheses—what is being tested?
- Then look at each finding, is it clearly stated, and does it relate to one of the hypotheses tested?
- Then consider how the finding is supported, for example, with statistics, tables, or figures.

Statistics EEG researchers are expected to have a basic understanding of common statistical measures and what they mean. Covering statistics in detail is beyond the scope of this chapter and the reader is encouraged to consult an introductory textbook in their field of study. Here, the focus is on why it's important to critically evaluate the statistical findings rather than the statistics themselves.

For our example experiment, we would expect statistical analyses that test the effects of stress on P300, compare patients and controls, and compare stress effects in patients and controls. Each stated finding should be supported by statistical evidence. Suppose the author states that P300 was larger under stress but only in patients. However, they don't state what their significance threshold was and it's not clear whether they corrected for multiple comparisons. In this case, we cannot be sure that the finding is correct and might draw different conclusions based on the significance threshold used. Not reporting this information undermines confidence in the finding. As a reader, it's important to assess whether the statistical evidence supports the conclusions drawn by the author.

Tables Tables are an effective and concise way of summarising large amounts of data. You should check whether the numbers presented in the table support the description in the text.

In our example, we might expect measures such as population characteristics or scores on questionnaires and neuropsychological tests to be reported in a table. Consider our earlier concern about age matching. If only average values for the age of each group are given but no standard deviation, we might be less convinced that the groups are well matched. Without this information we can't make a sound evaluation.

Figures Figures are important in EEG papers because they summarise key findings in a visual format. The figures should support the findings stated in the results section.

In our example experiment, we would expect to see an overlay of averaged ERPs for the two groups and the two conditions, or perhaps a difference wave comparing groups or conditions. What if the figure showed data from only one 'representative' participant instead? In this case, we would not be able to draw conclusions about group effects.

Figures play such an important role in EEG papers they are given their own section below.

36.5 How to Interpret EEG-Specific Figures

A major part of reading a results section is interpreting the figures. While the text should contain all outcome measures of a study, many papers will include compelling figures to clearly summarise their main results.

Figures are an integral part of a paper as they help you understand and interpret the results. Drawing the right conclusions requires a careful look at the figures. This section focuses on common features to pay attention to in a selection of widely used EEG figures. You can also refer to this section when designing your own accurate and informative figures.

36.5.1 Common Features in Figures

Here, we cover general best practice for informative and accurate figures before moving on to what you should look for in EEG-specific figures.

Caption and Legend Each figure should come with a meaningful caption that explains everything you need to know to understand the figure. Generally, the more complex the figure, the more detailed the caption. Often, a legend will be placed next to the figure to provide details about what is displayed.

Axes There is usually at least one axis, but often two or more. Each axis provides either a qualitative (e.g. experimental groups or conditions) or a quantitative (a scale in a certain unit) description of the results. An axis mostly relates to a single dimension but, especially for complex datasets, it can also include more than one dimension (e.g. when conditions are nested into experimental groups).

Categories While categories, such as experimental group or pre-post treatment measurements, can be conveyed via the axes of a plot, they might be indicated through different colour, line style, or symbol. This should be clearly defined in the legend.

Colour In figures that already use multiple axes, a colour scale might be used for the magnitude of another dimension (e.g. in a heatmap used with time-frequency domain data).

Scaling Each dimension in a figure has some further attributes like a unit, a range of displayed (or allowed) values, and a scale that defines the polarity and the distance between values. These details are crucial as the range of the scale and unit define the magnitude of the presented effects. Keep an eye on scaling when interpreting a figure.

Location It is important to know the spatial location where the EEG results were quantified. This can be a single recording electrode, or a *virtual* location like a region-of-interest that is generated by aggregating data across multiple channels. The spatial location should be clearly indicated by naming the electrode position or by displaying the location on a map. The electrode layout used should be given in the Methods section and if a non-standard layout is used (i.e. different from the classical 10–20 system), it should be explained in detail. However, if standard positions were used for the recording, the location is often simply indicated by the electrode label without further details. It is assumed that the standard positions and their nomenclature are commonly known.

Individual or Group Results Most EEG studies are done with a large sample, and the figures represent group data. Sometimes, the authors might present data from a single participant to highlight something specific like outlier values, or for a case study. For group data there are a few important aspects that should be mentioned somewhere: which groups are displayed (e.g. patients vs. controls), how many datasets are included in the plot, are specific datasets excluded from the plot, etc.? It might also be the case that a figure contains the data of the individual participants and an overlay of the group average for comparison.

Statistical Features Figures often contain data that has been aggregated in some way, for instance, averaged over time points (e.g. time intervals or trials), over observations (e.g. channels), or measurements (e.g. between-subjects or within-subjects). Consequently, statistical metrics are often reported together with the data to enrich the information in a figure, for example, the standard deviation or the standard error of the mean. These can be plotted in different ways, for example, as accompanying lines in different colour or line style, as differently coloured shadings, or as a simple error bar. Note that it is important to indicate what metric the error bar represents. The results of statistical tests might also be included in the figure and the areas of significant differences might be highlighted. The significance level should also be specified in such cases.

36.5.2 Examples of EEG Figures

The general principles outlined in the previous section should always be considered when looking at figures in a results section. Below we highlight features of commonly used figures for EEG results.

36.5.2.1 Waveforms

The most common way to display EEG data is through waveform plots. This can be raw EEG traces from one or multiple channels or processed EEG data. For results, the data is often integrated over time, like an ERP (averaged over trials), or over space, like a region-of-interest (averaged over channels).

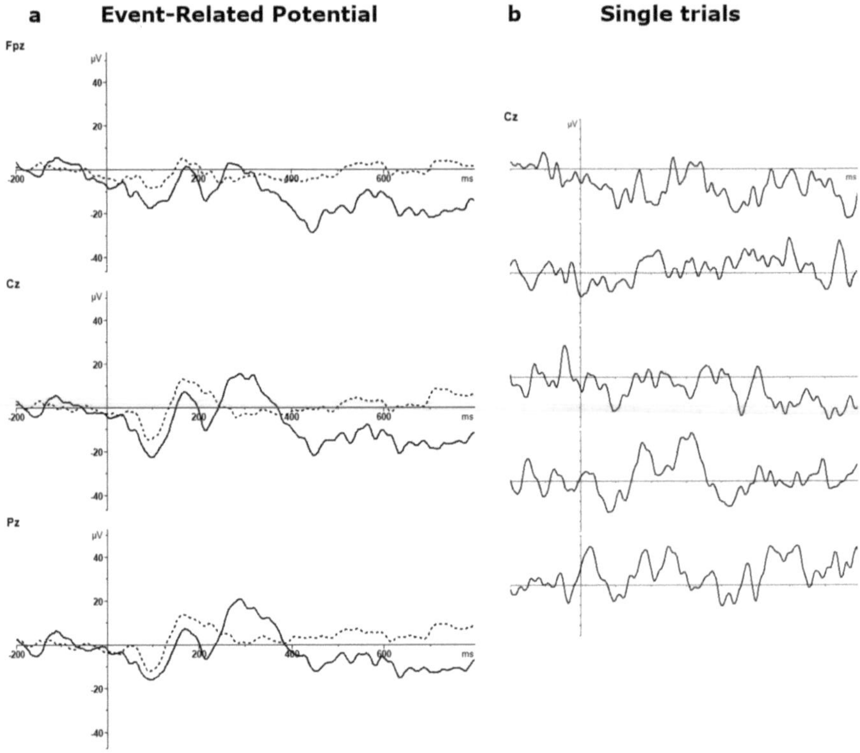

Fig. 36.1 Example plots for event-related time domain data. The left panel (**a**) shows an ERP at three midline electrodes (Fpz, Cz, Pz). Two conditions of an oddball task are displayed as an overlay, *targets* as a solid line, and *standards* as a dashed line. The right panel (**b**) shows a selection of five individual trials of the target condition at electrode Cz. The single trials are noisier and show more variability

What is displayed?

Continuous EEG Raw or processed EEG can be shown as a continuous data section from one or several channels. This is less common for analysis results, but it might be done when certain features of a recording should be highlighted. For example, for an epilepsy study, the continuous EEG might be displayed to demonstrate a specific spiking behaviour.

ERPs Many EEG studies investigate ERPs, so it is very likely that you will come across some kind of ERP result in a paper (see Fig. 36.1a). The results can be presented in different ways: a single condition, multiple conditions overlayed, multiple groups can be in the same graph or separated in adjacent graphs, a single ERP across all trials, or multiple ERPs for blocks of trials. The ERPs might be presented for a single or several channels. In the latter case, the results could be overlayed in a single plot. However, it is more likely that each channel is shown in a separate panel. If many or all channels are shown, they are often arranged like the electrodes on the scalp. The ERP could also be plotted for a region-of-interest. In that case, it should be indicated which channels were averaged. As ERPs result from averaging across trials and participants, statistical features are often added to the figures and should be considered together with the data.

Single Trials Sometimes the focus might be on single trials rather than ERPs, for instance, when the variability of the evoked response over trials is important. Figures might look as for ERPs but the data is not averaged (see Fig. 36.1b). The single trials might be displayed in different ways, for example, as overlay of all trials for a single channel (butterfly plot), as a 2D trial by time plot (with amplitude colour coded), or as a 3D trial by time by amplitude plot.

What to look for?

Check for displayed channels, time range, amplitude scale, data unit, information on displayed conditions and/or groups, use of colour and line style, statistical features (like standard deviation or standard error of the mean), the number of averaged trials and participants (for ERPs). Are there any features highlighted, e.g. artifacts, clinical features, etc.?

36.5.2.2 Topographies

Since EEG is mostly recorded from a sensor array covering the whole scalp (with variable density), topographical maps of the voltage distribution on the scalp are very common figures (see Fig. 36.3 for examples). Many ERPs or oscillatory effects have known spatial distributions and looking at topographies is very useful to judge whether an effect appears in the expected location. As the spatial distribution can change due to the experimental manipulation, age, or psychiatric conditions, topographies are often prominently displayed to show the difference in results between conditions or groups.

What is displayed?

Scalp Maps in 2D The voltage is plotted on a scalp map whereby each electrode represents one data point and the space in between electrodes is filled by calculating the voltage gradient (e.g. by spherical splines interpolation). Maps can show the scalp from multiple viewpoints; however, the most common representation is a top view with a projection of all electrodes onto a 2D plane (see Fig. 36.2a). Amplitudes are usually coded using a colormap (or grey scale) with the rainbow colormap being the most typical. Topographies can be plotted for a single time point, or as the average over a time interval. Figures commonly show the time interval around an ERP or frequency peak.

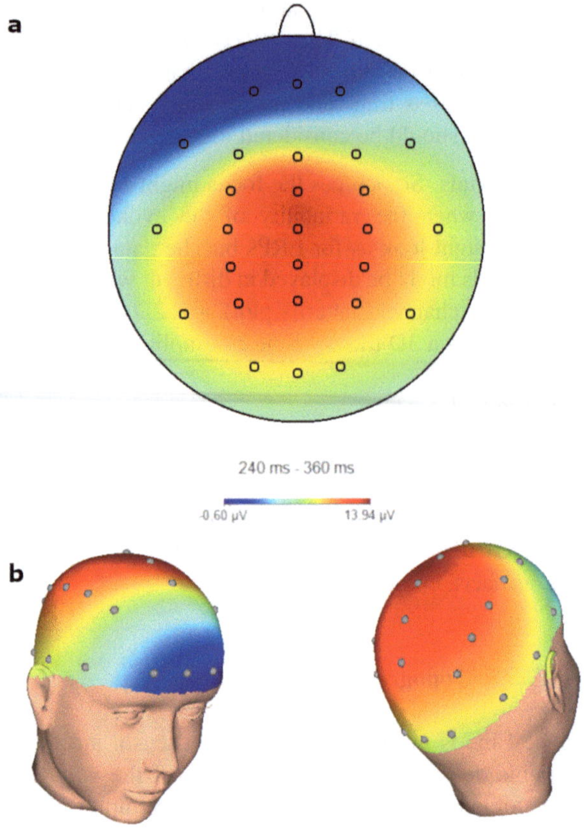

Fig. 36.2 Examples for topographies of EEG data based on the difference between two ERPs (targets minus standards) in an oddball experiment. The data is averaged within the time window 240–360 ms. The top panel (**a**) shows a 2D topography as a view from the top in which all scalp electrodes are projected on the same plane. The bottom panel (**b**) shows the same data projected onto a 3D head model from two different angles. Positive values are represented by warmer colours

Scalp Maps in 3D A 3D topography follows the same principles as its 2D counterpart, but the voltage distribution is plotted onto a head model. The 3D head can usually be rotated to focus on a specific view angle (see Fig. 36.2b).

What to look for?

Check for displayed channels, displayed time point or interval, amplitude scale, colour or grey scale, data unit, information on displayed conditions and groups, number of averaged trials.

36.5.2.3 Frequency Spectra and Time-Frequency Plots

If a paper is investigating oscillatory effects, it is likely that a plot of spectral activity will be provided (see Fig. 36.3).

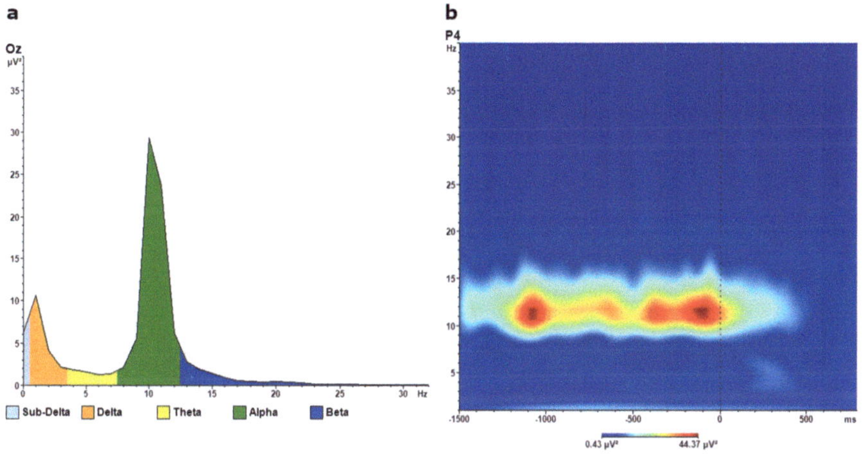

Fig. 36.3 Classical example plots for (time-)frequency domain analyses. The left panel (**a**) shows a typical frequency spectrum during a resting task with closed eyes. The power in the alpha band (in green) is elevated at occipital electrodes (Oz is displayed). The right panel (**b**) shows a time-frequency representation based on wavelet analysis. The x-axis refers to time, the y-axis to frequency. The example shows ongoing activation in the alpha band that is suppressed when a stimulus is presented (at 0 ms). Higher power values are represented by warmer colours

What is displayed?

Frequency Spectra The main difference to the waveform plots is that data is no longer represented as time points but as frequency bins. Each frequency bin has a spectral amplitude or power value (power = amplitude^2) assigned and the spectrum is usually plotted as a histogram or as a spectral curve (see Fig. 36.3a). When using a histogram style, the different frequency bands are often colour-coded. Either the full spectrum is shown (that is from 0 to the Nyquist frequency), or only a specific frequency range is displayed. Spectra mostly represent some type of averaged data, but single-trial representations are also possible.

Time-Frequency Plots This plot is a combined time and frequency domain representation. Usually plotted in 2D, time is commonly on the x-axis and frequency on the y-axis. The spectral values can be amplitude, power, or phase angles and they are mapped onto a colour scale. Values are plotted for each time-frequency point, so that a heatmap results (see Fig. 36.3b).

What to look for?

Check for displayed channels, displayed time (and frequency) range, frequency resolution, scale, data unit, information on displayed conditions and groups, number of averaged trials.

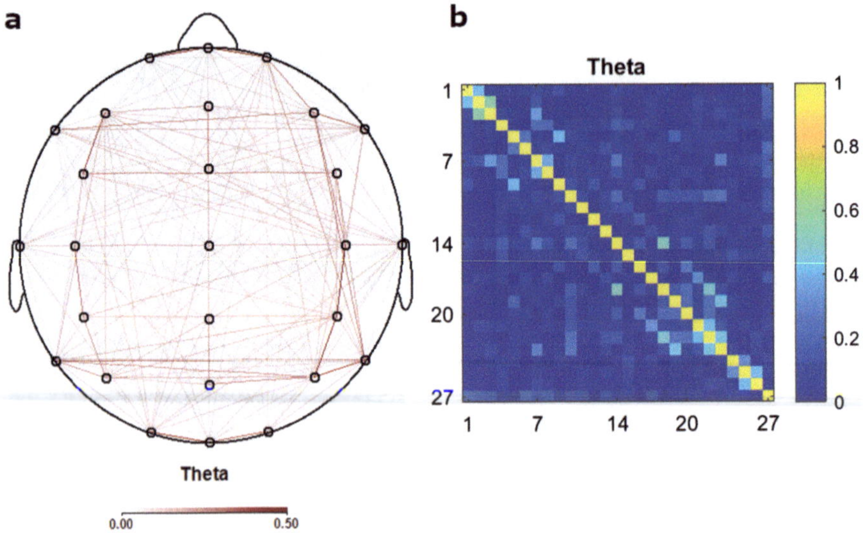

Fig. 36.4 Example plots for connectivity results. The left panel (**a**) shows connectivity values plotted onto a 2D scalp map. Each value is represented by a line between the respective electrode locations. Darker line colour represents higher connectivity. The right panel (**b**) shows the same data displayed as a connectivity matrix. Both x and y-axis represent the channels and the connectivity value for each channel pair is represented by colour, whereby yellow means higher connectivity

36.5.2.4 Connectivity Plots

Connectivity analyses have become common in EEG research, and thus figures of connectivity results appear regularly in papers. Note that many connectivity metrics are normalised and limited to a specific range of values (e.g. between 0 and 1) and are unitless.

What is displayed?

Connectivity metrics can be based on time or frequency measures (usually derived from the cross-correlation or cross-spectrum). Therefore, connectivity figures can also look like any of the waveform or spectral plots—with the exception that the results are not referring to a single location but to the relationship between two locations. Another very common way to show connectivity results is to map them onto *nodes* and *edges*, whereby the nodes represent the locations (e.g. electrodes) and the edges represent the connectivity metric between them. The edges can code the underlying value through colour or line width (see Fig. 36.4a). In cases where only a few selected connections are displayed, the value might just be written next to the edge. The results can be plotted onto a scalp map, EEG cap layout or as overlay on a brain sketch. More abstract representations also exist, for example as a connectivity matrix (see Fig. 36.4b) or circular graph.

What to look for?

Check for displayed channels, displayed connectivity metric and range, frequency resolution (if applicable), scale, information on displayed conditions and groups, number of averaged trials.

36.5.2.5 Brain Maps

Brain maps are commonly used for the presentation of fMRI results, but you might find them in EEG papers too (see Fig. 36.5). Examples are source-localised EEG or when EEG has been recorded simultaneously with fMRI and the results are combined. In the latter case, what is plotted onto a brain model is not EEG, but rather some kind of correlation measure between EEG and fMRI.

What is displayed?

The results are usually mapped on a standard brain model (e.g. the MNI template). Results can be displayed as sectional images in the three perpendicular orientations (coronal, sagittal, axial), or as overlay on 2D or 3D brain models.

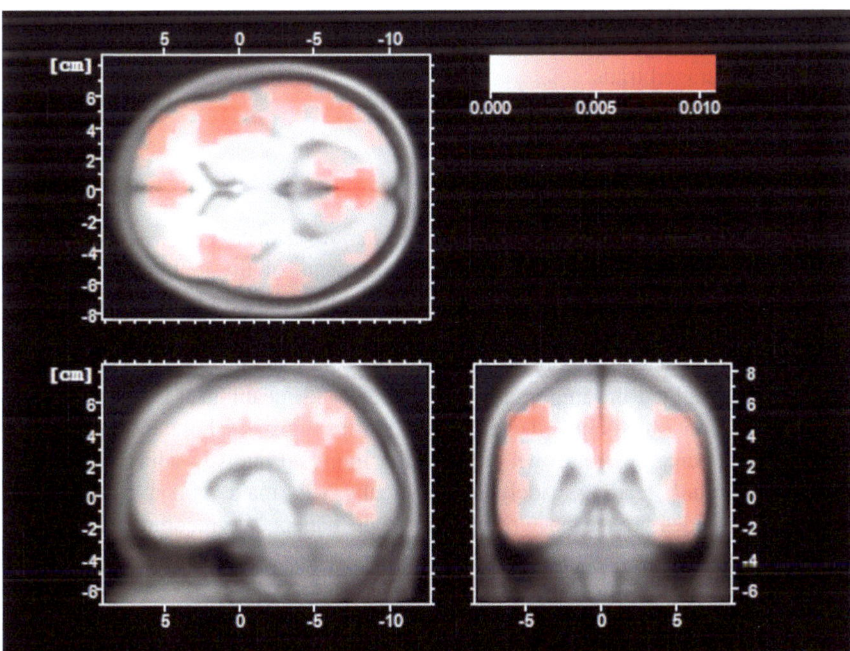

Fig. 36.5 Example plot from an EEG source analysis with the LORETA algorithm. Three slices from different directions of the MNI 305 standard template are shown with the source activation displayed as overlay. Higher activation is shown in more intense red

What to look for?

Check for displayed region or plane, coordinates in standard space, displayed metric and range, scale, data unit, information on displayed conditions and groups, number of averaged trials.

36.5.2.6 Statistical Plots

Statistical features might be part of EEG result figures, or they might even be the focus.

What is displayed?

Descriptive statistical values like the mean and standard deviation can be represented as histograms, line charts, or box plots. As in other types of plots, several conditions or groups can be contrasted. The mean (or a similar metric) might be highlighted, and error bars are commonly added (e.g. showing standard deviation). It is also quite likely that some indication of statistical significance is included.

What to look for?

Check for displayed channels, time or frequency range, scale, statistical metrics used, details on significance, information on displayed conditions and groups, number of averaged trials.

36.6 Conclusion

In this chapter, we have covered the main elements of the methods and results sections of a research paper and provided strategies for interpreting and evaluating them. An important skill for reading scientific papers is critical analysis. We encourage you to actively apply the advice given here to hone your evaluation skills. We have highlighted the importance of accurate and meaningful figures. The tips provided here not only help you evaluate the figures in papers, but they also serve as a guideline for creating your own high-quality figures. Finally, reading and writing go hand in hand; reading widely and critically will make you a better writer. Chap. 35 provides advice on writing your own papers, we recommend that you read both chapters since they are intended to be complementary.

References

Baker, M. (2016). 1,500 scientists lift the lid on reproducibility. *Nature, 533*, 452–454.

Carey, M. A., Steiner, K. L., & Petri, W. A., Jr. (2020). Ten simple rules for reading a scientific paper. *PLoS Computational Biology, 16*, e1008032.

Hubbard, K. E., & Dunbar, S. D. (2017). Perceptions of scientific research literature and strategies for reading papers depend on academic career stage. *PLoS One, 12*, e0189753.

Keil, A., Debener, S., Gratton, G., Junghofer, M., Kappenman, E. S., Luck, S. J., Luu, P., Miller, G. A., & Yee, C. M. (2014). Committee report: Publication guidelines and recommendations for studies using electroencephalography and magnetoencephalography. *Psychophysiology, 51*, 1–21.

Kozeracki, C. A., Carey, M. F., Colicelli, J., Levis-Fitzgerald, M., & Grossel, M. (2006). An intensive primary-literature-based teaching program directly benefits undergraduate science majors and facilitates their transition to doctoral programs. *CBE Life Sciences Education, 5*, 340–347.

Page, M. J., Mckenzie, J. E., Bossuyt, P. M., Boutron, I., Hoffmann, T. C., Mulrow, C. D., Shamseer, L., Tetzlaff, J. M., Akl, E. A., Brennan, S. E., Chou, R., Glanville, J., Grimshaw, J. M., Hrobjartsson, A., Lalu, M. M., Li, T., Loder, E. W., Mayo-Wilson, E., Mcdonald, S., et al. (2021a). The PRISMA 2020 statement: An updated guideline for reporting systematic reviews. *BMJ, 372*, n71.

Page, M. J., Moher, D., Bossuyt, P. M., Boutron, I., Hoffmann, T. C., Mulrow, C. D., Shamseer, L., Tetzlaff, J. M., Akl, E. A., Brennan, S. E., Chou, R., Glanville, J., Grimshaw, J. M., Hrobjartsson, A., Lalu, M. M., Li, T., Loder, E. W., Mayo-Wilson, E., Mcdonald, S., et al. (2021b). PRISMA 2020 explanation and elaboration: Updated guidance and exemplars for reporting systematic reviews. *BMJ, 372*, n160.

Pernet, C., Garrido, M. I., Gramfort, A., Maurits, N., Michel, C. M., Pang, E., Salmelin, R., Schoffelen, J. M., Valdes-Sosa, P. A., & Puce, A. (2020). Issues and recommendations from the OHBM COBIDAS MEEG committee for reproducible EEG and MEG research. *Nature Neuroscience, 23*, 1473–1483.

Picton, T. W., Bentin, S., Berg, P., Donchin, E., Hillyard, S. A., Johnson, R., Jr., Miller, G. A., Ritter, W., Ruchkin, D. S., Rugg, M. D., & Taylor, M. J. (2000). Guidelines for using human event-related potentials to study cognition: Recording standards and publication criteria. *Psychophysiology, 37*, 127–152.

Sun, T. T. (2020). Active versus passive reading: How to read scientific papers? *National Science Review, 7*, 1422–1427.

Van Noorden, R. (2023). More than 10,000 research papers were retracted in 2023—A new record. *Nature, 624*, 479–481.

Index

T. Warbrick (ed.), *The EEG Handbook*, https://doi.org/10.1007/978-3-032-20450-9

J

L

M

N

O

P

R

MIX
Papier aus verantwortungsvollen Quellen
Paper from responsible sources
FSC® C105338

If you have any concerns about our products,
you can contact us on
ProductSafety@springernature.com

In case Publisher is established outside the EU,
the EU authorized representative is:
Springer Nature Customer Service Center GmbH
Europaplatz 3, 69115 Heidelberg, Germany

Printed by Libri Plureos GmbH
in Hamburg, Germany